T0189153

The Handbook of Medical Image Perception

This state-of-the-art book reviews key issues and methods in me
research through associated techniques, illustrations, and examples
in the field, the book covers a range of topics including the his
perception research, the basics of vision and cognition, and dedic
especially those concerned with the interface between the clinici
medical image data. It summarizes many of the basic techniques
analyze medical image perception and observer performance researc
understand basic research techniques so they can adopt them for use

Written for both newcomers to the field and experienced researche
a broad overview of medical image perception, and will serve as a
years to come.

EHSAN SAMEI is a Professor of Radiology, Biomedical Enginee
Duke University, where he serves as the Director of the Carl E. Ravir
Laboratories (RAI Labs) and the Director of Graduate Studies for Medi
rent research interests include medical image formation, analysis, disp
with particular focus on quantitative and molecular imaging.

ELIZABETH KRUPINSKI is a Professor at the University of Arizona
of Radiology, Psychology, and Public Health. She is the Associate Dir
and Assessment for the Arizona Telemedicine Program, President of t
Perception Society, and serves on the Editorial Boards of a number o
radiology and telemedicine.

THE HANDBOOK OF MEDICAL IMAGE PERCEPTION AND TECHNIQUES

Edited by

EHSAN SAMEI

Duke University Medical Center

ELIZABETH KRUPINSKI

University of Arizona

CAMBRIDGE
UNIVERSITY PRESS

University Printing House, Cambridge CB2 8BS, United Kingdom

Published in the United States of America by Cambridge University Press, New York

Cambridge University Press is part of the University of Cambridge.

It furthers the University's mission by disseminating knowledge in the pursuit of education, learning and research at the highest international levels of excellence.

www.cambridge.org
Information on this title: www.cambridge.org/9781107424630

First published 2010
First paperback edition 2014

A catalogue record for this publication is available from the British Library

ISBN 978-0-521-51392-0 Hardback
ISBN 978-1-107-42463-0 Paperback

Dedicated to M^5
(Maija, Mina, Mateen, Mitra, and Maryam),
without whose love, understanding, and sacrifice
this project would have not been possible,
and to my mentors, Mike Flynn and Perry Sprawls,
who set examples before me of dedication, ingenuity, and professionalism.
E.S.

Dedicated to my parents Carole and Joseph Krupinski
who instilled in me the appreciation of life-long learning and hard work,
to my medical image perception mentors and friends Harold Kundel, MD, and Calvin Nodine, PhD,
and to my husband Michel Rogulski, PhD,
who supports and stands by me every day.
E.K.

CONTENTS

CONTRIBUTORS

CRAIG K. ABBEY
Department of Psychology
Building 429, Room 205a
University of California, Santa Barbara
Santa Barbara, CA 93106–9660
USA

EUGENIO ALBERDI
Centre for Software Reliability
City University
Northampton Square
London EC1V 0HB
UK

PETER AYTON
Department of Psychology
City University
Northampton Square
London EC1V 0HB
UK

ARTHUR BURGESS
Radiology Department
Brigham & Women's Hospital
308–1012 Pakington St.
Victoria, BC V8V3A1
CANADA

ROBERT CALDWELL
Department of Radiology
University of Iowa
3170 Medical Lab
Iowa City, IA 52242
USA

DEV CHAKRABORTY
Department of Radiology
University of Pittsburgh
3520 Forbes Avenue Parkvale Building
Pittsburgh, PA 15261
USA

WEIJIE CHEN
Center for Devices and Radiological Health
US Food and Drug Administration
10903 New Hampshire Avenue
Silver Spring, MD 20993–0002
USA

MIGUEL P. ECKSTEIN
Department of Psychology
Psychology East (Building 251), Room 3806
University of California, Santa Barbara
Santa Barbara, CA 93106–9660
USA

EDMUND FRANKEN
Department of Radiology
University of Iowa
3890 JPP
Iowa City, IA 52242
USA

MATTHEW FREEDMAN
Lombardi Building, S150
Box 20057–1465
3800 Reservoir Road NW
Washington, DC 20057–1465
USA

MARYELLEN GIGER
Department of Radiology
University of Chicago
5841 S. Maryland Avenue MC 2026
Chicago, IL 60637
USA

STEPHEN HILLIS
VA Iowa City Health Care System
CRIISP (152)
601 Highway 6 West
Iowa City, IA 52246–2208
USA

WALTER HUDA
Radiology
Medical University of South Carolina
169 Ashley Avenue
PO Box 250322
Charleston, SC 29425
USA

YULEI JIANG
Department of Radiology
University of Chicago
5841 S. Maryland Avenue MC 2026
Chicago, IL 60637
USA

ELIZABETH KRUPINSKI
Department of Radiology Research
University of Arizona
1609 N. Warren Building 211 Rm 112
Tucson, AZ 85724
USA

HAROLD KUNDEL
Department of Radiology
University of Pennsylvania
3400 Spruce St.
Philadelphia, PA 19104
USA

MATTHEW KUPINSKI
University of Arizona
Optical Sciences
1630 East University Boulevard
Tucson, AZ 85721
USA

XIANG LI
Duke University Medical Center
2424 Erwin Road
Suite 302 (DUMC) Box 2731
Durham, NC 27705
USA

DAVID MANNING
School of Medical Imaging Sciences
St Martin's College
Lancaster
Lancashire
LA1 3JD
UK

CLAUDIA MELLO-THOMS
University of Pittsburgh
Department of Radiology and
Training Program of Biomedical Informatics
3362 Fifth Avenue
Pittsburgh, PA 15213
USA

CALVIN NODINE
Department of Radiology
University of Pennsylvania
3400 Spruce St.
Philadelphia, PA 19104
USA

KENT OGDEN
Radiology Department
SUNY Upstate Medical University
750 E. Adams St.
Syracuse, NY 13210
USA

TERESA OSICKA
ISIS Center
Georgetown University
2115 Wisconsin Avenue NW,
Washington, DC 20057
USA

MARIA PETROU
Communications and Signal Processing Research Group
Department of Electrical and Electronic Engineering
Imperial College
South Kensington Campus
London SW7 2AZ
UK

ANDREY POVYAKALO
Centre for Software Reliability
City University
Northampton Square
London EC1V 0HB
UK

HANS ROEHRIG
Department of Radiology Research
University of Arizona
1609 N. Warren Building 211 Rm 112
Tucson, AZ 85724
USA

EHSAN SAMEI
Departments of Radiology, Physics, and Biomedical Engineering
Duke University
2424 Erwin Rd, Suite 302
Durham, NC 27710
USA

ROBERT SAUNDERS
Department of Radiology
Duke University
2424 Erwin Rd, Suite 302
Durham, NC 27710
USA

KEVIN SCHARTZ
Department of Radiology
University of Iowa
3170 Medical Lab
Iowa City, IA 52242
USA

JEFFREY H. SIEWERDSEN
Department of Biomedical Engineering
Johns Hopkins University
Baltimore, MD 21205
USA

LORENZO STRINGINI
Centre for Software Reliability
City University
Northampton Square
London EC1V 0HB
UK

RONALD SUMMERS
Radiology and Imaging Sciences Department
National Institutes of Health
Building 10 Room 1C660
10 Center Drive MSC 1182
Bethesda, MD 20892–1182
USA

GEORGIA TOURASSI
Department of Radiology
Duke University
2424 Erwin Rd, Suite 302
Durham, NC 27710
USA

MICHAEL J. ULISSEY
Director of Breast Imaging
Parkland Hospital
The University of Texas Southwestern Medical Center at Dallas
5323 Harry Hines Blvd.
Dallas, TX 75390–8896
USA

RICHARD VANMETTER
252 Walnut St. NW
Washington, DC 20012–2157
USA

ROBERT WAGNER
FDA/CDRH
HFZ-140
Silver Springs, MD 20993
USA

MARGARITA ZULEY
University of Pittsburgh
Director of Breast Imaging
Magee Womens Hospital
300 Halket St.
Pittsburgh, PA 15213
USA

CARL ZYLAK
Henry Ford Health System
Department of Radiology
2799 W. Garnd Blvd.
Detroit, MI 48202
USA

1

Medical image perception

EHSAN SAMEI AND ELIZABETH KRUPINSKI

1.1 PROMINENCE OF MEDICAL IMAGE PERCEPTION IN MEDICINE

Medical images form a core portion of all the information a clinician utilizes to render diagnostic and treatment decisions while a patient is under his/her care. As such, medical imaging is a major feature of modern medical care. An important requirement in using medical images is to understand what an image indicates; there is therefore a need to perceive (i.e. interpret) medical images and an associated need to have physicians subspecialized in medical image interpretation. The goal of this chapter is to provide a broad picture of the importance of medical image perception from a general healthcare perspective.

Medical imaging has been primarily ascribed to the subspecialty of radiology, with about a billion radiological imaging exams performed worldwide every year. The images include many types of examinations – single projection X-rays used in musculoskeletal, chest, and mammography applications; dynamic X-ray exams such as fluoroscopy, three-dimensional computed tomography (CT), and magnetic resonance (MR) exams; nuclear medicine emission images; and ultrasound. With the advent of digital imaging and multi-detector CT, the type and number of radiology examinations have been changing as well. The range of image types is also expanding rapidly with new modalities such as tomosynthesis and molecular imaging, which is being investigated for numerous applications, from identifying lesion margins during surgical removal to identifying cancer cells in the blood. Imaging technologies are extremely varied. Medical images can be grayscale or color, high-resolution or low-resolution, hardcopy or softcopy, uncompressed or compressed (lossy or lossless), acquired with everything from sophisticated dedicated imaging devices to off-the-shelf digital cameras.

While imaging is the central technology behind the subspecialty of radiology, during the past several years, imaging has also expanded beyond radiology to embrace other subspecialties including cardiology, radiation oncology, pathology, and ophthalmology, to name a few. Study of pathological specimens used to be limited to glass slide specimen "images" rendered by the microscope for the pathologist to view. With the advent of digital slide scanners in recent years, however, virtual slides are becoming more prevalent not only in telepathology applications but in everyday reading (Weinstein, 2001). In many medical schools and pathology residency programs, students are no longer required to purchase a microscope and a box of glass specimen slides. Students now learn from a CD with directories of virtual slides to view as softcopy images. Ophthalmology has relied on images for years (mainly as 35 mm film prints or slides) for evaluating such conditions as diabetic retinopathy. With the advent of digital images and high-performance color displays, screening raters are increasingly using softcopy images. Telemedicine has opened up an entirely new area in which medical images are being acquired, transferred, and stored to diagnose and treat patients (Krupinski, 2002). Specialties such as teledermatology, teleophthalmology, telewound/burn care, and telepodiatry are all using images on a regular basis for store-and-forward telemedicine applications. Real-time applications such as telepsychiatry, teleneurology and telerheumatology similarly rely on video images for diagnostic and treatment decisions.

One way to demonstrate the pervasiveness of medical imaging is to examine the amount of money spent each year on healthcare and then portion out the amount devoted to medical imaging (Beam, 2006). Relying on 2004 data from the Centers for Medicare and Medicaid Services (CMS), approximately 16% of the gross domestic product (GDP), or $1.6 trillion, is allotted to national healthcare expenditures (http://www.cms.hhs.gov/home/rsds.asp). Medicare expenditures represent 17% of national healthcare expenditures, of which Part B (43%) accounts for the non-facility or physician-related expenditures. Approximately 8% of Part B (or nearly $10 billion) constitutes physician-based imaging procedures. Imaging also accounts for over 40% of all hospital procedures reported in the discharge report according to the Agency for Healthcare Research and Quality (AHRQ) (http://www.ahrq.gov/data/hcup/). Based on Medicaid Part B spending, one can conservatively assume that imaging procedures comprise only 8% of non-Medicaid Part B health spending. Therefore, medical imaging in the USA is estimated to amount to $56 billion ($10 billion/17%/43%), or 0.5% of GDP.

With the pervasiveness of imaging in modern medicine, there has been significant attention and interest in the technology of imaging operations, ranging from hardware features to software functionalities. What is less appreciated is the perceptual act underlying the interpretation of these images (Manning, 2005). In order to impact patient care, an image must be *perceived and interpreted* (i.e. understood in the context of patient care) (Figure 1.1). If one assumes each of the one billion imaging examinations performed worldwide annually involves an average of four individual images per exam, one could compute that on the average, 120 medical image perception events take place every second! This astounding frequency speaks further of the pervasiveness of medical image perception in healthcare enterprise.

The Handbook of Medical Image Perception and Techniques, ed. Ehsan Samei and Elizabeth Krupinski. Published by Cambridge University Press.

Figure 1.1 As a fundamentally visual discipline, medical imaging requires psychophysical interpretation of the images to draw "meaning" from the imaging information and understand their clinical relevance.

Figure 1.2 The detection of a subtle abnormality is somewhat similar in difficulty to identifying the dog in a popular visual demonstration.

The need for interpretation of medical images comes from the fact that medical images are not self-explanatory. In the popular culture, "a picture is worth a thousand words," a phrase that reflects the power and utility of images. Ironically however, the interpretation of a medical image involves summarizing a multi-dimensional image into a few words because medical images *by themselves* do not deliver the certainty that they promise (Figure 1.2). This uncertainty, which necessitates interpretation, stems from the nature of medical imaging. Imaging is ultimately a visual discipline, impacted by psychophysical processes involved in the interpretation of images. For example, medical images can contain anatomical structures that can camouflage a feature of clinical interest that is not prevalent (in the case of screening). This uncertainty impacts the psychology of interpretation. Added to this complexity are notable variations from case to case and a multiplicity of compounding abnormalities and related factors that the interpreter needs to be mindful of.

There are clearly a significant number of images being viewed and interpreted by clinicians today in a variety of clinical specialties. As such, diagnostic accuracy cannot be defined independently of the interpretation, and any limitations or suboptimality in terms of how the images are used can significantly influence the diagnostic and therapeutic clinical decisions that they enable. Given a one-to-one link between an image and its interpretation, imaging technology alone can offer little in terms of patient care if the image is misinterpreted. The complexities of image interpretation can lead to interpretation errors and clinicians do make mistakes in the interpretation of image data (Berlin, 2005, 2007). Estimates in radiology alone suggest that in some areas there may be up to a 30% miss rate and an equally high false-positive rate. Errors can also occur in the recognition of an abnormality (e.g. whether a lesion is benign or malignant). Such errors can have a significant impact on patient care due to delays or misdiagnoses. What is less well appreciated is the prominent contribution of the inherent limitations of human perception to these errors. Image perception is the most prominent yet least appreciated source of error in diagnostic imaging. The prominence of imaging reading errors in malpractice litigation is an example of this ignorance.

The likelihood of error in the interpretation of images emphasizes the need to understand how the clinician interacts with the information in an image during interpretation. Such an understanding enables us to determine how we can further improve decision-making. That brings us to the science of medical image perception. Error is one reason to study medical image perception.

1.2 THE SCIENCE OF MEDICAL IMAGE PERCEPTION

First and foremost, it is important to understand the nature and causes of interpretation error. For that objective, one needs to distinguish between visual errors (estimated to amount to about 55% of the errors) because the clinician does an incomplete search of the image data (Giger, 1988); and cognitive errors (45%), where an abnormality is recognized but the clinician makes a decision-making error in calling the case negative (Kundel, 1978). Visual errors are further subdivided into errors where the clinician fails to look at the territory of the lesion (30%) (Kundel, 1975, 1978), and those when he/she does not fixate on the territory for an adequate amount of time to extract the lesion's relevant features (25%) (Carmody, 1980).

Contributing to interpretation errors are a host of psychophysical processes. Camouflaging of the abnormality by normal body features (so called anatomical noise) is one of the main contributors to interpretation error. Masking of subtle lesions by normal anatomical structure is estimated to affect lesion detection threshold by an order of magnitude (Samei, 1997). The visual search process, necessitated by the limited angular extent of the high-fidelity foveal vision of the human eye, is another important contribution to image interpretation. Preceded by a global impression or gist, a visual search of an image involves moving the eye around the image scene to closely examine the image details (Nodine, 1987). Studies on visual search have highlighted the prominent role of peripheral vision during the interpretation, where there is an interplay between foveal and peripheral vision as the observer scans the scene (Kundel, 1975). As a result, there are characteristic dwell times associated with correct and incorrect decisions that are influenced by the task and idiosyncratic observer search patterns (Kundel, 1989). Satisfaction of search – once an abnormal pattern is recognized, it takes additional diligence on the part of the clinician to look for other possible abnormalities within an image – is yet another contributing factor to errors (Berbaum, 1989; Smith, 1967; Tuddenham, 1962, 1963). Studies have explored the impact of expertise and prior knowledge in that behavior.

Image quality is yet another topic of interest. While intuitively recognized, image quality has been more elusive than image interpretation to characterize in such a way that it would directly relate to diagnostic accuracy (or its converse, diagnostic error). In that regard, it is important to understand how best to assess image quality and its impact on perception in order to optimize it and minimize error (Krupinski, 2008). Studies have focused on the impact of image acquisition, imaging hardware, image processing, image display, and reading environment on image quality and diagnostic accuracy.

Ergonomic aspects of interpreting medical images also play a role in the perception process. There is a need to understand the impact of ergonomic and presentation factors to minimize error (Krupinski, 2007), including determination of the causes of fatigue and how they can be minimized, the contribution of fatigue to error, the environmental distractions, the impact of the viewing interface, especially with softcopy images, and the impact of the color tint of the image.

Though we hope and aim for consistent and correct clinical decisions with every case, that aim is hard to achieve. The likelihood of two clinicians rendering two different interpretations of the same image is unsettlingly high and the expertise of the clinician plays an important role in this problem. Medical expertise is the ability to *efficiently* use contextual medical knowledge to make accurate and consistent diagnoses. Medical imaging expertise further involves perceptual and cognitive analysis of image features and manifests itself in a rich structured knowledge of normalcy and "perturbations" from the normal, an efficient hypothesis-driven search strategy, and an ability to generalize visual findings to idealized patterns. Achieving such expertise requires talent further honed by motivated effortful study, preferably supervised, and dedicated work, where accuracy is roughly proportional to the logarithm of the number of cases read annually (Nodine, 2000). Topics of interest in this line of investigation include the impact of the clinician's experience, age, and visual acuity on accuracy, toward better training and utilization of medical imaging clinicians.

Considering the impact of image perception on diagnostic accuracy, it is often necessary to test various imaging technologies and methods in terms of the associated impact on image perception. Such studies require the use of experienced clinicians, which is an expensive undertaking. Thus, there is a great need for accurate computational programs that can model visual perception and predict human performance. A host of such perceptual models have been developed, including the ideal human observer model, non-prewhitening models, channelized models, and visual discrimination models (Abbey, 2000). These models naturally require a reasonably accurate understanding of the image interpretation process. As our knowledge of the process is limited, so is the accuracy of these models. As such, their use often requires certain assumptions, verifications of their accuracy and relevance in pilot experiments, and certain calibrations, e.g. adding internal noise to make the model predictions fit human results. Nonetheless, these models have demonstrated valuable, though limited, utility in many applications, and their advancement continues to shed light on the image interpretation process.

By and large, image interpretation is currently a human task. However, increasingly, artificial intelligence tools are being used to aid in interpretation or to replace the radiologist altogether. The most common technology currently used is computer aided diagnosis (CAD), computer algorithms that examine the image content for certain abnormal features of clinical interest and then prompt the clinician for a closer examination of those features (Doi, 2007). CAD is becoming an important tool for interpreting medical images, considering the exponential growth of imaging and the shortage of specialized expertise. There is currently a need to understand the impact of CAD on diagnosis by investigating issues such as how best to integrate the human and the machine in such a way that the strength of both can be fully utilized towards improved diagnosis. For example, an experienced clinician might ignore the CAD prompts or be distracted by them if the system indicates too many false-positives. On the other hand, an inexperienced clinician might overly depend on CAD, initiating unnecessary follow-up procedures or dismissing an abnormality that might not have been picked up by the CAD algorithm. Such patterns might also change over time as a clinician gets used to a system, and such "getting used to" might not necessarily lead to improved diagnosis or efficiency. Thus, there is a need to understand the impact of CAD on the clinician's psychology, expertise, efficiency, and specialization paradigms.

Fundamental to the discussion above is the need to measure diagnostic accuracy itself (Metz, 2006; Wagner, 2007). There are a number of measures of performance such as fraction correct, sensitivity, and specificity. However, such simple measures do not adequately reflect accuracy as they can be dependent on disease prevalence or the criteria applied by the clinician, e.g. a clinician who calls all cases abnormal will have a perfect sensitivity but poor specificity, and vice versa. Seeking an overall performance measure independent of disease prevalence and criterion, receiver operating characteristic (ROC) analysis has emerged as the current gold standard for measuring diagnostic accuracy. However, ROC analysis has a number of limitations, including being limited primarily to single tasks, nonbinary confidence ratings, and location-independent decisions. In recent years, a number of advances of the ROC methodology have been developed, a welcome expansion which has shown continued progress.

1.3 WHY A CLINICIAN SHOULD CARE ABOUT MEDICAL IMAGE PERCEPTION

Medical image perception is a mature science that continues to be advanced by expert scientists. When over-specialization causes specialized "territories" to be left to the experts, one may ask why a clinician who interprets medical images needs to care about medical image perception. Needless to say, no one expects a clinician to also be a medical perception scientist. However, some knowledge of perception issues and concerns can provide vital advantages for the clinician who interprets medical images. Those advantages can be grouped into five categories.

1. Patient care-related: Understanding perceptual issues could help a clinician to improve his/her performance. Knowledge of key perceptual factors such as satisfaction of search, the relevance of prolonged dwell time, search strategies, and psychological impacts of CAD can affect the way he/she interprets medical images. Awareness of these issues enforces a greater care about the way the images are created, a greater appreciation for image quality and its effect on accuracy and efficiency, an appreciation for the influence of fatigue and the proper ergonomic design of the working environment, and higher confidence in the use of new display technologies.
2. Science-related: Being better informed about key perceptual factors enables a more proper design of projects involving medical images, develops an ability to better answer perceptual questions that inevitably arise in the review of imaging-related papers and grant applications, and increases proficiency in the reviewing of such papers and grants.
3. Teaching and learning-related: Knowledge of perceptual factors can help clinicians better communicate their expertise to trainees and help clinicians hone their perceptual skills.
4. Consumer-related: Understanding the importance of perceptual factors enables a clinician to be a better shopper of medical image-related products and services. For example, he/she will be more mindful of the image quality performance of acquisition and display devices, and the importance of the graphical user interface of picture archiving and communication system workstations.
5. Profession-related: Awareness of image perception issues enables a clinician to better educate patients, other medical professionals, and the public about the statistical nature of medical image interpretation, and to play a more effective role in related malpractice litigations.

1.4 ABOUT THIS BOOK

As outlined above, medical image perception is a frequent clinical task and a notable component of modern medicine. With perceptual error as one of the major sources of medical decision errors, our knowledge of perceptual issues gives us resources to minimize these errors and to educate future medical imaging clinicians and scientists. This book aims to provide a comprehensive reflection of medical perception concepts and issues within a single volume. Chapters in this text deal with a variety of perceptual issues in detail.

The first part of the book offers chapters by four prominent scientists, reflecting on historical developments of the field and its theoretical foundations. This part includes some reflections of the late Robert Wagner, the legendary perception scientist whose work and impact has been paramount in shaping the field as we know it today. The second part of the book includes six chapters discussing the science of medical image perception. Main topics include visual and cognitive factors, satisfaction of search, and the role of expertise. This part concludes with the perceptual relevance of image quality and reflections on the limitations of the human visual system. Part three focuses on perception metrology, with chapters on the logistics of designing perception experiments, and ROC methodology and its variants. This part ends with discussion of perceptual observer models and their implementation. Part four focuses on decision support and CAD, with topics ranging from the design of CAD studies to perceptual factors associated with the use of CAD in interpreting chest, breast, and volumetric images.

The last major part of the book offers six additional chapters about specific optimization considerations from a perceptual standpoint. Applications include radiography, CT, mammography, image processing, and image display. This part further offers a perspective on ergonomic design of workplaces for radiologists. The book ends with an epilogue outlining future possible directions for medical image perception science.

REFERENCES

Abbey, C.K., Bochud, F.O. (2000). Modeling visual detection tasks in correlated image noise with linear model observers. In Beutel, J., Van Metter, R., Kundel, H. (eds). *Handbook of Medical Imaging, Vol. 1: Physics and Psychophysics*. Bellingham, WA: SPIE Press, pp. 655–682.

Beam, C.A., Krupinski, E.A., Kundel, H.L., Sickles, E.A., Wagner, R.F. (2006). The place of medical image perception in 21st-century health care. *JACR*, **3**, 409–412.

Berbaum, K.S., Franken E.A., Dorfman, D.D., *et al.* (1989). Satisfaction of search in diagnostic radiology. *Invest Radiol*, **25**, 133–140.

Berlin, L. (2005). Errors of omission. *AJR*, **185**, 1416–1421.

Berlin, L. (2007). Accuracy of diagnostic procedures: has it improved over the past five decades? *AJR*, **188**, 1173–1178.

Carmody, D.P., Nodine, C.F., Kundel, H.L. (1980). An analysis of perceptual and cognitive factors in radiographic interpretation. *Perception*, **9**, 339–344.

Doi, K. (2007). Computer-aided diagnosis in medical imaging: historical review, current status and future potential. *Comput Med Imag & Graphics*, **31**, 198–211.

Giger, M.S., Doi, K., MacMahon, H. (1988). Image feature analysis and computer-aided diagnosis in digital radiography. 3. Automated detection of nodules in peripheral lung fields. *Med Phys*, **15**, 158–166.

Krupinski, E.A., Jiang Y. (2008). Evaluation of medical imaging systems. *Med Phys*, **35**, 645–659.

Krupinski, E.A., Kallergi, M. (2007). Choosing a radiology workstation: technical and clinical considerations. *Radiol*, **242**, 671–682.

Krupinski, E.A., Nypaver, M., Poropatich, R., *et al.* (2002). Clinical applications in telemedicine/telehealth. *Telemed J e-Health*, **8**, 13–34.

Kundel, H.L. (1975). Peripheral vision, structured noise and film reader error. *Radiol*, **114**, 269–273.

Kundel, H.L., Nodine, C.F., Carmody, D. (1978). Visual scanning, pattern recognition and decision-making in pulmonary nodule detection. *Invest Radiol*, **13**, 175–181.

Kundel, H.L., Nodine, C.F., Krupinski, E.A. (1989). Searching for lung nodules: visual dwell indicates locations of false-positive and false-negative decisions. *Invest Radiol*, **24**, 472–478.

Manning, D.J., Gale, A., Krupinski, E.A. (2005). Perception research in medical imaging. *Br J Radiol*, **78**, 683–685.

Metz, C.E. (2006). Receiver operating characteristic analysis: a tool for the quantitative evaluation of observer performance and imaging systems. *JACR*, **3**, 413–422.

Nodine, C.F., Kundel, H.L. (1987). Using eye movements to study visual search and to improve tumor detection. *RadioGraphics*, **7**, 1241–1250.

Nodine, C.F., Mello-Thoms, C. (2000). The nature of expertise in radiology. In Beutel, J., Van Metter, R., Kundel, H. (eds). *Handbook of Medical Imaging, Vol. 1: Physics and Psychophysics*. Bellingham, WA: SPIE Press, pp. 859–894.

Samei, E., Flynn, M.J., Kearfott, K.J. (1997). Patient dose and detectability of subtle lung nodules in digital chest radiographs. *Health Physics*, **72**(6S).

Smith, M.J. (1967). *Error and Variation in Diagnostic Radiology*. Springfield, IL: Charles C. Thomas.

Tuddenham, W.J. (1962). Visual search, image organization, and reader error in Roentgen diagnosis: studies of psychophysiology of Roentgen image perception. *Radiol*, **78**, 694–704.

Tuddenham, W.J. (1963). Problems of perception in chest roentgenology: facts and fallacies. *Radiol Clin North Am*, **1**, 227–289.

Wagner, R.F., Metz, C.E., Campbell, G. (2007). Assessment of medical imaging systems and computer aids: a tutorial review. *Acad Radiol*, **14**, 723–748.

Weinstein, R.S., Descour, M.R., Liang, C., *et al.* (2001). Telepathology overview: from concept to implementation. *Human Path*, **32**, 1283–1299.

PART I

HISTORICAL REFLECTIONS AND THEORETICAL FOUNDATIONS

2

A short history of image perception in medical radiology

HAROLD KUNDEL AND CALVIN NODINE

"Offering an account of the past, in disciplinary histories as in ethnic and national ones, is in part a way of justifying a contemporary practice. And once we have a stake in a practice, we shall be tempted to invent a past that supports it."

K. A. Appiah (Appiah, 2008)

2.1 FOREWORD

Medical radiology is a practical field in which images are produced primarily for the purpose of making inferences about the state of health of people. Research in radiology is also practical. Historically, imaging physicists have concentrated on developing new ways to visualize disease and on improving image quality. They have worked on the development of psychophysical models that express mathematically how observers respond to basic properties of displayed images such as sharpness, contrast, and noise. Radiologists have concentrated on image interpretation, which is using images for diagnosis, follow-up, staging, and classification of disease. Image perception, which is the process of acquiring, selecting, and organizing visual information, has generally been neglected perhaps because radiologists take for granted their ability to make sense of the patterns in images. Research in image perception has been motivated by two factors: first, the realization that human factors are a major limitation on the performance of imaging systems and second, the appreciation of the extent of human error and variation in image interpretation. Radiologists certainly are surprised when they discover that they either missed a lesion or saw one that really wasn't there.

This chapter will trace the study of perception and psychophysics as it has unfolded in books and journal articles. We will concentrate on observer error and variation, which has been a major stimulus for the development of a body of statistical methodology known as receiver operating characteristic (ROC) analysis and for attempts at understanding the perceptual basis for image interpretation and reader error. The chapter is based in part on material used by one of us (HK) for a talk given in 2003 at the Medical Image Perception Society (MIPS) in Durham, NC. It reiterates material already published in the *Journal of the American College of Radiology* (Kundel, 2006). Manning, Gale, and Krupinski (Manning, 2005), as well as Eckstein (Eckstein, 2001), also have published histories of medical image perception. A review of research and development in diagnostic imaging by Doi (Doi, 2006) contains observations about image perception and Metz (Metz, 2007) has written a tutorial review of ROC analysis that is considerably more detailed and authoritative than the material presented here.*

2.2 FLUOROSCOPES AND FLUOROSCOPY: A LESSON IN OPTIMIZING IMAGE SYSTEM PERFORMANCE

One of the earliest articles about visual perception in radiology was a discussion of visual physiology and dark adaptation in fluoroscopy by Béclère (Béclère, 1964). Béclère's article was published in 1899 but it was not until 1941 that the impact of dark adaptation on the visibility of details at fluoroscopy was seriously studied. A radiologist, W. Edward Chamberlain, working with the medical physicist George Henny, used the phantom developed by Burger and Van Dijk (Burger, 1936) to measure contrast-detail curves for fluoroscopic screens. They came to the conclusion that although the fluoroscopic screens in use at the time were technically almost equal in sharpness and contrast to images on X-ray films, the decrease in visual acuity and intensity discrimination of the retina at low brightness levels "render the available sharpness and contrast more or less invisible." Chamberlain presented the results in the Carman Lecture at the Radiological Society of North America (RSNA) (Chamberlain, 1942) and suggested that a device called an image intensifier that had been patented recently by Irving Langmuir of the General Electric Research Laboratories could provide a technological solution to the visibility problem. The subsequent development of the image intensifier (Coltman, 1948) vastly improved fluoroscopy and facilitated the development of cineradiography, cardiac catheterization, and interventional radiology.

2.3 THE PERSONAL EQUATION: OBJECTIVELY EVALUATING IMAGE SYSTEM PERFORMANCE

Chamberlain was also involved in the first extensive study of error and variation in radiology. In 1946 the United States

* The history of perception research in radiology as seen through the eyes of two participants is biased by our own experiences and by our tunnel vision. We apologize in advance to those participants in the historical events recorded here whose contribution was slighted, misinterpreted, or not mentioned. Remember there are errors of omission and commission in image interpretation and in recalling history. Feel free to inform us about our error since that is the way that we learn and eventually become experts.

The Handbook of Medical Image Perception and Techniques, ed. Ehsan Samei and Elizabeth Krupinski. Published by Cambridge University Press.

Table 2.1 *Between-observer disagreement. The number of cases read as negative for tuberculosis (neg) by the first reader that were read as positive (pos) by the second reader. The percentage between-observer disagreement is calculated as 100*neg/pos.*

Readers	Neg/pos readings	Percentage inter-observer disagreement
N/M	21/62	34
O/M	19/62	31
P/M	27/62	43
Q/M	11/62	18
Average	19/62	31

Table 2.2 *Within-observer disagreement. The number of cases read as negative for tuberculosis (neg) on a second reading after receiving an initial positive reading (pos). The percentage within-observer disagreement is calculated as 100*neg/pos.*

Readers	Neg/pos readings	Percentage intra-observer disagreement
M	18/118	10
N	4/59	7
O	14/83	17
P	39/96	41
Q	22/106	21
Average	19/92	21

Public Health Service (USPHS) and the Veterans Administration (VA) initiated an investigation of tuberculosis case finding by the newly developed technique of photofluorography. The VA had the responsibility of evaluating induction and discharge chest radiographs on millions of World War II veterans and wanted to use the best of four imaging techniques available for chest screening: 14 by 17 inch celluloid films, 14 by 17 inch paper negatives, 35 mm photofluorograms, and 4 by 10 inch stereophotofluorograms. A "Board of Roentgenology" chaired by Chamberlain and consisting of two radiologists, three chest specialists, and a statistician was convened to address the issue. They designed a study in which five readers independently interpreted four sets of 1,256 cases radiographed using each of the techniques. After a lapse of at least two months, the 14 by 17 inch films were interpreted a second time. The results published in 1947 (Birkelo, 1947) in the *Journal of the American Medical Association* (JAMA) were inconclusive because the variation among readers was greater than the differences among the techniques. The disagreement between pairs of readers averaged about 30% and within pairs of readers about 20%. Tables 2.1 and 2.2 contain brief extracts of the extensive results.

Tables 2.1 and 2.2 illustrate not only the extent of reader disagreement but also the awkward method for summarizing the results. The investigators lacked statistical tools to characterize these data. The JAMA article was accompanied by an editorial (Editorial, 1947) with the title "The personal equation in the interpretation of a chest roentgenogram," which expressed astonishment at the magnitude of observer disagreement and stated: "These discrepancies demand serious consideration." Indeed, the publication of the USPHS–VA study led to a flurry of activity that is succinctly described by two of the major participants, the radiologist L. Henry Garland (Garland, 1949) and the project statistician Jacob Yerushalmy (Yerushalmy, 1969).

The phrase "the personal equation" goes back to 1796 when the British Astronomer Royal, Nevil Maskelyne, found that his observations of the time that a certain star crossed the meridian were different from those of his assistant (Stigler, 1968). The transit time was used to set ship navigational clocks and although the error of eight-tenths of a second only translated into one-quarter of a mile at the equator, it was important to an astronomer. Maskelyne and his assistant tried to get their measurements to agree but after repeated attempts they failed. He fired the assistant! Twenty years later, while writing his *Fundamenta Astronomiae*, Friedrich Bessel found Maskelyne's account and did some experiments that also showed observational variation among astronomers. He tried, unsuccessfully, to develop "personal equations" to adjust for the differences between observers.

$$\text{John's value} = \text{Jane's value} + \text{bias correction} \quad (2.1)$$

It is ironic that in 1994 an editorial (Editorial, 1994) in the *New England Journal of Medicine* (NEJM) accompanying an article with the title "Variability in radiologists' interpretation of mammograms" (Elmore, 1994) expressed similar sentiments to those in the JAMA editorial. It is distressing that the authors either ignored or were unaware of 50 years of research on observer variation in radiology.

The results of the VA chest screening study eventually were expressed as under-reading and over-reading. A number of follow-up studies were designed "with the hope of discovering the components responsible for this variability" (Yerushalmy, 1969). Two groups of the radiologists that participated in the studies were designated CRN for Chamberlain, Rigler, and Newell and GMZ for Garland, Miller, and Zwerling. The CRN results (Newell, 1954) were published in a paper titled "Descriptive classification of pulmonary shadows: a revelation of unreliability in the roentgen diagnosis of tuberculosis." The GMZ results (Garland, 1949) were summarized by L. Henry Garland in his presidential address to the RSNA in 1948 titled "On the scientific evaluation of diagnostic procedures." His 1959 update of error in radiology and medicine in general is often quoted to support a statement that radiologists disagree with each other 30% of the time (Garland, 1959). Diagnostic unreliability has not gone away. Similar observations about disagreement (Felson, 1973; Gitlin, 2004; Goddard, 2001) are made whenever it is specifically studied.

Garland could not explain the observed variability. He classified the errors using a taxonomic approach that was later elaborated by Smith (Smith, 1967) and updated by Renfrew *et al.* (Renfrew, 1992; see Table 2.3 in Section 2.5). The GMZ group also studied reading strategies, which included dual reading and trying to control the attitude of the reader. The use of dual reading as an error-compensating mechanism was the major practical suggestion that resulted from the USPHS–VA study (Yerushalmy, 1950). Garland wrote the following about the attitude studies: "In studying the problem, the group was very conscious of the penalty in the form of over-reading which must be paid for the advantage of a reduction in under-reading." They had hit upon attitude or bias toward a particular outcome as a source of variability and recognized that it influenced the ebb and flow of true and false positives but they could not deal with it because they lacked an adequate model. That model was found in signal detection theory and a radiologist, Lee Lusted (Lusted, 1968), was largely responsible for its introduction into both radiology and the entire medical community.

2.4 RECEIVER OPERATING CHARACTERISTIC (ROC) ANALYSIS

2.4.1 The introduction of signal detection theory into radiology

The theory of signal detectability was developed by mathematicians and engineers at the University of Michigan, Harvard University, and the Massachusetts Institute of Technology partially as a tool for describing the performance of radar operators. Lee Lusted was exposed to the concepts of signal and noise in 1944 and 1945 while working in the radio research laboratory at Harvard University (Lusted, 1984). In 1954, as a radiology resident at the University of California in San Francisco (UCSF), he was introduced to the problem of observer error when he participated in one of the film reading studies supervised by Yerushalmy and Garland. Apparently his mind was prepared for a logical leap when in 1956 and 1957 he encountered a plot of percentage false negatives against percentage false positives in the laboratory of W. J. Horvath, who was responsible for optimizing the performance of the Cytoanalyzer, which was a device for automatically analyzing Papanicolaou smears (Horvath, 1956). At that time Lusted plotted a "performance curve" for chest X-ray interpretation. In 1959 he showed the curve reproduced in Figure 2.1 in the Memorial Lecture at the RSNA and published it in 1960 (Lusted, 1960). This was the first published example of an ROC curve for performance data from radiology.

Although Figure 2.1 shows a plot of false negatives against false positives, the usual convention, shown in Figure 2.2, is to plot true positives against false positives.

Lusted (Lusted, 1969) saw the ROC curve as a useful tool to accomplish two things: first, to use a parameter such as the area under the curve (AUC) as a single figure-of-merit for an imaging system and second, to decrease the observed variability in reports about images by separating the intrinsic detectability of the signal, which is a sensory variable, from the decision criteria, which is a matter of judgment. He stated: (Lusted, 1969) "It is very difficult for a human observer to maintain a constant decision attitude over a long period of time. This is a possible explanation for the finding that a radiologist will disagree with his own film interpretation about one out of five times on a second reading of the same films." He wrote a very influential book, *Introduction to Medical Decision Making* (Lusted, 1968), and went on to become a founding member of the Society for Medical Decision Making and the first editor of the journal, *Medical Decision Making*.

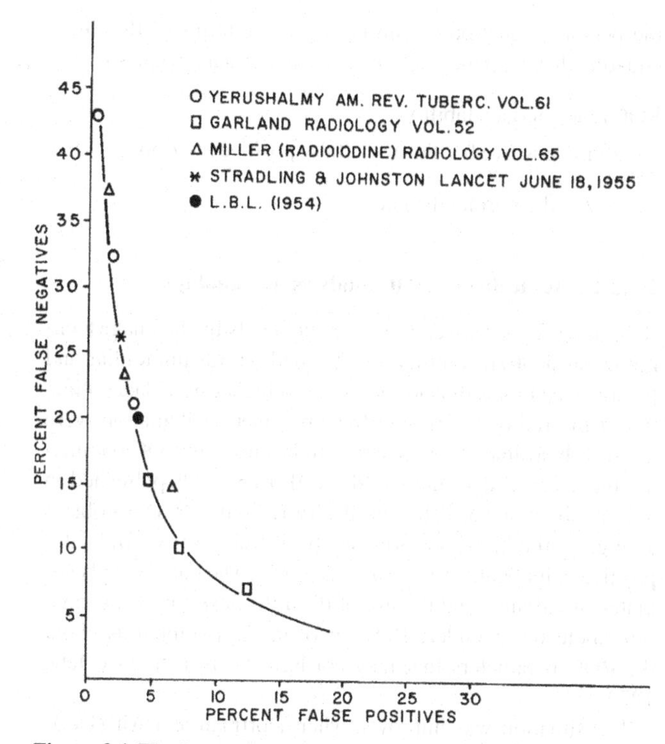

Figure 2.1 The "operating characteristic" curve showing the reciprocal relationship between percentage false negatives and percentage false positives. Most of the data were from studies of the interpretation of photofluorograms for tuberculosis. From Lusted (Lusted, 1960) with permission.

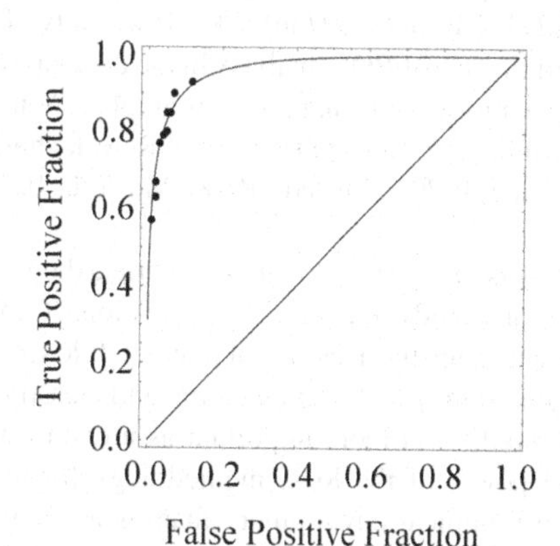

Figure 2.2 A conventional receiver operating characteristic curve showing the reciprocal relationship between the fraction of true positives and the fraction of false positives. The data points are the same as those in Figure 2.1. The curve is a binormal curve with an area under the curve (AUC of Az) of 0.87 that was fit by inspection to the data points.

Signal detection theory is a psychophysical model that describes an observer's response in terms of some known or estimated distribution of a signal and noise in the stimulus. The theoretical foundations of signal detection theory laid out in a book by Green and Swets was originally published in 1966 (Green, 1966) and reprinted with revisions in 1974 (Green, 1974). ROC analysis, which is derived from the theory of signal detectability, has become a powerful tool in visual systems evaluation. It turns out that some of the variability among observers can be reduced by applying a signal detection theory model. Perhaps the personal equation should be written in terms of the linear equation that describes an ROC curve in normal deviate space.

$$\text{ROC index of detectability} = z\,(\text{true positive fraction}) - z\,(\text{false positive fraction}) \quad (2.2)$$

where z is the normal deviate.

2.4.2 Early studies of ROC analysis in radiology

ROC analysis was not embraced immediately by the image technology evaluation community. The method was unfamiliar and practical examples illustrating experimental design, data collection using rating scales, and ROC parameter calculation were not readily available. Some early studies used an ROC parameter, the index of detectability, d', read from tables published in a book about signal detection theory (Elliott, 1964), to obtain a single estimate of performance from true positive and false positive pairs (Kuhl, 1972; Kundel, 1968). The data lacked estimates of variance and the use of d' in the absence of information about the complete ROC curve made assumptions about the ROC parameters that may not have been justified (Metz, 1973a).

The situation was improved when Dorfman and Alf (Dorfman, 1969) at the University of Iowa published a method using maximum likelihood for estimating the parameters of the ROC curve. Much of the subsequent development of statistical methodology was based on this work. In the 1970s David Goodenough, Kurt Rossmann, and Charles Metz (Goodenough, 1972, 1974; Metz, 1973b) at the University of Chicago demonstrated the use of ROC analysis in the evaluation of film-screen combinations for standard radiography. A number of articles describing the value and the use of ROC technique were published in the 1970s (Lusted, 1978; McNeil, 1975; Metz, 1978).

In 1979 Swets *et al.* (Swets, 1979) reported the results of a multi-institutional study, supported by the National Cancer Institute (NCI), comparing the accuracy of radionuclide scanning and computed tomography (CT) for detecting and classifying brain tumors. The study was more important as a demonstration of the potential power of the ROC methodology for technology evaluation in a clinical environment than as a comparison of two imaging methods. The methodology for evaluating "diagnostic systems" using ROC analysis was described in a book by John Swets and Ronald Pickett (Swets, 1982) that laid the groundwork for future developments in statistical methodology. A FORTRAN version of the Dorfman–Alf computer program was published in the book as an appendix. This became a prototype for the subsequent development in the 1980s by the group at the University of Chicago of a very influential ROC analysis software package called ROCFIT. It was superseded in the 1990s by a new package called ROCKIT.

2.4.3 Developing methods for the statistical analysis of ROC data

A test of whether the difference between two values of the area under the ROC curve is due to a real difference in the imaging techniques that yielded the values or just due to chance can be done by calculating a critical ratio (CR), denoted z, which is the ratio of the observed difference ($AUC_1 - AUC_2$) to the standard error ($SE_{(diff)}$) of the difference. The CR is then used to estimate the probability that the difference is real (or statistically significant) (Hanley, 1983).

$$z = \frac{AUC_1 - AUC_2}{SE_{(diff)}} \quad (2.3)$$

The AUC can be calculated using the procedure of Dorfman and Alf (Dorfman, 1969). Calculating the $SE_{(diff)}$ is not as straightforward. Swets and Pickett (Swets, 1982) proposed a model that included estimates of the variability due to case sampling, reader sampling (between reader variability), reader inconsistency (within reader variability), and the multiple correlations between cases and readers. They also presented a methodology for approximating the estimates. In 1992 Dorfman and Berbaum from the University of Iowa and Metz from the University of Chicago published a method for analyzing rating scale data using a combination of the Dorfman–Alf method for calculating ROC parameters and the classical analysis of variance (ANOVA) (Dorfman, 1992). The so called multi-reader multi-case (MRMC) or Dorfman, Berbaum, and Metz (DBM) method separates case variance from within and between reader variance, providing a method for deciding whether any observed differences are due to the readers or to the cases.

Recent work on methodology has focused on improving techniques to account for variance (Beiden, 2001) and on accurate estimates of statistical power (Chakraborty, 2004; Hillis, 2004, 2005; Obuchowski, 2000a).

2.4.4 ROC analysis becomes a standard method for technology evaluation

In 1989 four articles that reviewed the state of the art of ROC analysis were published by the groups that were most active in methodological development: Berbaum *et al.* from the University of Iowa (Berbaum, 1989), Gur *et al.* from the University of Pittsburgh (Gur, 1989), Hanley from McGill University (Hanley, 1989), and Metz of the University of Chicago (Metz, 1989). The fact that four reviews were published indicates the growing interest in ROC analysis as a methodology for imaging technology evaluation.

A count of the articles in six radiology journals indexed by PubMed that use the phrase "ROC" in either the title, the abstract, or the keywords is shown in Figure 2.3.

There is a steady increase in the number of citations since 1974. Note that the jump in 1988 may be due to an increase in publications but may also be an artifact caused by the addition to the database of abstracts and the keyword "ROC".

There has been increased use of ROC analysis for technology evaluation and a steady development of the methodology. There is now a new generation of review articles (Metz, 2007; Obuchowski, 2005) and ROC analysis is even beginning to appear in statistics textbooks (Zhou, 2002).

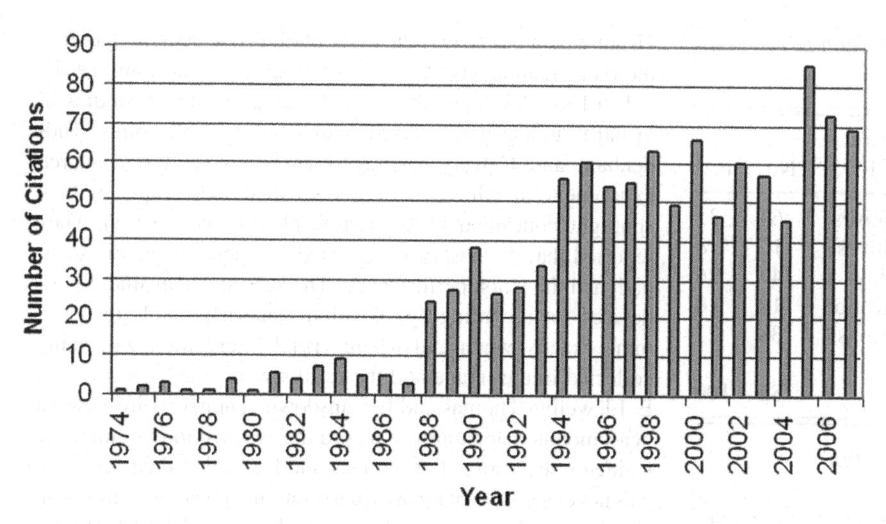

Figure 2.3 Results of a survey of the indexing in PubMed of six radiology journals: *Academic Radiology*, *Acta Radiologica*, *American Journal of Roentgenology*, *British Journal of Radiology*, *Investigative Radiology*, and *Radiology*. The journal title, abstract, and keywords were queried for "ROC." The sudden jump in 1988 may be due to a change in the indexing but the general upward trend is evident.

2.4.5 Problems with ROC analysis that have not been solved

When ROC analysis was first being studied, a few problems were identified that at the time proved to be intractable. They included: (1) analyzing the results from cases with multiple lesions on each abnormal image; (2) accounting for responses given in the wrong location on an abnormal case; and (3) dealing with situations where the diagnostic truth was unknown.

Egan *et al.* (Egan, 1961) had addressed the issue of multiple signals in 1961 and called it "the method of free response." In 1978, Bunch *et al.* (Bunch, 1978) applied the free response method, now called "free response ROC" (FROC), to images and proposed a method for analyzing the data that was not entirely satisfactory. In 1989 Chakraborty (Chakraborty, 1989) tackled the problem of data analysis and proposed a model for the FROC curve and a more satisfactory method for analyzing the data. A number of other groups have worked on the problem (Edwards, 2002; Obuchowski, 2000b) and the methodology may now be sufficiently mature so as to be useful for practical applications (Chakraborty, 2004).

Incorrect location data basically results in an image with two possible decision outcomes: a false positive for the incorrectly located lesion and a false negative for the lesion that was missed. Which one should be used in the ROC analysis? The so-called "location ROC" (LROC) was first tackled in 1975 by Starr, Metz, Lusted, and Goodenough (Starr, 1975) and picked up by Swensson (Swensson, 1993, 1996, 2000) in 1993, who made considerable progress. Chakraborty (Chakraborty, 2002) has proposed a model that accounts for location errors.

The *diagnostic truth* remains a problem, especially for those who wish to assemble large verified image databases for use by the research community (Dodd, 2004). The original expert panel method used by the GMZ group for studying performance is still in general use despite serious limitations (Revesz, 1983). In order to characterize detection, the GMZ group needed to be able to specify whether an image was truly positive or negative. They decided to use a "roentgenographic criterion" and to define positive and negative "not in terms of disease, but in terms of the presence or absence of significant shadows on the roentgenogram" (Yerushalmy, 1969). The procedure was to obtain a large number of interpretations on each image and to call back any individual with a suspicious lesion. The repeat exams were then interpreted again and determined to be either roentgenographically positive or roentgenographically negative. Once the images were dichotomized in this way, reports could be grouped into *under-reading* or missing a positive film (false negatives) and *over-reading*, or calling a negative film positive (false positives) (Garland, 1949). We have not come very far since then. Consensus or agreement of an expert panel is still one of the major methods for determining ground truth. A few investigators have suggested approaches either for establishing an ROC curve without knowledge of the truth (Henkelman, 1990) or for estimating reliability rather than accuracy (Kundel, 1997, 2001).

2.5 CLASSIFICATION OF ERROR

The development of methods for describing variability in statistical terms provided a powerful analytical tool but did not explain why the variability existed in the first place. Apparently readers are unable to maintain consistent decision criteria. Lusted (Lusted, 1969) stated: "It is very difficult for a human observer to maintain a constant decision attitude over a long period of time. This is a possible explanation for the finding that a radiologist will disagree with his own film interpretation about one out of five times on a second reading of the same films." Inconsistency plays only a small part in the generation of error. There are other sources.

From 1964 to 1966 Marcus Smith, a radiologist in New Mexico, with the cooperation of other radiologists and physicians in the area, collected and classified 437 errors using the categories shown in Table 2.3 (Smith, 1967). He did a thorough literature review and actually related the classification of errors back to the seventeenth-century work of Francis Bacon and Thomas Brown. Renfrew *et al.* (Renfrew, 1992) reviewed and classified errors in 182 cases that were presented at problem case conferences from 1986 to 1990. They used a classification that was similar to that of Smith and found about the same percentage of cases in each category.

Table 2.3 *Classification of medical errors by Smith in 1967 and Renfrew in 1992.*

Cause of error (Smith, 1967)	No	Pct	Cause of error (Renfrew, 1992)	No	Pct
Under-reading	209	48	False negatives	64	35
Complacency*	60	14	False positives	15	8
Lack of knowledge	14	3	Classification	47	26
Faulty reasoning	43	10	Communication	18	10
Communication	66	15	Complications	38	21
Unknown	45	10			
Total	437	100		182	100

*A mixture of false positives and misinterpretations

It is of interest that in his description of errors of under-reading, Smith explicitly includes satisfaction of search as a cause.

2.6 PERCEPTION OF THE MEDICAL IMAGE: UNDERSTANDING THE OBSERVER

"The process of roentgen diagnosis comprises three basic steps: the recording, the perception and the interpretation of critical roentgen shadows. Volumes have been written on the recording and interpretation of these shadows but their perception is so spontaneous that radiologists have largely taken it for granted."

William Tuddenham (Tuddenham, 1962)

2.6.1 Studies of visual search

In 1961 a radiologist, William Tuddenham, gave the Memorial Lecture at the RSNA (Tuddenham, 1962) with the title: "Visual search, image organization and reader error in roentgen diagnosis." He discussed visual physiology, visual search and some of the psychological principles that govern what we see in images. He suggested that errors of perception might arise from "incomplete coverage" or "unpatterned search" and that "our quest for meaning may lead us to abandon search prematurely." He also suggested that improvement in teaching methods could result in decreasing errors. He was not the first radiologist to discuss visual search but he was the first radiologist to actually measure the visual scanning behavior of radiologists (Tuddenham, 1961).

Until the development of eye tracking apparatus, studies of visual search generally used search time as end points. Tuddenham and Calvert performed an ingenious visual search experiment in which observers scanned radiographs with a spotlight controlled by a joystick (Tuddenham, 1961). They reported that the observer who used the most systematic scan-path had the worst performance. The subsequent availability of devices for recording eye position made it possible to determine exactly when and where visual information was being collected in the image and the results were equally surprising. E. Llewellyn Thomas and E. Lansdown (Thomas, 1963) used a head-mounted apparatus to record the eye position of radiology residents searching chest images and found that their scanning was not exhaustive but that visual fixations were concentrated on boundaries in the images. Kundel and Wright (Kundel, 1969b) recorded the eye position of radiologists and radiology residents viewing chest radiographs that were either normal or contained a solitary pulmonary nodule. They found that some of the scan-paths were like surveys with the eyes moving circumferentially around the lungs, some were concentrated on suspicious regions, but most were too complicated to characterize by inspection. The fixations tended to be concentrated on the lungs when the task was lung-specific, "search for nodules," as opposed to more distributed over the chest image when the task was general, "search for any abnormality." A study by Kundel and LaFollette (Kundel, 1972) recorded the eye position of medical students, residents, and radiologists. One of the chest radiographs in the test set showed a large right upper lobe opacity and small left lower lobe opacity. The scanpaths of six observers at different experience levels are shown in Figure 2.4. By the time students were in the third year of medical school they recognized the lesions and fixated them within four seconds.

Figure 2.5 shows the complete scan of the observers shown in Figure 2.4, which averaged about 20 seconds. Note that the fixations were concentrated on the abnormalities, leaving parts of the image without coverage.

The two notable findings that were followed up and verified by this and other laboratories were the concentration of fixations on the lesions to the exclusion of other areas of the image and the speed with which an obvious and subtle opacity on a chest radiograph was fixated by observers with appropriate training.

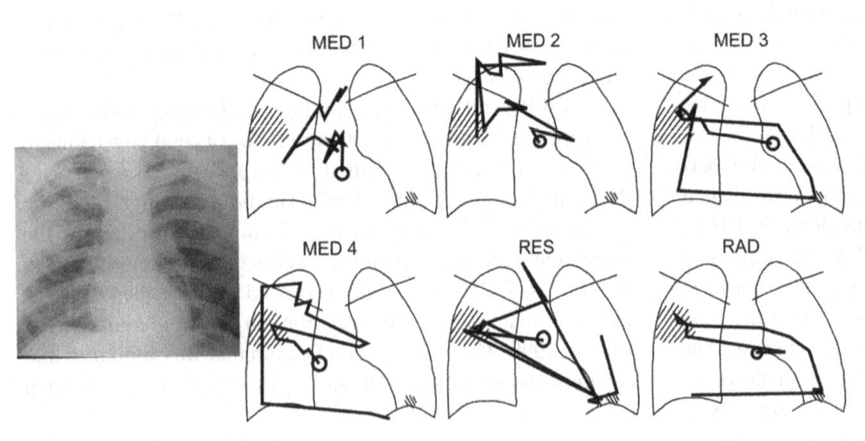

Figure 2.4 The chest radiograph is on the left side and on the right side are the scanpaths recorded during the first four seconds of viewing from four medical students (MED 1, MED 2, MED 3, and MED 4), a radiology resident (RES), and a radiologist (RAD). Notice how the pattern of inspection of the lesions changes and the prompt fixation of the lesions starting with the third-year medical student. Redrawn from Kundel and LaFollette (Kundel, 1972) with permission.

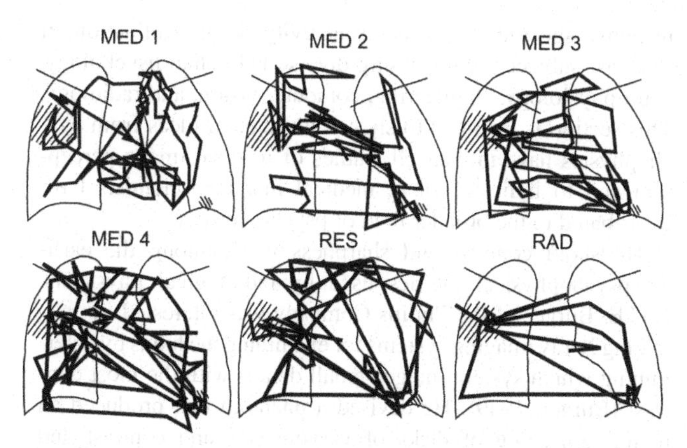

Figure 2.5 The complete scanpath of first- to fourth-year medical students, radiology residents, and radiologists viewing a chest film with a large right upper lobe and a small left lower lobe opacity. Note the concentration of fixations on the lesions and the more efficient appearance of the pattern of the radiologist. Redrawn from Kundel and LaFollette (Kundel, 1972) with permission.

2.6.2 Studies of gaze dwell time on lung nodules, breast cancers, and extremity fractures

A lot of the work on analyzing error concentrated on lung nodule detection. It quickly became apparent that sorting through clinical cases to find lung nodules was both labor-intensive and yielded nodules with a variety of characteristics that were difficult to quantify. Nodules could be simulated optically without much difficulty, providing an endless supply of chest images with nodules in known locations having uniform, mathematically definable characteristics (Kundel, 1968, 1969a). Kundel, Nodine, and Carmody (Kundel, 1978) studied the eye position of observers searching for lung nodules. They measured the location of the axis of the gaze and the time (gaze dwell time) that a five-degree useful visual field centered on the axis of the gaze included a nodule. They found that of 20 missed nodules, 30% were never fixated by the useful visual field, 25% were fixated only briefly, and 45% received prolonged visual attention but were not reported. These misses were classified as scanning, recognition, and covert decision errors respectively.

Further studies using chest radiographs (Kundel, 1989), mammograms (Krupinski, 1996), and skeletal radiographs (Hu, 1994; Krupinski, 1997) correlated gaze dwell time with lesion location or suspected lesion location and with decision outcome: true positive, false positive, false negative, and true negative. Fixation survival curves, examples of which are shown in Figure 2.6, show the percentage of fixations that remain after the elapsed gaze duration. The true positive and false positive outcomes had the longest fixation dwell times. The false negatives on the lung nodules and the mammograms also showed significant fixation dwell times. Kundel, Nodine, and Krupinski (Kundel, 1990) used fixation dwell time information to provide perceptual feedback about the location of potential lesions, much in the manner of computer aided diagnosis (CAD). The procedure required gaze tracking while the observer viewed a chest radiograph. Immediately after viewing, the observer pointed out positive locations. All positive locations eventually would be scored as true positive or false positive. Then regions with clusters of fixations that had long dwell times and were not pointed out could be identified as potential false negatives and shown to the observer with an appropriate prompt. The work on perceptual feedback was reviewed by Krupinski, Nodine, and Kundel (Krupinski, 1998) in 1998.

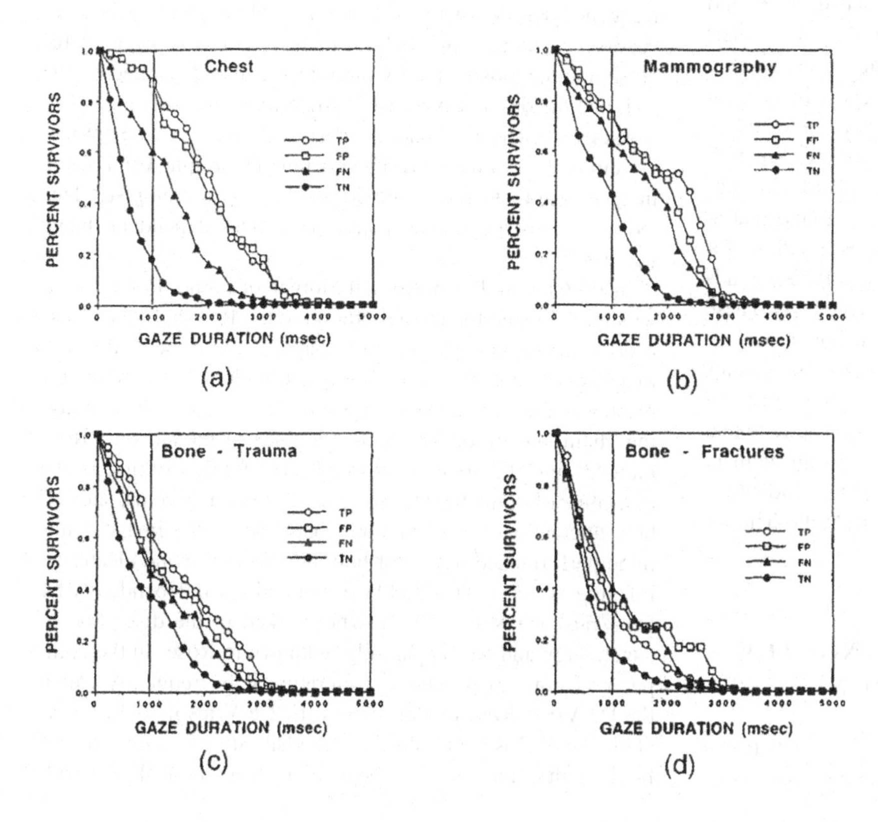

Figure 2.6 Survival function curves associated with TP, FP, TN, and FN decision outcomes for nodules in chest radiographs, tumors in mammograms, traumatic bone injuries, and fractures of the extremities. The survival function is a plot of the percentage of the total fixations located on the lesion that remained on the lesion after the elapsed gaze time. From Krupinski *et al.* (Krupinski, 1998) with permission.

The curves are taken from Figure 3 in Krupinski, Nodine, and Kundel (Krupinski, 1998) with permission. The survival functions indicate the probability of survival of a cumulative fixation cluster as a function of gaze duration. The vertical lines indicate the percentage of fixations remaining after 1000 ms.

2.6.3 Satisfaction of search (SOS)

To our knowledge, Marcus Smith (Smith, 1967) first used the term "satisfaction of search" to describe one possible mechanism for missing lesions and Tuddenham (Tuddenham, 1962) used the more general term "satisfaction of meaning." Neither of them presented any objective evidence for the existence of a phenomenon in which the presence of one abnormality on an image blocked the perception of a second abnormality. In a series of elegant experiments, Kevin Berbaum and the group at the University of Iowa showed that the SOS phenomenon did really exist (Berbaum, 1990, 1991, 1994). The SOS phenomenon was verified independently by Samuel *et al.* (Samuel, 1995). Most of the missed lesions are fixated but not recognized (Berbaum, 2000, 2001; Samuel, 1995) suggesting that SOS is not strictly a search or scanning problem but rather a suppression of recognition. The final chapter on SOS still has to be written.

2.6.4 Development of the concepts of global analysis and holistic perception

In 1975, Kundel and Nodine (Kundel, 1975), following up the observation of the speed with which experienced observers fixated abnormalities, showed radiologists chest images with a variety of straightforward lung and cardiac abnormalities for 200 ms, equivalent to the duration of one visual fixation, and found that the average performance as measured by ROC analysis was surprisingly good. The readers achieved an average AUC of 0.76 on flash viewing on a set of chest images on which they had achieved an average of 0.96 with an unrestricted viewing time. Oestmann *et al.* (Oestmann, 1988) did a flash viewing study using chest radiographs: 40 normal, 40 with subtle nodules, and 40 with obvious nodules. They found that at a false-positive fraction of 20%, the true-positive fractions for subtle and obvious cancers were 30% and 70% at 0.25 seconds and 74% and 98% at unlimited viewing time, respectively. These experiments reinforced the notion that medical image perception has a major "global" component (Kundel, 2007). It seems that most of the visual pattern recognition that goes on occurs at the very onset of viewing and much of the visual activity that follows is largely confirmatory, although there is an element of discovery search. Current research is focusing on identifying image properties that attract visual attention (Mello-Thoms, 2003; Perconti, 2007).

2.7 PSYCHOPHYSICAL MODELING: THE OBSERVER IMAGE INTERACTION

Psychophysics is the scientific study of the relationship between a stimulus, characterized in physical terms, and an observer's response specified by either sensitivity or discrimination. It is a vast subject encompassing diverse fields like psychology, neurophysiology, engineering, computer vision, and radiology. This section is intended to show how seminal ideas from psychophysics have influenced studies of medical image perception and to indicate where medical imaging scientists have contributed to the broader field of psychophysics.

Threshold contrast and sharpness were among the earliest psychophysical metrics used by radiological physicists. G.C.E. Burger of the Philips Company was interested in optimizing X-ray imaging systems for examining the lungs by determining which system imaged small details with the most contrast (Burger, 1949). He devised a phantom that produced an image consisting of disks of varying size and contrast and determined the smallest disk that could just be detected at each contrast (Burger, 1950). This produced a unique contrast–detail curve, which is usually plotted as threshold contrast against size and was the prototype for many modern test phantoms.

Contrast–detail curves are a relatively simple way to determine the sensitivity of the observer but they are not adequate descriptors of image system performance. In the early 1960s radiological physicists began to adopt methods for characterizing imaging systems based on the work of an RCA engineer, Otto Schade (Schade, 1964), who introduced the concept of the *modulation transfer function* (MTF). Simplistically, the MTF is a plot of contrast sensitivity expressed as a percentage of a reference value against spatial resolution expressed as cycles/mm. The advantage of using the MTF is that total system MTF can be expressed as the product of the MTF of all of the system components including the eye. Although necessary, the MTF is still not sufficient to define image quality; image noise had to be considered and expressed in terms of spatial frequency. Kurt Rossmann, a radiological physicist at the University of Chicago, pointed out that even after accounting for noise, the diagnostic quality of an image also depended on the task that the observer was asked to perform (Rossmann, 1970). Rossmann stated (Rossmann, 1969): "Parameters of the optimal system will depend on the object being radiographed and on the image detail that needs to be detected by the radiologist." This led to the now popular catchphrase of "task-dependent image quality."

In 1966 a radiologist, Russell Morgan of Johns Hopkins University, delivered the annual oration to the RSNA, "Visual perception in fluoroscopy and radiography," about the analysis of imaging systems (Morgan, 1966). He included the frequency response of the human visual system in his analysis of the imaging chain and introduced a psychophysical model, the "Rose model," into the analysis. Rose (Rose, 1948), working in the discipline of optical engineering on television systems, showed how fluctuations in the photons that produce the image determine performance limits for both human vision and electronic imaging systems. This had been pointed out independently by De Vries (De Vries, 1943), who worked in the discipline of psychology and was apparently unknown to Rose. In the computer vision and psychology literature it is frequently called the De Vries–Rose model (or just the De Vries model by psychologists). Rose asserted that the visibility of a target that is brighter than the surround depends upon random fluctuations in the number of photons that arrive at the sensor. He further assumed that the arrival of photons with time followed a Poisson distribution and that the standard deviation of the distribution was equivalent to noise. He asserted that the noise limited the ability to detect contrast and indicated that a minimal signal-to-noise ratio (SNR) was required for signal detection. He also showed that imaging systems can be evaluated using an absolute scale based on quantum efficiency, which involves counting the number of incoming photons per unit area in a given time. Wagner and Brown (Wagner, 1985) showed how the application of SNR models could be used to evaluate and compare imaging systems. The Rose model also has been used to determine the statistical efficiency of human contrast discrimination in the presence of noisy backgrounds (Burgess, 1981). Burgess (Burgess, 1999) has described the implications and limitations of the Rose model for modern signal detection theory.

In the 1979 RSNA New Horizons Lecture, Kundel (Kundel, 1979) pointed out that existing psychophysical models relating physical image properties to observer responses were inadequate for use in images as structurally complicated as anatomical radiographs. He suggested that "surround complexity" had to be included as a noise component in the psychophysical equation. Detection experiments performed with George Revesz of Temple University had shown that ribs and vascular shadows interfered with the detection of lung nodules (Kundel, 1976). They used the term *conspicuity*, a concept developed by the psychologist Engel (Engel, 1971) to express the visibility of a target embedded in a structured surround. They attempted unsuccessfully to develop a psychophysical equation to quantify conspicuity (Revesz, 1974, 1985).

The effect of anatomical structures on nodule detection has been verified by Ehsan Samei and his colleagues (Samei, 1999, 2003) and imaging scientists are beginning to incorporate terms that express the effect of structured noise into ideal observer models (Eckstein, 1995). An ideal observer is one who can utilize all of the available information and perform a task with minimal cost or error. Actual performance can be compared with ideal observer performance to gain insight into the efficiency of an imaging system. Kyle Myers (Myers, 2000) has reviewed progress in the development of ideal observer models, and Harry Barrett and Kyle Myers (Barrett, 2003) put the models into the wider context of imaging science. Reviews of the contribution of medical imaging scientists have been written by Chesters (Chesters, 1992), Eckstein (Eckstein, 2001), and Burgess (Burgess, 1995). In 2007, a special issue of the *Journal of the Optical Society of America* featured papers about image quality and observer models from authors in a number of different fields including radiology (Kupinski, 2007). Psychophysical modeling has improved our understanding of perception and still seems to be the most fruitful avenue to the development of criteria and metrics for image quality that reflect human performance.

2.8 SUMMARY AND SPECULATION

Research in medical image perception and psychophysics has been driven by the awareness of the extent of human error and variation in the imaging process and by the need to use human beings for image technology evaluation. The displayed image is a meaningless grayscale pattern unless viewed and analyzed by an intelligent observer and, luckily for radiologists, computers are not as intelligent as people. Recent reviews show that error is about the same as it was when first systematically measured 60 years ago (Goddard, 2001; Robinson, 1997). The need for objective comparison of imaging modalities is especially important with the development and commercialization of CAD.

In this historical survey, we have tried to show how the need for technology evaluation led to the development and refinement of ROC analysis by medical imaging scientists and how the physics community used the concepts of statistical decision theory to help develop psychophysical models for image system performance that could be used to evaluate image quality independently of human judgment. We have also showed how a few stalwart workers have tried to improve our understanding of perception itself, the mysterious process by which the eye-brain converts the patterns in light into meaningful representations of the world around us. Surely this is a daunting task, understanding the workings of the brain, but it leads us to understanding how people learn to recognize patterns and it holds the promise of showing us how to improve our teaching of image interpretation.

In 1949 Garland (Garland, 1949) enumerated three objectives for future research that are still very relevant:

(1) Determine reliable methods for measuring the relative number of lesions missed by a reader.
(2) Study the probable reasons for missing lesions and their characteristics.
(3) Investigate methods of interpretation that might lead to a reduction in the number of lesions missed.

We have come very far in the area of performance measurement and although still in their infancy studies of psychophysical models offer the hope of eliminating or at least minimizing the need for observer performance studies. We understand the sources of error a little better but still need a lot of work in that area. Finally, interpretation methods for improving accuracy are still only dreams. As a field that is steeped in technology, we have turned to technology in the form of CAD to aid interpretation, but that is not enough. Recently there has been an increase of interest in defining the nature of expertise and the methods for attaining it (Lesgold, 1988; Nodine, 1996; Norman, 1992; Proctor, 1995; Wood, 1999). This may be the path to follow to improve our performance and we believe that further study of expertise should be encouraged because it has the promise of leading to better diagnosis and better patient care.

REFERENCES

Appiah, K.A. (2008). *Experiments in Ethics*. Cambridge, MA: Harvard University Press.
Barrett, H.H., Myers, K.J. (2003). *Foundations of Image Science*. Hoboken, NJ: John Wiley and Sons.
Béclère, A. (1964). A physiologic study of vision in fluoroscopic examinations. In Bruwer, A. (Ed.) *Classic Descriptions in Diagnostic Roentgenology*. Springfield, IL: Charles C. Thomas.

Beiden, S.V., Wagner, R.F., Campbell, G., *et al.* (2001). Components-of-variance models for random-effects ROC analysis: the case of unequal variance structures across modalities. *Acad Radiol*, **8**, 605–615.

Berbaum, K.S., Dorfman, D.D., Franken, E.A., Jr. (1989). Measuring observer performance by ROC analysis: indications and complications. *Invest Radiol*, **24**, 228–233.

Berbaum, K.S., Franken, E.A., Jr., Dorfman, D.D., *et al.* (1990). Satisfaction of search in diagnostic radiology. *Invest Radiol*, **25**, 133–140.

Berbaum, K.S., Franken, E.A., Jr., Dorfman, D.D., *et al.* (1991). Time course of satisfaction of search. *Invest Radiol*, **26**, 640–648.

Berbaum, K.S., El-Khoury, G.Y., Franken, E.A., Jr. (1994). Missed fractures resulting from satisfaction of search effect. *Emerg Radiol*, **1**, 242–249.

Berbaum, K.S., Franken, E.A., Jr., Dorfman, D.D., *et al.* (2000). Role of faulty decision making in the satisfaction of search effect in chest radiography. *Acad Radiol*, **7**, 1098–1106.

Berbaum, K.S., Brandser, E.A., Franken, E.A., *et al.* (2001). Gaze dwell times on acute trauma injuries missed because of satisfaction of search. *Acad Radiol*, **8**, 304–314.

Birkelo, C.C., Chamberlain, W.E., Phelps, P.S., *et al.* (1947). Tuberculosis case finding. A comparison of the effectiveness of various roentgenographic and photofluorographic methods. *JAMA*, **133**, 359–366.

Bunch, P.C., Hamilton, J.F., Sanderson, G.K., *et al.* (1978). A free-response approach to the measurement and characterization of radiographic observer performance. *J Appl Photogr Eng*, **4**, 166–171.

Burger, G.C.E., Van Dijk, B. (1936). Über die physiologischen Grundlagen der Durchleuchtung. *Fortschr Rontg*, **54**, 492–496.

Burger, G.C.E. (1949). The perceptibility of details in roentgen examinations of the lung. *Acta Radiol Diag*, **31**, 193–222.

Burger, G.C.E. (1950). Phantom tests with X-ray. *Philips Technical Review*, **11**, 291–298.

Burgess, A.E., Wagner, R.F., Jennings, R.J., *et al.* (1981). Efficiency of human visual signal discrimination. *Science*, **214**, 93–94.

Burgess, A. (1995). Image quality, the ideal observer, and human performance of radiologic detection tasks. *Acad Radiol*, **2**, 522–526.

Burgess, A.E. (1999). The Rose model, revisited. *J Opt Soc Am A*, **16**, 633–646.

Chakraborty, D.P. (1989). Maximum likelihood analysis of free response operating characteristic (FROC) data. *Med Phys*, **16**, 561–568.

Chakraborty, D.P. (2002). Statistical power in observer performance studies: comparison of receiver operating characteristic and free-response methods in tasks involving localization. *Acad Radiol*, **9**, 147–156.

Chakraborty, D.P., Berbaum, K.S. (2004). Observer studies involving detection and localization: modeling, analysis, and validation. *Med Phys*, **31**, 2313–2330.

Chamberlain, W.E. (1942). Fluoroscopes and fluoroscopy. *Radiology*, **38**, 383–412.

Chesters, M.S. (1992). Human visual perception and ROC methodology in medical imaging. *Phys Med Biol*, **37**, 1433–1476.

Coltman, J.W. (1948). Fluoroscopic image brightening by electronic means. *Radiology*, **51**, 359–367.

De Vries, H. (1943). The quantum character of light and its bearing upon threshold of vision, the differential sensitivity and visual acuity of the eye. *Physica*, **10**, 553–564.

Dodd, L.E., Wagner, R.F., Armato, S.G., 3rd, *et al.* (2004). Assessment methodologies and statistical issues for computer-aided diagnosis of lung nodules in computed tomography: contemporary research topics relevant to the lung image database consortium. *Acad Radiol*, **11**, 462–475.

Doi, K. (2006). Diagnostic imaging over the last 50 years: research and development in medical imaging science and technology. *Phys Med Biol*, **51**, R5–27.

Dorfman, D., Alf, E.J. (1969). Maximum likelihood estimation of parameters of signal-detection theory and determination of confidence intervals – rating method data. *J Math Psych*, **6**, 487–496.

Dorfman, D.D., Berbaum, K.S., Metz, C.E. (1992). Receiver operating characteristic analysis. Generalization to the population of readers and patients with the jackknife method. *Invest Radiol*, **27**, 723–731.

Eckstein, M., Whiting, J. (1995). Lesion detection in structured noise. *Acad Radiol*, **2**, 249–253.

Eckstein, M.P. (2001). The perception of medical images 1941–2001. *Opt Photonics News*.

Editorial (1947). The "personal equation" in the interpretation of a chest roentgenogram. *JAMA*, **133**, 399–400.

Editorial (1994). The accuracy of mammographic interpretation. *N Engl J Med*, **331**, 1521–1522.

Edwards, D.C., Kupinski, M.A., Metz, C.E., *et al.* (2002). Maximum likelihood fitting of FROC curves under an initial-detection-and-candidate-analysis model. *Med Phys*, **29**, 2861–2870.

Egan, J., Greenberg, G., Schulman, A. (1961). Operating characteristics, signal detectability, and the method of free response. *J Acoust Soc Am*, **33**, 993–1007.

Elliott, P. (1964) Tables of d'. In Swets, J. (Ed.) *Signal Detection and Recognition by Human Observers*. New York: John Wiley.

Elmore, J.G., Wells, C.K., Lee, C.H., *et al.* (1994). Variability in radiologists' interpretation of mammograms. *N Engl J Med*, **331**, 1493–1499.

Engel, F.L. (1971). Visual conspicuity, directed attention and retinal locus. *Vision Res*, **11**, 563–567.

Felson, B., Morgan, W.K., Bristol, L.J., *et al.* (1973). Observations on the results of multiple readings of chest films in coal miners' pneumoconiosis. *Radiology*, **109**, 19–23.

Garland, L.H. (1949). On the scientific evaluation of diagnostic procedures. *Radiology*, **52**, 309–328.

Garland, L.H. (1959). Studies on the accuracy of diagnostic procedures. *AJR Am J Roentgenol*, **82**, 25–38.

Gitlin, J.N., Cook, L.L., Linton, O.W., *et al.* (2004). Comparison of "B" readers' interpretations of chest radiographs for asbestos related changes. *Acad Radiol*, **11**, 843–856.

Goddard, P., Leslie, A., Jones, A., *et al.* (2001). Error in radiology. *Br J Radiol*, **74**, 949–951.

Goodenough, D.J., Rossmann, K., Lusted, L.B. (1972). Radiographic applications of signal detection theory. *Radiology*, **105**, 199–200.

Goodenough, D.J., Rossmann, K., Lusted, L.B. (1974). Radiographic applications of receiver operating characteristic (ROC) curves. *Radiology*, **110**, 89–95.

Green, D.M., Swets, J.A. (1966). *Signal Detection Theory and Psychophysics*. New York: John Wiley and Sons.

Green, D.M., Swets, J.A. (1974). *Signal Detection Theory and Psychophysics*. Huntington, NY: Krieger.

Gur, D., King, J.L., Rockette, H.E., *et al.* (1989). Practical issues of experimental ROC analysis. *Invest Radiol*, **25**, 583–586.

Hanley, J.A., McNeil, B.J. (1983). A method of comparing the areas under receiver operating characteristic curves derived from the same cases. *Radiology*, **148**, 839–843.

Hanley, J.A. (1989). Receiver operating characteristic (ROC) methodology: the state of the art. *Crit Rev Diag Im*, **29**, 307–355.

Henkelman, R.M., Kay, I., Bronskill, M.J. (1990). Receiver operating characteristic (ROC) analysis without truth. *Med Decis Making*, **10**, 24–29.

Hillis, S.L., Berbaum, K.S. (2004). Power estimation for the Dorfman-Berbaum-Metz method. *Acad Radiol*, **11**, 1260–1273.

Hillis, S.L., Obuchowski, N.A., Schartz, K.M., *et al.* (2005). A comparison of the Dorfman-Berbaum-Metz and Obuchowski-Rockette methods for receiver operating characteristic (ROC) data. *Stat Med*, **24**, 1579–1607.

Horvath, W., Tolles, W., Bostrom, R. (1956). Quantitative measurements of cell properties on Papanicolaou smears as criteria for screening. First International Cancer Cytology Congress. Chicago, IL: American Cancer Society.

Hu, C.H., Kundel, H.L., Nodine, C.F., *et al.* (1994). Searching for bone fractures: a comparison with pulmonary nodule search. *Acad Radiol*, **1**, 25–32.

Krupinski, E.A. (1996). Visual scanning patterns of radiologists searching mammograms. *Acad Radiol*, **3**, 137–144.

Krupinski, E.A., Lund, P.J. (1997). Differences in time to interpretation for evaluation of bone radiographs with monitor and film viewing. *Acad Radiol*, **4**, 177–182.

Krupinski, E.A., Nodine, C.F., Kundel, H.L. (1998). Enhancing recognition of lesions in radiographic images using perceptual feedback. *Opt Eng*, **37**, 813–818.

Kuhl, D.E., Sanders, T.D., Edwards, R.Q., *et al.* (1972). Failure to improve observer performance with scan smoothing. *J Nucl Med*, **13**, 752–757.

Kundel, H.L., Revesz, G., Stauffer, H.M. (1968). Evaluation of a television image processing system. *Invest Radiol*, **3**, 44–50.

Kundel, H.L., Revesz, G., Stauffer, H.M. (1969a). The electro-optical processing of radiographic images. *Radiol Clin N Am*, **7**, 447–460.

Kundel, H.L., Wright, D.J. (1969b). The influence of prior knowledge on visual search strategies during the viewing of chest radiographs. *Radiology*, **93**, 315–320.

Kundel, H.L., LaFollette, P.S. (1972). Visual search patterns and experience with radiological images. *Radiology*, **103**, 523–528.

Kundel, H.L., Nodine, C.F. (1975). Interpreting chest radiographs without visual search. *Radiology*, **116**, 527–532.

Kundel, H.L., Revesz, G. (1976). Lesion conspicuity, structured noise, and film reader error. *AJR Am J Roentgenol*, **126**, 1233–1238.

Kundel, H.L., Nodine, C.F., Carmody, D.P. (1978). Visual scanning, pattern recognition, and decision making in pulmonary nodule detection. *Invest Radiol*, **13**, 175–181.

Kundel, H.L. (1979). Images, image quality and observer performance. *Radiology*, **132**, 265–271.

Kundel, H.L., Nodine, C.F., Krupinski, E.A. (1989). Searching for lung nodules: visual dwell indicates locations of false-positive and false-negative decisions. *Invest Radiol*, **24**, 472–478.

Kundel, H.L., Nodine, C.F., Krupinski, E.A. (1990). Computer displayed eye position as a visual aid to pulmonary nodule interpretation. *Invest Radiol*, **25**, 890–896.

Kundel, H.L., Polansky, M. (1997). Mixture distribution and receiver operating characteristic analysis of bedside chest imaging using screen-film and computed radiography. *Acad Radiol*, **4**, 1–7.

Kundel, H.L., Polansky, M., Phelan, M. (2001). Evaluating imaging systems in the absence of truth: a comparison of ROC analysis and mixture distribution analysis using CAD in mammography. *Proc Soc Photo-Opt Instrum Eng*, **4324**, 153–158.

Kundel, H. (2006). History of research in medical image perception. *J Am Col Radiol*, **3**, 402–408.

Kundel, H.L., Nodine, C.F., Conant, E.F., *et al.* (2007). Holistic component of image perception in mammogram interpretation: gaze-tracking study. *Radiology*, **242**, 396–402.

Kupinski, M.A., Watson, A.B., Siewerdsen, J.H., *et al.* (2007). Image quality. *J Opt Soc Am A*, **24**, B198.

Lesgold, A., Rubinson, H., Feltovich, P., *et al.* (1988). Expertise in a complex skill: Diagnosing X-ray pictures. In Chi, M., Glaser, R., Farr, M. (Eds.) *The Nature of Expertise*. Hillsdale, NJ: Erlbaum.

Lusted, L.B. (1960). Logical analysis in roentgen diagnosis. *Radiology*, **74**, 178–193.

Lusted, L.B. (1968). *Introduction to Medical Decision Making*. Springfield, IL: Charles C. Thomas.

Lusted, L.B. (1969). Perception of the Roentgen image: applications of signal detection theory. *Radiol Clin N Am*, **7**, 435–445.

Lusted, L.B. (1978). General problems in medical decision making with comments on ROC analysis. *Semin Nucl Med*, **8**, 299–306.

Lusted, L.B. (1984). Editorial: ROC recollection. *Med Decis Making*, **4**, 131–134.

Manning, D.J., Gale, A., Krupinski, E.A. (2005). Perception research in medical imaging. *Br J Radiol*, **78**, 683–685.

McNeil, B.J., Keeler, E., Adelstein, S.J. (1975). Primer on certain elements of medical decision making. *N Engl J Med*, **292**, 211–215.

Mello-Thoms, C., Dunn, S.M., Nodine, C.F., *et al.* (2003). The perception of breast cancers – a spatial frequency analysis of what differentiates masses from reported cancers. *IEEE T Med Imaging*, **22**, 1297–1306.

Metz, C.E., Goodenough, D.J. (1973a). Letter: On failure to improve observer performance with scan smoothing: a rebuttal. *J Nucl Med*, **14**, 873–876.

Metz, C.E., Goodenough, D.J., Rossmann, K. (1973b). Evaluation of receiver operating characteristic curve data in terms of information theory, with applications in radiography. *Radiology*, **109**, 297–303.

Metz, C.E. (1978). Basic principles of ROC analysis. *Semin Nucl Med*, **8**, 283–298.

Metz, C.E. (1989). Some practical issues of experimental design and data analysis in radiographic ROC studies. *Invest Radiol*, **24**, 235–245.

Metz, C.E. (2007). ROC analysis in medical imaging: a tutorial review of the literature. *Radiol Phys Tech*, **1**, 2–12.

Morgan, R.H. (1966). Visual perception in fluoroscopy and radiography. Annual oration in memory of John D. Reeves, Jr., M.D., 1924–1964. *Radiology*, **86**, 403–416.

Myers, K.J. (2000). Ideal observer models of visual signal detection. In Beutel, J., Kundel, H.L., Van Metter, R.L. (Eds.) *Handbook of Medical Imaging*. Bellingham, WA: SPIE Press.

Newell, R.R., Chamberlain, W.E. & Rigler, L. (1954). Descriptive classification of pulmonary shadows. A revelation of unreliability in the roentgen diagnosis of tuberculosis. *Am Rev Tuberc*, **69**, 566–584.

Nodine, C.F., Kundel, H.L., Lauver, S.C., *et al.* (1996). The nature of expertise in searching mammograms for masses. *Proc Soc Photo-Opt Instrum Eng*, **2712**, 89–94.

Norman, G.R., Coblentz, C.L., Brooks, L.R., *et al.* (1992). Expertise in visual diagnosis: a review of the literature. *Acad Med*, **67**, S78–S83.

Obuchowski, N.A. (2000a). Sample size tables for receiver operating characteristic studies. *AJR Am J Roentgenol*, **175**, 603–608.

Obuchowski, N.A., Lieber, M.L., Powell, K.A. (2000b). Data analysis for detection and localization of multiple abnormalities with application to mammography. *Acad Radiol*, **7**, 516–525.

Obuchowski, N. (2005). Fundamentals of clinical research for radiologists, ROC Analysis. *AJR Am J Roentgenol*, **184**, 364–372.

Oestmann, J.W., Greene, R., Kushner, D.C., *et al.* (1988). Lung lesions: correlation between viewing time and detection. *Radiology*, **166**, 451–453.

Perconti, P., Loew, M.H. (2007). Salience measure for assessing scale-based features in mammograms. *J Opt Soc Am A*, **24**, B81–B90.

Proctor, R.W., Dutta, A. (1995). Perceptual skill/the development of expertise. *Skill acquisition and human performance*. Thousand Oaks, CA: Sage Publications.

Renfrew, D.L., Franken, E.A., Jr., Berbaum, K.S., *et al.* (1992). Error in radiology: classification and lessons in 182 cases presented at a problem case conference. *Radiology*, **183**, 145–150.

Revesz, G., Kundel, H.L., Graber, M.A. (1974). The influence of structured noise on the detection of radiologic abnormalities. *Invest Radiol*, **9**, 479–486.

Revesz, G., Kundel, H.L., Bonitatibus, M. (1983). The effect of verification on the assessment of imaging techniques. *Invest Radiol*, **18**, 194–198.

Revesz, G. (1985). Conspicuity and uncertainty in the radiographic detection of lesions. *Radiology*, **154**, 625–628.

Robinson, P.J.A. (1997). Radiology's Achilles' heel: error and variation in the interpretation of the Roentgen image. *Br J Radiol*, **70**, 1085–1098.

Rose, A. (1948). The sensitivity performance of the human eye on an absolute scale. *J Opt Soc Am*, **38**, 196–208.

Rossmann, K. (1969). Image quality. *Radiol Clin N Am*, **7**, 419–433.

Rossmann, K., Wiley, B.E. (1970). The central problem in the study of radiographic image quality. *Radiology*, **96**, 113–118.

Samei, E., Flynn, M.J., Eyler, W. (1999). Detection of subtle lung nodules: relative influence of quantum and anatomic noise on chest radiographs. *Radiology*, **213**, 727–734.

Samei, E., Flynn, M.J., Peterson, E., *et al.* (2003). Subtle lung nodules: influence of local anatomic variations on detection. *Radiology*, **228**, 76–84.

Samuel, S., Kundel, H.L., Nodine, C.F., *et al.* (1995). Mechanism of satisfaction of search: eye position recordings in the reading of chest radiographs. *Radiology*, **194**, 895–902.

Schade, O.S. (1964). Modern image evaluation and television (the influence of electronic television on the methods of image evaluation). *Appl Optics*, **3**, 17–21.

Smith, M.J. (1967) *Error and Variation in Diagnostic Radiology*. Springfield, IL: Thomas.

Starr, S.J., Metz, C.E., Lusted, L.B., *et al.* (1975). Visual detection and localization of radiographic images. *Radiology*, **116**, 533–538.

Stigler, S.M. (1968). *The History of Statistics*. Cambridge, MA: Harvard University Press, 240–242.

Swensson, R.G. (1993). Measuring detection and localization performance. In Barrett, H.H., Gmitro, A.F. (Ed.) *Information Processing in Medical Imaging*. New York, NY: Springer-Verlag.

Swensson, R. (1996). Unified measurement of observer performance in detecting and localizing target objects on images. *Med Phys*, **23**, 1709–1725.

Swensson, R. (2000). Using localization data from image interpretations to improve estimates of performance accuracy. *Med Decis Making*, **20**, 170–185.

Swets, J.A., Pickett, R.M., Whitehead, S.F., *et al.* (1979). Assessment of diagnostic technologies. *Science*, **205**, 753–759.

Swets, J.A., Pickett, R.M. (1982). *Evaluation of Diagnostic Systems. Methods from Signal Detection Theory*. New York, NY: Academic Press.

Thomas, E.L., Lansdown, E.L. (1963). Visual search patterns of radiologists in training. *Radiology*, **81**, 288–291.

Tuddenham, W.J., Calvert, W.P. (1961). Visual search patterns in roentgen diagnosis. *Radiology*, **76**, 255–256.

Tuddenham, W.J. (1962). Visual search, image organization, and reader error in roentgen diagnosis. *Radiology*, **78**, 694–704.

Wagner, R.F., Brown, D.G. (1985). Unified SNR analysis of medical imaging systems. *Phys Med Biol*, **30**, 489–518.

Wood, B.P. (1999). Visual expertise. *Radiology*, **211**, 1–3.

Yerushalmy, J., Harkness, J.T., Cope, J.H., *et al.* (1950). The role of dual reading in mass radiography. *Am Rev Tuberc*, **61**, 443–464.

Yerushalmy, J. (1969). The statistical assessment of the variability in observer perception and description of roentgenographic pulmonary shadows. *Radiol Clin N Am*, **7**, 381–392.

Zhou, X.-H., Obuchowski, N.A., McClish, D.K. (2002). *Statistical Methods in Diagnostic Medicine*. New York: John Wiley & Sons.

3

Spatial vision research without noise

ARTHUR BURGESS

The human visual system (HVS) is remarkably complex and sophisticated. To date there is no man-made system that can rival its dynamic luminance range – a ratio of about 10^5 for lowest to highest luminance in a single scene and a ratio of 10^{10} over all luminance adaptations. There is no machine vision system that can rival its pattern recognition capabilities. In this chapter, I will describe some of the past work on absolute thresholds of simple aperiodic signals as well as work using periodic signals (mostly windowed sine waves). This will be a short description of studies of human vision without noise – except for studies investigating photon-capture fluctuations at the retina. The chapter starts in the nineteenth century with very simple investigations designed to determine a few "laws" that describe human capabilities. A few important milestones are as follows. Blackwell (Blackwell, 1946) started the era of modern work using rigorous psychophysical methods and aperiodic test patterns such as discs. One very important aspect of his work was to establish that human observers could perform in a consistent manner with similar results from different observers. This was very surprising to many psychologists. The next dramatic change was Schade's (Schade, 1956) measurements of contrast sensitivity using one-dimensional (1D) sine wave test patterns. Perhaps the most important step of all was Campbell and Robson's (Campbell, 1968) hypothesis of independent detection of spatial frequency components and subsequent psychophysics experiments to test their hypothesis. The term "multiple visual channels" came later.

3.1 NINETEENTH CENTURY WORK

The earliest investigations of spatial vision concentrated on acuity (the ability to distinguish details in compact signals), luminance increment discrimination and spatial summation effects. One example of acuity measurement is use of the familiar Snellen letter chart, named after the Dutch ophthalmologist Hermann Snellen, who developed it in 1862. Ernst Weber (1795–1878) was one of the first to study the human response to a luminance increment stimulus in a quantitative fashion. He found that the relationship between stimulus magnitude and perceived intensity is logarithmic, with a discrimination just-noticeable difference (JND) given by a constant value, $JND = \Delta L/L$, where L is luminance. One application, for example, is that the brightness of stars is described on a logarithmic (magnitude) scale. Gustav Fechner (1801–1887) later offered an elaborate theoretical interpretation of Weber's findings, which he called Weber's law, though his admirers made the law's name a hyphenate, the Weber–Fechner law. Fechner related the sensation, Y, and the physical intensity, X, using $Y = a \log(X) + b$. The Italian astronomer Annibale Ricco (1844–1911) investigated spatial summation for compact signals. He found that with a small area near the fovea, about 0.1 degrees^2, there was complete summation described by the equation $CA = k$, where C is contrast threshold, A is signal area, and k is a constant. This is now known as Ricco's law. For larger signals, spatial summation for compact signals is often described by Piper's law, $C\sqrt{A} = k$. The equation is often valid for aperiodic signals subtending an angle between 0.1 and 20 degrees. It is named after the German physiologist Hans Edmund Piper (1877–1915) who formulated it. In the temporal domain, there is also Bloch's law, $R = It$, where R is the response, I is intensity, and t is time; the law seems to hold for presentations under 100 ms.

3.2 BLACKWELL – THE MODERN ERA BEGINS

Blackwell (Blackwell, 1946) was one of the first to do psychophysics experiments to evaluate contrast thresholds over a wide range of conditions. This was part of a World War II program under contract to the US Office of Scientific Research and Development. Aside from the data provided, the major contribution of Blackwell's work was the demonstration that such measurements could actually be done in a reliable and reproducible manner. Because of this situation, I will spend some time discussing it. At the outset of his work, there was considerable skepticism about its feasibility. Therefore he proceeded in a very careful manner. The experimental viewing conditions were meticulously designed and controlled. There were 19 observers who took part in a variety of preliminary experiments over six months to a year. So they were veterans of between 35,000 and 75,000 observations (decision trials) before the data collection reported in the publication. Blackwell found in his preliminary experiments that the precision of the initial physical photometric measurements was less than the precision of responses to visual stimuli. Five (and occasionally eight) contrasts were used for each stimulus and were presented in random order using an eight-alternative forced choice (8-AFC) procedure. The contrasts had been selected to give probability of detection between 10% and 95%. A small red indicator point was projected on the viewing screen (62 feet from the observers) and the stimulus was superimposed at one of eight locations on the circumference of a circle of three degree radius. After a six second viewing period,

The Handbook of Medical Image Perception and Techniques, ed. Ehsan Samei and Elizabeth Krupinski. Published by Cambridge University Press.

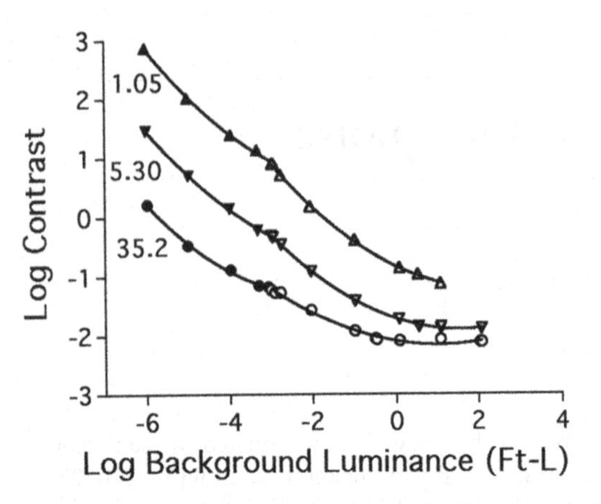

Figure 3.1 Results of Blackwell's experiments for detection of a disc signal over a range of background luminances. The associated numbers indicate the disc angular diameter in milliradians (one degree equals 17.5 mRad). The cusps in the curves occur near the luminance where cones begin to be activated. The data were obtained from Table II of Blackwell's paper.

observers selected a location using an eight-position electrical indicator. Average cumulative normal detection probability versus contrast curves were constructed and thresholds were defined as 50% correct after allowance for guessing. One of the figures in the paper shows such a curve based on 450,000 observations! It should be carefully noted that Blackwell's work was done before the development of signal detection theory so the proper analysis of forced-choice experiments lay in the future. He did mention that upon interrogation, observers stated that they did not feel confident of having "seen" a stimulus unless the probability of detection was greater than 90%! This is consistent with my own experience in 2AFC experiments with signals in noise – where observers usually have little confidence when the probability of detection is less than 80%. The stimuli were discs of various diameters and luminances, L, superimposed on uniform backgrounds of lower luminance, L_0. Contrast was defined as $C = (L - L_0)/L_0$. Results for three of the five signal diameters (in milliradians) that were investigated are shown in Figure 3.1. The data were obtained from Table II of Blackwell's paper. There is a cusp in the curves near a background luminance of 10^{-3} foot-lamberts (Ft-L) for all disc diameters. This corresponds to the point where cones are no longer activated and signals are detected using rods (scotopic vision).

Blackwell also did contrast threshold experiments using signals darker than the background but they will not be discussed here. The Blackwell paper represented a historical moment in vision research. It demonstrated that psychophysical measurements could be done in a very reliable and reproducible manner.

3.3 CONTRAST SENSITIVITY MEASUREMENTS

The next dramatic change in vision research (Schade, 1956) was the measurement of contrast sensitivity using 1D sine wave test patterns. The patterns were 1D because of hardware limitations. The 2D displays that we now use did not become generally available until the 1970s and even then were very expensive. In 1981, my first 320 × 256 image display system cost about $30,000 – not including the minicomputer system (which cost another $70,000).

Otto Schade (1903–1981) published in 1956 the first measurement of the visual contrast sensitivity function (CSF) as a function of spatial frequency. Schade, a colleague of Albert Rose, worked at Radio Corporation of America (RCA) from 1931 until his retirement in 1968 and, among many other things, developed the concept of the modulation transfer function (MTF), which has had wide application in imaging. Schade made a number of contributions to vision research and was invited by Bob Wagner to speak at an SPIE medical X-ray conference in 1974 on the topic of evaluation of noisy images. I had the pleasure of meeting him there. He was then retired but still had the enthusiasm of a graduate student for his subject – showing us an advance copy of his wonderful book on image quality (Schade, 1975). Wagner subsequently wrote a number of papers translating Schade's work (based on a sampling aperture concept) into the more modern notation with which we are familiar.

Schade's CSF measurements for detection of static 1D sine wave patterns on a special TV monitor (60 Hz frame rate and 500 to 800 vertical lines) are shown in Figure 3.2. He gave measurements for a wide range of luminances with the monitor mounted behind an opening in a large white adapting screen whose luminance was adjusted to match the average monitor luminance. He used a method of adjustment. The sine wave amplitude was increased steadily from zero at a fixed rate until the observer subjectively "saw" the test pattern. This ascending threshold value was recorded. The procedure was repeated until the observer achieved an approximately steady threshold criterion and these "learning" results were discarded. Then 10 to 15 more trials were done and the results averaged for the threshold curve determination. Each curve in the figure is based on 200 to 600 trials.

Schade interpreted his CSF results as a combination of the low pass characteristic of the MTF of the eye optical system followed by the effects of neural interaction at the retina. The low frequency falloff is due to the physiology of the retina and the way the brain interprets visual information. From early times

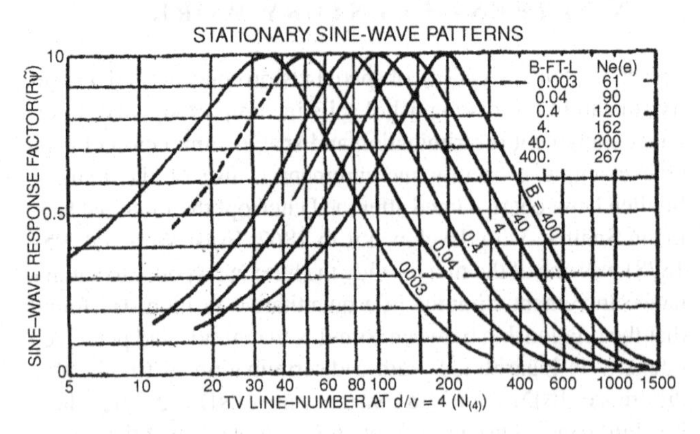

Figure 3.2 The CSF measured by Schade (Schade, 1956) using a 1D sine wave pattern at six average luminance values. The spatial frequency conversion factor is 1 cycle/degree = 28.6 TV line-numbers. The peak for B of 400 Ft-L occurs at 7 cycles/degree. (Reproduced with permission of the Optical Society of America.)

in vision research there had been interest in the phenomenon of Mach bands. This is an illusion that appears when there is a narrow gradient between a large dark gray region in an image and a large light gray region. We perceive a narrow darker band in the dark gray region adjacent to the gradient and a narrow lighter band in the light gray region. This can be interpreted as an indication of a decrease in the CSF at low frequencies. This view is consistent with the physiological discovery that the retinal ganglion cells of the frog (Barlow, 1953) and cat (Kuffler, 1953) respond to contrast rather than absolute brightness. Note that Kuffler had earlier suggested this possibility and was Barlow's PhD thesis supervisor. They found receptive fields with antagonistic regions – an "ON" (excitatory) center and an "OFF" (inhibitory) surround. If this receptive field is interpreted as a point spread function (PSF), then the corresponding Fourier transform has a band pass form. Note there is a marked dependence on mean luminance in Schade's CSF results due to retinal adaptation.

3.4 CHANNELS – INDEPENDENT DETECTION OF SOME FOURIER COMPONENTS

DePalma and Lowry (DePalma, 1962) investigated the frequency dependence of both sine wave and square wave detection. They claimed that, at high spatial frequencies, sine wave gratings were seen at lower contrast than square waves of the same frequency – with the implication that there was something very special about the smoothly changing luminance. This indirectly led to the spatial frequency channels concept discussed below. The following is John Robson's version of the story (personal communication). He had been a graduate student of Fergus Campbell at Cambridge University. Fergus asked John for his opinion about the Lowry and DePalma results. John said that, unless the laws of physics had recently changed, their results could not be correct and he didn't think there was any point in trying to repeat their observations, which must simply be wrong. This was due to the fact that the first harmonic of the square wave would be at three times the frequency of the fundamental and at high enough frequencies it would be outside the optical pass band of the lens. The fundamental of the square wave would have a frequency the same as the sine wave, but as is shown in equation (3.1) below, the fundamental would have an amplitude that was higher by a factor of $4/\pi$. So one would expect that the square wave would be detected at lower contrast. This is the opposite of the Lowry and DePalma results. They did a preliminary experiment and found that the relative visibility of sine and square wave gratings of high spatial frequency was, as well as they could measure it, exactly what would be expected if the optics filtered out the higher harmonic components of the square wave when they fell above the optical resolution limit. They then did investigations of the relative visibilities at frequencies below the CSF peak. This was based on the view that the harmonic components were not removed (or might even be relatively enhanced within the neural structures), so the square waves might become relatively more visible. John said that, making some assumption about the neural MTF (i.e. interpreting the contrast sensitivity function as an overall MTF of the optics + nervous system), it ought to be possible to predict the deviation of the threshold contrast ratio from its asymptotic high-frequency value. So he tried to make the predictions assuming various detection criteria for the effective internal image (e.g. peak-to-peak, maximum luminance slope, etc.) but couldn't get any predictions to fit their data. The contrast threshold ratio only deviated from the high-frequency value when the spatial frequency was quite a lot lower than would be predicted by any of those detection criteria. It looked as though the harmonic components were of no consequence until they could be independently detected. They then did some experiments with two-component compound gratings, which gave results consistent with this somewhat surprising notion that "components of the retinal image" might be detected independently. This ultimately led to the "multiple channels" hypothesis, which makes sense since the HVS obviously needed many channels in order to see – a single-channel device would simply represent a spatial frequency filter and a photometer.

Campbell and Robson (Campbell, 1968) described a number of psychophysics experiments to test their hypothesis of independent detection of spatial frequency components. The term "multiple channels" came later – but they understood that was what their work was all about. After developing a high-quality apparatus, their first experiment involved repeating the comparison of thresholds using sine wave and square wave test signals (gratings, as they are called in spatial vision research). A square wave, $F_{SQ}(x)$, of period X, can be described by its Fourier series components according to

$$F_{SQ}(x) = \frac{4}{\pi}\left(\sin\frac{2\pi x}{X} + \frac{1}{3}\sin 3\frac{2\pi x}{X} + \frac{1}{5}\sin 5\frac{2\pi x}{X} + \cdots\right). \quad (3.1)$$

The human CSF was known to have a peak at low frequency so if the channels hypothesis was correct, the threshold for detection of a square wave at higher frequencies should be determined by the threshold for a sine wave at the fundamental frequency. Furthermore, the ratio of square wave versus fundamental sine wave thresholds should be $4/\pi$ at higher frequencies. As is shown in Figure 3.3, this is precisely what they found. This suggested that frequency components separated by twice the fundamental frequency of the test pattern were detected independently.

Campbell and Robson did several other experiments to test their hypothesis. One was a comparison of thresholds for sine and rectangular waves of a wide range of duty cycles. Again, a rectangular wave can be described by a Fourier series. The contribution of the fundamental will depend on the duty cycle (the fraction of time the intensity is in its upper state). They found that the ratio of thresholds agreed extremely well with results predicted by their theory. Another experiment was based on suprathreshold discrimination. The idea is that a square wave and a sine wave should be indistinguishable until the first harmonic of the square wave is above threshold. The sine/square contrast ratio was set to $4/\pi$ and the subject was provided with a control that enabled him to adjust the contrast of both gratings together while maintaining a constant ratio of their contrasts. The display was switched repeatedly between the sine and square patterns until the observer was able to discriminate between the patterns.

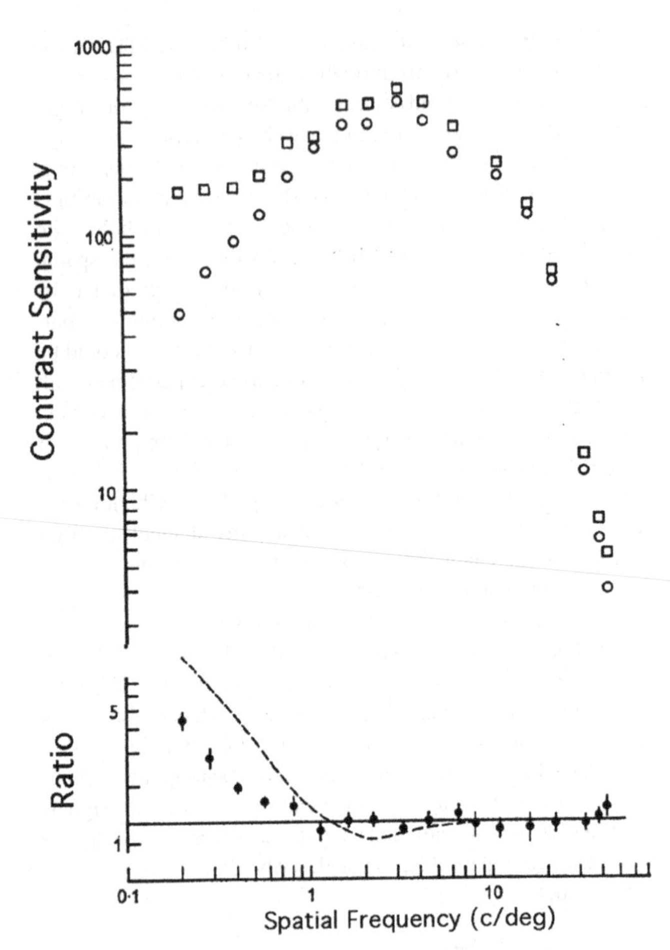

Figure 3.3 A plot of one subject's contrast sensitivity versus spatial frequency from the Campbell and Robson paper (Campbell, 1968), which introduced the concept of independent detection of spatial frequency components (visual channels). The open squares are for a one-dimensional square wave test pattern while the open circles are for a 1D sine wave test pattern. The filled circles at the bottom show the ratio of the contrast sensitivities at each frequency – the bars show +/– one standard error of the mean. The solid line is for a ratio of 1.273 ($4/\pi$), the amplitude of the fundamental sine wave in a unit-contrast square wave test pattern. The dashed line is a theoretical result for a simple peak detector mechanism. (Reproduced with permission from *The Journal of Physiology*.)

The results were precisely those predicted by their channels hypothesis. They repeated this experiment by pairing a sawtooth (triangular wave) and a sine wave at the fundamental frequency of the sawtooth. Again the results agreed with the predictions by their hypothesis. The publication of these experimental results set the stage for much of spatial vision research for the next two decades.

A conceptual representation of channels is shown in Figure 3.4. The CSF and the channels are assumed to be isotropic, so the dependence is on radial frequency. The isotropic assumption is in fact incorrect; it is known that orientation dependence exists but it will be neglected. This channels plot is a very simplified version of the real status of human visual channels. This will be discussed below.

The next step was an attempt to determine the bandwidths of the channels. A channel can crudely be thought of as having its input limited by a filter with a center frequency, f_C, and

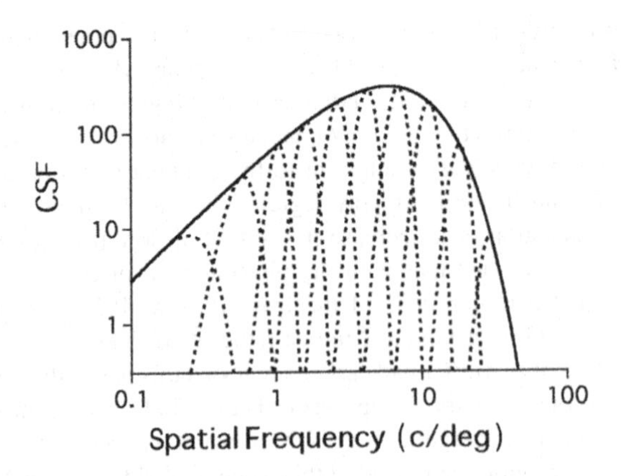

Figure 3.4 This is a conceptual illustration of the view of channels as it existed in the 1970s. The solid envelope curve represents the CSF in radial frequency while the dashed curves represent individual channels – Gabor functions (Gaussian-windowed sine functions) as an example.

50% transmission frequencies at af_C and bf_C, where a is less than unity and b is greater than unity. The "bandwidth" was usually estimated by determining how far apart in frequency two sine waves must be for the interaction between them to drop below a specified level. This estimation was done using a variety of methods. One involved having the subject adapt to a high-contrast sine wave pattern followed by a CSF measurement. The hypothesis was that the sensitivity of the channel centered on the adaptation frequency would be reduced and a dip in the CSF would occur near that frequency. The width of the dip would give a measure of bandwidth. This effect was found and the channel bandwidths were estimated to be about 1.8 octaves (full-width at half-maximum) independent of frequency. A variety of other experiments gave similar results. A number of functional forms for the channels were proposed and investigated. These included Gabor functions (sine waves with Gaussian envelopes) and difference of Gaussians (DoG functions), among others. To the best of my knowledge there is no consensus on a functional form that is good for all situations. It was also found that the CSF was not isotropic; there was a relative decrease along the diagonals in the 2D frequency domain.

3.5 SUBSEQUENT WORK

Another notable concept was proposed by Hauske *et al.* (Hauske, 1976) – matched filtering by the visual system. They used the technique of adding subthreshold 1D sinusoidal patterns while evaluating the detection of edges, lines, and bars. They assumed that the CSF could be considered to be a transfer function and then interpreted their results as suggesting that they could describe the visual system as a matched filter, which extracts an input signal contaminated with noise of specified spectral energy density. Another finding was that the psychometric function (detectability index, d', versus signal amplitude) was nonlinear. This finding was often interpreted as an indication of the existence of a nonlinear transducer function after the visual filters. It was also found that the psychometric function for amplitude discrimination was linear with d' proportional to the amplitude of the difference signal. The lower amplitude signal was usually referred to as the "pedestal." The plot of the threshold for detection of the difference signal as a function of the pedestal amplitude had a dipper shape because over a considerable range the discrimination threshold was lower than the detection threshold (with zero pedestal amplitude). The next step was to use 2D oriented channel filters, usually with four orientations. Finally methods for combining (pooling) the outputs of channels were investigated. Let c_k be the output of the kth channel of a total of K, let R be the combined output, and let b be a parameter to be determined. Then the equation (a Minkowski metric) for R was given the form

$$R = \left[\sum_{k=1}^{K} c_k^b \right]^{1/b} \tag{3.2}$$

This is a simplistic equation but one can think of it representing energy detection when b equals two and representing a response to just the channel with the maximum output when b goes toward infinity. As the century closed, the models used by vision research became more elaborate.

A consortium of ten laboratories is now doing a collaborative data collection, known as ModelFest, of observer thresholds for 43 stimuli – including Gabor functions with a variety of spatial frequencies, orientations, and Gaussian envelope widths, as well as a few Gaussian discs and a windowed sample of noise. The data (16 observers as of 2005) have been deposited in a public database to allow any researcher to test models. This is a massive undertaking and should give valuable results. Watson (Watson, 2005) used the latest data to evaluate a number of models and proposed a parsimonious standard model that is simple to implement and gives a very good fit (for practical purposes) to threshold data for all stimuli. The surprising thing about Watson's findings is that introducing spatial frequency channels into the models had very little effect on predicted thresholds. So his standard model does not include any channel decomposition.

3.6 RELATIONSHIP TO MEDICAL IMAGING PERCEPTION RESEARCH

We use visual channels in some signal detection models used for medical imaging research, and the number of channels is small. Any visual channels model with a small number of channels should be considered to be a highly simplified model for computational convenience rather than an accurate description of how the visual system really works. The word "small" must be emphasized. By small, I mean on the order of six different center frequencies and six orientations. Each retina has about 10^8 rods and 10^6 cones feeding "data" through intermediary cells to about 10^6 ganglion cells – each with its own receptive field. There are other specialized cells (with their own receptive fields) also collecting data from rods and cones. The ganglion cells and the other specialized cells all feed data into about 10^6 optic nerve fibers, which carry data (by nerve pulses) to the visual cortex. In the cortex, more than 10^8 cells then analyze the input data in particular ways and respond to particular features. When referred back to the retina, each cortical cell has its own receptive field in the sense that it responds best only to very specific features of stimuli reaching a specific small area on the retina. These features include spatial frequency and orientation as mentioned above (but with much finer gradations), direction of motion, temporal variation profile, disparity between inputs to the two eyes from the same location in the visual scene (for stereo vision), and color, among many other features. All of these cortical receptive fields are in fact channels that represent aspects of the visual scene. So there are, in fact, millions of channels available. Hubel, Weisel, and coworkers began their physiological studies of receptive fields and organization of cells by function in the cat's visual cortex in the early 1960s. They provided much of the early understanding of visual cortex function. Hubel and Weisel were awarded the 1981 Nobel Prize in Physiology or Medicine for their pioneering work. The totality of this analysis by this extremely large number of cells is then coordinated by higher levels in the brain to utilize the input to our two eyes. There is still very much research needed to understand the workings of this system and how information is coded. Some researchers estimate that as much as 25% or more of the brain is devoted to the process of vision. For those who want to learn about current views on visual physiology I recommend the excellent book, *Foundations of Vision*, by Wandell (Wandell, 1995). So to repeat (in conclusion), the concept of channels is much more complex than the very simple concepts illustrated in Figure 3.4, and also much more complex than the simple channels models we use in medical imaging.

REFERENCES

Barlow, H.B. (1953). Summation and inhibition in the frog's retina. *J Physiol*, **119**, 69–88.

Blackwell, H.R. (1946). Contrast thresholds of the Human Eye. *J Opt Soc Am*, **36**, 624–643.

Campbell, F.W., Robson, J.G. (1968). Application of Fourier analysis to the visibility of gratings. *J Physiol*, **197**, 551–566.

DePalma, J.J., Lowry, E.M. (1962). Sine-wave response of the visual system. II. Sine-wave and square-wave contrast sensitivity. *J Opt Soc Am*, **52**, 328–335.

Hauske, G., Wolf, W., Lupp, U. (1976). Matched filters in human vision. *Biol Cybern*, **22**, 181–188.

Kuffler, S. (1953). Discharge patterns and functional organization of the mammalian retina. *J Neurophysiol*, **16**, 37–68.

Lowry, E.M., DePalma, J.J. (1961). Sine-wave response of the visual system, I. The Mach Phenomenon. *J Opt Soc Am*, **51**, 740–746.

Schade, O.H. (1956). Optical and photoelectric analog of the eye. *J Opt Soc Am*, **46**, 721–739.

Schade, O.H. (1975) *Image Quality: a Comparison of Photographic and Television Systems*. Princeton, NJ: RCA Laboratories.

Wandell, B. (1995). *Foundations of Vision*. Sunderland, MA: Sinauer Associates, Inc.

Watson, A.B. (2005). A standard model for foveal detection of spatial contrast. *J Vision*, **5**, 717–740.

4

Signal detection theory – a brief history

ARTHUR BURGESS

4.1 INTRODUCTION

I will first describe early investigations of the effects of noise in images, starting with Albert Rose's 1948 fluctuations theory approach to modeling of thresholds as a function of luminance (photon fluence). Sturm and Morgan (Sturm, 1949) used the Rose model to estimate the effects of X-ray image intensification in fluoroscopy and the contrast-detail (CD) diagram phantom approach to measure human performance. The CD diagram shows the variation of the contrast required for signal detection at some defined reliability as a function of signal size. The work by Sturm and Morgan is discussed in detail in the next chapter, on the history of signal detection applications in medical imaging. The methods described by these authors became the standard evaluation approaches in medical imaging for the next 25 years. Next will be a brief introduction to signal detection theory (SDT) that was developed for radar applications in the early 1950s and was then applied to research in audition (hearing) in the late 1950s. Some suggestions were put forward in the early 1970s of how SDT might be applied in radiology.

This will be followed by experimental investigations in the early 1980s of human signal detection in uncorrelated (white) noise to determine whether our performance can be close to the limits predicted by SDT (the ideal observer concept) and whether humans could meet certain SDT constraints and requirements, such as the ability to do cross-correlation detection. At the same time, observer performance was being evaluated for a variety of tasks using simulated computed tomography (CT) images. Once performance with white noise was reasonably well understood, investigations shifted in the early 1990s to tasks involving spatially correlated noise (also known as colored noise) and statistically defined (but simulated) correlated backgrounds to determine whether humans could compensate for the spatial correlations (prewhiten) as would be done by the ideal observer. During this period, observer SDT models became more elaborate – based on image domain vector-matrix calculations including channels. There are very good chapters of mathematical reference material on SDT and applications in an SPIE handbook – see the introduction by Myers (Myers, 2000) and details about models by Eckstein *et al.* (Eckstein, 2000a). Background material on application of SDT to psychophysics can also be found in earlier classic books – see a collection of articles reprinted in Swets (Swets, 1964) plus detailed explanations by Green and Swets (Green, 1966). One can access brief discussions of some other material online using "detection theory" in Wikipedia.

4.2 ALBERT ROSE – THE BEGINNING

Early theoretical analysis of signal detectability in noisy images was based on the Rose model, first described in 1948 for analysis of human vision. Albert Rose (1910–1990, Figure 4.1) was a physicist who joined the staff at Radio Corporation of America (RCA) in 1935 after gaining his PhD from Cornell University. His first assignment was to design a new television camera with greatly increased sensitivity. His work led, in 1939, to the image orthicon television camera tube that was several hundred times more sensitive than previous cameras. It was first used for military purposes during World War II before becoming the commercial television broadcast camera of choice until 1965. He was widely honored and wrote several books including *Vision: Human and Electronic* (Rose, 1973). In 1942 Rose became interested in the relative limitations in the light sensitivity of photographic film, television pickup tubes, and the human eye (Rose, 1946). In this paper he presented two important proposals: (a) the use of an absolute scale (quantum efficiency) for evaluating imaging system performance, and (b) a simple model for evaluating signal detectability by human observers – now known as the Rose model (Rose, 1948). Quantum efficiency is the fraction of incident photons that are actually used by the imaging system. The Rose model allowed calculation of a signal-to-noise ratio (SNR) based on a particle fluctuation theory approach. This ratio describes the strength of the signal to be detected relative to a measure of the random fluctuations that interfere with signal detection. Details about SNR interpretation are discussed below. A few years later Peterson, Birdsall, and Fox (Peterson, 1954) presented a rigorous (ideal observer) SDT based on Bayesian probabilistic analysis. The Rose model is useful for simple calculations but has a very narrow range of validity. The Rose model will be outlined in this section. Burgess (Burgess, 1999a) presented details of the model and its relationship to the ideal observer approach.

Rose (Rose, 1948) used a particle-based fluctuations theory to evaluate human data from detection measurements at a variety of light levels. The data came from measurements published in 1928 by Cobb and Moss for luminances between 10^{-4} and 10^{-1} footlamberts (Ft-L, where 1 Ft-L $= 3.43$ candelas/m^2) plus measurements published in 1935 by Connor and Ganoung for luminances between 1 and 100 Ft-L. Note that the upper limit for scotopic (completely dark-adapted) vision is about 3×10^{-3} Ft-L and the transition mesopic region is from about 3×10^{-3} to about 0.3 Ft-L where both rods and cones contribute. The

The Handbook of Medical Image Perception and Techniques, ed. Ehsan Samei and Elizabeth Krupinski. Published by Cambridge University Press.

Figure 4.1 Photograph of Albert Rose (reproduced with permission from the IEEE).

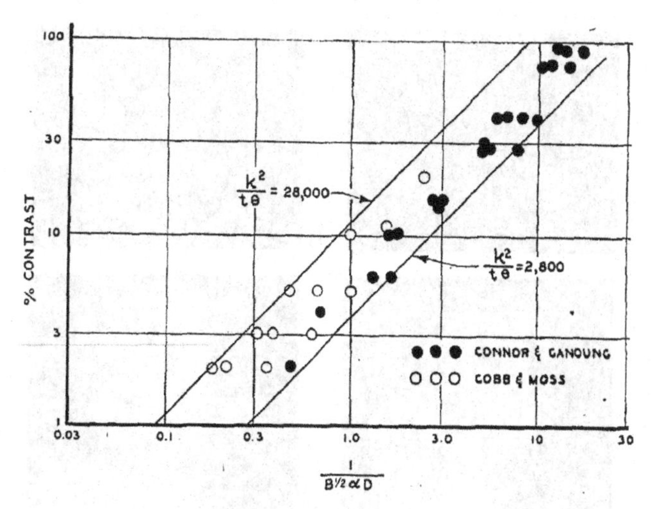

Figure 4.2 Contrast thresholds for signal detection in noiseless backgrounds under a range of luminances, B, from 10^{-4} to 100 Ft-L. The dark symbols are for scotopic vision and the open symbols are for photopic vision. The highest quantum efficiency is toward the lower right corner. The parameters are explained in the text. The diagonal lines indicate theoretical predictions for an ideal observer with the parameter products indicated. (Reproduced from Rose (1948) with permission from the Optical Society of America (OSA).)

photopic region is above the mesopic region and cone vision dominates. Rose assumed that observer vision was limited by photon fluctuation and used the following contrast threshold equation to compare to the data:

$$C_T^2 = \frac{5k^2 \times 10^{-3}}{tB\Theta\alpha^2 D^2} \quad (4.1)$$

The parameters are k, the threshold SNR which he had found to be about 5 based on observer experiments, the visual integration time t (in seconds), the scene brightness B (in Ft-L), the quantum efficiency of the eye Θ (with a maximum value of unity), the angle subtended by the test object α (in minutes of arc), and the eye lens aperture D (in inches). Rose produced a reduced plot to combine all data. He plotted contrast threshold as a function of the product, $\alpha D\sqrt{B}$, and the results (Figure 8 from his paper) are shown in Figure 4.2. The lines bound the data with the product $k^2/t\Theta$ between 2.8×10^3 and 2.8×10^4.

One can make an estimate of the quantum efficiency using reasonable estimates of 5 for the threshold SNR and 0.2 seconds for the integration time of the eye. This gives estimates of Θ approximately 0.5% for photopic vision and 5% for scotopic vision. These values are in not bad agreement with later results obtained by more sophisticated methods. They are obviously dependent on the selection of the values for k and t.

Later Rose (Rose, 1953) used the images shown in Figure 4.3 to demonstrate the maximum amount of information that can be conveyed by various known numbers of photons. Bob Wagner, Bob Jennings, and I had the pleasure of having dinner with Albert Rose in 1984, when he mentioned that he had had great difficulty in convincing the journal reviewers that the photographs in the figure were legitimate.

4.3 WHAT IS SIGNAL-TO-NOISE RATIO?

Noise is a ubiquitous problem in science and engineering. Anyone attempting to define an SNR must somehow characterize the strengths of both signal and noise. In this chapter I will assume that the term "noise" refers to truly random noise – as a result of a random process. The term "noise" is also used in contexts. This issue is discussed in my chapter on signal detection in medical imaging (Chapter 5), which includes patient structure backgrounds. The theory of signal detectability provides a definition of SNR that is not arbitrary (see Chapter 17 by Abbey and Eckstein). SNR calculations can be very misleading if not done correctly. The key point is that attempts to minimize the effect of noise on task performance, by filtering for example, must take into account the nature of the task. This leads to very different optimization strategies for different tasks. The Rose model approach, as we have used it in medical imaging, is valid for evaluation of detectability of sharp-edged signals at known locations in white noise added to a known mean background. This is referred to as the signal-known-exactly/background-known-exactly (SKE/BKE) task. The development of the Rose model is as follows. Let the mean signal (with area A) be described by an increment of $E[\Delta N_s]$ equal to $E[A\Delta n_s]$, photons above the background mean, where E[. . .] denotes expectation value. The background has a mean of $E[N_B]$ expected photons within the signal boundary and $E[n_B]$ expected photons per unit area. The background photon fluctuations have a standard deviation

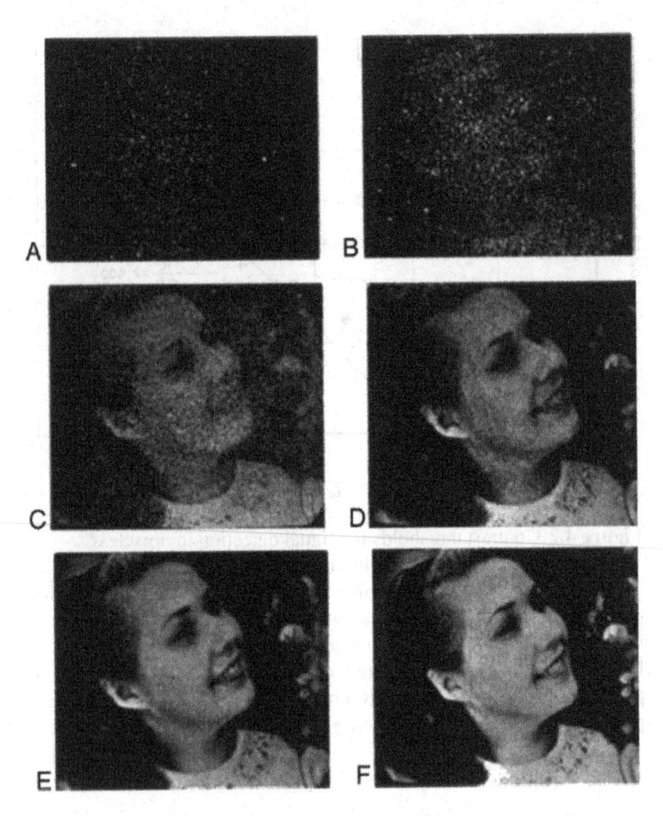

Figure 4.3 Picture used by Rose (1953) of woman with flowers, to demonstrate the maximum amount of information that can be represented with varying numbers of photons. The photon counts change by roughly a factor of seven between images. Each photon is represented as a discrete visible speck. The inherent statistical fluctuations in photon density limit one's ability to detect or identify features in the original scene. (Reproduced with permission of the OSA.)

per unit area of σ_{NB}. Let the signal contrast, C, be defined as $C = E[\Delta n_s]/E[n_B]$. Rose defined SNR as

$$SNR = \frac{mean}{\sigma_{NB}} = \frac{E[\Delta N_S]}{\sqrt{E[N_B]}} = \frac{AE[\Delta n_S]}{\sqrt{AE[n_B]}} = C\sqrt{AE[n_B]} \quad (4.2)$$

A number of assumptions are needed for the Rose model to be equivalent to modern SDT based on SKE/BKE cross-correlation detection in Gaussian noise:

(1) The Rose model neglects the fact that noise in the potential signal location has unequal variances for the signal-present and signal-absent cases. Hence the Rose model is an approximation that is valid only in the limit of low-contrast signals.
(2) The photon noise has a Poisson distribution, whereas the above SDT approach was based on a Gaussian distribution. We need to assume that photon densities are large enough that Poisson noise can be approximated by Gaussian noise with the same mean and variance.
(3) Rose used completely defined signals at known locations on a uniform background for his experiments and analysis, so the SKE and BKE constraints of the simple SDT analysis were satisfied.
(4) He assumed perfect use of prior information about the signal.
(5) He used a flat-topped signal at the detector plane (which is only possible when there is no imaging system blur). This assumption meant that integration of photon counts over the known signal area is equivalent to cross-correlation detection.

Rose used his model to assess the detectability of signals, in that he asked the question, "What SNR was required in order to detect a signal?" His approach was to use a constant, k, defined as the threshold SNR, and to suggest that the value of k must be determined experimentally. The signal was expected to be reliably detectable if its SNR were above this threshold. Once k is selected, the corresponding contrast threshold C_T is given by

$$C_T = \frac{k}{\sqrt{AE[n_b]}} \quad (4.3)$$

Using this definition of threshold, SNR has the unfortunate effect of mixing the measure of relative signal strength (SNR) with the proposed decision accuracy and the observer's decision criterion. However, it followed the convention of the day. People unfamiliar with SDT sometimes still use the threshold SNR concept. Some early publications had used a value of k equal to unity. Rose performed experiments and concluded that k was somewhere between 3 and 7, with a number around 5 being the most likely value. Subsequently, people in radiology using the Rose model discussed values in the range of 3–5. Given the confusion about the definition of a "threshold" and the subjective nature of the experiments (more on this below), it is not worth pursuing this topic further except to make one point. In 1980, I did experiments in collaboration with several other people and found human observer efficiencies near 50% for a variety of tasks (Burgess, 1981). Free response experiments described by Burgess (Burgess, 1999a) gave results consistent with a value of 5 for k when an arbitrary number of known discs were located anywhere within a defined region of about 1000 times the signal area. This immediately suggested that one could have used Rose's estimate of k of about 5 to predict about 50% human signal detection efficiency in white noise in the late 1950s when observer efficiency was first defined (Tanner, 1958).

4.4 INTRODUCTION TO SIGNAL DETECTION THEORY

4.4.1 The likelihood ratio approach

I will first give a brief summary of the main principles of SDT and then describe some history. A more complete description of the mathematics is given by Myers (Myers, 2000) and by Abbey *et al.* (Abbey, 2000) in the same book. The analytical approach depends on the decision task (the intended use of the image) – which is usually either classification or estimation based on data. A simple binary classification task will be discussed here since it is the easiest to understand. Assume that we are given a noisy digital image described by the column vector, **g**. In a binary decision task it is known that one of two situations (events, j, with j = 0 or 1) is true. For example it may be that the alternative events are: a signal is present or absent, or alternatively one of two signals (A or B) is present. The observer will consider two hypotheses, h_j, about which event actually occurred given the conditional *a priori* probability density of the image data, $p(\mathbf{g} \mid h_j)$, and the *a priori* probability density, $p(h_j)$, for each hypothesis being correct. The observer's problem is to decide which hypothesis is most likely to be correct. SDT gives a method of determining the best solution to this decision problem, which is used by the ideal observer. The analysis is based on Bayes' theorem. It can be shown that the *a posteriori* probability density, $p(h_j \mid \mathbf{g})$, of a particular hypothesis, h_j, being true is described by

$$p(h_j \mid \mathbf{g}) = \frac{p(h_j)p(\mathbf{g} \mid h_j)}{p(\mathbf{g})} = \frac{p(h_j)p(\mathbf{g} \mid h_j)}{\sum_{\text{All } j} p(h_j)p(\mathbf{g} \mid h_j)} \tag{4.4}$$

Frequently one wants to evaluate the evidence provided by the event independently of the *a priori* probability of the hypotheses or it may be that the two events have equal *a priori* probability. Then the evidence for the correctness of the hypotheses can be summarized by the likelihood ratio, defined as

$$L(\mathbf{g}) = \frac{p(\mathbf{g} \mid h_0)}{p(\mathbf{g} \mid h_1)} \tag{4.5}$$

Any monotonic function of the likelihood ratio can also be used to evaluate the evidence, since such transforms will not change decisions. It is often convenient to use the log(likelihood ratio), denoted by $\lambda(\mathbf{g})$, as a decision variable. Since noise is present, the decision variable will be a random variable. The value of the decision variable is then compared to a criterion value to arrive at a choice of hypothesis. The criterion value will affect the false alarm and signal miss rates. So the decision maker will select a criterion that gives desired rates for the two types of errors, usually based on *a priori* probability of the presence of a signal and the costs associated with the two kinds of errors.

The calculation of the likelihood ratio is only straightforward for a limited range of tasks. The simplest task concerns a decision about the presence or absence of a signal under SKE/BKE conditions. The calculation is particularly tractable if the noise is additive and uncorrelated with a Gaussian probability distribution. For this situation, analysis based on the log(likelihood ratio) indicates that the optimal strategy is to use the signal itself as a template to cross-correlate with the data to obtain a decision variable. This procedure is also known as matched filtering.

The likelihood ratio approach was first described by V.A. Kotelnikov (1908–2005), a Russian communications engineer and information theory pioneer. He is best known for having discovered, independently of others, the Nyquist–Shannon sampling theorem. Kotelnikov began investigating signal detection in noise in 1946 but was unsuccessful in getting his work published in the Russian literature (apparently the editors claimed lack of interest to communications engineers). Finally his dissertation was published in 1956 (in Russian), and subsequently translated from the Russian by R.A. Silverman and published in English (Kotelnikov, 1959).

4.4.2 The matched filter

The prewhitening matched-filter model for SKE signal detection comes up very frequently in the medical imaging literature. It was first developed by D.O. North (North, 1943, 1963) of RCA Laboratories to analyze optimization of radar system detection performance during World War II. Van Vleck and Middleton at the Harvard Radio Research Laboratory nearly simultaneously discovered it. North's main activity was developing electronic tubes. I was told in the 1980s by someone at RCA that he was informed about the detection optimization problem, solved it in one afternoon and never regarded it as a particularly significant achievement. His paper is a model of clarity of thought and a pleasure to read. North's work was first described in an RCA internal document, which was subsequently distributed to generations of graduate students in mimeographed form. The copies eventually became almost unreadable, so the report was finally published in a refereed journal in 1963 so a clean copy would be available for reference. Related radar reception work was done at the MIT Radiation Laboratory during World War II.

North asked the question, what is the best filter to use at the radar receiver so that its output gives the best contrast between signal and noise? In other words, if the signal is present, then the filter should give a sharply peaked output, whereas, if the signal is not present, the filter should give as low an output as possible. It was assumed that the filter designer knew the temporal profile and arrival time, t_0, of the signal (a short radar echo pulse of a particular shape). So the problem is to detect a known signal, $s(t)$, in a waveform, $x(t)$, with additive random noise, $n(t)$, by use of a filter with an impulse response, $h(t)$. The filter input and output functions are

$$\begin{aligned} &\text{input:} \quad x(t) = s(t - t_0) + n(t) \\ &\text{output:} \quad y(t) = s_o(t - t_0) + n_o(t) \\ &\text{where} \\ &\quad s_o(t - t_0) = h(t)^* s(t - t_0) \quad \text{and} \quad n_o(t) = h(t)^* n(t) \\ &\quad \text{and } {}^*\text{denotes convolution.} \end{aligned} \tag{4.6}$$

North chose to select the optimum filter by maximizing the ratio of the signal amplitude to the noise amplitude of the output at some instant in time, based on the known arrival time of the signal. Since the noise is a random variable, it was characterized by the mean-square value of its amplitude, $E[n_o^2(t)]$. This gave the following quadratic ratio, $\rho(t_m)$, to characterize the relative strength of output signal and noise at t_m:

$$\rho(t_m) = \frac{s_0^2(t_m)}{E[n_o^2(t)]} \tag{4.7}$$

Note that use of this equation does not consider or guarantee that the output signal will in any way resemble the input signal. The only concern is to maximize a scalar quantity at t_m. The analysis is easiest to do in the frequency domain, so let capitalized terms represent the Fourier transforms of the lower case terms:

$$\begin{aligned} S_o(f) &= H(f)S(f) \quad \text{and} \quad P_o(f) = E[\mid H(f) \mid^2 N(f)^2] \\ &= \mid H(f) \mid^2 P(f) \end{aligned} \tag{4.8}$$

$P(f)$ and $P_o(f)$ are the power spectra of the input and output noise. The power spectra can have arbitrary form; there is no need for the noise to be uncorrelated (white). The goal is to determine the particular filter that maximizes the ratio $\rho(t_m)$. It can be shown that

$$\rho(t_m) = \frac{\left[\int H(f)S(f)df\right]^2}{\int H^2(f)P(f)df} \quad \text{and for } \rho_{max}(t_m), \qquad H_{MF}(f) = \alpha S^*(f)/P(f) \tag{4.9}$$

The parameter α is an arbitrary scaling factor that can be taken into consideration when selecting a decision criterion. The function $S^*(f)$ is the complex conjugate of $S(f)$. Furthermore, the optimum measurement time equals the known arrival time. This is a description of the prewhitening matched filter (PWMF), by direct (one-step) filtering of the input. Prewhitened matched filtering can also be understood by consideration of the following two-step procedure. The first step involves a filter, $H(f)$, that converts the input fluctuations to white noise. This filter is described by $H_1(f) = 1/\sqrt{P(f)}$. When the filter is applied to the input, the result is $x_2(t) = h_1(t)^*s(t - t_0) + w(t)$, where $w(t)$ is white noise. This is the so-called "prewhitening" procedure. The signal, $s_2(t)$, in $x_2(t)$ has a different profile than the one in $x(t)$ because of the filtering. The optimum strategy is now to filter $x_2(t)$ using a second filter, $h_2(t)$, optimized to detect the revised signal profile. It can be shown that the optimum second filter and two-step filtering results are

$$H_2(f) = \alpha S^*(f)/\sqrt{P(f)} \quad \text{so}$$
$$H_{MF}(f) = H_1(f)H_2(f) = \alpha S^*(f)/P(f) \qquad (4.10)$$

So the one-step and two-step total matched filters are the same. The two-step description makes the notion of "prewhitening" directly observable. It is hidden in the one-step procedure. The term "observer template" is frequently encountered in the medical imaging signal detection literature. What it means is this. Rather than doing frequency domain matched filtering, the observer can use a template, $h(t)$ in the notation of this section, to cross-correlate with image data. The optimum template is the inverse Fourier transform of the prewhitening matched filter. One other point is worth noting. In receiver operating characteristic (ROC) or "yes/no" decision tasks where a signal may or may not be present, the signal peak amplitude (or profile scaling factor) and the template (or filter) scaling factor α must be known to the decision maker in order to select an appropriate decision criterion. This is not the case in M-alternative forced-choice (MAFC) experiments where only one (known) signal is present together with M–1 known and equally detectable alternatives. The optimum decision strategy is to select the alternative with the highest likelihood ratio value. The template scaling parameter does not come into play and can be any non-zero value. It does not need to be known to the decision maker since it is applied to all M alternatives.

4.4.3 The ideal observer

More complete ideal observer analysis based on the likelihood-ratio approach was developed by Peterson *et al.* (Peterson, 1954), also to evaluate optimization of radar systems. It was subsequently adapted for analysis of human auditory and visual signal detection and discrimination analysis. This approach is also described in some detail by Burgess (Burgess, 1999a). For simple situations the matched-filter approach and the likelihood-ratio approach give identical results. For more complex situations, the likelihood-ratio approach brings the full power of probability theory analysis to the problem and is much superior. Woodward and Davies (Woodward, 1952) did a related development of SDT using Jeffrey's theory of inverse probability.

Tanner and Swets (Tanner, 1954) did visual signal detection experiments, which gave results that were inconsistent with the previously common view that visual thresholds represented some "hard" lower detectability limit. This latter theory is known as "high threshold" theory – the view was that detection probability remained at zero until some particular signal amplitude was reached. Tanner and Swets found that visual signal detection probability increased monotonically from zero as signal amplitude was increased from zero. Tanner and Swets also showed the equality of visual signal detectability values, d′, determined by "yes/no" ROC and multiple location forced-choice methods. These results were regarded as an important validation of the application of SDT to human subjects and a demonstration that results of different psychophysical tests can have meaning in spite of different procedures.

4.4.4 Observer efficiency

The next classic paper in SDT was the introduction of the concept of observer efficiency by Tanner and Birdsall (Tanner, 1958). Let E_I be the signal energy required for the ideal observer to detect a signal at a given performance accuracy (90% correct in a two-alternative forced-choice [2AFC] task, for example). Let E_T be the signal energy required for the observer under test to detect the same signal at the same accuracy under the same conditions. They calculated efficiency at the selected decision accuracy (or performance level, P) using a ratio of these two energies. It should be carefully noted that their definition could be applied to tasks where the signal is only known in the statistical sense (SKS) as well as for SKE conditions. Barlow (Barlow, 1962) calculated efficiency in a different manner, using d' values for the two observers at a selected signal amplitude, A. It should be carefully noted that the two definitions are only equivalent in the case where the values of d′ are proportional to signal amplitude. This situation is rarely true for human observers and is not true for the ideal observer if the signal is not known exactly but is only defined statistically. The two definitions of efficiency are

$$\text{(Tanner \& Birdsall)}\ \eta = \left(\frac{E_I}{E_T}\right)_P, \quad \text{(Barlow)}\ \eta = \left(\frac{d'_T}{d'_I}\right)_A \qquad (4.11)$$

In the context of radiological or nuclear medicine imaging, the Tanner and Birdsall definition can be thought of as comparing how many photons the two observers need to perform the same task with the same accuracy. For MRI it would be the ratio of image acquisition times, all else being equal. The efficiency approach was used for evaluation of a few auditory signal detection and discrimination experiments in the late 1950s. This was possible because auditory signals and noise were easily generated and recorded. Efficiencies were found to be very low (1% or less) so the ideal observer model was never considered as a starting point for modeling of hearing. The low auditory efficiency is not surprising because the observer must make precise use of a known waveform and arrival time to do cross-correlation. We are not able to do this. There were no investigations of visual efficiency because the required display hardware was not available.

The efficiency approach was first applied to vision by Barlow (Barlow, 1978) using computer-generated random dot images. He was attempting to measure cortical spatial tuning curves that would be analogous to the contrast sensitivity curves for luminance (grayscale) patterns. The basic idea was that the small dots were reliably transduced by the retina and that any spatial tuning that was found might be due to receptive fields in the visual cortex. Barlow used Poisson statistics to generate patterns and chose observer efficiency as the measure of pattern detectability. He consistently found efficiencies of approximately 50%, independent of pattern size or shape. Subsequently, Burgess and Barlow (Burgess, 1983) used random dot patterns with Gaussian statistics to allow independent adjustment of the mean and the variance. This, in turn, separated effects due to sampling efficiency (which is an estimate of the accuracy with which the appropriate signal detection template is used and positioned) from effects due to intrinsic observer variability (internal noise). Their measurements suggested that virtually all of the reduction in human dot detection efficiency was due to centrally located internal noise. Sampling efficiency appeared to be approximately 100% for a wide range of random dot patterns. The word sampling was selected to be consistent with the original formulation of decision accuracy by Fisher. The work by Burgess and Barlow was actually done in 1979 and then submitted for publication. Unfortunately the journal moved its editorial office and all records of the manuscript submission and referee's comments were lost. This oversight was subsequently corrected and the paper appeared in print several years later.

4.5 ONE-DIMENSIONAL WHITE NOISE EXPERIMENTS

Pollehn and Roehrig (Pollehn, 1970) were the first to measure the contrast sensitivity function (CSF) in the presence of visual noise. They generated 1D sinusoidal signals with added 1D white noise in each raster line for display on a linearized television monitor. The noise will be white in 2D because noise in adjacent raster lines is uncorrelated. They made a number of measurements using two methods. One was the method of limits where the signal amplitude was decreased from a large value until the observer (subjectively) reported that it could no longer be "seen." The amplitude was lowered further and then increased until the observer reported that it could be "seen" again. The descending and ascending thresholds were then averaged. They used a second method for three selected frequencies. The observer was shown a randomly selected test pattern of one of these frequencies with amplitudes near the previously estimated thresholds and asked which pattern was present. Thresholds were defined as the amplitude that gave 50% correct responses. There was good agreement between the results for the two methods. They did a variety of experiments with six observers. CSF values for five noise levels are shown in Figure 4.4. They did CSF measurements four months apart to evaluate reproducibility. They also checked for consistency by using several viewing distances and varied the number of cycles in the sinusoidal signals. When the image noise level was

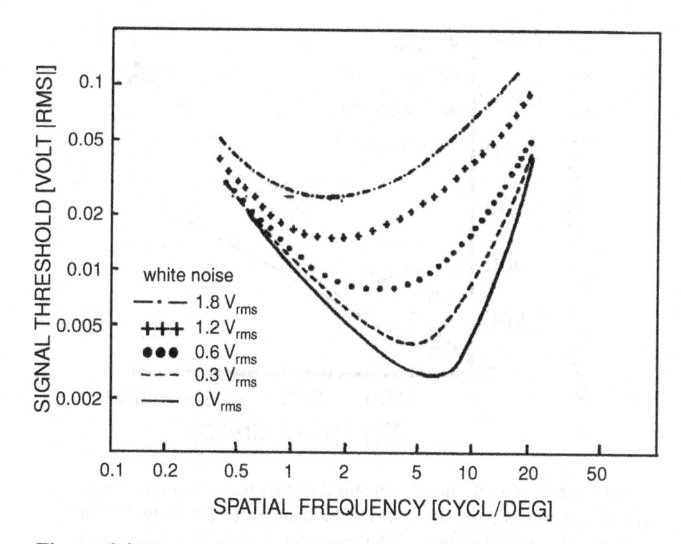

Figure 4.4 Measured amplitude thresholds from Figure 7 of Pollehn and Roehrig (1970). The data are averages (for six observers) for detection of 1D sinusoidal signals in added 2D white noise. Such a plot is an unnormalized inverse of the 1D CSF. Note that as the RMS noise level increases, the minima shift to lower frequencies and the curves flatten. (Reproduced with permission of the OSA.)

varied, they found that the signal threshold values S_{TH} as a function of the root mean square value of the image noise, N, could be described using

$$(S_{TH})^2 = a\left(N^2 + R^2\right) \tag{4.12}$$

They interpreted the parameter R as internal noise. The values of the parameters a and R varied with spatial frequency. They also measured the CSF with 1/f noise in the horizontal direction of the television scan lines. The noise would be uncorrelated (white) in the vertical direction so its influence is hard to interpret.

4.6 TWO-DIMENSIONAL WHITE NOISE EXPERIMENTS

Research with 2D images only became possible with the advent of image "frame buffers" (now called display memory) in the 1970s. But they were very expensive at first. A 640 × 512 display system cost about $100,000 in 1980 ($270,000 in 2007 dollars), so very few university laboratories had one. Burgess *et al.* (Burgess, 1981) did the first investigations of decision efficiency with computer-generated grayscale images using a 640 × 512 system. We used uncorrelated Gaussian noise and demonstrated that humans could perform amplitude discrimination tasks with efficiencies well over 50% with sampling efficiencies as high as 80%. We used amplitude discrimination tasks because we had found in preliminary experiments that the detection psychometric function (d′ versus signal amplitude) was nonlinear at low amplitudes and the effect became increasingly important for sine wave signals as spatial frequency increased. We found that for amplitude discrimination tasks the psychometric functions were both linear and proportional to signal amplitude. A discrimination threshold was defined as the signal energy required to obtain d′ equal to 1 (76% correct in a 2AFC task). Four

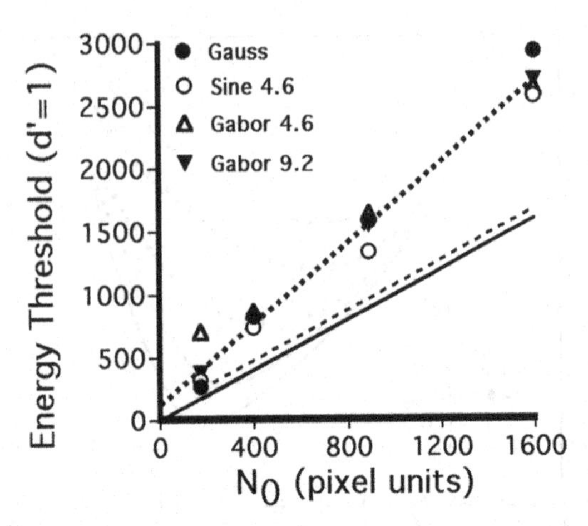

Figure 4.5 A plot of energy threshold results for signal amplitude discrimination as a function of white noise spectral density, N_0, in comparable units. The human data are averages for three observers and 1024 trials per observer per datum. The periodic signal frequencies (4.6 and 9.2) are given in cycles/degree. The solid line shows results expected for the ideal observer. The dashed line parallel to it includes additive internal noise. The dotted line through the human data has a slope of 1.61 indicating a sampling efficiency of 62%. The highest absolute discrimination efficiency in the plot is 59% (for the sine wave at N_0 equal to 1600). The highest sampling efficiency is 80% (for the 4.6 cycles/degree Gabor function). (Plot recreated using data obtained from work related to Burgess *et al.* (1981).)

signals (a compact 2D Gaussian disc, a 4.6 cycles/degree two-cycle sine wave, and two Gabor functions) were used and four noise spectral densities were used. The signal energy threshold results for amplitude discrimination as a function of image noise spectral density, N_0, are shown in Figure 4.5. Observer efficiencies depended on the noise spectral density. The highest values were about 50% obtained at the highest noise spectral density. For this spectral density the standard deviation of the image pixel noise was 15.6% of the 8-bit display range. We found efficiencies of about 50% for detection of aperiodic signals such as sharp-edged discs and compact 2D Gaussian functions. Efficiency for detecting sinusoidal signals was in the 10–25% range. The decrease in sinusoidal detection efficiency is probably due to an inability to make precise use of absolute phase information. A phase uncertainty (jitter) of as little as a quarter of a cycle dramatically reduces sine wave detection accuracy. At the peak of the contrast sensitivity curve (roughly four cycles/degree), a quarter of a cycle corresponds to approximately the full width at half maximum of the optical point-spread function of the eye.

The fact that the regression lines through the observer data in Figure 4.5 do not pass through the origin indicates that observers have internal noise. It is not likely to be neural noise. The more probable explanation is that we cannot do mechanical calculations of likelihood ratios, which leads to noisy decision-making. The fact that the regression lines are not parallel to the ideal observer line indicated that we do not do perfect sampling of the data. For example, our template may vary in size, shape, or position from trial to trial even though we are given complete information about the signal and its possible positions. These variations would introduce additional noise into the decision process as discussed below.

The measures used in the above plot will now be discussed. White noise digital images with a nominal sampling distance of unity have a Nyquist frequency of 0.5 and a two-sided spectral density, N_0, equal to the pixel variance. The energy threshold concept arises as follows. The task is done under SKE and BKE conditions with white noise. Centered on a known location (x_0, y_0), let the image data be described by $g(x - x_0, y - y_0) = As(x - x_0, y - y_0) + n(x - x_0, y - y_0)$ where the signal is described by $As(x - x_0, y - y_0)$ and the white noise is described by $n(x - x_0, y - y_0)$. The optimum ideal observer detection strategy on a trial-by-trial basis in a 2AFC task with white noise is to use the signal as a template, $t(x, y)$, and cross-correlate it with the data at the potential signal locations, $g(x - x_0, y - y_0) \otimes t(x, y)$. Recall that convolution of two functions in the space domain can be described by multiplication of their Fourier transforms in the frequency domain. When cross-correlation is done in the space domain, the Fourier domain multiplication involves the complex conjugate of one of the functions. So cross-correlation using the signal as a template can be described in the frequency domain as $G(u, v)T^*(u, v) = A^2S(u, v)S^*(u, v) + N(u, v)T^*(u, v)$, where * indicates the complex conjugate. The noise is white with spectral density N_0 and statistics that are independent of location. The noise can be assumed to have zero mean without loss of generality. It can be shown using equation (4.18) below that d′, or equivalently the SNR for the ideal observer with white noise, is given by

$$(d')^2 = \left(\frac{1}{N_0}\right) \iint [As(x, y)]^2 \, dxdy = \left(\frac{1}{N_0}\right) \iint A^2 \mid S(u, v) \mid^2 dudv = \frac{E_S}{N_0} \tag{4.13}$$

The integral over the quadratic content of the signal is known as its energy, E_S. One important observation is that for the ideal observer doing an SKE task, d′ will be proportional to signal amplitude. This is also true for all linear models of signal detection.

One point on terminology should be noted. In SDT-based work, signals are described by amplitude profiles rather than contrast. In radiology, the term "contrast" can have a number of meanings and therefore can be ambiguous. It is often calculated by normalizing the signal profile by the local image mean value, L_0, in its vicinity. So the signal contrast profile is defined by $c(x - x_0, y - y_0) = s(x - x_0, y - y_0)/L_0$. If the contrast representation is used for the signal, then it must also be used to normalize the noise fluctuations. Since N_0 describes the squared noise fluctuations, the $(L_0)^2$ terms in both the numerator and denominator of the equation cancel. This is true for all model observers so there is no benefit obtained from using the contrast representation in SDT calculations. This is a useful property since "contrast" values in digital images are arbitrary; they are determined by details of the display window and level settings.

After the brief Burgess *et al.* (Burgess, 1981) paper was published, we unsuccessfully submitted a much more detailed paper to another journal. The associate editor for vision (a physiologist) was not a fan of the application of linear systems theory or SDT to visual perception. The journal editor also suggested that, in spite of the high efficiencies, we had not demonstrated

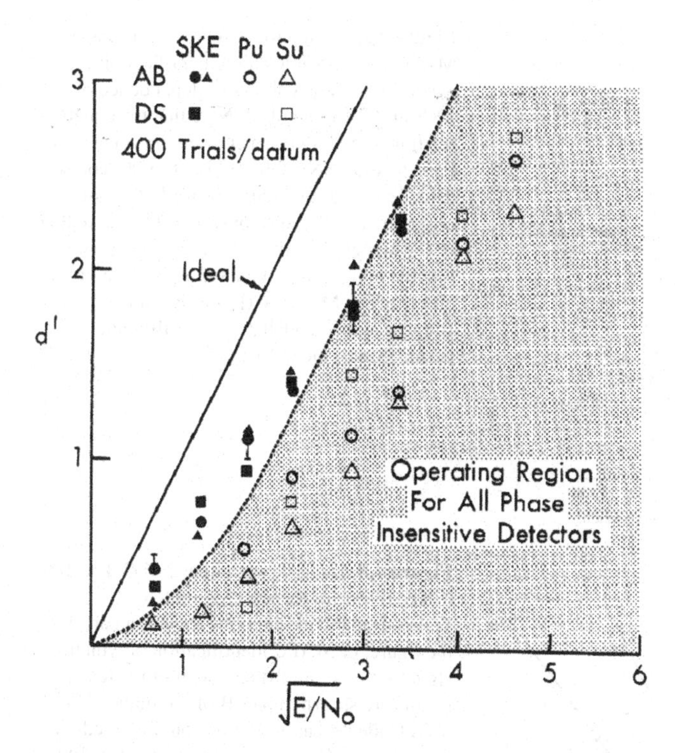

Figure 4.6 Detectability of a two-cycle sine wave (4 cycles/degree) in white noise as a function of SNR [sqrt(E/N_0)]. The shaded area indicates performance that is worse than the model observer that is ideal except for the fact that it cannot use sine wave phase information for detection (known as an autocorrelator). The data points are human results explained in the text. Note that humans do better than the autocorrelator for the SKE condition when the SNR is below about 2. (Reproduced from Burgess and Ghandeharian (1984a) with permission of the OSA.)

that humans could do cross-correlation detection. At the same time I had applied to the Medical Research Council of Canada for funding to continue the work. One of the grant reviewers took exception to the fact that I had ignored the autocorrelation theory of vision, which had been proposed in the early 1970s. An autocorrelator is an energy detector, which is not able to use sine wave phase information, and therefore performs worse than a cross-correlator under SKE conditions. On the next submission of the grant, I discussed this point and got funded. However, because of the above objections, I undertook a program of investigations about human observer abilities: use of phase information for sine wave detection and discrimination, detection in multiple-alternative locations, detection of Walsh–Hadamard signals with and without reference signals, and identification of these signals from an M-alternative menu. These investigations (all done using white noise) convincingly demonstrated that we performed better than the best possible autocorrelator and were consistent with the view that we were able to perform as suboptimal cross-correlators using a Bayesian decision strategy.

The sine wave detection study by Burgess and Ghandeharian (Burgess, 1984a), was designed to determine whether we could do better than the very best possible autocorrelator performance. The results are shown in Figure 4.6. Two human observers did experiments with three conditions. The first condition was based on a signal known exactly but variable (SKEV) paradigm. The profile of the two-cycle sine wave (4.6 cycles/degree) was made known to the observer but varied from trial to trial. In one case, the signal phase was selected randomly for each trial and the reference signal shown to the observer was simultaneously changed. In the other two conditions the signal was only defined statistically for the human observers (the SKS paradigm). In one case (points labeled PU), the observer was uncertain about signal phase because it varied randomly from trial to trial but the phase of the available reference signal never changed. In the final condition there was signal starting position uncertainty (points labeled SU). The performance of the autocorrelator was determined by Monte Carlo experiments. It is clear in Figure 4.6 that the human observers, under SKE conditions, do better than the best possible phase-insensitive detector for SNRs below about 2. There was another experiment involving discrimination of two-component sine wave signals that also clearly demonstrated that we could use phase information. These phase experiments were the first steps in demonstrating that we can do Bayesian cross-correlation signal detection similar to the ideal observer.

The second study by Burgess and Ghandeharian (Burgess, 1984b) was designed to see if human observer performance for signal detection with M-alternative and statistically independent locations could be predicted from 2AFC detection results using the standard SDT calculation. The additional locations gave additional opportunities for a noise-only location to yield a cross-correlation result that was higher than that obtained at the actual signal location. This effect could be accounted for using standard probability theory methods. This means that the value of d′ in an MAFC experiment depends on both SNR and the value of M. The signal was a disc (eight pixel diameter) in white noise. The number of locations ranged from 2 to 98. The results are shown in Figure 4.7. Part A of the figure shows the values of d′(M) as a function of SNR. The fact that the data fall on a common line indicates that M-alternative results can be predicted from 2AFC results. There is also a small offset; the line does not go through the origin so human d′ results are not proportional to SNR. This feature is found in all human experiments. Two observers did this experiment but only the results for one are shown. Part B of the figure shows SNR threshold variation for 90% correct decisions as M increases. The human results all fall on a curve that is $\sqrt{2}$ higher than the ideal observer for all values of M. This is just another way of presenting the data in part A, but makes the point about performance prediction from SDT more clearly. The ratio of performances indicated that the human absolute efficiency is 50%. The results of this study indicate that humans are able to behave in a Bayesian manner, making use of prior knowledge about possible signal locations in a way that can be predicted by SDT.

The third study by Burgess (Burgess, 1985) was also based on the Bayesian use of prior knowledge issue. The signal set was a collection of 2D Walsh–Hadamard functions, which are checker-board-like as shown in Figure 4.8. The functions are orthogonal so alternative decisions are statistically independent. A number of experiments were done but only two will be discussed here. One experiment was 2AFC detection using an SKE but variable paradigm. The signal was randomly selected from the ten highlighted choices in part A of the figure and a reference version was shown to the observer during the decision trial. The second experiment was M-alternative signal identification. One

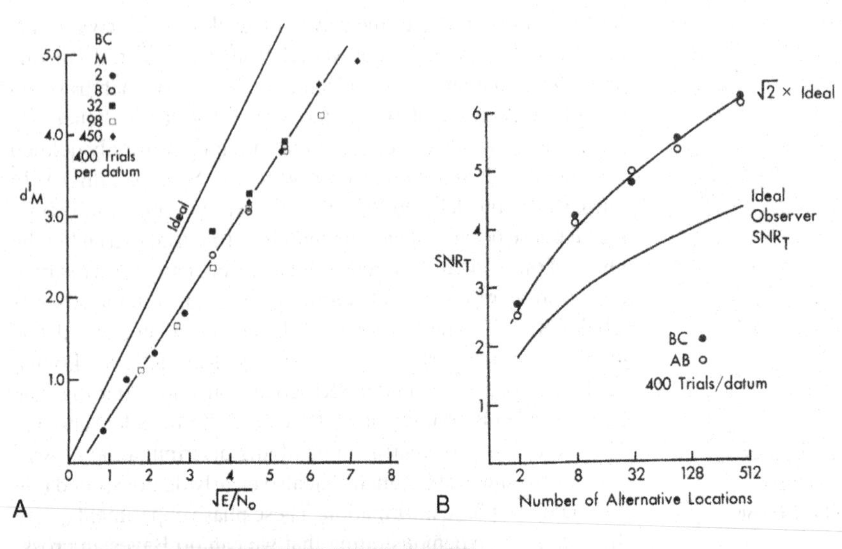

Figure 4.7 Results for observer BC doing an MAFC disc signal location identification experiment. Part A shows d' dependence of SNR and the value of M. Note that the equation relating d' and SNR depends on the value of M. Part B shows SNR thresholds (for 90% correct detection) as a function of M for both human observers and the ideal observer. The curve that is $\sqrt{2}$ higher than ideal observer data shows that the humans are maintaining 50% detection efficiency as M varies. (Reproduced from Burgess and Ghandeharian (1984b) with permission of the OSA.)

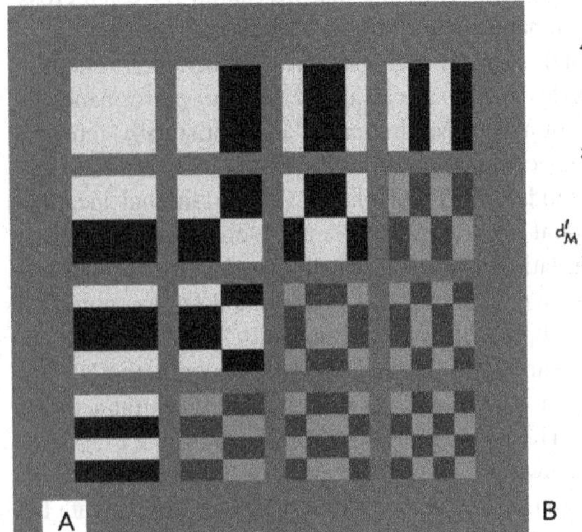

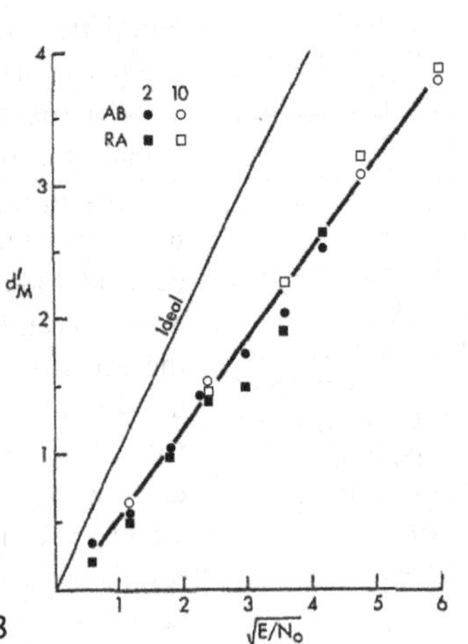

Figure 4.8 The task was either 2AFC detection (SKE) of one of the ten highlighted Walsh–Hadamard functions shown in part A of the figure, or MAFC identification of which signal was present from a menu of the ten. The results are shown in part B of the figure. The MAFC identification results could be predicted from the 2AFC results using the standard SDT calculation. (Reproduced from Burgess (1985) with permission of the OSA.)

signal was randomly selected from the set and the observer's task was to identify the signal and indicate his choice using the reference menu of all possibilities.

Human efficiencies were consistently high for all three studies: about 50% for detection of simple aperiodic signals, about 35% for Walsh–Hadamard signal detection and identification, and in the vicinity of 10–25% for SKE detection of sine waves. The results of the series of studies presented clear evidence that our ability to do signal detection tasks in white noise could be described by an observer model that could do cross-correlation detection and make use of prior information in the manner that a Bayesian decision maker would. Our performance efficiency was not 100%, but 50% for a simple task is quite good.

The final experiments in this series, by Burgess and Colborne (Burgess, 1988), were designed to study one possible source of inefficiency – internal noise. This was done in two ways. In the first method (two-pass), observers viewed the sequence of 2AFC image pairs twice with an interval of many weeks between observation trials. The signal would be in the same field in identical pairs of noise fields each time. This method allowed for analysis of observer consistency – a mechanical observer with no data acquisition or decision noise would make identical decisions on the two viewings of the image pairs. Note that the mechanical observer need not be ideal. An estimate of observer internal noise could be obtained using the probability of correct response and the probability of decision agreement. The second method was based on twinned noise. The noise fields for the two alternatives were identical. The mechanical observer would subtract the two image fields and the signal would be revealed in a noiseless background 100% of the time. A second experiment was then done using different noise backgrounds in the two alternative images. Relative performance results for the two cases could be analyzed to yield estimates of internal noise. All experiments were done using the four image white noise variances shown in Figure 4.5. The results were surprising. We found that there were two noise components. The first was the internal noise indicated by the nonzero intercept in the above Figure 4.5. This was independent of the noise level in the displayed images and was referred to as static noise with spectral density, N_s. The value of the static noise spectral density can be estimated using the intercept of the regression lines with the abscissa in Figure 4.5. The other noise component had a variance that was proportional to the variance of the image noise (with a constant fraction of about 64%). Since this component depended on noise in the image, it was referred to as induced noise. When image noise is white with spectral density, N_0, this induced noise can be included in observer models by using an effective image noise spectral density, $N_{eff} = (1 + \rho^2)N_0$, where ρ^2 equals 0.64. This means that observer absolute efficiency will be at most about 60%. This is consistent with a number of results. A simple model for human performance with white noise can be obtained by modifying the ideal observer equation, (4.13), to include these two sources of internal noise. The result is

$$(d'_H)^2 = \left(\frac{1}{(1+\rho^2)N_0 + N_s}\right) \iint A^2|S(u,v)|^2 dudv \quad (4.14)$$

Another point about this induced internal noise component is that it influences the slope of the regression line in Figure 4.5. Induced internal noise cannot be distinguished from suboptimal sampling efficiency on the basis of experiments to the date of this investigation (Burgess, 1988). In fact, induced noise and suboptimal sampling may actually be two ways of talking about the same phenomenon – inconsistent use of a template for image data collection. The topic of induced noise would be revisited about a decade later using both low-pass and high-pass noise (Burgess, 1998). This will be discussed below.

Researchers in spatial vision research also began to use noise-limited images and observer efficiency to investigate human performance. Pelli (Pelli, 1981, 1985) used 1D sine waves in noise to measure observer efficiency as a function of noise spectral density. He interpreted the nonlinear psychometric function for detection as an indication of channel uncertainty. As the price of 2D image display systems dropped dramatically in the 1980s, vision scientists also began to investigate noise-limited detection of more complex signals. These included 2D spatial sine-wave detection by Kersten (Kersten, 1983), detection of 2D visual noise by Kersten (Kersten, 1986), American Sign Language communication by Pavel *et al.* (Pavel, 1987), recognition of 3D objects in noise by Tjan *et al.* (Tjan, 1995), discrimination of fractal images by Knill *et al.* (Knill, 1990), and estimation of fractal dimension by Kumar *et al.* (Kumar, 1993). Geisler and collaborators in 1985 began an extensive program of investigations using the ideal observer concept in order to evaluate mechanisms through stages of the visual system. Geisler (Geisler, 2003) describes their work, which is based on the idea of introducing known properties of the visual system as constraints to the ideal observer and then comparing human and ideal performance of detection, discrimination, and identification tasks subject to these constraints. Geisler referred to this as sequential ideal observer analysis.

4.7 ESTIMATION OF OBSERVER TEMPLATES

The high human observer efficiencies for decision tasks with white noise suggest that we are able to do a reasonably good job of matched filtering. That is to say we are able to effectively use the signal (under SKE conditions) as a template. This requires that the template not only has the correct functional profile, but during a large number of decision trials it must be consistently of the correct size and it must be consistently placed in each of the possible signal locations. It would be nice if we had a more direct method of estimating the template used by the observer.

Classification image analysis provides such a method. The basic idea of the method is to take the average difference (over a large number of trials) between the image chosen in a 2AFC trial (whether it is a correct choice or not) and the image that is not chosen. The method was first used in audition research because signals and noise for decision trials could be recorded and replayed – see Ahumada and Lovell (Ahumada, 1971). For white noise, this average can be computed analytically. A transformation of the average result provides an estimate of the linear template used by the observer. Abbey *et al.* (Abbey, 1999) described the theoretical background of the method. They did Monte Carlo validation experiments (10,000 trials) using a 2D Gaussian signal and a model (non-prewhitening with an eye filter [NPWE]) observer applying a known template to the images. They then did preliminary human observer experiments (two observers and 2000 trials each after 840 training trials). There was very good agreement between estimated human and NPWE templates. For the human experiments, the results are easiest to explain using the filters obtained by Fourier transformation of the templates. At low frequencies the human filter fell in between those expected for an ideal observer (using the signal as a template) and the NPWE observer model. The human filter results extended to higher frequencies than expected for either observer model – suggesting that the human observers were incorporating some spatial frequencies with little or no signal content into their observer templates.

The mathematical derivation of the method is somewhat involved so it will only be outlined here. Let the signal be described by the vector, **s**, in white noise described by the vector, **n**, added to a known background described by the vector, **b**, added together to create an image, **g**. The observer is assumed to use a template described by the vector, **w**. The observer's internal responses, λ, included internal noise, a Gaussian random sample value for each trial, ε. Using the superscripts + and – to indicate the images with the signal present and absent, the images and the observer's internal responses are given by

$$\begin{aligned} \mathbf{g}^+ &= \mathbf{s} + \mathbf{n}^+ + \mathbf{b} \quad \text{and} \quad \mathbf{g}^- = \mathbf{n}^- + \mathbf{b} \\ \lambda^+ &= \mathbf{w}^t\mathbf{g}^+ + \varepsilon^+ \quad \text{and} \quad \lambda^- = \mathbf{w}^t\mathbf{g}^- + \varepsilon^- \end{aligned} \quad (4.15)$$

The same deterministic background is used in both images so it disappears when differences are calculated. However, the method can be extended to include stochastic backgrounds. The human observer's internal response cannot be observed, but the observer's score, o, on a given trial, j, can be used as a surrogate, giving $o = \text{step}(\lambda^+ - \lambda^-)$, where $o = 1$ if the decision is correct and $o = 0$ if the decision is incorrect. The probability of correct response for N trials is the mean (expectation) value $P_C = E(o)$. Let σ be the standard deviation of the white noise and $\Delta\mathbf{n}$ be the difference between the noise vectors for the two images in a 2AFC trial, $(\mathbf{n}^+ - \mathbf{n}^-)$. Then the classification image estimate, $E(\Delta q)$, can be described by

$$E(\Delta q) = \frac{1}{(N-1)\sigma^2} \sum_{j=1}^{N} (o_j - P_C)\Delta \mathbf{n}_j \quad (4.16)$$

This classification image is used as an estimate of the template used by the observer. It should be noted that a very large number (many thousands) of 2AFC decision trials are needed to obtain a reliable estimate.

Abbey and Eckstein (Abbey, 2002) used this method to estimate the templates used by two very experienced observers

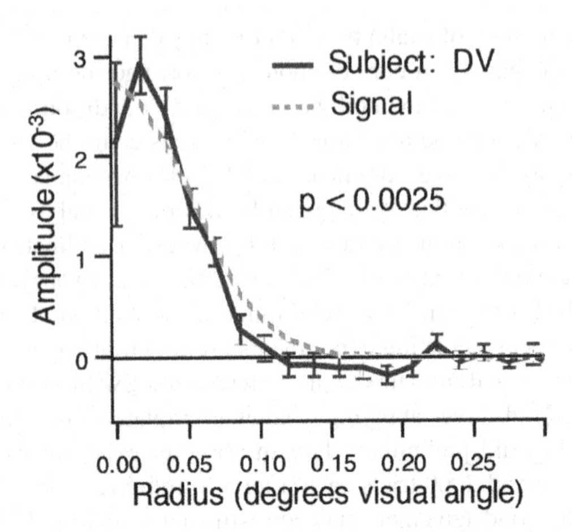

Figure 4.9 Radial average of the observer template estimate for one very well-trained observer. The radial profile of the signal is also plotted. The error bars are one standard error for averages in each radial distance bin. (Reproduced from Abbey and Eckstein (2002) with permission of the Association of Researchers in Vision and Opthalmology (ARVO).)

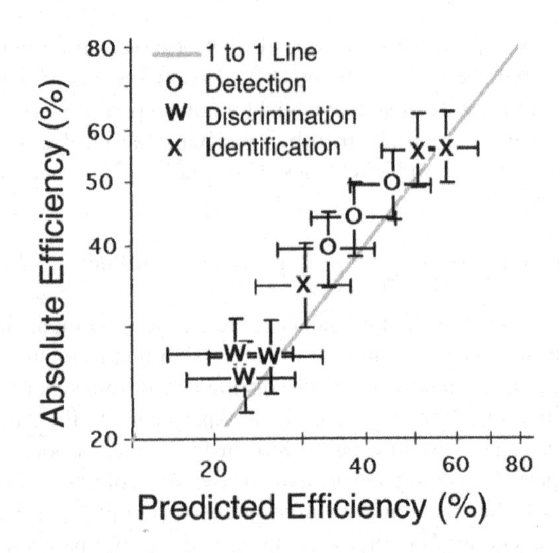

Figure 4.10 A comparison of human observer efficiencies that were calculated by Abbey and Eckstein (2006). The absolute efficiencies were determined from the percentage of correct responses and the predicted efficiency was determined in a Monte Carlo experiment using the individual estimated linear templates of the three observers. The error bars are 95% confidence limits. (Adapted from Figure 7 of Abbey and Eckstein (2006) with permission of the ARVO.)

during 2AFC detection of a Gaussian signal. The results obtained for one observer are shown in Figure 4.9. They found that the classification image estimates were isotropic, so they averaged over angle. To reduce the noise in the results, the data were binned by radial visual angle. The estimates at small angle have larger errors because fewer angular values were available for averaging. The scaled template (the signal) that would be used by the ideal observer is overlaid on the human data. There is reasonably good agreement between the human and ideal templates except that the humans present a negative amplitude over one range of radial distance. They suggested that this indicates the possibility of an inhibitory surround in the template (this is consistent with the low frequency drop in the human CSF used in the NPWE model).

Abbey and Eckstein (Abbey, 2006) investigated the template used for three 2AFC tasks in Gaussian white noise. The first was detection of a Difference of Gaussians (DoG) signal. The second task was discrimination between DoG signals that were identical except for contrast – the different signal for this task is also a DoG signal with the same spatial form. The third task was identification (actually form discrimination) between two DoG signals of different spatial extent, selected so that the different signal was also a DoG signal with the same spatial form as that present for the other two tasks. The ideal observer would use the same DoG filter to perform all three tasks. They selected the DoG signal because it has a band-pass profile in the spatial-frequency domain. They could tune the signal to be detected to spatial frequencies of interest by adjusting the DoG parameters. They selected a peak spectral amplitude at about four cycles/degree and a bandwidth (full-width at half-maximum) of approximately 1.8 octaves. The observers' classification of images across the three tasks were significantly different. Greatest variability appeared to be at low spatial frequencies (less than five cycles/degree). In this range they found frequency enhancement in the detection task, and frequency suppression and reversal in the contrast discrimination task. In the identification task, observer template estimates agreed reasonably well with the ideal observer template. They evaluated the effects of hypotheses of a nonlinear transducer and intrinsic spatial uncertainty as explanations of the human observer divergence from the ideal observer results. They found that the variation in classification images that they obtained for the human observers could not be explained as a transducer effect alone. They found that the effects of nonlinear spatial uncertainty could act as a mechanism for low-frequency enhancement of the classification images for the detection task. However, none of the models investigated fully explained the observed human data. They compared the human observer efficiencies that were calculated from the percentage of correct responses with the efficiency determined on the basis of estimated linear templates of the three observers. The results are shown in Figure 4.10.

4.8 CAN WE PREWHITEN?

4.8.1 Matched filters

In the late 1980s, visual signal detection research shifted to the issue of trying to understand how the form of the random noise power spectrum affects human signal detection performance. When the image noise (or stochastic background fluctuations) is spatially correlated, there are two main classes of models that we use in medical imaging. One is based on the ideal prewhitening (PW) matched filter strategy while the other class is based on suboptimal non-prewhitening (NPW) matched filters (Wagner, 1972). The PW model is an observer that is able to modify its template (or filter) to compensate for correlations in the noise – a process known as prewhitening. The NPW model in an observer is unable to modify its template (or filter) to prewhiten correlated noise. The PW and NPW models are obviously identical for white noise and uniform backgrounds. As will be seen later, the NPW model does very poorly for stochastic (spatially varying) backgrounds. The problem arises because the model uses the signal as a template for cross-correlating and local variations in background are included in the decision variable. This problem can be overcome by using a template with an antagonistic center-surround structure similar to retinal receptive fields so that there is (on average) zero response to local background variations. The Fourier transform of the appropriate template will be zero at zero spatial frequency ("DC"). This can be achieved by multiplying the signal amplitude spectrum by an eye filter with a "zero DC value." One isotropic form could be $E(f) = f^b \exp(-cf)$, where f is radial frequency, and b and c are adjustable parameters. This modified model is referred to as the NPW model with an eye filter (NPWE), which was first proposed by Ishida *et al.* (Ishida, 1984). The models are discussed in detail in Chapter 17.

The SNR (or d′) equations for the models performing an SKE/BKE task will now be discussed. The imaging system modulation transfer function (MTF) will not be included for simplicity. Assume that the signal to be detected has amplitude spectrum S(u, v). Assume that the colored noise has a power spectrum N(u, v). Then the SNR equations are given by

$$\begin{aligned}
&[\text{PW}] \quad (d')^2 = \int \frac{|S(u,v)|^2 dudv}{N(u,v)} \\
&[\text{NPW}] \quad (d')^2 = \frac{\int |S(u,v)|^2 dudv}{\int S^2(u,v)N(u,v)dudv} \\
&[\text{NPWE}] \quad (d')^2 = \frac{\int |S(u,v)E(u,v)|^2 dudv}{\int S^2(u,v)E^4(u,v)N(u,v)dudv}
\end{aligned} \quad (4.17)$$

With reference to equation (4.18) below, the relationship between the PW and NPW models is illustrated as follows. Suppose the model observer performs detection by cross-correlating the image data with an arbitrary template, t(x, y), which has Fourier transform, T(u, v). The signal may not be symmetrical so one must be careful about the distinction between cross-correlation and convolution. In the space domain convolution integral, a reversed version of the template function is used. In the Fourier domain, one uses the product S(u, v)T(u, v). In the space domain cross-correlation integral, the template function is not reversed. In the Fourier domain, one uses the product $S(u, v)T^*(u, v)$. The Fourier domain SNR (or d′) integral calculation goes as follows. The first integral gives the $(d')^2$ result for the filter corresponding to the arbitrary template. The second line (NPW) shows what happens if the observer is not ideal and uses the signal as a template. The third line shows what happens if the PW matched filter, $T_{MF}(f) = \alpha S^*(u, v)/N(u, v)$, is used:

$$\begin{aligned}
\text{Arbitrary.} \quad (d')^2 &= \frac{\left[\int^2 S(u,v)T^*(u,v)dudv\right]^2}{\int T^2(u,v)N(u,v)dudv} \\
\text{NPW.} \quad (d')^2 &= \frac{\left[\int^2 S(u,v)S^*(u,v)dudv\right]^2}{\int S^2(u,v)N(u,v)dudv} \quad (4.18) \\
\text{PW.} \quad (d')^2 &= \frac{\left[\int^2 [S(u,v)S^*(u,v)/N(u,v)]dudv\right]^2}{\int [S^2(u,v)N(u,v)/N^2(u,v)]dudv} \\
&= \int^2 \frac{|S(u,v)|^2 dudv}{N(u,v)}
\end{aligned}$$

The first observer uses signal data as "seen" through the template filter and is affected by noise as "seen" through that template filter. So the numerator is the "perceived" signal energy and denominator is the "perceived" noise power. The interpretation of the other models is similar. A particular version of the first model was used by Barrett and Swindell (Barrett, 1981). They assumed that the template (filter) would be based on the point-spread function (MTF) of the imaging system. Thus, all of the noise within the bandwidth of the system would be involved in the SNR calculation. Based on the first measurements of human efficiency in white noise, that point of view could be dismissed. High efficiency cannot be obtained without using a filter bandwidth fairly close to the signal bandwidth.

4.8.2 Linear discriminant analysis models

There will be situations where the Bayesian ideal observer will use nonlinear procedures to calculate a decision variable. The performance calculations could then become mathematically intractable. Fiete *et al.* (Fiete, 1987) proposed a linear-discriminant-analysis-based model (which they called the Hotelling observer) adopted from the pattern-recognition literature, where it is called the Hotelling trace criterion (HTC) and it achieves the best possible performance (maximizes d′) using linear procedures. This is named in honor of Harold Hotelling (1895–1973), a mathematical statistician and very influential economic theorist. The Hotelling approach is very similar to the Fisher linear discriminant function approach as described below. They are both concerned with calculating a measure of class separability – in our context, whether a known signal is present or not in a noisy image. Ronald Fisher (1890–1962) was an English statistician who created many of the foundations for modern statistical methods. In the context of signal detection, both their approaches involve a vector representation of the "signal" and a covariance matrix representation of the fluctuations (noise). The difference between the Fisher method (Fisher, 1936) and the Hotelling method (Hotelling, 1931) lies in how the covariance matrix is determined. The Hotelling method assumes that the covariance matrix is known *a priori* (which could be the case for simulated images) while in the Fisher method the covariance matrix is calculated from sample data (which would be the case for natural images, mammograms for example). For the case of 2D images, one can think of the approaches as calculating whitening signal detection performance using image domain data rather than spatial frequency domain data. For some situations, this image domain approach is preferable for digital images. However, as will be seen later, there are digital image situations where frequency domain methods are satisfactory.

Barrett and coworkers refer to their model as the Hotelling observer. I have always preferred the name Fisher–Hotelling (FH) observer for two reasons. One is that the term "Fisher linear discriminant function" is well known in many fields and no further explanation is needed, while the Hotelling term is somewhat obscure. The other reason is that the FH term covers both methods of obtaining the covariance matrix, **K**. Let the image domain signal be represented by a number of samples formed into a column vector, **s**. For example, digital image pixel values within a region of interest (ROI) can read in raster scan (lexicographic) order to obtain the signal vector. The SNR equation, with superscript t indicating vector transpose, is

$$(d')^2 = \mathbf{s}^t\mathbf{K}^{-1}\mathbf{s} \tag{4.19}$$

Note the form of this expression in comparison with the Fourier domain equation for the PW model. Crudely speaking, they both include the signal description squared (once as a signal and once as a template) and an inverse description of the noise covariation.

The difficulty with using the FH model is if the ROI under consideration has size $N \times N$ then the covariance matrix must be of size $N^2 \times N^2$, which can become very large. This causes two problems; one is that the matrix will become very difficult to invert. Secondly, if the covariance matrix must be estimated then at least N^2 ROIs must be available (preferably $3N^2$ or more). If the signal has very compact support, a small window, say 8×8 or 12×12, centered on possible signal locations, can be used so the covariance matrix will be 64×64 or 144×144. There are methods that will allow inversion of such matrices. For larger signals, Yao and Barrett (Yao, 1992) proposed a simpler approach than can be used in many medical imaging situations. Since the signals that concern us are frequently compact and smoothly varying, it is feasible to represent the image data by coefficients of a set of smoothly varying basis functions centered on possible signal locations. The basis set actually consists of vectors for digital images, but I will call them functions for the moment. It also would be convenient if the basis set consisted of orthogonal functions that are normalized but these are not necessary conditions. The effect of this approach is to dramatically reduce the dimensionality of the calculation.

For example, the signal may have compact support of 32×32 pixels, so the signal vector has 1024 entries and the covariance matrix will be 1024×1024. Usually it is possible to represent a compact and smooth signal satisfactorily by a set of 5 to 8 basis functions with support about 3 times the signal support size. The dot product of each basis vector with the signal vector yields an element in the reduced vector description of the signal – which now has 5 to 8 entries. Similarly, the dot product of each basis vector with noise and background image vectors is used to determine a 5×5 to 8×8 reduced covariance matrix. Note that, in principle, there are an infinite number of basis function sets that could be used. The choice is usually based on the relevance of the basis function set to the situation under study or alternatively some aspect of mathematical convenience. For example, the sine and cosine (complex exponential) basis set is used for Fourier analysis. This is a poor set for image domain analysis because the functions are infinite in extent. A better choice would be wavelets or another spatially compact set. The view is sometimes put forward that the basis functions should represent templates associated with visual system channels. As was discussed in Chapter 3, the idea that we have a small number of channels centered on discrete center frequencies is very naive. There are two benefits from the use of a small number of basis functions: a reduction in dimensionality for calculations and a limitation on the ability of observer models to do prewhitening. It will be seen later that experimental results suggest that we have such a limited capability. In addition the linear discriminant function method allows for decorrelation of outputs of arbitrary basis functions that are not orthogonal.

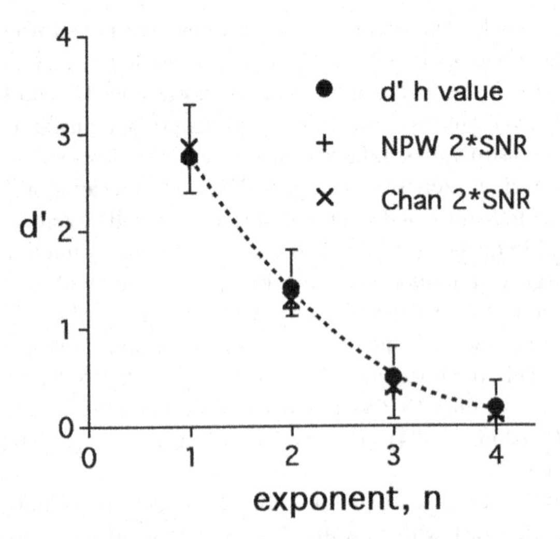

Figure 4.11 The results from observer experiments by Myers *et al.* (1985) for detection of a known signal in band-pass noise with the spectrum form $N(f) = Af^n \exp(-bf^2)$. The results would be independent of n for a PW observer. The solid circles are human results (with error bars estimated by the authors). Also shown are results (scaled by a factor of two) for the NPW model and for a simple Rect function channels model (averaged for four combinations of channel parameters). The dashed line is only a guide for the eye with no theoretical significance. (Adapted from data in Figure 1 of Barrett *et al.* (1993).)

4.8.3 Human observer experiments with correlated noise

4.8.3.1 High-pass noise

The main question in this section is, how effective are humans at compensating for noise correlations? That is, can we prewhiten the noise? The first study was done by Myers *et al.* (Myers, 1985) who had the goal of comparing human performance with the predictions of the PW and NPW models. They used a known signal (SKE) and isotropic band-pass noise spectra of the form $N(f) = Af^n \exp(-bf^2)$, where n ranged from 1 to 4. These spectra produce noise with short-range negative correlations. They produced images with the noise level adjusted so that d' for the ideal observer would remain constant as the exponent was varied. Their results are shown in Figure 4.11 together with predictions for both the NPW model and a channels model described below. Human results decreased dramatically as the exponent increased showing that we do poorly compared with the ideal prewhitening observer. There is very good agreement between the human results and predictions (after scaling) of the nonideal models as the exponent increases. One should note that, except for CT and SPECT images that have small exponent values, such band-pass noise is an artificial (experimental) strategy to force large differences between PW and NPW model predictions. Our poor performance should not be a surprise. Fluctuations with short-range negative correlations are never found in natural images or scenes, so perhaps the visual system has not been optimized through evolution to cope with such statistics.

Subsequent research using noise with low-pass power spectra showed that observer performance often gave efficiencies in the vicinity of 50%, just as was found for white noise. For low-pass noise power spectra, prewhitening models are better predictors of human performance than are non-prewhitening models. This work will be discussed below. So the performance with high-pass must be regarded as an anomaly. It now appears that the effects of low-pass and high-pass noise are very different.

Myers *et al.* (Myers, 1987) later suggested that the reduction in performance with the band-pass noise might be due to spatial-frequency channels in the visual system. The channels would irreversibly combine noise contributions from a range of spatial frequencies and would preclude any subsequent attempts to prewhiten the noise. They evaluated the Myers *et al.* (Myers, 1985) data using an artificial channels model consisting of nonoverlapping Rect functions in the frequency domain. This strategy avoided correlation between channels and thus simplified calculations. The channels were of constant "Q" in form; that is to say the fractional bandwidth, $(f_{max} - f_{min})/f_C$, was always proportional to the center frequency, f_C. They tried four combinations of initial channel center frequency and fractional bandwidth. There was very little difference in results for the four combinations so the average values are shown in Figure 4.11.

4.8.3.2 Random backgrounds

Rolland and Barrett (Rolland, 1992) did the first detection experiments with low-pass stochastic backgrounds. They used a Gaussian disc signal and white noise superimposed on statistically defined (lumpy) backgrounds that simulated nuclear medicine images. The lumps were randomly placed Gaussian blobs that had a standard deviation three times that of the signal. They used three different spectral densities for the lumpy backgrounds and assumed (for mathematical convenience) that the model nuclear medicine imaging system had an aperture with a 2D Gaussian transmission profile. The independent variable for one investigation was the ratio of aperture to signal size and in the other it was exposure time. They calculated the performance of both the Hotelling and NPW (without an eye filter) observer models for all the experimental conditions. After scaling of model predictions, there was very good agreement between the Hotelling model and human results. The NPW model predictions were dramatically different from human results. In fact, this would be expected because the basic NPW model has no way of eliminating the local variations in background and would include these variations in its decision variable values. They concluded that humans are able to use prewhitening. However, the prewhitening issue was not settled. Burgess (Burgess, 1994) subsequently showed that the modified NPWE model (with an eye filter) gave predictions that were also in very good agreement with the Rolland and Barrett human observer data.

It should not be assumed that the above findings indicate that the NPWE model is always reliable because that is not the case. Subsequent investigations by Burgess *et al.* (Burgess, 1997, 1999b) demonstrated that NPWE model predictions do not agree with human results for a number of situations. The results in Burgess *et al.* (Burgess, 1999b) will be discussed here. The experimental strategy was to include two image noise components – white noise, N_1, filtered noise, N_2, and an estimate of human observer internal noise. In addition, two different

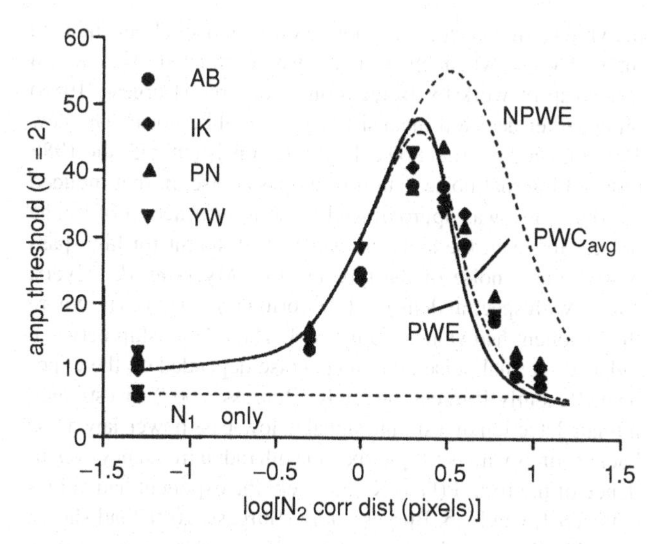

Figure 4.12 Amplitude thresholds for detection of a 2D Gaussian signal with two-component image noise and an estimate of human observer internal (white) noise. The lower dashed line is a fit through the average human results when only the white image noise component, N_1, is present. The other set of data are results when the N_2 component, filtered by a low-pass Gaussian filter, is present. The results are for the four human observers (filled symbols), the NPWE model, a PWE model with an eye filter, and a PWC model with spatial frequency channels. All the models included observer internal noise. (Reproduced from Burgess *et al.* (1999b) with permission of the OSA.)

signals were used, a 2D Gaussian and a simulated 2D nodule. Two forms of filtered noise were produced from white noise with a Gaussian probability distribution: (a) using filters with isotropic low-pass Gaussian transfer functions of varying bandwidths and (b) using isotropic low-pass power-law filters to give spectra of the form $N(f) = A/f^{\beta}$. Since the filtering process was linear, the filtered noise would have spatially constant (stationary) statistics and Gaussian probability distributions. This allowed exact calculation of the performance of all observer models because they would use linear procedures. Figure 4.12 shows results for four human observers and three model observers detecting a Gaussian signal for the case where the N_2 component had a low-pass Gaussian power spectrum. The model results with N_2 present were scaled to fit the average human data at the minimum correlation distance. Obviously the NPWE model is in very poor agreement with human results. Note that the PW model with an eye filter would be pointless if observer internal noise had not been included. Since the eye filter would then operate on both the signal and noise, its effects would cancel in the d′ calculation. The PWE model gave a better fit and the prewhitening model with channels (PWC) gave the best fit. Note that the amplitude threshold has a maximum value when the correlation distance of the N_2 noise is approximately equal to the standard deviation of the 2D Gaussian signal (three pixels). This is not surprising since at that point the signal is most easily confused with the "mottled" appearance of fluctuations in the N_2 noise background. The results for all the other three experimental conditions were similar.

Some related papers might be of interest. Burgess (Burgess, 1998) discussed evidence for the human capability to compensate for spatial noise correlations (prewhiten) or partially prewhiten. In particular I reported on new data about induced internal noise when image noise was correlated. This was a follow-up of work by Burgess and Colborne (Burgess, 1988) on observer decision inconsistency with white noise. The new 1998 induced internal noise data were consistent with the 1988 induced internal noise data for low-pass noise, in that induced internal noise was approximately a constant fraction of image noise. However, the situation was very different for high-pass power image noise of the type used by Myers *et al.* (Myers, 1985), with spectral density of the form $P(f) = Kf^n H^2(f)$, where the exponent had values from 1 to 4. The relationship between induced internal noise and image noise depended on the exponent. Recently Burgess and Judy (Burgess, 2007) investigated human detection of a simple signal in low-pass power-law noise backgrounds with isotropic spectra with radial frequency dependence of the form $P(f) = K/f^{\beta}$, where the exponent had values between 1.5 and 3.5. Burgess *et al.* (Burgess, 2001) had shown that CD diagrams for a prewhitening matched filter observer performing such detection tasks had a slope, m, with a linear dependence on the power-law exponent, $m = (b - 2)/2$. This gives the unusual prediction that the CD diagram will show *no variation* in contrast threshold with signal size for an exponent of 2 and a positive CD diagram slope for larger exponents. The results of Burgess and Judy (Burgess, 2007) agreed with this prediction. This does not prove that humans can prewhiten since other observer models also give similar CD slopes, but their results are consistent with a prewhitening capability.

4.8.3.3 Human templates with correlated noise

Abbey and Eckstein (Abbey, 2007) estimated the linear templates used by human observers for a variety of tasks with correlated noise. This was done using the classification image analysis method that they had previously used for tasks with white noise. The signal tasks were: (a) detection of a 2D Gaussian and (b) discrimination between two different Gaussian profiles. The ideal observer would base its decision on the difference signal. The Gaussians had different widths and the amplitudes adjusted so that the Fourier transform of the difference signal would be zero at zero frequency. They used four noise spectra: white, low-pass, high-pass, and band-pass. For these tasks, performance is generally enhanced by an observer's ability to "prewhiten" correlated noise as part of the formation of a decision variable. They also did observer experiments to determine psychometric function (d′ as a function of signal amplitude for each task) to assess the linearity of responses. For each experimental condition, they compared the Fourier transform of the estimated human observer template (i.e. the filter) with the ideal observer matched filter and the signal spectrum (which would be used by the NPW model without an eye filter). There was no point in including an eye filter because the backgrounds were flat and constant. They found that observer efficiency in these tasks was well-represented efficiency determined using the estimated templates in Monte Carlo experiments. The human results showed strong evidence of adaptation to different correlated noise textures. This adaptation is captured in the frequency weighting of the classification images.

4.9 SIGNAL KNOWN STATISTICALLY TASKS

Most signal detection calculations and experiments have been done using the SKE paradigm because of ease of analysis. For human observers, a high-contrast, noiseless reference copy of the signal is always displayed. In most cases the same signal is used over a complete block of trials. Less frequently the signal is varied from trial to trial, using the "signal known exactly but variable" (SKEV) paradigm (Burgess, 1984a, 1985). The reference signal is changed on each trial to indicate the signal that might be present. These paradigms are not very close to clinical situations where limited information is available about properties of potential signals. The next step in realism is to investigate observer performance for tasks with the signal known statistically (SKS). For this task, some signal parameter or parameters are defined by probability distributions. The signal detectability calculation for the SKS case has two stages – a linear template stage (one per signal option) to give with likelihoods and a nonlinear stage to combine likelihoods to obtain a decision variable. The first stage is analogous to that for SKE detection except that a template for each possible signal is used or cross-correlation with image data at each possible location. This gives a likelihood, λ_{jk}, of the jth signal option being present at the kth location option. On each trial, the likelihood, l_k, of any of the signals being present at the kth location and not being present at the M–1 other locations is calculated. The observer then selects the location with the highest location likelihood l_k. For Gaussian distributed noise, this calculation is done using the nonlinear equation

$$l_k = \sum_{j=1}^{J} \left[\frac{1}{\sqrt{2\pi\sigma_j^2}} \right]^M \exp\left(\frac{-(\lambda_{j,k} - \mu_{j,s})}{2\sigma_j^2} \right) \prod_{m=1}^{M} \times \left[\exp\left(\frac{-(\lambda_{j,m} - \mu_{j,N})}{2\sigma_j^2} \right) \right]^{(1-\delta_{k,m})} \quad (4.20)$$

The parameters in the equation are: J possible signal types, σ_j is the standard deviation of the response of the jth template, $\mu_{j,s}$ is the expected response of the jth template to the jth signal type, $\mu_{j,N}$ is the expected response of the jth template to signal-absent (noise-only) locations, M is the number of signal locations, and $\delta_{k,m}$ is the Kronecker delta function (which equals unity when k equals m and is zero otherwise). When the likelihood of the signal being present at the kth location is being evaluated, the exponent $(1 - \delta_{k,m})$ allows evaluation of the probability of the signals being absent (i.e. noise only) at other locations since when k equals m the exponent becomes zero. The SKS likelihood equation cannot, in general, be solved in closed form for correlated signals but model observer performance can be obtained by direct evaluation of the equation by applying templates to the image ensemble.

Judy *et al.* (Judy, 1997) did ROC experiments to measure the effect of signal size uncertainty on human observer detection performance in white noise and CT-like noise. They used smoothed disc and Gaussian signals and compared human performance with that of the NPW observer model. They defined a loss ratio as d'_{human}/d'_{NPW}. The signals had a wide range of sizes with effective areas ranging from 19 to 3414 pixels for the disc signal and 24 to 9069 pixels for the Gaussian signal. They also did control SKE experiments using the same signal sizes. The observers always knew the location of the center of the signal. The very surprising result was that size uncertainty had no effect on signal detectability. The best human performance was in the effective area range between approximately 100 and 1000 pixels where the loss ratio was about 0.62. The loss ratios decreased at smaller and larger areas.

Eckstein and Abbey (Eckstein, 2001) investigated performance for tasks under two paradigms, SKEV and SKS, using a set of signals that varied in size and shape. They used a signal that was a projected ellipsoid with major and minor axes randomly varying from 2 to 20 pixels giving 361 possible signals. There were two alternative locations centered on separate ROIs. The image noise had two components, white and power-law with an exponent of three. They used four different white noise standard deviations. They found, as would be expected, that human detection performance was better for the SKEV task than for the SKS task. However, the differences were small. When compared to ideal observer performance, average human efficiency was lower for the SKEV task (about 23% versus 36% for the SKS task). They also compared human detection performance with those of the NPW and NPWE observer models. There was very poor agreement with the NPW model results as the white noise standard deviation was varied and for small amounts of white noise human performance was better. The variation of human performance with the amount of white noise was similar to that of the NPWE model and in general the NPWE model performed better than humans. They concluded that, to a first approximation, human SKEV results could be used as a surrogate for SKS results.

Eckstein *et al.* (Eckstein, 2002) compared human and NPWE model observer performance for SKEV and SKS detection tasks. The image sets were simulated arterial filling defects embedded in real coronary angiographic backgrounds in images that had undergone different amounts of JPEG and JPEG 2000 image compression. They used a signal set that was simulated filling defects in simulated vertical arterial segments added to coronary angiograms. The filling defects were generated using projected ellipsoids with the vertical axis randomly varying from 3 to 25 pixels and the horizontal axis randomly varying from 3 to 10 pixels. This gave 184 possible signals available for random selection on a given decision trial. They determined observer performance for uncompressed images and 5 compression ratios between 5 and 30. Their results showed that although the model and to a lesser extent human performance improved when it was known which signal was present in each decision trial (SKEV task), conclusions about which compression algorithm is better (JPEG vs. JPEG 2000) for the task would not change whether one used an SKEV or SKS task. These findings suggested that the computationally more tractable SKEV models could be used as a good first approximation for automated evaluation of the more clinically realistic SKS task.

Eckstein *et al.* (Eckstein, 2003) again evaluated the optimization of a compression algorithm using model and human observers doing SKEV and SKS tasks. The purpose was to determine whether the benefits of the optimization procedure could be determined using the SKEV task rather than the SKS task – which would be more difficult to perform. In this work, they used three models: the NPWE observer, and two versions of the channelized Fisher–Hotelling observer (with Gabor and Laguerre–Gauss basis function sets). They compared observer performance for lesion detection in three classes of images – uncompressed, a default JPEG 2000 compression coding setting, and JPEG 2000 compression optimized using a genetic algorithm. Again, there was good agreement between optimization results for the SKEV and SKS tasks.

4.10 CORRELATED ALTERNATIVE DECISIONS

The analysis of observer performance is nearly always based on the assumption of statistically independent alternative decisions. Usually it is straightforward to design experiments so that this assumption is met, by using a signal placed in one of M independent background images for example. However, there are situations where it is not possible to do this. One example experiment is search for a signal that can be anywhere within a defined area. Alternative signal locations can overlap and alternative decisions about signal location will be correlated. The situation is also complex if the background image noise is spatially correlated with slowly varying luminance changes. The alternative signal locations could be selected so that signal overlap never occurs but the noise would still introduce correlations between alternative decisions. Eckstein *et al.* (Eckstein, 2000b) proposed that a modified figure of merit, d'_r, could be calculated from the standard MAFC detectability index, d′, using $d'_r = d'/\sqrt{1-r}$, where r is the correlation between decision variable values (responses). They showed theoretically that this expression was exact for 2AFC experiments. Burgess (Burgess, 1995) described the method of calculating the standard MAFC detectability index. Eckstein *et al.* showed theoretically that their above expression was exact for 2AFC experiments.

They then did 4AFC Monte Carlo experiments with model observers detecting simulated compact filling defects in simulated vertical arterial segments added to coronary angiograms. The models were the NPWE observer and a windowed Fisher–Hotelling observer. The filling defect signal was a 5-pixel diameter disc smoothed by a Gaussian function with a standard deviation of one pixel. The window (12 × 12 pixels) used to calculate the covariance matrix for the Fisher–Hotelling observer was large enough to include the complete signal area. They determined the correlation between model observer responses for the four signal locations and compared measured and predicted detectability indices. There was very good agreement between measured values and those predicted by the above relationship. Abbey and Eckstein (Abbey, 2000) did a more detailed theoretical analysis and showed that the modified figure of merit, d'_r, equation was correct for all MAFC experiments. This relationship allowed calculation of the percentages of correct responses, P_c, with and without taking correlations into account.

4.11 NON-STATIONARY BACKGROUNDS

Most signal detection investigations in medical imaging have used backgrounds that actually are statistically stationary or are assumed to be stationary. (The topic of statistical stationarity will be discussed in section 5.5.2.) However, clinical medical images usually are statistically non-stationary to varying degrees. Bochud *et al.* (Bochud, 1999) evaluated the spatial variation of statistics in mammograms – see their Figure 6. They concluded that there is considerable non-stationarity. It should be noted that they used images that were digitized films and were evaluated with data values proportional to X-ray exposure whereas images displayed to radiologists have data values proportional to log(X-ray exposure). Statistical variations of images evaluated in the exposure domain will be exaggerated. I recently evaluated statistical variations for a large collection of images in my database of digital mammograms that used the log(X-ray exposure) domain. In brief for my images (within the compressed breast region), the relative spatial variation of mean values, mean(x, y)/average [mean(x, y)], is about 1%, compared to 20% in the Bochud *et al.* paper. I also evaluated the relative spatial variation of the coefficient of variation, COV(x, y) = standard deviation (x, y)/mean(x, y). I found a maximum variation in the COV(x, y) of about 9% compared to 45% in the Bochud *et al.* paper. The variation of the mean that I found for the log(exposure) is probably low enough to allow acceptance of the assumption of wide-sense stationarity. The situation is not so good for the changes in the coefficient of variation across the ensemble at particular points (x, y). So in summary, I saw a considerable reduction in the spatial variation in the second-order statistics using my log(exposure) scale ROIs.

Bochud *et al.* (Bochud, 2000) investigated calculation of figures of merit for performance by model observers doing signal detection with non-stationary backgrounds. Their discussion of different types of background statistics and alternative figures of merit is particularly useful. They investigated two categories of medical backgrounds (mammography and coronary angiography) to see if they possess the qualities of semi-stationarity or local stationarity. Note that the mammograms for this study were the same as those used for Bochud *et al.* (Bochud, 1999) with a data scale proportional to X-ray exposure. They concluded that neither category of background could be considered to be stationary. They presented a closed-form method for d′ calculation based on the noise covariance matrix. They concluded that the most appropriate way of computing d′ for a model observer applied on patient structure backgrounds remains either the direct application of the model template or the computation of a closed-form expression based on the noise covariance matrix. Since that matrix for an ROI of size $N \times N$ has dimension $N^2 \times N^2$, the covariance matrix approach is only feasible for a very small signal s so a small ROI can be used (10×10 for example). They suggested that the error in d′ introduced by the stationarity assumption most likely depends on the model observer template. Zhang *et al.* (Zhang, 2006a, 2006b) measured human performance for detecting a signal in both statistically stationary and non-stationary noise. The test images were designed so that performance would be constant for model observers that assumed statistically stationary noise and did not use variations in local statistics. Zhang *et al.* (Zhang, 2006a) found that the performance of an ideal observer that could take such variations into account was about 140% higher with the non-stationary backgrounds while human performance was 30% higher. They concluded that humans can adapt their strategy to the local statistical properties of non-stationary backgrounds. Zhang *et al.* (Zhang, 2000b) compared human performance for globally statistically stationary and globally statistically non-stationary noise, both computer-generated. The latter was oriented filtered noise that changed as a function of spatial position in the image. Results showed that human performance was better when the signal was detected in the non-stationary images than when detected in the stationary images. The performance of the two models they tested that did not adapt to the local statistics of the noise (NPWE and non-adaptive PW) did not differ across the two test image conditions. They concluded that humans are using templates that adapt to the local statistics of backgrounds (a locally adaptive PW or Hotelling). However, human observers' ability to adapt their template to the local noise properties is very suboptimal when compared with an ideal observer that uses specific templates optimally individualized for the local statistics in the non-stationary background.

4.12 THE *A-CONTRARIO* OBSERVER

This observer model was proposed as an alternative to the standard SDT approach – where *a-contrario* means "by contradiction." The initial development was by Desolneux *et al.* (Desolneux, 2001) for a task with no noise or background structure present. Grosjean *et al.* (Grosjean, 2006) extended the work to signal detection in medical images with image noise and patient structure. The *a-contrario* observer approach is statistical and based on the Gestalt principle of vision. According to this principle an observer builds an *a priori* model of normal image contents – a texture for example. The observer then makes decisions based on the presence of departures (a potential signal) from this model. With analogy to SDT, this observer does make use of a hypothesis, H_0, about the properties of the texture. The *a-contrario* approach, in the most general case, does not make any assumptions about the size, shape, or pattern for the potential signal.

Grosjean *et al.* investigated the task of detecting a "spot" (a compact suspicious deviation in intensity values) in a texture image. They did not make any assumptions about the size, shape, or pattern for the spot. In effect, they are not using a hypothesis, H_1, about the nature of the signal. They do make use of a hypothesis, H_0, about the properties of the texture. For the purposes of analysis, they limited analysis to cases where textures could be described by colored noise. The basic idea of the *a-contrario* observer analysis is the detection of a "spot" in the image that is unlikely to occur by random chance. I will first describe the method in the abstract and then outline a specific example. Suppose that in order to detect spots, we make local measurements of some image property that yield values, m_j, of a random variable, M. We select a measurement procedure so that as m increases, the probability that the measurement is due to a random fluctuation of the texture decreases. The next step is to select a decision criterion, μ, so that we conclude that a significant departure from random fluctuations (a "spot") exists at some image location if m is greater than μ. We need to take into consideration the fact that the number of false alarms (NFA) per image will depend on the value of μ and select a value of μ that gives a sufficiently low value of NFA (a threshold, ξ, of

0.01 per image for example). The measurement procedure will depend on the nature of the abnormal structure to be detected in the images.

One specific example of the *a-contrario* observer is similar to the non-prewhitening matched filter. Suppose we are trying to detect masses or microcalcifications in mammograms. These are compact structures, which are localized and approximately round, that will give increased local intensity in image data values. So, as a reasonable approximation, we can use circular templates to make measurements. The measurement, m, at a particular image location is determined by cross-correlating the template with image data at that location. Since we are interested in local changes, one example template could have two components, a central flat-topped disc of diameter, D, that gives a positive weight (+1, for example) to data underneath it and a surrounding annulus of equal area that gives a negative weight (–1) to data underneath it. So, on average, the template measurements will be zero if no abnormality is present. Assume that we have no prior knowledge about the location of the abnormality. The observer strategy is to scan the image and make measurements at sufficiently closely spaced grid points – with spacing small compared to the disc diameter. We conclude that an abnormality is present at any location (x, y) if the value of m at that location exceeds the criterion for the desired NFA threshold. Assume that we have no prior knowledge about the particular size of the abnormality but do know that it lies within some range. Then we repeat the scanning procedure using a number of template diameters that sample the size range in sufficiently small increments – for example with a ratio of 1.2 for adjacent template diameters. Grosjean *et al.* (Grosjean, 2006) evaluated detection of masses in digital mammograms and found that contrast thresholds increased as mass size increased in the same manner as found for human and SDT-based models (this unusual contrast threshold elevation effect will be discussed in Chapter 5).

4.13 VISUAL DISCRIMINATION MODELS

The visual discrimination model (VDM) approach potentially provides another alternative to SDT. The method is also sometimes referred to as the visual difference model approach or the perceptual difference approach. The development of visual (or image) discrimination models was motivated originally (Carlson, 1980) in the 1980s and 1990s by the desire to apply advances in psychophysics and the neurophysiology of human visual perception to practical benefit in the design and evaluation of technologies for image processing and display. The goal of these models was to predict the effects of imaging system hardware and/or processing parameters on the perceived quality of rendered images. Image quality was generally defined in terms of the fidelity of displayed images relative to a standard reference image. The Carlson and Cohen approach was a very simple just-noticeable difference (JND) model. Subsequent models were more sophisticated and included a large variety of human visual system features including ocular MTF, luminance adaptation, spatial and temporal contrast sensitivity, tuned channels, and contrast masking. Some references include Lubin (Lubin, 1993) concerning the Sarnoff model, Daly (Daly, 1993), and Watson (Watson, 1993) concerning the NASA-Ames model. It should be carefully noted that until recently, the VDM approach was not designed to deal with signal (or lesion) detection in random backgrounds or noise. It was only capable of dealing with deterministic backgrounds – even though these backgrounds could be complex. Ahumada and Beard (Ahumada, 1997) demonstrated experimentally that image discrimination models predict detection in fixed noise but not in random noise – "fixed noise" means that the same noise ROI is used repeatedly while random noise means that each ROI is a new sample from a random process.

Jackson *et al.* (Jackson, 1996) were the first to investigate the use of the VDM approach in medical imaging – for design of an imaging system. Application of the VDM method to medical image investigations did not require fundamental changes in the underlying model features but did require input image pairs to be generated with identical (twinned) backgrounds, only one of which contained the signal to be detected. Jackson *et al.* (Jackson, 1997) evaluated VDM observer detection performance for images with and without a breast tumor and reported signal detectability that correlated well with human performance in ROC experiments as a function of X-ray detector parameters. Eckstein *et al.* (Eckstein, 1997) evaluated human and VDM model performance for images with and without lesions added to coronary angiograms. They used both single and multiple channel models with several other human visual system features and found that multiple channels with broadband masking correlated best with human results. Of course, the use of twinned input images is a limitation that made the simulated detection task very different from signal detection experimental tasks and normal reader tasks in diagnostic radiology.

It is possible that the VDM approach could be used for optimization in some applications such as testing image compression algorithms. Eckstein *et al.* (Eckstein, 2004) investigated automated optimization of JPEG 2000 compression parameters using (a) the NPWE observer model and (b) the Watson visual discrimination model. They used test images that combined real X-ray coronary angiogram backgrounds with simulated filling defects of 184 different size/shape combinations. They then did human 4AFC detection experiments to compare filling defect detection performance with compressed images with the JPEG 2000 parameter settings obtained by the two optimization methods. They found that human detection performance was better with the compressed images created using the settings from NPWE model optimization.

Nafziger *et al.* (Nafziger, 2005) and Johnson *et al.* (Johnson, 2005) used two different versions of the Sarnoff model to evaluate human CD diagram performance for a simulated mass added to $1/f^3$ power-law noise. One goal was to determine whether their results agreed with the positive CD slope found by Burgess *et al.* (Burgess, 2001). They did find positive CD slopes but they were much lower than those found by Burgess *et al.* Also, the predicted contrast thresholds of the discrimination models were much higher than the contrast thresholds for human observers doing detection tasks with random $1/f^3$ power-law noise backgrounds. Johnson *et al.* (Johnson, 2005) had used a VDM with spatial frequency channels. They compared the model predictions with human performance of mass detection in mammograms under a variety of image display conditions and monitors. Their VDM-channelized model observer did predict the same rank ordering as found for human observer performance over the range of test conditions found experimentally. So, one is led to conclude that the VDM approach is not, at present, useful for predicting both trends and magnitudes of human detection thresholds in random noise backgrounds. However, it does appear that the VDM approach can predict trends in a qualitative manner.

4.14 WHERE ARE WE NOW?

In summary, the evidence from experiments has shown that human signal detection in low-pass filtered random noise and statistically defined random backgrounds can be reasonably well described by a suboptimal SDT-based Bayesian observer model, which includes partial prewhitening. This implies that humans can partially compensate for positive spatial correlations in noise. It is clear the NPW model without an eye filter performs very badly if the background is not known exactly; that is to say, the background is only defined by its statistics. The other conclusion is that while the NPWE model can be useful on occasion with statistically defined backgrounds, it can also be a very poor predictor of human performance. In fact Burgess *et al.* (Burgess, 1997) showed that, for their experiments with two-component noise, the elevation of the signal energy required for threshold detection was not proportional to the scaling factor of the spectral density of the N_2 component and therefore human performance could not be predicted by any choice of the eye filter in the NPWE model. So, while it is often convenient to use because its performance is easy to calculate, it must be used with great care. There is also the issue of a choice between spatial domain and Fourier domain analysis.

There are a number of considerations concerning use of the PW and linear discriminant analysis models. For simulated images that include one- or two-component wide-sense stationary Gaussian noise, the Fourier domain PW model is quite satisfactory. For more complicated situations, for example the Rolland and Barrett (Rolland, 1992) lumpy background approach or with Poisson noise, the Fisher–Hotelling (LDA) approach is the method of choice. The use of models with clinical backgrounds, such as coronary angiograms and mammograms, is more complicated still and is discussed in Chapter 5 on signal detection in medical imaging. For example, Burgess *et al.* (Burgess, 2001) found that, for mammogram backgrounds, results for the Fourier domain PW model and a channelized Fisher–Hotelling model were very similar.

The final issue is stationarity. This condition is easy to meet for simulated images but is quite likely to be not completely met with backgrounds from clinical images. In that event, the use of spatial domain analysis is the safest approach. The methods for analysis of model performance include simulation of 2AFC tasks using NPWE or NPW templates and two linear discriminant analysis methods; one is use of a small-window ROI (12 × 12 pixels for example) if the signals are small or basis functions if the signals are larger. As an alternative to simulation of 2AFC decision tasks, one can use the set of individual decision variables, $\{k \mid k = 1, \ldots, K\}$, that were determined by cross-correlation. The means, $E(\lambda)$, and variances, σ_λ^2, are then used to calculate detectability indices using the equation

$$(d')^2 = \frac{[E(\lambda_S) - E(\lambda_N)]^2}{\left[\sigma_{\lambda S}^2 + \sigma_{\lambda N}^2\right]/2} \tag{4.21}$$

where the subscripts S and N indicate signal present and absent cases. Many publications of Eckstein and collaborators provide guidance on use of Monte Carlo simulation methods and analysis of models – for both simulated and clinical backgrounds.

REFERENCES

Abbey, C.K., Eckstein, M.P., Bochud, F.O. (1999). Estimation of human-observer templates for 2 alternative forced choice tasks. *Proc SPIE Med Imag*, **3663**, 284–295.

Abbey, C.K., Eckstein, M.P. (2000). Derivation of a detectability index for correlated responses in multiple alternative forced-choice experiments. *J Opt Soc Am*, **A17**, 2101–2104.

Abbey, C.K., Eckstein, M.P. (2002). Classification image analysis: estimation and statistical inference for two-alternative forced-choice experiments. *J Vision*, **2**, 66–78.

Abbey, C.K., Eckstein, M.P. (2006). Classification images for detection, contrast discrimination, and identification tasks with a common ideal observer. *J Vision*, **6**, 335–355.

Abbey, C.K., Eckstein, M.P. (2007). Classification images for simple detection and discrimination tasks in correlated noise. *J Opt Soc Am*, **A24**, B110–B124.

Ahumada, A.J., Lovell, J. (1971). Stimulus features in signal detection. *J Acoust Soc Am*, **49**, 1751–1756.

Ahumada, A.J., Beard, B.L. (1997). Image discrimination models predict detection in fixed but not random noise. *J Opt Soc Am*, **A14**, 2471–2478.

Barlow, H.B. (1962). A method of determining the overall quantum efficiency of visual discriminations. *J Physiol (London)*, **160**, 155–168.

Barlow, H.B. (1978). The efficiency of detecting changes in density of random dot patterns. *Vision Res*, **18**, 637–650.

Barrett, H.H., Swindell, W. (1981). *Radiological Imaging: Theory of Image Formation, Detection and Processing*. New York, NY: Academic Press.

Barrett, H.H, Yao, J., Rolland J.P., Myers, K.J. (1993). Model observers for assessment of image quality. *Proc Nat Acad Sci USA*, **90**, 9758–9765.

Bochud, F.O., Abbey, C.K., Eckstein, M.P. (1999). Further investigation of the effect of phase spectrum on visual detection in structured backgrounds. *Proc SPIE Med Imag*, **3663**, 273–281.

Bochud, F.O., Abbey, C.K., Eckstein, M.P. (2000). Visual signal detection in structured backgrounds. III. Calculation of figures of merit for model observers in statistically nonstationary backgrounds. *J Opt Soc Am*, **A17**, 193–205.

Burgess, A.E., Wagner, R.F., Jennings, R.J., Barlow, H.B. (1981). Efficiency of human visual discrimination. *Science*, **214**, 93–94.

Burgess, A.E., Barlow, H.B. (1983). The efficiency of numerosity discrimination in random dot images. *Vision Res*, **23**, 811–819.

Burgess, A.E., Ghandeharian, H. (1984a). Visual signal detection. I. Ability to use phase information. *J Opt Soc Am*, **A1**, 900–905.

Burgess, A.E., Ghandeharian, H. (1984b). Visual signal detection. II. Signal location identification. *J Opt Soc Am*, **A1**, 906–910.

Burgess, A.E. (1985). Visual signal detection. III. On Bayesian use of prior knowledge and cross correlation. *J Opt Soc Am*, **A2**, 1498–1507.

Burgess, A.E., Colborne, B. (1988). Visual signal detection. IV. Observer inconsistency. *J Opt Soc Am*, **A5**, 617– 627.

Burgess, A.E. (1994). Statistically defined backgrounds: performance of a modified nonprewhitening matched filter model. *J Opt Soc Am*, **A11**, 1237–1242.

Burgess, A.E. (1995). Comparison of receiver operating characteristic and forced choice performance measurement methods. *Med Phys*, **22**, 643–655.

Burgess, A.E, Li, X., Abbey, C.K. (1997). Visual signal detectability with two noise components: anomalous masking effects. *J Opt Soc Am*, **A14**, 2420–2442.

Burgess, A.E. (1998). Prewhitening revisited. *Proc SPIE Med Imag*, **3340**, 55–64.

Burgess, A.E. (1999a). The Rose model, revisited. *J Opt Soc Am*, **A16**, 633–646.

Burgess, A.E. (1999b). Visual signal detectability with two-component noise: low-pass filter effects. *J Opt Soc Am*, **A16**, 694–704.

Burgess, A.E., Jacobson, F.L., Judy, P.F. (2001). Human observer detection experiments with mammograms and power-law noise. *Med Phys*, **28**, 419–437.

Burgess, A.E., Judy, P.F. (2007). Signal detection in power-law noise: effect of spectrum exponents. *J Opt Soc Am*, **A24**, B52–B60.

Carlson, C., Cohen, R. (1980). A simple psychophysical model for predicting the visibility of displayed information. *Proc Soc Info Display*, **21**, 229–246.

Daly, S. (1993). The visible differences predictor: an algorithm for the assessment of image fidelity. In *Digital Images and Human Vision*. Ed. Watson, A.B., Cambridge, MA: MIT Press, 179–206.

Desolneux, A., Moisan, L., Morel, J.M. (2001). Edge detection by Helmholtz principle. *J Math Imaging Vision*, **14**, 271–284.

Eckstein, M.P., Ahumada, A.J., Watson, A.B. (1997). Image discrimination models predict visual detection in natural medical image backgrounds. *Proc SPIE Human Vision, Visual Processing, and Digital Display VIII*, **3016**, 44–56.

Eckstein, M.P., Abbey, C.K., Bochud, F.O. (2000a). Practical guide to model observers in synthetic and real noisy backgrounds. In *Handbook of Medical Imaging Vol. I: Physics and Psychophysics*. Eds. Beutel, J., Kundel, H.L., Van Metter, R.L., Bellingham, WA: SPIE Press, 593–628.

Eckstein, M.P., Abbey, C.K., Bochud, F.O. (2000b). Visual signal detection in structured backgrounds. IV. Figures of merit for model performance in multiple-alternative forced-choice detection tasks with correlated responses. *J Opt Soc Am*, **A17**, 206–217.

Eckstein, M.P., Abbey, C.K. (2001). Model observers for signal known statistically tasks. *Proc SPIE Med Imag*, **4324**, 91–102.

Eckstein, M.P., Abbey, C.K., Pham, B.F. (2002). The effect of image compression for model and human observers in signal known statistically tasks. *Proc SPIE Med Imag*, **4686**, 13–24.

Eckstein, M.P., Zhang, Y., Pham, B., Abbey, C.K. (2003). Optimization of model observer performance for signal known exactly but variable tasks leads to optimized performance in signal known statistically tasks. *Proc SPIE Med Imag*, **5034**, 123–134.

Eckstein, M.P., Zhang, Y., Pham, B.T. (2004). Metrics of medical image quality: task-based model observers vs. image discrimination/perceptual difference models. *Proc SPIE Med Imag*, **5372**, 42–52.

Fiete, R.D., Barrett, H.H., Smith, W.E., Myers, K.J. (1987). Hotelling trace criterion and its correlation with human observer performance. *J Opt Soc Am*, **A4**, 945–953.

Fisher, R.A. (1936). The use of multiple measurements in taxonomic problems. *Ann Eug*, **7**, 179–188.

Geisler, W.S. (2003). Ideal observer analysis. In *The Visual Neurosciences*. Eds. Chalupa, L., Werner, J., Boston, MA: MIT Press, 825–837.

Green, D.M., Swets, J.A. (1966). *Signal Detection Theory and Psychophysics*. New York, NY: John Wiley & Sons Ltd.

Grosjean, B., Muller, S., Souchay, H. (2006). Lesion detection using an a-contrario detector in simulated digital mammograms. *Proc SPIE Med Imag*, **6146**, 61460S.

Hotelling, H. (1931). The generalization of Student's ratio. *Ann Math Stat*, **2**, 360–378.

Ishida, M., Doi, K., Loo, L.-N., Metz, C.E., Lehr, J.L. (1984). Digital image processing: effect on detectability of simulated low-contrast radiographic patterns. *Radiol*, **150**, 569–575.

Jackson, W.B., Beebee, P., Jared, D.A., *et al.* (1996). X-ray image system design using a human visual model. *Proc SPIE Med Imag*, **2708**, 29–40.

Jackson, W.B., Said, M.R., Jared, D.A., *et al.* (1997). Evaluation of human vision models for predicting human-observer performance. *Proc SPIE Med Imag*, **3036**, 64–73.

Johnson, J.P, Lubin, J., Nafziger, J.S., Krupinski, E.A., Roehrig, H. (2005). Channelized model observer using a visual discrimination model. *Proc SPIE Med Imag*, **5749**, 199–210.

Judy, P.F., Kijewski, M.F., Swensson, R.G. (1997). Observer detection performance loss: target size uncertainty. *Proc SPIE Med Imag*, **3036**, 39–47.

Kersten, D.A. (1983). Spatial summation in visual noise. *Vision Res*, **24**, 1977–1990.

Kersten, D.A. (1986). Statistical efficiency for the detection of visual noise. *Vision Res*, **27**, 1029–1040.

Knill, D., Field, D., Kersten, D. (1990). Human discrimination of fractal images. *J Opt Soc Am*, **A7**, 1113–1123.

Kotelnikov, V.A. (1959). *The Theory of Optimum Noise Immunity*. New York, NY: McGraw-Hill.

Lubin, J. (1993). The use of psychophysical data and models in the analysis of display system performance. In *Digital Images and Human Vision*. Ed. Watson A.B., Cambridge, MA: MIT Press, 163–178.

Myers, K.J., Barrett, H.H., Borgstrom, M.C., Patton, D.D., Seeley, G.W. (1985). Effect of noise correlation on detectability of disk signals in medical imaging. *J Opt Soc Am*, **A2**, 1752–1759.

Myers, K.J., Barrett, H.H. (1987). Addition of a channel mechanism to the ideal-observer model. *J Opt Soc Am*, **A4**, 2447–2457.

Myers, K.J. (2000). Ideal observer models of visual signal detection. In *Handbook of Medical Imaging Vol I: Physics and Psychophysics*. Eds. Beutel, J., Kundel, H.L., Van Metter, R.L., Bellingham, WA: SPIE Press, 558–592.

Nafziger, J.S., Johnson, J.P., Lubin, J. (2005). Effects of visual fixation cues on the detectability of simulated breast lesions. *Proc SPIE Med Imag*, **5749**, 566–571.

North, D.O. (1943) and (1963). Analysis of the factors which determine signal–noise discrimination in pulsed carrier systems. *RCA Tech Rep* PTR6C (1943), reprinted in *Proc IRE*, **51**, 1016–1028.

Pavel, M., Sperling, G., Reidl, T., Vanderbeek, A. (1987). Limits of visual communication: the effect of signal-to-noise ratio on the intelligibility of American Sign Language. *J Opt Soc Am*, **A4**, 2355–2365.

Pelli, D.G. (1981). *Effects of visual noise*. Doctoral thesis, Cambridge University.

Pelli, D.G. (1985). Uncertainty explains many aspects of visual contrast detection and discrimination. *J Opt Soc Am*, **A2**, 1508–1530.

Peterson, W.W, Birdsall, T.G., Fox, W.C. (1954). The theory of signal detectability. *IRE Trans Info Theory*, **PGIT-4**, 171–212.

Pollehn, H., Roehrig, H. (1970). Effect of noise on the MTF of the visual channel. *J Opt Soc Am*, **60**, 842–848.

Rolland, J.P., Barrett, H.H. (1992). Effect of random background inhomogeneity on observer detection performance. *J Opt Soc Am*, **A9**, 649–658.

Rose, A. (1946). A unified approach to the performance of photographic film, television pickup tubes, and the human eye. *J Soc Motion Picture Eng*, **47**, 273–294.

Rose, A. (1948). The sensitivity performance of the human eye on an absolute scale. *J Opt Soc Am*, **38**, 196–208.

Rose, A. (1953). Quantum and noise limitations of the visual process. *J Opt Soc Am*, **43**, 715–716.

Rose, A. (1973). *Vision – Human and Electronic*. New York, NY: Plenum Press.

Sturm, R.E., Morgan, R.H. (1949). Screen intensification systems and their limitations. *Am J Roentgenol*, **62**, 617–634.

Swets, J.A. (1964). *Signal Detection and Recognition by Human Observers*. New York, NY: John Wiley & Sons Ltd.

Tanner, W.P., Swets, J.A. (1954). A decision-making theory of visual detection. *Psychol Rev*, **61**, 401–409.

Tanner, W.P., Birdsall, T.G. (1958). Definitions of d′ and η as psychophysical measures. *J Acous Soc Am*, **30**, 922–928.

Tjan, B.S., Legge, G.E., Braje, W.L., Kersten, D. (1995). Human efficiency for recognizing 3-D objects in luminance noise. *Vision Res*, **35**, 3053–3069.

Wagner, R.F., Weaver, K.E. (1972). An assortment of image quality indices for radiographic film-screen combinations – can they be resolved? *Proc SPIE Med Imag*, **35**, 83–94.

Watson, A.B. (1993). DCTune: a technique for visual optimization of DCT quantization matrices for individual images. *Soc Info Display Digest*, **24**, 946–949.

Woodward, P.M., Davies, I.L. (1952). Information theory and inverse probability in telecommunications. *Proc IEE (London)*, **99** Pt. III, 37–44.

Yao, J., Barrett, H.H. (1992). Predicting human performance by a channelized Hotelling observer model. *Proc SPIE Med Imag*, **1768**, 161–168.

Zhang, Y., Abbey, C.K., Eckstein, M.P. (2006a). Observer performance detecting signals in globally nonstationary oriented noise. *Proc SPIE Med Imag*, **6146**, 292–301.

Zhang, Y., Abbey, C.K., Eckstein, M.P. (2006b). Adaptive mechanisms for visual detection in statistically non-stationary oriented noise. *J Opt Soc Am*, **A23**, 1549–1558.

5

Signal detection in radiology

ARTHUR BURGESS

5.1 TRIBUTE TO ROBERT F. WAGNER

This chapter is dedicated to the memory of Robert F. Wagner (1938–2008; Figure 5.1), whose untimely death was a great shock to his multitude of friends and admirers in the medical imaging community. He contributed to many areas related to quantitative analysis and assessment of image quality and was active to the end. Bob grew up in Philadelphia, PA, and was an outstanding student during his school years. In 1959, Bob obtained a BS in electrical engineering from Villanova University (Philadelphia) and was class valedictorian. Bob served in the Order of St. Augustine from 1959 to 1965. He obtained an MA in theology from Augustinian College (Washington, DC) and an MS in physics from Catholic University (Washington, DC) in 1965. He received his PhD in theoretical nuclear physics from Catholic University in 1969 and then did a post-doctoral fellowship at Ohio State University (Columbus). Bob sent hundreds of applications for a permanent position while he was a post-doctoral fellow. Unfortunately the timing was very bad. The supply and demand curves for people with a physics PhD degree crossed orthogonally in the early 1970s. Many of us who finished graduate school at that time also entered medical imaging because of the poor job prospects in other areas of university and industrial research. Fortunately, around the same time, diagnostic radiology departments in medical schools had decided that physicists and engineers would be useful people to have around doing imaging research. For all of us looking for work in those days, including Bob, this was fortunately the beginning of a new era in medical imaging and led to a lifetime of very interesting research possibilities. Bob finally got a job in 1972 with the Division of Electronic Products of the Bureau of Radiological Health (BRH), now the Center for Devices and Radiological Health (CDRH) in the US Food and Drug Administration (FDA). He was Chief of the Diagnostic Imaging Section from 1976 to 1992.

Bob's initial role was to help elucidate the relationship between the information content of X-ray images and patient radiation dose. His first assignment was to do a survey of possible metrics to quantify projection X-ray image quality, with the plan to present the results six months later at the first Society of Photo-optical Instrumentation Engineers (SPIE) Symposium on Applications of Optical Instrumentation in Medicine (Chicago, Illinois, November 1972). The result was a wide-ranging paper (Wagner, 1972) that is still useful reading today. Two very important points made in the paper were: image quality *must* be defined in terms of the task the image is destined to perform, and the analysis must include the human image perception stage. During the 1970s, Bob and collaborators made a number of contributions to accurate characterization of X-ray imaging system components including film-screen systems and noise power spectra – see Wagner *et al.* (Wagner, 1974) and Wagner (Wagner, 1977) for early work. Around the same time Bob translated into modern notation the ideas of Otto Schade (Schade, 1987) on noise equivalent apertures and bandwidths as scalar metrics to characterize the modulation transfer functions (MTFs) of imaging system components. Schade had made many legendary contributions to electronic and photographic image quality assessment. Bob had noticed that Schade's signal-to-noise ratios (SNRs) – including display for human vision – were excellent approximations to the SNRs that derive from formal signal detection theory (SDT) for an observer which we later called the non-prewhitening (NPW) matched filter.

In the late 1970s, Bob and coworkers began to consider applications of information theory to assessment of medical imaging systems – see Wagner *et al.* (Wagner, 1979). In another line of investigation, Bob and coworkers investigated the spatial frequency dependence of the noise equivalent quanta metric, NEQ(f), which has now become a standard in our field. NEQ(f) is simply a scaling of the measured noise power back to the axis of the exposure quanta through the system transfer (MTF and gain) characteristics.

Wagner and Brown (Wagner, 1985b) proposed to use information theory and statistical decision theory methods to create a "grand unified theory" for analysis and evaluation of all medical imaging systems. The initial name for their theory was Grand Unified ANalysis Of (GUANO) Medical Imaging. This attempt at light-heartedness did not seem to meet the approval of the journal editorial staff. The paper became a classic since it brought order to the assessment of detectability of simple structures across imaging systems based on a large variety of physical processes. The high efficiency of human observer experimental results for tasks performed with white noise (described in section 4.6 of this book) suggested that human observer models could be incorporated into this overall system SNR analysis approach for image quality assessment purposes. If measured human efficiency is very low with a particular system for a particular task then the system designer should consider a redesign to allow humans to extract the necessary information. One example extensively investigated by Bob and his collaborators was our low efficiency for texture discrimination tasks in ultrasound (Smith, 1983; Wagner, 1985a, 1990). This work plus many published theoretical analyses of ultrasonic imaging

The Handbook of Medical Image Perception and Techniques, ed. Ehsan Samei and Elizabeth Krupinski. Published by Cambridge University Press.

Figure 5.1 Photograph of Robert F. Wagner (1938–2008), who was a giant among researchers in medical imaging. This photo was taken at the 1994 Newport Beach SPIE meeting; some of us were enjoying a picnic on the beach. Over a career spanning 36 years in medical imaging, Bob and his large number of collaborators made very many seminal contributions to image quality analysis and the development of the use of SDT concepts in these image quality applications. (Reproduced with permission of K. Hanson.)

system properties and interaction with tissue led to development of improved medical ultrasonic data display approaches. Alternatively, if human efficiency is very high with a particular system for a particular task then one can be confident that there is not much room for system improvement. Bob continued to work on image quality-related problems throughout his career, while also working on other topics.

For example, in the late 1980s, he became interested in image reconstruction issues in collaboration with Ken Hanson (Wagner, 1992). They used the SDT approach to assess the effects on signal detection of image artifacts produced by random high-contrast background features in objects to be imaged. The first motivation for evaluation of system and human performance with statistically defined backgrounds (discussed in Chapter 4) came from surprising and counterintuitive differences between predictions for systems optimized for signal known exactly (SKE)/background known exactly (BKE) decision tasks and clinically successful systems. This led to many friendly but heated debates with Harry Barrett and significant advances (Barrett, 1990, 1995) in the theoretical approach to image quality analysis using statistically defined backgrounds. Another significant publication related to image quality was by Tapiovaara and Wagner (Tapiovaara, 1993). Finally, extensive details of image quality assessment theory and methods were provided in ICRU Report 54 (Sharp, 1996).

In the late 1990s Bob also became interested in reader variability in ROC studies of clinical images together with the role of reader responses correlated by use of the same image evaluation set. There were related issues with computer systems for aiding diagnosis. This was directly related to the FDA's role in approval of new diagnostic equipment. This led to many studies of multireader multicase (MRMC) ROC analysis issues (Obuchowski, 2004; Wagner, 2007).

Bob's research was characterized by imagination, an unerring sense of what was important in the long term and the ability to probe in depth. Bob had the ability to patiently explain subtleties to those of us that were not blessed with his profound understanding of topics. This was always done with a very appealing sense of humor. He was the recipient of many honors including fellowships of the Institute of Electrical and Electronics Engineers (IEEE), the Optical Society of America (OSA), the Society for Imaging Science and Technology (SPSE), and the SPIE, as well as several FDA awards. Bob Wagner's chapter in this book (Chapter 6) is about his dinners with the giants of modern imaging science. This would be a fitting description for his own role in medical imaging even when he dined alone.

5.2 INTRODUCTION

This description of signal detection research and applications in radiology will span six decades. Work began in the late 1940s when Sturm and Morgan used the simple Rose model and contrast-detail (CD) diagram methods to investigate the possible benefits of image-intensified fluoroscopy. The Rose model is described in Chapter 4. An example image of a Rose–Burger phantom used for CD diagram estimation is shown below in Figure 5.3. This CD diagram method was used frequently during the 1950s and 1960s to evaluate the relative merits of X-ray imaging systems (image quality) and for system quality assurance. This CD diagram approach has several defects. First, the observer knows exactly what the distribution of signals looks like and makes a subjective decision as to whether a particular signal can be "seen". There is no way of obtaining an objective estimate of signal detectability – such as the probability of correct response. Different observers may have different biases. Also, the CD phantom evaluations are usually done using a very small number of images so statistics are poor. Wagner and Weaver (Wagner, 1972) were the first to consider the use of metrics-based SDT to evaluate image quality. Wagner made many seminal contributions to both applications of SDT in medical imaging as well as theoretical and experimental methods of image quality investigation. Some of these will be mentioned below. A few notable beginnings of work on detection problems in medical imaging are as follows. The first attempts to quantify the effects of anatomical structure in projection radiographs were by Kundel and coworkers – see Revesz *et al.* (Revesz, 1974). Judy *et al.* (Judy, 1981) were the first to use SDT methods to investigate signal detection in simulated computed tomography (CT) image noise. Kundel *et al.* (Kundel, 1985) described the first quantitative study of chest structure effects in a paper with the delightful title "Nodule detection with and without a chest film." Judy *et al.* (Judy, 1992) compared human performance with several SDT-based models for detecting simulated signals in liver

CT images. Bochud *et al.* (Bochud, 1995) were the first to investigate detection of simulated signals in mammogram backgrounds. Eckstein *et al.* (Eckstein, 1998) investigated detection of simulated filling defects in coronary angiograms. Samei *et al.* (Samei, 1999) investigated the relative effect of quantum noise and patient structure fluctuations on detection of subtle simulated nodules in chest radiographs. Burgess *et al.* (Burgess, 2001a) obtained the surprising result of a positive slope for the CD diagram for mass detection in mammogram backgrounds. Abbey *et al.* (Abbey, 2002) estimated human observer templates for detection of a simulated mass in mammographic backgrounds. Castella *et al.* (Castella, 2007a and 2007b) extended this work. Bath *et al.* (Bath, 2005a) described a collaborative investigation of the variable components that limit nodule detection in chest radiographs. Other topics to be discussed in this chapter include the nature of random noise and power spectrum measurements. The first topic is important because one must often distinguish between variable patient structure in images and true random noise. The second is important because patient structures often have power-law spectra, which are much more difficult to estimate than image noise spectra. Finally, the chapter contents will be summarized and the present status of SDT-related investigations in medical imaging will be discussed.

5.3 STURM AND MORGAN – THE BEGINNING

Russell H. Morgan (Figure 5.2) and his collaborator Ralph E. Sturm of Johns Hopkins Medical School were the first in radiology to make use of Albert Rose's model concepts and equations for signal detection in noise (see Chapter 4 on the history of SDT). Their work (in the late 1940s) was motivated by the impending introduction of electronic image intensifiers for use in X-ray fluoroscopy. Morgan was an outstanding radiologist who was born in London, Ontario, Canada. He received his MD in 1937 from the University of Western Ontario. He did his radiology residency at the University of Chicago and developed an automatic photoelectric X-ray timer there in the early 1940s. Morgan went to Johns Hopkins in 1946 as professor and chairman of the newly created Department of Radiology in the School of Medicine. In 1971 he became dean and vice president for the health divisions of the Johns Hopkins University School of Medicine, a position he held until 1975. He had an excellent understanding of mathematics, physics, and engineering principles and made a number of research contributions to X-ray imaging.

The image intensifier was invented in 1934 for use with infrared light. It was further developed during World War II for military use. Coltman (Coltman, 1948) of Westinghouse invented the X-ray electronic image intensifier. It subsequently went into clinical use in 1953. Before then, radiologists looked directly at the output of a very dim X-ray intensifying screen under very low light conditions in a darkened room – using scotopic (dark-adapted) vision. By convention, fluoroscopy was done in the mornings and radiologists spent any time outside the dark X-ray rooms wearing red goggles to preserve their dark-adapted vision. Radiologists realized that the low light

Figure 5.2 Photograph of portrait of Russell H. Morgan, MD, painted by Herbert Abrams, MD, 1982, oil on canvas. Herbert Abrams was formerly Chairman of Radiology at Brigham and Women's Hospital, Harvard Medical School. (Reproduced with permission from the Alan Chesney Medical Archives of the Johns Hopkins Medical Institutions.)

conditions reduced their visual acuity and also that the most light-sensitive part of their retina was about 20 to 30 degrees above and to the right of the fovea. Try looking at the star cluster "Pleiades" in the constellation of Taurus during the summer if you happen to be outside a city. If you look directly at the cluster you can make out five or six stars but if you carefully shift your foveal gaze away you will get the fuzzy impression of more stars. When the announcement was first made concerning development of the X-ray electronic image intensifier, radiologists expected that they would soon be able to "see" anatomy as well using fluoroscopy as they could when looking at film radiographs. Sturm and Morgan began their research to investigate what might be expected and soon realized that they had to deliver some bad news to the radiological community (as well as some good news of course).

The reasons for the good and bad news were as follows. Before electronic image intensification the detection of objects was limited by the photon fluence at the observer's retina. The number of light photons reaching the retina from an object of 1 mm diameter might be as low as 100 within the estimated storage time of the visual system (0.2 sec). This meant that the root mean square (RMS) fluctuations within the object would be 10 photons (or 10% of the mean). Radiologists did not notice fluctuations at this scale because their scotopic visual acuity was so low when directly viewing a phosphor screen (20/200 or worse, which is in the legally blind region). Normal (good) visual acuity under daylight (photopic) conditions is 20/20.

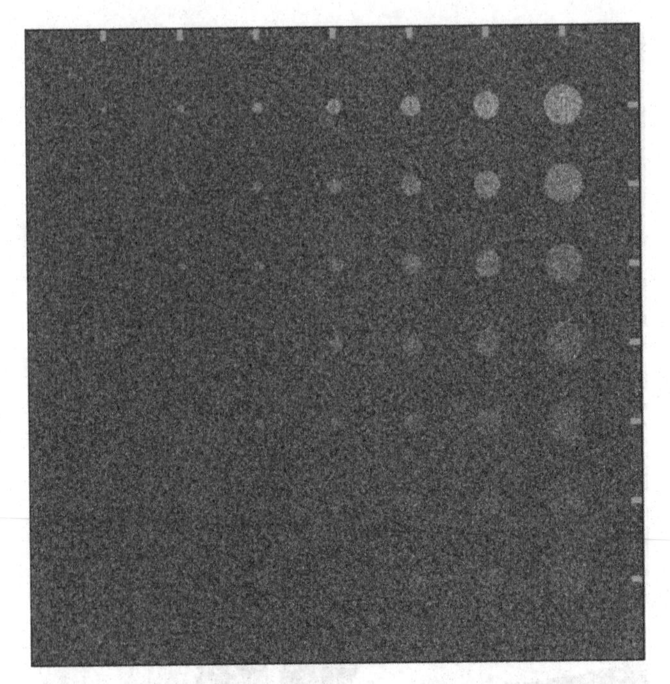

Figure 5.3 Simulation of the Rose–Burger phantom using white noise added to a calculated test pattern. Object diameter varies by steps of about a factor of $\sqrt{2}$ along each row and contrast varies by steps of about a factor of $\sqrt{2}$ down each column. The SNR list for the largest radius disc is 57.0, 39.9, 28.5, 19.9, 14.2, 10.4, and 7.6. The SNR values along the diagonal from the lowest-contrast large disc to the highest-contrast small disc are approximately 7.4 ± 0.5.

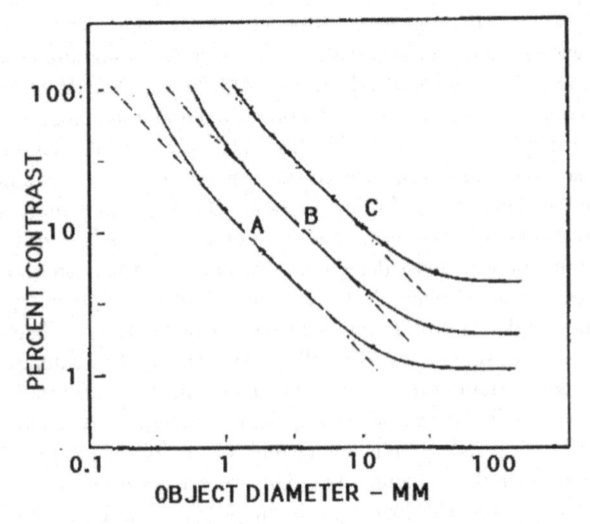

Figure 5.4 Experimental contrast thresholds for fluoroscopy. Conditions correspond to: (A) anterior-posterior (AP) chest, (B) AP abdomen, (C) lateral abdomen. (Reproduced from Figure 7 of Sturm and Morgan (1949) with permission from the *American Journal of Roentgenology*.)

In modern terminology the quantum sink of the combined system, with an X-ray screen viewed directly by the eye-brain system, was at the retina. The quantum sink is the point where the quantum flux per unit area (measured at the input to the detector system) is the lowest (Cunningham, 2000). Sturm and Morgan carefully designed and executed an observer experiment using their own creation of what is now know as a Rose–Burger phantom (see Figure 5.3). They did very careful calculations of the expected quantum fluence at each stage in the imaging system/eye-brain chain and estimated expected contrast thresholds for three different situations. These were: (a) fluoroscopy of the average anterior-posterior (AP) chest, (b) the average AP abdomen, and (c) the average lateral abdomen. Six observers viewed the very low light level fluoroscopy test images in a darkened room after one hour of adaptation. They identified the smallest diameter test object that they could subjectively "see" at each contrast level. The theoretical results and the experimental results (averages for the observers) are shown on Figure 5.4. There is very good agreement over the object diameter range from about 0.5 to 10 mm. Their explanations for the disagreements outside this range would be described as follows in modern terms. They suggested that disagreement at small diameters is probably due to retinal blurring of the signals, which reduces contrast and edge sharpness, which mixes noise fluctuations from the area around the signal with noise fluctuations within the signal boundary. They suggested that the disagreement at large diameters is probably due to incomplete spatial integration.

Now consider the situation with electronic image intensification. The Sturm and Morgan work was published four years before any units were available for clinical use. However, radiologists had known for a considerable time that intensifiers were under development. Sturm and Morgan stated that there was a widespread impression among radiologists that a limitless improvement in visual acuity and contrast discrimination would be available by just increasing the luminance of the image. They pointed out that with electronic image intensification, the quantum sink would be somewhere within the electronic intensifier system. In this case, the detectability of objects or lesions in the patient would be limited by quantum fluctuations at the point of the quantum sink. They presented results of calculations for an idealized electronic intensifier chain and concluded that no improvement in visual performance could be expected once the output luminance of the intensifier was about 30 to 50 times greater than that available for systems without electronic intensification. However, they did point out that if the gain of the electronic intensifier system were large enough, radiologists would be able to view the fluoroscopic images using daylight vision. This would turn out to be the major benefit of the electronic intensifier. The red goggles would be eliminated and visual acuity would increase. Unfortunately, the quantum fluctuations would become visible given the low tube currents used in fluoroscopy (e.g. 3 to 5 mA). Given the short storage time of the visual system (about 0.2 sec), the effective current-time product (mA) per "perceived image frame" would be about 1. So the effective quantum noise levels would still be much higher than in radiography – which was done using many tens of mAs per image.

Sturm and Morgan also did human perception experiments using film-based radiographs of their phantom, determined the CD diagrams, and presented CD diagram results for nine conditions. They included the three imaging cases (chest and abdomen) mentioned above together with three systems – radiography, idealized electronic intensified fluoroscopy, and the standard direct-viewing fluoroscopy of that day. Their CD diagram results demonstrated very clearly that X-ray imaging is a photon-noise-limited process. They stated in their conclusions that "it must be emphasized again, however, that the fundamental limitation of visual performance during screen intensifications is the number of roentgen-ray photons absorbed by the intensifying device."

The CD diagram approach has been used up to the present for imaging system evaluation. The phantoms are of varying degrees of sophistication. Early phantoms had a fixed test disc signal array with contrast decreasing along columns and disc diameter varying along rows. A simulation of such a phantom is shown in Figure 5.3. Perception thresholds were based on subjective judgment of the lowest contrast disc "seen" at each size. Such judgments can suffer from large variability and bias, so they have limited reliability. More recently phantoms have been developed based on the objective 4AFC decision task. The CDMAM phantom (Karssemeijer, 1993), for example, can be evaluated automatically to eliminate human limitations. This can give much more reliable results if a large enough number of images are used – to represent the statistical variation due to imaging noise.

5.4 DETECTION IN CT AND SPECT NOISE

5.4.1 CT power spectrum

Riederer *et al.* (Riederer, 1978) pointed out that uncorrelated statistical noise in the projection data would lead to unique correlations in the CT reconstruction noise. The power spectra will be of the form $P(f) = fH^2(f)$, where H(f) is a low-pass filter function – referred to as an apodization or regularization filter. H(f) will depend on the reconstruction algorithm used in a specific scanner. The same form of power spectrum occurs for all methods of image reconstruction from projections, including single photon emission CT (SPECT). Note that this would apply to image acquisition noise. It will not apply to patient structure. The idealized CD diagram equation (Rose model type) for a flat-topped signal of area A and a difference, $\Delta\mu$, in attenuation coefficient of a signal from the local background is given by

$$d' = \Delta\mu A^{0.75}\sqrt{NEQ} \qquad (5.1)$$

Therefore the slope of the standard log-log CD diagram plot of contrast threshold versus signal diameter would be expected to be –3/2. This result is also obtained for prewhitening matched filter observers (subject to some constraint on signal profile); see equation (5.4) in the mammography structure section below. A number of CD diagram investigations were done using a Burger–Rose type phantom but human results did not agree with this predicted result – see Cohen and DiBianca (Cohen, 1979) and Muka *et al.* (Muka, 1995). Cook *et al.* (Cook, 1996) measured the CD diagram slope for detection in a Rose–Burger type phantom in a study of the effects of image compression algorithms on signal detection. They used a subjective threshold estimation method and found a slope of –0.83 ± 0.05, which is in poor agreement with theory. The issue of anomalous CD slopes for CT images is still an unsolved problem.

5.4.2 CT – humans versus the NPW observer

Judy *et al.* (Judy, 1981) were the first to investigate signal detection in simulated CT image noise. They compared human observer results to an NPW observer model. Their data can be used to estimate a human observer of about 50% relative to the NPW observer, which is in very good agreement with estimates from other work. The signal was a disc located in one of eight possible positions (nonoverlapping) within an annular ellipse (to provide a local background) placed on a uniform general background. The ellipse was brighter than the background and the signal had the same brightness as the background. In one version of the experiments, the task was MAFC selection of the most likely signal location. The NPW model observer did cross-correlation using the known signal as a template. Let the digital template (consisting of weighting values for each pixel) for the kth possible signal location be described by a column vector, $\mathbf{w}_k$, with elements $[w_{k1}, w_{k2}, ..., w_{kN}]$. Let the image data be described by a column vector, $\mathbf{g}$, with elements $[g_1, g_2, ..., g_N]$. The cross-correlation result for the kth possible signal location is then determined by the following inner product (where t indicates vector transpose):

$$X_k = \mathbf{w}_k^t\mathbf{g} = \sum_{j=1}^{N} w_{kj}g_j \qquad (5.2)$$

For the MAFC task the best strategy is to just select the location that gives the lowest cross-correlation result. Judy *et al.* refer to this as a minimum detector for their negative contrast signal detection task. They calculated the signal-to-noise ratio using the average difference, S, between the cross-correlation results with and without a signal and the standard deviation, σ, of the cross-correlation results (corrected for correlations between adjacent pixels). Because of nonlinearities in their image production, σ depended on whether the signal was present, σ_A, or absent, σ_R. There was, on average, about 15% difference between the two standard deviations, with σ_R being larger. One of their experimental data plots with standard error estimates is shown in Figure 5.5. The data are for two observers (PFJ and RGS) and ten experimental conditions – a baseline with a selection of signal contrast, signal size, and CT noise level; as well as nine other combinations of those parameters. The abscissa of the figure is the SNR for the NPW model calculated using σ_R and the ordinate is the observer detectability index. Note that the detectability index does not equal the SNR because the task is 8AFC detection. Using the data in the figure, one can estimate the average human efficiency at d_ϵ equal to 2.5 to be about 0.5 using the Tanner and Birdsall definition in equation (4.11). This is in very good agreement with other estimates of efficiency for detection of small aperiodic signals.

5.4.3 Liver structure in CT

Judy *et al.* (Judy, 1992) produced hybrid liver CT images by superimposing both bright and dark disc signals of various sizes (1.5 to 13 mm diameter) and contrasts on 92 normal liver CT images. The CD diagram slope flattened significantly for discs larger than 7 mm diameter. The CD log-log slope was less than predicted by SDT. Manipulating the display window (adjusting

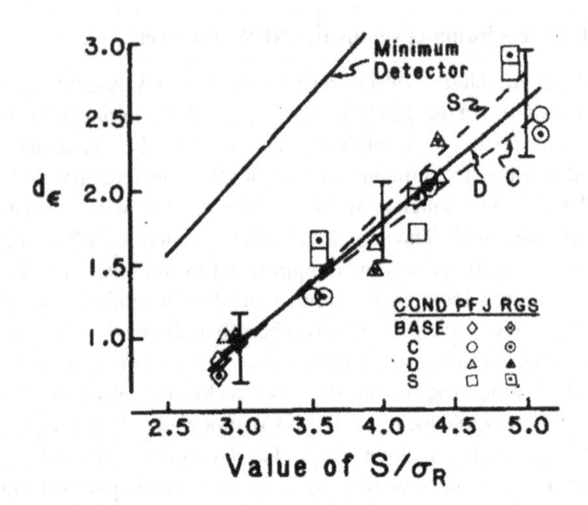

Figure 5.5 Detectability indices (d_ϵ) for 8AFC detection of a negative contrast disc in CT noise. The SNR calculation is based on the noise standard deviation, σ_R, for the case when no signal is present. They assumed that the NPW model would base its decision on the minimum cross-correlation result. The results for two human observers (PFJ and RGS) are given and estimated efficiency is about 0.5, based on SNR^2 ratios at selected performance levels. (Reproduced from Judy and Swensson (1981) with permission of the American Association of Physicists in Medicine.)

image contrast) had little impact on the results. Observer performance was determined using a rating scale ROC method. The regression line through the results for the first two diameters (1.5 and 3 mm) agreed with the expected slope of −1.5 based on the SDT estimate for CT noise. The CD slope became about −0.3 above 7 mm. This is a dramatic departure. Seltzer *et al.* (Seltzer, 1994) did another investigation. They measured spectra of the liver structure images' CT structure and compared human performance with the NPW and NPW with an eye filter (NPWE) models (both containing observer internal noise). The internal noise explained part, but not all, of the reduction in CD slope. They suggested that the reduction might be due to the fact that liver images have a fractal-like structure (Cargill, 1988), which has a low-pass power-law spectrum and contributes substantial low-frequency components.

5.4.4 SPECT imaging

Abbey and Barrett (Abbey, 2001) investigated human and model observer performance in a simulation of SPECT imaging – with both isotropic low-pass filtered ramp image noise, $N(f) = AfH^2(f)$, and low-pass power-law random anatomical background fluctuations $B(f) = K/f^n$. There were 15 experimental conditions including a baseline parameter set. The evaluations were done using three independent variable sets: seven choices of cut-off frequency for the regularizing filter H(f), five values of image noise scale, and five values of the power-law exponent (from 2 to 4). They compared human results for 2AFC/SKE detection of an isotropic, compact signal with predictions for a number of observer models. The two NPW models without an eye filter can be discounted because they did worse than humans. The NPWE model performances as a function of each of the independent variable sets had variations that resembled human variation. For some reason the performances of an NPWE model with internal noise were not evaluated. The performance of the ideal (PW) observer model was much better than humans, as would be expected. The only case where its performance variation resembled humans was in the study of the effect of the power-law exponent variation. Again, this model was not tested with internal noise. The final three models were based on linear discriminant analysis (LDA) with basis functions (which they referred to as channelized Hotelling models). After they included internal noise there was very good agreement with human performance.

5.5 EFFECTS OF PATIENT STRUCTURE – INTRODUCTION

5.5.1 What is "noise"?

Radiologists are very aware that confounding patient structure limits their ability to detect details in medical images. This is particularly true for the case of projection radiography, hence the interest over the years in developing subtraction and tomographic methods. These can dramatically reduce the effects of patient structure. Samei *et al.* (Samei, 2000) discussed some aspects of the structure problem.

The first attempts to quantify the effects of anatomical structure in projection radiographs were by Kundel and coworkers. Revesz *et al.* (Revesz, 1974) coined the term "conspicuity," to describe an estimate of the "visibility" of an abnormality (nodule in a chest image for example). Conspicuity for a nodule was defined using the ratio of: (a) signal strength (contrast or amplitude difference between the peak signal value and the mean local value) and (b) a measure of local radial gradients. So conspicuity depends on two things, signal strength and the complexity of the local background. Unfortunately, there is no way of incorporating such a measure into the SDT approach. The rest of this section will deal with methods that could only be used after the advent of digital imaging techniques.

The basic idea is to create hybrid images by adding realistic or simulated lesions (signals) to images of normal patient structure to serve as backgrounds. The patient structure images can also be manipulated to create related random noise images by randomizing the phases of the clinical images. Alternatively, random noise images with the same power spectrum as the clinical images can be produced. This hybrid image method ensures that the properties of the signal are completely controlled and can be made known to the observer (the SKE condition), and also ensures that the statistical properties of the backgrounds can be accurately estimated. Also the statistical properties of the backgrounds can be accurately estimated. Experiments can then be done using MAFC or ROC methods. The word "noise" will come up a number of times and different authors will sometimes have different definitions. In previous sections where simulated backgrounds were used, we were on safe ground because the images were always generated using random processes where the outputs are random noise and SDT could be safely used. I will first discuss some of the properties of random noise, and then I will discuss lesion detection (mass and microcalcification) in mammograms, and finally nodule detection in chest images.

Wikipedia has a number of entries concerning "noise." The first is in the context of sound where one finds, "*In common use, the word noise means unwanted sound or noise pollution. In electronics noise can refer to the electronic signal corresponding to acoustic noise (in an audio system) or the electronic signal corresponding to the (visual) noise commonly seen as 'snow' on a degraded television or video image. In signal processing or computing it can be considered data without meaning; that is, data that is not being used to transmit a signal, but is simply produced as an unwanted by-product of other activities. In Information Theory, however, noise is still considered to be information. In a broader sense, film grain or even advertisements in web pages can be considered noise. Noise can block, distort, or change the meaning of a message in both human and electronic communication. In many of these areas, the special case of thermal noise arises, which sets a fundamental lower limit to what can be measured or signaled and is related to basic physical processes at the molecular level described by well known simple formulae.*" There are other entries related to TV, radio, digital photography, etc. So, in the general sense, loud music played by a person in a nearby room can be "noise" to you since it can interfere with speech communication or even thinking.

In medical imaging, we are concerned about image structure that limits the ability of radiologists to detect abnormalities (lesions, alternatively referred to as signals). Visual scientists use the term "masking" to describe anything that reduces signal detectability – including other deterministic signals as well as random noise. Medical imaging people had tended to go the other way and call everything noise, such as "anatomical noise" or "structure noise," even if it had identifiable structure. It will be seen later in this section that patient structures in mammograms and chest images have some of the characteristics of a random process, but not completely. So the structure can be considered to consist partly of random noise and partly of recognizable features that can also make lesion detection more difficult. The recognizable features will mostly vary from patient to patient – examples are ducts and blood vessels in mammograms and ribs in chest images. One point that I want to emphasize very strongly here is that *the prewhitening procedure described earlier in the context of non-white random noise has nothing to do with taking the recognizable features into account when making signal detection decisions.* Prewhitening is a procedure used to optimize signal detection in the presence of spatially correlated random noise. This was discussed above in the introduction to SDT. The mathematical issues related to random noise will now be briefly discussed.

5.5.2 Random noise characteristics

Random noise arises from random processes. In contrast to deterministic processes, which give predictable results, random processes give results that can only be described by probability distributions. Random processes are also known as stochastic processes. A random process, in principle, produces an infinite number of sample functions which all differ in a probabilistic sense. Let V(t) be a real and continuous random process with sample functions, $v_k(t)$, where t is time and k is the index number (from 1 to infinity). Upper case letters indicate the process and lower case letters indicate sample functions. An individual sample function is also referred to as a realization. The collection of sample functions is known as an ensemble. A feature of a sample function is that it is, in principle, infinite in extent. One must be rather careful in thinking about random processes, as the following example shows. Suppose that you have a very large number of identical sine wave generators in your laboratory. You trip a circuit breaker at 5 pm each afternoon to shut them all off and close the circuit breaker at 9 am each morning to turn them all on at exactly the same time, t_0. The random process is such that the voltage output of the kth generator is a sample function defined by $v_k(t) = A\sin[2\pi(t - t_0) + \phi_k]$, where the phase term, ϕ, is the only random variable; the other parameters are deterministic – A, f, and t_0 are the same for all signal generators. If a stranger enters and looks at an oscilloscope monitoring the output of one signal generator, it will seem that he is viewing a deterministic sine wave. But there are two considerations. One is that the phase for the kth generator cannot be predicted given the phases for all the other generators. The second point is that the kth phase is not exactly known to anybody and, given the laws of physics, could never be exactly determined.

Strictly speaking, since images are two-dimensional (2D), one should be referring to random fields rather than random processes, but I will not use that term. There are a number of important classes of random processes. The first, known as "ergodic," is the least common and has the stringent requirement that all the statistical properties of the random process can be determined using one sample function – in the limit as its size goes to infinity. The second class, known as "strictly stationary," is a random process for which the joint probability distributions are the same for all locations. Then the probability distributions tell you all you need to know about the random process. Recall that a joint probability distribution can be described using all of its joint moments. The third class, "nth order weakly stationary," has the same joint moments for all locations up to nth order. SDT can be used for the subclass of "second order weakly stationary" processes, which are also known as wide-sense stationary (WSS) processes. The only requirement for WSS is that the first and second joint moments do not vary with respect to location or across the ensemble. This process will have a constant mean and variance, and the covariation of values at different locations will depend only on the distance between the locations. There is one subtlety to consider since some medical image structures have been found to have power-law spectra. A random process with a spectral density function of the power-law form $P(f) = K/f^\beta$ will have P(0) equal to infinity at zero frequency. So the variance will be infinite. So, strictly speaking, a random process with a power-law spectrum cannot be stationary. In practice, however, images have finite extent so one can set the value of P(0) approximately equal to the value of the first non-zero frequency spectral density. As was shown above, SDT can be used in a straightforward manner even if noise has spatial correlations. Another consideration is the probability density function (PDF) of noise amplitude fluctuations. The case of noise with a Gaussian PDF (known as Gaussian noise) is simple; if the first two moments are known then all higher order moments can be calculated. Unfortunately SDT with non-Gaussian PDFs is much more difficult.

The covariance issue will be discussed in one spatial dimension for simplicity. Let V(x) be a real and continuous random process with mean μ, variance σ^2, and sample function values v(x), where x is a location. The autocovariance function (ACVF) for two locations, x_1 and x_2, separated by a distance, ξ, is obtained using the expectation value, E[...], of the departure from the mean for the two locations

$$K_v(x_1, x_2) = E[\{v(x_1) - \mu\}\{v(x_2) - \mu\}] = E[v(x_1)v(x_2)] - \mu^2 = E[v(\xi)v(0)] = K_v(\xi, 0) \tag{5.3}$$

The function $K_v(\xi, 0)$ is usually abbreviated as $K_v(\xi)$. The autocorrelation function is defined by $R_v(\xi) = K_v(\xi)/\sigma^2$. The power spectrum (or power spectral density function) of the random process is the Fourier transform of the autocovariance function. It describes the frequency content of the amplitudes of the random process. The ACVF of white noise is a δ-function and the power spectrum is flat.

Note that the power spectrum involves squared quantities and it is always a real function. The phase spectrum of the random process is of no consequence. It will always be flat because the sample functions will have random and uncorrelated phase. This is a key point, which will arise later when various authors investigate the question as to whether a particular type of patient structure can be considered to be noise arising from a random process. If the patient structure has a non-uniform phase spectrum, it is not pure random noise.

5.5.3 Power spectrum analysis issues

Much of the work on effects of patient structure will involve power spectrum analysis so it is worthwhile considering it briefly here. The radial averages of patient structure spectra have been found to have power-law form, $P(f) = K/f^\beta$, where f is radial frequency. Power-law spectra are difficult to estimate accurately. The average exponents are typically around 3 so a linear-linear plot of P(f) versus f will be sharply peaked at low frequency. The results are much easier to interpret if the spectra are plotted in log[P(f)] versus log[f] form since a pure $1/f^\beta$ spectrum yields a straight line. It is not recommended to use linear-linear or log-linear plots because of the difficulty of interpretation. The sharp peak has another important implication. But first it is necessary to deal with the issue of data preparation.

Recall that the ensemble mean value was subtracted in the above autocovariance equation (5.3) calculation. The implication is that when spectral analysis is done using the method of discrete Fourier transformation of digital data, the ensemble mean should be subtracted from each region of interest (ROI) used – *not the mean of that ROI*. The latter approach will lead to serious error in estimation of spectral coefficients at low frequency. This is discussed below.

The next issue has to do with windowing (tapering) the ROI data so that it decreases smoothly to zero at the edges of the data array. If it is known that the spectrum is nearly isotropic then an isotropic window function should be used. The discrete Fourier transform (DFT) is implicitly periodic so an ROI of width W will be implicitly repeated with period W. An ROI image, r(x, y), is a subset of a larger image, g(x, y), obtained in essence by windowing the large image using a 2D Rect function w(x, y), so r(x, y) = g(x, y)w(x, y). So there will be large discontinuities in the implicitly periodic DFT representation. The DFT of the data from one ROI will be R(u, v) = G(u, v)**W(u, v), where ** indicates 2D convolution. Spectral estimation then involves averaging $|R(u, v)|^2$ over a large collection of ROIs. The convolution of "true" spectral data, $|G(u, v)|^2$ with the Fourier transform of the window function ensures that the spectral coefficient at frequency (u, v) will influence neighboring coefficients. This is called "spectral leakage" and is a bad thing, since the observed spectrum can be quite different from the true spectrum at some frequencies. The effect is particularly important for power-law spectra because the Rect energy spectrum $|W(u, v)|^2$ decays slowly, as $1/f^2$, along the (u, v) coordinate axes. Any window function has an important effect at low frequencies where spectra are changing very rapidly with frequency and spectral leakage will have a very strong influence at all angles. Now we return to the issue of which mean value to subtract. If this is done by inappropriately subtracting individual ROI means, the altered ROIs will have zero mean. The zero frequency coefficient of the DFT of each ROI will then be zero and the "forced" spectral density estimate at frequency (0, 0) will be near zero. Then spectral estimates at neighboring low frequencies will have inappropriately low values due to leakage. Leakage effects can be reduced by selection of a suitable window function, but the errors in coefficients are not completely eliminated.

The next step is estimation of the parameters describing the power spectrum. It is usually found that the 2D structure spectrum is reasonably close to being isotropic. So the radial average spectrum is calculated by averaging over angle. This is best done using log(spectral density) values averaged into log(radial frequency) bins of constant width. The results at low frequencies are systematically unreliable and must not be used in estimating the exponent of the power-law. This is due both to spectral leakage and the fact that a very small number of coefficients have been available in the 2D spectrum estimate. Similarly, high-frequency results where image quantum noise becomes important should not be used. Therefore a considerable fraction of the spectral data is not available for estimating the parameters K and β in $P(f) = K/f^\beta$.

There is a method of doing spectral analysis without a window function. I shall refer to it as 2D quadrant mirror replication. It has been used for many years but is rarely mentioned in the literature because it only works for spectra that are peaked at the origin. Aguilar (Aguilar, 1993) and Anguinano (Anguinano, 1993) described the method in companion papers. The basic idea is this. Assume the ROI data in an $N \times N$ array consists only of the representation of the letter "R." A larger array, $(2N - 2) \times (2N - 2)$, is created, the original data are placed in one quadrant, and mirrored versions are inserted (without duplication of boundary values) into the remaining quadrants. The boundary value duplication issue is not particularly important, so $(2N - 1) \times (2N - 1)$ or $N \times N$ arrays do not give significantly different results. The 2D mirror replication procedure is illustrated in Figure 5.6. The implicit periodic replication of the 2D mirrored ROI data no longer has discontinuities in data values at the unit cell boundaries so no window function is needed. This eliminates spectral leakage. There is still a small artifact along the spectrum coordinate axes due to the discontinuities in

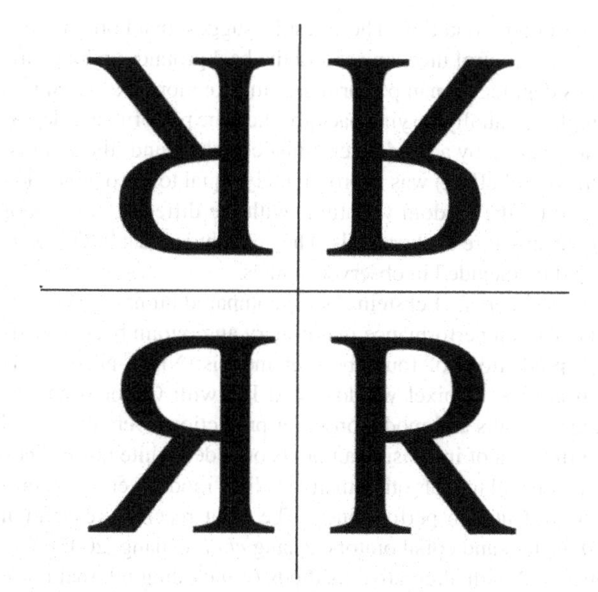

Figure 5.6 An illustration of the quadrant replication method of data preparation for DFT spectral analysis.

the gradients at the edges of the unit cells of the implicit periodic replication. This is very minor and easily corrected since the on-axis spectral coefficients are consistently twice as large as their nearest off-axis neighbors.

5.6 SIGNAL DETECTION WITH PATIENT STRUCTURE

5.6.1 Coronary angiogram backgrounds

Since the early 1990s Eckstein and a number of collaborators have investigated detection of simulated lesions added to highly variable backgrounds obtained from X-ray coronary angiograms. They were the first to use such "hybrid" images – where simulated lesions are added to real clinical backgrounds. The hybrid image approach ensures that the ground truth about image content is known when evaluating model and human observer detection performance. There is complete control of lesion properties such as size, shape, and contrast (and hence detectability) and lesion location in the backgrounds. Some of the experimental methods were very similar for the papers discussed below so the methods will be outlined here. Arterial segments were generated by mathematically projecting computer-generated 3D right circular cylinders. The signal was a hemi-ellipsoidal filling defect embedded within one of the 3D arterial segments. As a result, in the 2D projection the lesion (signal) appears as a brighter disc. The simulated artery and embedded signal were blurred with a Gaussian point spread function and then added to the patient structure backgrounds. They used a 4AFC detection task where the signal appeared at the center of one of four identical, clearly visible, simulated cylindrical artery segments added to different backgrounds that had been randomly selected from a database of angiogram ROIs.

Eckstein and Whiting (Eckstein, 1995) described the first investigation – which was to evaluate improvements in human and model observer performance due to edge enhancement of coronary angiograms. They did perceptual experiments using images with and without imaging processing. The detectability index, d′, for humans detecting the simulated filling defects increased by 20% after processing. Edge enhancement improved the d' value for the NPW model by 70%, which is not too surprising. The simple NPW model has no way of compensating for local background variations near the signal. The edge enhancement processing would reduce intensity variations due to slowly varying backgrounds and thus acts somewhat like the eye filter used in the NPWE model – which is far superior to the simple NPW model with random backgrounds. They found no improvement for two Fisher–Hotelling (FH) models that they tested – the local model with a small window (e.g. 12 × 12 pixels) and a channelized model with Gabor basis functions. As was mentioned in section 4.8.2, there are many names associated with linear discriminant analysis. The two of concern in medical imaging are Fisher and Hotelling. Their analyses only differ in how the covariance matrix is determined. The Fisher approach obtains an estimate from a finite collection of measured data. The Hotelling approach assumes that the covariance matrix is known *a priori.* So the Fisher approach is needed with data obtained from clinical images and the Hotelling approach can be used for simulated images. I will use the expression "Fisher–Hotelling (FH) model" through this entire chapter since both types of statistically defined backgrounds will be encountered.

Eckstein and Whiting (Eckstein, 1996) did experiments to determine whether two fundamental aspects of signal detection that were found to hold for human experiments with filtered noise could be reproduced when the backgrounds were angiogram structure. The first effect is that SDT makes specific predictions about the effect of changing the number, M, of alternative decisions in a forced-choice experiment when the underlying probability density functions are equal-variance Gaussians. The percentage of correct responses (PC) changes with M, for fixed signal and noise properties, but the value of d′ does not change because the transformation from PC to d′ takes the value of M into account. They used five values of M (2, 4, 8, 16, and 32) and found that the d′ results fell on a common line as a function of the square root of signal energy. This is in agreement with the prediction of SDT. The second effect is that d′ should be proportional to the square root of signal energy. They found that human results did not quite agree with this prediction. The relationship was linear but there was a small offset; the best-fit regression line did not pass through the origin. This is a common finding in all human signal detection experiments and is usually attributed to a small observer uncertainty about some signal property such as precise alternative locations – even though the observer is given cues to indicate possible locations. The offset is also found in experiments with white noise so no change had occurred when going to clinically realistic backgrounds.

The next experiment, by Whiting *et al.* (Whiting, 1996), was designed to compare the relative effects of increasing amounts of white noise when added to a uniform background in one condition and a randomly selected coronary angiogram background in the second condition. This was done as a function of signal contrast, added white noise variance, and feature motion. All four arterial segments in the four-alternative forced-choice (4AFC) task moved identically relative to the background in

32-frame image sequences displayed at 15 frames per second. Two naive human observers took part in the experiment. They found that, for all conditions, the threshold signal energy increased linearly as a function of added white noise variance. The regression lines had positive y-intercepts, as in Figure 4.5. The presence of the structured background increased both the y-intercept and the slope of the regression line. Thus, the presence of the structured background had a multiplicative effect, as well as an additive effect, on the degradation of performance. They concluded the multiplicative effect might be modeled by an increase in induced internal noise – see Burgess and Colborne (Burgess, 1998), as discussed in section 4.6.

The aim of the next investigation, Eckstein *et al.* (Eckstein, 1997a), was to determine the relative effects of two possible sources of human signal detection performance degradation due to the coronary angiogram background structure. The first effect is performance degradation due to random variation within an image and between images – that is to say the backgrounds act as noise. The second effect is performance degradation by a process called "contrast masking" which occurs even when the same distracting (deterministic) background is repeatedly used, so there is no random variation between images. Their experimental results showed that both effects contributed approximately equally to degradation of human signal detection performance. They concluded that both effects should be included in models of human observer performance. It should be noted that this (repeated) use of patient structure backgrounds is similar to the "twinned" background method mentioned in two discussions in Chapter 4. The "twinned" background method was used to estimate human observer internal noise and the conclusion was that, for white and low-pass noise, the backgrounds produced an "induced" internal noise whose spectral density was approximately equal to the image spectral density; see Burgess and Colborne (Burgess, 1988) and Burgess (Burgess, 1998). The induced internal noise effect is easy to include in the SDT models.

Eckstein *et al.* (Eckstein, 1997b) investigated whether modifying SDT models to include several attributes of spatial vision models would improve agreement between SDT model performance predictions and human performance. Spatial vision studies of detection of a signal superimposed on one of two identical (twin) backgrounds show performance degradation when the background has high contrast and is similar in spatial frequency and/or orientation to the signal. To account for this finding, spatial vision models include a contrast gain control mechanism that pools activity across spatial frequency, orientation, and space to inhibit the response of the receptor (channel) sensitive to the signal. In SDT-oriented tasks, in which the observer has to detect a known signal added to random-noise-like backgrounds, the main sources of degradation are the stochastic noise in the image and suboptimal human visual processing.

Eckstein *et al.* investigated how the noise and masking effects interact. They used three background conditions: (1) a uniform noiseless background, (2) repeated use of the same sample of a coronary angiogram background, and (3) different, randomly selected samples of coronary angiogram backgrounds. Their human observer results showed that detection degrades when going from the uniform background condition to the repeated background condition and degrades even further in the different backgrounds condition. These results suggest that both the contrast gain control mechanism and the background random variations degrade human performance in detection of a signal in a complex, spatially varying background. The performance degradation caused by a fixed deterministic background (the contrast gain control effect) was approximately equal to the degradation caused by the random variations with the different samples of patient structure backgrounds. They concluded that both effects should be included in observer models.

Eckstein *et al.* (Eckstein, 1998) compared human 4AFC signal detection performance in coronary angiogram backgrounds with predictions of four observer models: NPW, NPWE, FH with a 12 × 12 pixel window, and FH with Gabor channels. Human results and model observer predictions were compared as a function of increasing amounts of added white noise. They found that all models other than the NPW model were good predictors of human performance. The most recent investigation by Eckstein and collaborators, Zhang *et al.* (Zhang, 2007), was concerned with alternative methods of including internal noise in channelized FH models. This has the benefit of making the models more realistic and also reduces model performance so that it is more in line with human performance.

Zhang *et al.* (Zhang, 2007) evaluated two ways of inserting internal noise into Hotelling model observers. The first approach was to add internal noise to the output of individual channels, while in the second internal noise was added to the decision variable arising from the combination of channel responses. Details of the internal noise were also varied. Three FH model observers were studied: the FH observer with a small square window (FH_W), a channelized FH observer (FH_G) with orientation and spatial frequency tuned Gabor function channels, and a channelized FH observer with Laguerre–Gauss basis functions (FH_{LG}). Results were compared to human observer performance doing a 4AFC task of detecting known but variable simulated signals (the signal known exactly but variable (SKEV) paradigm described in Chapter 4). The signals were added to X-ray coronary angiogram backgrounds. They found that the internal noise method that led to the best prediction of human performance differed across the model observers. The FH_G model gave the best prediction of human performance when internal noise was added to channels. The FH_W and FH_{LG} models gave the best predictions of human performance when the internal noise was added to the decision variable. Their results can provide guidance to researchers about the best methods of including internal noise in FH model observers when evaluating and optimizing medical image quality.

5.6.2 Mammographic structure effects

Bochud and coworkers were the first to quantitatively investigate the effect of patient structure in mammography. In their first paper, Bochud *et al.* (Bochud, 1995) described a study of simulated microcalcification detection in hybrid images. They used the 2AFC method with five observers and the SKE paradigm. There were three kinds of backgrounds: (a) ROIs selected from digitized mammograms (sampling distance of 31 microns), (b) matched random noise images (i.e. random phase images) with the same power spectrum, and (c) images having only system noise. The mammogram images will provide backgrounds that are partly noise and partly identifiable structure. The other images will be random noise. The simulated microcalcification signal was the projection of a sphere (240 micron diameter) smoothed by the system MTF. They used four different signal amplitudes for each background type. Performance averaged across observers was significantly worse with the random phase images. Using 80% correct as a threshold criterion, the threshold contrast for detection of the signal in the random phase images was 0.08 optical density units. Detection thresholds for the mammographic and system noise backgrounds were both about 0.06 optical density units. They reached the conclusion that mammographic structure played essentially no role in detection of the small, simulated microcalcifications.

Bochud *et al.* (Bochud, 1997) continued the study of the relative effect of anatomical variation and system noise by including an 8 mm diameter simulated mass as well as a 220 micron diameter simulated microcalcification. The background ROIs were obtained from two mammograms and images of system noise alone produced using a Lucite phantom. The radial averaged spectral densities of the mammograms differed by a factor of about five at 1 cyc/mm but became identical above 1 cyc/mm. They added variable amounts of extra system noise to the mammogram ROIs. The results for the microcalcification detection were not consistent. System noise was the limiting factor in the case of mammographic anatomy with the lower spectral density. The situation was reversed for the mammographic anatomy with the higher spectral density – in that case the anatomical effect dominated. The results for simulated mass detection demonstrated that system noise was unimportant in this case. They compared human performance with that of an NPWE observer model. They had used four different amplitudes for each and prepared psychometric function plots (human d′ versus model SNR). The human functions were linear but had slopes below 1; they were about 0.7 for the microcalcification and 0.4 for the mass.

Bochud *et al.* (Bochud, 1999a) did 2AFC detection experiments with three different signals (an 8 mm simulated mass and two different simulated microcalcifications) added to two sets of mammogram backgrounds with different average power spectrum magnitudes (small and medium anatomical variations). The other experimental parameter was a variable amount of imaging system noise. They used an NPWE model to evaluate human results. One goal was to estimate the fraction of anatomical variation that could be considered to have arisen from a random process (they referred to this as the noisy component, and also "pure noise"). They compared human results to model predictions for the cases (a) where anatomy was considered to be deterministic and only system noise reduced detectability, and (b) where anatomy was considered to be completely random – so both it and system noise reduced detectability. They concluded that, for the mass, a small fraction of the anatomy acted as pure noise (14% and 5% for the cases of small and medium anatomical variations, respectively). The situation was dramatically different for the microcalcifications – small anatomical variations had very little effect on detectability while medium anatomical variations had the effect expected from pure noise with the same power spectrum.

Bochud *et al.* (Bochud, 1999b) did more sophisticated 2AFC observer experiments with four types of backgrounds – original mammograms, the mammograms after phase randomization, filtered random noise, and clustered lumpy backgrounds (CLB) with the same statistics (mean, variance, and spectrum) as the mammograms. They provided evidence suggesting that the statistics of their mammogram backgrounds were non-stationary (spatially varying). However, part of this finding may have been due to the fact that peripheral (uncompressed) parts of the breast may have been included. The average human d′ values for the randomized phase and filtered noise images were nearly identical and significantly lower (20%) than for the origin mammograms. The CLB backgrounds were produced using a complicated method of randomly placing clusters about each image. Each cluster contained randomly distributed Gaussian blobs (with randomly defined 2D properties). The individual 2D blobs were not isotropic and had random orientations. The scales of the clusters and blobs were selected to be quite different. It has been demonstrated that this method produces realistic looking simulations of mammograms. The average d′ results for the CLB case were significantly (30%) higher. This suggested that humans could make use of phase information in mammograms when doing the 2AFC tasks. They compared human performance to two observer models, NPW and NPWE. The performance of the two model observers were computed in two different ways – using the Fourier domain detectability index equations and by Monte Carlo experiments where calculated model templates were cross-correlated with the image pairs to make decisions. The template method has the advantage that it does not rely on an assumption of stationary background statistics. The NPW model did very poorly, as one would expect for statistically defined backgrounds. The NPWE template observer had slightly higher d′ values than predicted by the Fourier domain calculation and the d′ value for the mammographic backgrounds was slightly higher than those for the other three backgrounds. Both the human and NPWE template observer results are consistent with the view that mammographic structure is not pure random noise.

Burgess *et al.* (Burgess, 2001a) studied signal (lesion) detection in mammograms as a function of lesion size. They first determined the spectral properties of a large database of digitized normal images using ROIs in the compressed part of the breast. The images were carefully selected to be free of microcalcifications and artifacts. The power spectra were approximately isotropic (small angular dependence) and the radial averages of spectra had the form $P(f) = K/f^{\beta}$, with a mean exponent of about 3 (range from 2 to 4). Then they calculated the expected CD diagram for a PW observer model detecting a simulated mass in random noise with the average power spectrum. The diagrams' results were very surprising – the predicted slope was positive with a slope of +0.5 in log-log coordinates.

They then investigated human observer performance using the 2AFC hybrid image method. Scaled images of real tumor masses extracted from breast tissue specimen radiographs were added to normal tissue ROIs. The CD diagram slopes for these observer experiments were 0.3 for all four extracted masses when mass size was greater than 1 mm. The CD slope became negative for mass size less than 1 mm, suggesting that imaging system noise had become the dominant factor. All previous CD diagrams had been similar to Rose model results in white noise; the contrast threshold decreased as signal size increased and flattened at large signal sizes. They showed that the novel mammography CD slope is a consequence of the power-law dependence of the projected breast tissue structure spectral density. Under certain constraints of signal properties, the slope, m, of the CD diagram for detection in power-law noise can be described by the equation

$$m = (\beta - 2)/2 \tag{5.4}$$

One signal property constraint is that the integral for the PW observer model must converge. This means that the energy spectrum of the signal must decrease with frequency more rapidly than the power-law noise spectrum. The other constraint is that the signal spectrum (plus system MTF if included) must scale as signal size is increased. If this does not occur, then different CD slopes will be obtained for both human and model observers, as was shown by Chakraborty and Kundel (Chakraborty, 2001) and by Burgess (Burgess, 2001b). For example, suppose the signal is a sharp-edged disc smoothed by a Gaussian filter and the standard deviation of the Gaussian filter is fixed as disc diameter increases. The CD diagram slope for detection in $1/f^3$ noise in this case has a negative slope of –0.5.

The following is a simplistic explanation of this threshold amplitude versus size effect. As signal size increases, with amplitude held constant, its energy spectrum is increasingly concentrated at low spatial frequencies. The spectral density of patient structure is also increasing rapidly as frequency decreases. If the power-law exponent is large enough (greater than 2), then as lesion size increases the ratio of signal energy density and structure spectrum power will actually decrease. So lesion amplitude must be increased to maintain constant detectability. Another feature of power-law noise is that the pixel variance estimate depends on the size of the measurement area. This is very different from image noise – where variance becomes constant when the measurement area is considerably larger than the correlation distance of the noise. For the case of $1/f^3$ noise and mammograms, we found that the ensemble average variance increases linearly with the area of the measurement ROI. The relationship between variance and area changed dramatically from ROI to ROI.

Burgess *et al.* (Burgess, 2001a) did three other CD diagram investigations – human observer 2AFC detection of a simulated mass added to power-law noise images with a $1/f^3$ spectrum matched to the average mammogram spectrum, plus search for the simulated mass in known regions of mammogram backgrounds and the power-law noise. The results are shown in Figure 5.7. The CD diagram for the power-law noise images had a positive slope of 0.44, but thresholds were significantly higher than for the mammogram backgrounds. This was consistent with the Bochud *et al.* findings discussed above – the mammogram backgrounds were not acting as pure noise. The difference in 2AFC CD slopes was probably due to the fact that the mammogram ROIs had a wide range of power-law exponents (standard deviation of 0.35) so the CD slope for that case was an average across exponents while the filtered noise ROIs had nearly constant exponents (standard deviation of 0.03). The CD diagram slopes determined for an FH model were also different for the two types of background – they were 0.4 and 0.5 for the mammographic and filtered noise backgrounds, respectively. The average human absolute efficiency for detection in the filtered noise was 40% except for the 1 mm size, which had 24% efficiency. The average relative efficiency (compared to the PW observer model) for the mammographic background case is 57% with a range from 38% to 91%. The CD diagram comparison for the search experiments was completely different – mammogram backgrounds and matched power-law noise gave identical results. This suggests that for the search case mammogram backgrounds can be considered to be pure noise. Furthermore the CD diagrams for the search experiments could be predicted from the power-law 2AFC results using standard SDT analysis methods. An equivalent number of alternative locations, M*, was defined as the ratio of the search area and the signal area. This value of M* was used together with the 2AFC results for filtered noise to obtain the dashed curve through the search data in Figure 5.7.

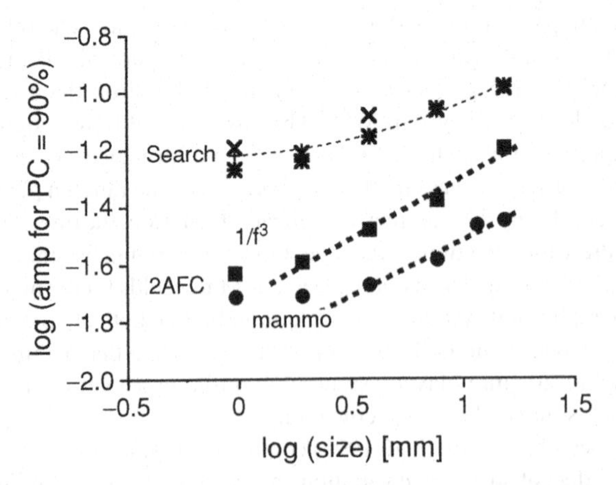

Figure 5.7 CD diagram results for detection of simulated mass signals in 2AFC and search experiments. The backgrounds were ROIs from the compressed part of mammograms and filtered power-law noise with the same exponent as the average for the collection of mammograms. The data points are averages for three observers. They had lower thresholds with the mammographic backgrounds for 2AFC detection. There was no significant difference in background dependence for the search task. The dashed curve through the search results is not a fit to the data. It was calculated from the 2AFC filtered noise results using an MAFC estimation method described in the text.

We also investigated detectability index results and CD slopes for a number of model observer procedures. One was the PW Fourier domain detectability calculation by integration. The second was a spatial domain determination using an FH model with channels and data acquisition using templates cross-correlated with the images used for the human experiments. The latter method does not require stationarity. There was very good agreement between the results for the two approaches – suggesting that statistical non-stationarity is not an important consideration. The third model, NPWE, was evaluated in two ways; detectability was calculated by numerical integration in the Fourier domain as well as by the spatial domain Monte Carlo method using the observer templates. The NPWE model gave good fits to the CD slopes but did not perform as well as the human observers for mammographic backgrounds, so it can be excluded.

Bochud *et al.* (Bochud, 2004) used MAFC experiments to investigate the effect of lesion location uncertainty on detectability of a microcalcification and a mass in normal mammographic backgrounds. In the simulated microcalcification (a projection of a 0.2 mm diameter sphere) experiment, there were M alternative locations in each image. The M possible locations were known to the observer and separation between locations decreased as M increased. The situation for the analysis of the performance of the MAFC task as it was performed in the microcalcification experiment is complex. The long-range correlations of the patient structure in each image will prevent statistical independence. In spite of this expected complexity, Bochud *et al.* found that their microcalcification results showed that performance changes with an increasing M could be approximated by MAFC SDT with the usual assumptions. This is consistent with the view that detectability of microcalcifications is mainly limited by image noise (and internal noise) rather than patient structure. However, a detailed analysis of the microcalcification results did show that there were statistically significant departures from the MAFC detectability index estimated using the SDT model. They investigated this effect in two ways. They found that the departures did not vary significantly as signal separation increased. This suggested that the departures did not arise from the violation of the assumption of statistically independent decisions because of background spatial correlations.

They also found that performance (percentage correct) diminished less steeply with number of possible signal locations than expected from decision variables with Gaussian PDFs. They then estimated the decision variable PDFs used by the observers and found very significant departures from Gaussian distributions. The PDFs were more compact than Gaussians. The design of the mass (10 mm diameter) experiment was very different. There were M independent background images (2 or 9) for each trial, with the signal present in one image. This ensures that structure is not correlated between alternative images. They found no statistically significant departure from the MAFC detectability index estimated using the SDT model. Therefore, for mass detection, they could not rule out the hypothesis of non-Gaussian decision variable PDFs. They also did a free search experiment where the microcalcification signal was located randomly anywhere within a defined area in the image. The observer used a cursor to select the most probable location. They determined the histograms of responses as a function of the separation between the actual and selected signal locations. The histograms of the two observers were very well fitted by assuming a 2D uniform distribution of incorrectly selected locations for distances greater than about one signal diameter. They also defined an effective number of alternative locations, M*, using the ratio of the area of the acceptable response circle and the search area (559 in their case). They found that there was very good agreement between the percentage of correct responses for the observers and with the prediction of SDT using the M* value.

It is well known that masses are harder to detect in dense breast tissue than in fatty tissue. Burgess *et al.* (Burgess, 2005b) determined the CD slopes of human and model observers for the two cases and estimated the size of the smallest masses that would be detectable in the two background types. They acquired a database of 6234 ROIs from the compressed part of normal and artifact-free full-field digital mammograms. The percentage of glandular tissue was estimated using a modified and semi-automated version of a quantitative method developed by Kaufhold *et al.* (Kaufhold, 2002). Spectral analysis was done

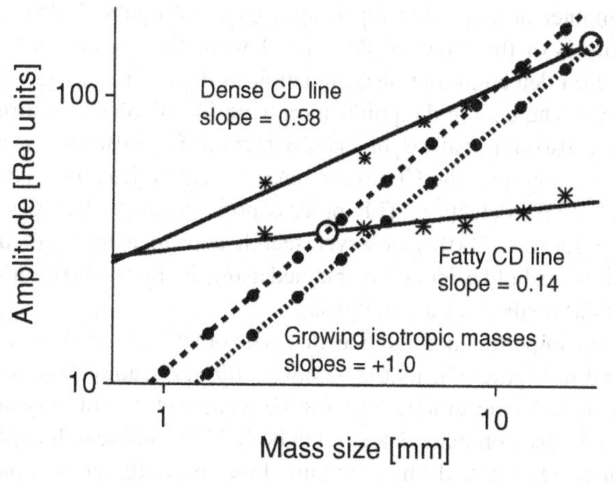

Figure 5.8 The CD diagram results (averaged across three observers) for detection of a real mass in fatty and dense tissue are indicated by * symbols and the best-fit regression results are indicated by solid lines. There is very good agreement between CD log-log slope measurements and theoretical predictions based on spectral analysis of the collections of images from the two density classes. The dashed line is for a growing mass in fatty tissue while the dotted line is for a mass in denser (25% glandular) tissue. The large circles where the lines cross indicate the mass sizes when they would become detectable with a 90% probability of correct response in a search experiment (d′ will depend on mass size).

on a collection of 673 fatty ROIs (12% glandular or less) and 867 dense ROIs (40% glandular or greater). The mean values and standard deviations of the power-law exponents were 2.42 ± 0.3 and 3.16 ± 0.3 for the fatty and dense ROI collections. The signal was a mass that had been extracted from a patient breast biopsy image and used previously (Burgess, 2001a). The CD diagram results (averaged across three observers) from 2AFC experiments are shown in Figure 5.8. There is very good agreement between expected CD slopes; 0.14 measured and 0.16 theoretical for the fatty case, and 0.58 for both measured and theoretical for the dense case. The growing mass lines are described below.

Now consider the effect of these CD diagram results on detection limits for an isotropically growing mass. By "isotropically growing mass," I mean that the three dimensions of the mass have constant ratios as the mass size increases. As can be seen in Figure 5.7, the CD lines for 2AFC and search detection are approximately parallel for mass size greater than 3 mm. However, search detection with 90% probability of correct responses required an increase in amplitude by a factor of about 3, a log(amp) difference of about 0.5. Since the amplitude units in Figure 5.8 are relative, the CD detection lines in that figure can also be interpreted as search CD lines with 90% probability of correct responses. If the mass is embedded in fatty tissue, the difference in linear X-ray attenuation coefficient between the mass and surrounding tissue is about 0.5 cm^{-1}, compared to the attenuation coefficient of 0.38 cm^{-1} for fatty tissue alone. The left-most dashed line with unit slope indicates the increasing amplitude (or contrast) of the mass signal as it grows surrounded by fatty tissue. The dashed line is positioned so that a 3 mm mass becomes detectable by a human observer with 90% probability in search throughout the entire compressed part of an average-sized breast. The growing mass data were obtained using a computer program developed for a paper (Burgess, 2005a) on analysis of the effect of detector element size on microcalcification detection and discrimination in digital mammography (DM). The data in the publication were for microcalcifications but I also did analysis of mass detection. It is also necessary now to interpret the CD line as one for search detection rather than 2AFC detection. This process normalized the Burgess *et al.* (Burgess, 2001a) observer search data to agree with the mass threshold calculation obtained using the Burgess (Burgess, 2005a) method for a 3 mm mass.

The important point to note that is no other normalization was done in Figure 5.8; that is to say *no further* scaling was done so the figure accurately portrays the relative effects of fatty and dense tissue on mass detection in both 2AFC and search experiments. The dotted line with unit slope shows the growth path of the mass when it grows embedded by 25% glandular tissue that gives a difference in linear X-ray attenuation coefficient of 0.25 cm^{-1}. This path crosses the dense tissue CD line at a mass size of about 20 mm. Therefore there is a dramatic effect of breast tissue density on mass detectability and most of the size threshold difference between fatty and dense tissue is due to the breast tissue statistics, as represented by the CD lines. The separation of the growth lines represents the lesser effect of intrinsic tissue contrasts. It should be noted that radiologists of course do not know the size or shape of a mass that might be present, so detection thresholds would be higher than for those in Figure 5.8. The radiological literature on size distributions of masses detected in mammograms indicates that the lower limit is about 5 mm. However, I have discussed the above results with some radiologists. They say that 3 mm masses can occasionally be "seen" but they are not definitive enough to be reported. Instead the patients are placed on a callback list to monitor changes.

As was mentioned above, the slope of the log-log CD diagram should depend on the power-law exponent as given by equation (5.4). Also, mammogram backgrounds have a range of exponents. My estimates based on digital mammogram ROIs from the compressed breast region have given values between 1.5 and 3.5. Burgess and Judy (Burgess, 2007) did experiments using filtered noise to determine whether the equation was valid over this range for a human observer. They found very good agreement for all exponents except for 3.5 where the CD slope was too low.

5.6.2.1 Observer templates with mammograms

Details of experimental and mathematical methods for estimation of the templates that human observers use when doing 2AFC SKE tasks with filtered noise are given in section 4.7. Abbey and colleagues did most of that work. The next step was to estimate templates in tasks involving clinical image backgrounds. Abbey *et al.* (Abbey, 2002) estimated human observer templates for detection of a 5 mm round simulated mass in mammographic backgrounds. They used a method that constrained the observer template to be represented by a limited number of linear features. This was done for five observers and then an overall average template was estimated. All the estimated templates had very little weight at very low spatial frequency and contained a band of large positive weight frequencies from 0.08 to 0.3 cyc/mm. The template weights then oscillated less

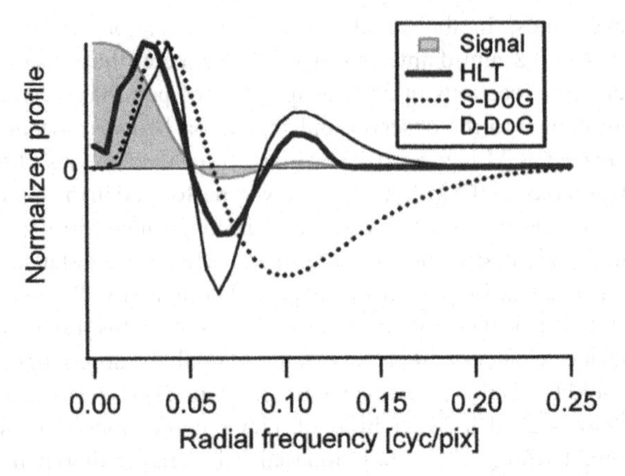

Figure 5.9 The Fourier transforms of human observer linear template (HLT) results – averaged over angle and across all five observers. From Castella *et al.* (Castella, 2007a). These are for signal detection in real mammogram backgrounds. The HLT results are compared to frequency domain results for the signal and templates used by two channelized FH models (sparse and dense DoG channels). (Reproduced with permission from the Optical Society of America.)

strongly between positive and negative values at higher frequencies. They compared these human estimates with templates for seven observer models. Several models gave templates in agreement with the human average for the first positive band but no models agreed with human results over the full frequency range.

Castella *et al.* (Castella, 2007a) estimated human observer templates associated with the detection of a realistic mass signal superimposed on real and simulated but realistic synthetic mammographic backgrounds. The signal was a projected spherical mass (4 mm diameter) extracted from a mammographic phantom. The simulated structure images were produced using the CLB method with parameters selected to match the power spectrum of the mammographic backgrounds. The nature of the mammographic backgrounds prevents the use of a direct linear-weighted sum of images to estimate the observer's template that had been used by Abbey and Eckstein (Abbey, 2006). So a more complex method was used. No attempt will be made to explain details here. The general idea is the same. Data from images selected by the observer to contain the signal (whether or not the 2AFC decision is correct) were combined. Similarly the data from images that were not selected were combined. An estimate of the template was then obtained using an iterative genetic algorithm. There were a lot of data. Five observers took part and the total number of trials per observer was 1400 for the real backgrounds, and 4000 for the CLB case. The Fourier transforms of the human observer linear template (HLT) results (averaged across all five observers) are shown in Figure 5.9. The figure also includes model observer results from Castella *et al.* (Castella, 2007b).

These results clearly demonstrate that an NPW detection model is not tenable. That model would use a template that is matched to the signal. The average observer template frequency response has a definite suppression of low spatial frequencies. But note that there is not complete suppression at zero frequency. The template result does not rule out an NPWE model because the eye filter would also show decreased response at low spatial frequencies. Castella *et al.* found that the observer performance results were much better for the mammographic backgrounds than for the CLB case. The 2AFC detectability indices would be 1.4 and 0.9, respectively. This is consistent with previous findings that mass detection in mammograms is easier than in power-law noise with matched statistics.

Castella *et al.* also did a twin-noise 2AFC experiment and used a new method to relate features in the backgrounds to observers' trial-to-trial decisions. In a twin-noise experiment, the same background is used in both images of the pair. A mathematically based observer would subtract the images and the stochastic background would disappear – so detection would be done with 100% accuracy. Human performance with twin noise is not perfect (as has been shown in a number of experiments), but it is much better than when the two backgrounds are different. They determined the scalar values of a number of statistical properties of the background image data. They found that no features were significant if measured over a large area centered on the two potential signal locations. The mass had a diameter of about 30 pixels and the only measurement region that gave significant features was 20 × 20 pixels. The four statistics that had the largest impact on detectability were the local standard deviation of pixel values and three measures based on local texture properties: coarseness, contrast, and strength.

Castella (Castella, 2007b) compared the human results described in Castella (Castella, 2007a) to results of a number of observer models for mammographic backgrounds and CLBs. Spatial domain templates for the models were determined and then Monte Carlo 2AFC decision trials were performed using the templates to obtain a percentage of correct responses. The two types of NPW models did worse than humans for both background types, so they can be dismissed. Two versions of the NPWE model were used; one had internal noise and one did not. The NPWE model without internal noise did much better than humans for both types of backgrounds. This result is not consistent with previous results of Burgess *et al.* (Burgess, 2001a), who found that humans did better than the NPWE model for mammographic backgrounds. When they added enough internal noise to the NPWE model to match the human CLB results, the surprising result was that the model performance with mammograms became the same as for CLBs. This was the only situation (human or model) in their study where performance with mammographic backgrounds was not significantly better than for CLBs.

Finally, the FH observer was evaluated using three different versions of channels: non-overlapping Rect functions with bandwidth proportional to centre frequency, sparse-DoG (with only three channels), and dense-DoG (with eight channels). The Fourier transforms of the templates used by the FH model with sparse and dense DoG channels are shown in Figure 5.9. Performance with dense-DoG channels and internal noise was also evaluated. The FH model results with Rect function channels were virtually identical to human results. The results with sparse-DoG channels were slightly worse than human results. These results suggest that the two models are not useful because humans are known to have internal noise – which was not included in the models. The FH model with dense-DoG channels performed better than humans. When internal noise was added to the model to fit the human CLB results, model results for mammograms determined using the same internal noise also matched human mammogram results. The results suggest that humans use much the same templates for 2AFC signal detection in the two types of backgrounds in spite of the fact that the mammographic backgrounds have recognizable structure.

5.6.2.2 Can mammography radiation dose be reduced?

The investigations of CD diagrams with mammographic backgrounds by Burgess *et al.* (Burgess, 2001a) suggest that patient structure is the limiting factor for detecting lesions larger than 1 mm in size while image noise dominates for smaller lesions. This, in turn, suggests that it might be useful to reconsider the technical factors (including tube voltage) that are used with the goal of reducing patient dose. Of course, image noise will increase as dose is reduced so the effect of dose reduction on detection and discrimination of microcalcifications and spiculations must be assessed. Observer models and human observer experiments have been used to evaluate the effect of radiation dose reduction on lesion detection in mammography. The results below suggest that a dose reduction of a factor of two may be feasible.

Chawla *et al.* (Chawla, 2007) added four types of simulated lesions to clinical mammograms. The lesions were benign and malignant masses plus two types of microcalcification clusters. They then added simulated image noise to mimic reduced dose. Performance of the channelized FH model observer was evaluated at three dose levels – full (normal) dose, half dose, and quarter dose. Two types of channels were used, Gabor and Laguerre–Gauss. Observer detection performance was evaluated by the ROC method using the SKEV paradigm. They concluded that 50% dose reduction had marginal or no significant effects on lesion detection. Samei *et al.* (Samei, 2007) used the same image sets as Chawla *et al.* (Chawla, 2007) to evaluate detection and discrimination performance by five experienced mammographers. They found that the detection of malignant masses was not significantly affected by dose reduction. The detection of benign masses was only significantly affected by dose reduction to one quarter of the normal dose. There was a clear reduction of detection of microcalcifications and discrimination between benign and malignant masses as dose was decreased.

Ruschin *et al.* (Ruschin, 2007) did human observer experiments using a free-response ROC (FROC) method to evaluate the effect of radiation dose reduction. They added two types of simulated lesions to normal digital mammograms. The five simulated malignant masses had been evaluated for realism and ranged in size from 8.7 to 9.4 mm. The five simulated microcalcification clusters had an average of 36 microcalcifications with average size 0.26 mm. They added two levels of simulated image noise so that the resulting images corresponded to 50% and 30% of the original dose level. Four radiologists experienced in mammography evaluated the images by searching for lesions and marking and assigning confidence levels to suspicious regions. There was no significant difference in results as a function of noise level for the mass detection task. There was a very significant decrease in microcalcification cluster detectability with increasing noise – indicating that a decrease in dose of a factor of two had an important effect.

Saunders *et al.* (Saunders, 2007) took the evaluation of technique factors in mammography one step further by also varying resolution. They investigated model and human observer performance for detection of masses plus detection and discrimination of microcalcification distributions at three dose levels. They modified clinical digital mammograms in two ways – changing the quantum noise level to simulate reduced dose and changing the resolution. They used three noise levels by adding correlated noise with spectra matched to the quantum noise in the original images. The effect of resolution on human performance was determined by using three different image display systems, one of which had far superior resolution while the other two were very similar. The resolution of the images used for model observer performance evaluation was varied by filtering of clinical digital mammograms. They produced a total of 2200 images for the experimental conditions. Five experienced breast-imaging radiologists took part in the experiments. The radiologist would view a mammographic region 5.12 × 5.12 cm and rate it into one of four categories: microcalcifications present in the center of the region, a benign mass present in the center of the region, a malignant mass present in the center of the region, or no lesion present. Overall lesion detection accuracy was computed as the average of sensitivity and specificity in detecting any lesion. They also evaluated performance of two observer models doing the same tasks. One model was the Sarnoff JNDMetrix visual discrimination model (VDM; Lubin, 1995), which is discussed in Chapter 4. The other model was the NPWE. They found that the performance of the radiologists was largely unaffected by decreasing resolution but decreased at higher noise levels. Specifically, noise had the greatest effect on their performance in microcalcification detection and mass discrimination. For the visual discrimination model, resolution affected performance strongly while noise had a more modest effect. The NPWE model was little affected by either resolution or noise. These results suggested that the observer models investigated in this study do not yet fully emulate human performance.

5.6.2.3 *Mammography tomosynthesis*

Digital breast tomosynthesis (DBT) is undoubtedly the next big thing in X-ray breast imaging. The procedure is essentially limited-angle CT with a 2D detector and a relatively small number of views. Many manufacturers are testing prototype systems. There are many proprietary reconstruction algorithms at the moment as well as methods proposed by university researchers. The algorithms have ad hoc aspects and many make nonlinear use of the data. This means that SDT models will be difficult to use to evaluate system performance and to compare lesion detectability results to those of projection mammography. Intuitively, one would expect better mass detectability with tomosynthesis images because of the reduced degrading effect of overlapping breast structure. However, it is quite possible that microcalcification detectability will be reduced. The effect of breast structure on mass detectability can be described mathematically.

One aspect of the tomosynthesis images is that the power spectrum will have a lower exponent on average. Patient structure in projection mammograms has an approximately isotropic power-law spectrum, $P(f) = K/f^{\beta}$, with an average exponent of about three. For very thin tomographic slices produced by standard CT methods with many views, one would expect an average exponent of about two. The reduction of one is predicted by application of the Fourier projection-slice theorem – see Gaskill (Gaskill, 1978). It is likely that the power-law exponent reduction will be somewhat different in tomosynthesis images because of the cruder data collection methods. The CD diagram slope will be zero for an exponent of two – see Burgess and Judy (Burgess, 2007). The other factor is that the scaling factor, K, of the average power spectrum should be lower. These two effects will make masses more detectable. However, it is also necessary to consider the effect of mass contrasts in multislice presentation. The study of lesion detectability in tomosynthesis images has been limited.

Zhou *et al.* (Zhou, 2006) did a task-based comparison of three non-proprietary mammography tomosynthesis reconstruction methods. The task was detection of a very low-contrast mass embedded in very dense fibro-glandular tissue. They used an anatomically realistic 3D digital breast phantom whose normal anatomic variability limited lesion detectability. They used an ensemble of 3D phantoms with randomly varying components: tree-like ductal structures, fibrous connective tissue, Cooper's ligaments, skin, pectoralis muscle, and power-law structural noise for small-scale object variability. Half the 3D phantoms had irregular masses of about 7 to 8 mm in size at variable locations in the breast volume. They simulated 11-angle data acquisition using realistic X-ray techniques and attenuation coefficient values. Three reconstruction algorithms were used: two deterministic methods – simple back-projection (BP) and simultaneous algebraic reconstruction technique (SART) – plus one statistical method, the expectation maximization (EM) algorithm. They generated 200 data acquisition realizations for reconstruction, half with signal present. Five human observers took part in 2AFC experiments with 200 image pairs per reconstruction algorithm. The potential mass locations were indicated but mass shape was variable. There were significant differences in percentages of correct responses (averaged over the observers). The images produced using the EM algorithm gave the best results (PC = 89%), the SART method was next (PC = 83%), while the results for the BP algorithm were very poor (PC = 66%). So it is clear that lesion detectability can be highly variable as reconstruction methods are changed.

Ruschin *et al.* (Ruschin, 2007) evaluated and compared the visibility of simulated tumors in 2D DM and breast tomosynthesis (BT) images of patients. Their method involved using two simulated 3D tumors with irregular margins. They were projected at appropriate angles to create 2D representations and then were added to each DM image as well as each BT projection image prior to 3D reconstruction. The same beam quality and approximately the same total absorbed dose were used for acquisition of each image type. Nine observers participated in a series of 4AFC human observer experiments designed to determine what signal intensities (contrasts) of the tumors were needed to achieve the same detectability, d′, equal to 1.96 (81.5% correct), with the two imaging methods. The required signal intensity was 0.010 (relative logarithmic units) in the central BT projection while a signal intensity of 0.038 was used for the DM images. Equivalent levels of tumor detection in BT images were thus achieved at around one-quarter the signal intensity in DM images, indicating that the use of BT may lead to earlier detection of breast cancer.

Good *et al.* (Good, 2008) assessed the ergonomic and diagnostic performance-related issues associated with the interpretation of BT-generated examinations. They used 30 selected cases and 9 experienced radiologists acted as observers. There were 3 image presentation conditions: projection DMs alone, the 11 low-dose projections acquired for the reconstruction of BT images, and the reconstructed BT examinations. Observers rated cases in three ways: using the FROC method, a case screening paradigm, and by providing subjective assessments of the relative diagnostic value of the two BT-based image sets as compared with projection digital mammograms. They found that their results were not statistically significant, mainly due to the small sample size.

5.6.3 Chest structure effects

It is well known that patient structure interferes with nodule detection in projection chest images. One aspect is bone – the chemical compositions of bone and tissue are very different, as is the dependence of X-ray attenuation coefficients on photon energy. Therefore there has been continuing study of dual tube voltage (kVp; sometimes referred to as dual energy) imaging methods to provide separate bone and soft tissue images. The work began with Lehmann *et al.* (Lehmann, 1981) and continues to this day. At the time of writing, the most recent paper is by Williams *et al.* (Williams, 2007). Another important aspect is the complexity due to vascular and bronchial structures.

Kundel *et al.* (Kundel, 1985) described the first quantitative study of chest structure effects in a paper with the delightful title "Nodule detection with and without a chest film." They used a simulated nodule radiographed first on a constant thickness phantom and then radiographed while it was on a chest phantom. Their data can be interpreted to show that the effect of the normal chest structure on nodule detection was about 25 times the effect of the radiograph quantum noise. As will be seen later in this section, this value is in quite good agreement with more recent estimates obtained using digital image methods.

Burgess (Burgess, 1985) did a 98AFC disc location identification experiment with white noise added to spatially varying but known backgrounds. Two backgrounds were sine waves with periods of twice and four times the disc diameter; the other was a square wave background with a period of four times the disc diameter. The disc locations were uniformly distributed over the phases of the periodic backgrounds. This experiment could be considered to be a crude simulation of the bone effect problem. Two different background amplitudes were tested: 20% and 40% of the mean background level. SNRs required for threshold detection ($d' = 2$) were elevated by about 40% on average but the elevation depended to some degree on disc location relative to the phase of the periodic background. The discs were consistently most difficult to detect when located at the transition between the bright and dark regions. The observers all reported that the task was quite different from detecting a disc in noise on a uniform background. With the uniform background, one or more candidate locations could usually be immediately "seen" in the image. The observer then considered the options and selected the most probable. With the varying backgrounds, the observers reported that they had to examine the bright, dark, and transition regions separately. The threshold elevations were much less than those seen with chest anatomy images but demonstrate the masking effect of even the simplest non-uniform backgrounds.

Samei *et al.* (Samei, 1999) investigated the relative effect of quantum noise and patient structure fluctuations on detection of subtle simulated nodules in chest radiographs. They did an ROC rating scale study with five experienced radiologists as observers using hybrid images and the SKE paradigm (with the known nodules located at the center of each ROI). The simulated nodules were digitally superimposed on 45 of 60 ROIs obtained from normal chest radiographs – by subtraction of the nodule profile from image data on a log(exposure) scale. The same numbers of quantum noise images for a uniform Lucite phantom were used. The X-ray exposure for the phantom images was carefully selected to give a quantum noise level equal to the average for the clinical images. Three nodule diameters were used for each image type. Contrasts were adjusted to ensure that the nodules were subtle. They used the nodule contrast-diameter product as the independent variable for data analysis. The product values for the quantum noise study ranged from 8.4×10^{-3} to 17×10^{-3}. The product values for the patient structure study ranged from 0.14 to 0.28. The nodules were much more detectable in the quantum noise – in the sense that a much lower contrast-diameter product was needed to reach a selected A_z value. Using a threshold value of the area under the ROC, A_z, equal to 0.8 and a fit to averaged observer data for the three contrast-diameter products, a product value of 0.2 would be needed for the patient structure case but only 0.014 for the quantum noise case, a ratio of 14. The (less accurate) estimate of 25 that one can obtain from the Kundel *et al.* (Kundel, 1985) experiment is in fair agreement with this ratio of 14. They measured the power spectra of the patient ROIs and plotted the radial average data on a log(NPS) versus linear frequency scale. I converted these results to an estimated log-log plot. The result appeared to be power-law to a good approximation for frequencies below 1.5 cyc/mm. The spectrum flattened out at higher frequencies where quantum noise becomes dominant. I estimated that the power-law exponent was about 3, which is in good agreement with exponents found later by Bath *et al.* (Bath, 2005b). Samei *et al.* compared human observer d' average results for all experimental conditions with calculated results for the PW and NPWE matched filter observer models. Both models performed better than the human observers. Note that they refer to the PW model as a Hotelling model but their appendix equation (A1) is actually the PW matched filter model. Also, both models gave similar results for both anatomic and quantum noise backgrounds.

Keelan *et al.* (Keelan, 2004) did experiments that were complimentary to those of Samei *et al.* Their conclusions were similar; imaging system noise plays a very small role in limiting nodule detectability. The difference was that they varied noise level rather than nodule contrast. This allows estimates of detectability over a range of system noise spectral density. They obtained an extremely low-noise image of a detailed anthropomorphic chest phantom and added varying amounts of system noise. The image acquisition was done using a digital detector system with 139-micron pixel pitch. They obtained ROIs for three selected chest regions, which they referred to as the rib, heart, and lung scenes. This would represent a range of anatomical structure. They superimposed a subtle version of a 10 mm simulated nodule at the center of the ROIs and did 4AFC experiments with 36 observers. They did analysis of detection thresholds (80% correct) as a function of system noise and extrapolated to zero system noise. This gave an estimate of the effect of anatomical structure alone. They compared the results to the effects of the variation in system noise that would be expected for the range of systems found in clinical use. They concluded this full range would affect the nodule contrast thresholds by only about 8% for the case of overlapping ribs and less in the other two anatomical regions.

A collaborative investigation of X-ray imaging optimization has recently been started by a team of European medical imaging scientists – the Radiological Imaging Unification Strategies (RADIUS) Group. Their initial findings – related to nodule detection in digital chest imaging – were published in a series of papers in 2005, which will be discussed here. Note that the first authors of the papers will vary. The aims of the studies were to determine the effects of the following on nodule detectability: (1) nodule location in the image, (2) several aspects of anatomical variation, and (3) imaging system noise. An introduction to the work is given in Bath *et al.* (Bath, 2005a). They assembled a database of 30 normal digital chest images. They then selected 5 regions for evaluation – with the goal of having nodule detectability approximately constant within each region. The regions were the lateral pulmonary, retrocardium, lower mediastinum, hilum, and upper mediastinum. ROIs were collected from the database and a 1 cm simulated nodule was added to half of the ROIs. The displayed images, g(x, y), were in the (inverted) detector dose domain. So the presence of a nodule would decrease the dose behind it and increase display luminance. The normalized signal profile, f(x, y), had a peak value of unity. The hybrid image, $g_H(x, y)$, with the nodule superimposed at location (x_0, y_0), was (before transformation for display) $g_H(x, y) = [1 - Cf(x - x_0, y - y_0)]\, g(x, y)$. They referred to C as the image contrast of the nodule. Other authors have produced additive mammographic hybrid images in the log(detector dose) domain so the displayed hybrid image was $g_H(x, y) = As(x - x_0, y - y_0) + g(x, y)$. The two hybrid image definitions are equivalent to first order – using a Taylor series expansion for the logarithmic grayscale transformation. However, some authors have described the signal-scaling factor as amplitude rather than contrast because it is less ambiguous. Contrast sometimes includes amplitude divided by local mean value. In each study, three contrast values were selected for each region to give appropriate detection difficulties. They did rating scale ROC experiments with five observers and the image sets from the five regions. Smooth curves were fitted to the average A_z data for the three contrasts to determine the image contrast required to obtain A_z equal to 0.8 (their threshold definition).

The first study was an investigation of location effects, described in Hakansson *et al.* (Hakansson, 2005a). First they determined the image contrast thresholds for each anatomical region. The results are shown in column 2 of Table 5.1. They also estimated the average scatter-to-primary (S/P) ratio to allow estimation of the object contrast inside a patient that would give

Table 5.1 *Contrast thresholds for the five regions under study and threshold ratios (relative to the clinical image values) for five other image versions. The regions are: lateral pulmonary (LP), retrocardium (RC), hilum (H), lower mediastinum (LM), and upper mediastinum (UM). The results in columns 2 to 5 are based on observer experiment results. The remaining columns (6–8) show results estimated using equation (5.5), a similar equation for $C(g_R)$ in Hakanson et al. (Hakanson, 2005c), and equation (5.6). The second column data are clinical image thresholds, $C(g_P)$, which are determined before correction for the S/P ratio. Columns 3 to 7 show threshold ratios relative to $C(g_P)$ for: random phase versions of the clinical images, g_R; identifiable patient structure (de-noised images), b_I; image noise, n_Q; estimated random patient structure, n_A; and random phase versions of the de-noised images, b_R. The values in the final column, labeled n_{FACTOR}, are estimates of the fractions of the thresholds in the clinical images that can be attributed to the sum of noise components, n_Q and n_A, using the quadratic threshold method.*

Region	Threshold $C(g_P)$	Threshold ratios for components					
		g_R	b_I	n_Q	n_A	b_R	n_{FACTOR}
LP	0.10	2.54	0.94	0.05	0.36	2.51	0.12
RC	0.10	1.35	0.87	0.08	0.49	1.26	0.24
H	0.26	1.20	0.87	0.02	0.49	1.10	0.24
LM	0.06	1.84	0.81	0.18	0.58	1.76	0.34
UM	0.14	1.60	0.71	0.06	0.71	1.42	0.5

A_z equal to 0.8. Finally they estimated the tissue thickness parallel to the X-ray beam that would be needed to give A_z equal to 0.8. The required contrast (by all definitions) varied by about a factor of 2.5 among the five regions but the order of values for the regions depended on which contrast threshold measure was used. However, the hilar region consistently had the highest threshold values, which should not be surprising. It is complex, with blood vessels and bronchial structures that have random orientations. When these structures are seen on end, they can mimic nodules. Before correction for scatter radiation, the lower mediastinum region required the least image contrast. It went to the middle of the set for the object contrast threshold measure (after S/P correction). The lateral pulmonary regions had the lowest object contrast threshold. They considered possible explanations for their findings. The explanation that they found most convincing is that the hilum contains structures that can be similar in appearance to subtle nodules and hence increase the difficulty of the decision task.

In the next study, Bath *et al.* (Bath, 2005b), they evaluated the possibly random component of patient structure using a technique that had been previously used by Bochud *et al.* (Bochud, 1999b) – who had compared the detectability of simulated lesions in mammograms with detectability in random phase images (i.e. random noise images) with the same power spectra. Bath *et al.* measured the power spectra of ROIs from the five regions of the set of 30 patient images that had been used for the location effect study. They used the quadrant mirror replication method described above. They found that the spectra (after averaging over angle) could be described by a weighted sum of power-laws, $P(f) = a/f^b + c/f^d$. The first term dominated at low frequency with exponents between 3.1 and 3.4 for the five regions. The second term exponents are in the vicinity of 0.5. This component dominates above about 1 cyc/mm and is probably due to system noise (it will be seen later that system noise had little effect on nodule detectability).

They then produced a large number of random noise images with the same power spectra for each region and added the simulated nodule to half. They did an ROC study to determine detection thresholds for these images and compared the results to those obtained with the patient images. The thresholds for the random noise images were significantly higher than those of the clinical images. The results were described using threshold ratios (random noise/clinical images). The ratios are given in column 3 of Table 5.1. These might be considered to be one measure of the extent to which the anatomical structure acts as random noise. This will be discussed later in the summary of all studies. A number close to 1 indicates that the structure is close to random. The average ratio was 1.59 across the five image regions. The largest ratio was for the lateral pulmonary regions (2.54). The smallest was for the hilum (1.20). The ratio for the case with nodules behind the heart was also small (1.35). The pulmonary area results are interesting. The ribs make a large contribution to the local noise power spectrum but do not seem to interfere that strongly with nodule detection.

The third study was an investigation into the effect of a particular component of structure on nodule detectability. In the paper by Bath *et al.* (Bath, 2005c), they made an interesting distinction between two types of structures. They defined *anatomical background* as identifiable structure whose appearance would not vary much with small changes in imaging geometry – such as bones and the heart. They defined *anatomical noise* as patterns that can change dramatically when small imaging geometry changes are made. They obtained two posterior-anterior (PA) chest images with slightly different geometry from nine volunteers – by moving the X-ray tube during one patient breath hold. They then used a "de-noising" method (Tischenko, 2003) based on the above distinction about anatomy.

The images were analyzed in the wavelet domain to determine the similarity of the images at various spatial scales. The noise component, which has a low correlation between the images, was then removed – this would include imaging system noise and anatomical noise (as they defined it). The remaining "de-noised" image was based on structure that was highly correlated between the two source images. The de-noised images were then segmented into the five regions as before. The subsequent procedures were the same as for the location effect study. The contrast thresholds for these hybrid images were compared to the thresholds that had been obtained for 30 patients in the location experiment. The thresholds were lower than for the original clinical images but not dramatically so. The ratios (de-noised/clinical) varied from 0.94 in the lateral pulmonary regions to 0.71 in the upper mediastinum. A high value indicates that there is very little difference in nodule detectability between the original images and the de-noised images. The value of 0.94 was barely significant given their estimates of experimental error. The average ratio across all regions was 0.83. When the comparison was limited to the images from the nine volunteers, the average ratio was 0.85. They were able to discount learning effects to some degree because two of the five observers had done the detection tasks with the de-noised images before doing the detection tasks with the original images.

The final study, Hakansson *et al.* (Hakansson, 2005b), was an investigation of the role of system noise. A large number of images of system noise alone were produced using a Lucite phantom selected to give the same scatter radiation and beam quality as the clinical images. Three dose levels were used, adjusted according to the image receptor doses estimated for the five regions in the clinical images. This ensured that system noise in the phantom images was the same as system noise in the corresponding regions of the clinical images. The simulated nodule was inserted at a random location in half the ROIs. The contrast threshold ratios (system noise only/original clinical images) are shown in column 5 of Table 5.1. The values ranged from 0.02 (for the hilum) to 0.18 (behind the mediastinum). The low value for the hilar region is consistent with previous results in that patient structure had a much greater effect on nodule detectability in that region. Hakansson *et al.* concluded that their results were consistent with those of Samei *et al.* (Samei, 1999), whose results they interpreted to indicate a contrast threshold ratio (system noise/clinical image) of about 0.07. The main conclusion was that imaging system noise played only a small role in limiting nodule detection. This will be discussed below.

Hakansson *et al.* (Hakansson, 2005c) summarized the results of the studies and discussed the role of the various aspects of noise and structure in limiting nodule detectability. They assumed that the clinical patient images, g_P, could be decomposed into additive components so $g_P = b_I + n_A + n_Q$, where b_I is identifiable (nonrandom) patient structure, n_A is patient structure that is not identifiable (considered to be random noise), and n_Q is quantum noise (which is assumed to dominate the system noise). They then made the interesting assumption that the contrast thresholds for images of individual components would contribute in a quadratic manner to the clinical image contrast threshold, $C(g_p)$, as uncorrelated processes so that

$$C^2(g_P) = C^2(b_I) + C^2(n_A) + C^2(n_Q) \quad (5.5)$$

The values of $C(g_p)$, $C(b_I)$, and $C(n_Q)$ were determined experimentally, so they used this equation to estimate values of $C(n_A)$, which had not been determined experimentally. The ratios, $C(n_A)/C(g_p)$, are given in column 6 of Table 5.1. They vary from 0.36 for the lateral pulmonary region to 0.71 for the upper mediastinum. Similarly, they decomposed the random noise images, g_R, into three components with the substitution of a random component, b_R, that would correspond to a random phase image obtained from the identifiable patient background (de-noised) component, b_I. The other two components, n_A and n_Q, were unchanged. The quadratic contributions to the random noise image threshold, $C(g_R)$, were described by an analogous combination equation. This equation was used to estimate the contrast threshold, $C(b_R)$, which had not been measured experimentally. The ratios, $C(b_R)/C(g_p)$, are given in column 7 of Table 5.1. A value of the ratio for $C(b_R)$ very close to the ratio for $C(g_R)$ would indicate that the identifiable patient background was the dominant factor in limiting nodule detectability. This occurs for the lateral pulmonary and lower mediastinum regions.

Hakansson *et al.* made the interesting argument that it might be better to base estimates of the contributions of the components to nodule detectability on squared threshold values rather than the threshold ratios themselves. Using this idea, one can calculate a noise contribution measure, n_{FACTOR}, to describe the relative role of system noise plus estimated patient random structure noise in reducing nodule detectability. The measure is defined as $n_{FACTOR} = \text{ratio}(n_Q)^2 + \text{ratio}(n_A)^2$, where the ratio values are given in columns 3 and 6 of Table 5.1. By reorganizing equation (5.5), one obtains

$$n_{FACTOR} = 1 - \left[\frac{C(b_I)}{C(g_P)}\right]^2 \tag{5.6}$$

As n_{FACTOR} goes from 0 to 1, the importance of the sum of noise components goes from none to completely dominant.

In summary, the results of the studies suggested the following. Imaging system noise plays only a very small role. There is considerable variation in contrast threshold among the selected regions of the chest images – a factor of 2.6. This variation appears for both contrast in the clinical images and object contrast estimated after correction for scatter radiation. Scatter radiation has very little effect on the object contrast threshold – for a given nodule size. Perhaps it will be necessary to determine CD diagrams to see if this is true in the sense of estimating when a nodule will become detectable as it grows. This is due to the fact that the object contrast and the cross-sectional area of the nodule increase during growth. The power spectra of the structures are not white, so the Rose model cannot be applied. The question of what fraction of patient structure can be viewed as random noise is harder to deal with and will now be discussed.

In this discussion I will assume that the random (non-recognizable) component of patient structure plays the major role in limiting nodule detectability to an expert who is very familiar with normal anatomy. Obviously this assumption may not be true and its violation could be the source of the inconsistencies mentioned below. It may be that the identifiable component plays a very important role but that would depend on the natures of both that component and the decision task. One could base the estimate on some function of the threshold ratios found by experiment or on estimates obtained using an equation such as equation (5.5) – with thresholds raised to some power. The choice of a quadratic form was ad hoc and it may be that some other exponent for summation would be superior. It would be expected that no matter which method is used, the rank order of the fraction estimate should be preserved. The rank orders are shown in Table 5.2.

There is no consistent order. The lower mediastinum had the lowest image contrast threshold and the random phase image experiment suggested a low importance for random structure in the clinical image. But the de-noised experiment suggested a relatively high importance. The hilum had the highest image contrast threshold and the random phase image experiment suggested the highest importance for random structure in the clinical image. But the de-noised experiment suggested an importance in the middle of the five regions. This discrepancy may be due to one aspect of the random noise image production procedure. The random images were based on power spectra averaged over angle while the 2D power spectra are unlikely to be isotropic. An alternative method would have been to randomize the phases of individual clinical images and use these to obtain random

Table 5.2 *Rank orders of the estimated importance of the random component of patient structure. A low number indicates a high importance. The columns g_R and b_I are based on measured thresholds relative to $C(g_P)$ for that chest region. The ranks for b_I, n_A, and n_{FACTOR} are identical because they are all related through the use of equation (5.5).*

Region	Threshold $C(g_P)$	Relative rank order g_R	b_I	n_A	n_{FACTOR}
LP	0.10	5	5	5	5
RC	0.10	2	3 (tie)	3 (tie)	3 (tie)
H	0.26	1	3 (tie)	3 (tie)	3 (tie)
LM	0.06	4	2	2	2
UM	0.14	3	1	1	1

background ROIs. This might have produced different results. The fact that the ranks for b_I, n_A, and n_{FACTOR} are identical is of no help since they are completely correlated. The values are all derived from the de-noised experimental results (since quantum noise was of little consequence) and the estimates, n_A and n_{FACTOR}, are based on equation (5.5). The inconsistency between results for the random phase images and de-noised images might also be due to some inexactness in the de-noising procedure. Finally, equation (5.5) may be based on an incorrect assumption about the exponent used for summation. This could be investigated by a threshold determination experiment using the noisy component that had been removed during the de-noising procedure. However, in spite of these inconsistencies, this series of very novel studies has produced interesting results and has produced useful progress in understanding the roles of the random and non-random components of chest structure.

5.7 WHERE ARE WE NOW?

5.7.1 Simulation experiments and models

There has been an amazing amount of progress in the understanding of medical image perception over the 35 years since I entered medical physics. Our ability to predict human performance of simple tasks under controlled conditions is vastly improved. In the early 1970s the Rose model represented the limit of our analysis capabilities. Today we can estimate human performance for a wide variety of tasks and even estimate the templates that we use. There are cases where the models have been successful enough to be used for optimization of image quality. Two examples are automatic task-based optimization of compression – see Zhang *et al.* (Zhang, 2004, 2005). There has also been a very large increase in the number of people working on perception problems and we have our own society with biannual conferences. Many very talented people have entered the field, they work on a number of topics, and many points of view are represented. I will try to summarize the current state of affairs in some areas. This is a personal view and not a consensus view, but I do hope it is close.

The modern work began with studies of task performance using white, Gaussian noise. The human results for the early 1980s (see section 4.6) showed that we could perform simple tasks with surprisingly high efficiency. At least the results seemed surprising at the time. Various experiments suggested that we are able to use the strategies employed by the Bayesian ideal observer. This allowed us to use a "top-down" SDT analysis approach using the detectability index equation of the ideal model as a starting point. We could then do experiments to investigate sources of inefficiency and modify the ideal observer model accordingly. This was a major departure from the "bottom-up" approach that was used at the time in spatial vision research – starting with known or assumed properties of the visual system beginning at the retina and using simple strategies for using information at higher levels in the brain. There was limited understanding of spatial vision psychophysics in 1980, so the "bottom-up" models were very simplistic. For simple tasks with white noise, it turned out that the modifications that were needed to the ideal observer model were quite minor. For example, with compact aperiodic signals, all that was needed was internal noise. When induced and static internal noises were included, the maximum predicted efficiency was about 60%. This was in good agreement with the value of 50% found in many experiments. Since the noise was white, there was no distinction to be made between prewhitening and non-prewhitening models.

Once the utility of the "top-down" approach was established, researchers began to investigate human performance with spatially correlated random noise and simulated random backgrounds (see section 4.8.3). The question was: can we compensate for random noise correlations? Or equivalently, can we take the correlations into account when making decisions? If this is done in an optimal manner, in the SDT terminology, we can do prewhitening. The selected random backgrounds were of two types: one was the "lumpy background" method of creating a discrete number of randomly placed Gaussian blobs with randomly selected parameters; the other was use of filtered noise. Experimental results were mixed. In some situations both the NPWE and the prewhitening models provided equally successful methods of describing human performance trends. Scaling of model performance was, of course, needed to fit human data (and hence account for other sources of inefficiency). On the other hand, the NPWE model did not work well for other situations. The message was that the NPWE model must be used with great care since it can sometimes give incorrect predictions. It does have the merit of simplicity and ease of performance estimates (doing Monte Carlo experiments using its template and image data). In my view, many people who are not sufficiently knowledgeable about observer modeling do not understand the dangers of casual use of the NPWE model.

The answer to the prewhitening question was a firm "somewhat." Two decades of research results suggested that we are only able to do partial prewhitening, but not completely compensate for spatial correlations in image noise. There are probably many explanations for this deficiency. In any event a relatively simple method was developed to include this phenomenon in a modified observer model. Myers and Barrett (Myers, 1987) suggested the use of a small, discrete set of spatial frequency channels. The equivalent image domain approach is to use a small set of spatially compact basis functions. One can think of the model as consisting of a bank of filters operating in parallel on input data. The important property of a filter is that it irreversibly mixes the noise within its frequency pass band and its output cannot be decorrelated. If the number of basis functions is small, a decorrelator after the filter bank can only do partial prewhitening. If the number of filters in the 2D frequency domain is extremely large and center frequencies are closely spaced, a decorrelator placed after the filter bank can still do a very good job of prewhitening the input noise. In fact, as the number tends toward infinity, the prewhitening tends toward completeness. The strategy in medical imaging has been to use a relatively sparse basis function set to control the amount of prewhitening and produce a better fit to human results.

The basis function approach can be difficult to implement in the frequency domain. The solution, developed by Barrett and coworkers, is to do calculations in the image domain using a vector-covariance matrix (quadratic form) approach adapted from pattern recognition analysis. The approach yields an LDA estimate of class separability (the two classes could be signal present and signal absent, for example). The model performance calculation then proceeds according to equation 4.19 on the historical aspects of SDT and related experiments. LDA separates the classes by a flat surface in a multivariate distribution function space. This linear classifier calculation goes by a number of names: Fisher's linear discriminant, Hotelling trace coefficient, and Mahalanobis distance, among others. The Hotelling trace is also called the Lawley–Hotelling or Hotelling–Lawley trace. It is unfortunate that the Hotelling name has come to dominate in medical imaging because the distinction between Fisher and Hotelling analysis only depends on how the covariance matrix is determined. As was mentioned above, Hotelling analysis is based on prior knowledge of the covariance matrix while Fisher linear discriminant analysis depends on determining it from image data. So, from a practical point of view, the difference depends on whether one is using simulated images with known statistics or using ROIs from clinical images whose statistics must be estimated. This distinction is of little consequence in medical imaging research – we do both.

In addition to allowing partial prewhitening, a major benefit of a relatively sparse basis function set is that it dramatically reduces the dimensionality of the LDA. This is an extremely important consideration. We normally use pixels as the basis set for images. As was mentioned in Chapter 4, pixels can be a good basis set when signals are small (say on the order of 10 pixels in a side or less) so analysis can be done using small ROIs – the windowed FH model approach. An accurate matrix estimate could be obtained using a few hundred images. Pixels are not a good choice for signal detection analysis for large signals. The covariance matrix will be very large and difficult to invert. Also, if it were necessary to estimate the matrix, a large number of images would be needed. A smooth and compact basis set with K elements makes a large difference. One can use an ROI that is about three or four times the support of the signal. But more importantly a vector with K entries can represent the essential features of the local image data. The covariance matrix is then $K \times K$. Typically a value of K of about 6 is adequate for isotropic and smooth signals and the reduced mathematical calculation is straightforward. So, the word "dramatic" is justified. However, ROI size and the value of K will depend on the nature of the background statistics. One must often do some investigation of alternative choices to see if there is any difference in the results.

Human CD diagram results with CT noise (with a ramp-like power spectrum for quantum noise) are still problematic.

As best I could determine by a literature search, no study has ever shown human results with the slope of –3/2 predicted from the power spectrum. The work that showed the closest agreement between human results and those of models was by Abbey and Barrett (Abbey, 2001) using simulated SPECT images with photon noise and simulated structure.

Probably the most interesting development in the last ten years has been direct estimation of human observer templates for a variety of tasks – by the method of classification images (see section 4.7 for details). Ahumada first used this method in acoustical signal detection research. He then (Ahumada, 1996) introduced it to spatial vision research for the study of vernier acuity. Abbey *et al.* (Abbey, 1999) extended it to other template estimations in noisy images. The previous psychophysical methods, comparing human and model performance as a function of some independent variable among a number of variables in a study, allows one to reach certain conclusions about which features of the ideal observer approach might be reasonably included in a human observer model and which human deficiencies might be included. Unfortunately, this indirect approach does not provide an unambiguous idea of what we are really doing. The template estimates for detection of a compact signal with a low-pass spectrum in white noise showed very good agreement between the template used by humans and that used by the ideal observer.

A subsequent study using discrimination tasks using a signal with a band-pass spectrum did not show such a clear relationship. The signal to be detected had zero spectral amplitude at zero frequency (DC) so any model using the signal as a template would have zero response at DC. Human observer template spectra did consistently show low-frequency suppression, but they were not consistently close to zero at zero frequency. However, the template estimates were good predictors of human task performance efficiency. Abbey and Eckstein (Abbey, 2007) also estimated human templates for a number of tasks with correlated noise. The human results showed strong evidence that we can adapt to different correlated noise textures. This was a very important point since it is desirable to be able to distinguish between the NPWE and PW models since the NPWE model is not able to adapt to noise power spectrum changes. However, there were again some significant departures from the templates that would be used by an ideal observer. Some templates for detection in white noise and high-pass noise suggested a suppression of human response at low frequency of the type found in the NPWE model that would not be present for the ideal observer. The other tasks were cases where the ideal (PW) observer would completely or nearly completely suppress the lowest frequencies but human data did not indicate quite as much suppression. In spite of the discrepancies in some details, the results were very impressive and consistent with the view obtained from other work that we can do partial prewhitening.

5.7.2 Current status of models

I will now summarize the current state of affairs with models. First, I will discuss the approximations that we make in order to deal with the complex reality of medical images. The use of simple linear systems theory requires that the imaging system be isoplanatic. That means that MTFs must not change over the image plane. This is definitely not the case for the X-ray focal spot MTF so blur of signals and patient structure will be a function of position. In spite of this fact, the standard practice is to use the focus MTF that is measured along the so-called central axis of the system. The second issue has to do with the Poisson nature of the photon noise. This is inconsistent with the requirement of additive Gaussian noise for use of the simple version of SDT. In one sense, this is not a big problem because the photon fluence at the detector plane is usually large enough that the distinction between the two PDFs at a particular location is of no consequence. Hence, for that aspect, the Gaussian assumption is a reasonable approximation. There is, however, an important distinction due to the fact that the photon fluence at the detector varies spatially in a given image – with attenuation changes along ray paths. So, the mean and variance of the Poisson noise will vary over the image. Secondly, all patients are different, so X-ray fluence will vary in the sample function index number direction of the space of the random process outputs. This means that there is no possibility that the image noise can be a WSS process. In spite of this fact, the system noise is usually assumed to be locally WSS for the purposes of SDT calculations. We make some of the same assumptions about the statistical properties of patient structure when it seems reasonable to model it as a Gaussian WSS random process – to a good approximation. So in summary, we make a number of assumptions in order to ensure that SDT calculations are tractable. This is standard modeling practice in science. Being reasonable scientists, we also take care to check as to whether our simplifying assumptions are a satisfactory representation of reality.

The next topic is the hierarchy of models. The ideal observer is *the gold standard* for top-down SDT analysis. In my view, we need to be very careful not to muddy the waters by sloppy terminology and misuse of the definition of the ideal observer. These will inevitably lead to confusion in our field. *The ideal observer is ideal and it is always ideal – full stop. Any model that includes deficiencies can never be ideal.* Barlow (Barlow, 1962) made this point in the context of evaluation of retinal use of photon fluence at extremely low light levels. A number of researchers doing similar research had started to modify the ideal observer model (as we do) to account for assumed human deficiencies, but they continued to use the term "ideal." It seemed that communication would soon be almost impossible because of the variety of meanings of the degraded "ideal observers." For example, in our field, a term such as "ideal channelized Hotelling observer" will only cause confusion. A model with a relatively small number of basis functions (channels) cannot be ideal so it seems inappropriate to imply that it is.

The other point about terminology is this. There are some situations where observer model X gives the same results as the ideal observer. That is because the full capabilities of the Bayesian ideal observer are not needed. For example, the PW matched filter model may be equivalent to the ideal observer for some task. It is confusing to reverse the sentence and say, "the ideal observer is equivalent to the PW matched filter model." That can suggest to a naive reader that the ideal observer is subordinate to the more general PW model, when in fact the opposite is true. The same applies to LDA models. They are used in situations where ideal observer performance is difficult or impossible to calculate (with known mathematical methods).

Sometimes LDA results are the same as ideal observer results, so the LDA model is equivalent to the ideal observer. At other times the LDA performance is simply the best possible using linear procedures. The superior ideal observer would use nonlinear procedures. When LDA and ideal observer analysis give identical results, it is confusing to a naive reader to suggest that the ideal observer is equivalent to the LDA model. It is the reverse that is true – the LDA model is subordinate to the more general ideal observer model. So one should say: "for this case, the LDA model is equivalent to the ideal observer."

It becomes increasingly difficult to calculate ideal observer performance as the complexity of the imaging situation increases. Unless otherwise stated, I will assume that the alternative signals are known exactly and that any background fluctuations can be treated as Gaussian WSS random noise. If the random noise is white then the ideal observer is simply a nonprewhitening matched filter observer because noise decorrelation is unnecessary. If the noise is correlated but WSS, the ideal observer uses the PW matched filter strategy. The NPWE observer model is sometimes a convenient approximation to a PW model with low-pass noise since it uses a bipolar center-surround template to remove local background fluctuations from the determination of decision variable values. However, in some situations humans do better than this model so it must be carefully tested. If the noise is not WSS, it is more convenient, and accurate, to do SDT analysis in the image domain using the model observer template and Monte Carlo methods with the noise ROIs. But there are some occasions where Fourier domain methods give comparable results.

If the goal is to take into account our limited capability to prewhiten then the LDA approach with basis functions is the best solution – to obtain partial prewhitening capability. The next step in complexity is the case in which the ideal observer would use nonlinear decision strategies. Ideal observer performance calculations then usually become mathematically intractable. In that case the LDA method will give an estimate of the best possible performance using linear strategies. This approach can be extremely difficult to implement using image samples (pixels) as a basis set. So the best approach is to use a basis set that is smoothly varying and localized in the image domain. This is usually referred to as a "channelized" observer model. Recall that a small (sparse) collection of "channels" is not an accurate representation of the extreme complexity of true human visual system channels. It is simply a method of reducing the dimensionality of the covariance matrix to facilitate calculations. It also has the benefit of being partially prewhitening, as humans seem to be. Since we are at present evaluating detection and discrimination of simple signals a sparse collection of "channels" is adequate. As we go to investigation of more complex signals the number of 2D "channels" that are needed will surely increase.

5.7.3 Patient structure

In the mid-1990s a number of people began to investigate signal detection in ROIs selected from clinical images. This immediately raised some important issues. SDT development was predicated on the assumption that the noise is Gaussian and WSS. There was no reason, *a priori*, to assume that either of these constraints is satisfied by ROIs from clinical images. For the case of mammograms, spatial variation of statistics is dramatically reduced by selecting ROIs that are from completely within the constant breast thickness part of the mammogram where the breast is in contact with the compression plate. ROIs that include any part of the peripheral part of the breast have systematic thickness variations, which severely confound both spectral analysis and signal detection analysis. In general, spatial variations will be due to a combination of system noise (nearly all quantum) plus two components of patient structure – identifiable anatomy and structure that might be considered random. In fact, as it turned out, mammograms came closest to being outputs of a random process.

The earliest experiments with mammogram ROIs were done by Bochud and coworkers, who found that system noise limited detectability of small objects like microcalcifications. The situation for masses in mammograms was much more interesting. Detectability of intermediate-sized objects like spiculations remains to be investigated. The first aspect of mass detectability was simple and well known to radiologists; it is completely limited by patient anatomical variation. In particular, superposition of normal structure in the projection images can mimic masses. Bochud *et al.* (Bochud, 1999b) did experiments with one mass size and compared human results with mammogram backgrounds to those with three kinds of statistically matched random backgrounds. Human performance with mammographic backgrounds was better than with two types of random backgrounds: (a) mammograms whose phases had been randomized, and (b) filtered noise with a spectrum matched to that of the mammographic backgrounds. However, the situation was the opposite for the third type – random lumpy backgrounds having matched second-order statistics. Human performance with these backgrounds was better than with mammographic backgrounds (I will return to this point below). A number of other 2AFC experiments demonstrated that we could detect masses more easily in mammogram ROIs than in filtered random noise with matched second-order statistics. The implication of this result is that we can, in a 2AFC experiment, discount some of the identifiable structure. Burgess *et al.* (Burgess, 2001a) found a very novel CD diagram with positive slope in a log-log plot. They showed that this was due to a $1/f^{\beta}$ power-law spectrum.

The results to date suggest that it is not unreasonable to assume that patient structure in mammograms is approximately due to a WSS random process. This suggests a simple method for modifying the PW observer model by correcting for the fact that the noise that *effectively* limits performance is not equal to that suggested by the measured mammogram power spectrum. Scaling factors are needed. The first factor will account for the detectability index ratio ($\rho = d'_R/d'_M$, random noise/mammogram) for mammogram and matched random noise backgrounds. The results of Bochud *et al.* (Bochud, 1999b) for a 10 mm mass gave the following ratios: 1.3 for clustered lumpy backgrounds and 0.75 for both the random phase images and filtered noise. Castella *et al.* (Castella, 2007a) found the opposite effect for CLBs using a more sophisticated simulation approach. The d' ratio for a 4 mm mass was about 0.6. This difference from the Bochud *et al.* results is not a cause for concern. It almost certainly indicates an improvement in the realism of the lumpy backgrounds. The CD diagram data of Burgess *et al.* (Burgess, 2001a) can be used to estimate d′ ratios for a range of sizes; the ratio was 0.67 for an 8 mm diameter mass.

Mammogram structure has less effect on mass detectability than the random noise with the same radial-averaged power spectrum. One can describe the mammographic power spectrum, $P_M(f)$, using an *equivalent* random noise spectrum, $P_{EQ}(f)$; that is a fraction of the noise spectrum, $P_R(f)$, which is matched to the second-order statistics of the mammogram ROIs. The equivalent random noise spectrum is given by $P_M(f) = P_{EQ}(f) = (1 - \rho^2)P_R(f)$. For $d'_R/d'_M = 0.7$, the value of $\rho^2 = 1 - (d'_R/d'_M)^2 = 0.5$. However, this is not the *effective* power spectrum to use in the SDT estimate of human performance. The *effective* power spectrum can be obtained by including an induced internal noise correction factor, $(1 + b^2)$, to obtain $P_{EFF}(f) = (1 + b^2)P_{EQ}(f)$. Burgess and Colborne (Burgess, 1988) estimated the factor $(1 + b^2)$ to be about 1.6 for white noise. Burgess (Burgess, 1998) obtained several estimates for low-pass noise; the most reliable was about 1.4. The effective power spectrum to use in 2AFC observer calculations for an 8 mm mass for a mammographic background would then be approximately

$$P_{EFF}(f) = (1 + b^2)P_{EQ}(f) = (1 + b^2)(1 - \rho^2)P_R(f) \approx 0.8P_R(f) \quad (5.7)$$

Human mass detection performance has been compared to that of a number of observer models using mammogram ROIs and a variety of random backgrounds. There has been variation in model results. The largest has occurred for the NPWE model. Most researchers evaluate NPWE model performance in two ways – numerical integration of the Fourier domain d′ equation and image domain Monte Carlo experiments using the NPWE template. Burgess *et al.* (Burgess, 2001a) found very good agreement between results. Bochud *et al.* (Bochud, 1999b) obtained different results – NPWE model results obtained by the Monte Carlo method were significantly better than those obtained by Fourier domain integration. This is not a very significant issue – it may be due to minor differences in the nature of the mammogram databases; for example, restriction on where the ROIs might be located.

Several methods have been used to simulate mammographic backgrounds. These include filtered noise with the same power spectrum, mammograms with randomized phase, simple lumpy backgrounds, and more complex CLBs. The Bochud (Bochud, 1999b) finding of better human detection of simulated masses in simple lumpy backgrounds than in mammograms suggests that simple lumpy backgrounds are not a good model to use. More recent results by Castella *et al.* (Castella, 2007b) using CLBs suggest that this problem has been solved. Castella *et al.* (Castella, 2008) developed a method of optimizing CLB parameters using a genetic algorithm. They did objective evaluations of radiologists' estimates of the realism compared to mammographic backgrounds. The results showed that the optimized CLB versions are significantly more realistic than the previous CLB model. Anatomical structures are well reproduced. It would be interesting to compare search experimental results for mammographic backgrounds and CLBs.

In summary, it seems that mammographic anatomy does not have a strong influence on detectability of small objects like microcalcifications. The opposite is true for large objects like masses. A number of 2AFC mass detection experiments have shown that mammographic backgrounds cannot be treated as truly random noise – *for that type of experiment.* This is not really surprising for 2AFC experiments. A detailed look at good-quality images shows clearly identifiable patient structure – blood vessels and ducts that are randomly oriented (to some degree), and skin pores. When radial average spectra are calculated, the orientation effects are also averaged – which can lead to a simple spectrum. So model performance calculated using mammographic power spectra per se is not really a good predictor of human 2AFC performance with clinical images. It is probable that this discrepancy can be accounted for in SDT analysis with mammogram backgrounds by a simple rescaling of the power spectrum. The situation is different in search experiments (with mass location uncertainty). Burgess *et al.* (Burgess, 2001a) found that mass detectability was identical for the mammogram and filtered random noise backgrounds with matched statistics. Furthermore they could predict the search experiment data using the 2AFC results for the filtered noise. It remains to be seen whether that is also the case for more complicated signal uncertainty tasks and search in complex CLBs.

The European RADIUS group began a study of nodule detection in chest images. They defined the image background as consisting of three components: (a) imaging system noise, (b) identifiable but variable patient structure, and (c) possibly random patient structure. They analyzed five regions separately: lateral pulmonary area, retrocardium, lower mediastinum, hilum, and upper mediastinum. One of their goals was to determine the relative contributions of the background components in limiting nodule detectability. This is a very difficult problem, since SDT, as it stands, has no way of dealing with non-random structure. They did make considerable progress toward their goal. They found that imaging system noise played a very small role and could usually be neglected. However, it is not yet clear how one should precisely calculate the relative effects of the two structure components. They tried two methods and the results do not appear to be consistent, but there was one exception. The only consistent result was that, by all measures, the apparently random structure in the lateral pulmonary region had much less effect than the identifiable but variable patient structure. The estimates of relative effects of apparently random structure for the other four regions depended on the estimation method. So, even though a start has been made, there is a considerable distance to go. It is an important problem, since it would be very nice to be able to modify our observer model calculations to include the effect of variable but identifiable patient structure in reducing nodule detectability.

REFERENCES

Abbey, C.K., Eckstein, M.P., Bochud, F.O. (1999). Estimation of human-observer templates for 2 alternative forced choice tasks. *Proc SPIE*, **3663**, 284–295.

Abbey, C.K., Barrett, H.H. (2001). Human and model-observer performance in ramp-spectrum noise: effects of regularization and object variability. *J Opt Soc Am*, **A18**, 473–487.

Abbey, C.K., Eckstein, M.P., Shimozaki, S.S., *et al.* (2002). Human observer templates for detection of a simulated lesion in mammographic images. *Proc SPIE Med Imag*, **4686**, 25–36.

Abbey, C.K., Eckstein, M.P. (2006). Classification images for detection, contrast discrimination, and identification tasks with a common ideal observer. *J Vision*, **6**, 335–355.

Abbey, C.K., Eckstein, M.P. (2007). Classification images for simple detection and discrimination tasks in correlated noise. *J Opt Soc Am*, **A24**, B110–B124.

Aguilar, M., Anguinano, E., Pancorbo, M.A. (1993). Fractal characterization by frequency analysis. II. A new method. *J Microscopy*, **172**, 233–238.

Ahumada, A.J., Jr. (1996). Perceptual classification images from Vernier acuity masked by noise. *Perception*, **25**(ECVP '96 suppl.), 18.

Anguinano, E., Pancorbo, M.A., Aguilar, M. (1993). Fractal characterization by frequency analysis. I. Surfaces. *J Microscopy*, **172**, 223–232.

Barlow, H.B. (1962). A method of determining the overall quantum efficiency of visual discriminations. *J Physiol (London)*, **160**, 155–168.

Barrett, H.H. (1990). Objective assessment of image quality: effects of quantum noise and object variability. *J Opt Soc Am*, **A7**, 1266–1278.

Barrett, H.H., Denny, J.L., Wagner, R.F., Myers, K.J. (1995). Objective assessment of image quality: II. Fisher information, Fourier crosstalk, and figures of merit for task performance. *J Opt Soc Am*, **A12**, 834–852.

Bath, M., Hakansson, M., Borjesson, S., *et al.* (2005a). Nodule detection in digital chest radiography: introduction to the RADIUS chest trial. *Radiation Protection Dosimetry*, **114**, 85–91.

Bath, M., Hakansson, M., Borjesson, S., *et al.* (2005b). Nodule detection in digital chest radiography: part of image background acting as pure noise. *Radiation Protection Dosimetry*, **114**, 102–108.

Bath, M., Hakansson, M., Borjesson, S., *et al.* (2005c). Nodule detection in digital chest radiography: effect of anatomical noise. *Radiation Protection Dosimetry*, **114**, 109–113.

Bochud, F.O., Verdun, F.R., Hessler, C., Valley, J.F. (1995). Detectability on radiological images: the influence of anatomical noise. *Proc SPIE Med Imag*, **2436**, 156–165.

Bochud, F.O., Verdun, F.R., Valley, J.F., Hessler, C., Moeckli, R. (1997). The importance of anatomical noise in mammography. *Proc SPIE Med Imag*, **3036**, 74–80.

Bochud, F.O., Valley, J.F., Verdun, F.R., Hessler, C., Schnyder, P. (1999a). Estimation of the noisy component of anatomical backgrounds. *Med Phys*, **26**, 1365–1370.

Bochud, F.O., Abbey, C.K., Eckstein, M.P. (1999b). Further investigation of the effect of phase spectrum on visual detection in structured backgrounds. *Proc SPIE Med Imag*, **3663**, 273–281.

Bochud, F.O., Abbey, C.K., Eckstein, M.P. (2004). Search for lesions in mammograms: non-Gaussian observer response. *Med Phys*, **31**, 24–36.

Burgess, A.E. (1985). Detection and identification efficiency: an update. *Proc SPIE*, **535**, 50–56.

Burgess, A.E., Colborne, B. (1988). Visual signal detection. IV. Observer inconsistency. *J Opt Soc Am*, **A5**, 617–627.

Burgess, A.E. (1998). Prewhitening revisited. *Proc SPIE Med Imag*, **3340**, 55–64.

Burgess, A.E., Jacobson, F.L., Judy, P.F. (2001a). Human observer detection experiments with mammograms and power-law noise. *Med Phys*, **28**, 419–437.

Burgess, A.E. (2001b). Evaluation of detection model performance in power-law noise. *Proc SPIE Med Imag*, **4324**, 123–132.

Burgess, A.E. (2005a). Effect of detector element size on signal detectability in digital mammography. *Proc SPIE Med Imag*, **5745**, 232–242.

Burgess, A.E., Jacobson, F.L., Judy, P.F. (2005b). Effect of breast tissue density on mass detection. Oral presentation. Presented at Medical Image Perception Society Conference XI, Windermere, UK.

Burgess, A.E., Judy, P.F. (2007). Signal detection in power-law noise: effect of spectrum exponents. *J Opt Soc Am*, **A24**, B52–B60.

Cargill, E., Barrett, H.H., Fiete, R.D., Kur, M., Patton, D.D. (1988). Fractal physiology and nuclear medicine scans. *Proc SPIE*, **914**, 355–361.

Castella, C., Kinkel, K., Verdun, F.R., *et al.* (2007a). Mass detection on real and synthetic mammograms: human observer templates and local statistics. *Proc SPIE Med Imag*, **6515**, 65150U.

Castella, C., Abbey, C.K., Eckstein, M.P., *et al.* (2007b). Human linear template with mammographic backgrounds estimated with a genetic algorithm. *J Opt Soc Am*, **A12**, B1–B12.

Castella, C., Kinkel, K., Descombes, F., *et al.* (2008). Mammographic texture synthesis: second generation clustered lumpy backgrounds using a genetic algorithm. *Optics Express*, **16**, 7595–7607.

Chakraborty, D., Kundel, H.L. (2001). Anomalous results for signal detection in mammograms. *Proc SPIE Med Imag*, **4324**, 68–76.

Chawla, A.S., Samei, E., Saunders, R., Abbey, C., Delong, D. (2007). Effect of dose reduction on the detection of mammographic lesions: a mathematical observer model analysis. *Med Phys*, **34**, 3385–3398.

Cohen, G., DiBianca, F.A. (1979). The use of contrast detail dose evaluation of image quality in a computed tomographic scanner. *J Comput Assist Tomogr*, **3**, 189–195.

Coltman, J.W. (1948). Fluoroscopic image brightening by electronic means. *Radiol*, **51**, 359.

Cook, L.T., Cox, C.G., Insana, M.F., *et al.* (1996). Contrast-detail analysis of the effect of image compression on computed tomographic images. *Proc SPIE Med Imag*, **2712**, 128–137.

Cunningham, I.A. (2000). Applied linear systems theory. In *Handbook of Medical Imaging, Vol. 1. Physics and Psychophysics*, eds. Beutel, J., Kundel, H.L., Van Metter, R.L., Bellingham, WA: SPIE Press, 79–159.

Eckstein, M.P., Whiting, J.S. (1995). Lesion detection in structured noise. *Acad Radiol*, **3**, 249–253.

Eckstein, M.P., James, S., Whiting, J.S. (1996). Visual signal detection in structured backgrounds. I. Effect of number of possible spatial locations and signal contrast. *J Opt Soc Am*, **A13**, 1777–1787.

Eckstein, M.P., Ahumada, A.J., Watson, A.B., Whiting, J.S. (1997a). What is degrading human visual detection performance in natural medical image backgrounds? *Proc SPIE Med Imag*, **3036**, 50–63.

Eckstein, M.P., Ahumada, A.J., Watson, A.B. (1997b). Visual signal detection in structured backgrounds. II. Effects of contrast gain control, background variations, and white noise. *J Opt Soc Am*, **A14**, 2406–2419.

Eckstein, M.P., Abbey, C.K., Whiting, J.S. (1998). Human vs. model observers in anatomic backgrounds. *Proc SPIE Med Imag*, **3340**, 16–26.

Gaskill, J.D. (1978). *Linear Systems, Fourier Transforms, and Optics*. New York, NY: John Wiley & Sons.

Good, W.F., Abrams, G.S., Catullo, V.J., *et al.* (2008). Digital breast tomosynthesis: a pilot observer study. *Am J Roentgen*, **190**, 865–869.

Hakansson, M., Bath, M., Borjesson, S., *et al.* (2005a). Nodule detection in digital chest radiography: effect of nodule location. *Radiation Protection Dosimetry*, **114**, 92–96.

Hakansson, M., Bath, M., Borjesson, S., *et al.* (2005b). Nodule detection in digital chest radiography: effect of system noise. *Radiation Protection Dosimetry*, **114**, 97–101.

Hakansson, M., Bath, M., Borjesson, S., *et al.* (2005c). Nodule detection in digital chest radiography: summary of the RADIUS chest trial. *Radiation Protection Dosimetry*, **114**, 97–101.

Judy, P.F., Swensson, R.G., Szulc, M. (1981). Lesion detection and signal-to-noise ratio in CT images. *Med Phys*, **8**, 3–23.

Judy, P.F., Swensson, R.G., Nawfel, R.D., Chan, K.H., Seltzer, S.E. (1992). Contrast-detail curves for liver CT. *Med Phys*, **19**, 1167–1174.

Karssemeijer, N., Frieling, J.T., Hendriks, J.H. (1993). Spatial resolution in digital mammography. *Invest Radiol*, **28**, 413–419.

Kaufhold, J., Thomas, J.A., Eberhard, J.W., Galbo, C.E., González-Trotter, D.E. (2002). A calibration approach to glandular tissue composition estimation in digital mammography. *Med Phys*, **29**, 1867–1880.

Keelan, B.W., Topfer, K., Yorkston, J., Sehnert, W.J., Ellinwood, J.S. (2004). Relative impact of detector noise and anatomical structure on lung nodule detection. *Proc SPIE Med Imag*, **5372**, 230–241.

Kundel, H.L., Nodine, C.F., Thickman, D., Toto, L. (1985). Nodule detection with and without a chest film. *Invest Radiol*, **20**, 94–99.

Lehmann, L.A., Alvarez, R.E., Macovski, A., *et al.* (1981). Generalized image combinations in dual KVP digital radiography. *Med Phys*, **8**, 659–667.

Lubin, J. (1995). A visual discrimination model for imaging system design and evaluation. In *Visual Models for Target Detection and Recognition*, ed. Peli, E., Singapore: World Scientific Publishers.

Muka, E., Blame, H., Daly, S. (1995). Display of medical images on CRT soft-copy displays: a tutorial. *Proc SPIE Med Imag*, **2431**, 341–359.

Myers, K.J., Barrett, H.H. (1987). Addition of a channel mechanism to the ideal-observer model. *J Opt Soc Am*, **A4**, 2447–2457.

Obuchowski, N.A., Beiden, S.V., Berbaum, K.S., *et al.* (2004). Multireader, multicase receiver operating characteristic analysis: an empirical comparison of five methods. *Acad Radiol*, **11**, 980–995.

Revesz, G., Kundel, H.L., Graber, M.A. (1974). The influence of structured noise on the detectability of radiological abnormalities. *Invest Radiol*, **9**, 479–486.

Riederer, S.J., Pelc, N.J., Chesler, D.A. (1978). The noise power spectrum in computed x-ray tomography. *Phys Med Biol*, **23**, 446–454.

Ruschin, M., Timberg, P., Svahna, T., *et al.* (2007). Improved in-plane visibility of tumors using breast tomosynthesis. *Proc SPIE Med Imag*, **6510**, 65101J.

Samei, E., Flynn, M.J., Eyler, W.R. (1999). Detection of subtle lung nodules: relative influence of quantum and anatomic noise on chest radiographs. *Radiol*, **213**, 727–734.

Samei, E., Eyler, W., Baron, L. (2000). Effects of anatomical structure on signal detection. In *Handbook of Medical Imaging, Vol. 1. Physics and Psychophysics*, eds. Beutel, J., Kundel, H.L., Van Metter, R.L., Bellingham, WA: SPIE Press, 655–682.

Samei, E., Saunders, R.S., Baker, J.A., Delong, D.M. (2007). Digital mammography: effects of reduced radiation dose on diagnostic performance. *Radiol*, **243**, 396–404.

Saunders, R.S., Baker, J.A., Delong, D.M., Johnson, J.P., Samei, E. (2007). Does image quality matter? Impact of resolution and noise on mammographic task performance. *Med Phys*, **34**, 3971–3981.

Schade, O.H. (1987). Image quality: a comparison of photographic and television systems. Reprinted in *SMPTE J*, **100**, 567–595.

Seltzer, S.E., Judy, P.F., Swensson, R.G., Chan, K.H., Nawfel, R.D. (1994). Flattening of the contrast-detail curve for large lesions on CT liver images. *Med Phys*, **21**, 1547–1555.

Sharp, P.F., Metz, C.E., Wagner, R.F., Myers, K.J., Burgess, A.E. (1996). *ICRU Report 54, Medical Imaging: the Assessment of Image Quality*. Bethesda, MD: International Commission on Radiological Units and Measurements.

Smith, S.W., Wagner, R.F., Sandrik, J.M., Lopez, H. (1983). Low-contrast detectability and contrast/detail analysis in medical ultrasound. *IEEE Trans Son Ultrason*, **SU-30**, 164–173.

Sturm, R.E., Morgan, R.H. (1949). Screen intensification systems and their limitations. *Am J Roentgenol*, **62**, 617–634.

Tapiovaara, M.J., Wagner, R.F. (1993). SNR and noise measurement for medical imaging. I. A practical approach based on statistical decision theory. *Phys Med Biol*, **3**, 71–92.

Tischenko, O., Hoeschen, C., Effenberger, O., *et al.* (2003). Measurement of the noise components in the medical x-ray intensity pattern due to overlaying non-recognizable structures. *Proc SPIE Med Imag*, **5030**, 422–432.

Wagner, R.F., Weaver, K.E. (1972). An assortment of image quality indices for radiographic film-screen combinations – can they be resolved? *Proc SPIE*, **35**, 83–94.

Wagner, R.F., Weaver, K.E., Denny, E.W., Bostrum, R.G. (1974). Towards a unified view of radiological imaging systems. Part I: noiseless images. *Med Phys*, **1**, 1–24.

Wagner, R.F. (1977). Towards a unified view of radiological imaging systems. Part II: noisy images. *Med Phys*, **4**, 279–296.

Wagner, R.F., Brown, D.G., Pastel, M.S. (1979). Application of information theory to the assessment of computed tomography. *Med Phys*, **6**, 83–94.

Wagner, R.F., Insana, M.F., Brown, D.G. (1985a). Progress in signal and texture discrimination in medical imaging. *Proc SPIE*, **535**, 57–64.

Wagner, R.F., Brown, D.G. (1985b). Unified SNR analysis of medical imaging systems. *Phys Med Biol*, **30**, 498–518.

Wagner, R.F., Insana, M.F., Brown, D.G., Garra, B.S., Jennings, R.J. (1990). Texture discrimination: radiologist, machine and man. In *Vision: Coding and Efficiency*, ed. Blakemore, C., London: Cambridge University Press, 310–318.

Wagner, R.F., Myers, K.J., Hanson, K.M. (1992). Task performance on constrained reconstructions: human observers compared with suboptimal Bayesian performance. *Proc SPIE*, **1652**, 352–362.

Wagner, R.F., Metz, C.E., Campbell, G. (2007). Assessment of medical imaging systems and computer aids: a tutorial review. *Acad Radiol*, **14**, 723–748.

Whiting, J.S., Eckstein, M.P., Morioka, C.A., Eigler, N.L. (1996). Effect of additive noise, signal contrast and feature motion on visual detection in structured noise. *Proc SPIE Med Imag*, **2712**, 26–38.

Williams, D.B., Siewerdsen, J.H., Tward, D.J., *et al.* (2007). Optimal kVp selection for dual-energy imaging of the chest: evaluation by task-specific observer preference tests. *Med Phys*, **34**, 3916–3925.

Zhang, Y., Pham, B.T., Eckstein, M.P. (2004). Automated optimization of JPEG 2000 encoder options based on model observer performance for detecting variable signals in X-ray coronary angiograms. *IEEE Trans Med Imag*, **23**, 459–474.

Zhang, Y., Pham, B.T., Eckstein, M.P. (2005). Task-based model/human observer evaluation of SPIHT wavelet compression with human visual system-based quantization. *Acad Radiol*, **12**, 324–336.

Zhang, Y., Pham, B.T., Eckstein, M.P. (2007). Evaluation of internal noise methods for Hotelling observer models. *Med Phys*, **34**, 3312–3322.

Zhou, L., Oldan, J., Fisher, P., Gindi, G. (2006). Low contrast lesion detection in tomosynthetic breast imaging using a realistic breast phantom. *Proc SPIE*, **6142**, 61425A.

6

Lessons from dinners with the giants of modern image science*

ROBERT WAGNER

* Adapted with permission of Oxford University Press from the original version in *Radiation Protection Dosimetry*, **114**, 4–10 (2005).

The organizers of this project have given me the honor and the opportunity to look back on some of the highlights of my career in the field of image science. When I started in 1972, many of the pioneers in this field were still active, so these reminiscences will take us back more than a half-century. I spent many delightful conferences, discussions, and dinners with all of the people I'll be referring to here. This review is just a sample of some of these memorable moments involving topics of continuing interest to many of us today. The chapter is organized according to the four categories of limitation to imaging performance: (1) Quantum-limited imaging; (2) Anatomical-background-limited imaging; (3) Artifact-limited imaging; and (4) Reader- or doctor-limited imaging.

We trace some critical moments in the history of image science in the last half-century from first-hand or once-removed experience. Image science used in the field of medical imaging today had its origins in the analysis of photon detection developed for modern television, conventional photography, and the human visual system. Almost all "model observers" used in image assessment today converge to the model originally used by Albert Rose in his analysis of those classic photo-detectors. A more general statistical analysis of the various "defects" of conventional and unconventional photon-imaging technologies was provided by Rodney Shaw. A number of investigators in medical imaging elaborated the work of these pioneers into a synthesis with the general theory of signal detectability and extended this work to the various forms of computed tomography (CT), energy-spectral-dependent imaging, and the further complication of anatomical-background-noise limited imaging. We call here for further extensions of this work to the problem of under-sampled and thus artifact-limited imaging, which will be important issues for high-speed CT and magnetic resonance imaging (MRI).

6.1 QUANTUM-LIMITED IMAGING

For most of us, the dominant figure from the mid-century was Albert Rose of RCA (called the Radio Corporation of America in my childhood). In the 1930s and 1940s Rose developed several advances in the sensitivity of pick-up tubes used in early television. He co-invented the Image Orthicon pick-up tube, which was used in guided missiles in World War II and served for several decades after the war as the major source of live television broadcasts.

Al Rose was doing fundamental research in photon detection and its conversion to electrical signals, and also had a keen interest in solar energy conversion. In the course of his many technological developments, he investigated the fundamental limits to image and signal detection using optical and electronic systems, including human vision and displays (Rose, 1946; 1948; 1953; 1973).

Rose realized that when imaging Poisson-distributed photons there are absolute limits on low-contrast detail detectability imposed by the statistical fluctuations in the finite number of detected photons. It is remarkable to observe today – given the fact that Rose's work was taking shape just a few years into the quantum age – how he appreciated the fact that the granularity of light quanta imposed limits to practical imaging system performance. An ideal detector was one that was limited only by that quantum noise – and this served as the baseline against which the so-called detective quantum efficiency (DQE) of real detectors with greater noise could be measured. Rose thus put many phenomena on a common footing, including human and electronic vision and photography.

Rose used his classic sequence of six images of the woman with flowers (Rose, 1953) – at different levels of exposure and therefore different signal-to-noise ratios (SNRs) – to estimate the DQE of practical imaging systems. He imaged that sequence of images with the system under test and observed how the results matched the originals with respect to the quality of what could be seen.

One of my favorite examples where he used that work was his analysis and prediction (almost 30 years ago) that home movies under "available light" using the photographic medium would not be able to compete with electronic movie-making for picture quality (Rose, 1976). He provided arguments that the DQE of a practical detector working with available light would have to exceed that of the human eye (a few percent) by one to two orders of magnitude. (One order of magnitude has to do with the reduction in lens aperture – and thus loss of light – needed for depth of focus of the camera; the other is related to the fact that the movies will be displayed at a higher brightness than the recorded brightness, which will make the eye more demanding.) The DQE of photographic film was stuck in the order of magnitude of one percent and he expected the DQE for electronic devices to approach 100% – and it is roughly halfway there now. In any case, I no longer know people who are shooting home movies on film, so the day he predicted has already arrived. (Photographic purists still prefer the conventional silver

The Handbook of Medical Image Perception and Techniques, ed. Ehsan Samei and Elizabeth Krupinski. Published by Cambridge University Press.

halide medium when high magnification is required, but digital technology is rapidly approaching their needs.)

Albert Rose also contributed a fundamental tool to our culture that we use every day, namely, what we all call "the Rose model" of signal detectability. In that model (Rose, 1946; 1948), the "signal" is the average integral over the region of interest on the image (assumed known *a priori*) minus the comparable measurement over the same area of the background. The "noise" is given by the fluctuations averaged over the comparable area of the background. We refer to this as the simplest form of "matched filter" since one integrates over an area *matched* to the expected signal. All forms of SNRs that arise from more formal approaches to signal detection theory degenerate into the Rose SNR in the limit of a simple disk signal in a background of uncorrelated noise. An excellent historical overview of all of these issues was recently given by Burgess (Burgess, 1999a; 1999b).

Otto H. Schade, Sr. – who also worked at RCA from the 1940s to the 1970s – was actually the first hero I discovered when I came into this field. Schade developed somewhat more elaborate models of signal detection than those of Rose. One of his monumental achievements was the analysis and consistent measurements of the imaging process of great interest to the TV viewing audience, namely, the chain from photographic movie capture on negative film to positive print and the optimal TV pick-up and coupling of that to a displayed version of the movie on a home TV screen (Schade, 1975).

I noticed that Schade's SNRs – including display for human vision – were excellent approximations to SNRs that derive from formal signal detection theory for an observer we later called the non-prewhitening matched filter. This observer does not take advantage of the optimal pre-processing of correlated noise known as "prewhitening," and so we started referring to it as the "non-prewhitening matched filter." Several key references to the relationship between the performance of the prewhitening matched filter, the non-prewhitening matched filter, and the incorporation of an eye filter and/or a finite number of receptive fields or channels are reviewed in Burgess, 1999a, Wagner, 1972, and Barrett, 1993.

For the 50th anniversary of the landmark work of Albert Rose, a special issue of the *Journal of the Optical Society of America* (Section A, Optics, Image Science, and Vision) was published in March, 1999. The editors of that issue (A. Burgess, R. Shaw, and J. Lubin) provide an anecdotal historical introduction (Burgess, 1999b). Then Burgess provides an elaborate review of the way in which the Rose model and signal-detection theory came together (Burgess, 1999a). Burgess himself played many pivotal roles in this history. He spent a sabbatical around the year 1980, split between the Cambridge (UK) labs of the vision researcher Horace Barlow and our own imaging research group here at the US Food and Drug Administration (FDA) outside of Washington, DC. He used the concept of the ideal (or Bayes) observer of signal detection theory as the standard against which to measure human performance for detection and discrimination tasks with noisy images and found human observer efficiencies in the range of 25–50%. These results encouraged us that the approach to signal detection analysis that we were using was indeed quite relevant to the perception of noisy images by human observers.

One of the central figures in the field of image science has been Rodney Shaw. In fact, I believe the expression "image science" was formally coined with the appearance of the book of that title by Dainty and Shaw (Dainty, 1974) in 1974. The first half of this book analyzes the photographic process from the point of view of statistical efficiency. Shaw greatly elaborated on the approach of Albert Rose and later Peter Fellgett (Shaw's Cambridge University (UK) mentor) to reach some conclusions regarding the statistical efficiency (DQE) of photography that were surprising to many investigators at the time – and are still surprising to our contemporaries who are not familiar with their work. Looking back over the DQE culture launched by Rose, Shaw remarked (Burgess, 1999b) that "the traditional photography community at first thought very little of having an absolute measure foisted on them by an outsider, especially since this measure made them custodians of a technology rated at less than 1% absolute efficiency."

In the book of Dainty and Shaw, the authors review the set of fundamental "defects" that lead to the low DQE of photographic film. Film has a so-called "threshold defect" – a minimum of about three photons are usually required to render a grain developable (for protection against fogging). So the maximum DQE one can expect from film is in the order of 33%. In the interest of providing photographic latitude, this number is designed to be a variable – but this variability is itself another source of noise which degrades the DQE. The variable grain size and variability of grain position are further sources of noise. Finally, film also has a "saturation defect" – that is (at least for general black-and-white photography), a grain is a binary or two-level (off-on) recorder. This coarse degree of quantization degrades the DQE further. In the end, the overall DQE comes down into the range of 1%.

In recent years Shaw has carried out an analysis of the process of digital photography (Shaw, 2003) and finds practical levels of DQE for that technology in the range of 10–30% (the latter for the black-and-white process).

An apparently subtle point was developed by Shaw in the course of all this work. Shaw spoke of the "inevitability" of the view that photographic noise will be thought of as amplified photon noise. His use of the word *inevitability* reflected the fact that a very long time was to elapse before this view became widely accepted in the photographic community (*decades* in the case of some investigators). The defects at each stage of the photographic process recounted in the previous paragraphs have an effect on the photographic SNR equivalent to that of a series of filters that would throw away corresponding fractions of the incoming photons – as in the famous family of six images of the woman associated with Al Rose's early work. The resulting noise or SNR is thus equivalent to that which would result from random capture of a number of photons much smaller than the number in the incoming photon stream, the fraction being in the order of 1%. Photographic granularity or noise is thus not some additive after-effect, but rather the result of this multiplicative cascade of random imperfect phenomena.

I always notice this phenomenon at work when I go to the movies and see a "Western" film and – presumably from now on – the reader will too! Whenever there is a scene of the wide-open sky, the film granularity is obvious not only because the scene is fairly uniform there but also because the noise is more perceptible when the scene is brighter. That is, the visual threshold for detecting small changes in contrast is lower when there is more light. I always think at this point, "I'm viewing amplified photon (or quantum) noise!" This may be among the few moments in life when the granularity of light photons is experienced so vividly. We rarely experience this with ordinary vision because the amplifier gain of the visual process appears to be set so that the noise is not quite visible (except at very low levels of light). (The reader who wonders why this noisy-sky phenomenon is not so obvious when watching Westerns on TV is encouraged to simply get up and turn off the room lights the next time such a scene comes onto the screen.)

Our appreciation of the work of Rose, Schade, and Shaw and our study of signal detection theory led my colleague David Brown and me in the mid-1980s to a formulation of the ideal observer performance for ten different modalities and geometries then used (or soon to be used) in medical imaging (Wagner, 1985). Kyle Myers, a protégé of Harry Barrett at the University of Arizona, soon joined our group and we were then invited by Professor Peter Sharp of Aberdeen to serve on the International Commission on Radiation Units and Measurements (ICRU) Report Committee #54. A synthesis of the fundamental principles of medical image science that derive from this and related works, together with many applications, were published in that report (ICRU, 1996).

An interesting historical sidelight was the fact that before the appearance of the ICRU report, there were some investigators in the medical physics community who claimed that the model underlying the so-called NEQ/DQE approach was not so obvious as its proponents claimed. Noise equivalent quanta (NEQ) is the numerator that combines the classical imaging measurements – including the modulation transfer function (MTF) and the noise power spectrum – in the modern formulation of DQE. Our response was that the NEQ/DQE approach *is not a model* (Metz, 1995). NEQ is simply a scaling of the measured noise power back to the axis of the exposure quanta through the system transfer characteristics. The comparison of the measured NEQ (what the image is "worth") with the measured level of the exposure quanta (what the image "cost") yields the DQE.

Several examples of the very good consistency between measurements and analysis following ICRU Report #54 compared to results of human observer experiments were given by Gagne *et al.* (Gagne, 1996). This consistency is our common experience when the task is simple signal detection against the background of a uniform test phantom.

One of the early applications of this approach makes an interesting historical point. When CT first became available it was not obvious why the exposures or doses were so high. The algorithm developers at the time thought it had something to do with the algorithms being suboptimal. My colleagues and I measured the physical characteristics of CT scanners – including the noise power spectrum of CT images on the absolute scale provided by CT numbers and their calibration to the attenuation of water – and compared them with what would be expected from ideal detection and reconstruction. We found that the two results were similar. (We are ignoring here the major problems with collimation in some early systems.) In our 1985 paper (Wagner, 1985) and the ICRU report (ICRU, 1996) we showed how the *noise multiplexing* intrinsic to the CT process was responsible for the high exposures required – a natural phenomenon that could not be finessed by better algorithms. The noise multiplexing arises during the fundamental process of strip or ray integration in CT; the signal of interest is collected from a region of a strip containing a lesion, say, while noise is collected from the entire strip (and ditto for every projection or view). This phenomenon is responsible for the high exposures necessary to achieve the great step in improved low-contrast sensitivity provided by CT. But the benefit is obviously not having to saw open the body for the task that CT addresses.

At that same time, Markku Tapiovaara was visiting our labs from the Radiation and Nuclear Safety Authority (STUK) in Finland. As he was preparing to return to Finland he asked us if we had some further problems we were worrying about at that time. We did – we were wondering about the issues that arise from the fact that X-ray detectors essentially measure an energy or current rather than counting photons. Markku's work on that problem led to an elegant formulation of the solution that we call the matched filter in the energy domain (Tapiovaara, 1985). The idea is analogous to looking for an orange in a noisy black-and-white photograph of a "still life." Such an image suppresses the energy or spectral information in the original scene. However, if the spectral information is available – say for the task of discriminating various kinds of soft tissue and/or calcifications in medical imaging – that information can be optimally weighted just as one does when one looks for the orange in a *color* photograph that carries the spectral information. Optimal image detection based on these principles will soon become available with photon-counting detectors and optimal signal processing (Lundqvist, 2003).

The history of our understanding of photon detectors that both amplify the incoming primary events and scatter the resulting secondaries provides another fascinating chapter of modern image science. Classical literature on this subject dates to the 1930s – but was limited to the large-area response. Shaw had generalized the earlier results into the spatial frequency domain and fine-detail response using semi-quantitative arguments based on well-understood limiting cases. A landmark contribution to the classical literature in this field – which capitulates much of the earlier work – was the paper by Rabbani, Shaw, and Van Metter (Rabbani, 1987) that provided a more general approach to that problem. This work was reviewed and extended to complex multi-stage systems by Cunningham and Shaw (Cunningham, 1999) in the special Rose celebratory issue of the *Journal of the Optical Society of America* cited earlier. We encourage young investigators to become familiar with this literature because many are surprised to learn that the Poisson component of the radiation stream goes through an amplifier *uncolored* by the system transfer function. Only the non-Poisson component gets weighted by the square of the system transfer function or MTF. This has many practical implications; in particular, it does not allow one to measure the MTF from the color of the output noise in a system with a Poisson-distributed input. Harry Barrett "reminds" his students that *Poisson* is French for "*independent*"!

The work sketched in the previous paragraph depends on the assumption of noise stationarity and the appropriateness of noise power spectral analysis. More recently, Barrett and colleagues (Barrett, 1997; 2004) have provided a more general approach to this problem based on correlated point processes that does not require this assumption.

6.2 ANATOMICAL-BACKGROUND-LIMITED IMAGING

We traditionally justified the use of simple phantoms and Fourier analysis of low-contrast detectability (e.g. NEQ and DQE analysis) because historically gall-bladder imaging and angiographic imaging were considered to be quantum-limited tasks, and we always wished to be conservative vis-à-vis those tasks.

Nevertheless, all these years we knew from the work of Harold Kundel and colleagues that real medical images give rise to limits to detection – often the dominant limitation – from the very anatomy itself, chest imaging being the paradigmatic example (Kundel, 2000; Samei, 2000). A landmark step in the direction of a combined analysis of quantum-limited and anatomical-limited imaging has been taken by Arthur Burgess and colleagues in recent years (Burgess, 2001). They analyzed the limits to lesion detectability generated by the variable tissue background in mammography. Modeling that background as power-law noise, and using the kind of model observers discussed in ICRU Report #54 (ICRU, 1996) (especially the appendices) they found very good agreement between the performance of human observers and that of the model observers. A striking result of that work was the fact that lesion detectability degraded with increasing size of the lesion in mammographic backgrounds whereas lesion detectability improves with increasing size of the lesion in the elementary Rose-model-like SNRs and simple phantoms where the background is uniform. The problem of sorting out radiation-limited tasks from anatomical-limited tasks has emerged as one of the dominant themes in contemporary research (Tingberg, 2004).

6.3 ARTIFACT-LIMITED IMAGING

An image is always a *sampled* version of the input scene or object, and this is particularly obvious with digital imaging systems. It is well known that *undersampling* of the original image scene leads to artifacts such as streaks, rings, or other patterns in the image that do not correspond to the original scene or object. Perhaps the most familiar example of this phenomenon is the Moiré pattern seen on TV images when the TV raster lines "beat" against a striped pattern of the clothing of a person on camera, resulting in the annoying fringe pattern that varies with the movement or position of the person.

A very nice summary of the Fourier-based approach to the understanding of undersampling artifacts for linear shift-invariant systems – including the concept of aliasing – was presented by Albert and Maidment (Albert, 2000), by analogy with the analysis of periodic structures in solid-state physics. Barrett and colleagues (Barrett, 1995) have taken a more general approach to this problem that is free of the assumption of shift invariance, and have demonstrated the role in that analysis of the "cross-talk matrix" previously described by them. The diagonal elements of this matrix are analogs of the system MTF squared; the off-diagonal elements measure how much interference or "cross-talk" a spatial frequency component in one input channel generates in the other spatial frequency channels. It is the hope of these authors that someone might build a test tool to measure the cross-talk matrix someday, and we still maintain that hope. A remarkable curiosity is that the approach of Barrett *et al.* is based on a Fourier series analysis on a finite support, not Fourier integrals, and their Fourier series approach *does not* assume shift invariance. The papers by Albert and Maidment and by Barrett *et al.* show – for problems where the noise is stationary – that a form of shift invariance is induced by averaging over the continuous range of possible positions for the target or lesion of interest. Gagne has presented some nice examples of this phenomenon (Gagne, personal communication, 2004). He demonstrated how, for low-contrast targets in simple backgrounds with stationary noise, averaging over the position of the target allows one to use the NEQ-based SNRs of ICRU Report #54 even for digitally sampled imagery.

Artifacts are a major issue for CT and MRI. Thirty years ago Joseph and Schulz (Joseph, 1980) took a fundamental approach to the analysis of the artifacts in CT. We consider here only the special problem of the finite number of *views* (as opposed to the finite number of samples per blur spot, which affects the presence or absence of aliasing). For the special case of parallel-projection geometry, their analysis reduces to the following compact result for the minimum number of projections N_{min} required for artifact-free reconstruction over a region within a radius R when the maximum spatial frequency in the reconstruction is υ_{max}:

$$N_{min} = 2\pi R \upsilon_{max} \tag{6.1}$$

Joseph and Schulz gave very vivid practical demonstrations of this expression at work using 90 views. This work moved into the background as the number of projections used in two-dimensional X-ray CT was increased to 180 and 360 or more as the technology evolved. However, it is relevant again today in the field of MRI, as we mention next.

Some of the original MRI images were made using the projection-reconstruction (PR) geometry and reconstruction methods used in two-dimensional CT at that time (early to mid-1970s). Fairly soon, however, investigators discovered how to obtain MR images in two-dimensional Cartesian k-space, or spatial-frequency space on an x-y grid. Many methods were then explored for traversing k-space quickly and more sparsely in the interest of faster imaging. A contemporary approach to this problem has been pursued by Mistretta and colleagues (Peters, 2000; Vigen, 2000; Barger, 2002) by reverting to the original PR geometry of CT imaging and obtaining speed by undersampling the number of views, with particular emphasis on three-dimensional PR MRI. These investigators demonstrated a 35-fold undersampling and associated speed-up for doing blood vessel imaging with an MRI contrast agent. For many applications, their so-called vastly undersampled isotropic projection imaging (VIPR) provides visual image quality competitive with much slower contemporary Cartesian-based approaches.

We mention these issues because a major goal of investigators from the school that I represent here is to devise *model observers* in the spirit of those used in ICRU Report #54 and related literature that are able to address the complication of the artifacts that arise in digital imaging in the presence of under-sampling. Myers, Wagner, and Hanson (Myers, 1993) showed that it is indeed possible to devise such observers even for the extreme case of CT imaging using only eight views. These model observers were used to explore the dependence of image quality on the parameters of nonlinear algorithms used for fast under-sampled systems. They showed that it was possible to design model observers whose performance tracked well with that of human observers of their images. We look forward to the day when test images of anatomically realistic phantoms that challenge deterministic artifacts can be quantitatively analyzed as readily as images of uniform phantoms of low-contrast signals are analyzed today using the NEQ/DQE approach.

6.4 READER- OR DOCTOR-LIMITED IMAGING

In the last decade the general field of receiver operating characteristic (ROC) analysis for studying the performance of imaging systems in the clinical setting has evolved into a multivariate approach called the multiple-reader, multiple-case (MRMC) ROC paradigm. This situation was driven by the great variability observed in the measurement of performance of human readers for a wide range of clinically important imaging tasks. One of the more dramatic examples of this variability was that studied by Beam, Layde, and Sullivan (Beam, 1996) for the population of US mammographers. The MRMC approach is now the predominant method for assessing medical imaging and associated methods of computer-aided diagnosis, both among academic investigators as well as sponsors of new technologies submitted for review to the FDA. An overview of the issues and several public case studies has been published by myself and a number of colleagues, as part of our efforts in the process of consensus discovery and guidance development for further industry and FDA interactions (Wagner, 2002; 2007). Some further practical advice for study designers was published in 2003 (Wagner, 2003). One of the contributions of our group in all of this is an approach for analyzing an observer study into its components of variance, allowing us to separate the uncertainties that arise from the readers – including the effects of computer aids to reading (Beiden, 2001) – from that which arises from the finite sample of cases and the underlying physics.

6.5 THE FUTURE

In this overview I've tried to provide samples of some critical moments in the evolution of our understanding of the problems associated with assessing the performance of medical imaging systems. During the last decade a major project was undertaken to analyze this problem in a very formal way using the same level of rigorous mathematics that was used in the development of quantum mechanics. A contemporary landmark of this project is the recent publication of *Foundations of Image Science* by Barrett and Myers (Barrett, 2004). This work includes the results of several decades of investigations by the authors and their colleagues, including many PhD and post-doctoral projects. It is easy to forecast that this monumental work will have an indefinitely long shelf-life. There is material in the book for launching countless new PhD and post-doctoral projects – and hopefully further efforts toward consensus measurements and guidance for practitioners in this rapidly maturing field that we share.

REFERENCES

Albert, M., Maidment, A.D.A. (2000). Linear response theory for detectors consisting of discrete arrays. *Med Phys*, **27**, 2417–2434.

Barger, A.V., Block, W.F., Toropov, Y., Grist, T.M., Mistretta, C.A. (2002). Time-resolved contrast-enhanced imaging with isotropic resolution and broad coverage using an undersampled 3D projection trajectory. *Magn Reson Med*, **48**, 297–305.

Barrett, H.H., Denny, J.L., Wagner, R.F., Myers, K.J. (1995). Objective assessment of image quality. II. Fisher information, Fourier crosstalk, and figures of merit for task performance. *J Opt Soc Am*, **A12**, 834–852.

Barrett, H.H., Myers, K.J. (2004). *Foundations of Image Science*. Hoboken, NJ: John Wiley.

Barrett, H.H., Wagner, R.F., Myers, K.J. (1997). Correlated point processes in radiological imaging. *Proc SPIE Med Imag*, **3032**, 110–124.

Barrett, H.H., Yao, J., Rolland, J.P., Myers, K.J. (1993). Model observers for assessment of image quality. *Proc Natl Acad Sci*, **90**, 9758–9765.

Beam, C., Layde, P.M., Sullivan, D.C. (1996). Variability in the interpretation of screening mammograms by US radiologists. *Arch Intern Med*, **156**, 209–213.

Beiden, S.V., Wagner, R.F., Campbell, G., Metz, C.E., Jiang, Y. (2001). Components-of-variance models for random-effects ROC analysis: the case of unequal variance structures across modalities. *Acad Radiol*, **8**, 605–615.

Burgess, A. E. (1999a). The Rose model revisited. *J Opt Soc Am*, **A16**, 633–646.

Burgess, A.E., Jacobson, F.L., Judy, P.F. (2001). Human observer detection experiments with mammograms and power-law noise. *Med Phys*, **28**, 419–437.

Burgess, A.E., Shaw, R., Lubin, J. (1999b). Noise in imaging systems and human vision. *J Opt Soc Am*, **A16**, 618.

Cunningham, I.A., Shaw, R. (1999). Signal-to-noise optimization of medical imaging systems. *J Opt Soc Am*, **A16**, 621–632.

Dainty, J.C., Shaw, R. (1974). *Image Science*. New York, NY: Academic Press.

Gagne, R.M., Jafroudi, H., Jennings, R.J., *et al.* (1996). Digital mammography using storage phosphor plates and a computer-designed X-ray system. In *Digital Mammography '96*, ed. Doi, K., Giger, M.L., Nishikawa, R.M., Schmidt, R.A. Amsterdam, Netherlands: Elsevier, 133–138.

International Commission on Radiation Units and Measurements (ICRU). (1996). Report #54. *Medical Imaging: The Assessment of Image Quality*. Bethesda, MD: International Commission on Radiation Units and Measurements.

Joseph, P.M., Schulz, R.A. (1980). View sampling requirements in fan beam computed tomography. *Med Phys*, **7**, 692–702.

Kundel, H.L. (2000). Visual search in medical images. In *Handbook of Medical Imaging. Vol. 1. Physics and Psychophysics*, ed. Beutel, J., Kundel, H.L., Van Metter, R.L. Bellingham, WA: SPIE Press, 837–858.

Lundqvist, M., Danielsson, M., Cederstrom, B., *et al.* (2003). Measurements on a full-field digital mammography system with a photon counting crystalline silicon detector. *Proc SPIE Med Imag*, **5030–5031**, 547–552.

Metz, C.E., Wagner, R.F., Doi, K., *et al.* (1995). Toward consensus on quantitative assessment of medical imaging systems. *Med Phys*, **22**, 1057–1061.

Myers, K.J., Wagner, R.F., Hanson, K.M. (1993). Binary task performance on images reconstructed using MEMSYS 3: comparison of machine and human observers. In *Maximum Entropy and Bayesian Methods*, ed. Mohammad-Djafari, A., Demoment, G. Dordrecht, Germany: Kluwer Academic, 415–421.

Peters, D.C., Grist, T.M., Korosec, F.R., *et al.* (2000). Undersampled projection reconstruction applied to MR angiography. *Magn Reson Med*, **43**, 91–101.

Rabbani, M., Shaw, R., Van Metter, R. (1987). Detective quantum efficiency of imaging systems with amplifying and scattering mechanisms. *J Opt Soc Am*, **A4**, 895–901.

Rose, A. (1946). A unified approach to the performance of photographic film, television pickup tubes and the human eye. *J Soc Motion Pict Eng*, **47**, 273–294.

Rose, A. (1948). The sensitivity performance of the human eye on an absolute scale. *J Opt Soc Am*, **38**, 196–208.

Rose, A. (1953). Quantum and noise limitations of the visual process. *J Opt Soc Am*, **43**, 715–716.

Rose, A. (1973). *Vision – Human and Electronic*. New York, NY: Plenum Press.

Rose, A. (1976). The challenge of electronic photography. *J Appl Photographic Engineering*, **2**, 70–74.

Samei, E., Eyler, W., Baron, L. (2000). Effects of anatomical structure on signal detection. In *Handbook of Medical Imaging. Vol. 1. Physics and Psychophysics*, ed. Beutel, J., Kundel, H.L., Van Metter, R.L. Bellingham, WA: SPIE Press, 655–682.

Schade, O.H. (1975). *Image Quality: A Comparison of Photographic and Television Systems*. Princeton, NJ: RCA Laboratories.

Shaw, R. (2003). End-to-end linearity considerations for photon-limited detection and display systems. *Proc SPIE Med Imag*, **5030**, 414–421.

Tapiovaara, M.J., Wagner, R.F. (1985). SNR and DQE analysis of broad spectrum X-ray imaging. *Phys Med Biol*, **30**, 519–529.

Tingberg, A., Bath, M., Hakansson, M., *et al.* (2004). Comparison of two methods for evaluation of image quality of lumbar spine radiographs. *Proc SPIE Med Imag*, **5372**, 251–262.

Vigen, K.K., Peters, D.C., Grist, T.M., Block, W.F., Mistretta, C.A. (2000). Undersampled projection-reconstruction imaging for time-resolved contrast-enhanced imaging. *Magn Reson Med*, **43**, 170–176.

Wagner, R.F., Beiden, S.V., Campbell, G., Metz, C.E., Sacks, W.M. (2002). Assessment of medical imaging and computer-assist systems: lessons from recent experience. *Acad Radiol*, **9**, 1264–1277.

Wagner, R.F., Beiden, S.V., Campbell, G., Metz, C.E., Sacks, W.M. (2003). Contemporary issues for experimental design in assessment of medical imaging and computer-assist systems. *Proc SPIE Med Imag*, **5034**, 213–224.

Wagner, R.F., Brown, D.G. (1985). Unified SNR analysis of medical imaging systems. *Phys Med Biol*, **30**, 489–518.

Wagner, R.F., Metz, C.E., Campbell, G. (2007). Assessment of medical imaging systems and computer aids: a tutorial review. *Acad Radiol*, **14**, 723–748.

Wagner, R.F., Weaver, K.E. (1972). An assortment of image quality indexes for radiographic film-screen combinations – can they be resolved? *Proc SPIE Med Imag*, **35**, 83–94.

PART II

SCIENCE OF IMAGE PERCEPTION

Perceptual factors in reading medical images

ELIZABETH KRUPINSKI

The interpretation of medical images relies on a combination of many factors including perception, cognition, human factors, technology, and even to some extent innate talent. It is difficult if not impossible to consider any of these in isolation because they interact in a number of both clear and subtle ways. However, when we study the interpretation of medical images, by necessity we sometimes must examine only a single variable at one time in order to try to fully understand its contribution to the overall interpretation process. Such is the case with perception. When one considers how clinicians read medical images, three basic stages are typically regarded as being involved: seeing, recognizing, and interpreting. It may sound simple, but the potential for failure at any point in the interpretation process is actually quite high – errors are made.

Radiology has for many years been the most common clinical specialty in which medical images are used to render diagnoses. Because of this, most image perception research to date has been done in this area. With the advent of telemedicine, however, there are a number of other clinical specialties in which images are used routinely such as pathology, dermatology, and ophthalmology. Virtual pathology is based on the scanning of glass slides into "virtual slides" (Weinstein, 2004). Teledermatology uses off-the-shelf consumer-grade digital cameras to acquire photographs of skin conditions (see Figure 7.1) (Krupinski, 1999a), and teleophthalmology uses digitally acquired photographs of the retina and other eye structures using digitally based non-mydriatic cameras (Taylor, 2007). The clinical areas in which digital images are being acquired for remote consultations via telemedicine are increasing tremendously (Krupinski, 2002).

All of these image-based clinical applications have a number of common features. In every case, clinicians rely on the optimal presentation of the image data in order to render a complete and accurate diagnostic decision. However, no matter what the image acquisition and display technology, abnormalities can be missed or misinterpreted, and the very important question is how do we avoid such errors? It is certainly possible to improve the imaging systems or develop new systems for imaging and analysis that will provide new and important information that was not available with traditional methods. In radiology improvements are continuously being made and new technologies developed. The virtual slides in pathology are now being scanned at multiple depths, recreating virtually the way that pathologists would focus on different planes of interest within a traditional specimen. Automated image analysis and interpretation schemes (i.e. computer-aided detection and diagnosis (CAD and CADx)) are being used extensively in radiology and are slowly being adapted to pathology, dermatology, and other image-based specialties (Mete, 2007; Emre, 2007). However, technology development is not the ultimate answer to eliminating medical image interpretation errors.

Clinicians still need to review the images and see, recognize, and interpret the information that is there and render a decision that will affect patient treatment and care. We need to understand not only the images and the technologies used to acquire and display them; we need to understand the interpreter of those images – the clinician. As the clinical environment changes and more and more various types of images become a part of the patient record, this becomes even more critical. This chapter reviews some of the basic issues associated with medical image perception research, especially in the context of radiology, but with examples from other areas of medical imaging as well.

7.1 VISION BASICS

The human visual system has been studied for thousands of years and we know much about it in terms of physiology, neural connections to the brain, and where in the brain information gets processed. This chapter will not describe the visual system or vision theories in great detail (see Forrester, 1996 for an excellent overview), but some aspects are relevant to medical image perception. At a very basic level there are two main aspects of vision that are important for interpreting most medical images, especially radiographs – spatial resolution and contrast resolution. Spatial resolution, or the ability to see detail, is highest at the fovea and declines rather sharply as one moves out towards the peripheral regions of the retina. In terms of visual acuity, it drops by about 75% at about five degrees from the center of the fovea (see Figure 7.2). This is a reflection of the distribution of the rods, which are responsible for sensing contrast, brightness, and motion; and the cones, which are responsible for fine resolution, spatial resolution, and color. The rods (about 115 million in total or 30,000/mm^2) are located mainly in the retina's periphery, while the cone population increases towards the macula. In the fovea itself there are only cones and they number about 6.5 million in total or 150,000/mm^2. Also contributing to these resolution differences is the fact that in the fovea there is a convergence of rods and cods on the next layer of visual cells, called the bipolar cells. The ratio of cones to bipolar cells in the fovea is essentially 1:1. In the periphery each bipolar cell receives input from 50–100 rods. This means direct transmission of a single signal versus a combination of multiple signals into one neural pathway.

The Handbook of Medical Image Perception and Techniques, ed. Ehsan Samei and Elizabeth Krupinski. Published by Cambridge University Press.

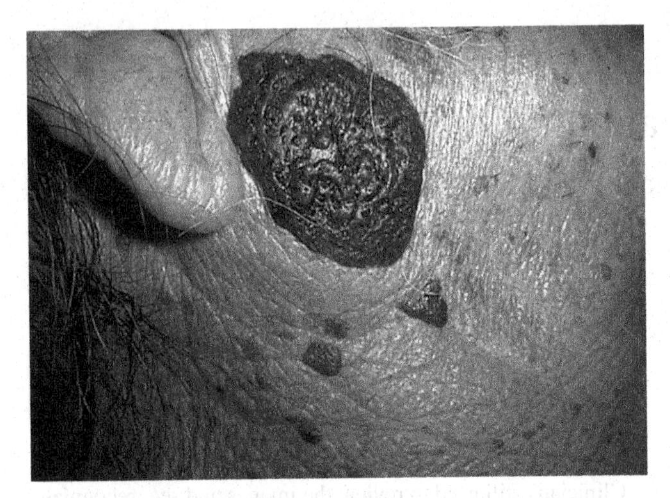

Figure 7.1 Typical teledermatology image acquired with a 1 Mpixel digital off-the-shelf consumer-grade digital camera.

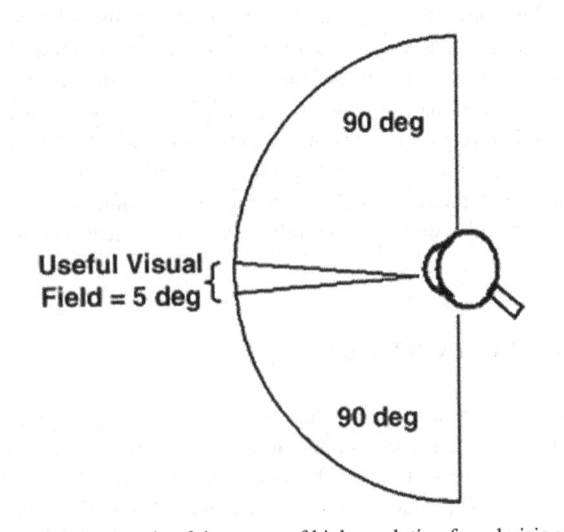

Figure 7.2 Schematic of the extent of high-resolution foveal vision and peripheral vision where the central foveal part represents the "useful visual field" that must be directed across an image during search to fixate on image detail.

The acuity or resolving power of the eye depends on physiological factors as well as the physical separation between the stimuli to be discriminated, the wavelength of the light, the illumination of the background on which the stimuli appear, and the level of dark/light adaptation of the viewer. The dependence of visual acuity on contrast (and vice versa) may be even more important. By testing observers' ability to discriminate a thin white line against a uniform background illumination, it has been found that the limits of resolution are about 0.5 minutes of arc. Due to diffraction effects, detecting this thin white line depends on the liminal brightness increment (lbi) or the point at which the bright and dark diffraction rings at the edges of the line are detected. If these rings are not different enough from the background, they will not be detected and neither will the line. Sinusoidal grating patterns of dark and light lines (where the average luminance is the same but the contrast between the lines differs) can be used to measure the lbi in a more direct and practical manner. The grating frequency (cycles per degree) and acuity, defined as 1/grating frequency, characterize grating discrimination. Quite a few psychophysical studies have been done with this method, and have shown that contrast sensitivity peaks in the mid-spatial frequency range around 3–5 cycles/degree. A number of other factors also affect contrast sensitivity, including luminance, the direction of the grating lines, the grating frequency, line width, motion, dark/light adaptation, and wavelength.

These spatial and contrast resolution limits can be readily characterized using discrete, generally geometric stimuli against fairly uniform backgrounds with classic psychophysical paradigms. With medical images the targets are not simple geometric shapes, the backgrounds are generally very complex, and the clinicians rarely know the type of target lesion they are looking for or where it is located *a priori*. Thus it is a bit simplistic to say that an abnormality of a given size and contrast (as rendered on a given display) should be detected if it falls within the limits of spatial and contrast resolution. Medical image perception is more complicated.

7.2 VISUAL SEARCH

It is clear that spatial resolution is greatest at the fovea, and what this means for perception research is that images need to be searched or scanned with the fovea, especially if small, low-contrast objects are going to be detected. There are some abnormalities that can be detected with peripheral vision (Kundel, 1975a), but normal anatomy (or structured noise) can significantly decrease detection the further away they are located from the axis of gaze (Kundel, 1975b). Peripheral vision contributes to the guidance of foveal vision (i.e. the "effective visual field") to abnormalities, but this is not the main mechanism for detection (Kundel, 1991). Search is required to cover an image adequately with foveal vision, so research into how radiologists search images is an active area of interest.

Tuddenham and Calvert in 1961 carried out one of the first studies of search patterns in radiology (Tuddenham, 1961). Radiologists were given a flashlight with an adjustable diameter and were asked to move the beam across a series of paper-based radiographic images as they searched for lesions. They adjusted the light's diameter to illuminate only the area that was required for accurate and comfortable interpretation. It was found that there is a significant difference between readers in their search patterns, and that readers tend to be rather non-uniform in their coverage. This non-uniformity in coverage might be a significant source of error.

The first study in radiology to record eye position of radiologists was by Llewellyn Thomas and Lansdown (Llewellyn Thomas, 1963). It demonstrated, in a more rigorous way, that search patterns are unique to an individual and non-uniformly cover the image. A number of studies have been done since then using eye-position analysis as a tool for understanding how the radiologist searches images and why errors tend to occur. The technology has changed of course, but eye-position recording is still used as a tool to observe and understand the perceptual processes involved in medical image perception. Most of the eye-tracking systems are based on the same general principles. They typically use a low-level infra-red light source that is reflected off the pupil and cornea back into a digital camera

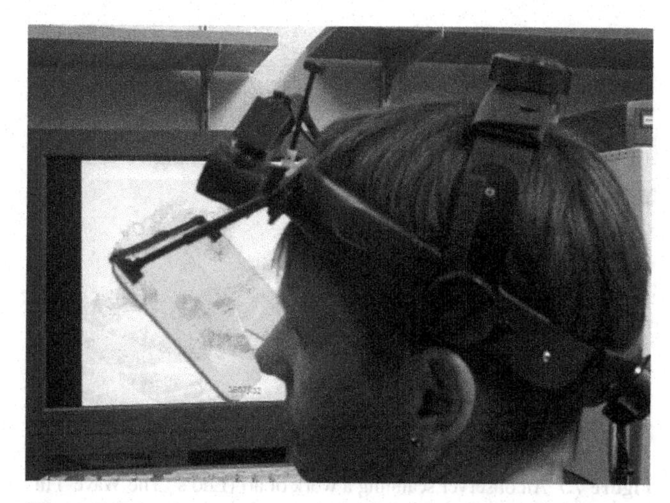

Figure 7.3 An observer setup to record eye-position during a search task. The optics of the recording system are located on the head band above the observer's eyes. The visor reflects the infra-red signal into and back out of the eyes to a digital camera also located above the eyes.

that samples this signal every 1/60 second. Figure 7.3 shows a typical observer setup for an eye-position recording system using a head-mounted system.

Since the system is calibrated to the scene (image) it is possible to correlate visual dwell information with locations in the image being viewed. Most of the head-mounted systems achieve at least one degree of accuracy (spatial error between true eye position and computed measurements) and remote systems are slightly less accurate although still quite useful. Both systems can use a magnetic head tracker in addition to the optical system to track any movements that the observer makes and incorporate these data into the eye-tracking record. Without a head tracker it is often necessary to restrict the observer's range of motion with a headrest. The raw x, y location data generated during search is typically processed using both spatial and temporal thresholds to sum the raw data into more meaningful units (Nodine, 1992). These thresholds can vary and are to some extent determined by the stimuli, the task, and the observers. Typically, the raw x, y coordinate data are transformed, using these thresholding algorithms, into fixations that generally last a minimum of 250 msec. Figure 7.4 shows a typical eye-position recording output file. Column 1 shows the temporal order of the fixations, column 4 shows the individual fixation durations, and columns 7 and 8 show the x, y locations of the fixations.

The fixations can be gathered together using spatial thresholds into clusters of fixations. This is often useful when determining how long someone looks at a given location. By adding in return visits to a given location (i.e. after scanning somewhere else in the image), cumulative clusters can be calculated to give a more accurate picture of the total time spent fixating a location. This is important in medical imaging because it is useful to know whether or not someone actually fixated a lesion. Figure 7.5 shows a typical search pattern of a radiologist searching for nodules in a chest image. Each small circle represents a fixation and the lines show the order in which they were generated. There is a nodule in the lower left lung with a single fixation on it. This single fixation was all the experienced radiologist needed to report the nodule's presence.

These types of search patterns are not unique to the radiologist's task of searching for lesions in X-ray images. The search patterns of pathologists searching virtual slides for cells indicative of cancer (see Figure 7.6) are quite similar.

One of the underlying assumptions in recording eye position is that the observer is not only directing high-resolution foveal vision to areas within the image, but they are directing their attentional and information-processing resources there as well in order to extract and process information about the scene in order to render some sort of decision about it. Even when observers look at works of art (see Figure 7.7) they are directing their perceptual and attentional resources to locations within the image that are interesting and/or informative – in this case to

```
Page:  1       EYENAL FIXATION PROGRAM

Fix Title:
Fix File : jj4a.fix        Code:
 ----------------------------- Fixation Criteria -----------------------------|
                                                                               |
| Vertical Degrees   :  0.50   1.00   1.50        1.00    Count:   3           |
| Horizontal Degrees:  0.50   1.00   1.50        1.00    Sample: 6    Blink:12 |
|----------------------------- EYEDAT Parameters -----------------------------|
| Seg:      1   Delay:  0.00 Secs   Max Dur:  600.00 Secs   PD Scale:  1.000   |
| Time Option : Real     Start Flag :       0       Stop Flag =        0      |
|-----------------------------------------------------------------------------|
| Fix |Scn|  Start Time  |  Dur.  |      saccade      |  Hor.  |  Ver.  |Pupil|Eye/Scn|
| no. |Pln| hh:mm:ss.fff|  secs  |  secs  |degrees|  pos   |  pos.  |diam.|inches |
|-----+---+-------------+--------+--------+-------+--------+--------+-----+-------|
|    1| 0 |15:43:10.333 | 0.100 | 0.000 |  0.00 |  11.09|  -9.00|235.3|   74.6|
| XDAT      0(    0h)   15:43:09.067
|    2| 0 |15:43:10.900 | 0.167 | 0.483 |  7.54 |  19.87|  -6.04|219.4|   65.3|
|    3| 0 |15:43:11.067 | 0.133 | 0.017 |  2.37 |  22.54|  -5.70|219.8|   65.1|
|    4| 0 |15:43:11.283 | 0.233 | 0.100 |  9.27 |  12.91|  -2.21|219.2|   60.4|
|    5| 0 |15:43:11.567 | 0.167 | 0.067 |  9.50 |  23.07|  -0.75|212.1|   62.2|
```

Figure 7.4 Typical output file from an eye-position recording trial. Column 1 shows the temporal order of the fixations, column 4 shows the individual fixation durations, and columns 7 and 8 show the x, y locations of the fixations.

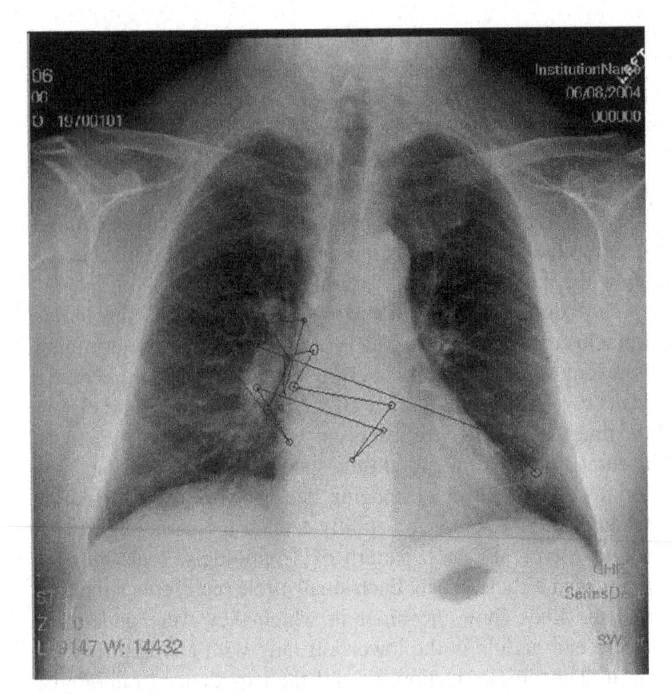

Figure 7.5 Typical eye-position pattern of a radiologist searching a chest image for nodules. Each small circle represents a fixation and the lines show the order in which they were generated. There is a nodule in the lower left lung with a single fixation on it. This single fixation was all the experienced radiologist needed to report the nodule's presence.

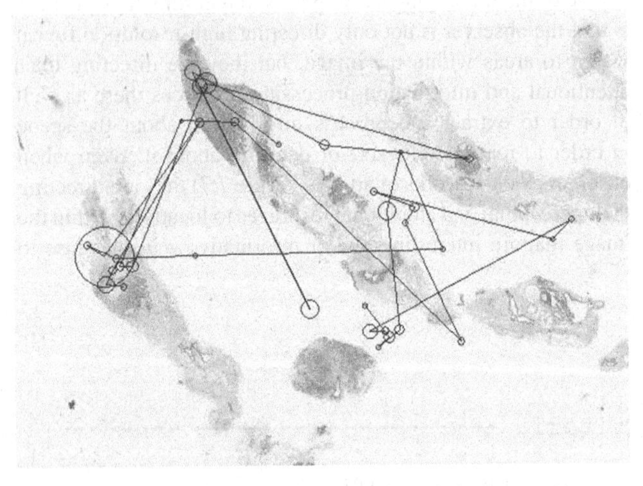

Figure 7.6 Typical eye-position pattern of a pathologist searching a virtual slide for cells indicative of breast cancer. Each small circle represents a fixation and the lines show the order in which they were generated.

Figure 7.7 An observer scanning a work of art (Edo's "The Wave") in order to decide if it is aesthetically pleasing.

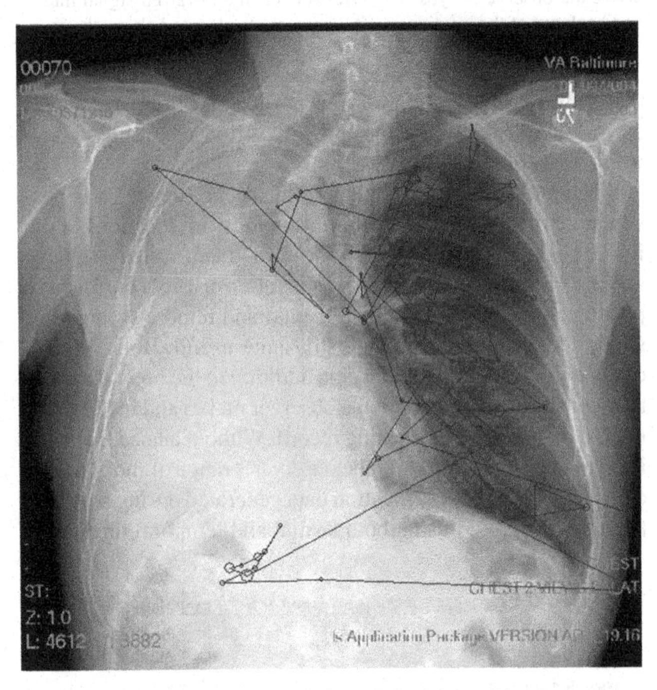

Figure 7.8 Eye-position pattern of a radiologist searching for nodules in a patient with only one lung. Since there is no relevant information in the right lung area (missing lung) the radiologist does not generate many fixations on that side. There is a nodule under the left clavicle that was fixated but not reported (false negative).

decide if the artwork is aesthetically pleasing or not (Locher, 2007).

Figure 7.8 shows the eye-position pattern of a radiologist searching for nodules in a patient with only one lung (the right lung is missing). Information-processing theory provides the basis for the current interpretations of visual search data such as these (Crowley, 2003; Haber, 1969; Krupinski, 1998; Nodine, 1987; Nodine, 1992; Spoehr, 1982). The initial glance at an image results in a global impression of the image that includes the processing and recognition of content such as anatomy, symmetry, color, and grayscale. The information gathered in this global impression is compared with information contained in long-term memory that forms the viewer's cognitive schema (or expectations) of what information is in an image. In some cases, the target of the search "pops out" in this global impression and the viewer makes a quick decision.

If there is no relevant information at some locations within the image (if there is no lung there can be no lung nodules), there is little need to direct high-resolution foveal gaze and attention to those areas. Thus the radiologist recognized within a split second of the image coming into view that there was no right lung present. This split-second recognition is related to what is known as the global percept, in which a massive amount of parallel processing occurs within the visual system, characterizing the image, its general layout, major features, boundaries etc. This early stage of scene perception takes place in less than 250 msec. The information processed in this global percept guides the subsequent search of the image with high-resolution foveal vision, where the extraction of feature details from complex image backgrounds takes place.

The early studies in medical image perception focused on trying to determine why errors were made – especially when lesions that were missed initially could often be easily detected when viewed another time. It is estimated that the miss rate in radiology is about 20–30% (false negatives) with a false positive rate of about 2–15% (Bird, 1992; Muhm, 1983; Robinson, 1997). To some extent, false positives are easier to understand than false negatives from a perceptual point of view. False positives often occur because the overlaying anatomic structures can mimic disease entities. A vessel on end or an ambiguous nipple marking can easily be mistaken for a nodule. For many false positives, there is clearly something in the image that attracts attention and leads to the false impression that a lesion exists. This is not always the case, however, because sometimes there does not seem to be anything clearly identifiable in the image but a lesion is still reported. Kundel *et al.* called these two types of false positives common and sporadic (Kundel, 1989). Common ones tend to be reported by more than one person and are associated with longer dwell times. Independent ratings of common false positives indicate that they indeed have more lesion-like features. Sporadic ones tend to be reported by only a single radiologist, are associated with few or no lesion-like features, and tend to have less visual dwell associated with them.

False negatives are a little more complicated, but probably more important to understand in terms of why they occur. Once technical reasons (e.g. patient positioning or exposure) are eliminated, the cause of such errors most likely resides in the reader. Tuddenham and Calvert suggested that lesions may be missed due to inadequate search of images and significant inter-observer variability in search strategies (Tuddenham, 1961). Kundel *et al.* followed up on these ideas and used eye-position recording to classify types of errors made during search (Kundel, 1978). Again, a key assumption is that the eye-position system records the axis of gaze and visual dwell is a reflection of information processing and allocation of attention and visual processing resources (Nodine, 1992).

Thus, false negatives have been classified into three categories (Kundel, 1978) based on how long they are dwelled on or fixated. About one-third of false negative errors fall into each category. The first error category is known as a search error because the observer never fixates the lesion with high-resolution foveal vision and thus cannot begin to process the information at that location in the image. Since there are no fixations directed here and no positive report of the lesion's presence, it is also assumed that the lesion was not detected in the global percept. The second type of error is called a recognition error, because these lesions are fixated with foveal vision, but not for very long. Inadequate time spent fixating reduces the likelihood that lesion features will be detected or recognized. Decision errors comprise the third group and these occur when the observer fixates the lesion for long periods of time, but either does not consciously recognize the features as those of a lesion or actively dismisses them. In Figure 7.8 there is a nodule located about 2 cm under the left clavicle. The radiologist fixated it but did not report it. The dwell time was less than 1000 msec (a typical cutoff point in chest viewing) so the false negative was considered to be a recognition error. The dwell time cutoff between recognition and decision errors may vary as a function of the type of image and type of lesion, but these types of errors have been noted in chest (Kundel, 1978; Kundel, 1989), bone (Hu, 1994; Krupinski, 1997), and mammography images (Krupinski, 1996a; Nodine, 2002).

Fixations are not the only relevant eye-position parameters used in understanding the perception of medical images. How the observer gets from one place to the next in an image is important as well. Scanpaths represent saccadic eye movements or movements of the eyes around a scene, locating interesting details and building up a mental "map" representing the scene. Again, the eyes move so that small parts of the scene can be centered in the fovea where visual acuity is the greatest. In one study of perception in pathology image interpretation, special attention was paid to eye movements, based on the hypothesis that eye movements may reflect meaningful differences in the cognitive abilities of novice and expert pathologists (Krupinski, 2006).

Medical students, residents, and experienced pathologists viewed a series of virtual slides on a color (9 Mpixel) liquid crystal display (LCD) as their eye position was recorded, with the task being to indicate where they would like to zoom to get a higher-resolution view of the image. The residents and medical students generated about one-third of all fixations with equal frequency on their first, second, and third preferred zoom locations, but the pathologists generated significantly fewer fixations (26%) on their first selected preferred zoom location and significantly more (44%) on the second preferred zoom location. This is consistent with the pathologists' ability to identify preferred zoom locations in their peripheral vision.

In terms of the saccades, the pathologists had significantly longer saccades (mean 0.500 sec) than the residents who, in turn, had significantly longer saccades (mean 0.244 sec) than the medical students (mean 0.129 sec). The pathologists generally fixated on a location and then made a long saccade to another fixation location and had few repeat fixations and small saccades around the same location. The residents and medical students had long saccades between locations as well, but had more short saccades and hence more fixations within a given area.

Total scanpath distance (in degrees of visual angle) was also calculated by summing all of the saccade distances or the lines between each fixation (see Figures 7.5–7.7). The pathologists had shorter total scanpath distances (mean 63.21°) than the residents (mean 176.37°) or the medical students (mean 205.74°). The saccades were also analyzed to examine whether there were differences in the number of saccades generated per image searched as a function of experience. There were, with the pathologists generating the fewest saccades followed by the residents, who were followed by the medical students who had the most saccades per image. There were also fewer saccades generated on the malignant cases than on the benign cases for all three groups.

7.3 USING VISUAL SEARCH TO STUDY THE READING ENVIRONMENT

Eye-position recording as a tool to understand medical image perception has been used to study other aspects of viewing than just errors. Chapter 29 of this book, by Krupinski and Roehrig on displays, details a number of studies that examined how the displays used to present medical images can influence not only diagnostic accuracy but search efficiency as well. Chapter 10, by Nodine and Mello-Thoms, discusses the body of data that we have on the impact that expertise has on visual search, perceptual processing, and decision-making accuracy.

From a purely perceptual standpoint, a couple of points regarding displays are warranted here. The Digital Imaging and Communications in Medicine 14 (DICOM-14) grayscale display function standard (Blume, 1996) was devised in part to provide a basis for providing similarity in grayscale perception for a given image when displayed on different display systems that may not have the same luminance. The emphasis on luminance and proper calibration of display monitors is important because it has been found that luminance levels drift over time (Evanoff, 2001; Groth, 2001; Parr, 2001). One study (Evanoff, 2001) monitored display parameters (including maximum and minimum luminance levels) of 98 monitors used over a five-day period for the oral portion of the American Board of Radiology (ABR) exam meeting and found that there was considerable luminance drift in even that short a period of time. It is clear that without regular calibration it is possible that these changes in luminance could affect detection and search performance (Krupinski, 1999b).

However, there is a significant amount of flexibility in the human visual system and it may be that those with more experience in radiology may actually become more perceptually sensitive to the crucial parameters of visual analysis required for detecting lesions in radiographic images (Sowden, 2000). In one study (Siegel, 2001) that assessed performance using displays with degraded image quality, there were no significant differences in sensitivity (although specificity varied a bit more), but the differences in performance were smaller for dedicated chest radiologists compared to general radiologists. It may take significant changes in display parameters to affect performance of experts significantly because of perceptual learning or adaptation.

The importance of perception and its dependence on the display medium are obviously crucial. In fact, the DICOM-14 display standard for digital radiology is based on the idea of perceptual linearization (Blume, 1996; Johnston, 1985; Pizer, 1981a, 1981b). Perceptual linearization capitalizes on the capabilities of the human visual system (i.e. threshold contrasts and just-noticeable differences) to optimize displays by producing a tone scale in which equal changes in digital driving levels yield changes in display luminance that are perceptually equivalent across the entire luminance range. This means that equal steps in brightness sensation represent equal steps in acquired image data, and studies have shown that it does matter (Blume, 1996; Johnston, 1985; Pizer, 1981a, 1981b).

One study (Krupinski, 2000) examined detection and search performance for detecting masses and microcalcifications in mammograms using a perceptually linearized DICOM-calibrated display versus a non-perceptually linearized display calibrated using the Society for Motion Picture and Television Engineers (SMPTE) pattern. It was found that there was statistically increased detection accuracy with the perceptually linearized display. Eye-position recording revealed significantly more efficient visual search patterns with the perceptually linearized display. The total viewing times, time to first fixate the lesions, decision dwell times, and number of fixation clusters generated during search were decreased with the perceptually linearized display. Proper calibration of the digital display in terms of luminance and tone scale clearly impacts efficient and accurate perceptual performance.

7.4 COLOR VISION

Medical image perception as applied to radiology is rather unique since radiology is essentially monochrome while most other clinical specialties that use images use color images. It is possible to add color to radiographic images to bring out certain details (e.g. color Doppler ultrasound (Blaivas, 2002; Beebe, 1999)), but the initial rendering is in shades of gray. Color may also be used when merging/registering images from different modalities to highlight the differences between images as an indication of disease or other changes (Rehm, 1994). Color is useful in medical image perception because the human visual system relies quite heavily on color perception. In fact, from an evolutionary perspective color may be more important for survival than spatial resolution (Chaparro, 1993)! Data from Levkowitz and Herman indicate that color has a dynamic range of at least 500 JNDs (just-noticeable differences or discriminable levels) (Levkowitz, 1992). Grayscale has a dynamic range of only about 60–90 JNDs – clearly more restricted. It's not that simple, however, because the visual system has lower spatial resolution in the color channels than in the luminance channels. In radiology at least, increasing dynamic range by going to color may actually be confounded by color-luminance dependencies, having a significant effect on contrast perception (Mullen, 1985; Granger, 1973). In other clinical specialties, where color is the rule rather than the exception, this not really an issue.

An interesting point to consider when discussing medical image perception and color images is that there are common deficiencies in color vision that might have to be considered when moving to color displays and developing methods to calibrate them. Very basically, color vision is based on the fact that there are three sets of cones in the retina. Each is sensitive over a range but has peak sensitivity at a specific wavelength; short at 419 nm (blue), medium at 531 nm (green), and long at 558 nm (red). There are relatively few short cones so overall photopic sensitivity is due largely to summed medium and long cones (555 nm). Perceptually, color is characterized by three properties. Saturation is the depth or purity of colors and is related inversely to the number of different incident wavelengths. Hue or color name (e.g. red) is a function of the wavelength(s) reaching the eye's photoreceptors. Brightness is the intensity of the color and is roughly proportional to the amplitude of the incident wavelength(s).

Color may be more relevant to pathologists, ophthalmologists, dermatologists, and other clinicians that utilize true-color images compared to radiologists who use grayscale and pseudo-color images only. Color vision deficiencies fall into two categories – acquired or inherited. One common acquired deficiency that occurs with age is the growth of cataracts, which not only affect acuity but lead often to a yellowing of the lens (xanthopsia) resulting in the lens absorbing blue and green, making perception in this range difficult. Corneal edema as a result of allergies, infections, and even contact lenses can result in the perception of halos and rainbows, especially around bright light sources (e.g. display devices). Age-related macular degeneration affects about 23% of people over 65 and is a loss of cones, resulting in a loss of both acuity and color perception. Migraines (and sometimes regular headaches) can cause a central blind spot (scotoma) in which colored zig-zags (like twisted pieces of yarn) shimmer or flash, without any actual external stimulus. Chromatopsia results in a distortion of color in which the scene is either infused with an illusory blue tint (cyanopsia) or an illusory yellow tint (xanthopsia). Clearly, if the diagnostic task requires proper perception of color within a medical image, these degradations in perceptual processing of color could affect diagnostic accuracy.

Much more common than the acquired color deficiencies are inherited deficiencies and these too can be important in the perception of color medical images. There are three types of congenital color vision deficiencies, referred to as monochromacy, dichromacy, and anomalous trichromacy. Most inherited color deficiencies are significantly more common in men than in women. Monochromacy or total colorblindness is the lack of all ability to distinguish colors. This occurs when two or all three of the cone pigments are missing and the result is that color vision is reduced to a single dimension. This is a relatively rare condition. Dichromacy occurs when only one of the cone pigments is absent or not functioning properly, reducing color to two dimensions. Protanopia is caused by the absence of red photoreceptors, making red colors appear dark; and in deuteranopia the green photoreceptors are missing, affecting red-green discrimination. Tritanopia is the rarest deficiency in this category, in which blue receptors are missing.

Anomalous trichromacy is rather common and occurs when one pigment is altered in its spectral sensitivity rather than missing altogether. Protanomaly is an alteration of red receptors (closer to green receptor response) resulting in poor red-green discrimination. Deuteranomaly affects the green receptors and is the most common color deficiency. It affects red-green discrimination, especially in males. Tritanomaly is rare and affects blue-yellow discrimination. There are really no significant studies that have examined color perception or color deficiencies and their impact on diagnostic accuracy in any field of medical imaging.

7.5 STEREOSCOPIC VIEWING

Most images (e.g. paintings or photographs), including medical images, are two-dimensional. They generally incorporate cues that lead to the perception of depth (e.g. interposition or linear perspective) but they do not truly incorporate depth information in the sense of seeing three-dimensional objects. Radiographic images are 2D images of 3D objects (the human body) and the third dimension is captured primarily as density differences and structural overlap. Dermatology images try to incorporate depth by acquiring one image of a lesion straight on and another from the side. Ophthalmology captures multiple views of the retina as well. In all of these cases, however, the clinician takes the 2D information and with knowledge of human anatomy "sees" the third dimension. The perception or interpretation of depth is done primarily in the mind of the clinician. Digital imaging, however, presents a unique opportunity for viewing medical images in 3D, potentially leading to better localization and characterization of lesions.

With proper acquisition devices and advances in display technologies, 3D is becoming much more feasible than in the past in many areas of medical imaging. The question of course is does it help? Stereotactic biopsy systems, for example in mammography or performing other procedures (Moll, 1998), have been shown to improve the accuracy and reduce the time spent performing these interventions. One study (Rosenbaum, 2000) that used rotating 3D images presented on a stereoscopic display showed that readers tended to have increased confidence in their perception of findings. Getty (Getty, 2007) has demonstrated similar success with stereoscopic viewing of mammography images in which standard 2D and stereoscopically presented 3D mammography images were presented to radiologists. Diagnostic accuracy was improved, especially for microcalcifications, with the stereo mammograms and a significant number of new lesions were actually detected compared to 2D viewing. It is likely that the improvement was due to better separation of over- and underlying tissues from the lesions. Manipulating images in a variety of ways to get better views or see things from a different angle also seemed to help in separating normal tissue from lesion tissue. Benign versus malignant characterization was also observed and attributed to the fact that the stereo images allow the observers to follow the extent of the lesions through the normal tissue more effectively and to visualize lesion characteristics obscured by overlying tissues in 2D images. There have been few studies in other medical imaging areas on potential benefits of 3D over 2D images.

7.6 PERCEPTION AND VISUAL FATIGUE

Clinicians in all specialties are viewing more medical images of all sorts (along with text data) from computer displays on a regular basis. Radiologists and pathologists are doing so at an extraordinary rate given the fact that image viewing comprises close to 100% of what they do. This long-term viewing of images at a relatively close distance for hours on end may however have a negative side. Computer vision syndrome (CVS) is a repetitive strain disorder characterized by eyestrain, blurred vision, double vision, dry eyes, tired eyes, headaches, neckaches, and backaches (Rechichi, 1996; Takahashi, 2001; Yunfang, 2000). According to the Occupational Safety and Health Administration (OSHA), about 90% of US workers using computers for more than three hours per day suffer from some form of CVS. Significant effects include perceptual errors (OSHA, 2001), performance errors, decrease in reaction time, fatigue, and even burnout. Up to about 15 years ago one could easily have said this has nothing to do with radiology or other medical image perception tasks, but today it has everything to do with medicine. Radiology in particular has changed from a predominantly film-based specialty to one in which images are read from computer monitors. CVS has not been studied to any degree in medicine, but it has been studied in other areas and the effects observed are likely to be observed in digital-based medical image interpretation as well. The aspect of most concern is the effect on perception and performance, since these two phenomena comprise the basics of what medical image perception is all about – viewing a diagnostic image and rendering a diagnostic interpretation.

Short-term visual fatigue has been often noted as one of the main symptoms of CVS, but there are long-term effects that could affect perception as well. Prolonged and repeated computer use can lead to vision problems (e.g. induced myopia or asthenopia) that are likely to need correction (Mutti, 1996; Sanchez-Roman, 1996). The cause of these problems seems to be the amount of effort required for vergence and accommodation responses with respect to fixating the computer display. Decreased vergence accuracy at far distances (Tyrrell, 1990) occurs with prolonged viewing and it is even more pronounced as task and stimulus complexity increase (Watten, 1994). This is especially relevant for medical imaging because interpreting medical images is a very complex task both perceptually and cognitively, and the stimuli themselves are very complex and vary considerably from case to case.

Objective factors can be monitored to determine if display users are fatigued and should take a break or stop altogether. Measures such as pupil diameter, eye movement velocity, dark focus of accommodation, and width of focal accommodation can be measured relatively easily to determine someone's relative state of fatigue and the results fed back to the viewer to act upon (Chi, 1998; Murata, 2001). Two factors that can affect fatigue in medical image viewing are display luminance and room illumination. Both medical-grade and off-the-shelf displays have higher and higher luminance values, and at least in radiology high luminance is associated with better diagnostic accuracy and more efficient visual search. There may be limits to how high we can go, however, in luminance before it starts to induce fatigue. One study (Saito, 1991) with rather simplistic stimuli showed that increased display luminance correlated highly with increased fatigue and the impact was lessened when display luminance and room illumination were kept constant. The less often the eyes had to dark- or light-adapt to changing light intensities from either the display or room lights, the less fatigue and effects on visual reaction time were observed. Other factors also can impact fatigue levels, including ambient noise (Takahashi, 2001), the height of the display (Sotoyama, 1996; Turville, 1998), experience with computers (Czaja, 1993; Krupinski, 1996b), and even user age.

Most of the factors are easily controlled for in the digital reading environment once users are aware of them. From a perceptual standpoint, room design, design of the immediate area clinicians work in, and the specifications of the display (e.g. luminance, contrast ratio, and pixel size) are all well within the control of the operator and should be calibrated and monitored on a regular basis to promote optimal reading conditions and avoid fatigue and stress.

7.7 CONCLUSIONS

There are many other aspects of perception that impact the interpretation of medical images. The aspects considered in this chapter, however, provide those interpreting medical images with an idea of the types of things they should be concerned with. Many things need to be considered in the optimization of medical image perception, including basic visual system physiology, the display, the room, the images, the task, and especially the clinician. Even something as simple as how often a clinician needs to take a break from viewing to avoid fatigue and decrease the probability of making errors is important to understand. The problem is not a static one that will be solved with "the" study on medical image perception.

Technology changes continually. The cathode ray tube (CRT) was replaced by the LCD, and true 3D projection displays, micro-displays, and even virtual reality displays are being explored. Film has been essentially replaced by digital acquisition and display in radiology and glass slides are being transformed into virtual slides that no longer require a light microscope to view in pathology. Digital images of the skin, the eyes, and many other organs are being acquired with simple digital cameras and transmitted around the world for quick and accurate interpretation. Each of these medical imaging applications presents new challenges to clinicians' perceptual and cognitive systems. Our continued exploration of medical image perception and the way the clinicians interact with the medical image will always be important.

REFERENCES

Beebe, H.G., Salles-Cunha, S.X., Scissons, R.P., *et al.* (1999). Carotid arterial ultrasound scan imaging: a direct approach to stenosis measurement. *J Vasc Surg*, **29**, 838–844.

Bird, R.E., Wallace, T.W., Yankaskas, B.C. (1992). Analysis of cancers missed at screening mammography. *Radiology*, **184**, 613–617.

Blaivas, M. (2002). Color doppler in the diagnosis of ectopic pregnancy in the emergency department: is there anything beyond a mass and fluid? *J Emerg Med*, **22**, 379–384.

Blume, H. (1996). Members of ACR/NEMA Working Group XI: the ACR/NEMA proposal for grey-scale display function standard. *Proc SPIE Med Imag*, **2707**, 344–360.

Chaparro, A., Stromeyer, C.F., Huang, E.P., *et al.* (1993). Colour is what the eyes see best. *Nature*, **361**, 348–350.

Chi, C.F., Lin, F.T. (1998). A comparison of seven visual fatigue assessment techniques in three data-acquisition VDT tasks. *Hum Factors*, **40**, 577–590.

Crowley, R.S., Naus, G.J., Stewart, J., *et al.* (2003). Development of visual diagnostic expertise in pathology: an information-processing study. *J Am Med Inform Assoc*, **10**, 39–51.

Czaja, S.J., Sharit, J. (1993). Age differences in the performance of computer-based work. *Psychol Aging*, **8**, 59–67.

Emre, C.M., Alp, A.Y., Stoecker, W.V., *et al.* (2007). Unsupervised border detection in dermoscopy images. *Skin Res Technol*, **13**, 454–462.

Evanoff, M.G., Roehrig, H., Giffords, R.S., *et al.* (2001). Calibration of medium-resolution monochrome cathode ray tube displays for the purpose of board examinations. *J Digit Imaging*, **14**, 27–33.

Forrester, J., Dick, A., McMenamin, P., *et al.* (1996). *The Eye: Basic Sciences in Practice*. Philadelphia, PA: WB Saunders.

Getty, D.J. (2007). Improved accuracy of lesion detection in breast cancer screening with stereoscopic digital mammography. Paper presented at the 93rd Annual Meeting of the Radiological Society of North America, Nov 25–30, Chicago, IL.

Granger, E.M., Heurtley, J.C. (1973). Visual chromaticity modulation transfer function. *J Opt Soc Am*, **63**, 1173–1174.

Groth, D.S., Bernatz, S.N., Fetterly, K.A., *et al.* (2001). Cathode ray tube quality control and acceptance testing program: initial results for clinical PACS displays. *Radiographics*, **21**, 719–732.

Haber, R.N. (1969). *Information-processing Approaches to Visual Perception*. New York, NY: Holt, Rinehart and Winston.

Hu, C.H., Kundel, H.L., Nodine, C.F., *et al.* (1994). Searching for bone fractures: a comparison with pulmonary nodule search. *Acad Radiol*, **1**, 25–32.

Johnston, R.E., Zimmerman, J.B., Rogers, D.C., *et al.* (1985). Perceptual standardization. *Proc SPIE Med Imag*, **536**, 44–49.

Krupinski, E.A. (1996a). Visual scanning patterns of radiologists searching mammograms. *Acad Radiol*, **3**, 137–144.

Krupinski, E.A., LeSueur, B., Ellsworth, L., *et al.* (1999a). Diagnostic accuracy and image quality using a digital camera for teledermatology. *Telemed J*, **5**, 257–263.

Krupinski, E.A., Lund, P.J. (1997). Differences in time to interpretation for evaluation of bone radiographs with monitor and film viewing. *Acad Radiol*, **4**, 177–182.

Krupinski, E.A., Nodine, C.F., Kundel, H.L. (1998). Enhancing recognition of lesions in radiographic images using perceptual feedback. *Opt Eng*, **37**, 813–818.

Krupinski, E., Nypaver, M., Poropatich, R., *et al.* (2002). Telemedicine/telehealth: an international perspective. Clinical applications in telemedicine/telehealth. *Telemed J E Health*, **8**, 13–34.

Krupinski, E.A., Roehrig, H. (2000). The influence of a perceptually linearized display on observer performance and visual search. *Acad Radiol*, **7**, 8–13.

Krupinski, E.A., Roehrig, H., Furukawa, T. (1999b). Influence of film and monitor display luminance on observer performance and visual search. *Acad Radiol*, **6**, 411–418.

Krupinski, E.A., Tillack, A.A., Richter, L., *et al.* (2006). Eye-movement study and human performance using telepathology virtual slides: implications for medical education and differences with experience. *Human Pathol*, **37**, 1543–1556.

Krupinski, E.A., Weinstein, R.S., Rozek, L.S. (1996b). Experience-related differences in diagnosis from medical images displayed on monitors. *Telemed J*, **2**, 101–108.

Kundel, H.L. (1975b). Peripheral vision, structured noise and film reader error. *Radiology*, **114**, 269–273.

Kundel, H.L., Nodine, C.F. (1975a). Interpreting chest radiographs without visual search. *Radiology*, **116**, 527–532.

Kundel, H.L., Nodine, C.F., Carmody, D.P. (1978). Visual scanning, pattern recognition and decision-making in pulmonary tumor detection. *Invest Radiol*, **13**, 175–181.

Kundel, H.L., Nodine, C.F., Krupinski, E.A. (1989). Searching for lung nodules: visual dwell indicates locations of false-positive and false-negative decisions. *Invest Radiol*, **24**, 472–478.

Kundel, H.L., Nodine, C.F., Toto, L. (1991). Searching for lung nodules: the guidance of visual scanning. *Invest Radiol*, **26**, 777–781.

Levkowitz, H., Herman, G.T. (1992). Color scales for image data. *IEEE Comp Graphics and Applic*, **12**, 72–80.

Llewellyn Thomas, E., Lansdown, E.L. (1963). Visual search patterns of radiologists in training. *Radiology*, **81**, 288–291.

Locher, P., Krupinski, E.A., Mello-Thoms, C., Nodine, C.F. (2007). Visual interest in pictorial art during an aesthetic experience. *Spat Vis*, **21**, 55–77.

Mete, M., Xu, X., Fan, C.Y., *et al.* (2007). Automatic delineation of malignancy in histopathological head and neck slides. *BMC Bioinformatics*, **8**, S17.

Moll, T., Douek, P., Finet, G., *et al.* (1998). Clinical assessment of a new stereoscopic digital angiography system. *Cardiovasc Intervent Radiol*, **21**, 11–16.

Muhm, J.R., Miller, W.E., Fontana, R.S., *et al.* (1983). Lung cancer detection during a screening program using four-month chest radiographs. *Radiology*, **148**, 609–615.

Mullen, K.T. (1985). The contrast sensitivity of human colour vision to red-green and blue-yellow chromatic gratings. *J Physiol*, **359**, 381–400.

Murata, A., Uetake, A., Otsuka, M., *et al.* (2001). Proposal of an index to evaluate visual fatigue induced during visual display terminal tasks. *Intl J Hum Comput Interact*, **13**, 305–321.

Mutti, D.O., Zadnik, K. (1996). Is computer use a risk factor for myopia? *J Am Optom Assoc*, **67**, 521–530.

Nodine, C.F., Kundel, H.L. (1987). Using eye movements to study visual search and to improve tumor detection. *Radiographics*, **7**, 1241–1250.

Nodine, C.F., Kundel, H.L., Toto, L.C., *et al.* (1992). Recording and analyzing eye-position data using a microcomputer workstation. *Behav Res Methods Instrum Comput*, **24**, 475–485.

Nodine, C.F., Mello-Thoms, C., Kundel, H.L., *et al.* (2002). Time course of perception and decision making during mammographic interpretation. *AJR Am J Roentgenol*, **179**, 917–923.

OSHA (2001). http://www.osha.gov/pls/oshaweb/owadisp.show_document?p_table=FEDERAL_REGISTER&p_id=15160. Last accessed May 20, 2007.

Parr, L.F., Anderson, A.L., Glennon, B.K., *et al.* (2001). Quality-control issues on high-resolution diagnostic monitors. *J Digit Imaging*, **14**, 22–26.

Pizer, S.M. (1981a). Intensity mappings to linearized displays. *Comput Graphics Image Process*, **17**, 262–268.

Pizer, S.M. (1981b). Intensity mapping: linearization, image-based, user-controlled. *Proc SPIE Med Imag*, **271**, 21–27.

Rechichi, C., Demoja, C.A., Scullica, L. (1996). Psychology of computer use: XXXVI. Visual discomfort and different types of work at videodisplay terminals. *Percept & Mot Skills*, **82**, 935–938.

Rehm, K., Strother, S.C., Anderson, J.R., *et al.* (1994). Display of merged multimodality brain images using interleaved pixels with independent color scales. *J Nucl Med*, **35**, 1815–1821.

Robinson, P.J.A. (1997). Radiology's Achilles' heel: error and variation in the interpretation of the Roentgen image. *Br J Radiol*, **70**, 1085–1098.

Rosenbaum, A.E., Huda, W., Lieberman, K.A., *et al.* (2000). Binocular three-dimensional perception through stereoscopic generation from rotating images. *Acad Radiol*, **7**, 21–26.

Saito, K., Hosokawa, T. (1991). Basic study of the VRT (visual reaction time): the effects of illumination and luminance. *Intl J Hum Comput Interact*, **3**, 311–316.

Sanchez-Roman, F.R., Perez-Lucio, C., Juarez-Ruiz, C., *et al.* (1996). Risk factors for asthenopia among computer terminal operators. *Salud Publica Mex*, **38**, 189–196.

Siegel, E.L., Reiner, B.I., Hooper, F., *et al.* (2001). The effect of monitor image quality on the soft-copy interpretation of chest CR images. *Proc SPIE Med Imag*, **4323**, 42–46.

Sotoyama, M., Jonai, H., Saito, S., *et al.* (1996). Analysis of ocular surface area for comfortable VDT workstation layout. *Ergonomics*, **39**, 877–884.

Sowden, P.T., Davies, I.R., Roling, P. (2000). Perceptual learning of the detection of features in x-ray images: a functional role for improvements in adults' visual sensitivity? *J Exp Psychol Hum Percep Perform*, **26**, 379–390.

Spoehr, K.T., Lehmkuhle, S.W. (1982). *Visual Information Processing*. San Francisco, CA: WH Freeman and Company.

Takahashi, K., Sasaki, H., Saito, T., *et al.* (2001). Combined effects of working environmental conditions in VDT work. *Ergonomics*, **44**, 562–570.

Taylor, C.R., Merin, L.M., Salunga, A.M., *et al.* (2007). Improving diabetic retinopathy screening ratios using telemedicine-based digital retinal imaging technology: the Vine Hill study. *Diabetes Care*, **30**, 574–578.

Tuddenham, W.J., Calvert, W.P. (1961). Visual search patterns in roentgen diagnosis. *Radiology*, **76**, 255–256.

Turville, K., Psihogios, J., Ulmer, T., *et al.* (1998). The effects of video display terminal height on the operator: a comparison of the 15" and 40" recommendation. *Appl Ergon*, **29**, 239–246.

Tyrrell, R.A., Leibowitz, H.W. (1990). The relation of vergence effort to reports of visual fatigue following prolonged near work. *Hum Factors*, **32**, 341–357.

Watten, R.G., Lie, I., Birketvedt, O. (1994). The influence of long-term visual near-work on accommodation and vergence: a field study. *J Hum Ergol*, **23**, 27–39.

Weinstein, R.S., Descour, M.R., Liang, C., *et al.* (2004). An array microscope for ultrarapid virtual slide processing and telepathology. Design, fabrication, and validation study. *Hum Pathol*, **35**, 1303–1314.

Yunfang, L., Wenjing, W., Bingshuang, H., *et al.* (2000). Visual strain and working capacity in computer operators. *Homeost Health Dis*, **40**, 27–29.

8

Cognitive factors in reading medical images

A survey of cognitive factors and models of medical image interpretation

DAVID MANNING

8.1 INTRODUCTION

We are very familiar with the task of observing visual scenes. We make judgments on what the scenes mean to us, and we use those judgments to take decisions on our actions. It is a natural part of how we interrogate the world and it gives us a personal framework of meaning and understanding. It also allows us to make sense of our environment by providing moment-to-moment information for a survival strategy as the environment changes. Seeing does not appear to interfere with the world in any way; we are confident in its efficiency and we feel we have complete control over it. In all these respects, we are justified in having the point of view that reading medical images is an extension, or just another version, of an activity we are all good at. But the truth about vision is much more difficult. It is inconsistent and unreliable and it is poorly controlled. Our eyes go to places we do not instruct them to go and lie to us on where they have been, and the brain does not always tell us the full story of what we have looked at (Elkins, 1996). So it is perhaps not surprising that studies of radiological performance over half a century have shown that there is a significant disagreement in the diagnostic decisions made by different radiologists given the same visual data, and this difference is remarkably resistant to change despite continual technical improvements in the quality of medical images (Samei, 2006). The phenomenon is referred to as inter-observer variance and is accompanied by an intra-observer variance in any single observer given the same task at different times. Another way of saying this is that observers make errors that are independent of the images. So, given that radiology strives for rigor and excellence in all aspects of the diagnostic process, technical and otherwise, we should ask if there are cognitive contributions to these visual errors that can be identified, better understood, and ultimately, reduced.

Radiology is to do with recognizing abnormality when it is present and normality when it isn't. But because we cannot recognize things we do not already know, this recognition must start from experience. Therefore vision must have a strong learning component to it (Gregory, 1990) and we use the learnt experiences to interpret every new visual event. This confirms an instinctive belief that we are by no means simple object-detectors. Rather, our eye-brain systems make interpretations of the visual data coming in. Signals are transmitting *information* or a message, and the purpose of a message is that when we have received it we should know more than we did before. Information theory (Shannon, 1948) is a method of measuring this process, and it has given rise to ideas that have connected what we believe our brains do with visual information to the development of artificial intelligence. It has allowed us to see how units of energy from the physical world can, somehow, become mental events. But as Frith (Frith, 2007) points out, although computers can now beat world champions in chess tournaments, humans are still much better than machines at, for example, recognizing natural objects and faces. Information theory and developments in computer science have shown us that although our brains have apparently solved such perceptual and cognitive problems, they are, in reality, very difficult problems to solve; consequently, we have a rather imperfect understanding of how cognition works in radiology.

What we do know, however, is that visual information processing and decision-making are key features of the activity and that experts perform better than non-experts. This is a very good starting point. We have a body of knowledge, explained in other chapters, on the principles of what it takes to achieve expert performance, and we know something of how it is developed in radiology. We know how relevant it is to clinical outcome and how computer aids to diagnosis can enhance human performance. Studies in these areas have provided an opportunity to construct some ideas on the cognitive factors involved in the image reading process.

Analysis of the radiological task is central to these themes and should provide an opportunity to identify why some operators perform better than others; but other inequalities and variables in physics and psychophysics have their contribution to make too and they have to be considered before the decision component can be isolated (see Beutel, Kundel, and Van Metter (Beutel, 2000) for a comprehensive treatment of this area of medical imaging). Other parts of this book address those variables by considering the physical, physiological, and perceptual influences on reader performance and they also deal with the techniques that have been developed for modeling and measuring performance outcomes. In this chapter we deal with the concepts involved in our attempts at understanding the cognitive features of medical image interpretation.

8.1.1 Plan of the chapter

The scope of this chapter is to look at the nature of the radiological task as a *problem-solving, visual-reasoning activity* and we consider the models which help in understanding how it is carried out. There are overlaps with many other sections of the book but particularly with the chapters on expertise and perception.

- In the first section we consider the radiological task in a brief overview that treats the image as a source of information.

The Handbook of Medical Image Perception and Techniques, ed. Ehsan Samei and Elizabeth Krupinski. Published by Cambridge University Press.

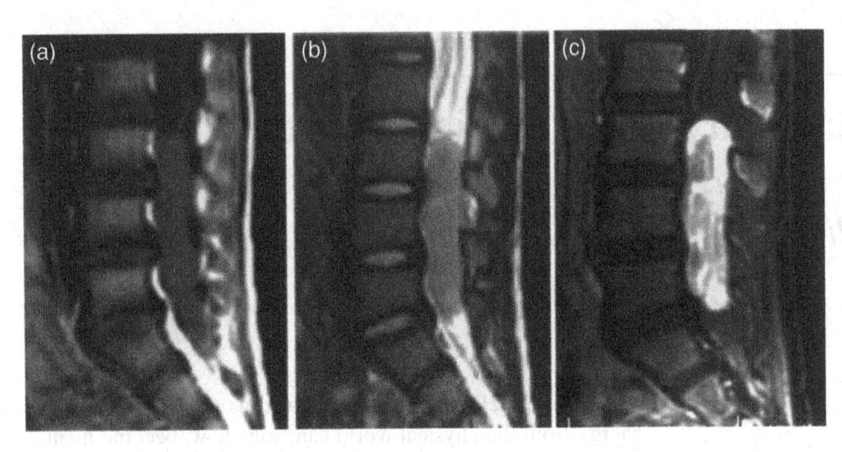

Figure 8.1 A magnetic resonance image (MRI) showing a sagittal section of the lumbar region. A primary malignant peripheral nerve sheath tumor is demonstrated of the cauda equina in a child. (a) T1-weighted sagittal view, (b) T2-weighted sagittal view, and (c) enhanced MRI with gadolinium contrast agent. Notice how the general image contrast and the visual appearance of the anatomical structures, particularly the tumor, are altered depending on the acquisition criteria. The reader can only interpret the meaning behind these images by drawing on a significant technical knowledge of how signals change with imaging method. Reprinted with permission from Macmillan Publishers Ltd: *Spinal Cord* 2004, **42**, 199–203. Copyright 2004.

All diagnostic procedures in medicine aim to reduce uncertainty by adding to the total sum of information the clinician has available for making a decision. The image reader must be clear about the precise nature of this information before deciding on its diagnostic inferences so we comment on how imaging modalities affect information properties.

- Next, we turn to the objects and features in medical images that radiologists use in the interpretation process. We take examples such as contrast, texture, and location to illustrate the perceptual characteristics used by readers to classify objects, and we hypothesize on the relative importance of these factors in understanding the image. This is followed by comments on how this understanding has to be communicated to the clinician who requested the examination.
- Through reference to cognitive theories, the third section deals with the way that readers recognize and name objects in the image. We consider how the theories provide insight into how readers select pertinent data and arrive at decisions. And we link these activities to radiology training, artificial intelligence, and inter-observer variance.
- Finally we deal with the issue of probabilistic reasoning in the image reading activity, how we update our decisions on the basis of new information, and how image readers are influenced by non-image information and biases in making their decision.

8.2 OVERVIEW OF THE INTERPRETATION TASK

Radiologists do not usually have the benefit of direct observation of their patients but, rather, have to rely on the interpretation of *images* that represent some structural or functional features of their patients' bodies. From these visual representations, inferences about states of health or disease are drawn. *Representation* is a key word here because the visual task of radiology is interpretation of the data, or information, recorded by the imaging method. This information is, of course, not the anatomy itself but is an abstraction of its physical or chemical features. For example, the shadow picture of an X-ray image is created from the spatial distribution of linear attenuation coefficients when an X-ray beam is transmitted through a person's body. Until the middle part of the 20th century this was the only form of medical image radiologists were faced with. So there was a limited range to the physical meanings behind the image features (or shadows). An observer could be certain, for example, that a dark shadow on a radiograph indicated transmission through tissue material of lower density and atomic number than that creating a lighter shadow. The meaning was consistent from one X-ray test to another and it meant that decision-making was constrained to the anatomical or disease significance of one type of shadow. But the image information can now be highly diverse and as new imaging modalities come along so the diversity increases. Technological developments over the last few decades have added many new imaging methods that have signals with entirely different information content concerning the human body and disease processes.

Before interpretations of the image can be made the observer needs first to have a clear notion of what the image features are physically representing.[1] This might seem an obvious statement to those familiar with the patterns demonstrated in radiographs but it is far from obvious to those first faced with the problem. Medical images are not self-explanatory and require substantial effort in their interpretation. Newcomers to radiology have to first establish a knowledge of how the image was formed and then use that knowledge as a mental engine running in the background as the diagnostic problem is negotiated (see Figure 8.1).

For skilled and experienced radiologists, however, the range of imaging methods is a positive benefit. It is commonplace for several imaging methods to be used in the investigation of the same diagnostic problem. Each method provides its own unique data contribution and decision-making is sometimes made easier as a result. There is often an additive effect of combining two or more imaging methods that outstrips the diagnostic performance of any one when used singularly. This advantage is down to enhanced decision-making and interpretation of the problem when greater amounts of more diverse data are made available to the reader.

So we have established that in medical imaging, the image features are representations of physical objects. They are not the objects themselves, and the precise character of the

[1] Even within the same imaging modality there can be differences in signal appearances, as shown in Figure 8.1. MRI is an example where, depending on the image acquisition sequence or algorithm, the same body tissue can appear to the observer as a black shadow, a white signal, or be absent altogether. Such technical complexities add to the interpretive – cognitive – challenge for those diagnosing from medical images.

representation must first be understood by the decision-maker. When this has been achieved, the informed observer can then go on to use the data in a problem-solving activity.

It would be misleading to imply that the image is the only patient data to which the radiologist has access. The clinical history of the patient will also accompany the images and this may be read in conjunction with the image interpretation to clarify and guide the process. Equally informative are previous radiological examinations. These are particularly important to the decision-making process when the imaging examination is a repeat of a previous study or is a follow-up on the progress of disease subsequent to treatment or some other intervention.

The availability of clinical information and previous examinations is generally assumed to be helpful, but some studies have questioned this and there appear to be some circumstances where it is not the case. Swenson *et al.* (Swenson, 1985), for example, showed that there were some benefits in scrutinizing medical images without any prior knowledge or preconceptions as to the possible clinical condition. The problem has been discussed in terms of the perceptual interference that extra data can have on the visual search process rather than in decision-making, although the end result will be diagnostic error with the same potential consequences. Curtailing search activity at the point where an expected lesion has been identified can have the effect of missing additional lesions. This has been characterized as a distracting effect of clinical information and is reported to be similar to the satisfaction of search (SOS) effect.

The phenomenon was also extensively studied with reference to the clinical severity of detected and missed lesions (Berbaum, 2007) and in checklist attempts to reduce its effects (Berbaum, 2006). The interpretation process for most radiologists tends to include attention to all available information for the task in hand, and although the distraction phenomenon cannot be denied, it is something of a contentious issue in the broad practice of radiology. In terms of decision bias it should be balanced against the reassuring effect it has. For example, prior data confirms for the radiologist that the correct procedure has been carried out, and it focuses attention at an early stage in the viewing process on the areas most likely to be implicated in the clinical problem. It ensures that time is not wasted looking at irrelevances in the image (Robinson, 1997). There can be little doubt that prior information will guide search and will also influence decisions but for the majority of practitioners the positive aspects of the effect are considered more important than its potential pitfalls.

8.3 INTERPRETATION PROCESS

8.3.1 Features in medical images

How radiologists recognize objects in medical images will be a topic we return to later when it will be related to cognitive theories on how we recognize objects in general; but at this stage we should at least consider some of the definable characteristics of radiological features, how they vary, and how they are organized into classes that can be agreed on by observers.

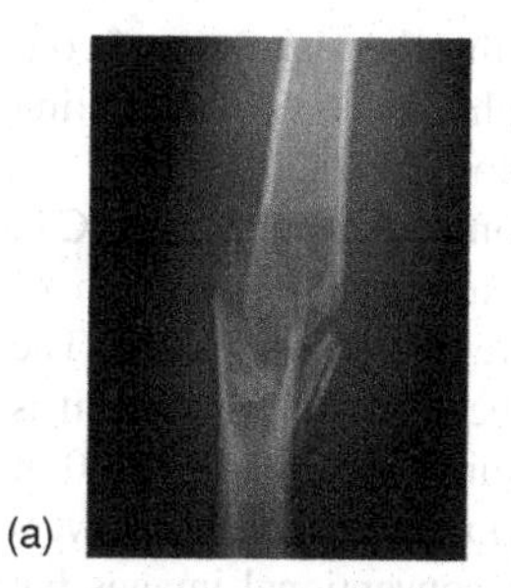
(a)

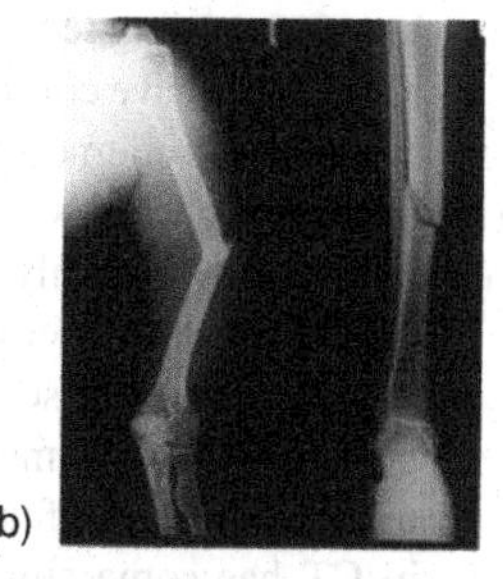
(b)

Figure 8.2 Shape is important in deciding what an image feature represents, but without contextual information on location these decisions are not useful. In (a), a fractured bone shaft is well demonstrated, but without additional information regarding its proximity to a joint as shown in the cases in (b) it has little value. Neighboring structures are used in radiology as frames of reference for location and extent of pathology.

Medical images portray structure and form (anatomy) and, through a deductive process in the observer, they infer function. The structural information is recognized and understood because of a pictorial knowledge base that the observer has compiled from studies of anatomy and function that normally precede radiology education. Recognition takes place when the observer is able to translate radiological objects into the "language" of anatomy learnt in a different visual form. But there are perceptual and cognitive obstacles to the translation that make medical images subject to variance in their interpretation. This variance is based on the reader's knowledge and experience. We can take two common imaging methods to illustrate.

(i) First, conventional radiography images. We will take the characteristic of *shape*. X-ray images record features of a 3D object (the body) on to a 2D receptor, and because depth cannot be represented in this process there is inevitable overlap and geometric reshaping or distortion of many structures. All this is normally represented in a grayscale of shadows. A basic principle used to help recognition is to have more than one view of the region, and these are sometimes termed "orthogonal projections." Views from different perspectives can then demonstrate structures as image features "en face" or "in profile" and as long as the observer is familiar with the real-world solid being represented, the structure can be recognized on the basis of shape mentally reconstructed from the views. A powerful support to the shape-based decision is the *location* characteristic of the structure (Figure 8.2). Radiology uses this as a principle in that all images of a region of interest tend to include contextual structures or location neighbors that can be used as frames of reference in a field of view (Herring, 2007).

But another characteristic of objects in medical images is "color." For the most part, the concept of color in medical images is restricted to black, white, and *shades of gray*; but these still have fundamental importance for feature recognition. The gray tone value of an observed feature will tell the observer a great deal about its make-up and also its congruence with its surroundings. It is this quality, known as contrast in medical images, which is so influential in the interpretation of their meaning. It is regarded with such importance that image recording technologies and the pharmacology of contrast agents have continuously sought new ways of exploiting this means of demonstrating radiological features.

(ii) Second, sectional X-ray images. Computed tomography (CT) images are a highly effective solution to the problem of 2D/3D overlap experienced in conventional radiography. The thickness of each section is between 1 and 7 mm and it is possible for some multi-slice examinations to have 1000 or more images. The feature recognition problem for observers in CT has some similarities with conventional images but there are important differences that relate to *location*. Slices remove the problem of structural overlap but they provide less information about the neighborhood relating to a feature in any single section. The observer must, therefore, inspect several contiguous slices to fully appreciate 3D relations and this activity adds to the cognitive demand of the immediate task. An advantage that sectional imaging brings to feature recognition, however, is the way it portrays 2D *shape* without any obscuring effect from adjacent structures above or below the slice. It requires the observer to draw on a different orientational knowledge base of anatomy which is arguably more complex, but presents fewer perceptual difficulties because normal shapes of structures are made invariant by the process.

The "color" issue is another feature of CT imaging that treats the characteristics of image features differently. *Grayscaling* is dependent on the same attenuation characteristics of objects as in conventional radiography but in this case the values are quantized, so allowing the viewer to "window" the image into the format that best demonstrates an item of interest. The end result is beneficial, but at a similar cost to the advantage given by CT's sectional slices in that the observer has to interact with the display and change settings to best demonstrate the item. The activity is additional to the interpretive task and requires extra cognitive resource.

8.3.2 Communicating findings: agreement on meanings

Irrespective of the imaging method used, radiological features and their variations have become well classified by practitioners and are documented extensively in standard textbooks on the subject. Radiologists have access to a wide source of knowledge in the form of databases of image features representing disease states and normal variations. Typically, these examples and feature descriptions are arranged in families and categories approaching the level of precision one would associate with histology. When medical images are read, what is learnt with respect to the diagnostic problem is then related to this knowledge in the language it uses. The knowledge is finally translated through various cognitive skills into decision processes and concepts.

Reading medical images, therefore, serves two purposes related to the diagnosis of disease:

1. Radiologists perform a decision-making process based on judgments of the nature of visual information. The greater the precision of these judgments, through being able to draw on an extensive and shared taxonomy of possibilities, the greater will be the specificity of the decision.

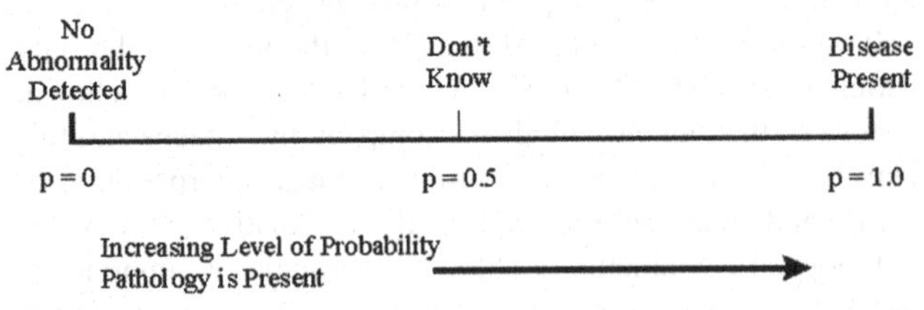

Figure 8.3 A visual analog scale to indicate confidence level in a diagnostic decision.

2. Decisions articulated to clinicians can be converted into actions to affect treatment based on the new diagnostic information. There needs to be an agreed understanding in the process of communicating what are sometimes arcane radiological findings in order to convey meaning and significance.

It is important to recognize the significance and essential contribution of these cognitive activities in the radiology reporting process. The clinician's diagnostic decision following an imaging test depends on his or her interpretations of the radiology report and although most doctors would review the images themselves, their assessment will be strongly influenced by what the radiologist says. However, radiology reports are the end points of judgments made in conditions of uncertainty. They are uncertain in the sense that all medical images are contaminated by some degree of noise that competes with the diagnostic signal, and when signals are subtle or noise presents ambiguities that mimic signals, decisions made are subject to error. Radiological reports tend, therefore, to be verbal expressions of probability. So if there is a mismatch between the ascribed probability given by the radiologist and that from the clinician, there is potential for misdiagnosis or inappropriate treatment. It is entirely possible that such a misunderstanding could spring from the *language* interpretation of the report rather than a mismatch of the *visual* interpretation or understanding of the image features.

Hobby *et al.* (Hobby, 2000) investigated the reliability of communication of levels of certainty in radiological reports and found large inter-observer and intra-observer differences in the meanings of expressions used. Phrases such as "unlikely," "probable," "no evidence of," "appears to be present," and "unable to exclude" are all in common use in radiology reports and they attempt to transfer to the recipient the concept of doubt or certainty that the radiologist holds for the visual decision. But in Hobby's study, terms like these were scored with very wide disagreement between report readers who were asked to give probability values to them.

Visual analog scales (VAS) have been used to see if they reduce differences in interpretation (Bryant, 1980; Maxwell, 1978; McCormack, 1988; Robinson, 1989) but although there are indications that numerical expressions of probability improve agreement between observers, they do not achieve complete congruence and have not been widely adopted in practice. Figure 8.3 can be used to illustrate the inter-rater differences with some terms such as "probably present." In simple experiments with tutorial groups of trainees one can demonstrate how, even with the help of these analog scales, there are still quite wide differences in people's minds in the interpretation of where the level of probability of disease is situated when such terms are used.

8.3.3 Mental models of meaning: schemata

Recognition of the relevance of perceived objects and decision-making regarding their significance are the important primary skills in reading radiographs, and it is thought that these cognitive factors are underpinned by mental models or *schemata*. Interpretation of meaning in the communication and classification of probability through language is the second stage. The mental models have meanings that can be expressed verbally in either concrete or abstract dimensions depending on the conditions of their use. For example, the same schema for a radiological object can be expressed as a *description* and the *name* of the structure (concrete) or as a *concept* that relates to the disease process it represents depending on the clinical situation (abstract). In studies of cognition in the interpretation of radiological information, verbal protocols are used as the means by which researchers can access the thought activities of observers (Lesgold, 1988). In investigations of this kind, readers are asked to tell the experimenters what thought processes they are using, how their decisions are being formed, and what significant factors they are taking into account based on the image information and external "prompts." The prompts referred to here are those generated by the reader in, for example, drawing on past experiences similar to the current problem. On the other hand, the approach taken by Nodine and Kundel (Nodine, 1987) and others has been biased towards recording visual search. These are perceptual experiments and gain a different insight into the interpretation process. They make a particularly valuable contribution because in many aspects of visual search, actions are unconscious to the observer who, therefore, cannot give valid verbal data on the activity. But both cognitive and perceptual approaches rely on *mental schemata* in their explanatory frameworks and they are the models most widely used in the concept of recognizing diagnostic clues in images.

8.4 COGNITIVE APPROACHES

8.4.1 Outlook

Problem-solving in medical image interpretation can be thought of as a selection process from a set of personally held options. The observer builds suitable new hypotheses and schemata, or activates existing ones, as possible solutions and then tests them against the available visual information. Neisser (Neisser, 1976) developed the idea of what he called anticipatory schemata that are used in the construction of a driver to the visual exploration of a scene as shown in Figure 8.4. This exploration samples visual information. It then feeds findings back to the schema, modifies and develops it; then the development updates and further drives the exploration. In this interpretation, the schema is a planning engine that keeps the perceptual cycle going to its conclusion in a diagnostic solution. Schemata may be multiple and it is likely that groups of schemata are activated and connected in many problem-solving tasks as we shall discuss in more detail in section 8.4.

But it seems that the activation process must be both fast and at the front of a sequence of events in searching a visual scene.

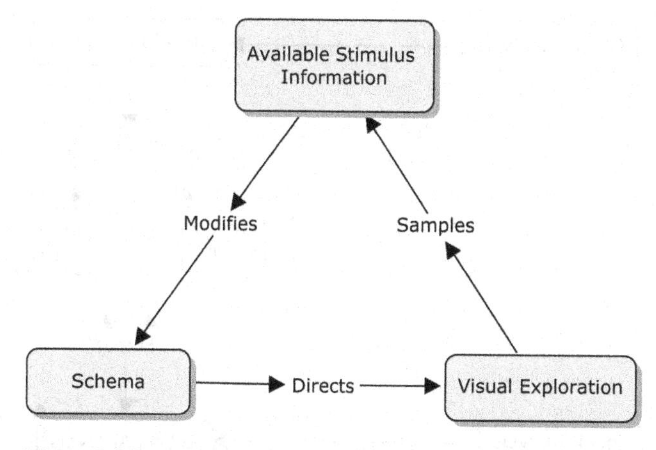

Figure 8.4 The adaptable schema as a cognitive planning engine for acquiring more visual data; after Neisser (1976).

Nodine and Mello-Thoms (Nodine, 2000) observed, in respect of the initial formation of a schema, that if it is to be effective in guiding search, then it must be generated very early in the viewing process.

Gregory (Gregory, 1970) and Rock (Rock, 1983) contributed to a first model of visual perception as a decision-centered process with a cognitive overlay. According to the model, the retinal signal is broken down by the visual system into primitive components such as color, form, and movement in a few hundred milliseconds. At this point all the necessary neural transformations are carried out and visual stimuli are processed into a literal perception. It is a very short, rapid mechanism during which the observer's visual attention is captured. It is stimuli-driven, but may also be modified from experience. The following stage of attention-selection uses knowledge of how the visual world is structured and organized to prompt the creation of a preferred perception. Gregory called this "applying the object hypothesis." During this stage, a goal-directed mechanism plays an important role based on the observer's expectations or intention, and it creates a preferred perception made up of definable objects (Figure 8.5). Conspicuous abnormalities are recognized rapidly and the observer finds those in a very short time, often without eye movements, but it is much slower than the initial phase and involves cognitive processes.

As Kundel (Kundel, 2000) points out, it is subject to change in response to added information about the implied meaning of the image. The point at which this change may take place is at the "covert decision" stage in Figure 8.5. "Covert" indicates the subconscious level of its operation and the decision to perceive the scene differently can be prompted by external or internal events. The external events may be a cue from a second reader, a colleague, or from a computer aid but the internal sources of a change in perception are less easy to identify. These may come from recall of a similar case experience in the past. When the preferred perception is reached, an overt, conscious decision can be made about the significance of the scene.

So-called "flash" experiments have investigated how much information can be gained from a single glance of short duration (200 msec). When chest radiographs are presented in this way, experienced radiologists are capable of recognizing up to 70% of the abnormalities that are subsequently picked up in free search. This indicates a rapid development of appropriate schemata or at least access to existing ones at the global phase of a model for

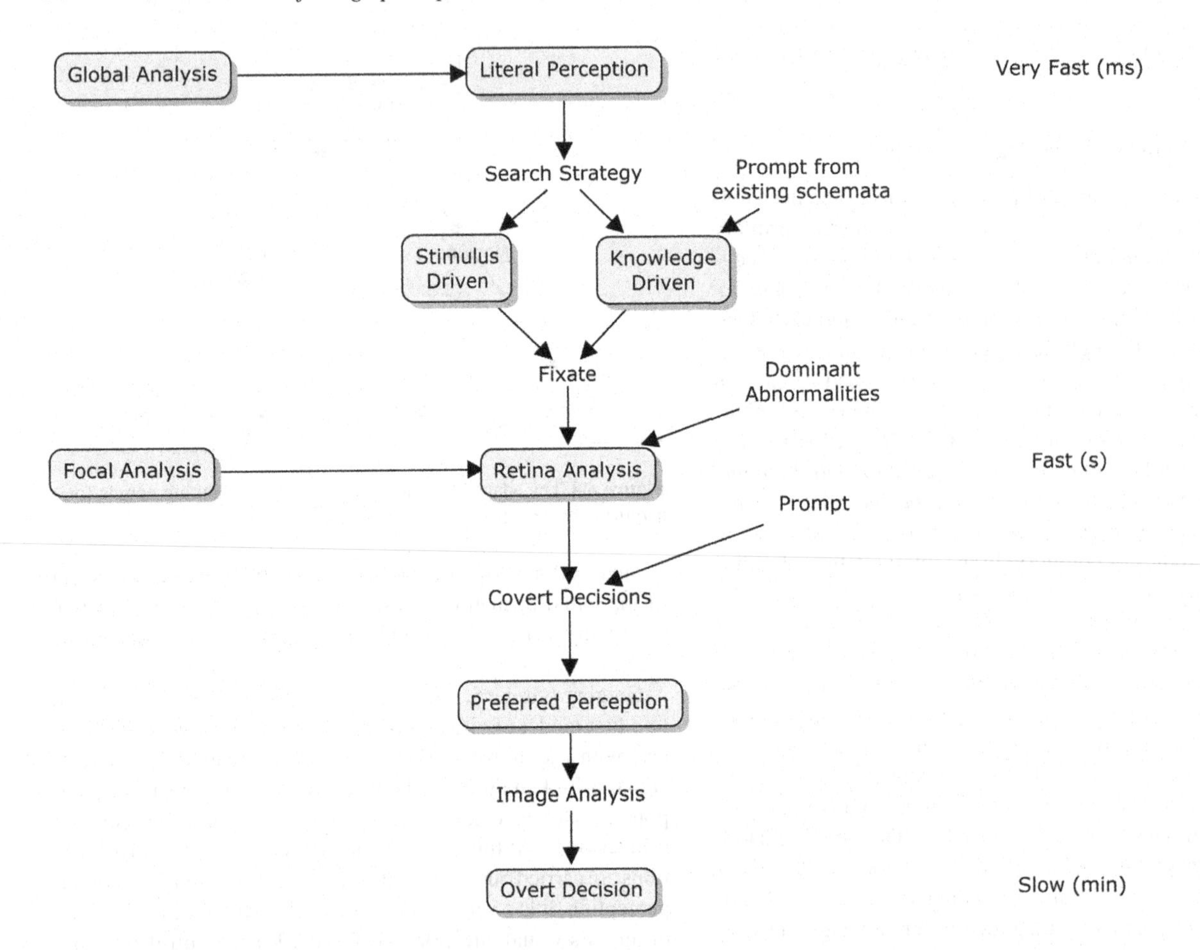

Figure 8.5 Decision-centered model of the development of perception based on the ideas of Gregory (1970) and Rock (1983).

visual search and detection. Kundel and Nodine (Kundel, 1975) used these flash experiments to develop such a model, based on the Gregory and Rock concept. The global initial response from the whole retina provides an overall, or schematic, impression at a very early stage and goes on to guide focal search (Kundel, 1983). This search process inspects image perturbations that, presumably, have been "flagged" at the global phase, and tests them against schemata of candidate objects for diagnosis. Figure 8.6 illustrates the process.

8.4.2 Recognition: cognition and perception

Identifying and discriminating specific objects in medical images can be thought of as a special form of object recognition. It is an analytical process. Recognition of radiological objects is only one part of a diagnostic problem-solving activity, of course, because there must also be a sense of relevance and applicability to the clinical picture. But it is a key stage in the process of building concepts concerning disease features. The observer attempts to determine the meaning behind objects perceived in the context of the medical problem that initiated the acquisition of the image (Kundel, 2006). Human perception and its relationship to cognition has been of interest to philosophers, psychologists, and physicists for decades, but more recent times have seen the subject also become a practical attraction to computer scientists because of its relevance to artificial intelligence. Its application to radiology is particularly topical because a machine intelligence that could faultlessly emulate the style and performance of radiology experts would be something of a holy grail in the medical applications of computers.

We recognize objects in the everyday world over a range of different scales and orientations and although these transformations occasionally affect our ability to identify items or figures, generally the brain's object recognition system is able to cope. Exceptions to this are illustrated by puzzles that are sometimes constructed to make familiar objects difficult to recognize by presenting them in an unusual context or from a strange perspective. Puzzles like this are usually in the form of photographs of the object rather than the object itself and they make use of some of the visual cues that are used in recognition of the real object but not all of them. In this way such puzzles are similar to the images of anatomical structures presented in radiology.[2]

It seems clear from this that for all the objects that we are able to recognize the brain must hold, in some sense, a representation of them that is made available for access so that we can identify ones we have met before. In many circumstances

[2] Nodine and Krupinski (1998) and Smoker *et al.* (1984), amongst others, have used the idea of visual puzzles as an instrument to investigate the existence of special perceptual skills or spatial abilities in radiology experts compared with lay people. Studies like these meet with less success than one might expect in demonstrating any such differences and so far, it seems the special expertise that radiologists have is highly specific to their professional task.

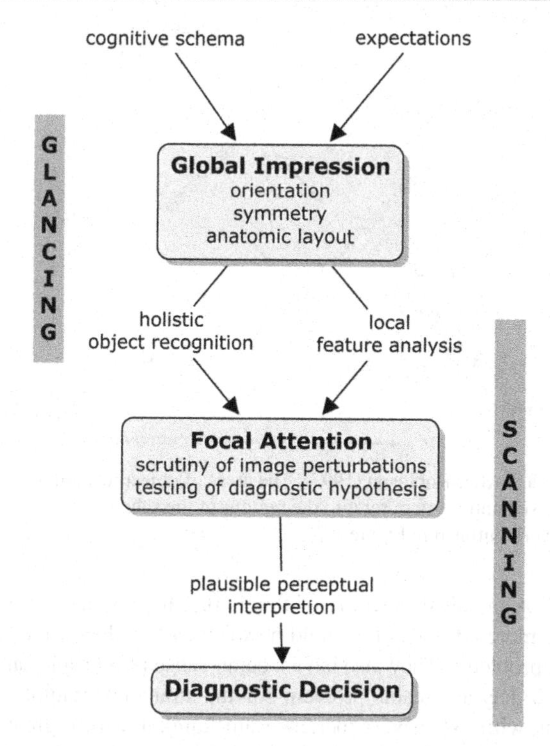

Figure 8.6 The global-focal model of perception in radiology developed by Kundel and Nodine (1983). Note the cognitive components in this model, represented by the prior expectations, hypothesis testing, and decision-making.

we do this even when they are presented in an obscure way or partially occluded. We can think of object recognition as a matching process where the object or feature of current interest is compared with an already held mental template that represents it. Such comparisons or matches are made quite easily and recognition takes place quickly when knowledge regarding the object is comprehensive and when we are presented with full 3D experience of it. In some cases we can even identify an object from only a general outline of its shape, but for this to happen we probably need to have a deep familiarity with the item (Ellis, 1996).

The recognition of objects probably involves the same dynamic and continuous exchange between perceptual data newly acquired from the visual scene and our stored knowledge of the world as pictured by Neisser's (Neisser, 1976) decision model. The interchange is also a cyclical updating process over the very short time-scales we have already discussed even though it is multi-staged and sometimes involves several iterations (Hendee, 1997). So cognition and perception should not be thought of as separate mechanisms, but rather as an interactive flow of information that leads to recognition in visual scenes. In the case of medical image interpretation, this is taken a stage further to the point of *skilled decision-making* involving the implications of the findings. Radiology is, therefore, a specialized task; it is not intuitive, and expert performance depends on knowledge acquired through specific, targeted learning. Skilled decision-making is a hallmark of expertise, and it is predominantly through an interest in the development of expertise that radiology has attracted attention from perceptual and cognitive directions of study (Nodine, 2000).

8.4.2.1 Recognizing objects in medical images: forming and retrieving schemata

Object recognition was analyzed most extensively by Marr (Marr, 1982), who took as his starting point the premise that vision is a computational process. Representations are made from light images formed on the retina and these are then stored in a symbolic form in the visual memory. There are three stages to the process in this computational model. The first stage is where the main features of the object are represented simply as intensity changes across the field of vision. In this way 2D information about the image is registered. Edge features are incorporated in this representation because they are usually examples of abrupt changes in intensity. Marr termed this stage the primal sketch. The second stage determines more subtle characteristics of the object through an assembly of cues concerning depth. These items of information are processed entirely from the point of view of the observer's position with respect to the object and Marr referred to this as the 2D sketch. In the case of objects in medical images, the depth cues will be quite different to the prompts we get due to stereopsis and shading from the world of solid objects. They are perhaps learnt or acquired by guided experience from looking at orthogonal projections or images of body sections referred to earlier. Marr's third stage is a full 3D model representation of the seen object independent of the viewer's position; a mental reconstruction of the solid object in a coordinated frame. So it specifies the shape of the real item, with all its surfaces and positioning with respect to its environment.

At the point of the 3D stage the feature is represented in a fairly standardized form that allows the observer to compare it to all known structures that have been stored from past experience in a sort of "look-up table" of candidate schemata. Ellis and Young (Ellis, 1996) have discussed the way in which objects might be recognized through a hierarchy of possibilities based on the situation. They point out that the level at which we need to recognize things can vary depending on the context of the decision goal. In radiology, for example, a rib in a chest radiograph might be identified, under different conditions and for different purposes, as: a piece of bone, a rib, a seventh left rib, or a seventh left rib that is fractured. There is clear inference from this approach that the cognitive system is very flexible. But Ellis and Young warn against overemphasizing the significance of this flexibility and although they acknowledge its existence, they doubt it is a property that we have in continuous use. While the facility is certainly available (as long as the observer has the required knowledge base), they comment that it does not follow that all the levels of recognition can be achieved with the same degree of ease. There are probably certain levels that are typical of everyday use and, because they are accessed frequently, they are recognized without conscious effort. This basic category level can be thought of as a visual representation, processed at the point of the object-recognition units in Figure 8.7 that simply describe what the item looks like. The semantic system, however, operates at a deeper level, perhaps not typical of everyday use, where the properties and attributes of the object are worked

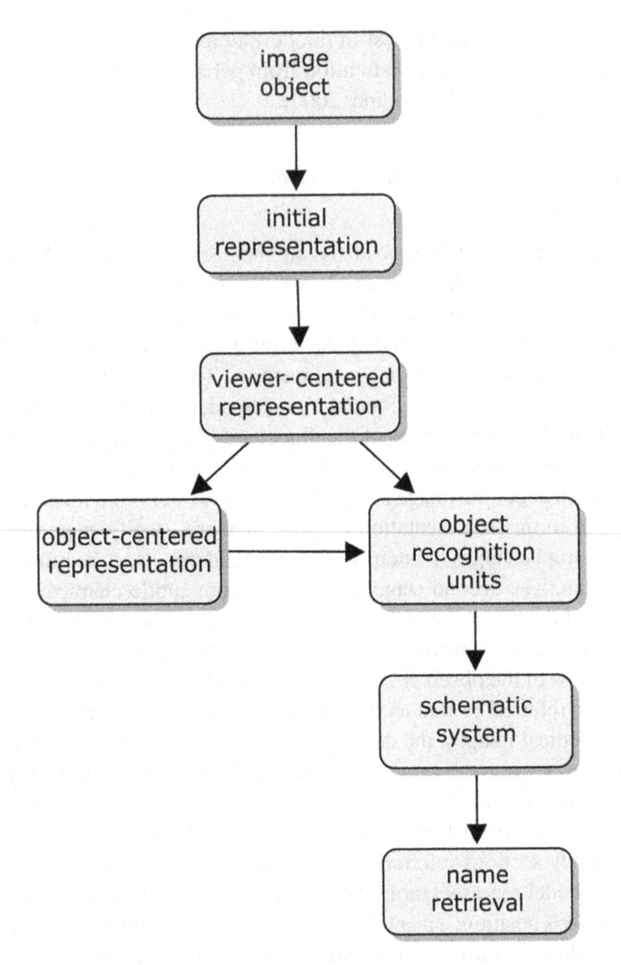

Figure 8.7 A model for object recognition based on the Marr (1982) framework. At the schematic (or semantic) level, the object is recognized with a depth of understanding that allows connection with a variety of other schemata of known objects. This gives the opportunity to develop broad concepts around the meaning of the object and not just a recall of its name.

on and where we extend recognition into areas of significance, relevance, and meaning.

8.4.2.2 Recognition and radiology performance

Levels of recognition have a clear relationship with learning and, by implication, with expertise and some sources of cognitive error. This might be particularly relevant in respect of the performance differences between readers with different levels of experience in their exposure to specific image appearances. Cognitive theories of recognition provide encouraging theoretical support to what we see in the development of expertise in radiological tasks and how it depends on extensive, specific experience accompanied by feedback. The importance of specificity of the experience has been emphasized by Gunderman *et al.* (Gunderman, 2001), who have observed that expertise in one domain in radiology does not necessarily transfer to another. However, greater relevant experience in experts allows their acquired knowledge to be better organized around integrated schemata (concepts). This gives them the facility to combine and recombine their factual knowledge and to ask "bigger questions." A practical outcome of this is that it provides experts with a more advanced threshold between easy and hard recognition problems. They also have a better sense of relevance and more widely applicable problem-solving strategies. Radiology experts with extensive experience are familiar with a greater number of image objects and have shifted more of them into the "everyday" category for recognition purposes. Expert decisions are therefore more certain. But for Gunderman, this also means that experts can not only point out and name more radiological objects but the structures they recognize give them wider concepts of diagnostic meaning. This is because integration of their schemata takes place at a higher semantic level.

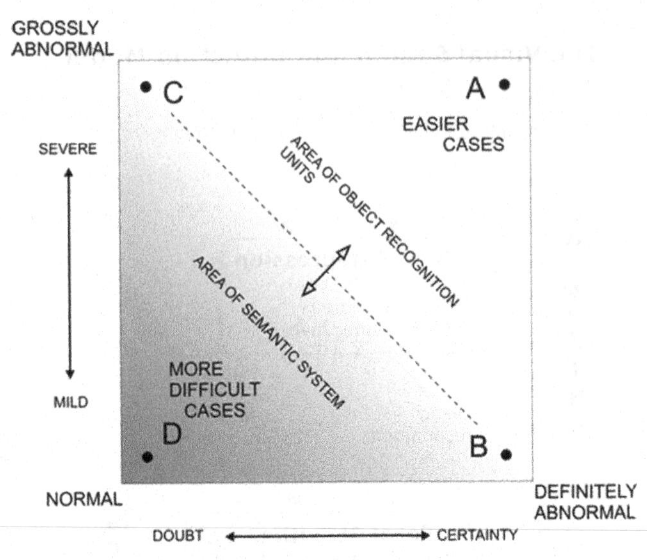

Figure 8.8 After Robinson (1997). The areas of object recognition and the semantic system refer to the regions of the schematic framework outlined in Figure 8.7.

Robinson (Robinson, 1997) argues that the importance of radiology as a medical specialty resides in the ability of its practitioners to make valuable decisions on what he calls "stress cases." These are the difficult cases in a framework that describes the relationship between degrees of abnormality, doubt, uncertainty, error, and variation as shown in Figure 8.8. The important feature of this framework from a cognitive perspective is the way it characterizes "easy" and "hard" cases.

The cases in A (Figure 8.8) are recognized with ease and with certainty because they are gross examples of typical or everyday cases where the reader has a stereotypical mental model of the abnormality in question. In B a minor abnormality is recognized with similar certainty because of its typical and clear appearance. But in C, despite the severity of the disease feature, the observer is uncertain because of his doubt over the significance of the finding. He is operating, without complete success, in the deeper semantic regions of his recognition schema. Finally in D the observer finds the case difficult because the disease feature mimics or is close to his stereotype of normality and from his knowledge base he is in doubt of the meaning of the appearance.

There will be factors that are apparently extrinsic to the cognitive object recognition model that will, nevertheless, have a powerful effect on whether a case is considered hard or easy by the observer and the subsequent level of his performance.

The question of how a difficult case comes to be difficult is worth comment. Relating back to the point made that we only recognize what we know, it is clear that anything that affects the caseload experience of the observer will have an effect on the level of recognition that the observer will hold for the feature in front of him. This effect might even be outside the context of formal instruction or training. An important example of this is the prevalence of the disease and its radiological appearance in the day-to-day experience of the observer. Where many examples of the feature are seen at a high level of frequency, the ease and speed with which they will be recognized will be equally high. But this obvious statement must be qualified by the understanding that *correct* easy, rapid recognition will only become commonplace in an observer who receives feedback on his decisions. Feedback on performance in decision-making is a very important feature in developing the knowledge base and schemata necessary for meaningful identification and recognition because without it there is no validation stage in the process of forming accurate mental models. Gunderman *et al.* (Gunderman, 2001) have cited Lesgold *et al.* (Lesgold, 1988) in giving some insight into this through studies of differences between experts and novices in Chi *et al.* (Chi, 1988). Their observation is that simply showing learners many examples of different lesions will eventually produce some results. But effective recognition with an associated concept of relevance is speeded by active instruction on the correctness (or otherwise) of their decisions and the alternatives that could be considered. Because of their scant experience novices tend to seize on an abnormal feature and grasp at a first explanation rather than considering wider possibilities. Faced with the same visual information, an observer who has benefited from wide experience and confirmatory feedback will bring to the problem more schemata for organizing the information into meaningful patterns and will suggest solutions beyond the perception of the learner.

One caveat must be added to the argument that frequent, extensive exposure and feedback will result in better radiological performance. It may also prime the observer to "recognize" features that are not there. This may be due partly to a bias that uses a "simplicity and likelihood principle" (van der Helm, 2000). Given a certain image feature which looks like a common disease sign, it will be *simpler* (in terms of the covert decision to take a preferred perception in Figure 8.5) to decide to call it something that is familiar. But there will also be an influence from the cognitive process that asks *how likely* it is that the feature represents disease. If recent experience has exposed the reader to a high prevalence of the disease in a routine caseload, an image that suggests the sign will be *more likely*, in the view of the primed decision-maker, to be a genuine presentation of the disease. We shall return to the issue of probability in decision-making a little later.

8.4.3 Recognition by chunking visual information

Descriptions of the strategies employed in object recognition have to take account of the evidence for the mechanisms that seem to be used. We used the term *perturbation* earlier in the context of the Kundel and Nodine model (Kundel, 1983, 2000) of an initial response that provides a schematic to guide focal search. Perturbations have been described by Nodine and Mello-Thoms (Nodine, 2000) as those parts of the image that radiologists find "odd." They speculate that what are recognized at the rapid first glance of an image are unexpected anomalies generated from the global impression. There are too many potential targets for this first glance to be a target search per se, but it is designed and intended to find something odd about the image on which search and attention can be focused. The reader must first know the difference between the "not-odd" and the "odd," and the term perturbations indicates the disturbance they create in the observer's schema for a "normal" or "not-odd" image. The level of disturbance from a perceptual point of view will be related to some physical characteristics of the object. The conspicuity of the object is a term sometimes given to this quality and includes characteristics such as size, contrast, and degree of edge-sharpness as well as the nature of its surroundings. It probably has an important influence on the preattentive phase of vision because it targets attention for subsequent search in the attentive phase. But there is less evidence to show that how conspicuous an individual object is influences its recognition. This tends to confirm the importance of the role that cognition plays in image interpretation. Detection is only a starting point in the decision-making process of whether an object is significant and worth reporting. Eye-tracking studies have shown that the measured conspicuity of small lung nodules in chest radiology, for example, has a bearing on their likelihood of being fixated but not on their identification as pathology (Manning, 2004). Detection and scrutiny of these perturbations does not always lead to a correct decision on their meaning. This decision feature of image interpretation is a particularly difficult one for computer aids to detection (CAD) and diagnosis (CADx) to access. Given that the majority of unreported lesions are detected by the eye-brain system but rejected at a cognitive level of recognition, it may be the case that computer aids to *detection* are not able to offer significant gains in performance to many experienced observers. Computer aids to improve radiological performance may eventually find their greatest success through ways of improving the decision-making component of the task.

Image perturbations may be limited to a local level for small objects, or they could be much more global in origin and in some cases they may be a combination of these where several small perturbations form a larger pattern; but in any event they provide a pop-out source of information for the plan that guides subsequent search of the image. The speed at which these perturbations are recognized at the global impression stage suggests a form of Gestalt, "whole-problem" approach that is distinct from a built-up step-wise construction. Some of these aspects of visual processing are used in art and in puzzles as conflicts and "surprise" features. Conflict and interest arises when the Gestalt, global impression suggests one visual decision but the subsequent focal attention suggests something else (Figure 8.9). This can occur in a medical image where the perturbation captured in the pop-out phase is a large, low-spatial-frequency object and gives the observer a convincing match to a well-established cognitive schema for a lesion. However, during search and focal attention, less obvious, smaller, high-spatial-frequency features make a contribution to the diagnostic decision and the original "flash" response is then modified (Figure 8.11).

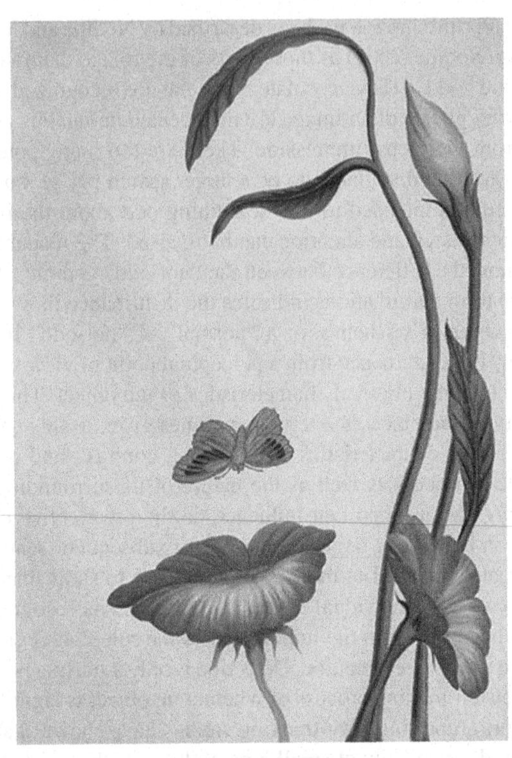

Figure 8.9 The global impression of the image gives, as a fast response, the woman's face. Moments later the observer realizes there is more to the scene, and that the face could be a coincidence of the shapes and positions of spurious objects. But our schemata for faces are so well established that we find the face image the most compelling. There are cognitive and perceptual models for the explanation of what is happening here. Included in these is the importance of the proximity of the image sub-units. This allows them to be actively grouped into face features by our semantic systems in the pre-attentive phase of vision. http://creativecommons.org/licenses/sa/1.0/

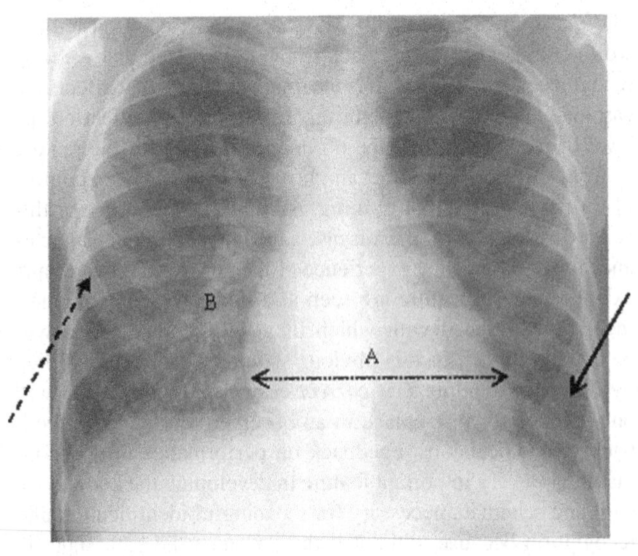

Figure 8.10 Radiologists will often arrive at a diagnostic decision very quickly through what is thought to be a chunking of the key features demonstrated in an image (www.bcm.edu). In this chest film there is: (i) mild cardiac enlargement (A) with (ii) pulmonary venous congestion (B), (iii) fluid within the horizontal fissure - - -→, and (iv) prominence of so-called Kerley's B lines which indicate lymphatic engorgement →. Collectively these signs indicate congestive cardiac failure to the experienced radiologist. Fast arrival at this decision probably involves little or no attention to irrelevant features, may be associated with priors, and can be thought of as an "Aha!" moment (Ramachandran, 2003).

Another aspect of the grouping phenomenon illustrated by Figure 8.9 is the way that it is not always necessary to take all the available information from an image to arrive at a conclusion. What Ramachandran (Ramachandran, 2003) has called "Aha!" moments occur when relevant "chunks" of information are merged to give a conclusive, confident decision even though some details may be absent from the scene. In these instances the large perturbation in the image may be incomplete but is made up of smaller components that collectively give a strong enough global message for a decision to be made – often very quickly (Figure 8.10). When radiologists are used as observers in experimental work their performance is noted as being fast and without apparent effort. The speed, typically only a few seconds for a single static image, with which inspection is carried out followed by an additional period for drawing the full diagnostic decision tends to suggest that this "chunking" of information plays an important part in the interpretation process (Manning, 2005).

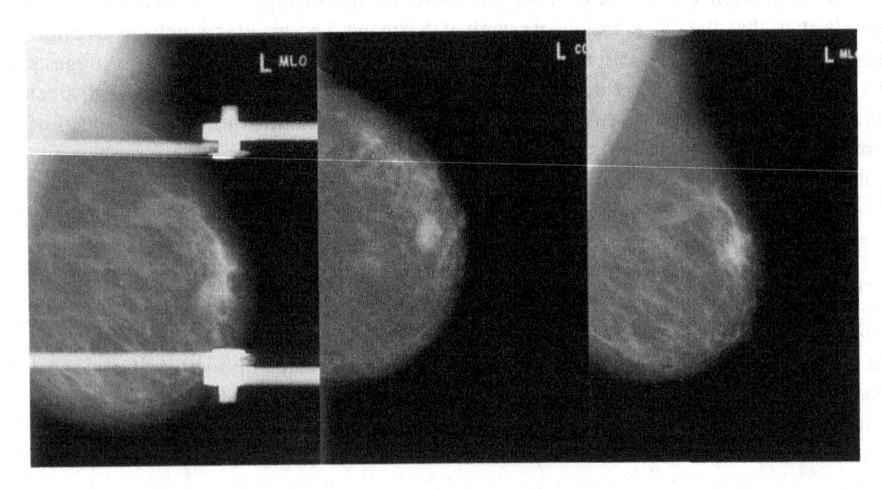

Figure 8.11 The mammogram views show a large mass lesion at the global stage of viewing. Closer focal attention reveals more information in the form of adjacent architectural distortion along with scattered suspicious microcalcifications to modify and refine this diagnosis.

8.4.3.1 Top-down and bottom-up strategies for recognition

Marr's approach to object recognition and his model for the development of schemata was taken on by Ullman (Ullman, 1996) to emphasize a so-called *top-down* view of recognition that applies to where the object is grasped in the whole-state, global event. This is in contrast to a *bottom-up* description of the recognition process more characteristic of the perceptual experimentalists. The differences in position taken by the cognitive and perceptual schools stems more from the methods they use to interrogate the problem rather than the theories themselves and, as we have seen, the approaches are not mutually exclusive. In fact they are closely similar in the sense that they both use the *object features* as the units of recognition and in medical image interpretation perceptual and cognitive schools agree that an important measure of expertise is speed and accuracy of the global recognition of normal from abnormal (Nodine, 2000).

The terms *top-down* and *bottom-up* are used in computer science to describe strategies of information processing and knowledge-ordering that can be useful in modeling object recognition in humans. The models implied by the terms are sometimes used in discussions on the global, intuitive, fast-response parts of the models as well as the slower focal attention stage. High-level familiarity with a visual problem, for example, may be associated with a prior expectation of what an image will contain. This "prior" has elements that are *top-down* and it complements the *bottom-up* primitive components of literal perception described in Figure 8.5. Search and focal attention are characterized by relatively slow scrutiny of image features and can be thought of as partly a *bottom-up* arrangement. Recognition at this stage seems to be based in part on fitting simpler component characteristics together to form a more complex object feature then matching the results to what is observed in the image. However, there are some clear *top-down* components to the focal attention stage when one considers the hypothesis testing and template fitting that is thought to occur.

In the *top-down* model of information processing, an overview of the system is constructed without going into detail for any part of it. Each part of the system is then refined by designing it in more detail. Each new part may then be refined again, in yet more detail until the entire specification is detailed enough to confirm the model as valid. *Top-down* models are often designed with the assistance of "dark boxes" that make it easier to bring the task to completion but provide little understanding of the component mechanisms.

In *bottom-up* design, individual parts of the system are first specified in great detail. The parts are then linked together to form larger components, which are in turn linked until a complete system is formed. This strategy often resembles a "seed" model, where the beginnings are small but eventually grow in complexity and completeness.

8.4.3.2 What's the object called?

Naming the object has an important part to play in the communication of decisions, ideas, and concepts in image reading so it should be closely allied to any cognitive theories we might have on object recognition. Naming can be incorporated into Marr's recognition schema but the assumption is that the semantic system does not, in itself, contain object names. The name-store or lexicon is separate from the semantic system but is accessed by it. We need to have a concept of what the object *means* to us, that is, to achieve the level of semantic representation in the schema, before we can retrieve its name from our lexicon. In radiology, pathological features are sometimes given names that are derived entirely from a description of what the feature looks like. These can be highly logical – for example "spiral fractures" (Figure 8.12) have that precise geometric appearance – but names occasionally verge on the exotic, as in "bat-wing" sign in acute alveolar pulmonary edema (Figure 8.13). Names derived from appearance, however, are probably very valuable in communicating meaning because of the level of their shared understanding and are preferable to those taken as eponyms from the doctor who first described the pathology; for example, Salter-Harris fracture (Figure 8.14). As with all language, irrespective of the system adopted, there is absolute importance in a common agreement within the culture of the names given to objects.

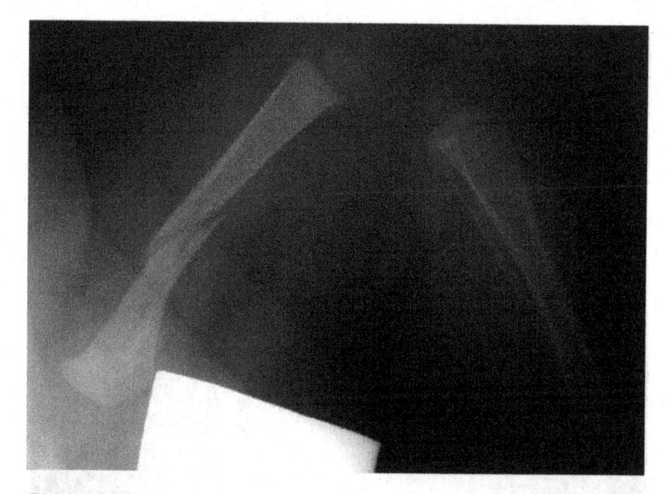

Figure 8.12 Names given to image features can be quite logical, as in this case of a *spiral fracture* of the femur of a two-year-old child. Communicating this information is easy because the description comes from a commonly understood term.

The arrangement of retrieving names from a lexicon separate from our semantic systems can account for the way we occasionally see an object, recognize it, fully comprehend and understand its conceptual meaning, but have difficulty in remembering its name. In fact there are circumstances where we recognize familiar objects without having a name for them at all, in which case we resort to rather convoluted descriptions of the items in trying to communicate verbally to others. When we eventually recall the name of an object whose name we know but have forgotten, it is always connected with a sense of meaning for the item or its context and never simply at the basic level of visual representation.

8.4.4 Deciding what is pertinent in the image

Medical images are rich in visual data, but only a small proportion of these data are relevant to the diagnosis. Problem-solving, generally, involves finding the most effective path through the gap between what is currently known in a situation, and what

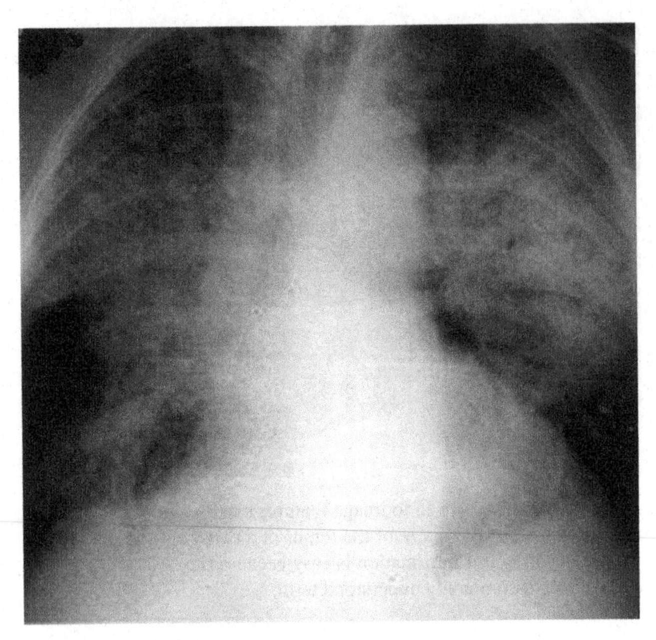

Figure 8.13 Radiologists call this appearance the "bat-wing" sign, characteristic of acute alveolar pulmonary edema. The term conveys a message that relates quite well to the general shape of the shadows seen in the lung fields. medcyclopaedia.com (GE UK).

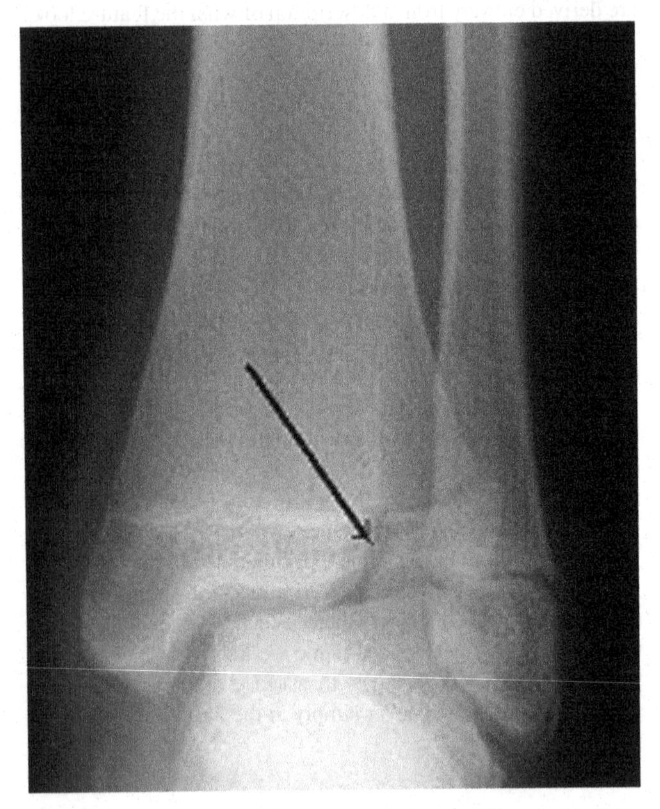

Figure 8.14 This is termed a Salter-Harris fracture. The nature of its appearance could not be communicated by its name unless the recipient was familiar with the language of radiology.

is targeted. One of the difficulties in medical problem-solving is that the target (diagnosis) is unknown and must be identified by the clinician. So the process is marked by the generation of a number of diagnostic hypotheses that are possible end-states or targets. These hypotheses are then tested to see which ones, if any, are viable solutions. This sort of logical medical reasoning applies to radiology where the hypotheses are developed on the basis of visual disease patterns observed in image features then tested against their congruence with other patterns of data and their likelihood of being right. The meaning behind the patterns is the key to successful decision-making in reading images, but there are various subjective judgments involved in this process that can bias the way observers evaluate not only immediate image information but also later evidence. For example, a primacy effect has been noted in clinical decision-making so that when initial hypotheses are formed, new data that conflict with the earlier hypotheses are more likely to be ignored or their importance relegated (Wood, 1999). If this is the case, there is something to be gained from readers being skeptical of their hypotheses until all the candidate ones have been vetted. Given the large number of possible diagnostic schemata and the fact that even irrelevant information is available to be utilized by radiologists, Raufaste, Eyrolle, and Mariné (Raufaste, 1998) have asked how it is that pertinent rather than non-pertinent hypotheses are selected.

One possibility is that each diagnostic concept is represented by a schema that acts as an information processor. Each of these "processors" is capable of being activated when a certain trigger threshold is crossed (Rumelhart, 1986). So when a sufficiently large cue from an image matches a schema, or even part of it, the schema is activated. If one assumes that (i) the most highly activated schemata are the ones that have the chance to influence decision behavior and (ii) the most appropriate schemata are usually the ones that match the cues and are the ones most likely to be selected, it provides a possible model for how pertinence is generated and relevant hypotheses are selected. This model has some support from experimental observations of radiologists. They carry out the very complex task with apparent ease, rapidity, and with repeated skill in familiar areas of work and they are adept at selecting both primary and differential diagnoses. But they perform less confidently, with lower accuracy, and with fewer alternative hypotheses when they are inexperienced (Wood, 1999), faced with unusual diagnostic problems, or have to interpret images from a new modality. This tends to support the idea that special expertise is associated with a well-developed armory of many ready-made templates. These templates are used frequently and so they give solutions quickly and accurately, but the formation of new ones as possible answers (when learning a new task) takes more time and effort.

What if a visual cue from a medical image does not exactly match a frequently used schema, but triggers one nearby that represents something similar? Could schemata communicate with each other and so enhance the range of objects we can recognize?

8.4.4.1 The connectionist approach

The idea of functional connections between schemata takes the idea of schematic triggering a stage further and suggests that activation above a threshold is not confined to the single schema but spreads its effect to the neighborhood. This has a priming effect on linked schemata and as a result, activation occurs at higher levels in those that match several cues. According to this model, schemata that are frequently activated undergo reinforcement and become dominant or "high-likelihood" solution templates. Other schemata, on the other hand, may have infrequent, low levels of activation, in which case they become eliminated (Balota, 1996).

This effect of competition would give fast access to hypothesis formation for disease processes that are more commonly experienced, and because of the relationship between case familiarity and expertise one would expect experts to arrive at their decisions very quickly and without the need to use all the available information.

The view of image interpretation represented by this description is termed connectionism and the models are often called "neural nets." They are an analogy with the artificial neural networks (ANNs) of parallel processing in computer science. According to this model, large numbers of simple processing elements make up pools of schemata and are connected to each other by excitatory or inhibitory links in various possible ways. An organized system of such pools of units (with each pool corresponding to specific representations and sensitive to specific levels of threshold stimulus) can form highly complex interaction patterns in this picture (Ellis, 1996). Changes to networks in the human cognitive system can be brought about by new experiences and feedback on decisions which, by analogy with the artificial systems, change the weightings of connections and pathways and therefore the significance of their relative contributions. This is a holistic view in neuroscience of how the brain is thought to work and it argues that the brain functions as a whole rather than as a set of discrete modules each with its own prescribed function. It tends to be defended by the fact that many areas of the cerebral cortex can be recruited to multiple tasks and everything is connected to everything else (Ramachandran, 2003).

ANNs can learn in either supervised or unsupervised ways. In supervised learning the system is given a set of examples where the truth is known and it maps each example to its category. Instant feedback on the truth in this arrangement means that the system learns success and failure immediately. This is a metaphor for the way in which radiology students learn against the truth fed back to them from their professors. But when the truth is unknown, unsupervised learning may be used instead. This method is nearer to many clinical situations where the ground truth is often difficult to determine. In unsupervised learning the ANN is allowed to create its own map between a "working truth" and the examples it is fed. It creates many more categories than in the supervised method and learning occurs by agreement rather than with a fundamental truth state. In many instances in clinical practice this is the way that radiology performance has to be measured because of the elusive nature of ground truth.

A problem with understanding the radiology interpretation task in ANN terms is that it can never be completely clear which processing elements are being used most heavily in the classification process. In ANNs the activity of the network is carried out at the unobservable level of the "hidden layer" in the multilayer system, giving little insight into how the decision output was generated (Figure 8.15).

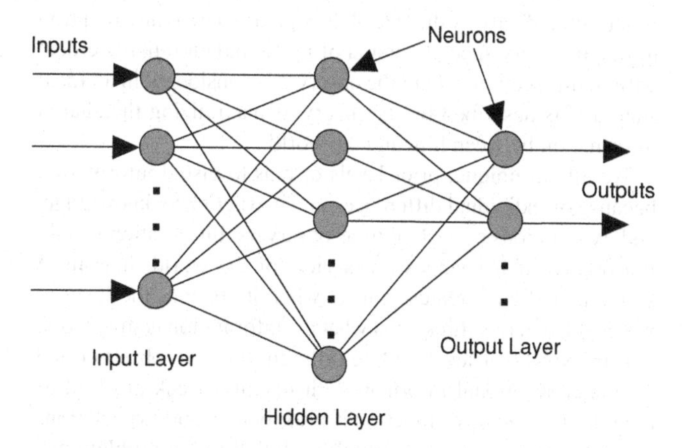

Figure 8.15 A typical multilayer perceptron (MLP) artificial neural network. The hidden layer processes information but is normally inaccessible – so it operates as a black box. All the layers are connected in this example; the input is binary data and the output layer conveys the network's decision as a *p*-value to the outside world. http://www.geocomputation.org/2000/GC016/Gc016.htm

However, as Nodine and Mello-Thoms (Nodine, 2000) have pointed out, this is remarkably similar to the way that radiology experts seem to operate. They are often unable to list all the steps they took or which factors weighed more heavily than others in the diagnostic decision. This hidden layer analogy is only relevant to the "stress cases" referred to in section 8.4.2.2 because in easy and obvious cases there is no doubt as to the decision trigger. But it should be appreciated that it is the difficult diagnoses, often based on borderline evidence, that are most valuable in medicine. These often represent the states of diseases at their earliest – the most important stage for good prognosis.

Artificial intelligence used in clinical diagnosis or as a teaching aid in radiology training is clearly an exploitable technology that can play a part in the interpretation process in medical imaging and there is little doubt it will grow in importance. Designs of these systems continue, quite rightly, to base themselves on how we think the brains of expert radiologists solve the problems. But there is currently a deficiency to this approach in one important area: artificial intelligence, unlike humans, has not yet been able to take fully into account *prior information and knowledge* in the formation or testing of hypotheses.

8.5 PROBABILISTIC DECISION-MAKING

8.5.1 Outlook

Information theory, referred to in section 8.1.1, provides a very valuable framework for describing how, when we sample the physical world with our visual systems, we convert incoming light energy into brain activity and experience it as a mental event telling us what is going on "out there." (I must stress here that the final part of the last sentence is not an attempt to sum up the nature of consciousness!)

But information theory, generally, takes no account of the observer and assumes that the amount of information transferred is determined only by the size of the signal. There is an exception to this in a very special version of its formulation that we deal with in this section, where the observer is considered in an ideal state and is described by the theory as maximizing the shared information between him and the world.

We all put unique, prior levels of bias to visual information because of individual differences in the experience, knowledge, and expectations of each of us as observers. Information is only informative if it changes one's view of the world. It is most valuable and has greatest impact when it changes the level to which the receiver thinks that what the information is suggesting is true. Now, that level will be different for each observer and it puts conventional information theory in a weak position as a model for cognitive function if prior beliefs and expectations are not accounted for. Clearly the solution to the problem can be found only if it is possible to determine the levels of belief before and after the information is received.

8.5.1.1 Bayes' Theorem

To illustrate how prior beliefs and probabilities are relevant to reading medical images, we need to take an example.

Example: probabilistic reasoning can be used to help a clinician analyze the possible causes of a patient's breathlessness and chest pain.

The purpose of diagnostic imaging is to provide information that will influence action along a treatment pathway. Eddy (Eddy, 1982) provides clear evidence for the claim that a natural thinking strategy for physicians is the use of "degrees of certainty." This strategy is used by them to select courses of action for treatment. In our example the clinician can have a degree of certainty about the patient's condition based on history, symptoms, and clinical examination. Further information will be gathered from imaging to refine the subjective probability of, let's say, a pulmonary embolus – a potential threat to life. Now the appropriate treatment for a pulmonary embolus is anticoagulation therapy and is not without risk, so the physician needs to be very sure that in all cases the treatment is only prescribed if it is needed; in this case, if there is definitely a clot in the patient's lung. The patient has an appropriate imaging test and the radiologist reports a positive result: that there is an indication of pulmonary embolus. How sure can the physician be that the patient should be treated with anticoagulation therapy for a pulmonary embolus?

Bayes' formula can be applied to this because it tells us the following:

$$P(\text{pe} \mid \text{pos}) = \frac{P(\text{pos} \mid \text{pe})P(\text{pe})}{P(\text{pos} \mid \text{pe})P(\text{pe}) + P(\text{pos} \mid \text{normal})\,P(\text{normal})} \tag{8.1}$$

$P(\text{pe} \mid \text{pos})$ is the probability that the patient has pulmonary embolus given that the imaging test has a positive report (this is called the *posterior probability*). $P(\text{pos} \mid \text{pe})$ is the probability that if a pulmonary embolus is present, the radiologist will correctly report it (this is the *sensitivity* or *true positive rate* for the test). $P(\text{pe})$ is the probability that the patient *has* a pulmonary embolus (this is the *prior probability*). $P(\text{normal})$ is the prior probability that the patient *does not have* a pulmonary embolus (his chest pain and breathlessness in this case are due to something less serious). $P(\text{pos} \mid \text{normal})$ is the probability that if the patient does not have a pulmonary embolus, the radiologist will *incorrectly* diagnose it as being present (this is the *false positive rate* for the test).

As a model of a probabilistic approach to clinical problem-solving, the Bayes system is helpful because it not only reflects what many physicians do but it throws light on the way that prior probabilities can affect the usefulness of the test decision (Phillips, 1995). In our example we are told how much we have to change our belief about the diagnosis of a life-threatening disorder given the new evidence from radiology (in our example the direction of change is that we are now *more certain that the condition is present* because the radiology report confirms our prior belief), and it also tells us what evidence we should have been expecting from radiology given our prior belief about the diagnosis (if the clinician felt confident that embolism was present before radiology, he now feels more strongly in this direction and has created a better belief).

There is an "updating" process going on here that actually benefits from error (Frith, 2007). Consider a reverse situation in our example where, before radiology, the physician felt mildly certain that the patient was *not* suffering from a pulmonary embolus. New information, even if it contradicts a previously held world view, allows the diagnostician to revise beliefs about the state of the world and to repeat the process as necessary in a loop of activity that continues to make refined predictions. At every loop, the error level gets smaller until the cognitive system "knows" what the reality is.

8.5.1.2 The ideal observer

An ideal observer is a concept that sets upper limits to the performance of a decision-maker in a task. The Bayesian ideal observer is the absolute upper limit and assumes the cognitive use of all the available evidence in the best possible way. It is useful as a benchmark against which to compare human performance because the model assumes incoming information is used interactively with previous visual experiences and current expectations (Donovan, 2007). Information is represented in Bayesian theory's statistical framework as a set of probability density functions (PDFs) where the decision output is a product of three such functions:

- the likelihood function associated with each cue
- the prior density function representing probabilities for the various possible states for the image before it was viewed
- the utility function representing the costs, risks, and benefits

These functions can be used in building realistic models of the cognitive task in image interpretation and identifying the relationship between the decision variables in human observers matched against the Bayesian ideal. Frith (Frith, 2007) points out that when it comes to rare events and large numbers, humans are not good at using evidence; we get it hopelessly wrong in estimating risks, benefits, and likelihoods from data of this kind. But our brains are near to ideal observers when we take evidence from our senses and we combine this information in a very Bayesian way; weak evidence tends to be ignored and strong evidence is emphasized.

There is good reason to conclude that in radiology the reasoning process can be explained quite successfully in terms of

Bayes' Theorem. It takes a hypothesis and compares it with one or more alternatives (the differential diagnoses); it includes, in all cases, the possibility that there is no disease present. It is carried out always on the assumption that the radiology report adds value to the diagnostic decision and the statistical significance of this added value can be measured in terms of its "diagnosticity" (Wood, 1999).

8.6 SUMMARY AND CONCLUSIONS

Reading medical images is an interpretation task that can be compared with other activities where humans take visual representations of objects and make sense out of their meaning. But radiology has some special requirements overlaying this everyday description that make the task a particularly difficult cognitive challenge. The nature of the information that creates diagnostic images is highly varied and if the reader is without a good understanding of how an image has been formed, it is difficult to identify or put meaning behind the objects and features it contains. This is quite different to the everyday world where our eye-brain systems have adapted entirely to information received from an environment lit by the visible spectrum.

Radiologists must also interpret images in two stages or levels; first in a description of what the objects are and then to make inferences on the state of health or disease in the context of their deviance or otherwise from the norm.

Since biological normality in a medical image is represented by a range of values and not a point, this requires a great deal of case experience and an encyclopedic knowledge base. But this is not to say that the cognitive activity of radiology is simply a look-up process from a table of facts. Image reading must be able to convey actionable meaning to the reader and the reader must in turn be able to articulate unequivocally this meaning to others who take responsibility in the treatment pathway.

The purpose of medical imaging is to provide diagnostic information through a decision-making activity and to reduce the uncertainty posed by a clinical problem. Interpretation and judgment are subjective and since all meaningful decisions are taken in conditions where the answer is uncertain, the process is subject to error. This means that there is some value in understanding the processes involved and the cognitive resources demanded of the practitioners during this activity. The better our understanding in this regard, the better will be our ability to take a "systems approach" to sources of error during the interaction of the cognitive systems of decision-makers with technology and working practices in the radiology reading room.

REFERENCES

Balota, D., Paul, S. (1996). Summation of activation: evidence from multiple primes that converge and diverge in semantic memory. *J Exptl Psych: Learning Memory and Cognition*, **22**, 336–345.

Berbaum, K.S., El-Khoury, G.Y., Ohashi, K., *et al.* (2007). Satisfaction of search in multitrauma patients: severity of detected fractures. *Acad Radiol*, **14**, 711–722.

Berbaum, K., Franken, E.A., Jr., Caldwell, R.T., Schartz, K.M. (2006). Can a checklist reduce SOS errors in chest radiography? *Acad Radiol*, **13**, 296–304.

Beutel, J., Kundel, H.L., Van Metter, R.L. (eds) (2000). *Handbook of Medical Imaging*, 1st edn. Bellingham, WA: SPIE.

Bryant, G., Norman, G. (1980). Expressions of probability. Words and numbers. *N Engl J Med*, **302**, 411.

Chi, M., Glaser, R., Farr, M. (1988). *The Nature of Expertise*. Hillside, NJ: Laurence Erlbaum.

Donovan, T., Manning, D. (2007). The radiology task: Bayesian theory and perception. *Br J Radiol*, **80**, 389–391.

Eddy, D. (1982). Probabilistic reasoning in clinical medicine. In *Judgment Under Uncertainty: Heuristics and Biases*, ed. Kahneman, D., Slovic, P., Tversky, A. Cambridge, UK: Cambridge University Press.

Elkins, J. (1996). *The Object Stares Back. On the Nature of Seeing*. New York, NY: Harcourt, Inc.

Ellis, A., Young, A. (1996). *Human Cognitive Neuropsychology: A Textbook With Readings*. New York, NY: Psychology Press; Hove, UK: Taylor Francis.

Frith, C. (2007). *Making Up the Mind – How the Brain Creates Our Mental World*, 1st edn. Oxford, UK: Blackwell.

Gregory, R. (1970). *The Intelligent Eye*, 1st edn. New York, NY: McGraw-Hill.

Gregory, R. (1990). *Eye and Brain. The Psychology of Seeing*, 4th edn. London, UK: Weidenfeld & Nicholson.

Gunderman, R., Williamson, K., Fraley, R., Steele, J. (2001). Expertise: implications for radiological education. *Acad Radiol*, **8**, 1252–1256.

Hendee, W., Wells, P. (1997). *The Perception of Visual Information*, 2nd edn. New York, NY: Springer.

Herring, W. (2007). *Learning Radiology. Recognizing the Basics*, 1st edn. Philadelphia, PA: Mosby Elsevier.

Hobby, J., Tom, B., Todd, C., Bearcroft, P., Dixon, A. (2000). Communication of doubt and certainty in radiological reports. *Br J Radiol*, **73**, 999–1001.

Kundel, H. (2000). Visual search in medical images. In *Handbook of Medical Imaging*, 1st edn., ed. Beutel, J., Kundel, H.L., Van Metter, R.L. Bellingham, WA: SPIE.

Kundel, H. (2006). History of research in medical image perception. *J Am Coll Radiol*, **3**, 402–408.

Kundel, H., Nodine, C. (1975). Interpreting chest radiographs without visual search. *Radiol*, **116**, 527–532.

Kundel, H., Nodine, C. (1983). A visual concept shapes image perception. *Radiol*, **146**, 363–368.

Lesgold, A.M., Rubinson, H., Feltovitch, P., *et al.* (1988). Expertise in a complex skill: diagnosing x-ray pictures. In *The Nature of Expertise*, ed. Chi, M.T.H., Glaser, R., Farr, M.J. Hillsdale, NJ: Lawrence Erlbaum.

Manning, D., Ethell, S., Donovan, T. (2004). Detection or decision errors? Missed lung cancer from the PA chest radiograph. *Br J Radiol*, **77**, 231–235.

Manning, D., Gale, A., Krupinski, E. (2005). Perception research in medical imaging. *Br J Radiol*, **78**, 683–685.

Marr, D. (1982). *Vision: a Computational Investigation into the Human Representation and Processing of Visual Information*, 1st edn. San Francisco, CA: W.H. Freeman.

Maxwell, C. (1978). Sensitivity and accuracy of the visual analogue scale: a psycho-physics classroom experiment. *Br J Clin Pharm*, **6**, 15–24.

McCormack, H., Horne, D., Sheather, S. (1988). Clinical applications of visual analogue scales: a critical review. *Psychol Medicine*, **18**, 1007–1019.

Neisser, U. (1976). *Cognition and Reality*, 1st edn. San Francisco, CA: W.H. Freeman.

Nodine, C., Krupinski, E.A. (1998). Perceptual skill, radiology expertise and visual test performance with NINA and WALDO. *Acad Radiol*, **5**, 603–612.

Nodine, C., Kundel, H. (1987). Using eye-movements to study visual search and to improve tumor detection. *Radiographics*, **7**, 1241–1250.

Nodine, C., Mello-Thoms, C. (2000). The nature of expertise in radiology. In *Handbook of Medical Imaging*, eds. Beutel, J., Kundel, H., Van Metter, R. Bellingham, WA: SPIE Press.

Phillips, C. (1995). *Logic in Medicine*, 2nd edn. London, UK: BMJ Publishing.

Ramachandran, V. (2003). *The Emerging Mind: 2003 Reith Lectures*. http://www.bbc.co.uk/radio4/reith2003/ (last accessed 27 March 2009).

Raufaste, E., Eyrolle, H., Marine, C. (1998). Pertinence generation in radiological diagnosis: spreading activation and the nature of expertise. *Cognitive Science*, **22**, 517–546.

Robinson, P. (1989). Lung scintigraphy – doubt and uncertainty in the diagnosis of pulmonary embolism. *Clin Radiol*, **40**, 557–560.

Robinson, P. (1997). Radiology's Achilles' heel: error and variation in the interpretation of the Roentgen image. *Br J Radiol*, **70**, 1085–1098.

Rock, I. (1983). *The Logic of Perception*, 1st edn. Cambridge, MA: MIT Press.

Rumelhart, D., McClelland, J. (1986). *Parallel Distributed Processing: Explorations in the Microstructure of Cognition*. Cambridge, MA: MIT Press.

Samei, E. (2006). Why medical image perception? *J Am Coll Radiol*, **3**, 400–401.

Shannon, C. (1948). A mathematical theory of communication. *Bell System Technical Journal*, **27**, 379–423, 623–656.

Smoker, W., Berbaum, K.S., Luebke, N., Jacoby, C. (1984). Spatial perception testing in diagnostic radiology. *Am J Radiol*, **143**, 1105–1109.

Swenson, R., Hessel, S., Herman, P. (1985). The value of searching films without specific preconceptions. *Invest Radiol*, **20**, 100–107.

Ullman, S. (1996). *High Level Vision: Object Recognition and Visual Cognition*. Cambridge, MA: MIT Press.

Van Der Helm, P.A. (2000). Simplicity versus likelihood in visual perception: from surprisals to precisals. *Psych Bulletin*, **126**, 770–800.

Wood, B. (1999). Decision making in radiology. *Radiol*, **211**, 601–603.

9

Satisfaction of search in traditional radiographic imaging[1]

KEVIN BERBAUM, EDMUND FRANKEN, ROBERT CALDWELL, AND KEVIN SCHARTZ

9.1 INTRODUCTION

The failure of observers to report abnormal radiographic findings (false negatives, FNs) accounts for about half of errors made by radiologists (Smith, 1967). In an attempt to explain one type of FN error, Tuddenham hypothesized that, with the discovery of one abnormality, the observer may be inclined to discontinue scrutiny of the image even if it has not been completely inspected (Tuddenham, 1962). In other words, a lesion is "missed" after detecting another lesion in the same image. Smith referred to such an error as "satisfaction of search" (Smith, 1967).

The operational definition of satisfaction of search (SOS) used in laboratory studies contains another criterion: for SOS to occur, the lesion missed when it appears with another lesion must be detected in the absence of the other lesion. In the laboratory, a known abnormality is defined as the *test* abnormality because detection of that abnormality is measured. The *test* abnormality is always presented twice to observers: once alone and once with another abnormality present within that same exam. SOS occurs when the *test* abnormality is missed in the presence of the *added* abnormality, but not in its absence. Although detection of any abnormality may affect detection of any other abnormality, SOS can only be measured rigorously on the *test* abnormality because it is the only abnormality that is presented by itself as a critical control condition. The long-term objective of our research is to understand and overcome SOS error.

The purpose of this chapter is to review the research on the SOS effect conducted using traditional radiographic imaging modalities. Although SOS effects have been studied in computed tomography (CT) (Berbaum, 2006b; 2007a; 2007b), our understanding of SOS in advanced imaging is just emerging. We focus on 14 studies of SOS that we conducted in our laboratory at the University of Iowa and a study conducted at the University of Pennsylvania (Samuel, 1995). We describe the impact of SOS on diagnostic accuracy in radiographic examinations, causes of SOS, and ways by which SOS can be reduced. Table 9.1 classifies these studies by the type of imaging and whether the goal of the study was to rigorously demonstrate the SOS effect using receiver operating characteristic (ROC) methods, investigate possible causes of SOS by studying visual search, or determine whether an intervention could reduce or eliminate the SOS effect. The reason for the empty cell in the table under multi-trauma radiology is that causes of SOS and interventions to prevent it in this modality are being studied in CT imaging rather than traditional radiography.

[1] Supported by US Public Health Service grants R01 CA-42453, R01 EB/CA00145, and R01 EB/CA00863 from the National Cancer Institute, Bethesda, MD.

9.2 SIGNIFICANCE OF SOS

"Missing a substantial finding on a radiograph is an event that escapes few radiologists during their professional careers, and when one of these errors is pointed out, most are astonished that they did not see what now (in retrospect) seems obvious to them."

Nodine, 2001

By far the most common errors in radiology are based on failure of perception where the radiologist simply does not "see" the abnormality (Berlin, 1996; Lev, 1999; Fidler, 2004; Halsted, 2004). Insight into the causes of these errors is limited for the simple reason that human consciousness does not have access to the workings of visual perception, only to its products, the percepts themselves. This is why introspection has long been rejected as a methodology for the serious study of perception in experimental psychology. While radiologists have experiences that inform them about error-prone reading situations, they have no special way of seeing into the neural processing that supports pattern recognition. Even with extensive research on perception using rigorous methods of experimental psychology, neuroscience, and related fields, current understanding of human pattern recognition does not allow full explanation of missed radiologic diagnosis. Progress in understanding the causes of error in diagnostic radiology has been achieved by experiments focusing on error-prone situations.

"One possible source of error had been pointed out by Tuddenham (1963), who proposed that when an observer was satisfied with the meaning of an image, active search was stopped. Smith (1967), in a delightful, anecdotal classification of observer errors, coined the term satisfaction of search. The phenomenon is real: observers do not report unexpected findings on images when they have found something suggested by the original search task (Berbaum, *et al.*, 1990, 1994a)."

Kundel, 2006

Errors attributable to SOS are common. Of the 182 cases presented at a "problem case" conference (Renfrew, 1992), 126 involved perceptual or cognitive error; of these over half were FN errors, and nearly one-fifth were due to an SOS effect. Anbari found an FN rate of 17% in 189 examinations of acute spinal injuries and attributed nearly a third of them (28%) to SOS (Anbari, 1997). SOS error in emergency medicine is sometimes referred to as premature closure. These errors are believed to be

The Handbook of Medical Image Perception and Techniques, ed. Ehsan Samei and Elizabeth Krupinski. Published by Cambridge University Press.

Table 9.1 *Fifteen studies of satisfaction of search in radiography*

Study goal	Chest radiology	Multi-trauma radiology	Contrast studies of the abdomen
SOS demonstration with ROC method	Berbaum (1990; 2000a); Samuel (1995)	Berbaum (1994; 2007b)	Franken (1994)
Causes of SOS	Berbaum (1991; 1998; 2000b)	Berbaum (2001)	Berbaum (1996)
Interventions to prevent SOS	Berbaum (1993; 2006b; 2007a)		Berbaum (2005)

The interested reader may obtain copies of all of the Iowa papers listed in the table from kevin-berbaum@uiowa.edu.

highly prevalent in emergency medicine, accounting for 91% of errors in that practice (Voytovich, 1985; Kuhn, 2002).

Appealing to the concept of SOS to explain misses does not provide rigorous proof that SOS caused the errors. SOS must be demonstrated using an appropriate control condition. To demonstrate that detection of one lesion causes another not to be detected, we must show that the other lesion was detected when it appeared alone. Further, when the distracting lesion is removed, that removal cannot be allowed to affect the physical nature of the second lesion and the background anatomy neighboring it.

We have studied SOS effects using four kinds of evidence: (1) changes in ROC curves, (2) gaze-dwell time relating to misses within images, (3) observer response times for inspection events, and (4) intervention strategies to reduce SOS effects. Our earliest investigations provided the first laboratory studies of SOS using three different types of traditional imaging examinations: chest radiography, trauma radiography, and contrast studies of the abdomen. ROC studies compared detection accuracy for *test* abnormalities that appeared on two types of radiographs: those with only the *test* abnormalities, and those with the *test* abnormalities and *added* abnormalities. All of these experiments demonstrated decreased reporting of *test* abnormalities when *added* abnormalities were present. Next, we used eye-position methodology to assess the causes of SOS. Gaze-dwell time (i.e. the time that the eye fixates a given point) was used to classify errors according to the Kundel–Nodine system (Kundel, 1978; 1987; Nodine, 1987). A failure to report an abnormality that has not been fixated is classified as a scanning error, a failure to report an abnormality that has been extensively fixated is classified as a decision-making error, and a failure to report an abnormality that has been fixated but not long enough to indicate decision-making activity is classified as a recognition error. In our ROC and eye-position experiments to provide a global characterization of the search process, we measured various response time intervals: (1) time to detect the *added* and *test* abnormalities, (2) time for false-positive (FP) responses, and (3) total search time. These studies revealed that SOS effects in chest and trauma radiology do not involve faulty scanning behavior, whereas the SOS in contrast studies of the abdomen is based on global changes in visual search behavior and faulty scanning. This suggests that the intervention most likely to reduce the SOS effect may depend on the cause – faulty scanning, faulty decision making, or faulty pattern recognition. Since gaze is partially under voluntary control, directing attention to neglected regions might reduce SOS caused by failure to fixate a missed abnormality. However, SOS based on faulty recognition may prove resistant to such an intervention. We studied whether clinical history and checklists can prevent SOS effects, and later, we studied whether image-based indications from computer-aided diagnosis can prevent SOS effects.

There are two primary ways to reduce diagnostic error in radiology. One method is to improve imaging technology so that abnormalities are more visible. The vast majority of technical research and clinical endeavor is concentrated in this activity (Beam, 2006; Samei, 2006). The other approach is to improve the interpretive process itself, the most crucial, yet least understood part of the diagnostic process.

"Research in the deeper aspects of image perception and in the interface between perception and analysis may hold the key to the problem of error and variability." Kundel, 2006

9.3 SOS IN CHEST RADIOGRAPHY

9.3.1 ROC experiments

9.3.1.1 Laboratory demonstration using the ROC method (Berbaum, 1990)

Our first experiment (Berbaum, 1990) attempted to determine whether an SOS effect, lesions remaining undetected following detection of an initial lesion, could be observed in a rigorous laboratory experiment. SOS had previously not been substantiated experimentally because of the difficulty in generating an appropriate control condition. It is not ordinarily possible, in a clinical image, to remove selectively a naturally occurring lesion leaving other lesions and background anatomy unchanged. To demonstrate that a lesion is not detected because another has been detected, the second lesion must be shown to be detectible in the absence of the first. Further, the physical nature of the second lesion and the background anatomy neighboring the second lesion must be identical in the presence and absence of the first lesion. Our primary goal was to develop a procedure to study SOS.

Our method compared detection accuracy for various native, subtle lesions (*test* abnormalities) with detection accuracy for those same lesions when a simulated pulmonary nodule (an *added* abnormality) was inserted into the radiograph. (The nodule was artificially created and photographically added to the radiograph.) The background anatomy and actual lesions were perfectly matched for the two conditions compared in our experiments. Simulated and native lesions were not spatially superimposed and the native abnormalities were physically identical

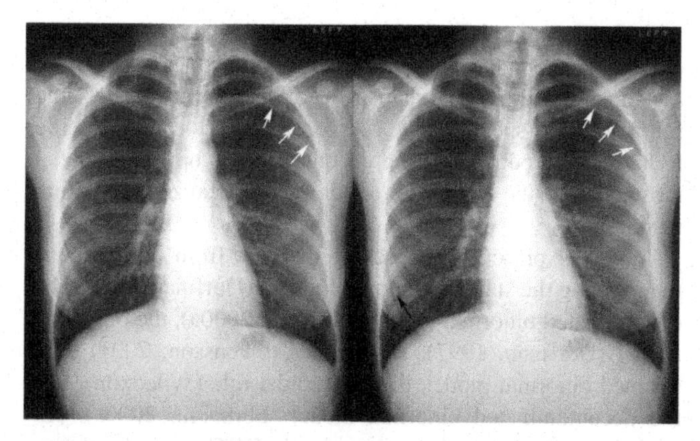

Figure 9.1 Example chest radiographs of a 25-year-old female patient with a left pneumothorax without (left) and with a simulated pulmonary nodule (right). The nodule, serving as the *added* abnormality, is indicated by a black arrow and the pneumothorax, serving as the *test* abnormality, by white arrows. No arrows were used when the images were displayed for observers.

with and without the nodules (see Figure 9.1). Use of a precise abnormality location response made it possible to sort out *test* and *added* lesion detections. Only responses related to the native lesion were analyzed.

Verification of diagnosis for abnormal cases was through follow-up imaging studies, surgery, clinical course, laboratory tests, and/or autopsy reports, which were part of the patient medical record. The 32 normal cases containing no indications of disease from any diagnostic source were selected to match as closely as possible the age and sex of the abnormal cases.

Eight volunteers from the Department of Radiology faculty at The University of Iowa served as observers. The case sample included 70 chest radiographs, 38 of which evidenced diverse, subtle, clinically significant abnormalities and 32 of which evidenced no native lesions. A diverse group of native abnormalities was used as test lesions. These included lesions of the lungs and pleura, heart, mediastinum, spine, abdomen, shoulder, and chest wall.

Observers were informed that each case might contain one, several, or no abnormalities and that the image set contained a wide range of abnormalities. No special mention was made of the frequency of nodules that might be seen. Before each trial in the experiment, the observer was informed only of the patient's age and sex. Three types of responses were collected. First, the observer was asked to use red ink to indicate any abnormal features on a photographic facsimile of a prototypic normal radiograph (the same for every trial). Because simulated and native lesions were not spatially superimposed, these graphic indications allowed determination of which lesion(s) actually had been detected. In addition, the observer was asked to provide a confidence rating and a disease category for each abnormality. The latter provided further evidence as to which lesion(s) had been detected. The abnormal/normal rating scale included: 1 = definitely, or almost definitely, an abnormal feature; 2 = probably abnormal; 3 = possibly abnormal; 4 = an anomalous feature present but probably normal; and 5 = definitely, or almost definitely, normal. If no radiographic feature was drawn, we assumed a rating of 5.

The experiment consisted of two parts separated in time by six months to reduce the likelihood that the study images and the responses to them would be remembered. Half of the cases presented in each part contained a simulated lesion and half did not. Thus, in the course of two parts, each radiograph in the case sample appeared twice, once with and once without a simulated nodule. Within each part, cases were presented in a pseudorandom order so that occurrence of native and simulated lesions and normal radiographs would be unexpected and balanced.

Detection of *test* lesions was substantially reduced in the presence of *added* nodules. ROC curves were derived from maximum likelihood estimates for each observer in our two experimental conditions (Dorfman, 1969). The average perceptual accuracy of the individual ROC curves as measured by A_z and d'_e was significantly reduced with addition of the nodules ($t(7) = 2.36$, $p < 0.05$; $t(7) = 2.65$, $p < 0.05$, respectively). The results indicated a substantial SOS effect, with diminished accuracy in perception of native lesions. Therefore, our first experiment demonstrated the existence of the SOS effect in a rigorous experiment and developed a paradigm that could be used for further investigation.

9.3.1.2 Replication using proper and joint ROC (Berbaum, 2000a)

At the time of the original demonstration, no methods were available to fit proper ROC curves. ROC plots in diagnostic radiology are constructed by extrapolation of fitted distributions to probability ranges wider than the range of the observed data from which those distributions' parameters are estimated. In the standard binormal model (Dorfman, 1969) this leads to the depiction of an impossible event in the ROC curve, a chance-line crossing such that images from diseased patients would have to look "more normal" than images from non-diseased patients. If the standard deviation of the actually-positive distribution is much larger than that of the actually-negative distribution, but the mean of the former distribution is not much larger than that of the latter distribution, the left tail of the actually-positive distribution will cut across the hump of the actually-negative distribution and lie above its left tail. This is what produces the inappropriate chance-line crossing by the ROC curve. It follows that the standard binormal model is not a correct account of the underlying decision process, at least as regards the area under the ROC curve, no matter how well it fits data. Because chance-line crossings greatly affect area under the ROC curve and hence conclusions from experiments, the fundamental flaw in the standard binormal model has motivated a decade-long search for proper models; that is, models of the decision process that inherently disallow chance-line crossings and still fit the data well.

The original demonstration of SOS in chest radiology is unlikely to be an artifact of inappropriate chance-line crossings (Berbaum, 1990) because the same result is shown in an analysis of d'_e. (ROC parameters like d'_e, sensitivity at a fixed specificity, and partial area, which index the ROC in ranges of specificity with ROC operating points, do not suffer from this limitation.) Nevertheless, we replicated the previous ROC study to test whether SOS in chest radiology could be demonstrated using proper ROC analysis and a new sample of readers.

Original data from Berbaum (Berbaum, 1990) and from a replication study with a new sample of 11 readers were analyzed with three proper ROC models (Swensson, 1996; 2001; Dorfman, 1997; 2000a; 2000b) and one ROC model for joint detection and localization. A joint ROC model curve has an ordinate equal to the probability of a true-positive (TP) detection response and a correct localization response, and an abscissa equal to the probability of an FP detection response (Swets, 1982). We extended the contaminated binormal model (Dorfman, 2000a) to joint ROC analysis through a straightforward assumption that the signal is correctly localized if – and only if – the latent observation is from the visible signal distribution. The contaminated binormal ROC model for detection data then becomes the decontaminated binormal ROC model for joint detection and localization data by leaving out the contaminated component for the signal distribution. The invisible (not correctly localized) component must be removed from the model, because detection responses that are not correctly localized have been removed from the data. Whereas the contaminated binormal model estimates the proportion of visible signals from the detection data with the assumption of no dependence on localization, the decontaminated binormal model estimates that proportion from correctly localized signals.

As noted above, Berbaum presented 70 chest radiographs to eight observers (Berbaum, 1990). The replication experiment presented 58 (30 abnormal) of the original cases to 11 radiologists. All cases were presented in two conditions: once with and once without a simulated pulmonary nodule. The detection accuracy for the native abnormalities was compared with and without the presence of the nodule. Areas under the ROC curves for the detection of native abnormalities were estimated for each observer in each experimental condition, and the average areas both with and without the *added* nodule were compared by using a paired *t*-test with degrees of freedom as determined by the number of readers.

Only the *test* abnormalities in these experiments were scored, not the *added* abnormalities, and location information was used to determine whether a response corresponded to the *added* abnormality, the *test* abnormality, or neither. Therefore, the scoring was essentially location-specific. The plots in Figure 9.2 show both the ROC points for three observers (rows) with ROC curves fitted with the contaminated binormal model (left column) and location-specific ROC (LROC) curves fitted with the decontaminated binormal model (right column). The contaminated binormal model produces a proper ROC curve that does not cross the chance line. There is no chance line for a location-specific ROC curve, and the location-specific ROC curve produced by the decontaminated binormal model terminates at the proportion of correctly localized TP rating responses.

Table 9.2 presents analyses of the data from both experiments using the standard binormal model (Dorfman, 1969), the contaminated binormal model (Dorfman, 2000a), the bigamma model (Dorfman, 1997), Swensson's (Swensson, 2001) constrained binormal model, the trapezoidal rule (Wilcoxon), and the decontaminated binormal model (Berbaum, 2000a). All statistics are based on the areas under the ROC curves for detection of native abnormalities estimated for each observer in each experimental condition. The first data column presents ROC or LROC area averaged across observers when the simulated nodule was absent; the second data column presents ROC or LROC area averaged across observers when the simulated nodule was present. The next two columns present the average difference in ROC area between the two conditions and the standard error of this average difference. The fifth column shows the correlation of the individual observer areas between the two conditions. The last three columns present details of the *t*-tests associated with the mean difference in individual ROC areas.

The contaminated binormal analysis, which offered the best fit to the data (according to likelihood-ratio chi-square statistics) without chance-line crossings, and the decontaminated binormal model found the highest levels of significant difference in area under the curve. The other proper ROC models and Wilcoxon also showed the SOS effect but with higher *p*-values. The standard improper binormal model gave a marginally significant SOS effect. The standard error of the difference in area was smaller for the contaminated binormal model and the decontaminated binormal model, thereby accounting for improved statistical power. Preventing inappropriate chance-line crossing reduces measurement error and provides more powerful statistical tests. With 19 observers, all models essentially demonstrated the SOS effect on area. Berbaum (Berbaum, 2000a) also analyzed the data of the original and replication experiments separately. The failure of some models to demonstrate an area effect with only 8 or 11 observers was probably the result of less valid measurement for those models. The analyses verify that the SOS effect on ROC area in chest radiology is not the result of an ROC curve-fitting artifact.

Table 9.2 *ROC analysis of the combined experiments of Berbaum (1990; 2000a) with 19 observers*

ROC model	ROC area nodules absent	ROC area nodules present	Mean difference	Difference standard error	*r*	*t*(18)	*p*
Standard binormal	0.65	0.60	0.05	0.02	0.67	2.06	0.06
Contaminated binormal	0.74	0.69	0.04	0.01	0.89	4.63	0.001
Bigamma	0.75	0.73	0.03	0.01	0.84	2.38	0.03
Constrained binormal	0.73	0.70	0.03	0.01	0.79	2.43	0.03
Wilcoxon	0.70	0.67	0.03	0.01	0.84	2.34	0.04
Decontaminated binormal	0.56	0.49	0.07	0.02	0.87	4.52	0.001

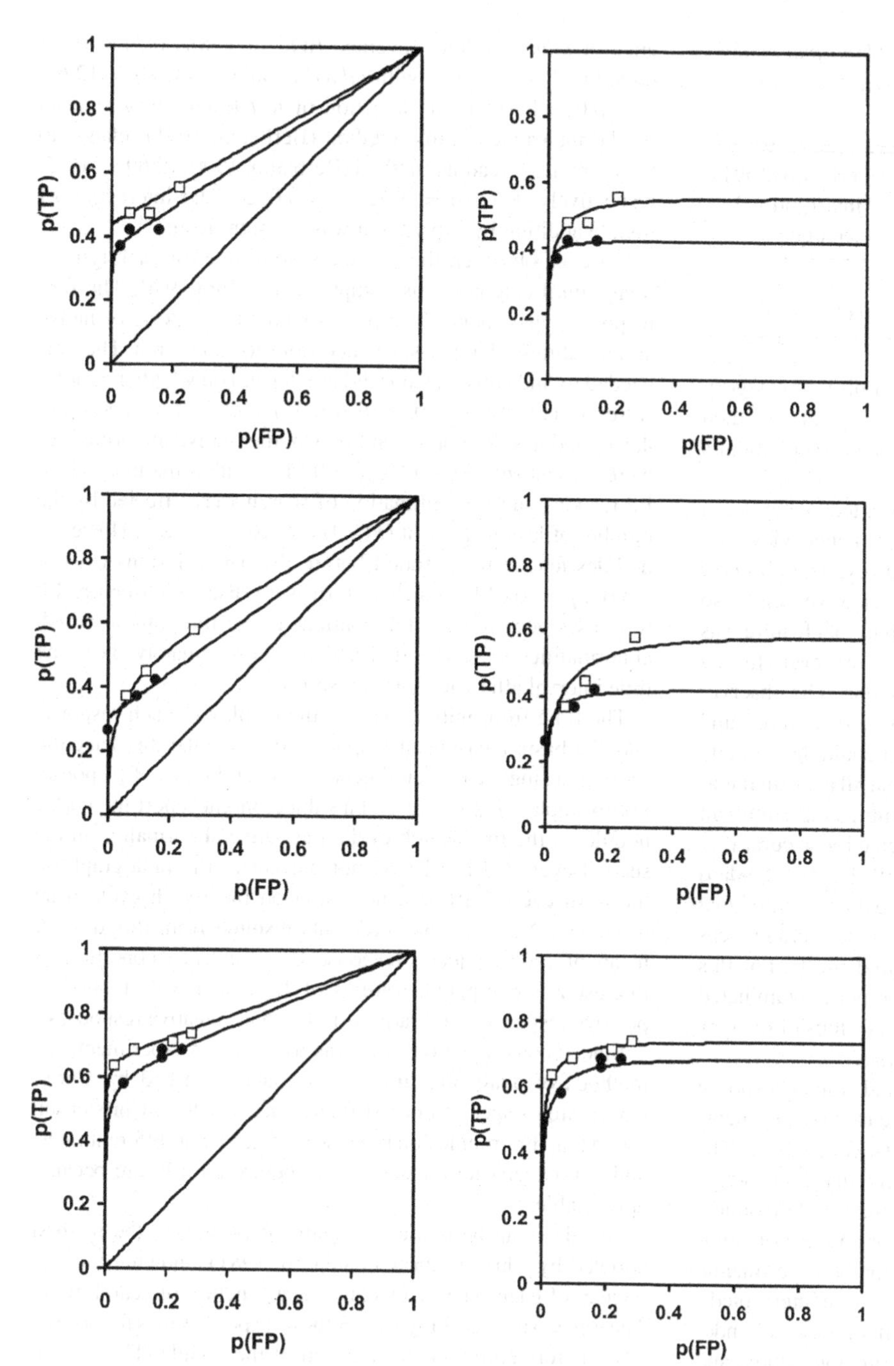

Figure 9.2 Data from three sample observers (rows) fitted by the contaminated binormal model (left column), and the decontaminated binormal model (right column). Plots show the ROC points and ROC curves for the detection of native abnormalities with the *added* nodule (open squares) and without the *added* nodule (solid circles).

9.3.2 Causes

9.3.2.1 Time course of search (Berbaum, 1991)

Tuddenham explained SOS error with the premise that once the need to find an abnormality is "satisfied," search is terminated before other lesions are detected (Tuddenham, 1962). The purpose of the Berbaum investigation was to determine whether this explanation accounts for the observed reduction in accuracy (Berbaum, 1991). This study attempted to relate SOS to total time of inspection and time intervals before, between, and after discovery of both types of lesions. Detection accuracy for native lesions in chest radiographs before and after the addition of a simulated nodular lesion was measured in ten new observers (volunteer faculty radiologists). As in the previous experiment, we compared detection of subtle, diverse lesions of the chest radiograph in the presence and absence of simulated pulmonary nodules. The case sample and nodule simulations of the first experiment were used.

The interruption technique used to measure visual search required the observers to (1) place their fingers on response keys, (2) listen to preliminary information read by an experimenter, (3) initiate the display of the patient images by pressing a key, (4) inspect the display, (5) press the right response key to indicate readiness to make a response (which terminated the display and recorded the inspection time), and (6) make a rating response (which was entered by the experimenter into the computer). If the rating indicated a normal examination, the experiment moved on to the next patient. If the rating indicated

Table 9.3 *Average areas under contaminated binormal model ROC curves derived show the impact of interrupting search and SOS (addition of nodule)*

Experimental condition	Berbaum (1990) SOS demonstration	Berbaum (1991) "interruption" technique
No nodule	0.78	0.70
Nodule	0.73	0.68

an abnormality, the observer also reported a disease category and the location of the abnormal feature or features, and then decided whether to continue inspection of the current patient's images or go on to the next.

Search was measured using an interruption technique. Each trial began with a radiograph hung on a darkened viewbox. An experimenter read the patient's age and sex. The observer pressed a telegraph key that turned on the viewbox lamps so that inspection could begin. As soon as an abnormal feature was discovered, the observer pressed another key; this event turned off the viewbox lamps and recorded search time. The observer rated the apparent abnormality of any feature discovered and described its location and type. The observer could then initiate further timed search cycles until confident that all abnormal features had been detected. This procedure permitted measurement of the various search intervals, as well as detection accuracy.

In the original publication (Berbaum, 1991), the standard binormal model had been fitted to the data of individual observers in each experimental condition and the average areas under the ROC curves compared using paired *t*-tests. For this review we re-analyzed the data using the proper contaminated binormal model (Dorfman, 2000a). The conclusions of the new analysis were the same as the original analysis.

Because we used the same case sample and similar response format, the data from both the previous and current experiments could be examined together in a single statistical analysis. The average ROC areas, shown in Table 9.3, were analyzed using a two-way analysis of variance (ANOVA) with a within-subject factor for the SOS manipulation (presence or absence of simulated nodule), and a between-subject factor for the experimental procedure (whether or not search was continuous or interrupted). This analysis tests whether a similar SOS effect occurred under the interrupted search paradigm, as in the previous study, and whether the interrupted search paradigm influenced accuracy. Detection of *test* lesions was substantially reduced in the presence of *added* nodules (average ROC area was 0.74 without *added* nodules vs. 0.70 with *added* nodules, $F(1,16) = 12.65$, $p < 0.01$). In addition, detection of *test* lesions was reduced by the interrupted search method (Berbaum, 1991) relative to the more usual reading method (Berbaum, 1990) (0.69 vs. 0.75, respectively, $F(1,16) = 4.02$, $p = 0.062$). The interaction of nodule addition by experiment was not significant.

For each observer, the average search time for each type of image until response was computed (see Table 9.4). The four response types included: reports of nodules, reports of native abnormalities, FP reports, and decisions to halt search. The time needed to find native lesions did not depend on whether nodules were present ($t(9) = -0.78$, $p = 0.46$), and the time needed to detect nodules did not depend on whether native abnormalities were present ($t(9) = -0.69$, $p = 0.51$). Neither the intervals to FP response nor to termination of search were affected by the number of lesions present ($F(1,9) = 0.20$, $p = 0.87$). However, nodules tended to be found before the native lesions ($t(9) = -3.03$, $p = 0.014$), which in turn were discovered before FP responses occurred. Search continued after the point at which abnormalities were reported and for approximately the same time interval after each initial response.

The time from initial onset of the display to each response was the basic measure of response time. Unlike the detection scoring, a single case could generate several types of response. For example, on a single trial a subject might report the *added* nodule on the first search cycle, the native abnormality on the second cycle (TP), a lesion not present in the radiograph on the third cycle (FP), and no lesion on the fourth cycle (true negative, TN). Four measurements resulted from this trial. A mean of each subject's response times in each condition of interest was computed. Ratings of 1, 2, or 3 were treated as positive responses, and ratings of 4 or 5 as negative responses.

Since choice reaction time was used in our experiment, the methods of analyzing time course data described by Christensen are of special interest (Christensen, 1981). Christensen found that different lesion types were detected at different rates, and that experts terminated search before the FP rate became appreciable.

Our data can be plotted in a similar fashion to the method adopted by Christensen (Christensen, 1981), showing the frequency of each type of response as a function of search time. The top two panels of Figure 9.3 show response times for normal cases before and after the addition of the nodule. The bottom two panels of Figure 9.3 show response times for abnormal cases before and after the addition of the nodule. The data in these

Table 9.4 *Mean response time in seconds for various case types for various response types. Different response types occur at different latencies, but different stimulus types do not require different latencies*

	Case type			
Response type	Normal exam	Abnormal exam	Normal with *added* nodule	Abnormal with *added* nodule
Nodule report			17.1	18.2
Native lesion report		23.9		25.5
False positive report	32.7	34.1	34.0	32.5
Termination of search	47.0	47.0	44.5	44.3

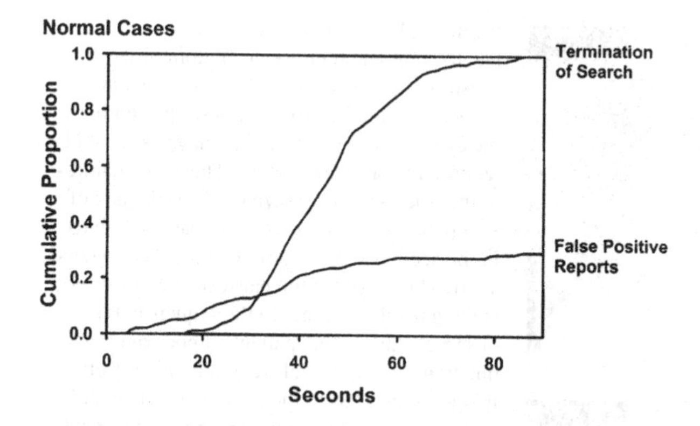

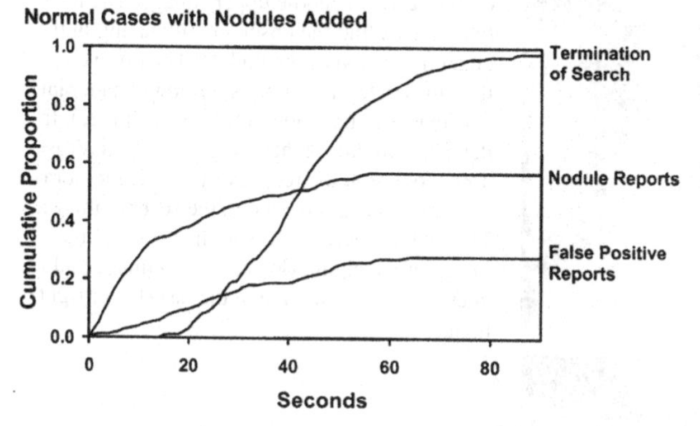

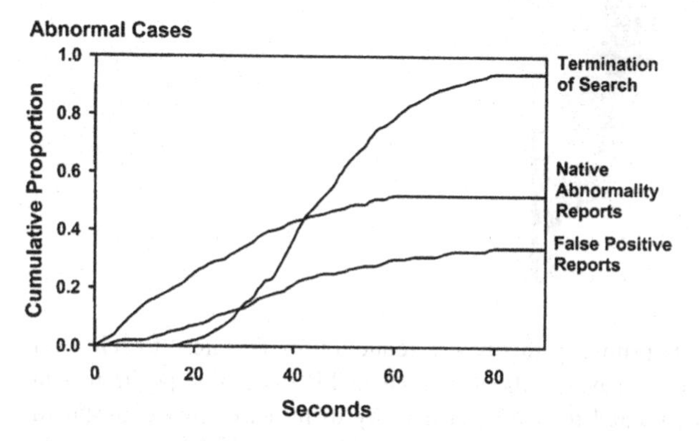

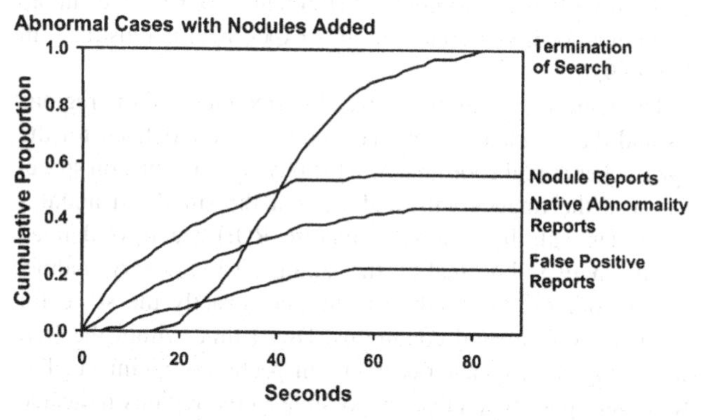

Figure 9.3 Cumulative frequency of each type of response is plotted as a function of search time. Each observer's response times were multiplied by a constant to set the total search time equal to the average of the ten observers. The top two panels show response times for normal cases before and after the addition of nodules. The bottom two panels show response times for abnormal cases before and after the addition of the nodule with more curves for detection of native abnormalities. Curves plot termination of search, FP reports, nodule reports, and native abnormality reports.

curves are scaled by multiplying each observer's response times by a constant to set individual search times equal to the average of the ten observers. The time-course functions presented here differ from those of Christensen (Christensen, 1981). Christensen's functions use the cumulative amount of reading time or "reading-seconds" pooled over readers and films. Later intervals of reading-seconds may represent various mixtures of the slowest readers and the longest searched films. Instead, our curves plot the cumulative frequency of each report type, as measured from the onset of film interpretation. The cumulative frequency distributions do not presuppose that radiologists generate TP and FP reports at the same rates for all readers and examinations, only that distributions from different readers can be equated by a multiplicative constant.

The reader may appreciate how the SOS effect is expressed in these time-course functions, although this is somewhat complicated because only one decision threshold is represented. Yet, a comparison of graphs for abnormal cases with and without *added* nodules (the bottom two panels) shows that the curve for native abnormalities is reduced with the introduction of the nodule, whereas the graphs for normal cases with and without *added* nodules (the top two panels) show that the FP rate for cases without nodules remains more or less constant. Note, however, that the FP curves in abnormal cases suggest a reduction in FP rate with the introduction of the nodule; FP rates for normal and abnormal cases appear to be affected differently by the introduction of the nodules. Perhaps FP rates are affected differently in normal and abnormal cases because the detection of a single abnormality has less impact than detection of two abnormalities.

When we planned this SOS paradigm, we assumed, based on the results of Kundel (Kundel, 1972) and Christensen (Christensen, 1981), that nodules would be detected rapidly. The curve for detection of the *added* nodules suggests that assumption was correct. Early in search, its slope is steeper than the function for detection of native abnormalities or the function for FP responses. The addition of the nodule has little impact on the FP and termination curves for normal cases.

For time-course functions, whether or not termination of search is considered "premature" depends on their slopes as they approach the terminal point. A slope near zero indicates that few additional lesions will be reported. If the slope of the FP function is increasing relative to that of the TP function late in search, cessation of search also may improve overall performance. Our curve for FPs is relatively flat for normal cases without nodules.

For abnormal cases, addition of nodules had little impact on the overall shape of the other curves. Where the functions for nodule, native abnormality, and FP responses reached a common slope (so that rates were equal), searches of many cases were terminated. This supports previous findings – observers halt search before responses are as likely to be false as true.

This experiment did not support Tuddenham's (Tuddenham, 1962) explanation of SOS that search is terminated after discovery of an abnormality. Observers inspect images for the same amount of time regardless of how many abnormalities are present, and appear to halt inspection before the TP and FP rates become unfavorable. The SOS effect occurs even though observers continue to inspect films after one abnormality is

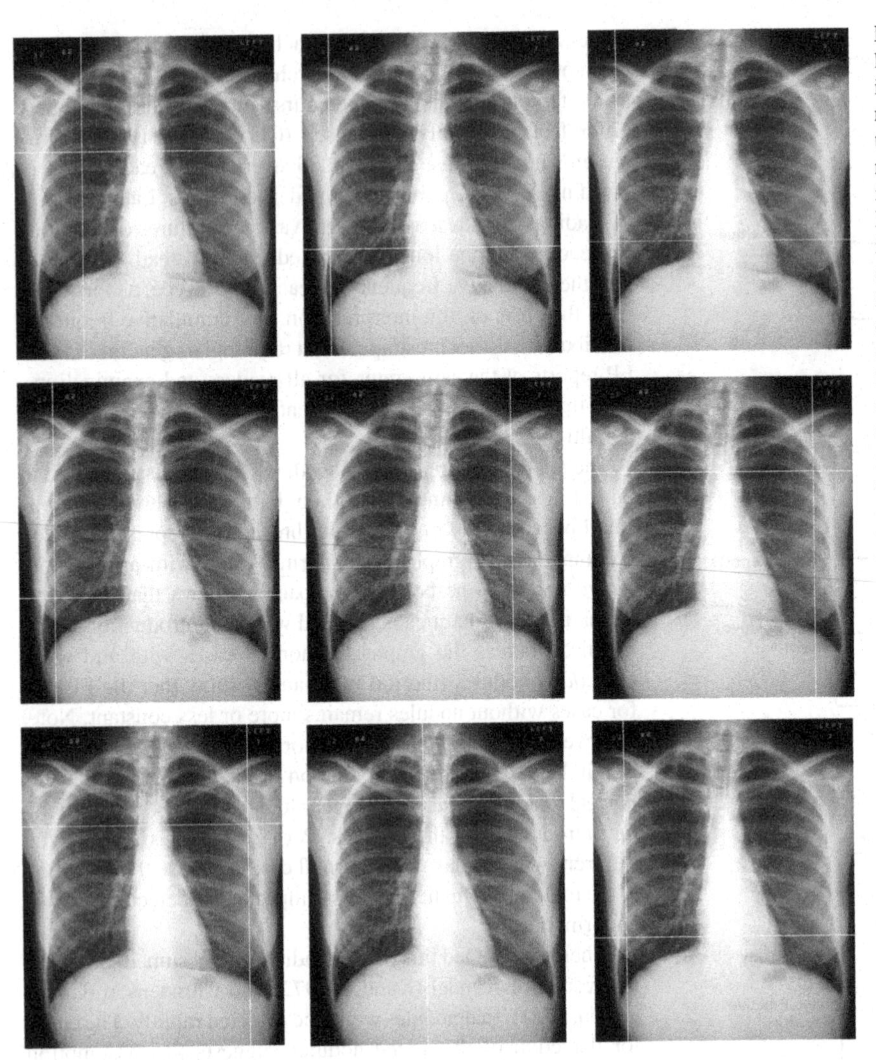

Figure 9.4 This sequence of images in left-to-right and then top-to-bottom order illustrates a few seconds of eye-movement recording with the crossed lines representing the eye gaze direction on the image as would be recorded by a scene camera. The abnormalities in these images are shown on the right side of Figure 9.1. One might assume that each image in the sequence is perhaps a quarter of a second apart. The length of the pause at each location is the gaze-dwell time. In the sequence, the observer starts in the patient's upper right quadrant (the upper left region in upper (left image), then goes to the position of the nodule on the second (middle upper) image. The next two images find the observer still gazing in the general location of the nodule. The observer then moves to the patient's left lower quadrant (the reader's right side of the center image). In the sixth (middle right) image, the observer is now looking in the region of the pneumothorax and continues doing so into the seventh image. Then the observer moves to the center of the image in the eighth (lower center) image and goes back to the nodule in the last (lower right) image.

reported. An alternative to the premature halting explanation of SOS is that perceptual resources available for a specific search task are limited. If the duration of search is relatively fixed even when multiple abnormalities are present, fewer perceptual resources are available for the detection of each one. Another factor may be perceptual set, a readiness (bias) to interpret images in particular ways generated by concluding that certain image features are indicative of a specific diagnosis. This perceptual set could hinder detection of features relevant to a different diagnosis.

9.3.2.2 Gaze-dwell time on missed abnormalities (Berbaum, 1996)

We tested the hypothesis that the SOS effect, which is associated with failure to detect native chest abnormalities in the presence of simulated nodules, is caused by reduced gaze on the native abnormalities. The gaze-dwell time of 19 radiologists was recorded for the region around abnormalities on images using an eye-tracking system (Applied Sciences Laboratory). Ten radiographs were reviewed, nine of which contained native abnormalities. Each image was seen with and without a simulated nodule. This proportion of abnormal to normal cases was used because SOS effect demonstrated in prior studies with these cases primarily involved a reduction in TP rates at each level of confidence without change in FP rates. Eye-position data was related to detection data by using cases that contributed most strongly to the SOS effect (Berbaum, 1990). An example of an eye-position record from this experiment is illustrated in Figure 9.4.

The regions containing native abnormalities and the regions around the simulated pulmonary nodules were defined for the eye tracker in real coordinates in the physical observation space. Because the images with and without the simulated nodules were identical, the regions of interest (ROI) which we defined for the native abnormality, the simulated pulmonary nodule, and all other areas of the image were exactly the same for our two experimental conditions. Dwell time within the ROI was computed for each observer's inspection of an image. The ROI were placed as close as possible to the regions indicated by a collaborating radiologist but were not placed closer than 1.25 cm, as this value represented the practical spatial-resolution limits of our eye-tracking system.

McNemar-like analysis. Our eye-position studies included far fewer cases than our detection experiments – ten cases in each experiment. This was because accurate recording of eye position with the equipment available at the time of these experiments required considerable calibration after each trial (case) of the

Table 9.5 *Before report gaze time on regions with native abnormality*

Class of detection outcome	Detection of native abnormality when the nodule was absent	Detection of native abnormality when the nodule was present	Frequency	Gaze time on native abnormality (in seconds)			
				Nodule absent	Nodule present	Difference in gaze time	*p*
1	No	No	61	6.5	7.1	+0.6	0.71
2	Yes	No	27	9.5	9.4	−0.1	0.95
3	No	Yes	13	11.2	19.2	+8.0	0.01
4	Yes	Yes	70	12.4	11.5	−0.9	0.53

experiment. Also, because the eye-tracker optics were head-mounted, observers would begin to feel discomfort after about 30 minutes of reading. Because ROC analysis could not be used, we needed a method that could be used with the sparse detection data of the eye-tracking experiments.

The confidence ratings were transformed into two categories: (1) reports of abnormality at any level of certainty were considered positive; and (2) a lack of a report of abnormality was considered negative. For each case with *test* abnormalities, the pair of responses were classified as follows: (a) the *test* abnormality was missed in both the presence and absence of an *added* abnormality, (b) the *test* abnormality was reported in the absence of an *added* abnormality, but was missed in its presence ("SOS outcome"), (c) the *test* abnormality was missed in the absence of an *added* abnormality, but was reported in its presence ("anti-SOS outcome"), or (d) the *test* abnormality was reported in both the presence and absence of an *added* abnormality. We evaluated the SOS effect over readers by testing the null hypothesis that the population probabilities of outcome categories b and c were equal (Berbaum, 1998). The McNemar test for non-independent proportions (McNemar, 1969) is inappropriate for counts summed over readers and cases because the counts are not identically distributed unless every reader has the same population probability of SOS. Nevertheless, we can evaluate the SOS over readers. Using the four classes of native-abnormality detection outcome, we constructed a separate McNemar-type four-fold table for each of the readers. We then computed the probability of each of the four joint events in the table. Under the McNemar null hypothesis (McNemar, 1969), the population probability (p) of the SOS outcome equals that of the anti-SOS outcome or, equivalently, $p(\text{SOS}) - p(\text{anti-SOS}) = 0$. We tested the McNemar null hypothesis for dependent-sample proportions taken over independent readers with a paired t-test generalizing to the population of readers. For discrete scales with a small number of distinct values, as was the case in the present study, the paired t-test is appropriate although somewhat conservative (Snedecor, 1989). A similar analysis was performed for the cases without a *test* abnormality for which an FP diagnosis was made. The confidence ratings were transformed into two categories, positive and negative. For each of the cases without native abnormalities, the pair of responses were classified into one of four categories: (a) no FP finding was reported either in the presence or absence of an *added* abnormality, (b) an FP finding was reported in the absence of an *added* abnormality, but not in its presence ("FP without-*added* abnormality event"), (c) an FP finding was reported in the presence of an *added* abnormality, but not in its absence ("FP with-*added* abnormality event"), and (d) an FP was reported in both the presence and absence of an *added* abnormality. We again tested the McNemar null hypothesis with a paired t-test.

Statistical analysis of the detection data of this study confirmed that the TP rate for native abnormalities decreased with the addition of simulated nodules, demonstrating the same SOS effect of our previous ROC studies. To relate detection results, particularly SOS, to gaze time, we classified the responses to each pair of trials, one with and one without the simulated nodule, into one of four classes as noted previously. Total gaze time within each region of interest was analyzed separately using a mixed model analysis of variance (Kirk, 1982; Dixon, 1992), in which case, readers and their interaction were treated as random effects while presence of the *added* nodule, detection outcome, and their interaction were treated as fixed effects.

Not surprisingly, observers looked longer at the region containing the nodule than they did at that same region when no nodule was present (9.9 vs. 1.5 seconds, $F(1{,}334) = 65.93$, $p < 0.0001$). Any gaze time after the observer began reporting the native abnormality was excluded from the total time on the native abnormality (see Table 9.5). The results showed marginally significant increased time on the region around native abnormalities when *added* nodules were introduced (9.8 seconds without vs. 10.2 seconds with nodules, $F(1{,}334) = 3.18$, $p < 0.076$). Tests of simple effects showed that when native abnormality detection responses were the same in the absence and presence of the nodule (i.e. both detected, or both missed), there were no changes in before-response dwell time on the native abnormality. There was also no difference in before-report dwell time on native abnormalities when SOS occurred (9.5 vs. 9.4 seconds, $p = 0.95$).

The threshold used to separate recognition from decision errors was defined by the 0.05 quantile of the distribution of dwell time associated with TP responses in both conditions. FN errors were categorized as 7% scanning errors, 35% recognition errors, and 58% decision errors when the nodules were absent, and as 5% scanning errors, 39% recognition errors, and 57% decision errors when the nodules were present. We also tested whether there was a general shift in dwell times associated with FN errors when the nodule was added. For dwell time on native abnormality regions in trials resulting in false negatives, those images without nodules averaged 7.45 seconds whereas those images with nodules averaged 7.80 seconds ($t(160) = -0.19$, $p = 0.85$, or Mann–Whitney rank sums, $p = 0.95$). *Without shifts in gaze-dwell times, changes in error classes with the introduction of the nodules are unlikely.*

For SOS in chest radiography, the *added* nodule did not affect the detection of the native abnormality through a change in dwell time on native abnormalities. Furthermore, gaze-dwell times associated with FN errors were quite long for images with and without the nodules (averaging 7 seconds), making it difficult to attribute the reduction in TP responses to the lack of gaze time. Reduction in gaze-dwell time on the missed abnormalities is not the cause of SOS errors in chest radiographs.

9.3.2.3 *Faulty decision making (Berbaum, 2000b)*

Our last investigation, in which we classified errors on the basis of gaze time according to the Kundel–Nodine method, indicated that the primary cause of SOS errors was faulty decision making (Berbaum, 1996). In contrast, another investigation of SOS (Samuel, 1995), in which errors were classified on the basis of gaze time according to the Kundel–Nodine method, indicated that faulty pattern recognition may be the primary cause of the SOS effect. The Kundel–Nodine method may have an unfortunate limitation when applied to the diverse abnormalities we used to study SOS in chest radiography. An unreported abnormality is assumed to have been recognized if the gaze time on the abnormality exceeds some sufficient value that is estimated from the gaze times on correctly reported abnormalities. A single temporal threshold for distinguishing recognition from decision-making errors may not work for the diverse abnormalities seen on chest radiographs because the recognition of different abnormalities may require different amounts of time. If so, use of a minimum estimated time from all abnormalities would overestimate decision errors. A more fundamental problem is that gaze time may not be a reliable indicator of the recognition of diverse abnormalities. Because multiple visual pattern analyzers may be needed to recognize diverse abnormalities, the eye can fall on an abnormality but a visual pattern may be analyzed that does not correspond to the abnormality. Gaze-dwell time would accumulate without recognition, and some errors would be incorrectly attributed to faulty decision making. These issues challenge the validity of our conclusion that SOS errors in chest radiology are caused by faulty decision making.

We performed this study to determine whether defective pattern recognition or decision making causes SOS errors in chest radiology. We used protocol analysis (Newell, 1972), a method borrowed from the study of human problem solving, to classify SOS errors into the same categories provided by an analysis of gaze-dwell time (Berbaum, 1998). The main question was whether the use of verbal protocols would result in the same proportion of error types as gaze-dwell time.

Twenty observers read 58 chest radiographs, half of which demonstrated diverse, native abnormalities. The radiographs were read twice, once with and once without the addition of a simulated pulmonary nodule. Observers verbally described their focus of attention during inspection, provided a separate account of abnormalities they would report, and gave a confidence of abnormality rating.

SOS effect on TP responses was studied with analysis of events in which the native abnormality was missed in one condition but not the other. An SOS event occurred when the native abnormality was reported in the absence of a simulated nodule but was missed in its presence. An anti-SOS event occurred when the native abnormality was missed in the absence of a simulated nodule but was reported in its presence.

Table 9.6 *Detection of native abnormalities: average area under the proper ROC curve*

	SOS experiment without verbal protocol procedure, Berbaum (2000a)	SOS experiment with verbal protocol procedure, Berbaum (2000b)
Without nodules (non-SOS condition)	0.74	0.75
With nodules (SOS condition)	0.69	0.74

We assessed the presence of the SOS effect by using proper ROC analysis applied to the confidence ratings. Areas under the ROC curve for the detection of native abnormalities on images with and without nodules were compared by using a paired t-test. We also compared the areas obtained during the current study with those obtained during a previous study (Berbaum, 2000a) by using two-sample t-tests.

No SOS reduction in the average area under the ROC curve was found for the detection of native abnormalities on images with simulated nodules (see Table 9.6; Berbaum, 2000b). The SOS effect on the ROC area in the previous study was statistically significant ($t(18) = 4.63$, $p < 0.001$; see Table 9.6; Berbaum, 2000a). A two-sample t-test was used to compare the average area under the ROC curve obtained in cases without simulated nodules both with and without the verbal protocol procedure. The results of this test demonstrated no statistically significant difference in ROC area between the two experiments when nodules were absent. Another two-sample t-test was used to compare the average ROC area obtained on images with simulated nodules evaluated both with and without the verbal protocol procedure. The results of this test demonstrated a statistically significant difference in ROC area between the two experiments when nodules were present ($t(37) = 2.18$; $p < 0.05$). *The requirement of providing a verbal protocol seems to have offered some protection against the SOS effect.* The SOS effect in the previous eye-position tracking experiment (Berbaum, 1996) was demonstrated by testing the McNemar null hypothesis. In order to determine whether a comparable SOS effect was also present in the FN errors of the current experiment with verbal protocols, we compared the frequencies of SOS and anti-SOS events for the 20 observers by testing the McNemar null hypothesis for dependent-sample frequencies taken over independent observers by using a paired t-test. The sample mean difference in frequency of SOS and anti-SOS events was statistically significant ($t(19) = 3.17$, $p < 0.01$), allowing us to reject the null hypothesis that the two population probabilities were equal and accept the alternative hypothesis that $p(\text{SOS}) > p(\text{anti-SOS})$. A similar analysis of FP responses occurring in one condition but not the other failed to find a statistically significant difference.

How does talking through the search prevent SOS? Based on neurophysiological and behavioral evidence, Ericcson (Ericcson, 1980; 1984) argued that speech can take place without any loss in information-processing capacity or commensurate performance reduction with regard to signals presented to the visual system. However, one would assume that talking about search can add no new information, nothing that the observer doesn't already know. Perceiving can be conceptualized as testing hypotheses about the world by performing tests on sensory data (von Helmholtz, 1867; Craik, 1943). Modern thinking adds the notion of "perceptual cycle" in which concepts, arising either from memory or an initial perceptual analysis, direct the sampling and testing of sensory data and the results modify the developing percept (Hochberg, 1970; Neisser, 1976). Recent thinking emphasizes that observer expectations determine whether a new focus of attention goes on to more sustained attention (Most, 2005). Discovery of an abnormality could influence subsequent perceptual cycles, biasing the perceptual hypotheses that will be tested in subsequent cycles to be more related to the found abnormality. Finding an abnormality may affect attention by changing the prevailing correspondence of sensory evidence and competing interpretations. Our procedure had two phases: verbalizing the current focus of attention, and summing up with a radiologic report. Perhaps noting each abnormality as it was found helped the observer to conclude the testing of one hypothesis in one perceptual cycle and then move on to a different diagnostic hypothesis in the next.

The SOS errors were classified using categories described by Kundel and his colleagues (Kundel, 1978; 1987; Nodine, 1987). But instead of basing the categories on length of gaze-dwell time, we used criteria applicable to verbal reports. The three categories of error were: (a) faulty image search (the observer did not indicate that the area of the native abnormality was inspected); (b) faulty pattern recognition (the observer noted the area containing the abnormality, but failed to note anything unusual); and (c) faulty decision making (the observer noted that the region might contain something unusual, but did not report a finding).

Table 9.7 shows the proportion of search, recognition, and decision errors averaged over observers according to the previous gaze-dwell time study and the current verbal protocol study. The two methods indicate different sources for the errors. A much higher proportion of search errors occurred with the verbal protocol study (0.37 vs. 0.08, $t(30) = 2.98$, $p < 0.01$), and a much higher proportion of decision errors occurred with the gaze-dwell study (0.51 vs. 0.12, $t(30) = 3.67$, $p < 0.001$).

Different proportions of error types were obtained when using gaze-dwell time and verbal protocol to classify SOS errors. The use of gaze-dwell time provided a lower proportion of search errors than did the use of a verbal protocol (8% vs. 37%). The verbal protocol method may overestimate search error because the voice and pointing finger cannot keep up with eye movement. Of course, incorrect attribution of errors to faulty visual search that are actually caused by faulty recognition does not affect the issue of whether SOS errors in chest radiology are caused primarily by recognition failure or decision-making failure (Berbaum, 1998; Samuel, 1995). The use of gaze-dwell time provided a higher proportion of decision errors than did the use of a verbal protocol (51% vs. 12%). For the diverse native abnormalities we used as *test* abnormalities, the gaze-dwell time method may overestimate decision-making error, incorrectly attributing some errors to faulty decision making when they were actually caused by faulty recognition. Gaze time on the abnormality can accumulate without recognition if the observer was attending to the wrong image features, which can happen with diverse abnormalities. Finding few decision-making errors (12%) with verbal protocol analysis directly contradicts the evidence that chest SOS effects are caused primarily by faulty decision making (Berbaum, 1998), supporting instead the conclusion of other studies that SOS errors in chest radiography are caused primarily by faulty pattern recognition (Berbaum, 1990; 1991; 2000a; Samuel, 1995).

Table 9.7 *Search, recognition, and decision errors according to gaze-dwell time and verbal protocol analysis*

	Gaze-dwell time experiment		Verbal protocol experiment
Search errors	8.3%	<	37.0%
Recognition errors	40.3%		50.5%
Decision errors	51.4%	>	12.4%

9.3.3 Interventions

9.3.3.1 *The influence of clinical history (Berbaum, 1993)*

In research spanning more than 15 years, two radiology perception laboratories attempted to determine whether clinical history affects perception by carefully controlling the information contained in clinical prompts (Swensson, 1977; 1979; 1982; 1985; 1988; 1990; Berbaum, 1986; 1988a; 1988b; 1989b; 1993; 1994b; 2006a). The general conclusion reached by Swensson and Berbaum is that Berbaum's results apply more to the effect of providing clinical history whereas Swensson's results apply more to the effect of providing a prior reader's conclusions (Swensson, personal communication, 1998; Berbaum, 2006a).

All of these studies separate effects of history on perception from effects on decision making by applying the same prompts to carefully matched normal and abnormal cases or locations (so that histories cannot lead *a priori* to better performance). If prompts were believed to the exclusion of radiographic findings, overall performance would be no better than would be obtained by guessing. Berbaum (Berbaum, 1986; 1988a; 1993; 1994b) found that categorical prompts that were correct for specific abnormalities in abnormal studies, but also presented with the normal studies, resulted in better detection accuracy than reading without prompts. Prompts that were plausible but incorrect for the abnormal studies as well as the normal studies led to few FP responses, and detection as accurate as reading without prompts. For these experiments, as well as Berbaum's other experiments on clinical history (Berbaum, 1988b; 1989b), the improvement in detection accuracy was based on an increased TP rate without increased FP rate, regardless of decision threshold.

A diagnosis completely missed without history, but detected with history, is evidence that history affects perception. In this situation, there are no radiographic findings available from

Table 9.8 *Experimental conditions*

	Berbaum (1991) (no clinical history)	Berbaum (1993) (clinical history provided)
Condition 1 (no *added* nodules)	No history, no nodules	History suggesting native abnormalities, no nodules
Condition 2 (*added* nodules present)	No history, nodules	History suggesting native abnormalities, nodules
Condition 3 (*added* nodules present)		History suggesting metastatic disease, nodules

perception that might be compared with the clinical prompt to improve decision making. Such evidence was provided by Berbaum (Berbaum, 1988b; 1989b), who found that knowledge of localizing symptoms and signs of trauma facilitate detection of fractures. Another type of evidence for an effect of history on perception comes from directly controlling whether the history is available at the time of image inspection. Presenting history before or after inspection should allow the same influence of history on decision making, but because the lifespan of visual memory is brief, perception can only be affected when the history is provided before inspection. Using categorical prompts with pediatric radiographs, Berbaum (Berbaum, 1994b) found that detection accuracy was better with history provided before inspection than inspection without history, but detection accuracy was no better with history provided after inspection.

If clinical information can substantially improve perception of abnormalities on radiographs and the SOS in chest radiology is caused largely by faulty pattern recognition (Berbaum, 1996; 2000b), perhaps clinical information can be used to counteract the SOS effect. Knowledge of clinical information could exact a penalty: increased SOS (reduced detection) of unprompted abnormalities. Search was measured using an interruption technique described previously (Berbaum, 1991) and the conditions of the new experiment were compared with those of the prior experiment that used the same technique to characterize the time course of search.

Table 9.8 lists the experimental conditions used in Berbaum (1991) and Berbaum (1993). Detection accuracy for native lesions was measured: (1) with histories suggestive of the native abnormality; (2) with these histories and simulated pulmonary nodules; and (3) with the same *added* nodules and histories suggestive of metastatic disease (i.e. for nodules). These conditions were also compared with those of the second experiment, which were similar but included no history (Berbaum, 1991). Ten additional volunteer radiologists from the faculty at The University of Iowa served as observers. The case set was the same as earlier studies (Berbaum, 1990; 1991) except that five cases used in previous studies were eliminated based on poor performance characteristics. It included 65 chest radiographs, 33 of which demonstrated native abnormalities.

Detection. In the original publication (Berbaum, 1993), the standard binormal model had been fitted to the data of individual observers in each experimental condition and the areas under the ROC curves analyzed using t-tests. For this review we re-analyzed the data (Berbaum, 1993) and the experiment with which it was compared (Berbaum, 1991) using the proper contaminated binormal model (Dorfman, 2000a). The conclusions of the new analysis were the same as the original analysis.

Given a history indicative of the *test* native abnormality, *test* abnormalities were detected with equal accuracy, whether or not an *added* nodule was present (0.82 with nodules present vs. 0.84 with nodules absent, $t(9) = -1.10$, $p = 0.30$). *In other words, there was no observable SOS effect in the presence of clinical history. Test* abnormalities were detected with significantly greater accuracy in both of these conditions than in the condition with *added* nodules and history indicative of metastatic disease (0.82 vs. 0.72, $t(9) = 4.88$, $p < 0.001$ and 0.84 vs. 0.72, $t(9) = 7.88$, $p < 0.0001$, respectively). No difference in detection accuracy was found between conditions with *added* nodules, one without history and the other with metastatic disease history (0.72 vs. 0.70, $t(9) = 0.83$, $p = 0.42$). A misdirecting history pointing to the *added* nodules did not degrade detection of other abnormalities to any greater extent than the unprompted nodules had degraded detection through SOS. Comparison between conditions without *added* nodules, one without history and the other with target history, revealed a substantial difference (0.73 vs. 0.84, two-group $t(18) = -4.08$, $p < 0.001$). *The results of these two comparisons suggest that test abnormality history is so important that SOS added lesions have no inhibitory effect.*

Detection time. A gamma distribution was fitted to each observer's response-time distribution for each type of response. Modal response times were computed from parameters of these curves. Various types of repeated measures analyses on modes were used to test specific hypotheses. The individual parameter estimates were averaged over observers to graph the average response-time distributions.

Figure 9.5 shows distributions of various responses for abnormal cases appearing with simulated nodules. Table 9.9 shows the average modal response times. Without history (top panel), nodules were detected before native abnormalities (Berbaum, 1991). As might be expected, history suggesting the native abnormality promotes rapid detection (middle panel), whereas history suggesting metastatic disease promotes more rapid detection of nodules (bottom panel). Nodules were reported before native abnormalities when both were presented along with a metastatic disease history (7.0 vs. 14.1 seconds, $t(9) = -3.45$, $p < 0.01$). However, with a native abnormality history, the order of detection was reversed: native abnormalities were reported before nodules (6.7 vs. 14.0 seconds, $t(9) = -4.62$, $p < 0.01$).

With history suggestive of the native abnormality, the average time to find native abnormalities was no different when nodules were added (6.7 seconds) than when they were absent (6.5 seconds). However, when nodules were present, the change of history from suggestive of native abnormality to suggestive of nodules increased time to detect native abnormalities (6.7 vs. 14.1 seconds, $t(9) = -6.67$, $p < 0.001$). Clearly, it is history rather than inclusion of an *added* nodule that controls time to detect native abnormalities.

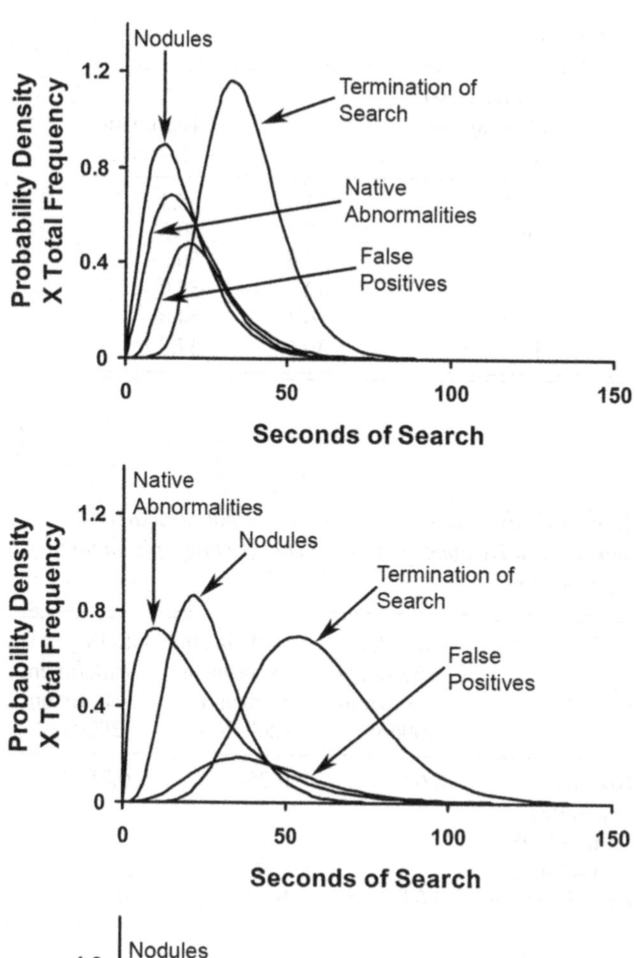

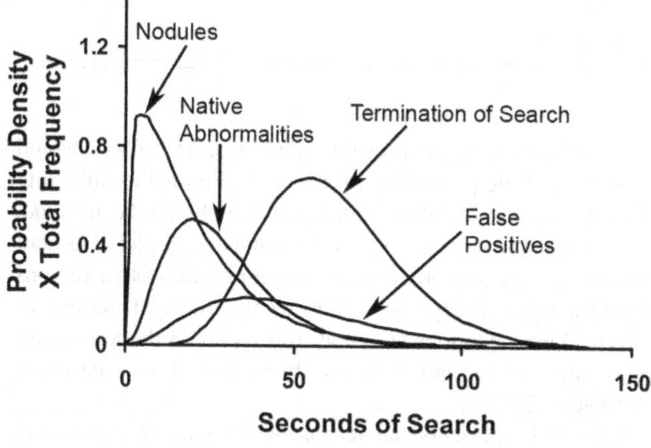

Figure 9.5 Response distributions as a function of search time for abnormal cases with simulated nodules presented with no history (top panel; from Berbaum, 1991), native abnormality history (middle panel), and metastatic disease history (bottom panel). Responses are nodule detection (true nodules), native abnormality detection (TPs), FP response, and termination of search. Curves are gamma distributions computed from parameter estimates of individual data estimated by the method of moments normalized to a mass equal to the total number of responses of each type.

Clinical prompts simultaneously influence accuracy and search. Even in the presence of simulated nodules, prompts improve detection for those abnormalities to which they refer, and they initiate early search for those abnormalities. A link between these two effects would be evidence that changed perception is the basis for improved accuracy. It is not certain that faster perception (or faster allocation of attention) is necessary for improved detection: if detection were somehow delayed, the prompts might still improve accuracy. Unlike the prompts, simulated pulmonary nodules affect detection latency very little. The impact of the *added* lesion may be less immediate than that of the prompt; detection of the nodule can only delay subsequent detection of other lesions, whereas history can promote detection at an earlier point in search.

Total search time was examined using separate ANOVAs with a within-subject factor for experimental condition, and another for cases with and without native abnormalities. The addition of the nodules produced a significant reduction in search time. For cases with native abnormalities, total search time was 52.9 seconds without nodules and 49.2 seconds with nodules present ($F(1,16) = 4.94$, $p < 0.05$), and for cases without native abnormalities, total search time was 60.8 seconds without nodules and 51.9 seconds with nodules present ($F(1,16) = 20.42$, $p < 0.001$). This was an unexpected outcome because the analysis of the previous study had demonstrated that the 2.5-second observed reduction in average mean search time was not statistically significant. The idea that discovery of the nodule triggers immediate termination of search remains doubtful. While total search time decreased with the addition of nodules, accuracy in detecting native abnormalities did not correspond to the length of search. Furthermore, the small reduction in total search time with the introduction of the simulated nodules is a rather late impact on search, occurring well after true and false responses.

Why should detection of lesions reduce search time? At some level, every radiologist must be aware that the working criterion of abnormality is relaxed as search proceeds. The increasing chance of FP response may be the more compelling factor. The simple passage of search time, however, may not be the most reliable indicator of when search should be halted. A good alternative may be the observer's level of confidence in his or her last report of abnormality.

9.3.3.2 Checklist (Berbaum, 2006b)

In radiology, checklists have been recommended to counteract SOS:

"Use of the worksheet helps to safeguard against one of the major hazards of both the free-search and the directed search when the 'direction' is by a specific clinical diagnosis – namely, that a search will be prematurely terminated because a significant signal or finding has been detected." Kinard, Orrison, and Brogdon, 1986

"There are a few recommendations that one can make to help prevent the more frequently detected abnormalities from decreasing vigilance for less frequently detected and unexpected abnormalities. First of all, if the abnormality is quickly interpreted, observers must be aware of the tendency to halt search early. If observers discover multiple abnormalities, they should make sure that they do not permit any one abnormality to capture all their visual attention. Another means to counteract the SOS effect is to adopt some heuristic method of self-prompting, such as an automatic checklist, that would increase vigilance for unexpected abnormalities."

Samuel, Kundel, Nodine, and Toto, 1995

Table 9.9 *Average modal response time in seconds – SOS with clinical history*

Clinical history suggests	*Added* nodule	Native (*test*) abnormality	Nodule detection	Native (*test*) abnormality detection	FP	Termination of search
Native (*test*)	Absent	Absent	*	*	34.3	65.3
abnormality	Absent	Present	*	6.5	37.9	60.5
Native (*test*)	Present	Absent	5.6	*	29.6	56.8
abnormality	Present	Present	14.0	6.7	42.4	55.9
Added nodules	Present	Absent	5.7	*	26.1	54.8
	Present	Present	7.0	14.1	36.6	57.8

Data from Berbaum (1993) – 10 observers

The leaders of these research teams, Gil Brogdon and Hal Kundel, are eminent radiologists and pioneers in medical image perception research. Their counsel should be considered.

We tested whether a checklist could reduce or eliminate SOS effects of *added* nodules on detection of native abnormalities in chest radiography. Fifty-seven chest radiographs, half of which demonstrated diverse, native abnormalities, were read twice by 20 observers, once with and once without the addition of a simulated pulmonary nodule. Area under a proper ROC curve for detecting the native abnormalities was estimated for each observer in each treatment condition. Radiologists used a checklist during the interpretation.

Each examination in the experimental session had a page of a response booklet dedicated to it. At the top of the page, there was a case number and a simple description of the age and sex of the patient. The checklist consisted of the following eleven items: 0: global Gestalt, 1: lungs, 2: heart, great vessels, 3: mediastinum, trachea, 4: chest wall, ribs, sternum, 5: pleural cavity, 6: humeri, scapulae, clavicles, 7: neck and cervical airway, 8: vertebrae, 9: abdomen, diaphragm, 10: other areas. For each item, the observer was required to check one of two boxes. One box was labeled "abnormal, report specifics below" and the other box was labeled "normal, nothing to report." When the observer checked the box "normal," they could then proceed on to the next element. When the observer checked the box "abnormal," they were directed to an area of the page where they noted the category of the abnormality (0–10), the abnormal feature, the specific diagnosis, and circled a confidence rating: 1 for suspicious for abnormality, 2 for possible abnormality, 3 for probable abnormality, or 4 for definite abnormality. Before returning to the checklist, the observer was asked to indicate the location of the abnormality on a photographic facsimile of a chest posteroanterior and lateral.

The response form for each case differed only in the patient information provided. The checklist begins with "global Gestalt." Many who have theorized about visual search suggest that there is an initial phase of inspection in which the observer orients to the general nature of the image (Nodine, 1987). Notice that the second item on the checklist is "lungs." Previous SOS experiments in chest radiology (Berbaum, 1991; 1993) suggest that, in the absence of clinical history, pulmonary nodules are found before non-pulmonary abnormalities. In addition, we did not want to preclude an SOS effect by directing search to the native abnormalities before the *added* nodules. There would seem to be little possibility of a second abnormality affecting detection of another abnormality that had already been found. Although many nodules seem to be found during global Gestalt anyway, a checklist order that places lungs after other regions would not give interpretable SOS results. Most of the rest of the checklist categories and order was based on the most common patterns of search found in the verbalization experiment (Berbaum, 2000b).

Table 9.10 *Average contaminated binormal ROC areas for detecting native abnormalities in chest radiographs in three experiments*

	Checklist experiment – Berbaum (2006b)	Verbalization experiment – Berbaum (2000b)	SOS experiment – Berbaum (2000a)
No *added* nodules (non-SOS condition)	0.67	0.75	0.75
Added nodules (SOS condition)	0.68	0.75	0.70

Table 9.10 presents the results of the current experiment and those of earlier experiments. In the checklist experiment, the paired t-test on ROC areas found no significant difference with and without *added* nodules ($t(19) = -0.56$, $p = 0.58$); the checklist eliminated the SOS effect. The SOS experiment (Berbaum, 2000a) previously reported a t-test on the data of the 19 observers, demonstrating an SOS effect on area. We repeated this analysis just on those 57 cases in common with the checklist experiment and achieved similar results ($t(18) = 5.11$, $p < 0.0001$). In the verbal protocol experiment, there was no SOS reduction in the average area under the ROC curve (Berbaum, 2000b). We repeated this analysis just on those 57 cases in common with the current checklist experiment and achieved exactly the same results ($t(19) = 0.20$, $p = 0.84$).

An analysis of covariance was used to compare the ROC areas across SOS conditions and verbal protocol and checklist experiments. In this analysis, SOS condition (non-SOS, SOS) was a within-subject factor (repeated measure), experimental procedure was a between-subject factor, and level of diagnostic experience (second-year resident, third-year resident, fourth-year resident, fifth-year resident, fellow, faculty) served as a covariate. This analysis adjusts differences in ROC area between experimental procedures to correct for any differences in reader experience between the two experiments. The difference in level of experience between the checklist and verbal protocol experiments was not statistically significant ($F(1,37) = 1.93$, $p = 0.17$). On average, ROC area was greater for the verbal protocol than the checklist (0.75 vs. 0.68, $F(1,37) = 17.26$, $p < 0.001$). Neither the main effect of SOS condition nor SOS condition by experiment interaction was significant.

The verbal protocol experiment, requiring observers to provide verbal descriptions of their search during the interpretation of cases, showed unexpected protection from SOS effects (Berbaum, 2000b). Describing visual search during radiologic interpretation was not studied as a way to prevent the SOS effect. The verbal descriptions were collected to help determine the cause of SOS. We thought that the protection from SOS might have been the result of observers generating their own internal checklist on the fly to help them to report their search behavior. The results of this actual checklist experiment show that this explanation cannot be correct. An actual checklist does not shield us from SOS in such a helpful way. SOS is not so much prevented as pre-empted. *The checklist disrupts perception even when the potentially distracting pulmonary nodule is not present.* Distraction by a nodule seems to be redundant rather than additive with disruption from the checklist. Gale proposed that a radiograph is selectively analyzed according to the reader's concept of it – its purpose, the anatomy it covers, how it ought to be inspected and interpreted – and the concept is then modified to fit what that analysis reveals (Gale, 1979). Following a preset pattern of search prevents what the analysis reveals from altering the ongoing interpretation and the subsequent hypothesis testing on the radiograph (Gale, 1979; 1983). Of course, we must consider that the problem may not be with the checklist per se, but with the fact that readers have to look away from the images to work through it (Sistrom, 2006).

9.3.3.3 Computer-aided diagnosis (Berbaum, 2007b)

A promising approach to prevent SOS errors in medical imaging diagnosis is computer-aided diagnosis (CAD). The rationale for CAD is that computer algorithms and human interpreters may have different, but complementary, strengths. By providing the human observer with the computer's findings, detection accuracy may be improved. Computer aids to help find cancer in mammograms, chest radiographs, and chest CT have been developed, where the observer's task is difficult but focused on a single type of disease.

CAD for nodule detection in chest radiography is well developed (e.g. Xu, 1997; MacMahon, 1999; Nakamura, 2000). Although CAD was not developed to counteract the SOS effect, it might prove an effective intervention to do so. The objective of this study is to determine whether an idealized CAD aimed at finding nodules can reduce SOS errors for other types of abnormality by easing the radiologist's burden of detecting nodules. Alternatively, there might be a penalty in reduced detection of unprompted abnormalities other than nodules.

To find out whether a future, perfectly performing CAD algorithm for nodule detection could prevent SOS in chest radiography, we simulated CAD prompts that pointed only to simulated nodules we had placed in the lung fields. We reasoned that if such a CAD could not prevent SOS, it would be unlikely that a less accurate CAD could do so. To test whether this idealized CAD prompt could alter the SOS effect, we used the same two conditions that were used in SOS demonstrations. Observers were asked to review an area of the film indicated by a CAD prompt for the location of a potential pulmonary nodule. CAD prompts pointed unerringly to the simulated nodules we placed in the lung fields and were present only when a nodule appeared in the radiograph. In this experiment, the simulated pulmonary nodule was always accompanied by a CAD prompt. This experiment was performed using 57 radiographic examinations of the chest that had been used in earlier SOS studies. Sixteen volunteer residents and faculty served as observers.

A response form for each patient in the experiment included photographs of the patient's images with a CAD prompt superimposed. A white circular disk was overlaid on a digitally acquired photograph of the radiograph and corresponded to the location of a simulated pulmonary nodule. The observer was first asked to review this area on the film and reject or accept this indication by checking one of two boxes. If they accepted the CAD indication, they would then circle their confidence that the feature was abnormal using the scale 1–4 where 1 = suspicious, but probably normal; 2 = possibly abnormal; 3 = probably abnormal; and 4 = definitely abnormal. Next, the response sheet contained a table for reporting other abnormal features on the film. The observer had been asked to continue describing other abnormal features, drawing their locations on the photographic facsimile, giving a likely diagnosis, and rating their confidence (using the same 1–4 scale) that the feature was abnormal. If no abnormal features were found, there was a box indicating "normal features, nothing to report."

Significantly more simulated nodules were reported in the SOS with CAD experiment than in the original SOS experiment (49.3 vs. 43.4 nodule reports, $t(33) = 3.31$, $p < 0.01$). The area under the contaminated binormal ROC curve was used to measure detection accuracy for native abnormalities. Areas were estimated for each observer in each experimental condition and compared by using a t-test for paired observations with degrees of freedom determined by the number of readers. There was a significant difference when nodules were added (average area 0.68 without nodules vs. 0.65 with nodules and prompts, $t(15) = 1.76$, p(one-tailed) < 0.05); thus CAD failed to affect the SOS effect. Further analysis showed that there was no difference in the magnitude of the SOS effect between the current experiment with CAD prompts and previous SOS experiments (Berbaum, 2000a).

ROC points are defined by sensitivities and specificities associated with boundaries between rating categories. Decision thresholds are revealed by fitting an ROC model to the response probabilities on either side of boundaries between the categorical ratings. Clearly, not all of the decision thresholds are equally consequential in patient care. To determine whether there are any shifts in response readiness associated with *added* nodules or CAD prompts, we analyzed the decision threshold between "possibly abnormal" and "suspicious, but probably normal." The decision threshold we studied was chosen to reflect the decision to report or not report abnormality.

A paired t-test on the FP fraction associated with the decision threshold for detecting native abnormalities estimated for each reader in each condition found a significant difference (FP fraction of 0.15 without nodules vs. 0.10 with nodules and prompts, $t(15) = 2.33$, $p < 0.05$), suggesting a threshold shift toward stricter reporting with the addition of nodules. By contrast, a similar analysis of the data from the previous SOS study without CAD prompts (Berbaum, 2000a) demonstrated no threshold shift.

A t-test for independent samples was used to test for a difference in decision threshold in control (non-SOS) conditions of the two experiments. This test demonstrated a marginally significant difference in the decision thresholds (FP rate 0.23 without CAD vs. 0.15 with CAD, $t(33) = 2.02$, $p = 0.052$). Therefore, there was a shift toward a more conservative threshold for reporting with the CAD prompt experiment, even in the condition without nodules or CAD prompts. We performed similar tests for a difference in decision threshold in experimental (SOS) conditions of the two experiments. This test demonstrated a significant difference in the decision thresholds (FP rate 0.19 without CAD vs. 0.10 with CAD, $t(33) = 2.60$, $p < 0.05$). Therefore, the same shift was evident in the experimental conditions as well. Thus, there was a shift toward more conservative thresholds for reporting native abnormalities with the addition of the nodules, which may actually be indicative of reduced visual search (see section 9.5.2).

Although CAD increased the recognition of the simulated nodules, it did not reduce the SOS effect of detecting nodules or the detection of other abnormalities.

9.4 SOS IN MULTI-TRAUMA PATIENTS

9.4.1 ROC experiments

9.4.1.1 Fracture detection in multi-trauma patients (Berbaum, 1994a)

Radiologists have long known that some abnormalities may draw and hold attention, diverting it from other injuries:

> "In the multiply injured patient there are obvious lesions that often overshadow other lesions, creating a significant possibility that they will be overlooked." Rogers, 1990

Common associations between lesions, called "clinical dyads," have been identified so that oversight may be avoided by seeking the second member of the pair when one is found. In our investigation, we studied perception of multiple lesions which are not commonly associated. Patients suffering multiple injuries often require a series of radiographs to examine all injured sites. The cases of our study were designed to simulate the radiographic examination of the multi-trauma patient. We undertook the study to determine whether SOS generated by finding a relatively obvious abnormality on one image would detract from recognizing a subtle lesion on another image of the same patient showing different anatomy.

Sixty-five simulated trauma patients were each depicted in series of radiographs that were assembled from radiographs of several actual patients. The radiographs were selected to simulate a single patient. Forty-six cases included a radiograph showing a subtle *test* fracture. In one experimental condition, none of the other radiographs in the patient's series contained a fracture. In a second experimental condition, a radiograph containing an *added* fracture was substituted for a radiograph that had no fracture in the first experimental condition. An example case is illustrated in Figure 9.6. In all, 215 radiographs were presented under each experimental condition. Of these 215 radiographs, 69 showed the 65 *added* fractures (or normal radiographs when the fracture was not added) and were not counted in scoring; 97 were normal radiographs, and 48 showed the 46 *test* fractures. Verification of the fractures was through follow-up studies, and clinical course, and under the supervision of a musculoskeletal radiologist. Normal images were from actual patients with no indication of abnormality in the anatomy involved from any diagnostic source.

Test fractures and dislocations included those of the **wrist**: scaphoid (3), navicular waist (1), distal radius (3), radial styloid (1), perilunate (dislocation, 1); the **hand**: metacarpals (10), fingers (2); the **chest and shoulder**: humeral greater tuberosity (1), clavicle (3); the **foot**: talar neck (1), cuboid (1), metatarsals (4), toes (3); the **ankle**: posterior (3) / lateral (2) / medial (1) malleoli; and the **leg, knee, and hip**: fibula (2), medial femoral condyle (2), tibial plateau (1), and acetabulum (1).

Added fractures and dislocations included those of the **wrist and elbow**: distal radius (3), radial head (3), ulnar styli (2), hamate/proximal phalanx (1), coronoid process of ulna (1); **hand**: metacarpals (5), metacarpal/carpal dislocation (1), fingers (2); **chest and shoulder**: humeral neck (1), clavicle (4), clavicle and scapula (3), scapula (4), acromion process (1), acromioclavicular separation (3), posterior shoulder dislocation (1), ribs (4); **foot**: calcaneus (1), cuboid (3), metatarsals (3), toes (2); **ankle**: lateral (3) / medial (3) malleoli; and **leg, knee, and hip**: fibula (1), intertrochantic femur (2), femoral neck (1), patella (2), acetabulum (2, one with femoral head dislocation), superior pubic ramus (1), pubis (2).

Figure 9.7 demonstrates how an example case would appear in each experimental condition and the use that would be made of observer responses to the images in scoring. In the control (non-SOS) condition, none of the other radiographs of the patient's series contained a fracture. In the experimental (SOS) condition, a radiograph containing an additional fracture (which we refer to as the *added* fracture) was substituted for a radiograph that had no fracture in the control condition. The substituted radiograph containing the *added* fracture and the normal radiograph that it replaced were carefully matched and depicted the same anatomy. This was possible because they were either of opposite extremities or they originated from different but carefully matched real patients.

The observer rated confidence that the acute fracture or dislocation was present as follows: 1 = definitely, or almost definitely, present; 2 = probably present; 3 = possibly present; 4 = probably not present, but some question; 5 = definitely, or almost definitely, not present. If a rating of 1–4 was given, the

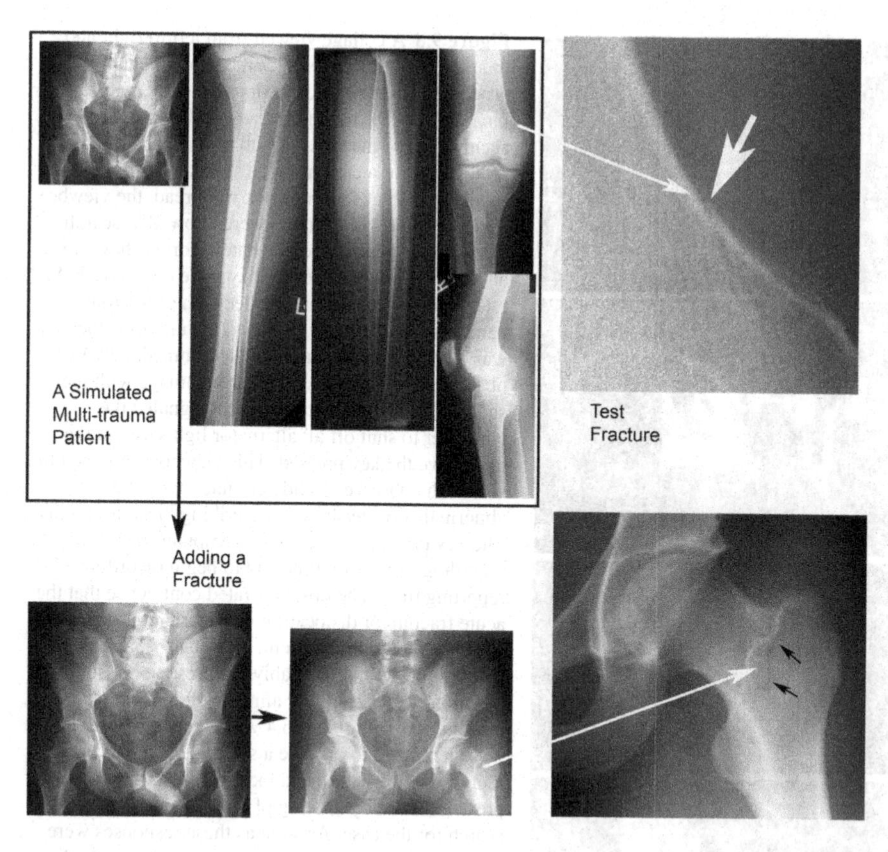

Figure 9.6 An example of a simulated multi-trauma patient is shown inside the box. It includes a normal anteroposterior (AP) radiograph of the pelvis, a normal AP and lateral radiograph of the left tibia and fibula, an AP radiograph of the right knee showing a subtle nondisplaced fracture of the medial femoral condyle, and a lateral radiograph of the right knee showing a fat-fluid level within the suprapatellar bursa. A more obvious fracture in the intertrochanteric region of the left hip is added by substituting a radiograph of the pelvis with this fracture for the one without a fracture.

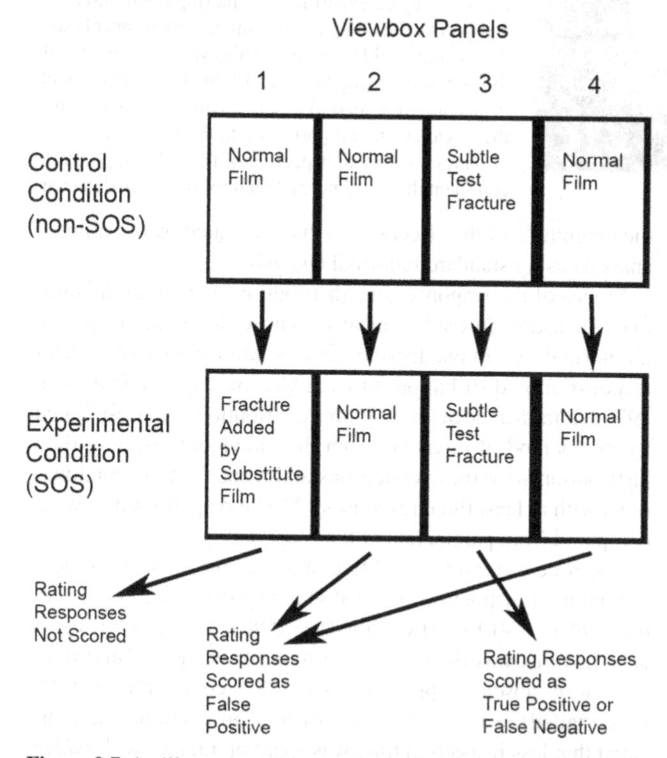

Figure 9.7 An illustration of how a simulated patient would appear in a four-panel viewer under each of the two experimental conditions of the experiment. In the SOS condition, a normal radiograph of the non-SOS condition is replaced with one showing the similar anatomy but containing a fracture. Responses to the radiograph with the *added* fracture and to the radiograph it replaced were not counted in scoring of accuracy.

observer was then asked to provide a specific diagnosis identifying the nature and location of the injury. The location and classification responses were used to determine whether rating responses referred to *added* or *test* fractures. A form of alternative free-response operating characteristic analysis described by Chakraborty was used to analyze the data (Chakraborty, 1990). Each case in the experiment supplied one FP rating: the most abnormal rating assigned to the normal radiographs of the case exclusive of the normal radiographs that replaced the *added* fracture radiograph. We limited analysis to cases in which the *added* fracture was reported with a certainty of at least 2.

The observers were instructed that the cases were of acute trauma and that some would be normal, while others would contain one or more clinically significant abnormalities. Figure 9.8 illustrates our experiment procedure within each experimental trial. The cases were examined twice by ten radiologists on occasions separated by four months. Half of the cases presented in each occasion were from each experimental condition so that any effect of reading occasion or order would be controlled. The presence of the *added* fracture was an experimental manipulation (independent variable), while detection of the *test* fractures was measured (dependent variable). Each part required two two-hour reading sessions to complete. All possible left-to-right orderings on the four panels of the film viewer of examinations showing no injuries, examinations containing a *test* fracture, and examinations containing an *added* fracture in the SOS condition (or no injury in the non-SOS condition) were used. The different orders were of approximately equal frequency and appeared at random across the course of the session.

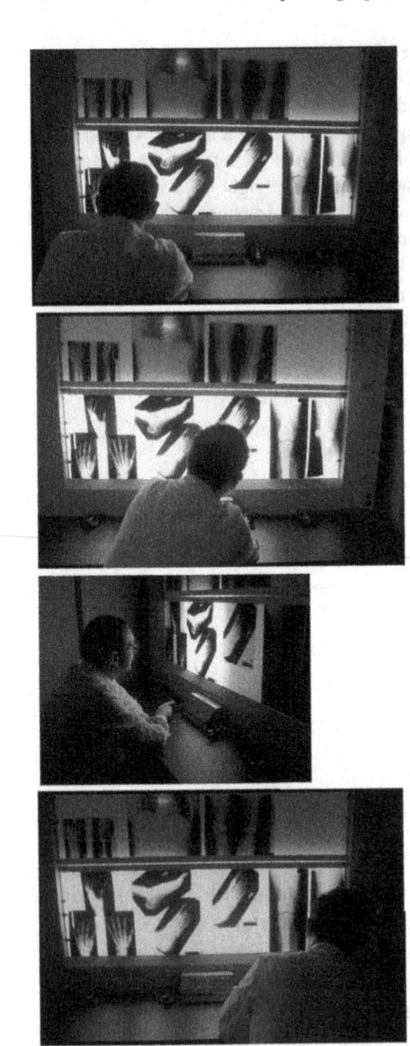

Figure 9.8 A collage viewed from left to right and then top to bottom illustrates the experimental procedure that permitted various time intervals during inspection of the radiographs as well as detection accuracy to be measured. Each trial began with the radiographs of a case hung on a darkened film viewer. After the patient's age and sex were read, the viewbox was illuminated. Observers were allowed to search the images of each case in whatever order they chose and report abnormalities as they were discovered. A telegraph key was located under each of the four panels in the alternator. When a fracture or dislocation was discovered, this was indicated immediately by the observer pressing the key under the image with the suspected abnormality. This event signaled the computer to shut off all alternator lights except the one above the key pressed. This viewbox remained lit so that the observer could continue to see the abnormality as he/she was describing the abnormality. The key press also caused the computer to stop recording inspection time and to begin recording reporting time. The observer rated confidence that the acute fracture or dislocation was present: 1: definitely, or almost definitely, present, 2: probably present, 3: possibly present, 4: probably not present, but some question, 5: definitely, or almost definitely, not present. If a rating of 1 to 4 was given, the observer was then asked to provide a specific diagnosis identifying the nature and location of the injury. A key press followed by a rating of 5 was used to terminate search for the case. As soon as these responses were recorded (by an experimenter entering them into the computer), the computer re-illuminated all panels so that search could continue. (This event coincided with the end of reporting time recording and resumption of inspection time recording.) The same procedure was then used again to report any additional injuries. The observer continued inspection of the radiograph until confident that all abnormal features had been detected.

In the original paper (Berbaum, 1994a), we analyzed the data only with the standard improper binormal model (Dorfman, 1969). Later, we analyzed the data using the proper contaminated binormal model of Dorfman (2000a), and the functionally proper constrained binormal model of Swensson (1996; 2001).

Table 9.11 presents the results from the standard binormal model, the contaminated binormal model, and the constrained binormal model. Note that the analysis based on areas from the standard binormal model and the constrained binormal model contain only nine observers because neither provides meaningful results for the observer-treatment with $p(\text{FP finding}) = 0$ for all operating points. The contaminated binormal model does provide a meaningful area measure for this situation, and so the analysis based on the contaminated binormal model included all ten observers. The two proper ROC models yield higher levels of statistical significance than the improper model even though they give smaller mean differences in observed ROC areas. This is because the standard error of the mean difference is much smaller for the proper models. The standard error of the mean difference is decreased by reduced variability about the means of the ROC areas in the two conditions and by increased correlation between areas. These results show that the SOS effect in skeletal radiology can be demonstrated with ROC models that do not allow inappropriate chance-line crossings, even though the magnitude of the effect is less than indicated by the original analysis using standard binormal analysis.

Modes of the response-time distributions were used as summary measures instead of means because the distributions are not normal. A gamma distribution was fitted to each observer's response-time distribution for each type of response (Johnson, 1970). Parameters of the gamma distribution were estimated by the method of moments from the individual response-time distributions (for those conditions and response type combinations with at least three responses). Modal response times were computed from parameters of these curves.

The time to report *added* fractures was the same regardless of whether *test* fractures were absent or present (12 vs. 12 seconds, $t(9) = -0.13$, $p(\text{two-tailed}) = 0.90$). The time to report *test* fractures was the same regardless of whether *added* fractures were absent or present (25 vs. 22 seconds, $t(9) = 1.31$, $p(\text{two-tailed}) = 0.22$). Analyses of total inspection time indicated that less inspection time was spent on images with *added* fractures (74 vs. 66 seconds, $F(1,9) = 12.55, p < 0.01$) and less inspection time was spent on images with *test* fractures (73 vs. 67 seconds, $F(1,9) = 10.12, p < 0.05$). These effects were small, amounting to 6 and 8 seconds, or 8% and 11% of total search time, respectively. However, because report time involved continued inspection of the radiograph being reported, reductions

Table 9.11 *Analysis with different ROC models*

Parameter	Standard binormal model	Contaminated binormal model	Constrained binormal model
Average ROC area without an *added* fracture ± standard deviation	0.85 ± 0.09	0.84 ± 0.09	0.87 ± 0.08
Average ROC area with an *added* fracture ± standard deviation	0.75 ± 0.16	0.80 ± 0.10	0.83 ± 0.12
Average difference in ROC area ± standard deviation of the difference	0.09 ± 0.04	0.03 ± 0.01	0.05 ± 0.02
Correlation between ROC areas	0.62	0.92	0.93
Paired *t*-test value	2.21	2.54	2.98
One-tailed *p* value <	0.05	0.02	0.01

in inspection time apparent from analysis of inspection time alone may be compensated for by inspection of the radiograph being reported during the report interval. Indeed, when report time was added to inspection time, a similar analysis found no differences in total search time.

This experiment was designed to simulate routine clinical interpretation; we used no restrictions on the order in which abnormalities were reported (e.g. left to right): the *added* and *test* fractures were randomly positioned on the viewer and observers were allowed to inspect the radiographs in any order they chose. Because of this, the observers may have sometimes come to the image with the *test* fracture before they came to the one with the *added* fracture. In this situation, a less robust SOS effect might be expected; a distracter would have less opportunity to affect perception or report of a target fracture on a radiograph already inspected. Nevertheless, we demonstrated SOS under these unrestricted viewing conditions. Detection of the *test* fractures was significantly reduced when *added* fractures were reported. However, when *test* fractures were detected, there was no difference in the time at which detection occurred in the two conditions. *SOS can occur where both added and test lesions are fractures and that SOS can propagate across a series of radiographs of a patient.*

9.4.1.2 Replication with new patients, readers, and technologies (Berbaum, 2007a)

The clinical practice of radiology has changed greatly since the last experiment was performed in 1994. Whereas the previous experiment used radiographs interpreted at a film viewer, current practice relies heavily on CT, magnetic resonance imaging, and direct digital radiography with interpretation at workstations equipped with high-resolution displays. In 2007, we attempted to replicate the SOS effect of finding *added* fractures on subsequent *test* fractures in a patient's multi-trauma series using modern images and displays. We describe only one of the three experiments reported by Berbaum (Berbaum, 2007a). Two experiments used CT studies of the spine and pelvis to investigate whether the clinical significance of *added* abnormalities controls the magnitude of SOS effects. Those studies are beyond the scope of this chapter because they used advanced imaging and because definitive answers to that question currently await further experiments.

In the current experiment, for each patient, the first examination displayed was normal for one experimental condition (the control condition) but included an *added* fracture in the other experimental condition (the SOS condition). Addition of fractures into the first examination was an experimental manipulation; we measured detection of the *test* fractures appearing in the second and third examinations, and gathered FP responses when both the second and/or third examinations were normal.

To simulate the radiologic examination of the multi-trauma patient, each patient in the experiments consisted of imaging studies of three different body parts. Seventy simulated multi-trauma patients were constructed. Although the examinations of each patient came from different sources, they were matched so that they would appear to belong to the same patient. Raw material came from digital radiographs from 304 actual patients presenting over 800 normal or abnormal examinations. To the extent possible, we used examinations from the same patient. Where this was not possible, we matched examinations by gender, age, and body type.

We presented radiographs examining three different body parts, excluding the spine. The *added* fracture was always presented as the first examination in the series. Detection accuracy was measured by scoring responses on the second and third examinations. The second and third body parts presented for each simulated patient were digital radiographs of extremities, chest, or pelvis and usually included multiple views. The same examinations of the second and third body parts were always presented for each patient in both control and experimental conditions of all three experiments. There were 27 simulated patients in which both the second and third examinations contained only normal digital radiographs, and 43 with a subtle *test* fracture. For each simulated patient, all responses to the second and third examinations were counted toward the patient rather than toward the examination so that there were 70 scored patients with 27 normals and 43 abnormals for the purposes of ROC analysis.

Cases were presented on two 3 Mpixel LCD monitors calibrated to the Digital Imaging and Communications in Medicine (DICOM) standard using the manufacturer's specifications. Tagged image format files (TIFFs) were generated from DICOM format digital radiographs and optimally resized to fill one display screen. Software was developed from ImageJ, a public domain image processing and analysis package written in the Java programming language (http://rsb.info.nih.gov/ij/, Rasband, 1997–2006). This was accomplished by modifying and extending ImageJ so that it had capabilities to display radiographic images. This software collected reader responses, not only their explicit judgments indicating detection and localization of abnormalities and associated confidence ratings, but also the display functions that they used to explore the image data including window and level adjustment.

Readers were instructed to search for all acute fractures and dislocations and to identify each abnormality by placing the mouse cursor over the abnormality and clicking with the right mouse button. This produced a menu box for rating their confidence that the finding was truly abnormal. The readers were directed to indicate their confidence that a finding was abnormal by using discrete terms such as "definitely a fracture or dislocation," "probably a fracture or dislocation," "possibly a fracture or dislocation," and "probably not a fracture or dislocation, but some suspicion." These discrete terms were transformed into an ordinal scale where 1 represented no report, 2 represented suspicion, 3 represented possible abnormality, 4 represented probable abnormality, and 5 represented definite abnormality.

Data for the experiment were collected in two sessions separated in time by several months. Half of the patients presented in each session were from the SOS condition and half from the non-SOS (control) condition. Thus, in the course of the two sessions, each patient appeared twice, once in each experimental condition. Within each session, patients were presented in a pseudorandom order so that the occurrence of fractures was unexpected and balanced. Before each trial of an experiment, the reader was always informed of the patient's age and sex. During the reading sessions, viewing distance was flexible, room lights were dimmed to about 5 foot-candles of ambient illumination, and there were no restrictions on viewing time.

The first examination in the series for each patient controlled the presence of the *added* fractures creating the non-SOS condition (without an *added* fracture) and the SOS condition (with an *added* fracture). The 70 *added* fractures were presented on digital radiographs of the foot (6), ankle (4), tibia/fibula (7), knee (5), pelvis (3), chest (3), shoulder (10), arm (5), elbow (4), wrist (11), and hand/fingers (12). The digital radiographs that were used in place of these examinations for the non-SOS condition had the same distribution of body parts.

Some of the radiographs appearing in the second or third examinations within each simulated patient presented a subtle *test* fracture; others presented no abnormalities. Whether the *test* fracture appeared in the second or third examination was random, but the same order was used for both experimental conditions in all experiments. The 47 *test* fractures were presented on digital radiographs of the foot (15), ankle (4), tibia/fibula (2), knee (3), shoulder (4), arm (3), elbow (1), wrist (3), and hand/fingers (8). The 97 normal examinations appearing in the second or third positions of the series (47 with the *test* fractures, and 54 as pairs to make 27 normal patients) were presented on digital radiographs of the foot (7), ankle (13), tibia/fibula (3), knee (20), pelvis (19), chest (15), shoulder (3), arm (3), elbow (5), wrist (2), and hand/fingers (7).

In this chapter, we limit ourselves to traditional analysis in medical image perception research, in which individual reader ROC parameters in each treatment are statistically analyzed to generalize to the population of readers. Accuracy parameters were estimated by fitting the contaminated binormal model to the rating data of individual readers in each treatment condition and the ROC areas, and sensitivities at specificity of 0.9 with and without the *added* fractures were compared using nonparametric Wilcoxon signed-rank tests (Dixon, 1992). Specificity of 0.9 was chosen as a convenient level at which to measure sensitivity because it was the even value that maximized the number of readers with operating points on both sides of the sensitivity value. A statistically significant reduction in detection accuracy for *test* fractures was found when the fractures were added (ROC area was 0.86 without *added* fracture vs. 0.81 with *added* fracture in the initial examination, $p < 0.01$; sensitivity at specificity of 0.9 was 0.71 without *added* fracture vs. 0.62 with *added* fracture in the initial examination, $p < 0.01$).

The experiment replicated the SOS effect in multi-trauma patients with modern digital acquisition and display methods using new images and readers, essentially doubling the evidence for SOS in musculoskeletal radiology.

9.4.2 Causes

9.4.2.1 Gaze-dwell times on missed fractures (Berbaum, 2001)

In these experiments, gaze time was used to determine whether fractures were missed because of misdirected attention. These studies were performed to determine whether SOS errors in patients with multiple fractures are caused by faulty visual scanning, faulty recognition, or faulty decision making. We also studied how the severity of the *added* fracture influenced visual search, postulating that fractures of greater clinical significance might induce visual neglect of other regions. Clinical necessities can be one reason for diagnostic oversights in the patient with multiple injuries (Rogers, 1982; 1984). We postulated that fractures of greater clinical importance might induce greater visual neglect of other regions. In the ROC trauma experiments already described (Berbaum, 1994a; 2001), the clinical importance of the *added* fractures was not very different from the *test* fractures being measured, and few of the detracting injuries involved *major* morbidity. The detection of some fractures has immediate implications for patient care.

Multi-trauma patients were depicted in a series of radiographs. Radiologists interpreted each series under two experimental conditions: when the first radiograph in the series included a fracture, and when it did not. In the first experiment, the initial radiographs showed nondisplaced fractures ("*minor*" extremity fractures). In the second experiment, the initial radiographs showed fractures of greater clinical significance ("*major*" fractures), with *major* morbidity. Each series also included a radiograph with a subtle *test* fracture and a normal radiograph on which detection accuracy was measured.

We selected cases from the previous ROC experiment that contributed to the SOS effect. The images consisted of radiographs of three different body parts. A representative arrangement is presented in Figure 9.9. The center panel of each case contained a radiograph with a subtle *test* fracture. The left panel had a radiograph of a specific body region. In the first experimental condition, that radiograph was normal; in the second, an *added* fracture was present. Addition of fractures into the left

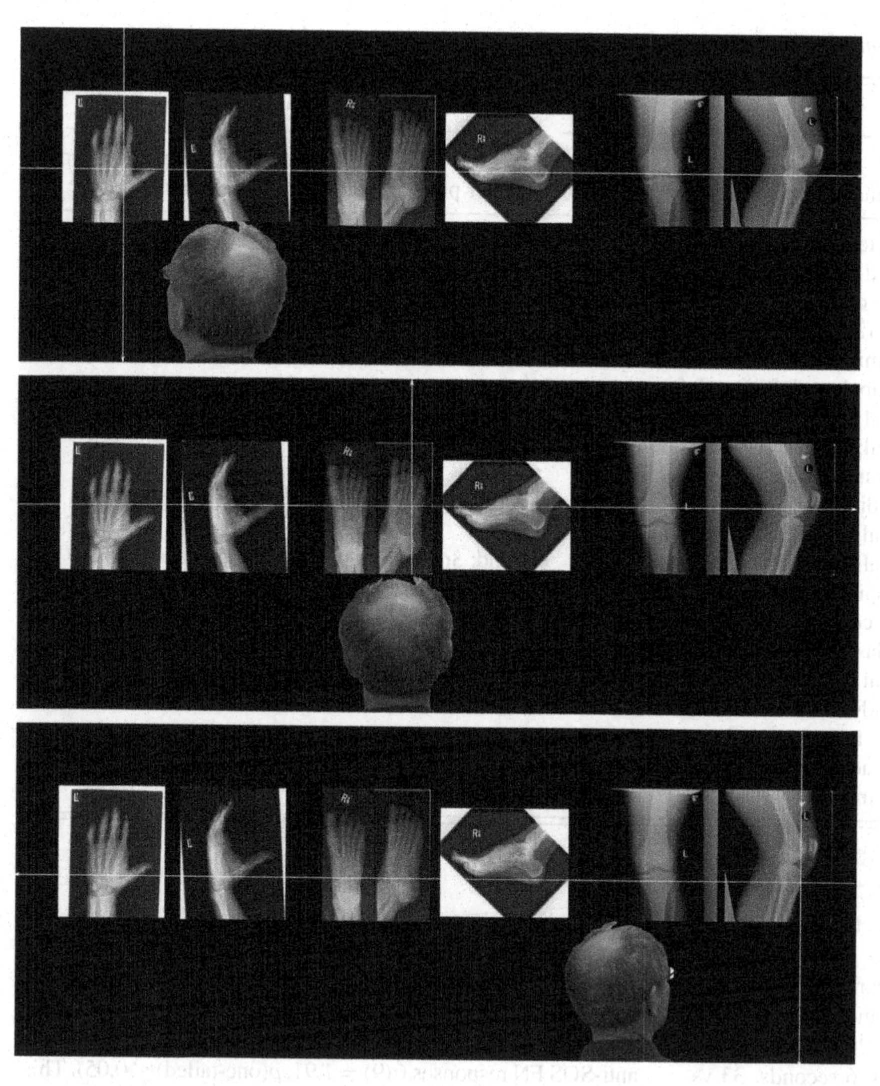

Figure 9.9 The arrangement of examinations in a simulated case: the PA and lateral radiographs of the hand show a fracture of the second metacarpal; the three views of the foot show a subtle fracture of the fifth metatarsal, and the PA and lateral radiographs of the knee are normal. The cross-hairs show the direction of gaze as mapped on to the viewing environment at three points during search.

panel was an experimental manipulation; we measured detection of the *test* fractures appearing in the center panel, and gathered FP responses only from the right panel containing a normal examination.

Table 9.12 lists our cases, the arrangement of the images in the three-panel bank of the film viewer, and the findings corresponding to *test* and *added* fractures. Note that Table 9.12 does not list the normal images appearing in the left panel for the first experimental (control) condition of each experiment. Occasionally, there were two views of the *test* or *added* fractures. In all, 30 radiographic examinations representing 10 patients were presented under each of the two experimental conditions.

Gaze-dwell time was recorded as ten senior staff radiologists reviewed ten simulated multi-trauma series in the first experiment and ten different senior staff radiologists reviewed ten in the second experiment. Observers were told these were images from a trauma series and were instructed to read the radiographs of each patient from left to right, to point out and describe the location of each abnormality that they wished to report, and to rate their confidence that the acute fracture or dislocation was actually present by using the words "definite," "probable," or "possible."

Data collection for the two experiments was completely independent. For each experiment, its data were collected in two sessions separated in time by several months. For each experiment, half of the cases presented in each session were from each experimental condition. Thus, in the course of the two sessions, each radiograph in the case sample appeared twice, once in each experimental condition. Within each part, cases were presented in a pseudorandom order so that occurrence of abnormal and normal cases was unexpected and balanced.

The global effect of the *added* fractures on search was assessed by studying dwell time within radiographs displayed on the left, center, and right panels. A separate analysis of variance was performed on the dwell time for each of the panels and the total time from all three panels. Each analysis of variance for repeated measures included factors for addition of fractures (absence or presence of fracture in the left panel), clinical importance of the *added* fracture (*minor* vs. *major fracture*), and cases. Generalization was to the population of observers.

Average overall search time was significantly reduced (9 seconds, 120 vs. 111 seconds, $F(1,18) = 8.42$, $p = 0.010$) when either a *minor* or *major* fracture was added regardless of the severity of the fracture. Conversely, average search time on the left panel (where fractures were added) increased by 3 seconds

Table 9.12 *Fractures presented in the ten simulated patients*

Left panel		Same images used in both experiments	
Experiment 1: *minor added* fracture	Experiment 2: *major added* fracture	Center panel: *test* fracture	Right panel: normal examination
Ankle: lateral malleolus Pelvis: nondisplaced linear pubis	Ankle: talar body and neck Cervical spine: extension tear drop C2; body of C7	Wrist: scaphoid Hand: 5th phalanx	Pelvis Tibia and fibula
Pelvis: hip dislocation with acetabular fracture	Pelvis: right superior pubic ramus, left inferior pubic ramus, left acetabulum, left iliac, left sacrum	Shoulder: clavicle	Wrist
Pelvis: acetabulum Hand: 4th metacarpal	Cervical spine: C2 ring fracture Cervical: displaced type 3 odontoid fracture of C2	Hand: 5th metacarpal Foot: 5th metatarsal	Tibia and fibula Chest
Wrist: ulnar styloid	Wrist: distal radius with dislocation carpal bones, ulnar styloid	Foot: posterior distal tibia	Shoulder
Chest: clavicle	Cervical spine: C6 burst fracture with C5 spinous process fracture	Hand: 3rd middle phalanx (intra-articular)	Tibia and fibula
Ankle: distal tibia	Ankle: comminuted distal tibia and fibula (Pilon fracture)	Hand: scaphoid	Knee
Shoulder: acromioclavicular separation	Cervical spine: traumatic spondylolisthesis of C2 (type III hangman fracture)	Knee: tibial plateau	Pelvis
Pelvis: linear fractures pubis	Pelvis: acetabulum, pubic ramus, sacrum	Foot: talar body	Hand

when a fracture was added, regardless of fracture severity. This added 3 seconds to the 9-second deficit to give a 12-second shortfall for the other images. Average search time was significantly reduced on both the center panel (*test* fracture) (7 seconds, 46 vs. 39 seconds, $F(1,18) = 20.26$, $p < 0.001$) and on the right panel (normal radiograph) (6 seconds, 33 vs. 27 seconds, $F(1,18) = 52.27$, $p < 0.001$) when a fracture was added to the left panel. Reduction was greater for *minor* fractures than *major* fractures (8 vs. 4 seconds). So about half of the reduced search time came from the *test* fracture radiographs and about half came from the normal radiographs. Although there seems to be a clear change in search behavior with less time devoted to other images when a fracture is introduced, the severity of the *added* fracture does not seem to make a great difference in search behavior.

Detection and eye-position data were analyzed using dichotomous categories: abnormalities were either reported (at any level of certainty) or not reported. For each observer, we counted the number of SOS events (the *test* fracture reported without an *added* fracture but missed when a fracture was added), and the number of anti-SOS events (the *test* fracture missed without an *added* fracture, but reported when a fracture was added). We tested for the presence of an SOS effect in the data by testing the null hypothesis that the population probabilities of SOS outcome and anti-SOS outcome are equal using a paired t-test, in which we generalized to the population of readers. When *minor* fractures were used as *added* abnormalities, there was no difference in the number of SOS and anti-SOS FN responses. Dwell time on regions of the *test* fractures for SOS errors and anti-SOS errors did not differ. When *major* fractures were used as *added* abnormalities, there were significantly more SOS FN responses than anti-SOS FN responses ($t(9) = 1.77$, p(one-tailed) $= 0.06$). When we limited the analysis of responses to those reader-case combinations in which the *added* fracture was reported when present, there were significantly more SOS FN responses than anti-SOS FN responses ($t(9) = 1.91$, p(one-tailed) < 0.05). This finding demonstrated an SOS effect in the TP data of the second experiment.

A Wilcoxon signed-rank test showed that median dwell time on the *test* fracture was greater for SOS errors than for anti-SOS errors (1.2 vs. 0.7 seconds, $p < 0.05$). Gaze time on *test* fractures suggests that the errors were not based on faulty scanning. Compared to the anti-SOS errors, SOS errors were comprised less of sampling (8% vs. 0%) and recognition (69% vs. 42%) errors and more of decision errors (23% vs. 58%). The paucity of sampling errors and predominance of decision errors suggest that although overall search time on other images was reduced when a fracture was added to the first image, this reduction did not cause the SOS errors.

An SOS effect could be demonstrated only in the second experiment with *major added* fractures. Analysis of dwell times showed that search on subsequent radiographs was shortened when the initial radiograph contained a fracture; however, the errors were not based on faulty scanning. The SOS effect in musculoskeletal trauma is not caused by faulty scanning. Demonstration of an SOS effect on *test* fractures with *major* but not *minor* additional fractures is sufficient to suggest the hypothesis that detection of other fractures is inversely related to the severity of the detected fracture. That hypothesis is under current investigation.

9.5 SOS IN CONTRAST STUDIES OF THE ABDOMEN

9.5.1 ROC experiments

9.5.1.1 SOS in the detection of plain-film abnormalities (Franken, 1994)

Anecdotal experience indicates that lesions detectable on plain films are often not identified in contrast studies of the same region. We studied the role of SOS in detection of plain-film abnormalities in contrast studies of the abdomen. In such studies, a nearly simultaneous plain film and a contrast film of the same patient are often available. Occasionally, abnormalities are demonstrated on both the plain-film and contrast examinations that are not the nominal subject of the contrast examination. These opportunistically discovered lesions presented an opportunity to study SOS. *Added* abnormalities included gastrointestinal or genitourinary contrast studies (both normal and abnormal), while *test* abnormalities were subtle lesions found anywhere in the radiograph outside the region of the contrast material.

There were two experimental conditions: (1) the cases of the sample (normal and abnormal) presented on a plain film of the abdomen; and (2) these same cases each presented with a contrast examination also recorded on a single abdominal film. The presence of contrast and contrast-demonstrated abnormalities was an experimental manipulation (an independent variable), whereas detection of the abnormalities separate from and visible outside of the contrast region was measured (a dependent variable). Figure 9.10 shows a pair of images from one case: the plain film without a contrast examination would be examined in one condition of the experiment, and the plain film with a contrast examination would be examined in the other condition.

Forty-three cases were presented, with each case having one plain film and one gastrointestinal or genitourinary contrast film. Twenty-two of these pairs included an abnormality unrelated to the lesion studied with contrast. Twenty-one were normal outside the contrast region. Verification of abnormalities was made

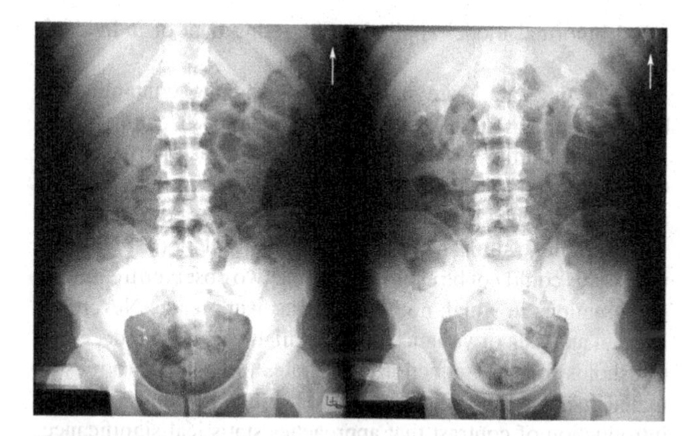

Figure 9.10 The *test* abnormality visible in the plain-film region on both non-contrast (left) and contrast (right) studies of a fracture of the left 8th and 9th ribs (white arrows). Note that a catheter in the bladder is also seen on both films. The excretory urogram on the right visualizes blood clots in the left pelvis and urinary bladder.

through follow-up studies, biopsy, clinical course, and similar material from the patient's medical record. Patients with no indications of plain-film abnormality who also had (normal or abnormal) contrast studies were selected to approximate the age and sex of the patients with plain-film abnormalities. Radiographs were selected so that the plain-film abnormalities were not superimposed by the contrast material.

Abdominal plain-film findings included subhepatic bubbles indicating abscess, a gastrostomy tube escaped into the small intestine, ankylosing spondylitis, gallstones, lesser trochanter fracture, nondisplaced fractures of the right ilium and left pubis, blastic bone metastasis of the rib, post-traumatic cyst on the ilium, fractures of the 10th and 11th ribs, a foreign body (tampon), Paget's disease of T-12 vertebral body and pelvis, calcified gallstones, blastic metastasis in the pelvis, Paget's disease in the hip, an absent femoral head, metastasis of the L-3 vertebral body, intraperitoneal barium (chronic), teeth in an ovarian teratoma, fractures of the 8th and 9th ribs, metastatic lung nodules, calcifications in the pancreas, and myostitis ossificans.

The contrast studies for patients with plain-film abnormalities included a normal endoscopic retrograde cholangiopancreatography (ERCP), a barium meal for upper gastrointestinal series (UGI) showing a "new" gastrostomy tube in the stomach, a T-tube cholangiogram demonstrating a biliary L-stent for pancreatic cancer, a barium enema revealing a rectovaginal fistula, a normal colostomy enema examination, a barium meal with small bowel follow-through indicating partial high-grade small bowel obstruction (2), an enteroclysis example of Meckel's diverticulum, a normal enteroclysis, a double contrast colon examination showing Crohn's disease, a barium meal for UGI with aspirated barium in left lower lobe bronchi, an excretory urogram of a horseshoe kidney, a ureteral stent injection revealing no ureteral obstruction, an enteroclysis showing chronic diffuse ischemic change, an excretory urogram showing an ileal conduit, a barium meal for small bowel follow-through showing partial obstruction at ileostomy, an excretory urogram of a normal urinary tract, a normal colostomy enema examination, a normal barium meal for UGI examination, an excretory urogram showing blood clots in the left ureter and pelvis, an excretory urogram showing hydronephrosis, and an ERCP with obstruction of the common bile duct.

The contrast studies for patients without plain-film abnormalities included antegrade barium meal for UGI showing Crohn's disease in the ileocolic anastomosis, an excretory urogram showing duplication in the right ureteral collecting system, an ERCP showing common hepatic duct stricture, enteroclysis showing Crohn's disease in the distal ileum (2), a post-colostomy barium enema, a barium meal for UGI post-subtotal gastrectomy and choledocho-jejunostomy, a hypaque UGI showing duodenal hematoma, a normal barium meal for small bowel follow-through examination, a barium meal for UGI showing Crohn's disease in the distal ileum, a normal excretory urogram (3), a barium enema showing ulcerative colitis, a normal barium enema examination (2), a normal enteroclysis examination, a cystogram showing left ureteral reflex, a barium enema showing colon diverticulosis, an enteroclysis showing partial small bowel obstruction, and a barium enema showing cancer of the transverse colon.

Ten volunteer radiologists from the Department of Radiology faculty or senior house staff at The University of Iowa served as observers. The experiment consisted of two parts separated in time by five months. Half of the cases presented in each part contained contrast images and half did not. Thus, in the course of two parts, each case appeared twice, once with and once without contrast. Within each part, cases were presented in a pseudorandom order. The observers were instructed that some of the cases could contain one or more abnormalities and others might be without any lesions. The time to find each abnormality and length of observation was measured. Confidence in abnormality was rated from 0 to 100% certainty. The classification and location responses were used to determine which ratings applied to contrast-demonstrated abnormalities and which applied to *test* abnormalities – those visible without contrast. The continuous ratings were grouped into six categories of rated abnormality defined as 0%, 1–59%, 60–79%, 80–89%, 90–99%, and 100% certainty.

A striking feature of our data was that there were far fewer responses indicating plain-film abnormality on the contrast examinations. Figure 9.11 shows the responses to plain-film and contrast studies in graphic form. The figure shows that observers missed plain-film abnormalities present on the contrast studies more often than they had on the plain film, indicating SOS for contrast examinations. However, they also had fewer FP responses of plain-film abnormalities related to the comparable plain-film regions on those films with contrast material. This reduction in FP rates differs from the SOS observed in previous studies. With the commensurate reductions in TP and FP rates, there may be no change in the ROC area but a systematic increase in decision criteria to report abnormality.

In the original paper (Franken, 1994), the rating data were analyzed using a reader jackknife method and the standard binormal ROC model (Dorfman, 1986) with the computer program, RSCORE-J (http://perception.radiology.uiowa.edu/). The reader jackknife method is much more likely to converge to a solution with few cases per reader than the maximum-likelihood method, which may fail to converge to a solution for some readers (Berbaum, 1989a). Interpretations of abdominal plain films and contrast films by radiologists showed no difference in area under the ROC curve (0.74 vs. 0.70, $t(9) = 0.60$, $p = 0.57$). Despite the presence of contrast, observers detected plain-film abnormalities on the contrast study as well as they had on the plain film. There was a systematic increase in the decision thresholds from plain film to contrast study ($F(1,9) = 8.80$, $p < 0.05$).

It has been argued that a disadvantage of pooling is that if individual observers do not use the rating scale in the same way in each experimental condition to produce approximately the same decision criteria, the pooled ROC curves will be biased downward, reducing their areas (Metz, 1987). Also, visual inspection of the pooled ROC curves showed that the curves for plain-film and contrast examinations crossed and suggested that they might differ in the region of the actual operating points to the crossover point, with subsequent differences canceling out to yield no difference in overall areas. Although the original paper included an analysis of partial areas under the ROC curves, which found no statistical evidence of a difference anywhere along the curves, we re-analyzed the data for this using the proper contaminated binormal model applied to the rating data of individual readers.

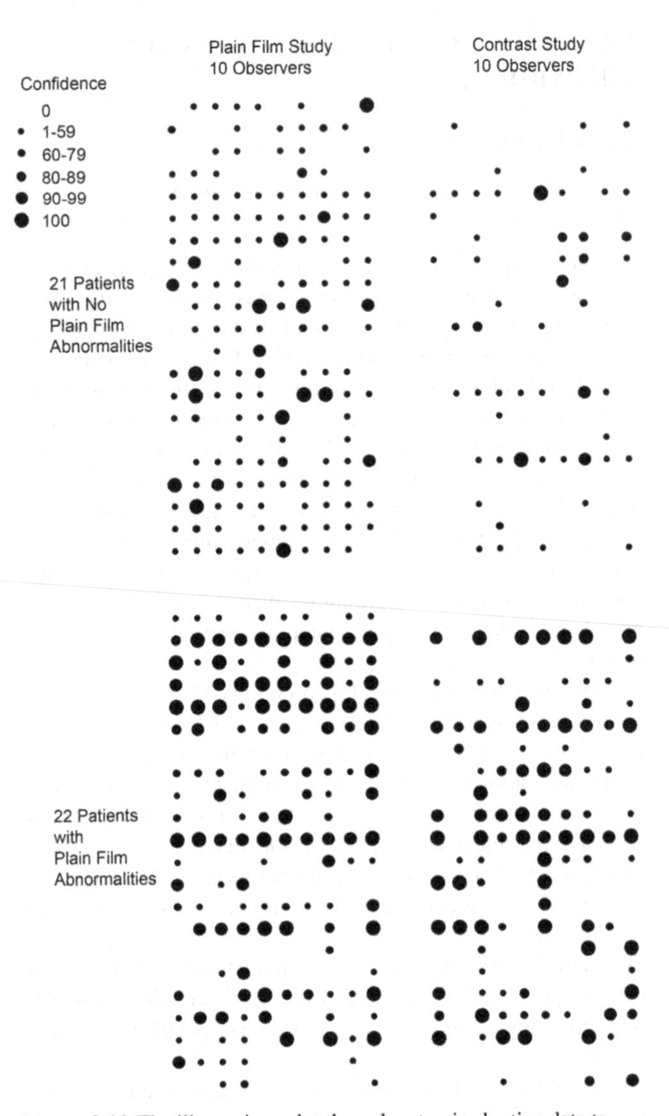

Figure 9.11 The illustrations plot the polycotomized rating data to demonstrate graphically the systematic decrease in TP and FP responses on contrast studies compared with plain film.

When the data were re-analyzed using the contaminated binormal model, no difference in ROC area for detecting plain-film abnormalities was demonstrated (0.70 for plain film vs. 0.68 for contrast studies, $t(9) = 0.41$, $p = 0.69$). This lack of an area difference agrees with the finding reported in the published article. Using the contaminated binormal model, decision thresholds could not be estimated for the two observer/treatment combinations in which no FP responses were made. Nevertheless, an analysis of variance of the available estimated decision thresholds found a reduction in FP rates (computed from estimated decision thresholds) associated with ROC points with the introduction of contrast that approaches statistical significance (average FP probability was 0.14 for plain films and 0.06 for contrast examinations, $F(1,7) = 5.25$, $p = 0.056$). Once again results from the proper ROC model agree with the finding using the reader jackknife method. This is shown in Table 9.13.

Table 9.13 *Average FP rate associated with fitted decision thresholds*

Threshold (category boundary)	Plain film	Contrast examination
0% vs. 1–100%	0.23	0.11
0–59% vs. 60–100%	0.18	0.08
0–79% vs. 80–100%	0.13	0.06
0–89% vs. 90–100%	0.10	0.03
0–99% vs. 100%	0.06	0.01

Analysis of decision times relied on medians. Total search times were not reduced by the introduction of *test* plain film or the contrast-demonstrated abnormalities: median total search time was 65 seconds when no lesions were present, 81 seconds when only a plain-film abnormality was present, 82 seconds when only a contrast-demonstrated abnormality was present, and 100 seconds when both were present. A statistically significant increase in total search time was found when either plain-film abnormality was present ($F(1,9) = 23.64, p < 0.01$) or when contrast was present ($F(1,9) = 19.04, p < 0.01$). These increases may in part be due to the time needed to report abnormalities. However, the search was not interrupted: the radiograph remained illuminated during verbal report so that inspection could continue. We assume that the observer continued to search for abnormalities during the reporting time as well.

The median time to detect *test* abnormalities without contrast was 21 seconds vs. 45 seconds to detect the same abnormality with contrast ($t(9) = -4.00, p < 0.01$). The median time for finding contrast-demonstrated abnormalities on radiographs without plain-film abnormalities was 19 seconds, and 20 seconds with plain-film abnormalities ($t(9) = -0.11$, $p = 0.91$). When both plain-film and contrast abnormalities were present, the median time for detecting contrast-demonstrated abnormality was 20 seconds vs. 45 seconds for detecting plain-film abnormalities ($t(9) = -2.52, p < 0.05$).

On average, the plain-film abnormality was detected after detection of the contrast-demonstrated abnormality. Detection of the plain-film abnormalities occurred later on the contrast examinations than on the plain films. However, the time to detect the contrast-demonstrated abnormality was little affected if a plain-film abnormality was present. The increase in cut points cannot be related to this delay because criteria for reporting an abnormality have been shown to be progressively relaxed during the course of search.

The nature of SOS in this study differs from SOS with chest and musculoskeletal radiographs. The response times indicate that observers probably interpreted the presence of contrast as indicative of the purpose of the imaging study. This expectancy also might have had a continuing effect by holding attention focused in the region of contrast, even when no abnormality was found. But our analyses of ratings provide strong evidence that the changes in decision criteria demonstrated by the ROC analysis were attributable to detection of the contrast-demonstrated abnormality, rather than simply a perceived difference in the task demands of contrast studies.

Reports of plain-film abnormalities are made with less confidence on the contrast study than on the plain film. This suggests that a plain-film abnormality must be more obvious to be reported on the contrast studies. There are two possible interpretations: if observers spent more of their attention and perceptual resources on the regions containing contrast, they would have less available for the non-contrast regions. On the other hand, they might recognize the plain-film abnormalities equally well on the contrast studies but discount them because they believe them to be irrelevant to the purpose of the study.

9.5.2 Causes

9.5.2.1 Gaze analysis of SOS effects in contrast studies (Berbaum, 1996)

In this experiment, we attempted to determine which of the explanations of SOS in abdominal contrast studies, presented at the end of the last section, was correct by recording gaze-dwell time as a measure of the degree of attention to various regions within the images. Gaze-dwell time was used to determine whether this is because observers fail to scan plain-film regions in contrast studies or because they discount plain-film abnormalities that were actually scanned. The gaze of ten radiologists was studied.

The case sample of this eye-position study was a subset of those from the previous study. We selected the ten cases that contributed most to the observed shift in operating points. Because the results of the previous study involved a reduction in both TP and FP rates at each ROC operating point, we included about an equal number of cases with (6) and without (4) plain-film abnormalities. Table 9.14 lists our cases and identifies the findings present in both plain-film and contrast images.

There were two experimental conditions: (1) the cases of the sample (normal and abnormal) presented on a plain film of the abdomen; and (2) these same cases each presented with a single film from a contrast examination. The presence of contrast and contrast-demonstrated abnormalities was an experimental manipulation or independent variable, whereas detection of the abnormalities separate from and visible outside of the region of contrast was the dependent variable. Figure 9.12 shows a pair of images from one case of the experiment. The plain film without contrast was examined in one condition of the experiment and the contrast examination in the other.

Regions of plain-film and contrast-revealed abnormalities were defined for the eye tracker. Since both our plain-film abnormalities and areas occupied by contrast were of various sizes and shapes, those regions had to be defined for the eye tracker in real coordinates in the physical observation space. Figure 9.13 illustrates how this was accomplished using a graphic representation of the radiographs shown in Figure 9.12. A gastrointestinal radiologist indicated the location of plain-film abnormalities in both plain-film and contrast examinations, and location of contrast in both the contrast examination and the corresponding area in the plain film (indicated by the curved lines in Figure 9.13). Boundaries of ROIs for plain-film abnormality and for contrast material were placed around the lines drawn by the radiologist. These ROIs compensated for size differences in area between plain-film and contrast examinations.

Table 9.14 *Plain film examinations and contrast studies*

Abdominal plain film findings	Type of contrast study	Contrast study findings
Normal	Upper gastrointestinal series	Duodenal hematoma
Blastic bone metastasis of left 11th rib	Barium meal for upper gastrointestinal series	Partial high-grade small bowel obstruction
Fracture 10th & 11th ribs on right	Enteroclysis	Normal
Normal	Barium enema	Ulcerative colitis
Ankylosing spondylitis	T-tube cholangiogram	Biliary L-stent for pancreatic cancer
Normal	Barium meal for UGI series	Partial small bowel with Crohn's disease
Gastrostomy tube escaped into distal small bowel	Upper gastrointestinal series	"New" gastrostomy tube in stomach
Paget's disease, left hip	Enteroclysis	Changes of small bowel ischemia
Normal	Cystography	Left ureteral reflex
Foreign body	Double contrast colon	Crohn's disease

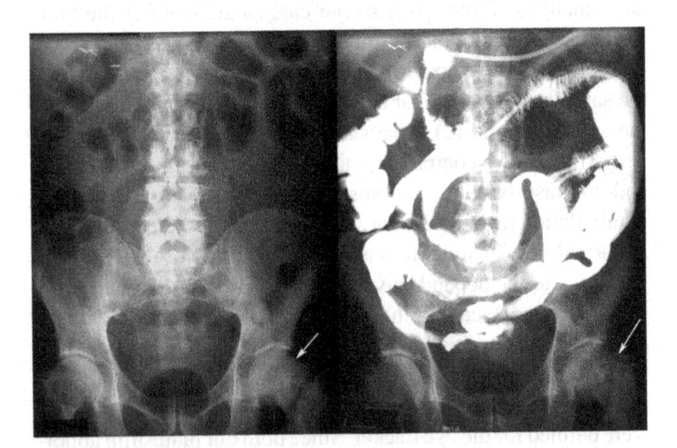

Figure 9.12 The *test* abnormality seen in the plain-film region on both non-contrast (left) and contrast (right) studies is Paget's disease in the left hip (white arrows). The enteroclysis contrast study on the right visualizes changes of small bowel ischemia.

We plotted cumulative dwell time of fixations in each of the three ROIs as a function of search time. Figure 9.14 is representative, showing data of one observer from one SOS case. The upper panel of Figure 9.14 shows that in the plain film, each of the three ROIs received considerable inspection. Non-contrast regions received attention not markedly different from that on the contrast region. The lower panel of Figure 9.14 shows that for the contrast examination, the slope of the function for the

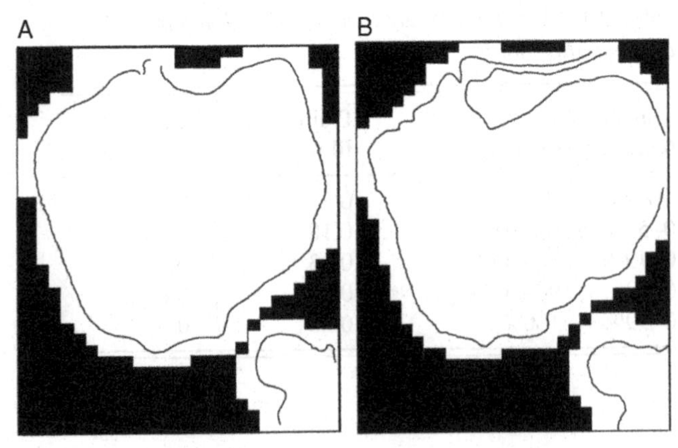

Figure 9.13 Graphic display of the regions of interest associated with the radiographs shown in Figure 9.12: A, representation of the radiograph; B, representation of the enteroclysis study. Irregular lines were drawn by a gastrointestinal radiologist to indicate the location of plain-film abnormality (Paget's disease in the left hip) and the region of contrast in the contrast study and to indicate the region of the same anatomic structures on the plain-film examination.

contrast region increased substantially, reflecting that more of the inspection time was devoted to the contrast region.

ANOVA on gaze time in all ROIs combined demonstrated that for cases without plain-film abnormalities, subjects spent more time searching contrast studies than plain film (97 vs. 82 seconds, $F(1,66) = 5.51$, $p < 0.05$). For cases with plain-film abnormalities, time spent searching contrast and plain-film studies did not differ (101 vs. 95 seconds, $F(1,106) = 0.23$, $p = 0.63$). Thus, there was no evidence for a premature halting of search in the contrast studies.

For cases without plain-film abnormalities, subjects spent more total time gazing at regions with contrast than the corresponding regions on plain films (80 seconds vs. 51 seconds, $F(1,66) = 19.71$, $p < 0.0001$). They spent less time on normal, non-contrast regions of contrast studies than on corresponding regions of plain film (17 seconds vs. 31 seconds, $F(1,66) = 8.31$, $p < 0.01$). For cases with plain-film abnormalities, subjects also spent more time gazing at regions with contrast than the corresponding regions on plain film (71 seconds vs. 45 seconds, $F(1,106) = 6.38$, $p = 0.013$). They spent much less time on normal, non-contrast regions of contrast studies than on corresponding regions of plain film (14 seconds vs. 26 seconds, $F(1,106) = 19.17$, $p < 0.01$). In none of these analyses was the effect of detection outcome or its interaction with the presence of the contrast statistically significant.

Each observer's shortest non-zero dwell time associated with a TP response for a plain-film abnormality was used as the minimum inspection time needed for the observer to recognize any plain-film abnormality. With these criteria on total cluster dwell time (which averaged about two seconds), the FN errors for plain-film examination were 14% scanning, 14% recognition, and 71% decision, whereas the FN errors for contrast examinations were 47% scanning, 30% recognition, and 23% decision. A test for significance of difference between two proportions (Bruning, 1987) on the data from total dwell time showed that the proportion of scanning errors for plain film was significantly different from that of contrast examinations (0.14 vs. 0.47, $z = 2.08$,

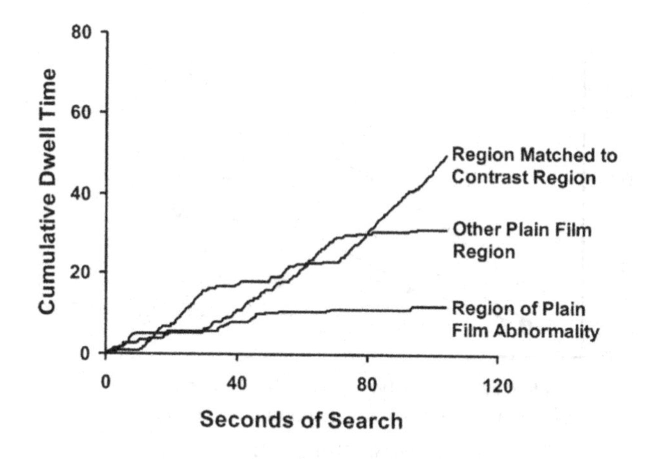

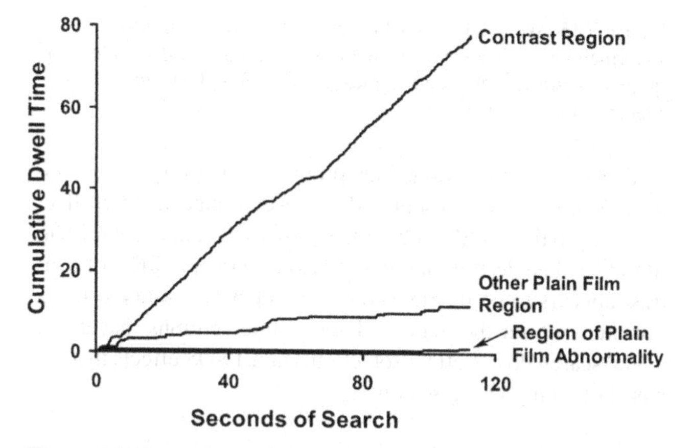

Figure 9.14 Data of one observer from one SOS case. Upper panel: data from the plain-film examination; lower panel: data from the contrast examination. Each of the three ROIs received considerable inspection on the plain film. In the contrast study, a dramatic drop in inspection time occurred for both the regions of abnormality and non-contrast area of the film, with a corresponding considerable increase for the contrast area.

$p < 0.05$). Using a similar test, we found that the proportion of decision errors for plain film was significantly different from that of contrast examinations (0.71 vs. 0.23, $z = 3.05, p < 0.05$).

Taken together, this eye-position experiment and the previous ROC experiment (Franken, 1994) present a consistent picture of the cause of the SOS observed in contrast studies of the abdomen. There was a global change in search behavior in the contrast studies, with altered distribution of foveal sampling across the image. The reduced dwell time on non-contrast regions led to increased scanning errors for plain-film abnormalities. The combined reduction in TP and FP responses appeared to be based on a kind of visual neglect.

It is doubtful that visual neglect explains SOS in which pulmonary nodules detract from detection of native abnormalities (Berbaum, 1990; 1991; 1993), or where one fracture detracts from detecting others (Berbaum, 1994a; 2007a). In those studies, SOS error involved an increased FN rate without a change in FP rate, and equally frequent FP rates in normal image regions with and without the distracting abnormality, suggesting equal search of those regions.

SOS induced by contrast appears to represent a different class of SOS. Within the framework of signal detection theory, we assume that absence of scanning is associated with absence of evidence on the latent evidence dimension. Reduced evidence means that thresholds on the latent non-negative decision dimension will always define few TP and FP responses.

If this "visual neglect" explanation of SOS in contrast examinations is correct, then focusing search on the neglected region would seem an obvious tactic for reducing SOS. Since the SOS effect in contrast examinations appears to be based on faulty scanning, and visual scanning is largely under voluntary control, a variety of actions to encourage more scanning of the plain-film regions might reduce the SOS. A more fundamental approach might be to separately inspect and report regions of contrast and non-contrast. Of course, the clinical impact of SOS in contrast studies is limited by the common practice, already part of current standards, of obtaining a preliminary film of the abdomen in those situations where there is a reasonably high prevalence of plain-film abnormalities.

9.5.3 Interventions (Berbaum, 2005)

We understand the cause of the SOS effect in contrast studies of the abdomen reasonably well: when contrast is introduced, observers focus on the contrast and neglect plain-film regions leading to reductions in both TP and FP responses to plain-film regions. Within the framework of signal detection theory, less inspection corresponds to less evidence of abnormality on the latent decision dimension both for abnormal and normal examinations. With reduced evidence, the estimated standardized decision thresholds increase because they are measured relative to the mean of the distribution of normal cases.

To test whether the order of reporting different types of abnormalities could alter the SOS effect, we included the same two conditions that were used in the original demonstration of that effect: presentation of each case with a plain radiograph and with a contrast examination. As in that experiment, there was no instruction about the order in which abnormalities were to be sought and reported. However, here we added a third experimental condition with each case presented with the contrast examination: observers were instructed to search for and report plain-film abnormalities before searching for and reporting abnormalities revealed by contrast. The presence of contrast material and contrast-demonstrated abnormalities were an experimental manipulation (an independent variable), while detection of the abnormalities separate from and visible outside of the contrast region was measured (a dependent variable).

In this study, several of the examinations used in the original study (Franken, 1994) were replaced. The following three pairs of plain-film and contrast studies were removed: a plain film showing a tampon in the vagina and a double contrast colon study also showing Crohn's disease, a plain film showing calcified gallstones and an excretory urogram showing a horseshoe kidney, and a plain film showing metastasis of the L-3 vertebral body and a barium meal for UGI showing no further abnormality. The following four pairs of plain-film and contrast studies were added: a plain film showing gallstones and a barium enema showing a "normal" post-colostomy colon, a plain film showing a foreign body (a needle) and a hypaque enema showing no further abnormality, a plain film showing a large calcified left-flank mass and a barium meal for UGI showing a nasogastric tube in the stomach, and a plain film showing posterior L4–5 bone fusion and an excretory urogram showing a normal urinary tract.

Thirteen volunteer radiologists served as observers. When an abnormality was detected, the observer was asked to provide a specific diagnosis, identify the feature, and circle its location on an accompanying diagram of the abdomen. The observer indicated a level of confidence that the feature was abnormal. Confidence ratings were discrete. If no abnormalities were detected, a box was checked indicating normal, no abnormalities to report.

Because this investigation concerned shifts in decision thresholds with SOS, an important methodological consideration was how the decision thresholds were defined. A small number of discrete categories defined using natural language terms such as *definitely abnormal, probably abnormal, possibly abnormal, probably normal,* and *normal, no abnormalities to report* provides the best chance of category boundaries having fixed boundaries in the observer's memory. Fitting the ROC data with the contaminated binormal model estimates decision thresholds, just as the standard binormal model does (Dorfman, 1969; 1995). We used the FP rates of the theoretical ROC points to measure decision thresholds. As a check on estimation of decision thresholds, we also fitted the data with the standard binormal model and studied the empirical operating points as well. All three methods produced the same results and conclusions, so we report only the results for the contaminated binormal model (Dorfman, 2000a; 2000b). Analysis focused on changes in decision thresholds among the treatment conditions.

FP probability for reporting plain-film abnormalities was significantly higher on the plain-film studies than the contrast studies (0.20 vs. 0.10, $F(1,11) = 18.15, p < 0.01$), and significantly higher on the plain-film studies than on the contrast studies with plain-film abnormalities reported first (0.20 vs. 0.14, $F(1,11) = 8.71, p < 0.05$). We postulate that the instruction to report plain-film abnormalities before contrast-demonstrated abnormalities ought only to increase FP reports based on an increased inspection of plain-film regions. By this test, a statistically significant difference in decision thresholds was found between contrast studies and contrast studies with plain-film abnormalities reported first (0.10 vs. 0.14, $F(1,11) = 4.20, p < 0.05$ by a one-tailed test). Reporting plain-film abnormalities before contrast-demonstrated abnormalities on contrast studies of the abdomen failed to eliminate the shift in decision thresholds when contrast is added, a shift that indicates reduced inspection of the plain-film regions. In addition, each pair of viewing conditions interacted with threshold definition ($p < 0.05$) with larger differences in FP probabilities for each pair of viewing conditions with more lenient decision thresholds. The nature of these interactions is illustrated in Figure 9.15. This significant statistical interaction, together with Figure 9.15, suggests that differences in average FP probabilities between viewing conditions were greater for more lenient threshold definitions.

If visual neglect of non-contrast regions causes SOS in contrast studies, then focusing search on the neglected regions ought to reduce errors. After all, visual scanning is largely under voluntary control. Separating the inspection of contrast and non-contrast regions is the simplest form of a checklist. Our

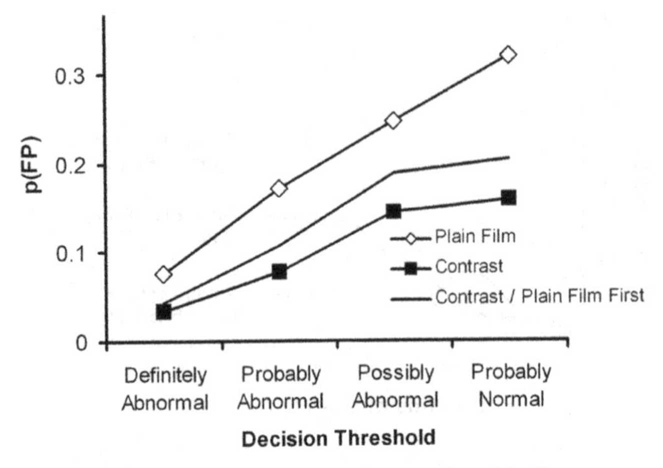

Figure 9.15 Average FP probabilities for reporting plain-film abnormalities at different decision threshold definitions for plain-film studies, contrast studies, and contrast studies with plain-film abnormalities reported first.

evidence shows that evaluating non-contrast regions before contrast regions in contrast studies does not eliminate the SOS effect on decision thresholds, although it reduces the effect somewhat. SOS visual neglect of plain-film regions in the presence of contrast appears to be resistant to our voluntary attempts to focus our attention on the neglected spots. Interventions that direct visual search do not offer protection against SOS effects that are based on faulty visual search.

9.6 SUMMARY OF SOS EFFECTS IN TRADITIONAL RADIOGRAPHIC IMAGING

SOS in chest radiography. Detection accuracy measured by ROC methodology for various subtle lesions in chest radiographs was compared with accuracy for detecting the same lesions when a simulated pulmonary nodule was added to the radiograph (Berbaum, 1990; 2000a). Detection accuracy was reduced with *added* nodules; this reduction in accuracy was based on fewer TP reports with no change in the rate of FP reports. In terms of signal detection theory (Green, 1962), a change in the rate of FP reports would indicate a shift in the threshold for reporting abnormality, which in turn would associate the SOS effect with faulty decision making. A lack of change in FP rate tends to rule out decision making as a causal factor.

The experiments in chest radiography did not support Tuddenham's (Tuddenham, 1962) hypothesis that SOS is caused by a termination of search after an abnormality has been found (Berbaum, 1991; 1993; 1998). In our experiments, observers continued to inspect films after reporting one abnormality. The time needed to find *test* lesions did not depend on whether *added* lesions were present, and the time needed to find *added* lesions did not depend on whether *test* lesions were present. These findings tended to rule out changes in search behavior as a cause of SOS error. This led us to hypothesize that SOS error in chest radiography might represent a failure in pattern recognition.

To test this hypothesis, eye-position recordings were made while observers interpreted chest radiographs that yielded SOS errors in the earlier ROC studies. The reduction in detection of native abnormalities produced by *added* nodules was not based on the length of time the observers looked at the native abnormalities (Berbaum, 1998). Instead, the long gaze times suggested that SOS errors may have been based on faulty decision making. This was clearly inconsistent with the finding of stable decision thresholds noted above. To resolve the issue, we used an analysis of verbal reports of attention taken during interpretation (Berbaum, 2000b). The results indicated that gaze time may not be a good discriminator of recognition and decision-making errors when applied to diverse abnormalities such as those in our SOS experiments. Verbal protocol analysis (Newell, 1972; Ericcson, 1980; 1984) suggested that SOS in chest radiology is based primarily on faulty pattern recognition (Berbaum, 2000b).

Based on this conclusion, we might expect that interventions to prevent SOS must improve the recognition of image features. In fact, we have evidence that clinical histories suggesting *test* abnormalities prevent SOS effects on them in chest radiology (Berbaum, 1993). However, the common recommendation to prevent SOS by using a checklist (Kinard, 1986; Samuel, 1995) was not found to be of value, because the checklist itself was a distraction (Berbaum, 2006b). Similarly, simulated computer-aided diagnosis prompts that unerringly indicated the locations of simulated pulmonary nodules failed to prevent the detection of those nodules from producing SOS effects on other abnormalities (Berbaum, 2007b). Thus, computer nodule checking will not prevent SOS in chest radiography when searching for diverse abnormalities.

SOS in multi-trauma patients. Trauma patients often require a series of radiographs to evaluate all potentially damaged organs. We constructed an experimental paradigm for studying SOS in multi-trauma patients by including a preliminary radiograph with or without a fracture in the patient's radiographic series (Berbaum, 1994a; 2001). Detection of subtle *test* fractures was substantially reduced when additional fractures were included. As in chest radiology, ROC area was reduced but decision threshold (FP rate) was unaffected, suggesting that decision making was not the cause of the SOS errors. Response time measurements did not support Tuddenham's premature termination of search explanation of SOS, suggesting that faulty scanning was not the cause of the SOS errors. Similarly, an experiment measuring gaze-dwell time in multiple orthopedic trauma cases indicated that the SOS errors were not caused by faulty scanning (Berbaum; 2001). The observer almost always fixated the fracture when an SOS error was made. This experiment suggested that the seriousness of the detected fracture may affect the number of SOS errors. We performed ROC experiments to test this (Berbaum, 2007a).

SOS in contrast studies of the abdomen. It often turns out that the simplest inference about SOS effects has been proven wrong on further study. In radiographic contrast studies of the abdomen, the presence of contrast material itself distracts from other non-contrast abnormalities (Franken, 1994). This experiment demonstrated a substantial change in decision thresholds estimated by ROC analysis, suggesting that observers became more conservative in reporting non-contrast abnormalities in the presence of contrast material. ROC area was unaffected by the SOS manipulation because the reductions in TP reports were accompanied by reductions in FP reports (at each ROC point). When there is no change in ROC area, the overlap between the distributions of signal and noise (abnormal and normal) does not change. This implies that the SOS misses were not caused by a degradation of pattern recognition. According to signal detection theory (Green, 1962), shifting thresholds point to decision-making error. In other words, the observers actually perceived the abnormalities, but discounted the percept and did not report them. Thus the process underlying SOS effects with contrast in abdominal radiographs is different from that of other types of examinations. It took the application of eye-tracking methodology in a second experiment to discover that the ROC threshold shift was not based on faulty decision making, but instead on a massive visual neglect of non-contrast regions (Berbaum, 1996). What appeared to be a shift toward more stringent decision thresholds in the ROC experiment (Franken, 1994) actually represented a shift in the underlying distributions of signal and noise from under fixed thresholds (Berbaum, 2005). The errors in the experiment were predominantly faulty scanning, according to Kundel–Nodine error classification. If this SOS is caused by failure to attend, and attention is at least partially under voluntary control, the solution seems obvious. But when tested in a third experiment, reading the non-contrast regions before the image regions with contrast offered no protection from SOS (Berbaum, 2005). At every step, without subsequent effort to answer the remaining question, we would have been misled. This is a painstaking, step-by-step process, but it is the only way to know more tomorrow than we do today.

9.7 CURRENT RESEARCH ON SOS IN ADVANCED IMAGING

Most medical perception studies have involved traditional radiography. Unfortunately, we still know relatively little about perception in advanced imaging. While most reports of SOS involve traditional radiography, clinical observations suggest a high frequency of SOS in CT imaging (Davis, 1996; Gurney, 1996; Kakinuma, 1999). White (White, 1996), in a retrospective analysis of patients with cancers overlooked at CT, noted that 43% of patients had a major distracting finding.

The thrust of our most recent research has been to bring the study of SOS effects into the realm of advanced imaging, looking for SOS effects in cervical spine and chest CT. We expect that substantial SOS effects in advanced imaging will yet be found experimentally, in correspondence with clinical observation.

There is wide agreement that a critical issue facing radiology is the growth in the volume of imaging studies being performed (Hillman, 2003; Rothenberg, 2005) and the increasing volume of data collected in each study (Andriole, 2004; 2006; Jacobson, 2006). As imaging technology changes, what we consider a diagnostic error also changes. The capability to find smaller and subtler lesions dramatically increases the volume of images that need to be reviewed.

> "The introduction of cross-sectional imaging and the maturation of computed tomography and magnetic resonance imaging have resulted in unprecedented increases in the number of images that must be interrogated to make clinical diagnoses."
>
> Jacobson, 2006

There is so much data that could be interrogated that an exhaustive search may not be possible in any practical time frame. The sheer number of images that are included in a single study may represent increased risk for failure of human perception (Armato, 2001). With increasing amounts of information available, we expect SOS to be a substantial contributor to observer error.

"What information consumes is rather obvious: it consumes the attention of its recipients. Hence a wealth of information creates a poverty of attention, and a need to allocate that attention efficiently among the overabundance of information sources that might consume it." Herbert A. Simon, 1971

REFERENCES

Anbari, M.M., West, O.C. (1997). Cervical spine trauma radiography: sources of false-negative diagnoses. *Emergency Radiology*, **4**, 218–224.

Andriole, K.P., Morin, R.L. (2006). Transforming medical imaging: the first SCAR TRIPTM conference: a position paper from the SCAR TRIPTM Subcommittee of the SCAR Research and Development Committee. *Journal of Digital Imaging*, **19**, 6–16.

Andriole, K.P., Morin, R.L., Arenson, R.L., *et al.* (2004) Addressing the coming radiology crisis—the Society of Computer Applications in Radiology Transforming the Radiological Interpretation Process (TRIPTM) initiative. *Journal of Digital Imaging*, **17**, 235–243.

Armato, S.G., Giger, M.L., MacMahon, H. (2001). Automated detection of lung nodules in CT scans: preliminary results. *Medical Physics*, **28**, 1552–1561.

Beam, C.A., Krupinski, E.A., Kundel, H.L., Sickles, E.A., Wagner, R.F. (2006). The place of medical image perception in 21st-century health care. *Journal of the American College of Radiology*, **3**, 409–412.

Berbaum, K.S., Brandser, E.A., Franken, E.A., Jr., *et al.* (2001). Gaze dwell times on acute trauma injuries missed because of satisfaction of search. *Academic Radiology*, **8**, 304–314.

Berbaum, K.S., Caldwell, R.T., Schartz, K.M., Thompson, B.H., Franken, E.A., Jr. (2007b). Does computer-aided diagnosis for lung tumors change satisfaction of search in chest radiography? *Academic Radiology*, **14**, 1069–1076.

Berbaum, K.S., Dorfman, D.D., Franken, E.A., Jr. (1989a). Measuring observer performance by ROC analysis: indications and complications. *Investigative Radiology*, **24**, 228–233.

Berbaum, K.S., Dorfman, D.D., Franken, E.A., Jr., Caldwell, R.T. (2000a). Proper ROC analysis and joint ROC analysis of the satisfaction of search effect in chest radiography. *Academic Radiology*, **7**, 945–958.

Berbaum, K.S., El-Khoury, G.Y., Franken, E.A., Jr., *et al.* (1988a). Impact of clinical history on fracture detection with radiography. *Radiology*, **168**, 507–511.

Berbaum, K.S., El-Khoury, G.Y., Franken, E.A., Jr., *et al.* (1994a). Missed fractures resulting from satisfaction of search effect. *Emergency Radiology*, **1**, 242–249.

Berbaum, K.S., El-Khoury, G.Y., Ohashi, K., *et al.* (2007a). Satisfaction of search in multi-trauma patients: severity of detected fractures. *Academic Radiology*, **14**, 711–722.

Berbaum, K.S., Franken, E.A., Jr. (2006a). Commentary: Does clinical history affect perception? *Academic Radiology*, **13**, 402–403.

Berbaum, K.S., Franken, E.A., Jr., Anderson, K.L., *et al.* (1993). The influence of clinical history on visual search with single and multiple abnormalities. *Investigative Radiology*, **28**, 191–201.

Berbaum, K.S., Franken, E.A., Jr., Caldwell, R.T., Schartz, K.M. (2006b). Can a checklist reduce SOS errors in chest radiography? *Academic Radiology*, **13**, 296–304.

Berbaum, K.S., Franken, E.A., Jr., Dorfman, D.D., *et al.* (1986). Tentative diagnoses facilitate the detection of diverse lesions in chest radiographs. *Investigative Radiology*, **21**, 532–539.

Berbaum, K.S., Franken, E.A., Jr., Dorfman, D.D., *et al.* (1990). Satisfaction of search in diagnostic radiology. *Investigative Radiology*, **25**, 133–140.

Berbaum, K.S., Franken, E.A., Jr., Dorfman, D.D., *et al.* (1991). Time-course of satisfaction of search. *Investigative Radiology*, **26**, 640–648.

Berbaum, K.S., Franken, E.A., Jr., Dorfman, D.D., *et al.* (1996). The cause of satisfaction of search effects in contrast studies of the abdomen. *Academic Radiology*, **3**, 815–826.

Berbaum, K.S., Franken, E.A. Jr., Dorfman, D.D., *et al.* (1998). Role of faulty visual search in the satisfaction of search effect in chest radiology. *Academic Radiology*, **5**, 9–19.

Berbaum, K.S., Franken, E.A., Jr., Dorfman, D.D., Barloon, T.J. (1988b). Influence of clinical history upon detection of nodules and other lesions. *Investigative Radiology*, **23**, 48–55.

Berbaum, K.S., Franken, E.A. Jr., Dorfman, D.D., Caldwell, R.T., Krupinski, E.A. (2000b). Role of faulty decision making in the satisfaction of search effect in chest radiography. *Academic Radiology*, **7**, 1098–1106.

Berbaum, K.S., Franken, E.A., Jr., Dorfman, D.D., Caldwell, R.T., Lu, C.H. (2005). Can order of report prevent satisfaction of search in abdominal contrast studies? *Academic Radiology*, **12**, 74–84.

Berbaum, K.S., Franken, E.A., Jr., Dorfman, D.D., Lueben, K.R. (1994b). Influence of clinical history on perception of abnormalities in pediatric radiographs. *Academic Radiology*, **1**, 217–223.

Berbaum, K.S., Franken, E.A., Jr., El-Khoury, G.Y. (1989b). Impact of clinical history on radiographic detection of fractures: a comparison of radiologists and orthopedists. *American Journal of Roentgenology*, **153**, 1221–1224.

Berlin, L. (1996). Malpractice issues in radiology: perceptual errors. *American Journal of Roentgenology*, **167**, 587–590.

Bruning, J.L., Kintz, B.L. (1987). *Computational Handbook of Statistics*, 3rd edn. Glenview, IL: Harper Collins, pp. 272–275.

Chakraborty, D.P., Winter, L.H.L. (1990). Free-response methodology: alternative analysis and a new observer-performance experiment. *Radiology*, **174**, 873–881.

Christensen, E.E., Murry, R.C., Holland, K., *et al.* (1981). The effect of search time on perception. *Radiology*, **138**, 361–365.

Craik, K.J.W. (1943). *The Nature of Explanation*. London: Cambridge University Press, p. 81.

Davis, S.D. (1996). Through the "retrospectroscope": a glimpse of missed lung cancer in CT. *Radiology*, **199**, 23–24.

Dixon, W.J. (1992). *BMDP Statistical Software Manual*, Vol 1. Berkeley, CA: University of California Press, pp. 155–174; 201–227; 521–564.

Dorfman, D.D., Alf, E., Jr. (1969). Maximum likelihood estimation of parameters of signal detection theory and determination of confidence intervals – rating method data. *Journal of Mathematical Psychology*, **6**, 487–496.

Dorfman, D.D., Berbaum, K.S. (1986). RSCORE-J: pooled rating method data: a computer program for analyzing pooled ROC curves. *Behavioral Research Methods and Instrumentation*, **18**, 452–462.

Dorfman, D.D., Berbaum, K.S. (1995). Degeneracy and discrete ROC rating data. *Academic Radiology*, **2**, 907–915.

Dorfman, D.D., Berbaum, K.S. (2000a). A contaminated binormal model for ROC data. II. A formal model. *Academic Radiology*, **7**, 427–437.

Dorfman, D.D., Berbaum, K.S. (2000b). A contaminated binormal model for ROC data. III. Initial evaluation with detection ROC data. *Academic Radiology*, **7**, 438–447.

Dorfman, D.D., Berbaum, K.S., Metz, C.E., *et al.* (1997). Proper receiver operating characteristic analysis: the bigamma model. *Academic Radiology*, **4**, 138–149.

Ericcson, K.A., Simon, H.A. (1980). Verbal reports as data. *Psychological Review*, **87**, 215–251.

Ericcson, K.A., Simon, H.A. (1984). *Protocol Analysis: Verbal Reports as Data*. Cambridge, MA: MIT Press.

Fidler, J.L., Fletcher, J.G., Johnson, C.D., *et al.* (2004). Understanding interpretive errors in radiologists learning computed tomography colonography. *Academic Radiology*, **11**, 750–756.

Franken, E.A., Jr., Berbaum, K.S., Lu, C.H., *et al.* (1994). Satisfaction of search in detection of plain film abnormalities in abdominal contrast examinations. *Investigative Radiology*, **29**, 403–409.

Gale, A.G., Johnson, F., Worthington, B.S. (1979). Psychology and radiology. In *Research in Psychology and Medicine*, Vol 1. London: Academic Press.

Gale, A.G., Worthington, B.S. (1983). The utility of scanning strategies in radiology. In *Eye Movements and Psychological Functions: International Views*. Hillsdale, NJ: Lawrence Erlbaum, pp. 169–191.

Green, D.M., Swets, J.A. (1962). *Signal Detection Theory and Psychophysics*. New York, NY: John Wiley & Sons, pp. 86–116.

Gurney, J.W. (1996). Missed lung cancer at CT: imaging findings in nine patients. *Radiology*, **199**, 117–122.

Halsted, M.J., Kumar, H., Paquin, J.J., *et al.* (2004). Diagnostic errors by radiology residents in interpreting pediatric radiographs in an emergency setting. *Pediatric Radiology*, **34**, 331–336.

von Helmholtz, H. (1867, 1963). *Handbook of Physiological Optics*. New York, NY: Dover.

Hillman, B.J. (2003). Economic, legal, and ethical rationales for the ACRIN National Lung Screening Trial of CT screening for lung cancer. *Academic Radiology*, **10**, 349–350.

Hochberg, J. (1970). Attention, organization and consciousness. In *Attention: Contemporary Theory and Analysis*. New York, NY: Appleton-Century Crofts.

Jacobson, F.L., Berlanstein, B.P., Andriole, K.P. (2006). Paradigms of perception in clinical practice. *Journal of the American College of Radiology*, **3**, 441–445.

Johnson, N.L., Kotz, S. (1970). *Continuous Univariate Distributions. 1: Distributions in Statistics*. New York, NY: John Wiley & Sons, pp. 186–187.

Kakinuma, R., Ohmatsu, H., Kaneko, M., *et al.* (1999). Detection failures in spiral CT screening for lung cancer: analysis of CT findings. *Radiology*, **212**, 61–66.

Kinard, R.E., Orrison, W.W., Brogdon, B.G. (1986). The value of a worksheet in reporting body-CT examinations. *American Journal of Roentgenology*, **147**, 848–849.

Kirk, R.E. (1982). *Experimental Design*, 2nd edn. Belmont, CA: Wadsworth, pp. 75; 429–455; 531–548.

Kuhn, G.J. (2002). Diagnostic errors. *Academic Emergency Medicine*, **9**, 740–750.

Kundel, H.L. (2006). History of research in medical image perception. *Journal of the American College of Radiology*, **3**, 402–408.

Kundel, H.L., LaFollette, P.S. (1972). Visual search patterns and experience with radiological images. *Radiology*, **103**, 523–528.

Kundel, H.L., Nodine, C.F., Carmody, D. (1978). Visual scanning, pattern recognition and decision making in pulmonary nodule detection. *Investigative Radiology*, **13**, 175–181.

Kundel, H.L., Nodine, C.F., Thickman, D., Toto, L. (1987). Searching for lung nodules: a comparison of human performance with random and systematic scanning models. *Investigative Radiology*, **22**, 417–422.

Lev, M.H., Rhea, J.T., Bramson, R.T. (1999). Avoidance of variability and error in radiology. *Lancet*, **354**, 272.

MacMahon, H., Engelmann, R., Behlen, F.M., *et al.* (1999). Computer-aided diagnosis of pulmonary nodules: results of a large-scale observer test. *Radiology*, **213**, 723–726.

McNemar, Q. (1969). *Psychological Statistics*, 4th edn. New York, NY: Wiley, pp. 54–58.

Metz, C.E. (1987). Current problems in ROC analysis. In *Proceedings of the Chest Imaging Conference 1987*. Madison, WI: The University of Wisconsin, pp. 315–336.

Most, S.B., Scholl, B.J., Clifford, E.R., Simons, D.J. (2005). What you see is what you set: sustained inattentional blindness and the capture of awareness. *Psychological Review*, **112**, 217–242.

Nakamura, K., Yoshida, H., Engelmann, R., *et al.* (2000). Computerized analysis of the likelihood of malignancy in solitary pulmonary nodules with use of artificial neural networks. *Radiology*, **214**, 823–830.

Neisser, U. (1976). *Cognition and Reality*. San Francisco, CA: W.H. Freeman.

Newell, A., Simon, H.A. (1972). *Human Problem Solving*. Englewood Cliffs, NJ: Prentice Hall Publishing.

Nodine, C.F., Kundel, H.L. (1987). Using eye movements to study visual search and to improve tumor detection. *RadioGraphics*, **7**, 1241–1250.

Nodine, C.F., Mello-Thoms, C., Weinstein, S.P., *et al.* (2001). Blinded review of retrospectively visible unreported breast cancers: an eye-position analysis. *Radiology*, **221**, 122–129.

Rasband, W.S. *ImageJ*. US National Institutes of Health, Bethesda, Maryland. http://rsb.info.nih.gov/ij/ (1997–2006).

Renfrew, R.L., Franken, E.A., Berbaum, K.S., Weigelt, F.H., Abu-Yousef, M.M. (1992). Error in radiology: classification and lessons in 182 cases presented at a problem case conference. *Radiology*, **183**, 145–150.

Rogers, LF. (1982). *Radiology of Skeletal Trauma*. New York, NY: Churchill-Livingstone, p. 1.

Rogers, L.F. (1984). Common oversights in the evaluation of the patient with multiple injuries. *Skeletal Radiology*, **12**, 103–111.

Rogers, L.F., Hendrix, R.W. (1990). Evaluating the multiply injured patient radiologically. *Orthopedic Clinics of North America*, **21**, 437–447.

Rothenberg, B.M., Korn, A. (2005). The opportunities and challenges posed by the rapid growth of diagnostic imaging. *Journal of the American College of Radiology*, **2**, 407–410.

Samei, E. (2006). Why medical image perception? *Journal of the American College of Radiology*, **3**, 400–401.

Samuel, S., Kundel, H.L., Nodine, C.F., Toto, L.C. (1995). Mechanism of satisfaction of search: eye position recordings in the reading of chest radiographs. *Radiology*, **94**, 895–902.

Simon, H.A. (1971). Designing organizations for an information-rich world. In *Computers, Communications, and the Public Interest*. Baltimore, MD: Johns Hopkins Press, pp. 37–52.

Sistrom, C. (2006). Radiology checklists, satisfaction of search, and the talking template concept. *Academic Radiology*, **13**, 922–923.

Smith, M.J. (1967). *Error and Variation in Diagnostic Radiology*. Springfield, IL: Charles C. Thomas, p. 27.

Snedecor, G.W., Cochran, W.G. (1989). *Statistical Methods*, 8th edn. Ames, IA: Iowa State University Press, pp. 146–147.

Swensson, R.G. (1988). The effects of clinical information on film interpretation: another perspective. *Investigative Radiology*, **23**, 56–61.

Swensson, R.G. (1996). Unified measurement of observer performance in detecting and localizing target objects on images. *Medical Physics*, **23**, 1709–1725.

Swensson, R.G., Hessel, S.J., Herman, P.G. (1977). Omissions in radiology: faulty search or stringent reporting criteria? *Radiology*, **123**, 563–567.

Swensson, R.G., Hessel, S.J., Herman, P.G. (1979). Detection performance and the nature of the radiologist's search task. In *Symposium on the Optimization of Chest Radiography* (Bureau of Radiological Health). Washington, DC: US Government Printing Office.

Swensson, R.G., Hessel, S.J., Herman, P.G. (1982). Radiographic interpretation with and without search: visual search aids the recognition of chest pathology. *Investigative Radiology*, **17**, 145–151.

Swensson, R.G., Hessel, S.J., Herman, P.G. (1985). The value of searching films without specific preconceptions. *Investigative Radiology*, **20**, 100–114.

Swensson, R.G., King, J.L., Gur, D. (2001). A constrained formulation for the receiver operating characteristic (ROC) curve based on probability summation. *Medical Physics*, **28**, 1597–1609.

Swensson, R.G., Theodore, G.H. (1990). Search and nonsearch protocols for radiographic consultation. *Radiology*, **177**, 851–856.

Swets, J.A., Pickett, R.M. (1982). *Evaluation of Diagnostic Systems: Methods from Signal Detection Theory*. New York, NY: Academic Press.

Tuddenham, W.J. (1962). Visual search, image organization, and reader error in roentgen diagnosis: studies of the psychophysiology of roentgen image perception. *Radiology*, **78**, 694–704.

Tuddenham, W.J. (1963). Problems of perception in chest roentgenology: facts and fallacies. *Radiologic Clinics of North America*, **1**, 227–289.

Voytovich, A., Rippey, R., Suffredini, A. (1985). Premature conclusions in diagnostic reasoning. *Journal of Medical Education*, **60**, 302–307.

White, C.S., Romney, B.M., Mason, A.C., *et al.* (1996). Primary carcinoma of the lung overlooked at CT: analysis of findings in 14 patients. *Radiology*, **199**, 109–115.

Xu, X.W., Doi, K., Kobayashi, T., MacMahon, H., Giger, M.L. (1997). Development of an improved CAD scheme for automated detection of lung nodules in digital chest images. *Medical Physics*, **24**, 1395–1403.

10

The role of expertise in radiologic image interpretation

CALVIN NODINE AND CLAUDIA MELLO-THOMS

10.1 INTRODUCTION

This chapter is about expertise in radiology. In the *Handbook of Medical Imaging, Volume 1* we reviewed a large segment of the literature on expertise in radiology (Nodine, 2000). Rather than provide the reader with an updated general review of research on expertise, we refer you to the recent *Cambridge Handbook on Expertise and Expert Performance* (Ericsson, 2006). The plan of this chapter is to concentrate on expertise in medical image interpretation. This will be accomplished by drawing primarily on image perception research in radiology in order to expand the reader's understanding of the perceptual and cognitive skills that define experts in radiologic image interpretation.

A hallmark of an expert in radiology is consistent and reliably accurate diagnostic performance in the task of interpreting medical images. Acquiring expertise in radiology requires not only specialized training and experience (both of which must involve some form of feedback) but also some degree of talent and motivation to reach a level of performance that would qualify as an expert. The amount and kind of training that it takes to make an expert in radiology is a key question in developing performance standards. Even after radiology residency training, the performance of novice radiologists rarely approaches that of more experienced radiologists. This is not surprising given a research framework that stresses the point that expert performance across a wide range of medical domains is rare (Ericsson, 2006).

The problem of studying expertise in radiology is further complicated by the fact that acquired expertise is domain-specific. Thus for example, acquiring expertise in interpreting chest radiographs does not directly transfer to interpreting mammograms. The structured knowledge learned from a massive amount of experience interpreting a wide range of variations in normal and abnormal chest patterns is largely non-transferable to other subdomains within radiology. There are some experts who possess unique skills, perhaps talents, and are able to cross the domain-specific divide. But, as a rule, this is atypical.

The subdomain-specific knowledge of expertise in radiology may have a plus side to it. Limiting knowledge to a specific standardized anatomic scene may facilitate tuning of specific perceptual and cognitive skills that give the expert a distinct advantage. Specifically tuned cognitive skills allow experts to capture the essence of a problem at first glance (Charness, 2001). Thus, subdomain-specific knowledge is a double-edged sword in both limiting and tuning perceptual-cognitive skills in radiology, as well as chess and other domains (e.g. visual arts).

Attempts to develop artificial intelligence systems (AI research) to simulate radiology expertise in interpreting medical images or assisting in the interpretation of medical images have so far failed to approach expertise standards in clinical practice, particularly in the context of mammography screening (Fenton, 2007; Gur, 2004). Perhaps this is due to the fact that these systems are often developed based upon the experts' answers to the developers' queries about how they performed a given task. Unfortunately experts are usually conscientiously unaware of the very fast perceptual/cognitive mechanisms that lead them to the decisions they make. Hence, in being questioned about their decision-making process, the experts may reason backwards, from the decision to the elements in the image that would have led to such a decision, regardless of whether or not such elements did in fact play a role in the actual decision-making process.

The new *Cambridge Handbook* is advertised as "the first handbook where the world's foremost 'experts on expertise' review scientific knowledge on expertise and expert performance," but research on expertise in radiology is mentioned only twice in 900 pages. Research on expertise in chess is characterized as representing the *Drosophila*, or model task environment, in which to study human cognitive processes (Ericsson, 2006). Certainly pioneering efforts by de Groot (de Groot, 1978), Chase and Simon (Chase, 1973), and Newell and Simon (Newell, 1972) brought a fresh view of the role of perceptual and cognitive processes of experts in playing the game of chess. However, there were also similar pioneering efforts going on in radiology by Llewellyn-Thomas and Landsdown (Llewellyn-Thomas, 1963), Tuddenham and Calvert (Tuddenham, 1961), and Kundel and Wright (Kundel, 1969) that brought fresh views about the role of perceptual and cognitive processes of experts interpreting radiographs that were running in parallel at about the same time.

10.2 EARLY PERCEPTUAL AND COGNITIVE RESEARCH: CONNECTIONS BETWEEN CHESS AND RADIOLOGY

There are important connections between cognitive psychology and AI research, which early on focused on expertise in the game of chess. Since both radiology and chess have a large perceptual skill component, research in radiology expertise can benefit from an understanding of chess expertise. As we will show, in both domains pre-existing knowledge about complex underlying problem-solving space and efficient search

The Handbook of Medical Image Perception and Techniques, ed. Ehsan Samei and Elizabeth Krupinski. Published by Cambridge University Press.

strategies (both visual search and memory search) are key theoretical issues. Furthermore, expertise in chess and radiology depend in similar ways on massive amounts of training and practice.

Arguably, the most striking research finding about expertise skill, whether it be in the domain of radiology or chess, is the expert's ability to extract the essence of a visual scene after a single glance or a few brief glances. This was a key finding in early research by Kundel and Nodine (Kundel, 1975), who tested experienced radiologists by flashing chest radiographs for 200 ms (to prevent visual inspection of the image through the use of visual scanning) and found that overall accuracy of interpretation was 70% correct. About the same time, following the lead of de Groot (de Groot, 1978), Chase and Simon (Chase, 1973) tested chess players by briefly presenting (5 sec) arrays of chess positions on a board and found that accuracy of recalling the test-game positions was greater for master chess players than for less skilled players.

Taken together, the main conclusion from both lines of research studying vastly different problems in different domains was that experts are able to search and grasp larger perceptual units by recognizing them faster as configurations, chunks, or templates than their less skilled counterparts. The mechanism behind this is largely due to perceptual/cognitive tuning. For example, in the context of radiology it entails a global response responsible for recognizing the radiologic scene as an anatomic pattern containing potentially relevant diagnostic information referenced against schematic forms of normal anatomic patterns stored in memory. This enhanced encoding of the radiologic scene facilitates perceptual discrimination and differentiation of abnormality and directs detailed focal visual search by checking fixations, according to Kundel and Nodine's model (first proposed in 1975). In the context of chess it has been shown that for chess masters, the mechanism is also related to perceptual/cognitive tuning, recognizing game configurations of chess pieces as chunks or templates, and indexing relevant moves represented in long-term memory (LTM), which triggers plausible counter-move strategies.

In both cases, rapid recognition is gained by long hours of perceptual learning studying configurations representing meaningful strategies and predicted outcomes in a problem-solving context. For the radiologist, variations in anatomic configurations take on diagnostic significance by identifying pathologic features. For the chess master, variations in configurations could also be interpreted as having diagnostic significance in terms of game plan by identifying plausible next "best" moves to a winning outcome. Thus, there is a lot to be learned about expertise from looking across these two research domains.

10.3 SEARCH ON DIFFERENT RESEARCH PATHS

The 30-plus-year-old radiology and chess studies were responsible for initiating a significant amount of research in expertise that continues today (Kundel, 1969; Llewellyn-Thomas, 1963; Tuddenham, 1961). Their significance lies in the similarities both in the problem-solving tasks and the mechanisms used to explain how the perceptual and cognitive processes of experts contribute to performance. Kundel and Wright (Kundel, 1969) was the first study of expertise by showing differences in scanning strategies as a function of levels of observers, from novices to experienced radiologists. Since inception, the two lines of research have taken different paths. Radiology research in the 1960s, when computers began to come into use in the medical field, focused on visual search because of the assumption that radiographic interpretation relies primarily on detailed visual sampling and discovery search for diagnostic interpretation and decision-making. Chess was also viewed as dependent on search, but the selection of game strategy and choice of chess moves was assumed to rely less on discovery search and more on memory search for recalling game patterns stored in LTM.

Because radiology was initially viewed as primarily a visual discipline and chess was initially viewed primarily as a cognitive reasoning discipline, the choice of research methodologies to study each was also different. Radiology research chose eye-tracking methodology from applied psychological research (e.g. detection of targets on radar screens (NAS, 1973)) to monitor eye fixations of radiologists during visual search of radiographs. Chess research chose a new methodology from AI research that used verbal protocol analysis to provide insights into memory and reasoning processes that were assumed to reflect game play. They represent different forms of complex human problem-solving, one medical and the other gaming.

Over the years different methodologies have been adopted by research in both domains. Because of the large perceptual component in study of expertise in both domains however, eye movement research has gained attention (de Groot, 1996). The search model in chess started out in the 1970s using think-aloud diagnostic verbal protocol analysis that accompanied chess game interpretation. This is contrasted with Kundel and others, who placed much emphasis on the development of effective visual search strategies for detecting lesions based on eye-fixation scanning patterns as a way of measuring perceptual skill of radiologists. Emphasis on importance of perceptual skills of chess players in choosing game-playing strategies led to eye movement studies for similar reasons.

10.4 COMPARING PERCEPTUAL AND COGNITIVE SKILLS IN CHESS AND RADIOLOGY

A review of expertise research in both radiology and chess domains was published in *The Handbook of Medical Imaging* (Beutel, 2000). Continuing research since that chapter was written has shed new light on the role of perceptual/cognitive skills in both radiology and chess. As already indicated, there is a lot of cross play between expertise research in radiology and chess (Wood, 1999). We will touch on some of this now.

In chess, level of expertise is quantified using the Elo rating system (named after the man who developed it), in which points are accrued based on games won at different levels of game difficulty and competition. An Elo rating of 2500 points is required to achieve Grand Master level, and there are at least three standard skill levels below this, where a standard skill level is 200 points based on nominal class intervals (standard deviations).

Only about 3% of competitive chess players reached Grand Master level in 2006. It takes an estimated 10 years of dedicated practice and tournament chess play to reach beginning Grand Master level. Garry Kasparov of Russia scored 2812 in the year 2000 to put him at the Super Grand Master level. At this level the probability of him winning a game against a 100th-ranked Grand Master competitor, who had a rating of 2616, was 75% wins and 25% losses (Ross, 2006).

In radiology, professionally prescribed medical training followed by practice and experience interpreting medical images are necessary to achieve a reliably acceptable skill level in performing diagnostic image interpretation. As with chess, years of continued practice are required to improve performance to a standardized level that maximizes diagnostic accuracy and minimizes error. For the task of accurately detecting breast cancer in mammograms, one study found that it takes about 25 years of post-residency image interpretation experience to achieve this level of expertise (Smith-Bindman, 2005).

With this much competition experience and practice, acknowledged chess master-level players would approach game performance standards roughly equivalent to diagnostic performance standards in radiology. However, diagnostic performance standards in radiology can be subjected to verifiable diagnostic outcomes that are quantitatively analyzable by decision theory, whereas win-loss game outcomes in chess can only be quantified probabilistically by sequential analysis of moves and countermoves against a hypothetical opponent having a known Elo rating.

A recent large-scale national study by Pisano *et al.* (Pisano, 2005) reported that the diagnostic accuracy of mammography screening performance of radiologists from 33 sites was 0.74 (for film) as measured by area under the curve (AUC) using receiver operating characteristic (ROC) analysis. Estimating the empirical operating point on the ROC curve of the sample of 42,760 women with verified cancer (Pisano, 2005, Figure 1A), this would translate into an approximate 75% true positive rate and a 15% false positive rate. Comparing win-loss performance of chess masters based on the Elo measure, which ranks win-loss game performance, with diagnostic accuracy performance of practicing radiologists based on an ROC measure is not actually possible given the different methods of quantifying performance in the two domains. Suffice to say, master-level chess players would probably perform at similar overall expertise levels in their domain as expert radiologists in their domain.

10.5 NEW CHALLENGES IN RADIOLOGY EXPERTISE RESEARCH

In the past, the study of expertise in radiology was largely limited to two subdomains, chest radiology and mammography. Both of these subdomains are composed of two-dimensional (i.e. flat) X-ray images of three-dimensional anatomic scenes (chest and breast, respectively). As new 3D imaging modalities come more into focus in radiology (such as computed tomography (CT) scans, magnetic resonance imaging (MRI), etc), research in developing expertise in these new subdomains is beginning to develop (Yang, 2002). Nonetheless, these new 3D imaging techniques present challenges to researchers designing experiments to study image understanding and expertise performance. New methodologies, particularly within an eye-tracking framework, will require precise synchronization between the eye tracker and the computer controlling the radiologist's scroll through a 3D image stack of chest or breast anatomy (Yang, 2002).

Expertise, as used in most of the research that we will be discussing, refers to performance of experienced radiologists that are used as observers under a variety of experimental conditions. Typically, these observers are full-time radiologists, often in academic settings, who have interpreted medical images for several years and thus have accrued a substantial amount of diagnostic image-reading experience. These "experts" enter experiments as part of a sample that includes less experienced observers such as community-based radiologists, radiology fellows, radiology residents, and other levels of radiology trainees. In the context of studying expertise, observers enter the typical radiology experiment at a level of skill to be contrasted with lesser skill levels based primarily on amount of training and image-reading experience. For example, Nodine *et al.* (Nodine, 1999) have shown that at the end of their residency novice radiologists who had read on average 650 mammograms during their mentor-guided training performed significantly below the average US radiologist with a five-year reading experience.

A similar research protocol exists in experiments studying expertise in chess. A key question is how does reading experience influence diagnostic performance in radiology? The answer to this question is clouded by the fact that not all image-reading experience in radiology is the same, and approaches to training are not consistent even across different subdomains of radiology. This is partly because, like in any other learning and mentor-guided training setting, the use, and therefore the effectiveness, of informative feedback or deliberate practice (Ericsson, 1996) is highly variable, especially when it takes place in the context of diagnostic medicine or chess tournament environments where opportunities for learning are very limited.

10.6 MODELING VISUAL SEARCH RESEARCH IN RADIOLOGY: IMAGE PERCEPTION REVISITED

The remainder of the chapter will attempt to outline major research findings since 2000 that have changed our perspective about modeling skilled performance in radiology. The original modeling came about from research carried out at Temple and Penn. Three aspects of modeling skilled performance of radiologists will be addressed. First is an important shift in how image perception initiates visual search. Second, a focus on object recognition, which we believe sets the stage for search by categorizing whether or not the overall scene is perturbed signaling abnormality. Third, some suggestions for training novices and less experienced radiologists to better acquire expertise skills in interpreting medical images.

Modeling visual search in radiology has been a primary goal of image perception research for many years. We and others (Gale, 1983) have emphasized the perceptual approach and primarily used eye tracking to study image perception. This methodology was chosen because we initially viewed image interpretation as primarily a visual-search problem.

The perceptual approach embedded in our modeling has perceptual learning as the basis for gaining expertise in what we consider to be primarily a visual problem-solving task. This approach is contrasted to the cognitive AI approach to the study of expertise in radiology (Lesgold, 1988). The cognitive AI approach viewed image interpretation within a logical reasoning perspective reflective of medical diagnostic problem-solving. As a result, the method of verbal protocol analysis was chosen to study what image features radiologists said they looked at, and how these identified features were translated into diagnostic statements during film reading. This cognitive AI approach, borrowed from chess research carried out at about the same time, drew primarily on human problem-solving and reasoning in modeling the radiologists' task.

One problem with using such a cognitively structured approach to describe the perceptual processing of a visual task (such as reading a medical image) is that some hypotheses are formed and discarded too quickly for storage in short-term memory, and this renders the observer unable to actually verbalize all hypotheses considered. In addition, particularly with expert observers, diagnostic impression on a case occurs very quickly. The perception of most medical images by experienced radiologists is instantaneous. Diagnostic interpretation takes somewhat longer. Even children as young as six years of age, when asked what they see, recognize the bones in a chest X-ray image (Kundel, 1969)! Hence, having made a tentative decision on a case a few seconds after image onset, the observer may then play the "feature-detection game" (Wolfe, 1989), whereby the observer draws logical conclusions from the image features that would have led to the decision and verbalizes these associations, regardless of whether or not they actually played a part in the decision-making process just because logic dictated it.

Both perceptual and cognitive approaches have recognized the importance of the interaction of perceptual and cognitive skills in expertise performance. The interplay of these two skills is difficult to untangle theoretically and experimentally. This is especially true when studying a complex task like radiographic interpretation and diagnostic decision-making that is vastly more complicated than many of the problems studied in the experimental psychology domain of human visual perception and cognition. This is also the case for studying chess play.

Kundel and Nodine started out studying the visual search strategies of radiologists by measuring what they looked at while scanning radiographic images to find abnormalities, typically lung nodules. Their modeling of the search process was largely influenced by experimental psychological research on visual search (Gregory, 1970, 1973; Hochberg, 1978; Neisser, 1976). But as research continued and different dependent eye-movement parameters were explored, the picture that evolved focused more and more on the global impression that initiated visual search and less on the scanning process functioning as a search and discovery strategy. Requiring search to discover whether an image did or did not contain an abnormality was inconsistent with what the eye movement data were indicating about how fast an abnormality was fixated after image onset. A detailed perceptual exploration and discovery-search process was just too cumbersome to fit the facts. Richard Gregory has made a similar point in reviewing physiological research on the eye-brain system by saying, "It seems implausible that the brain has to compute for vision... considering the cortex, processing speed is critical when there are combinatorial exponential explosions of possibilities (true in chess), or for searching through large ensembles (certainly true in interpreting medical images), thus SPEED can determine what is possible or impossible" (Gregory, 2001).

Kundel *et al.* (Kundel, 2007b) have found that recognition of abnormality in radiology is too rapid in most cases to have been caused by a search strategy requiring an exhaustive, or even optimally heuristic, search of the medical image, detection of potential targets, analysis for distinctive features stored in LTM, and selection of the best fitting target candidate. This perceptual/cognitive process emphasizes too much computation and too much of a bottom-up, parts-to-whole view of image perception. Rather, Kundel *et al.* (Kundel, 2007b) have come to the view that at least for experienced radiologists, image perception starts with a global or holistic impression designed to recognize whether the anatomic scene departs from an experience-based representation referenced in memory such as a normal anatomic schema. This prototype-like normal schema acts to bias perception by setting a perceptual threshold on how much the schematic normal prototype must be perturbed before signaling possible abnormality. The prototypic normal schema is developed as a result of massive numbers of negative case readings by radiologists relative to positive findings among which are a few truly abnormal cases (5–10% abnormal findings; see Pisano, 2005). Global or holistic recognition of a perturbed radiographic scene initiates discovery search by focusing visual attention immediately on the source of the perturbation. Regardless of the initial impression outcome, radiologists carry out inspection of the image by scanning and focally scrutinizing newly discovered findings. Recent evidence shows that in mammographic interpretation, experienced radiologists fixate true abnormalities very rapidly (within 1–2 sec of image onset), and most scanning that follows seems to be checking and confirming that no further hidden lesions are present (Mello-Thoms, 2005). This follow-up scanning typically takes 5–10 sec.

10.7 RECASTING THE ROLE OF VISUAL SEARCH IN RADIOLOGY

We have outlined a number of research findings in our discussion so far. We now turn to the research that has guided these ideas. In our chapter in the *Handbook of Medical Imaging* (Nodine, 2000), we reported on a number of studies that showed the importance of training and experience on the perceptual and cognitive skills of radiology experts.

An important factor in measuring expertise skill is time to decision. We reported that experts were faster than less experienced radiologists in making diagnostic decisions when interpreting mammogram test sets in an experimental laboratory setting (Nodine, 1999). In this study we developed a new method for measuring decision time during visual search. Observers viewed craniocaudal (CC) and mediolateral oblique (MLO) paired views of a single breast on a digital workstation.

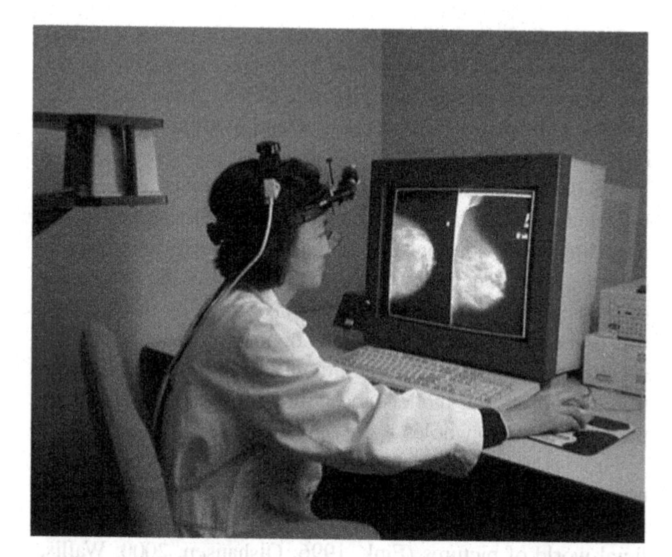

Figure 10.1 A mammographer scanning the display workstation containing CC (left) and MLO (right) images of one breast. Eye position was monitored during the initial interpretation phase with an eye-head tracker. The eye-tracker was worn on a headband by the viewer. It contained an infrared light source that imaged the viewer's left eye and a television camera that picked up limbus-reflection changes from the cornea as the eyes scanned the mammographic display. The eye-position data were sent to a control unit for analysis to determine the axis of gaze. The headband also contained a magnetic sensor that picked up changes in head movement within a magnetic field produced by a transmitter behind the viewer. Eye-head position data were integrated to determine the location of the axis of gaze on the mammogram.

Lesion identification and decision confidence were recorded when the observer clicked a mouse-driven pointer. This action also recorded the decision time, from image onset to lesion detection, that the lesion was localized.

Using this interactive computer display, we were able to measure decision times for detecting individual lesions in either CC or MLO views or both, or time to determine that the composite breast image was lesion-free. This method of measuring the time course of visual search indicated that experienced mammographers identified breast lesions rapidly and accurately. Experienced mammographers classified mammograms faster and more accurately than radiology residents and technologists. Less experienced observers depended more on scanning and discovery search to detect lesions. This led to longer search times that resulted in more errors than correct responses.

The new method of measuring the time course of detecting breast lesions in mammograms was further refined in a follow-up experiment (Nodine, 2001). In this experiment we attempted to test both initial impression and outcome of discovery search aspects of our image perception model by procedurally separating these two phases in the experiment. In addition, eye movements were measured during the initial classification phase.

Observers interpreted a two-view digitized breast image on a workstation, as shown in Figure 10.1, by first deciding whether the mammogram pair (CC and MLO views displayed together) showed findings suggestive of the presence of cancer. Eye movements were recorded during initial evaluation from image onset until the observer reported he/she was done, at which time the observer gave his/her initial impression, normal or abnormal, and confidence. The observers were allowed to change their initial decision, and take appropriate action consistent with their revised decision in the second phase. Regardless of the initial decision, normal or abnormal, a second phase followed, during which the observer inspected the images, marked lesion sites, if appropriate, on both CC and MLO views with a cursor, and indicated final decision confidence (for every lesion with an abnormal decision). The time course of both initial and final decisions was measured throughout both phases. The methodological advantage of this procedure was that in addition to obtaining decision-time measures for both phases, eye movement data obtained in the initial global overview phase could be related to lesion sites identified during the second confirm-and-localize phase.

We found that experienced mammographers detected 71% of true lesions within 25 sec of viewing the images, whereas it took less experienced trainees 40 sec to detect only 46% of true lesions, suggesting again that expertise performance is reflected in both speed and accuracy of decision-making. These findings were interpreted with a global impression/focal search model. Using positive predictive value (PPV = TP/(TP + FPn), where TP is the number of true lesions correctly reported and FPn is the number of lesions incorrectly reported on normal cases), performance began high and leveled off for both mammographers and trainees, but decision time was faster and PPV performance was higher for mammographers. This is shown in Figure 10.2.

Once again, as decision time increased, errors also increased, resulting in a significant decrease in diagnostic performance accuracy, particularly for trainees. Shorter median eye-fixation dwell times (which are viewed as a measure of information processing) for true positive decisions of trainees compared to

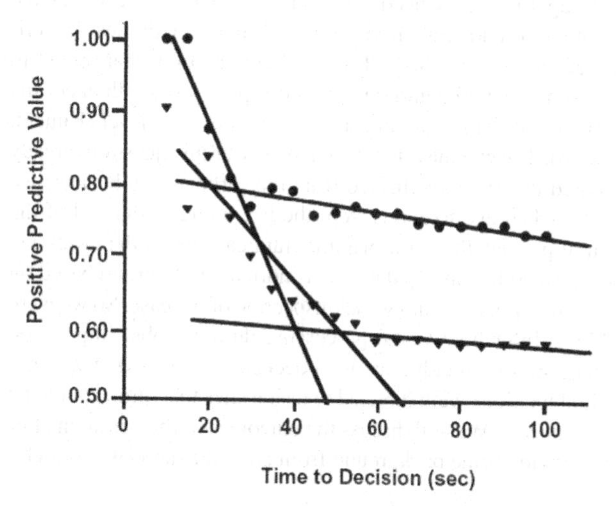

Figure 10.2 Positive predictive value for mammographers (circles) and trainees (triangles). Positive predictive value is a function of time to decision for final decision phase for cases with normal findings. Positive predictive value is calculated as TP/(TP + FPn). Positive predictive value performance begins high and levels off for both mammographers and trainees. Each set of positive predictive value data was fit by two linear regression lines. These lines cross at approximately 25 sec for mammographers and at approximately 40 sec for trainees.

mammographers indicated that trainees had difficulty perceptually differentiating true from false lesions. This was particularly true when the instances were in competition (as distractors) with true lesions.

These two studies begin to show a picture of the skills which experts in mammography use to perform image interpretation. By modeling image perception as global impression/focal search we have developed a methodological framework that helps to identify critical perceptual and cognitive components of expertise performance. Eye movement recording has provided a detailed view of the role that visual information processing plays in performing image interpretation. Our findings support and expand on earlier research by Christensen *et al.* (Christensen, 1981). So far we have seen that expertise skill contrasted with the skills that trainees acquire is closely related to the amount of image-interpretation practice and experience they acquire beyond basic resident training.

Eye-movement recording has enabled us to separate out the decision time required to categorize a test case containing both CC and MLO images of a single breast as either normal or abnormal, from the decision time required to examine and gather diagnostic evidence to confirm or disconfirm in favor of an alternative to the initial decision. The decision data were supported by localizing a potential abnormal finding, and coordinating this with eye-movement data during the initial decision phase. This required the observer to interact with pop-up menu displays and the mouse-cursor to point out lesion coordinates and indicate initial and final decisions with confidence levels. Measures of lesion detection supported by eye-fixation data and coordinated with cursor-localization provide verifiable evidence of the relationship between perceptual and decision-making aspects of the diagnostic interpretation performance of mammographers.

As the picture of how the time course of visual search affected decision-making became more clear, we realized that decision-making for experienced observers depended more on global impression and less on visual search than for the less experienced observers. This led to a follow-up study that aimed to determine how the initial impression specifically influences the ultimate decision outcome as a function of skill level (Kundel, 2007b). The eye movement data to answer this question already existed in the data collected from the earlier study because we had tracked eye fixations from the beginning to the end of the initial presentation of a mammogram case when the observers were asked to classify the case as normal or abnormal based on overall impression and visual inspection of the case. So we were able to infer how visual processing during global impression contributes to the ultimate final decision by looking at whether initial fixations were directed to lesion sites that turned out to be true cancers. We will discuss the outcomes of this research after we provide some background from both old and new research.

10.8 NEW PERSPECTIVES ON AN OLD PROBLEM: THE GESTALT VIEW OF PERCEPTION

There is a growing body of psychological literature that indicates that facial recognition occurs instantaneously and holistically, and that holistic recognition may extend to other stimuli such as objects and scenes (Peterson, 2003). This theoretical interpretation of research findings has a definite Gestalt flavor suggesting that the perceptual whole is recognized before analysis of any of its parts (Navon, 1977).

We have long supported this Gestalt view of perception, reflected best in James Gibson's theory (Gibson, 1979) in modeling how radiologists recognize abnormalities in medical images, although Gibson never talked about the perception of medical images. One is immediately struck by the fact that this global-focal Gestalt view runs opposite that advocated by much of the computer vision community (computer-aided detection/diagnosis) that models and builds image perception from parts (features) to wholes.

There is a great deal of speculation and controversy among neuroscientists about how the human eye-brain system recognizes objects in the visual-world environment, let alone the artificial world of pictures (Fink, 1996; Olshausen, 2000; Wallis, 1998). This neuroscience theorizing is stretched even further when extended to the recognition of abnormalities in medical images that have yet to be studied.

Presumably, a great deal of perceptual learning would be required to train the eye-brain system to recognize the various complex visual transformations of human anatomy produced by 2D imaging, not to mention 3D imaging. However, considering the length of training and extensive experience that radiologists acquire interpreting medical images, it is not unreasonable to assume that recognition of an abnormality, or lack thereof, may be carried out perceptually in the same rapid Gestalt or holistic manner as recognition of something as familiar as a face. An experienced radiologist may interpret, in just one year alone, 2500–4000 mammograms (Smith-Bindman, 2005). Is it possible, as we have suggested, that this perceptual experience could tune specific neurons in the radiologist's eye-brain system that would account for rapid holistic recognition of medical images?

There is tantalizing evidence that this may indeed be possible. In a recent unique neurological study, Haller and Radue (Haller, 2005) tested experienced radiologists and non-radiologists (neurologists) using functional MRI images of their brains and found selective enhancement of brain activation during visual scanning to detect changes in chest radiographs (albeit non-diagnostic in nature) vs. non-radiologic images (electron microscopic images). Neurological differences between radiologists and non-radiologists were associated with significantly faster responses although performance accuracy was low, which may have been due to the investigators' choice of atypical test–retest image manipulations.

Early modeling of image perception in radiology gave the global response an important role in visual search, namely that of analyzing visual input from the retinal image to determine the presence of gross deviations from normal, and selectively guide foveal scanning to examine and resolve these sources of concern in order to reach a diagnostic decision (Kundel, 1978). In this study they found that of 20 missed nodules, 30% were not fixated (scanning errors), 25% were fixated too briefly to be recognized as measured by short dwell (recognition errors), and 45% received sufficient dwell to be recognized but were not reported (decision-making errors). As image perception research

Table 10.1 *Results of mixture distribution analysis for time required to first fixate a cancer for all three research sites. Times are given in seconds*

	Fast component 1		Slow component 2	
n	Mean time (95% CI)	Proportion (95% CI)	Mean time (95% CI)	Proportion (95% CI)
400	0.72 (0.69–0.77)	0.57 (0.51–0.63)	4.75 (4.48–5.02)	0.42 (0.37–0.49)

progressed, we gave more and more credit to the importance of the global response. The modeling of Kundel and Nodine has now evolved to the point that we believe that global retinal analysis interacts with LTM to trigger a highly tuned holistic recognition of the overall radiologic scene capable of determining whether it is or is not perturbed, and directing visual attention to the source of the potential perturbation. The lesser emphasis on focal search, and the expanded role that the global response plays in image perception, can account for the speed/accuracy superiority of experienced radiologists.

10.9 HOLISTIC RECOGNITION BEGINS TO PLAY A ROLE IN RADIOLOGIC IMAGE PERCEPTION

If holistic recognition is responsible for the rapid and accurate performance of expertise in radiology, we should be able to test this from the eye-position data already acquired from our previous studies of breast imagers. Earlier studies contained eye-fixation data starting from onset of the mammogram case and continuing until an initial decision was made and rated for confidence.

Eye movement data from Nodine, Mello-Thoms, Kundel *et al.* (Nodine, 2002) formed the basis for testing the hypothesis that holistic recognition is an important perceptual/cognitive skill in radiology expertise (Kundel, 2007a). The data were obtained from a sample of three experienced mammographers, three mammography fellows, and three radiology residents undergoing mammography training, each viewing 40 difficult mammogram cases of which half contained lesions. The hypothesis was tested by determining at what point, in the course of decision time from image onset to initial classification of a mammogram case, the eye first fixated a true lesion and how this affected the final decision outcome.

Because we view the holistic recognition process as capable of characterizing the complete radiographic scene and pinpointing the spatial location of a perturbation within the scene, time to hit a true lesion with an eye fixation was assumed to require a rapid direct fixation to the target in less than 1 sec. This assumption was incorporated into the analysis of the eye-fixation time-to-hit data. Using this stringent operational definition of "hit time," it was found that time to fixate a true lesion was directly related to performance (index of detectability, d′).

The most experienced mammographer, also with the highest performance, recognized 55% of true cancers, and directly fixated the lesion within 1 sec of image onset. The worst performing and least experienced radiology resident recognized only 20% of true cancers with follow-up direct fixations to the lesion. This radiology resident actually fixated true lesions in 90% of the mammogram cases during subsequent discovery search without recognizing 70% of them.

Overall, 61% of true lesions were directly fixated by experienced mammographers; 50% directly fixated by fellows; and 42% directly fixated by residents within 1 sec of image onset. Thus, holistic recognition of cancers in mammograms was directly related to level of expertise as reflected in time (months, years) of experience interpreting mammograms for cancer detection. These findings, although limited to a small sample of observers, provide further support for the importance of perceptual skills in holistically recognizing perturbed radiographic scenes and rapidly focusing visual attention to the source of the perturbation.

In a follow-up study, these eye-tracking data on time to first fixate the true lesion were combined with eye-tracking data from two other sites investigating mammographic interpretation performance (Kundel, 2008). This increased the total number of readers to 19 and the mammographic case sample to a total of 400 cases from three research sites. Previous holistic modeling suggested a two-stage model with a fast and a slow component. In this model, the fast component represented the fixations that occurred as a result of the initial global impression, and the slow component represented fixations that occurred on cancers as a result of discovery search. Applying this holistic model to eye-tracking data from the three sites, it was possible to derive the time required to first fixate a cancer on a mammogram and examine the fit based on a two-component mixture distribution analysis. The results of the mixture distribution analysis for each of the two components of the model are shown in Table 10.1.

Overall, the pooled data from the three research sites show that 67% (297/443) of cancers were fixated within the first second of viewing. This means that within an estimated median average fixation duration of about 300 msec after image onset, the eye-brain system must have determined where, in image space, to launch the first fixation. The speed of this perceptual/cognitive process suggests parallel processing of the visual data from the entire radiologic scene acquired during the first few hundred milliseconds. Such rapid visual information uptake strongly supports a holistic analysis of the image scene. This finding defines the role of the fast perceptual component of the two-stage model. The remaining 33% of cancers (146/443) were first fixated later in search, defining the role of discovery search during the slow component.

Table 10.2 shows how the data fall within the two components of the mixture distribution model when broken down by percentage of fixations and percentage of true positives. Initial fixation of cancers does not guarantee that cancer recognition will take

Table 10.2 *Percentage of all cancers first fixated (Pct Fx) and the percentage of true positives (Pct TP) in each component given as mean and 95% confidence intervals for all readers and all research sites*

	Fast component 1		Slow component 2	
n (Readers)	Pct Fx	Pct TP	Pct Fx	Pct TP
19	60 (58–62)	63 (61–65)	40 (38–42)	52 (49–55)

place. During the fast component, of the 67% of cancers first fixated, 63% (186/296) were ultimately reported correctly as positive. Of the remaining 33% of cancers fixated during the slow discovery search component, only 52% (103/198) were correctly reported as positive. This difference in favor of the fast component over the slow component was consistent across all but 3 of the 19 observers. This finding suggests that holistic scene analysis plays a key role in early recognition of true lesions that are ultimately reported as true positive decisions. We interpret this to mean that the development of a highly tuned perceptual recognition skill is a critical factor in acquiring image interpretation expertise.

Twenty-eight years earlier, without the benefit of neuroscience research, Swensson, Hessel, and Herman (Swensson, 1982) proposed perceptual mechanisms of skilled observers that "function as a preliminary filter, automatically selecting specific pattern features for explicit attention and evaluation" (Swensson, 1980). This perceptual filter of the skilled radiologist "provides better differentiation between normal and abnormal radiographic findings than the explicit evaluation of the specific features directs his (the observer's) attention."

The findings reported above support those of Myles-Worsley *et al.* (Myles-Worsley, 1988), who earlier showed that expertise was related to faster and more accurate interpretation of chest X-ray images. They proposed two kinds of knowledge as being associated with radiology expertise: knowledge of "characteristic features of clinical normal exemplars," and knowledge of "a particular set of uncharacteristic features that signal pathology."

10.10 NEW MODEL OF VISUAL SEARCH

The modeling of the perceptual and decision-making processes that are reflected in the recent studies reviewed above has undergone several revisions since we last presented it in 2000. At that time we proposed a perceptual/cognitive framework for development of expertise in radiology as: SEARCH & DETECT–RECOGNIZE–DECIDE. This perceptual/cognitive framework emphasized the development of a heuristic search and detection strategy; fine-tuning of visual recognition of targets through practice; and decision-making that balanced the likelihood of being correct against the possibility and cost of being wrong.

A lot of research has been carried out since then, and the perceptual/cognitive framework has undergone a major shift in terms of the role that visual search plays in image perception. This is due to new evidence from several lines of research (radiology, cognitive science, neuroscience, and psychology) indicating that holistic recognition plays a key role in the perception of faces, objects, pictorial scenes, and now, as we have suggested, even perhaps radiographic images. Thus, a new framework in which to consider the development of radiology expertise is proposed in which search is de-emphasized and recognition becomes the lynchpin for visual search: RECOGNIZE & DETECT–SEARCH–DECIDE. This change suggests a new model, shown in Figure 10.3.

This new model is supported by psychological, neurophysiological, and computer vision research on object recognition. One research focus has been on visual recognition of objects, faces, and pictorial scenes of familiar objects (Peterson, 2003).

Carpenter *et al.* (Carpenter, 1998) developed a neural network model that employed what they refer to as a what-and-where filter for object recognition and scene understanding. In this computer vision model, object recognition is initiated preattentively top-down, by a learned prototype that primes target cells to selectively respond only when a match with target features of the visual input is identified. Their matching-rule modeling was based on current neurophysiological evidence (Carpenter, 1993).

More recent psychological and neurophysiological modeling of object recognition has provided strong support for the Carpenter *et al.* (Carpenter, 1998) model. In a reaction time experiment, Grill-Spector *et al.* (Grill-Spector, 2005) tested the three components of a model of object recognition: categorization, detection, and identification. This was done by briefly (17–167 ms) presenting objects from pictorial scenes containing animals, vehicles, or musical instruments. In separate experiments to test whether object detection precedes perceptual categorization and whether perceptual categorization precedes object identification, they contrasted each component against the other plus within-category identification by measuring response speed and accuracy. Their results showed that subjects responded as rapidly and accurately on categorization tasks as on detection-only tasks, and by the time objects were categorized the within-category identification seemed to have occurred.

These findings, plus evidence from their neurophysiological object recognition studies using functional MRI in human subjects, support the view that initial neuronal responses in specific cortical regions of the brain act by rapid initial categorization of the visual input to expedite identification by specifically tuning neuronal responses in testing object input against schematic representations in LTM (Carpenter, 1993). The tuning of these neuronal responses which act as so-called perceptual analyzers have been associated with selective enhancement of brain activation in radiologists during search and detection of changes in chest radiographs vs. non-radiologic images (Haller, 2005). Together, these studies support the theoretical position that rapid recognition of the global scene and detection of perturbations therein initiates image perception and visual search. The support for this model gives a clearer understanding of the fundamental perceptual/cognitive basis underlying visual expertise in radiologic image interpretation.

The holistic model shows a way that object recognition can work for radiologic interpretation by asking, in the words of Grill-Spector and Kanwisher (Grill-Spector, 2005) and the model of Carpenter *et al.* (Carpenter, 1998):

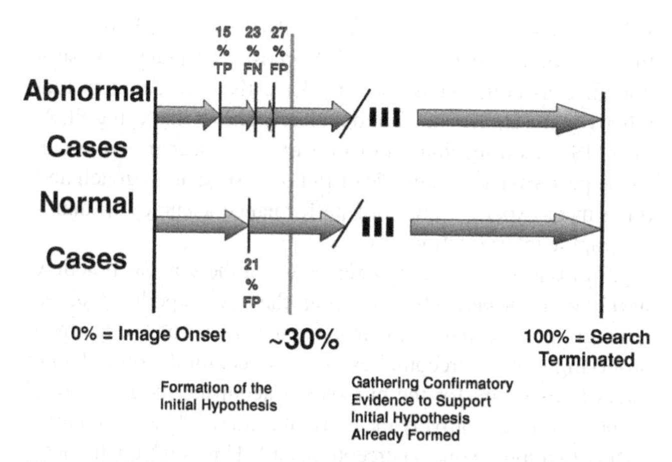

Figure 10.3 Schematic percentile representation of visual search behavior of expert breast imagers when reading mammograms, starting at image onset and terminating when the expert concludes his/her search. The figure shows both lesion-containing ("abnormal") and lesion-free ("normal") cases. For the abnormal cases, the experts' eyes first hit the location of true lesions that they correctly report (true positive responses), followed by the locations of true lesions that they do not report (false negative responses), followed by lesion-free locations of the background parenchyma that they interpret as containing a lesion (false positive responses). These events occur at, respectively, a little over one-tenth the overall duration of the search, around two-tenths the overall duration of the search, and a little short of three-tenths the overall duration of the search. For the normal cases, locations in the background parenchyma that are interpreted as containing lesions (false positive responses) are initially hit by the eyes around the first two-tenths of the overall duration of the search.

WHAT IS IT? (The "it" is the detected object such as a radiographic image, and categorizing the radiographic object as perturbed or not by indexing LTM.)

WHERE IS THE THING? (The "thing" is a perturbation that is either holistically identified as present or not, and perhaps whether or not it contains pathological features, followed by an appropriate search strategy to where the thing is.)

THAT MAKES IT WHAT IT IS? (The "what it is" is the holistically recognized radiographic object, with or without the identified perturbation, interpreted as a normal or abnormal radiographic image.)

Unfortunately, for the most part, neither the experimental psychology nor neurophysiological modeling of object recognition has touched base with image perception modeling in radiology. There are a few exceptions, such as the model presented in Figure 10.3, which has been the result of a long-term cross-disciplinary research effort between experimental psychologists and radiologists. Former graduate students such as Carmody, Krupinski, and Mello-Thoms, and researchers, primarily from the UK, have recently expanded on the model (Yang, 2002).

Part of the problem in experimental psychology has been a focus on non-applied problems and the testing of perceptual theory using very simplistic tasks. These simplistic tasks have limited generalizability to more complex applied problems such as those in radiology. Thus, progress in understanding the fundamental nature of image perception in radiology has been slow. In contrast, perceptual theorizing has exploded with advances in the "vision sciences" as evidenced by new journals and new investigators from many different disciplines entering the perceptual research arena. However, with all the expansion, very little spin-off has been assimilated into image perception research in radiology. More needs to be done to encourage the new generation of vision scientists to tackle complex applied research problems that bring perception, cognition, learning, vision science, neuroscience, and psychophysics all under one umbrella. It is almost as if newly trained vision scientists are afraid to tackle more complex applied problems. Perhaps the reach of image perception modeling summarized in Figure 10.3 is a case of "fools rushing in," but isn't this how scientific revolutions start (Kuhn, 1970)?

10.11 RADIOLOGY TRAINING: EXPERIENCE, EXPERIENCE, EXPERIENCE, BUT PRACTICE STILL MAKES PERFECT

If the goal of expertise research is to understand what role training plays in teaching new radiologists how to interpret radiologic images accurately and reliably, what have we learned so far? The answer to this the last time we asked the question in 2000 was that in the end, the key in developing perceptual/cognitive skill in radiographic interpretation is practice. The term practice was qualified, not as case-reading experience per se, but case-reading experience enriched by various forms of feedback in a question/answer format such as self testing, competitive testing, teaching colleagues, puzzling out difficult cases, participating in research both as subject and as investigator, etc.

Smith-Bindman *et al.* (Smith-Bindman, 2005) reported that physicians who interpreted a high volume of mammograms (2500–4000) annually, and had a higher screening focus as opposed to higher diagnostic focus, found more cancers and made fewer false positive responses than physicians who read fewer mammograms. This finding was interpreted as meaning that a select group of physicians, comprising only 8% of the total sample, were able to find a balance point (tradeoff) in making positive decisions that produced a reasonably high level of sensitivity without decreasing specificity. For this group, the tradeoff between sensitivity and specificity had no effect on overall performance, just different operating points on the same ROC curve.

These high-volume physicians begin to approach a criterion that would qualify them as experts as we have defined the term. The question remains, however, as to whether the performance of this sample reflects an increased confidence in the ability to recognize when a normal variant is uncharacteristically perturbed and draw upon knowledge of pathology to logically decide whether the perturbation signals abnormality. However, volume as used in the Smith-Bindman *et al.* study (Smith-Bindman, 2005) does not qualify as practice as defined by enriched case-reading experience of the question/answer type. What is lacking is some reference to whether the term volume included some form of feedback built in to the screening practice that informed the physicians about how accurately they were reading mammograms. Thus, in this study, the statistic of volume alone blurs the definition of reading experience into a meaningless concept.

In the Smith-Bindman *et al.* study (Smith-Bindman, 2005) one predictor of performance accuracy was the amount of case-reading experience derived from the Breast Cancer Surveillance Consortium and the American Medical Association Masterfile. Volume as a physician characteristic, along with years of experience, has been linked to accuracy in several large-scale statistical mammographic screening studies (Barlow, 2004; Beam, 2003). But as we have already said, volume per se leaves out critical information about the conditions under which case-reading experience is acquired. This makes it a hollow scientific term.

An investigation that looks at outcome relationships as a function of case-reading experience within a quasi-experimental framework is Esserman *et al.* (Esserman, 2002). This study compared UK with US radiologists using a standardized test set referred to as PERFORMS 2 (PERsonal PerFORmance in Mammography Screening) that was developed by Gale and Walker (Gale, 1991) and revised in subsequent years. This test is considered a teaching tool which is part of quality assurance in the National Health Service's Breast Screening Programme in the UK. It was designed specifically to provide feedback to radiologists each year as they developed image interpretation skills. Because of the emphasis on optimizing the effectiveness of mammography screening in the UK, sensitivity was defined as the percentage of cases correctly recalled, given those known to require recall, and specificity was defined as the percentage of cases correctly classified as normal, given all cases known to be normal. These terms are closer to being operationally defined than those used in the typical US volume studies cited above. Furthermore, required specificity is set at 0.90 in the UK to reduce the typical over-call-back rate that is more common in US mammography screening. We should point out that the call-back rate is higher in the US because in this country allegations of missed breast cancer continue to be the leading cause of malpractice litigation against radiologists and physicians with other specialties (Berlin, 2003).

The central hypothesis of the study was that case volume drives both sensitivity and specificity. The Esserman *et al.* study is different from the US volume studies considered above in a number of ways (Esserman, 2002). The performance of both UK and US radiologists was measured on the same test set to start. Data on reader volume were obtained from records provided by the facilities at the two observer sample sites. These data were used to assign three volume levels: low, medium, and high. The resulting breakdown of the final sample of 60 US radiologists by volume was 25% with a volume of 100 or fewer mammograms/month, 43% with a volume of 101–300 mammograms/month, and 32% with a volume of more than 300 mammograms/month. All 194 UK radiologists in the sample were dedicated high-volume readers from the national Breast Screening Programme (5000 mammograms/year).

The Esserman *et al.* results (Esserman, 2002) show that average sensitivity at specificity 0.90 (a stringent criterion for US radiologists) was about the same for high-volume US and UK radiologists. However, sensitivity was significantly lower for medium-volume and low-volume US radiologists compared to UK radiologists. Low-volume US radiologists also detected significantly fewer cancers than UK radiologists. Thus, when test conditions are held constant using the PERFORMS 2 test, and volume verified by facility records databases, the hypothesis that volume is a determinant of diagnostic accuracy was supported. This conclusion does not have the shortcomings that other volume studies have, and it clearly shows that using PERFORMS 2 as a teaching tool in a breast cancer program can shape performance standards of radiologists that approach and often meet expertise levels of performance accuracy in mammographic interpretation.

All of this argues that experience is not the same as practice. In the world of perceptual learning, the term experience alone refers simply to exposure of an image with no purpose other than observing it as a perceptual event. Practice, on the other hand, refers to exposure for the purpose of acquiring the perceptual event within a cognitively meaningful framework designed ultimately to improve one's perceptual skill. This is what Charness (Charness, 1996) implied by the term "deliberate practice" that was used to explain the acquisition of expertise in chess. All of these forms of enriched practice fit roughly into what leads to skilled performance. Or, to put the term "practice" into the cognitive framework of President George W. Bush:

> "See, in my line of work you got to keep repeating things over and over and over again for the truth to sink in, to kind of catapult the propaganda." George W. Bush, Greece, NY, May 24, 2005

The goal in acquiring expertise in image interpretation, as we have already outlined it, is the forming of a visual strategy for grasping the essence of a new medical image of some anatomic scene by referring to an embedded diagnostic schema (EDS) and deciding whether the perceptual impression of the case suggests that it is "normal" or "abnormal." This is a tentative hypothesis based on a global impression, but it is a good starting point for hypothetic-deductive diagnostic testing and decision-making. This EDS is the structural framework in LTM for matching the newly acquired radiologic scene to previously learned radiologic patterns or templates representing normal anatomic variants. This schema is perceptually primed by holistic recognition of the new radiologic scene that (a) enables the formation of a tentative diagnostic hypothesis (e.g. probably normal or probably abnormal) and (b) provides a plan for hypothesis testing that focuses visual attention on distinctive features (perturbations) that either confirm or disconfirm the initial tentative diagnostic hypothesis. Thus, perception, cognition, and LTM play key roles in diagnostic hypothesis testing in radiology, and the more practice the medical image interpreter acquires testing holistic recognition skills and enriching EDS, the better the result. The concept of visual perception as hypothesis testing has been the cornerstone of R.L. Gregory's perceptual theory for many years (Gregory, 1970). Hypothesis testing was also incorporated into the elementary perceiver and memorizer (EPAM) theory of memory and perception by Feigenbaum (Feigenbaum, 1962), and a new version of the memory theory called chunk hierarchy and retrieval structure (CHREST) by Gobet and Simon (Gobet, 2000) that simulates cognitive processes during the presentation and recall of chess positions.

We have shown over and over again that while both experienced and inexperienced radiologists gaze three times longer at false positive and negative decision sites, it is the true positive lesion sites that capture visual attention by a ratio of 2:1, and thus differentiate experienced from inexperienced observers. In most cases the ball game is over after 1–2 sec of viewing for an experienced radiologist, and scanning does not end here but continues to enrich interpretation and to satisfy search (Berbaum, 1990). The inescapable conclusion from all this is that experts are more perceptually sensitive to what constitutes a true perturbation in a medical image, and this perceptual sensitivity must come from seeing, classifying, storing, and receiving feedback on large numbers of normal and abnormal perturbation templates in LTM (not unlike the chess master).

The enhancement of sensitivity to simulated X-ray image nodules has been linked to perceptual learning by Sowden *et al.* (Sowden, 2000). Massed practice and feedback were proposed as critical factors in acquiring enhanced visual sensitivity. This is the basis of developing perceptual learning skills such as those described in this chapter. Without a repertoire of templates of image perturbations that have to some degree been verified, perceptual differentiation of true from false perturbations has no meaningful context within which to operate and error ensues. We have seen how accurately and reliably high-volume UK radiologists perform compared to low- and medium-volume US radiologists. Elmore and Carney (Elmore, 2002) criticize the findings of Esserman *et al.* (Esserman, 2002) on superficial grounds. These include limitations in the PERFORMS 2 testing situation when comparing US and UK radiologists because of the possibility that a "Hawthorne effect" may have unduly influenced performance. This seems unlikely since US radiologists performed *below* their UK counterparts who were simply taking the PERFORMS 2 as a routine part of the national Breast Screening Programme. The authors also cite legal and financial variations that may have an impact on sensitivity vs. specificity between the UK and US, but Elmore and Carney miss the most important point in the comparison: that UK film-readers from the beginning of training receive systematic feedback about their performance as the result of biannual testing by PERFORMS 2.

The PERFORMS 2 test is used in the UK as a teaching tool and feedback shows individual film-readers where they stand on performance relative to the UK performance goals, and more specifically where errors in decision-making occurred as part of a self-assessment exercise. This type of feedback is not only important within a perceptual learning framework of acquiring expertise skills, but is also important in motivating film-readers to become more accurate and reliable in interpreting medical images.

The lack of a teaching tool such as PERFORMS 2 in US mammography training and a required minimum volume of 480 mammograms per year may explain, in part, why high-volume UK film-readers generally outperformed US film-readers. As Elmore and Carney conclude, "practice probably does improve accuracy, but it may not make us perfect" (Elmore, 2002). No expert is perfect, but what Esserman *et al.* demonstrate is that film-readers from the UK's Breast Screening Programme can be trained to find a balancing point in performance that maximizes detection of true cancers while at the same time minimizing error. This tradeoff defines an established standard of expertise for screening mammography in the UK. Because the US has not developed national breast-screening standards, the performance variability of US radiologists is much greater than that of UK film-readers. Few would argue that reader variability of US radiologists could be reduced by utilizing a training program that includes a teaching tool like the PERFORMS 2 test. This type of training has a proven level of teaching effectiveness built in because it is specifically aimed at tuning the perceptual/cognitive skills of the image interpreter.

10.12 HUMAN EXPERTISE – HOW EXPERT RADIOLOGISTS SEE THE FOREST AND THE TREES

In the words of Anderson (Anderson, 1995), "one becomes an expert by making routine what to the novice requires creative problem-solving ability." In a variety of domains it has been shown that the acquisition of expertise involves working through a large number of cases, which then form a database that permits analogies and aids future problem-solving (Norman, 2006). This internal database allows experts to grasp the essence of a new image by referring it to deeply embedded anatomic schema held in LTM, and these schema serve as anchors for diagnostic hypothesis testing. In addition, a significant advantage of the hierarchically organized schemas that represent the experts' knowledge in their domain is that this arrangement allows for a marked reduction in use of working memory resources (Kalyuga, 2003). In contrast, trainees must acquire large amounts of information from the image and apply a problem-solving strategy in order to reach a diagnosis, a process that is slow due to the demands it places on the limited resources of working memory. It is widely accepted that any given subject can only retain about seven items (±2) in working memory at any given time (Miller, 1956), but these do not have to be simple items, such as single features. Hence, the speed advantage of the experts over the trainees is that, during the diagnostic reading of an image, experts can bring into working memory a single high-level representation of a schema associated with the perceived abnormality in the image, whereas the trainees must juggle several low-level representations of features associated with the same percept (Kalyuga, 2003).

This markedly different use of memory resources is a consequence of not only the limited exposure that trainees have had with normal and abnormal images but perhaps also of medical education itself, with its emphasis on clinical features and their association with disease. As pointed out by Kulatunga-Moruzi *et al.* (Kulatunga-Moruzi, 2004), a "forward-reasoning" strategy in which careful identification and consideration of all clinical features precedes any clinical diagnosis is advocated by numerous physicians involved in medical education. The main reason behind this choice is that it is believed that this strategy minimizes the risk that the trainee will prematurely choose an incorrect diagnosis before having examined all possible diagnoses supported by the clinical features present in the image. One problem with such an assumption is that clinical features can vary quite widely, and as such their detection in a given image may be dependent upon the trainees' overall perception of the case. In addition, extraction of too many irrelevant features may yield several competing diagnoses, which may worsen trainee performance. In contrast, experts rarely use more than four cues when making complex decisions (Harries, 2000). Furthermore, Norman *et al.* (Norman, 2000) have shown that when trainees were required to produce a list of clinical features before generating their diagnosis they produced a long list of competing diagnoses, obtained from the irrelevant features extracted, whereas when trainees produced the diagnosis first and then explained which features they used to generate this diagnosis, they were much more economical in their feature extraction strategy, and their performance reflected the better choice of features used to make the diagnosis.

The adoption of a "forward-reasoning" strategy in the training of radiology residents may be based upon the Dreyfus (Dreyfus, 1986) model of expertise development. According to this model, five stages separate trainees from experts. In the first stage, the "novice" shows a rigid observation of the taught decision-making rules. At this level, knowledge is treated without any reference to the actual context of the image, and the novice is unable to recognize the relevance of the perceived clinical features. Decision-making is based upon the analytical assessment of the problem, and demand on working memory resources is at its peak. At the next stage the "advanced beginner" is able to place information in context, but he/she is still unable to determine the relative importance of local and global attributes of the image. In the third stage the radiologist is deemed "competent." At this stage he/she can cope with the crowding in the image; that is, with the presence of several distractors in addition to any possible target. Recognition of relevance of clinical features is also observed at this level, and reading strategies become more standardized. At the next level, the radiologist is deemed "proficient" and holistic recognition of the context begins to occur. Together with holistic recognition, deviations from normal anatomic patterns become easier to identify, and decision-making requires fewer resources from working memory. At the final stage, the radiologist is deemed an "expert," and at this point no rules or guidelines are assumed to be necessary. The image is assessed holistically and decision-making is carried out intuitively in most situations. Analytical analysis of the image only occurs in novel situations or when holistic perception is somewhat impaired (for example, when the radiologist first encounters a new imaging technology or when the quality of the image is poor). Thus, the main characteristic of this model of acquisition of expertise is that automaticity is the defining characteristic of the final "expert" (Raufaste, 1998).

In contrast to the Dreyfus model, where cognitive processing characterizes the novices and perceptual processing characterizes the experts, the model of Lesgold *et al.* (Lesgold, 1988) suggests that perceptual processing is what defines the novices, whereas experts rely on a mix of both perceptual and cognitive processing. In Lesgold's model, at the first stage novices begin to develop some subsymbolic processing, as they are exposed to the first batch of images. Novices start the second stage by processing the image using only perceptual cues, and the outcome of this process is a set of hypotheses that are compatible with such perceptual cues. However, as Raufaste *et al.* (Raufaste, 1998) point out, a major problem with this strategy is discrimination insufficiency; that is, several diagnoses are often compatible with the data. Hence, the need to develop some type of cognitive-based processing becomes evident. In the later phase of the second stage of this model, this cognitive-based processing begins to develop, and decisions stop being based solely on perceptual cues and start to rely also on additional information obtained from the image. Finally, in the third stage, cognitive processing is fully developed, and although it is still applied to the output of the perceptual processing, now the two processes are in harmony, and the "expert" can select appropriate schemas from LTM and contrast them against perceived image elements. Lesgold *et al.* (Lesgold, 1988) also suggested that expert radiologists apply the chosen schema early, generally within 2 sec of image onset, and then continually test and modify the chosen schema in accordance with information acquired from the image.

Lesgold's finding that experts apply a chosen schema within 2 sec of image onset is in agreement with the Kundel *et al.* (Kundel, 2007b) eye-position track studies which showed that, when experts and novices visually scan two-view mammograms searching for breast cancers, the median time for the experts' eyes to hit the location of a cancer that they correctly report (i.e. true positive responses) is 0.96 seconds from image onset, whereas for the novices it is 2.15 seconds. In contrast, the median time for the experts' eyes to hit for the first time the location of a cancer that they visually inspect but fail to report (i.e. false negative responses) is 2.53 seconds from image onset, whereas for the novices this time is 4.12 seconds. The median times for the experts' eyes to first hit the location of a cancer found in the Kundel *et al.* study are remarkably similar to the median times found by Mello-Thoms *et al.* (Mello-Thoms, 2005), who also used two-view mammograms, but from a different patient population and using different experts to the Kundel *et al.* study.

Interestingly, all of these studies are in agreement with Mackworth and Morandi's (Mackworth, 1967) report that, when examining pictures, observers fixated locations deemed more informative within 2 sec of image onset. This led these authors to hypothesize that an initial analysis of the meaningful elements of an image can be made during early visual processing. Neisser (Neisser, 1967) took this a step further and suggested that image perception occurs in two stages: 1. a preattentive stage in which the entire image is processed in parallel and where a global, or holistic, view of what is being displayed is acquired; 2. a stage in which focal attention examines items or groups of items. Following Neisser's model, it has been suggested that the preattentive stage may bias the selection of the areas that will be subjected to further analysis using focal attention (Antes, 1981). This has been confirmed by Mello-Thoms *et al.* (Mello-Thoms, 2005), who modeled the behavior of expert breast imagers when reading mammograms as a two-stage process, as shown in Figure 10.4. In the first stage, after holistic recognition of the image as "perturbed" or "not perturbed" (in accordance with Kundel and Nodine's model; see Figure 10.4), the experts form an initial hypothesis about the case by selecting the appropriate schema in LTM. This stage entails the initial one-third of the total time allocated by the expert for visual search, and in this stage the locations of the majority of true lesions that the expert will be reporting (i.e. true positive decisions), as well as the locations of most lesion-free areas of the background that the expert has mistakenly identified as being a lesion and will be reporting (that is, false positive decisions), are fixated for the first time. This is followed by a second stage, which encompasses the remaining two-thirds of the time allocated for visual search, and in this stage the experts' eyes scan the image mostly to gather confirmatory evidence to support the initial hypothesis already

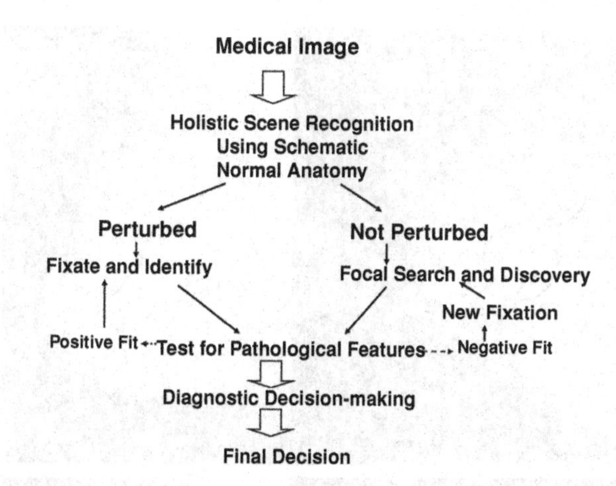

Figure 10.4 New image perception model of visual search in radiology. The new model is based on recent visual recognition research. The model shows the flow of visual information processing from presentation of a radiographic image to diagnostic decision. Initially, the radiographic scene is holistically recognized by indexing an encoded representation in LTM, referred to as schematic normal, to determine whether the radiologic scene matches the encoded representation or is in some way perturbed. If holistic matching determines that no uncharacteristic perturbations are present, focal discovery search for possible undetected perturbations ensues. If holistic matching determines that the radiologic scene is perturbed, the eyes are immediately directed to focus visual attention on the source of the perturbation. The model assumes that recognition of the holistic scene as perturbed or not perturbed initiates the appropriate follow-up focal search strategy. Identified perturbations are scrutinized for distinctive pathological features that signal abnormality. Decision-making follows.

formed. This is in agreement with Joseph and Patel (Joseph, 1990), who proposed that diagnostic hypotheses are selected early, and further processing is only used to eliminate some of the hypotheses initially generated, instead of for the generation of new hypotheses. It is in this second stage that the lesions detected as a result of the discovery search, which comprise the slow component of the Kundel *et al.* (Kundel, 2008) holistic model, are fixated.

Based upon these models, what is the main advantage in visual search that experts have over trainees? Cave and Batty (Cave, 2006) have suggested that experience may allow for better use of the information that is available in the preattentive representation of the image. This hypothesis is supported by Taylor (Taylor, 2007), who says that the real advantage of expertise is in the collection of information from the preattentive (or holistic) representation, rather than in the assessment of the features present in the image. In fact, recent studies using both magnetoencephalography and functional MRI (fMRI) to measure activity in the visual cortex have suggested that the early analysis of the preattentive representation of the image, which, by its very nature, is a low-spatial-frequency representation of the image contents, may serve as a top-down facilitator for the recognition of the image; in other words, this rapid process may be used by the visual system to generate predictions of what is being displayed, and these predictions are then matched with the results of the slower channels that process higher spatial frequencies (Bar, 2006).

According to this view, the bias generated by the processing of the low-spatial-frequency channels may influence the detailed processing of the high-spatial-frequency channels; that is, the radiologist's initial holistic representation of the image may in fact determine how successful he/she will be in detecting any true lesions that may be present in the case. This prediction has been confirmed in a study of breast cancer detection by expert breast radiologists using two-view mammograms (Mello-Thoms, 2005), which showed that the effect of the formation of the initial hypothesis is very strong. Namely, for the cases that contained a pathology-verified cancerous lesion, if that lesion was picked up as a "perturbation" in the expert's holistic representation of the image, then in 74% of those cases the expert correctly reported the true cancer and did not make any false alarms in the reading of the case. If, however, the initial hypothesis was incorrect and the area that attracted the expert's eyes initially was a lesion-free area of the parenchyma where the expert deemed a lesion to be present and reported this "perceived lesion" (i.e. a false positive response) then in only 5% of these cases was the expert able to revert course and detect and report the true cancer present in the images. In these cases, these true cancers were fixated very late in the course of search, and can thus be attributed to the discovery scanning suggested by Kundel *et al.* (Kundel, 2008). These findings also support Puri and Wojciulik's (Puri, 2008) view that expectation of the correct exemplar produces a benefit, whereas expectation of an incorrect exemplar results in a measurable cost.

In addition to being able to better assess the information gathered holistically, experts usually employ an elegant and efficient search strategy in which less time is devoted to fixating non-informative areas of the image (Krupinski, 1996; Kundel, 1972). Kundel and Wright (Kundel, 1969) suggested that "there may be strategies of search common to [experienced] radiologists." Their original insight was confirmed recently in a study that compared the visual scan strategy used by expert breast radiologists who had independently read a case set of digital mammograms (Mello-Thoms, 2008). The results showed very high agreement rates in the areas of the background fixated for at least 300 ms by these observers. For example, when the experts detected and reported a true cancerous mass, they agreed in 45% of the background areas fixated "before" first laying eyes on the cancer, and on 80% of the background areas fixated "after" first fixating the cancer. Moreover, the efficient search strategy employed by experts allows them to avoid unnecessary fixations in non-informative areas of the background, a strategy that speeds up their performance and that avoids the generation of too many conflicting diagnoses.

Furthermore, because of this economical strategy, it has been reported that radiologists cover on average 20% of the area of a chest radiograph with their high-resolution foveal vision (Kundel, 1972). Similar coverage rates have recently been found for expert breast radiologists reading two-view digital mammograms (Mello-Thoms, 2008). In this case it was shown that the experts cover, on average, 17% of the area of a two-view digital mammogram with their high-resolution, detailed foveal vision. Interestingly, coverage is a little smaller in the cases where the radiologists detect and report the true cancerous breast mass (15%) than it is in either the cases where they visually inspect but do not report the mass (19%) or where they report a lesion-free area of the background as being a malignant breast mass (19%).

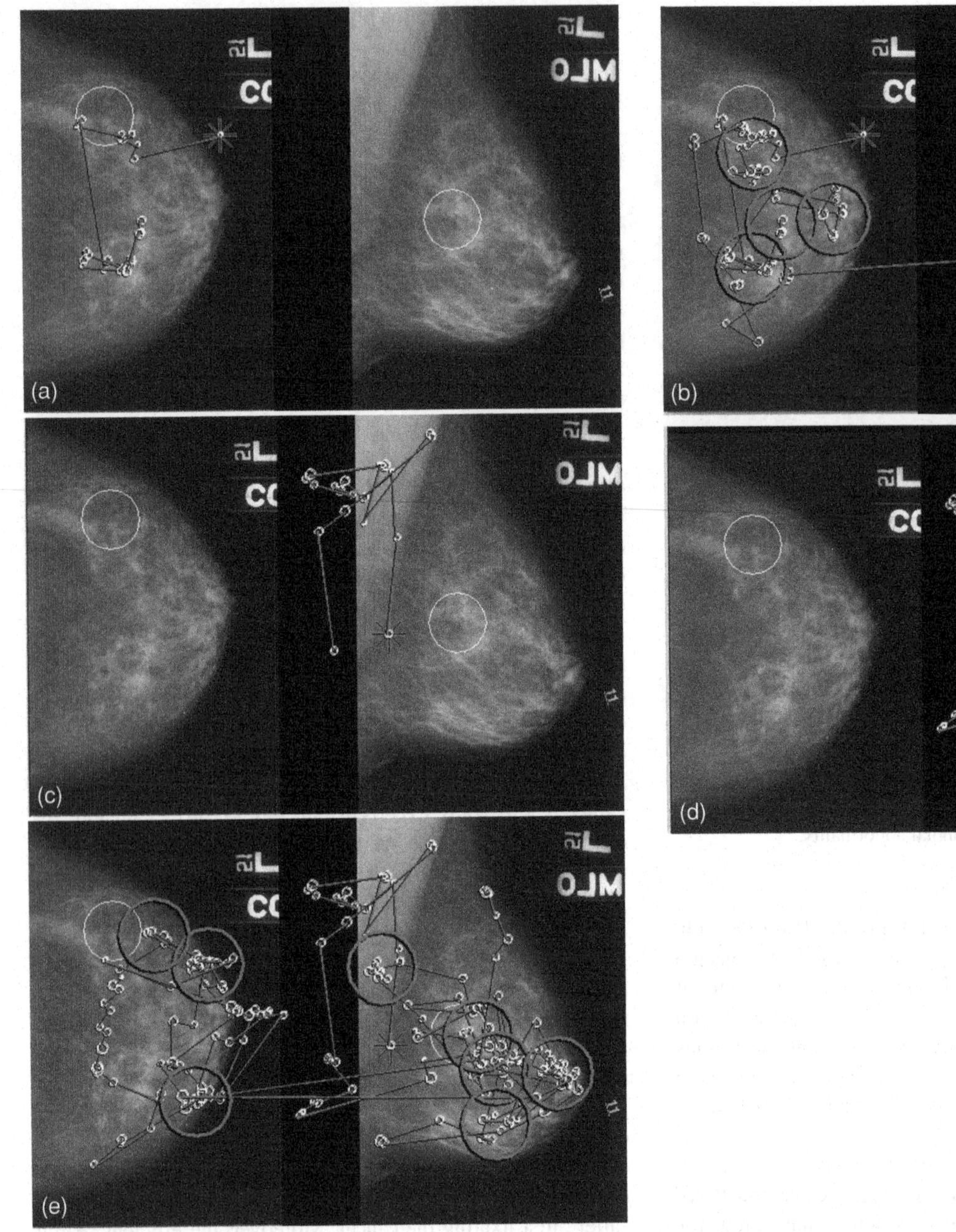

Figure 10.5 Comparative scanning of an expert breast imager and a radiology trainee. In this figure, the large light-colored circle marks the true lesion location in both views of the breast, whereas the large red circles mark the locations that attracted visual dwell for 1 sec or longer. The small white circles mark the locations fixated by the observers for 100 ms. (a) First 10 seconds of expert's search; note that by this time the lesion has already been fixated several times in the CC view, and the expert is seeking confirmation of the already established initial hypothesis. (b) Expert concludes search in 25 seconds. Lesion has attracted prolonged dwell in both views. (c) First 10 seconds of trainee's search; note that lesion has not attracted any visual attention in either view, and instead attention is concentrated in the pectoral muscle. (d) First 25 seconds of the trainee's search; attention is still solely concentrated in the MLO view, and the trainee seems to "look everywhere" but is unable to really find anything so far. (e) Trainee concludes search in 42 seconds. Although the lesion in the MLO view did attract prolonged visual dwell, there was no recognition of the finding, whereas the lesion in the CC view did not attract any amount of visual attention.

The evolution of the visual scan pattern between radiology trainees and expert radiologists has been previously reported (Kundel, 1972). In Figure 10.5 we show an example of this by contrasting the visual search strategy of an expert breast imager and a radiology trainee. Notice that within 10 seconds of image onset the eyes of the expert have already fixated several times the location of the lesion, whereas in the same time frame, the eyes of the novice are still wandering around the breast parenchyma, concentrating on the pectoral muscle. The expert's search is concluded within 25 sec, with long fixations (i.e. fixations that lasted 1 sec or longer) occurring in the location of the lesion in the two-views. Around the same time frame, the novice's eyes are still wandering around the MLO view of the breast, and no side-by-side comparative fixations have taken place, checking a possible "perceived finding" with anything in the CC view. The novice's search proceeds to 42 seconds, with no long fixations in the location of the lesion in the CC view, and scattered long fixations all through the breast parenchyma. From this example, it is clear that the expert was better at gathering information from the initial holistic representation, and as a result, the expert's search was faster and more efficient.

Direct comparison of the visual search strategy of an expert and that of a trainee, as shown in Figure 10.5, seems to suggest that the trainee indeed uses a "forward reasoning" strategy whereby all clinical features are examined in one view of the breast before proceeding to the second view. This behavior entails the use of a "mental protocol," where a list of features are checked and ruled in or out according to what the trainee perceives to be in the image. This begs the question, is this semi-automatic checking of features an example of "deliberate practice"? Or does it just constitute a natural follow-up to the passive, "trainee-as-a-receptacle-of-information" model used in most radiology residency programs (Gunderman, 2001)? The latter certainly seems to be the case, as "active" instruction using simulators has recently been shown to have a positive effect in the training of radiology residents (Monsky, 2002). In addition, Mueller *et al.* (Mueller, 2007) have recently shown that another type of "active" learning does improve performance of radiology trainees.

One may wonder why "active" learning may be more beneficial to radiology trainees than simple "passive" absorption of knowledge. Again, the answer to this question is centered on the process of transference of knowledge between the trainee's working memory, where information is transitionally stored, and their LTM, where truly absorbed information is kept. It is well known that when trainees can elaborate on knowledge at the time of learning, they are better able to later retrieve the learned information (Siamecki, 1978). Furthermore, as pointed out by Gunderman (Gunderman, 2000), a very effective process to transfer information to LTM is by relating the new information to be learned with some already learned information stored in LTM, in a process called encoding. In addition, an "active" learning strategy may help trainees better deal with the concept of uncertainty, because in this case learning goes beyond the quest for the "correct" (as opposed to the "wrong") answer and becomes more of a process where a given problem can be looked at from several different angles, where questions can arise more freely, and new approaches can evolve (Gunderman, 2001). Finally, in "active" learning, the trainee relies less on memorization than on understanding of what they need to know (Gunderman, 2003).

A significant part of the success enjoyed by expert radiologists seems to be based upon their capacity to perceptually learn how to identify potential lesion candidates. As Taylor (Taylor, 2007) points out, "this enhancement [in the expert's ability to recognize potential lesions] must reflect a change in the visual pathway in how low- and high-level representations are computed or processed." Dulaney and Marks (Dulaney, 2007) have recently determined that stimulus-specific learning can be associated with low-, local-level processing, whereas task-specific learning is related to high-, global-level processing. This finding lends support to the reverse hierarchy theory of perceptual learning (Ahissar, 2004), according to which changes in high-, global-level representations guide the selective tuning of low-, local-level perceptual processing. In other words, as the years of "deliberate practice" start to accumulate for the radiologists as they practice their craft, their ability to generate a more precise holistic representation of the image (i.e. a high-level representation) improves, and this more accurate holistic representation fine-tunes the perception of potential lesions (that is, low-level representations). In sharp contrast with this process, Kundel (Kundel, 2007a) has recently pointed out that very little time is spent instructing radiology trainees on how to gather the holistic impression of the image; instead, much of the instruction currently given is aimed at getting the trainees to agree on how to classify abnormalities. In other words, instruction in the high-level representation is substituted by instruction in the low-level processing, but this apparent attempt to improve the overall reading of the image (the high-level) with enhancements to the classification of lesions (the low-level) goes against the reverse hierarchy theory of perceptual learning (Ahissar, 2004).

How else do experts differ from novices? Novices tend to base their decisions mostly on the salient features of the image, whereas the experts use behavioral and functional information, in addition to the anatomical information obtained from the image, to activate schema that best fit their perception of the image (Hmelo-Silver, 2004). In addition, the acquired experience that experts have reading cases allows them to separate task-relevant from task-redundant information more effectively than novices, and to concentrate their efforts mostly on the relevant parts of the task (Haider, 1996). In radiology, it has been shown that experts zoom in on the location of abnormalities a lot faster than trainees, but at the same time, experts in this domain are much less aware of the details present in normal images (Myles-Worsley, 1988). Finally, as a result of having seen more cases, experts possess a larger internal database of "normal" and "abnormal" findings, and this aids them to recognize potential lesions quickly, as *a priori* predictions have been shown to expedite object recognition (Estes, 1994).

Hence, we may question whether all that is needed to produce expert radiologists is an appropriate learning environment, which can provide trainees with the necessary resources to read medical images, or whether expert radiologists are just naturally better at visual search tasks, and thus an innate degree of talent is necessary to acquire expertise in this field. In other words, are radiologists just "natural born searchers"? In our view this question has been elegantly answered in a study by Nodine and Krupinski (Nodine, 1998), in which radiologists and laypeople searched line drawings of Al Hirschfeld's "Nina" pictures and Martin Handford's "Waldo" ("Wally") pictures. The authors' hypothesis was that if radiologists were just inherently better searchers, they would find more Ninas and Waldos than laypeople. However, this is not what their results showed; there were no differences in overall detection performance between the two observer groups in either task, a finding that suggests that radiologists are not intrinsically better searchers. In light of these results, one may wonder whether radiologists are perhaps inherently better detectors of subtle signals. This has been examined by Rackow *et al.* (Rackow, 1987), who investigated whether radiologists were better at naturally detecting low-contrast signals embedded in anatomic structure. Again, there were no differences in performance between the radiologists and the laypeople. Finally, one may ask whether radiologists have higher innate visual acuity (such as that required to determine the 3D characteristics of objects presented as 2D projections) than radiology trainees and medical school students (Bass, 1990). Once more, the results showed no differences benefiting radiologists. In conclusion, the primary point made by these studies is that, while no one can deny the existence of true innate abilities in any area of expertise, this is not what mainly determines whether today's trainee will be tomorrow's expert.

REFERENCES

Ahissar, M., Hochstein, S. (2004). The reverse hierarchy theory of visual perceptual learning. *Trends in Cognitive Science*, **8**, 457–464.

Anderson, J.R. (1995). *Cognitive Psychology and Its Implications*, 4th edn. New York, NY: WH Freeman and Company.

Antes, J.R., Penland, J.G. (1981). Picture context effects on eye movement patterns. In *Eye Movements: Cognition and Visual Perception*, eds. Fisher, D.F., Monty, R.A., Senders, J.W., Hillsdale, NJ: Lawrence Erlbaum Publishers, pp. 157–170.

Bar, M., Kassam, K.S., Ghuman, A.S. *et al.* (2006). Top-down facilitation of visual recognition. *Proceedings of the National Academy of Sciences*, **103**, 449–454.

Barlow, W.E., Chi, C., Carney, P.A. *et al.* (2004). Accuracy of screening mammography interpretation by characteristics of radiologists. *JNCI*, **96**, 1840–1850.

Bass, J.C., Chiles, C. (1990). Visual skill: correlation with detection of solitary pulmonary nodules. *Investigative Radiology*, **25**, 994–998.

Beam, C.A., Conant, E.F., Sickles, E.A. (2003). Association of volume and volume-independent factors with accuracy in screening mammogram interpretation. *JNCI*, **95**, 282–290.

Berbaum, K.S., Franken, E.A., Dorfman, D.D. *et al.* (1990). Satisfaction of search in diagnostic radiology. *Investigative Radiology*, **25**, 133–140.

Berlin, L. (2003). Malpractices issues in radiology – breast cancer, mammography, and malpractice litigation: the controversies continue. *AJR*, **180**,1229–1237.

Beutel, J., Kundel, H.L., Van Metter, R.L. (eds.) (2000). *The Handbook of Medical Imaging, Volume 1: Physics and Psychophysics*. Bellingham, WA: SPIE Press.

Carpenter, G.A., Grossberg, S. (1993). Normal and amnesic learning: recognition memory by a neural model of cortico-hippocampal interactions. *Trends in Neuroscience*, **16**, 131–137.

Carpenter, G.A., Grossberg, S., Lesher, G.W. (1998). The what-and-where filter. Aspatial mapping neural network for object recognition and image understanding. *Computer Vision and Image Understanding*, **69**, 1–22.

Cave, K.R., Batty, M.J. (2006). From searching for features to searching for threat: drawing the boundary between preattentive and attentive vision. *Visual Cognition*, **14**, 629–646.

Charness, N., Krampe, R., Mayr, U. (1996). The role of practice and coaching in entrepreneurial skill domains: an international comparison of life-span chess skill acquisition. In *The Road to Excellence*, ed. Ericsson, K.A., Mahwah, NJ: Erlbaum Associates, pp. 51–80.

Charness, N., Reingold, E.M., Pomplun, M. *et al.* (2001). The perceptual aspects of skilled performance in chess: evidence from eye movements. *Memory & Cognition*, 1146–1152.

Chase, W.G., Simon, H. (1973). The mind's eye in chess. In *Visual Information Processing*, ed. Chase, W.G., New York, NY: Academic Press, pp. 215–281.

Christensen, E.E., Murry, R.C., Holland, K. *et al.* (1981). The effect of search time on perception. *Radiology*, **138**, 361–365.

de Groot, A.D. (1978). *Thought and Choice in Chess*. The Hague, Netherlands: Mouton Press.

de Groot, A.D., Gobet, F. (1996). *Perception and Memory in Chess*. Assen, Netherlands: Gorum Press.

Dreyfus, H.L., Dreyfus, S.E. (1986). *Mind Over Machine: the Power of Human Intuition and Expertise in the Era of the Computer*. New York, NY: The Free Press.

Dulaney, C.L., Marks, W. (2007). The effects of training and transfer on global/local processing. *Acta Psychologica*, **125**, 203–220.

Elmore, J.G., Carney, P.A. (2002). Does practice make perfect when interpreting mammography? *JNCI*, **94**, 321–323.

Ericsson, K.A. (1996). The acquisition of expert performance. In *The Road to Excellence*, ed. Ericsson, K.A., Mahwah, NJ: Lawrence Erlbaum Associates, pp. 1–50.

Ericsson, K.A., Charness, N., Feltovich, P.J. *et al.* (2006). *The Cambridge Handbook of Expertise and Expert Performance*. New York, NY: Cambridge University Press.

Esserman, L., Cowley, H., Carey, E. *et al.* (2002). Improving accuracy of mammography: volume and outcome relationships. *JNCI*, **94**, 369–375.

Estes, W.K. (1994). *Classification and Cognition*. New York, NY: Oxford University Press.

Feigenbaum, E.A., Simon, H.A. (1962). EPAM-like models of recognition and memory. *Cognitive Science*, **8**, 305–336.

Fenton, J.J., Taplin, S.H., Carney, P.A. *et al.* (2007). Influence of computer-aided detection on performance of screening mammography. *New England Journal of Medicine*, 1399–1409.

Fink, G.R., Halligan, P.W., Marshall, J.C. *et al.* (1996). Where in the brain does visual attention select the forest and the trees? *Nature*, **382**, 626–628.

Gale, A.G., Vernon, J., Millar, K., Worthington, B.S. (1983). Interpreting radiographs in a single glance. *Radiology*, **149**, 253.

Gale, A.G., Walker, G.E. (1991). Design for performance: quality assessment in a national breast screening programme. In *Ergonomics – Design for Performance*, ed. Lovesay, E.J., London, UK: Taylor & Francis.

Gibson, J.J. (1979). *The Ecological Approach to Visual Perception*. Boston, MA: Houghton Mifflin.

Gobet, F., Simon, H.A. (2000). Five seconds or sixty? Presentation time in expert memory. *Cognitive Science*, **24**, 651–682.

Gregory, R.L. (1970). *The Intelligent Eye*. New York, NY: McGraw-Hill.

Gregory, R.L. (1973). *The Confounded Eye in Illusion in Nature and Art*. New York, NY: Duckworth Co., pp. 49–95.

Gregory, R.L. (2001). Analog or digital? In *Looking at Looking*, ed. Parks, T.E., Thousand Oaks, CA: Sage, pp. 115–118.

Grill-Spector, K., Kanwisher, N. (2005). Visual recognition: as soon as you know it is there, you know what it is. *Psychological Science*, **16**, 152–160.

Gunderman, R.B. (2000). Illuminating the "black boxes" of learning and recall. *Academic Radiology*, **7**, 641–646.

Gunderman, R.B. (2001). Is technical school a good model for radiology residency? *AJR*, **177**, 1005–1007.

Gunderman, R.B., Williamson, K.B., Frank, M. *et al.* (2003). Learner-centered education. *Radiology*, **227**, 15–17.

Gur, D., Sumkin, J.H., Rockette, H.E. *et al.* (2004). Changes in breast cancer detection and mammography recall rates after the introduction of a computer-aided detection system. *JNCI*, **96**, 185–190.

Haider, H., Frensch, P.A. (1996). The role of information reduction in skill acquisition. *Cognitive Psychology*, **30**, 304–337.

Haller, S., Radue, E.W. (2005). What is different about a radiologist's brain? *Radiology*, **236**, 983–989.

Harries, C., Evans, J.S.B.T., Dennis, I. (2000). Measuring doctors' self-insight into their treatment decisions. *Applied Cognitive Psychology*, **14**, 455–477.

Hmelo-Silver, C.E., Pfeffer, M.G. (2004). Comparing expert and novice understanding of a complex system from the perspective of structures, behaviors, and functions. *Cognitive Science*, **28**, 127–138.

Hochberg, J.E. (1978). *Perception*, 2nd edn. Englewood Cliffs, NJ: Prentice-Hall.

Joseph, G.M., Patel, V.L. (1990). Domain knowledge and hypothesis generation in diagnostic reasoning. *Medical Decision Making*, **10**, 31–46.

Kalyuga, S., Ayres, P., Chandler, P., Sweller, J. (2003). The expertise reversal effect. *Educational Psychologist*, **38**, 23–31.

Krupinski, E.A. (1996). Visual scanning patterns of radiologists searching mammograms. *Academic Radiology*, **3**, 137–144.

Kuhn, T. (1970). *The Structure of Scientific Revolutions*, 2nd edn. Chicago, IL: University of Chicago Press.

Kulatunga-Moruzi, C., Brooks, L.R., Norman, G.R. (2004). Using comprehensive feature lists to bias medical diagnosis. *Journal of Experimental Psychology: Learning, Memory and Cognition*, **30**, 563–572.

Kundel, H.L., Wright, D.J. (1969). The influence of prior knowledge on visual search strategies during viewing of chest radiographs. *Radiology*, **93**, 315–320.

Kundel, H.L., LaFollette, P.S. (1972). Visual search patterns and experience with radiological images. *Radiology*, **103**, 523–528.

Kundel, H.L., Nodine, C.F. (1975). Interpreting chest radiographs without visual search. *Radiology*, **116**, 527–532.

Kundel, H.L., Nodine, C.F., Carmody, D.P. (1978). Visual scanning, pattern recognition and decision-making in pulmonary nodule detection. *Investigative Radiology*, **13**, 175–181.

Kundel, H.L. (2007a). How to minimize perceptual error and maximize expertise in medical imaging. *Proc SPIE Medical Imaging: Image Perception, Observer Performance and Technology Assessment*, **6515**, 1–11.

Kundel, H.L., Nodine, C.F., Conant, E.F., Weinstein, S.P. (2007b). Holistic component of image perception in mammogram interpretation: gaze-tracking study. *Radiology*, **242**, 396–402.

Kundel, H.L., Nodine, C.F., Krupinski, E.A., Mello-Thoms, C. (2008). Using gaze-tracking data and mixture distribution analysis to support a holistic model for the detection of cancers on mammograms. *Academic Radiology*, **15**, 881–886.

Lesgold, A., Rubinson, H., Feltovich, P., *et al.* (1988). Expertise in a complex skill: diagnosing X-ray pictures. In *The Nature of Expertise*, eds. Chi, M., Glaser, R., Farr, M., Hillsdale, NJ: Erlbaum, pp. 311–342.

Llewellyn-Thomas, E., Lansdown, E.L. (1963). Visual search patterns of radiologists in training. *Radiology*, **81**, 288–292.

Mackworth, N.H., Morandi, A.J. (1967). The gaze selects informative details within pictures. *Perception & Psychophysics*, **2**, 547–552.

Mello-Thoms, C., Hardesty, L., Sumkin, J., *et al.* (2005). Effects of lesion conspicuity on visual search in mammogram reading. *Academic Radiology*, **12**, 830–840.

Mello-Thoms, C., Ganott, M., Sumkin, J., *et al.* (2008). Different search patterns and similar decision outcomes: how can experts agree in the decisions they make when reading digital mammograms? In *International Workshop on Digital Mammography 2008, Lecture Notes on Computer Science 5116*, ed. Krupinski, E.A., Berlin, Germany: Springer-Verlag, pp. 212–219.

Miller, G.A. (1956). The magical number seven, plus or minus two: some limits to our capacity for processing information. *Psychological Review*, **63**, 81–97.

Monsky, W.L., Levine, D., Mehta, T.S. *et al.* (2002). Using a sonographic simulator to assess residents before overnight call. *AJR*, **178**, 35–39.

Mueller, D., Georges, A., Vashow, D. (2007). Cooperative learning as applied to resident instruction in radiology reporting. *Academic Radiology*, **14**, 1577–1583.

Myles-Worsley, M., Johnston, W.A., Simons, M.A. (1988). The influence of expertise on X-ray image processing. *Journal of Experimental Psychology*, **14**, 553–557.

National Academy of Sciences. (1973). Edited papers. *Symposium on Visual Search.* Washington, DC: NAS.

Navon, D. (1977). Forest before trees. The precedence of global features in visual perception. *Cognitive Psychology*, **9**, 353–383.

Neisser, U. (1967). *Cognitive Psychology*. Englewood Cliffs, NJ: Prentice-Hall, Inc.

Neisser, U. (1976). *Cognition and Reality*. San Francisco, CA: WH Freeman.

Newell, A., Simon, H. (1972). *Human Problem Solving*. Englewood Cliffs, NJ: Prentice Hall.

Nodine, C.F., Krupinski, E.A. (1998). Perceptual skill, radiology expertise, and visual test performance with NINA and WALDO. *Academic Radiology*, **5**, 603–612.

Nodine, C.F., Kundel, H.L., Mello-Thoms, C. *et al.* (1999). How experience and training influence mammography expertise. *Academic Radiology*, 575–585.

Nodine, C.F., Mello-Thoms, C. (2000). The nature of expertise in radiology. In *Handbook of Medical Imaging, Volume 1: Physics and Psychophysics,* eds. Beutel, J., Kundel, H.L., Van Metter, R.L., Bellingham, WA: SPIE Press, pp. 859–893.

Nodine, C.F., Mello-Thoms, C., Kundel, H.L. *et al.* (2001). Blinded review of retrospectively visible unreported breast cancers: An eye-position analysis. *Radiology*, **221**, 122–129.

Nodine, C.F., Mello-Thoms, C., Kundel, H.L., Weinstein, S.P. (2002). Time course of perception and decision making during mammographic interpretation. *AJR*, **179**, 917–923.

Norman, G.R., Brooks, L.R., Colle, C.L., Hatala, R.M. (2000). The benefit of diagnostic hypotheses in clinical reasoning: experimental study of an instructional intervention for forward and backward reasoning. *Cognition and Instruction*, **17**, 433–448.

Norman, G., Eva, K., Brooks, L., Hamstra, S. (2006). Expertise in medicine and surgery. In *The Cambridge Handbook of Expertise and Expert Performance*, eds. Ericsson, K.A., Charness, N., Feltovich, P.J., Hoffman, R.R., New York, NY: Cambridge University Press, pp. 339–353.

Olshausen, B.A., Field, D.J. (2000). Vision and the coding of natural images. *American Scientist*, **88**, 238–245.

Peterson, M.A., Rhodes, G. (2003). *Perception of Faces, Objects and Scenes: Analytic and Holistic Processes*. New York, NY: Oxford University Press.

Pisano, E.D., Gatsonis, C., Hendrick, E. *et al.* (2005). Diagnostic performance of digital versus film mammography for breast-cancer screening. *New England Journal of Medicine*, 1773–1783.

Puri, A.M., Wojciulik, E. (2008). Expectation both helps and hinders object perception. *Vision Research*, **48**, 589–597.

Rackow, P.L., Spitzer, V.M., Hendee, W.R. (1987). Detection of low-contrast signals: a comparison of observers with and without radiology training. *Investigative Radiology*, **22**, 311–314.

Raufaste, E., Eyrolle, H., Marine, C. (1998). Pertinence generation in radiological diagnosis: spreading activation and the nature of expertise. *Cognitive Science*, **22**, 517–546.

Ross, P.E. (2006). The expert mind. *Scientific American*, August, 64–71.

Siamecki, N., Graf, P. (1978). The generation effect: delineation of a phenomenon. *Journal of Experimental Psychology: Human Learning and Memory*, **4**, 592–604.

Smith-Bindman, R., Chu, P., Migloretti, D.L. *et al.* (2005). Physician predictors of mammographic accuracy. *Journal of the National Cancer Institute*, 358–367.

Sowden, P.T., Davies, I.R.L., Roling, P. (2000). Perceptual learning of the detection of features in x-ray images: A functional role for improvement in adults' visual sensitivity? *Journal of Experimental Psychology HPP*, **26**, 379–390.

Swensson, R.G. (1980). A two-stage detection model applied to skilled visual search by radiologists. *Perception and Psychophysics,* **27**, 11–16.

Swensson, R.G., Hessel, S.J., Herman, P.G. (1982). Radiographic interpretation with and without search: visual search aids the recognition of chest pathology. *Investigative Radiology,* **17**, 145–151.

Taylor, P.M. (2007). A review of research into the development of radiologic expertise: implications for computer-based training. *Academic Radiology*, **14**, 1252–1263.

Tuddenham, W.J., Calvert, W.P. (1961). Visual search patterns in roentgen diagnosis. *Radiology*, **76**, 255–256.

Wallis, G.M. (1998). Temporal order in human object recognition learning. *Journal of Biological Systems*, **6**, 299–313.

Wolfe, J.M., Cave, K.R., Franzel, S.L. (1989). Guided search: An alternative to the feature integration model for visual search. *Journal of Experimental Psychology, HPP*, **15**, 419–433.

Wood, B.P. (1999). Visual expertise. *Radiology*, **211**, 1–3.

Yang, G.Z., Dempere-Marco, L., Xiao-Peng, H. *et al.* (2002). Visual search: psychophysical models and practical applications. *Image and Vision Computing*, **20**, 291–305.

11

Image quality and its perceptual relevance

ROBERT SAUNDERS AND EHSAN SAMEI

The traditional methods physicists used for analyzing the performance of a digital radiography system were physical quantities like image resolution, noise, or the signal-to-noise ratio. These metrics were straightforward to measure and allowed for simple comparisons of two radiography systems. However, physicians were primarily interested in how well a system could help them perceptually. Would one system produce images that made a malignancy more conspicuous? Or did another mammography detector produce images that better highlighted architectural distortions? Therefore, this chapter reviews the physical metrics of image quality traditionally used and outlines how those might be extended toward measures of system performance that reflect the perceptual utility of an imaging system.

11.1 THEORY OF IMAGE QUALITY

The standard approach for defining image quality relies on linear systems theory (Blackman, 1959; Dainty, 1974; Dobbins, 2000). Linear systems theory decomposes an image into its spatial frequency components (Arfken, 1995). This decomposition is illustrated in Figure 11.1, where a relatively complicated one-dimensional signal is broken down into a small number of simple spatial frequencies.

11.1.1 Resolution

Linear systems theory explains how the detector acts on each individual spatial frequency. The modulation transfer function (MTF) explains the magnitude of the detector blur in terms of each of the frequency components (Giger, 1984a; Fujita, 1985; Workman, 1997; Bradford, 1999). Figure 11.2 illustrates this process for the signal from Figure 11.1. In this case, most of the low frequencies are present in the final signal, but the system suppresses the higher-frequency components. The effect of this is that the output signal is much smoother than the original input signal. This blurring effect can be seen in clinical images, as in Figure 11.3. This figure shows a simulated mammographic background with simulated microcalcifications. For an ideal detector (shown on the left), those microcalcifications are subtle but visible. A detector with a lower MTF (shown on the right) produces a blurrier image where the microcalcifications begin to fade into the normal parenchyma. This shows that it is important that the MTF can translate the spatial frequency components of abnormalities. For example, microcalcifications have very high-frequency details, due to their sharp edges, which means that a system must have a high MTF even at higher spatial frequencies in order to be visible on images. Other abnormalities may not have such high-frequency behavior and may not require as high an MTF. Therefore, the MTF needs for a particular system depend on the clinical task of interest.

11.1.2 Noise

Another factor that can obscure abnormalities is noise. That noise can be quantified using the noise power spectrum (NPS), which reveals the texture and magnitude of the noise (Giger, 1984b; Giger, 1986). Figure 11.4 illustrates how the NPS affects noise texture. White noise has a flat NPS where the intensity changes rapidly from pixel to pixel. In contrast, correlated noise has a differently shaped NPS; in this figure the noise has more low frequencies, which leads to a bumpier texture.

The relevance of the NPS for clinical images can be seen in Figure 11.5. This figure again shows a simulated mammographic background with simulated microcalcifications. White noise adds a grainy appearance to the image, but the microcalcifications remain visible. The same amount of correlated noise adds a more blobby appearance to the image and some of the microcalcifications are obscured by the noise. For clinical images, it is important not only to quantify the amount of noise but also its texture.

Both resolution and noise can obscure lesions. Which factor is more relevant for diagnostic performance: resolution or noise? If one system has better resolution while the other has better noise properties, which will produce the most clinically useful images?

11.2 IMPACT OF RESOLUTION AND NOISE ON DIAGNOSTIC ACCURACY

Before one is able to address the issue of perceptual image quality, one must decide what imaging task is important. Each imaging task presents different anatomy and different pathologies, each having different spatial frequency characteristics. For instance, searching for rib fractures will require looking for small high-frequency objects while searching for lung nodules means the physician is searching for small lower-frequency

The Handbook of Medical Image Perception and Techniques, ed. Ehsan Samei and Elizabeth Krupinski. Published by Cambridge University Press.

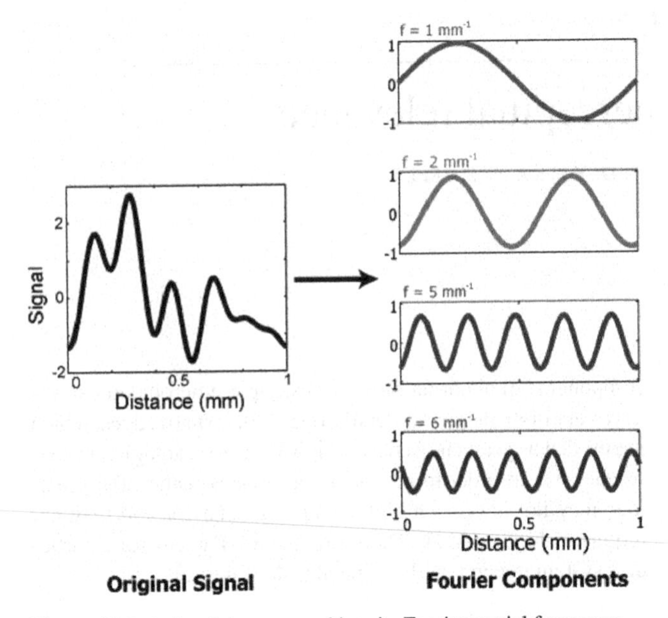

Figure 11.1 A signal decomposed into its Fourier spatial frequency components.

bumps. This section will discuss methods that were used to assess perceptual performance for two imaging tasks: the search for nodules in chest radiography and the search for masses and microcalcifications in mammography.

11.2.1 Chest imaging

This work examined how lung nodule detection was affected by the resolution and noise of two different types of flat-panel detectors: a direct selenium-based detector and an indirect cesium iodide-based detector (Saunders, 2003; Saunders, 2004). These resolution and noise properties are illustrated in Figure 11.6. The selenium-based detector offers higher resolution at the expense of higher noise, while the cesium iodide-based detector offers lower resolution with better noise properties. As well, the noise in the selenium system is approximately white (constant across spatial frequency), while the noise in the cesium iodide system is more correlated.

Figure 11.7 shows a lung image that had been modified to have the same resolution and noise as these two systems. The cesium iodide image has less noise but the anatomy appears more blurred, while the selenium image has sharper detail at the expense of greater noise.

To examine how these resolution and noise factors affected human performance, an observer experiment was conducted. Small simulated lung nodules were inserted into the centers of regions containing actual lung anatomy. This image set was then shown to experienced radiologists who scored the images based on whether they believed that image contained a lesion or not. Those scores were analyzed using receiver operating characteristic (ROC) analysis to compute a measure of overall accuracy (Metz, 1986; Metz, 2000).

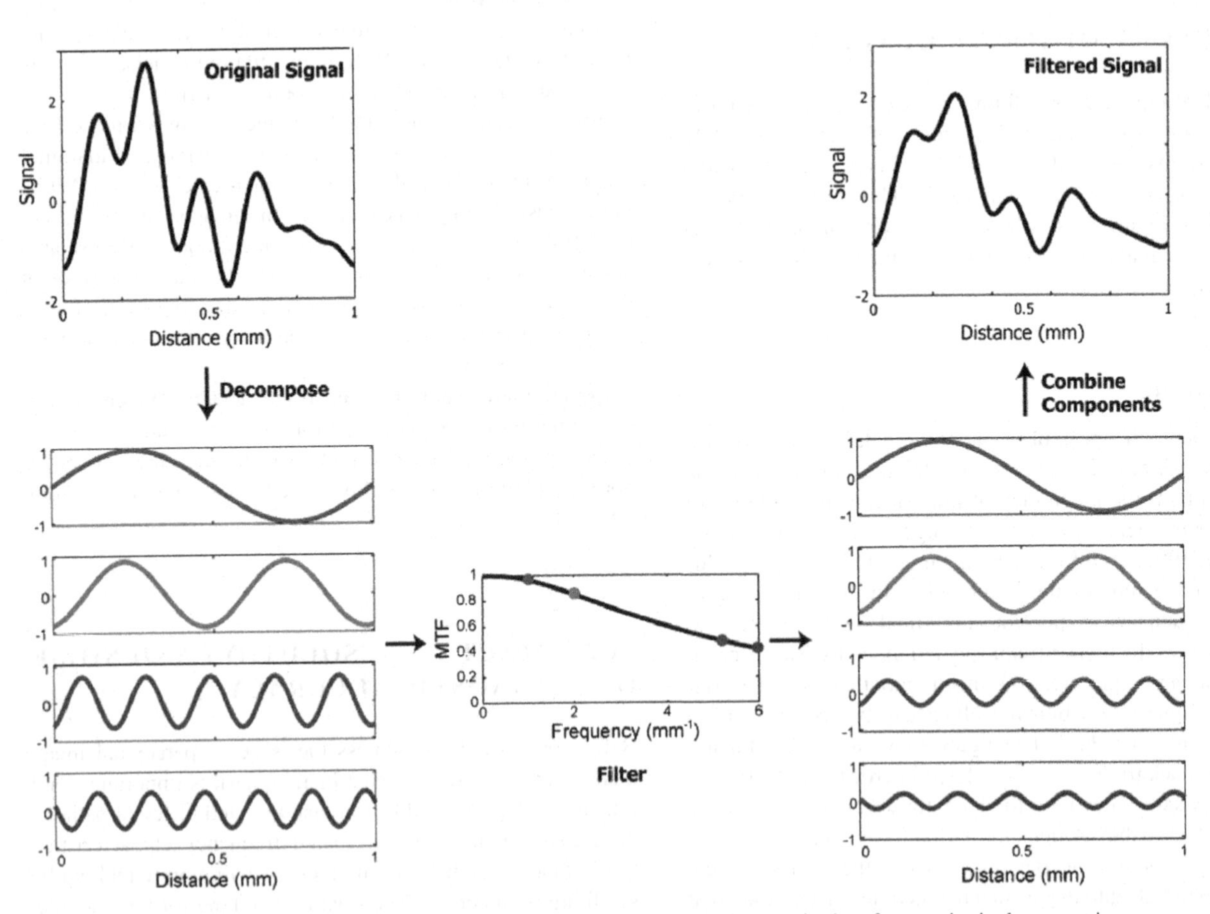

Figure 11.2 Illustration of how the modulation transfer function translates input frequencies into frequencies in the output image.

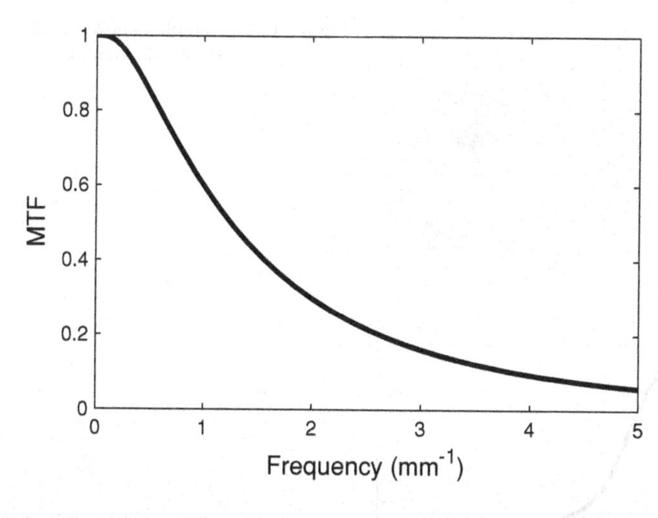

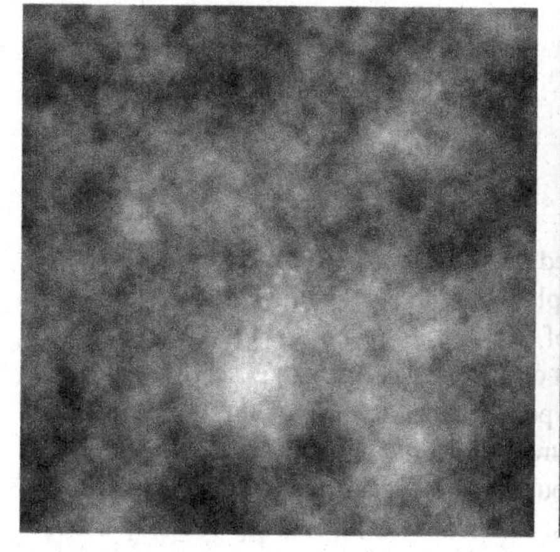

Figure 11.3 Example of how the MTF affects image quality in clinical images.

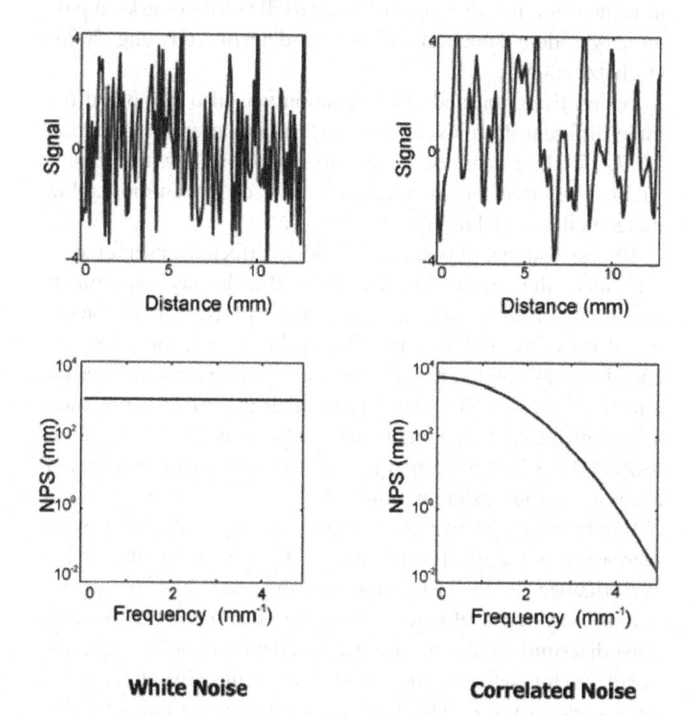

Figure 11.4 Illustration of how the noise power spectrum affects noise texture.

The results of this experiment are shown in Figure 11.8. As expected, accuracy, reflected by A_z, increased with increasing nodule size. Observers tended to perform slightly better on the selenium system than the cesium iodide system, although this difference was only statistically significant for the 2.75 mm case. This suggests that resolution appears to be slightly more important for lung nodule detection than noise, although this controlled experiment would need to be verified by clinical trials.

11.2.2 Mammography

The second imaging task investigated was the detection and discrimination of breast masses and microcalcifications in mammography. The detection and discrimination performance was first compared across different medical displays (Saunders, 2006c). These displays included a medical-grade liquid crystal display (LCD), a cathode-ray tube (CRT), and a CRT whose resolution had degraded like that of an aged CRT. The resolutions of these three displays are illustrated in Figure 11.9. The LCD has the highest resolution, then the normal CRT, and finally the degraded CRT (Saunders, 2006b). The effects of this resolution on a mammographic region can be seen in Figure 11.10.

A database of several thousand mammograms was created by inserting simulated masses and microcalcifications into real digital mammograms. Those images were then processed to

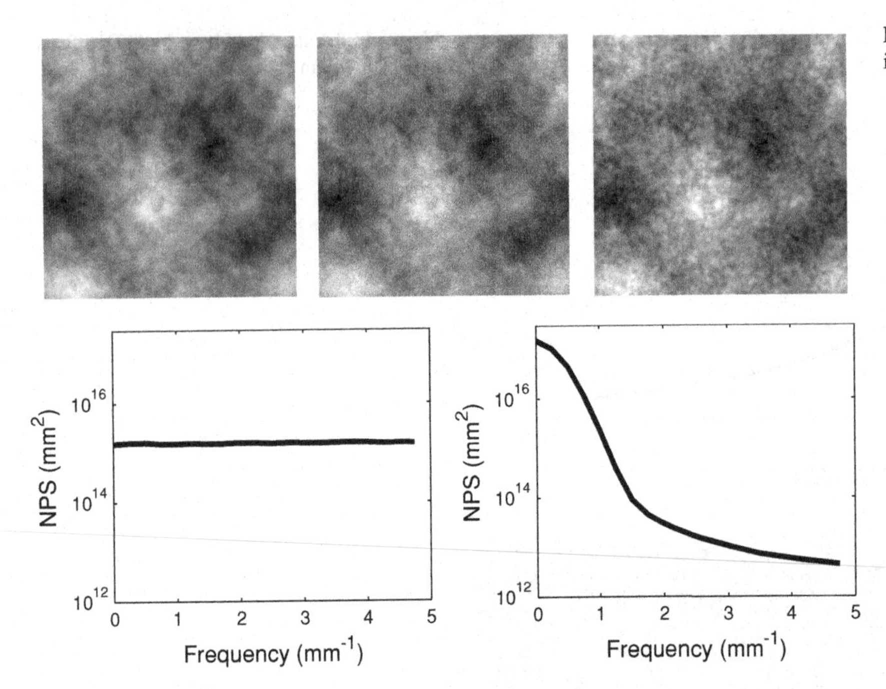

Figure 11.5 Example of how the NPS affects image quality in clinical images.

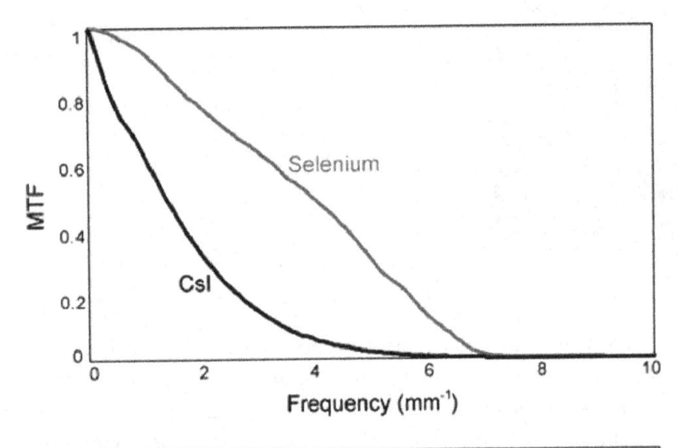

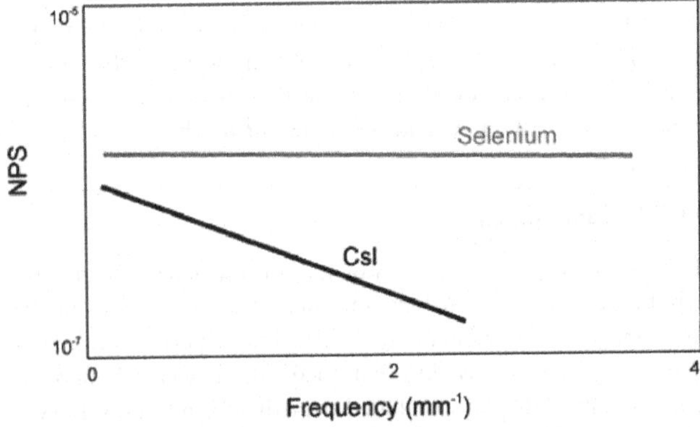

Figure 11.6 MTF and NPS of two flat-panel chest radiography detectors.

have similar appearance to clinical mammograms. Finally, experienced breast imaging radiologists viewed these images on each of the three displays and rated whether the images contained a benign mass, a malignant mass, microcalcifications, or no lesion. The results were analyzed for overall accuracy and accuracy at four different clinical tasks.

The results showed that observer performance was very similar for the three displays. Figure 11.11 illustrates overall accuracy as a function of display, while Figure 11.12 shows accuracy at four different clinical tasks. This suggests that observers would have similar performance on any of the three medical-grade displays. Figure 11.13 illustrates these results and compares them to previous work on breast mass detection (Krupinski, 2005; Saunders, 2006c). This also compares to previous work in chest radiography, which did not find statistically significant differences between LCDs and CRTs for the tasks of pulmonary nodule detection and catheter detection (Hwang, 2003; Oschatz, 2005).

Lesion detection and discrimination was also compared for three different dose levels: full, half, and quarter dose (Samei, 2007). As dose decreased, noise increased as shown in Figure 11.14. The effect of this increased noise on a mammographic image is illustrated in Figure 11.15.

To test whether this increased noise affected observer performance, the image database from the display experiment was used. In this case, the images were processed with additional noise to emulate the reduced dose conditions. Experienced breast-imaging radiologists then viewed these images on a medical-grade LCD and rated whether the images contained a benign mass, a malignant mass, microcalcifications, or no lesion. The results were again analyzed for overall accuracy and accuracy at four different clinical tasks.

Figure 11.16 shows that overall accuracy declined with increasing noise (decreasing dose). As seen in Figure 11.17, performance at the tasks requiring the most high-frequency details was most affected: microcalcification detection and mass discrimination. The task that relied most on low-frequency detail, malignant mass detection, was minimally affected by the increased noise. This implies that increased noise mostly affects the perception of high-frequency details, as those details represent the defining features of microcalcifications and distinguish malignant and benign masses (Samei, 2007). This

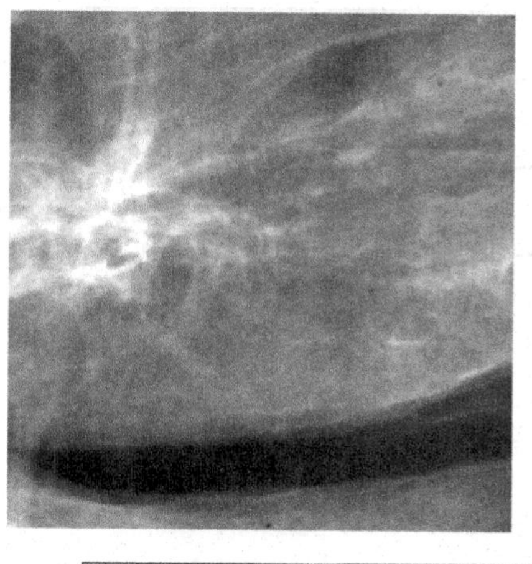

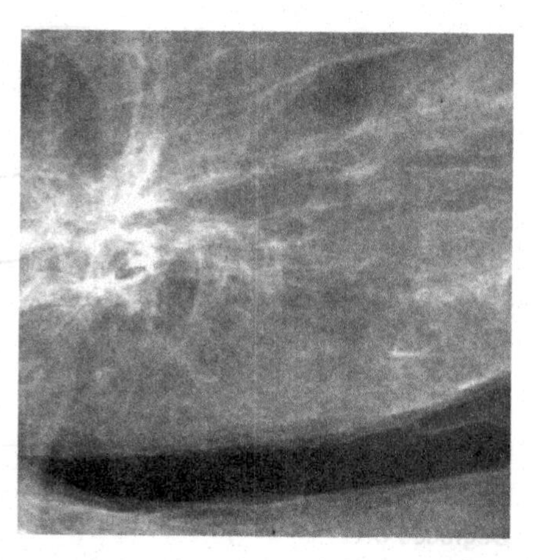

Figure 11.7 A lung region displayed with the resolution and noise corresponding to a cesium iodide-based system (left) and a selenium-based system (right).

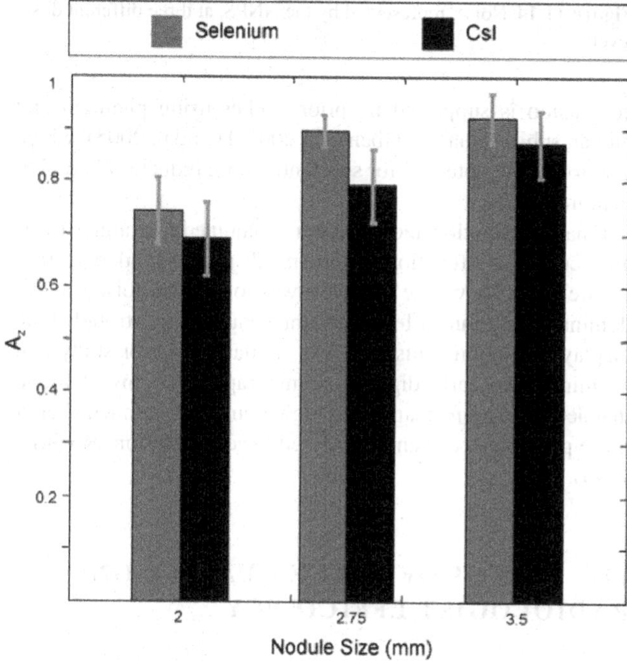

Figure 11.8 Observer performance (A_z) as a function of nodule size for two different chest radiography systems.

Figure 11.9 Resolution of three different mammography displays (Saunders, 2006b).

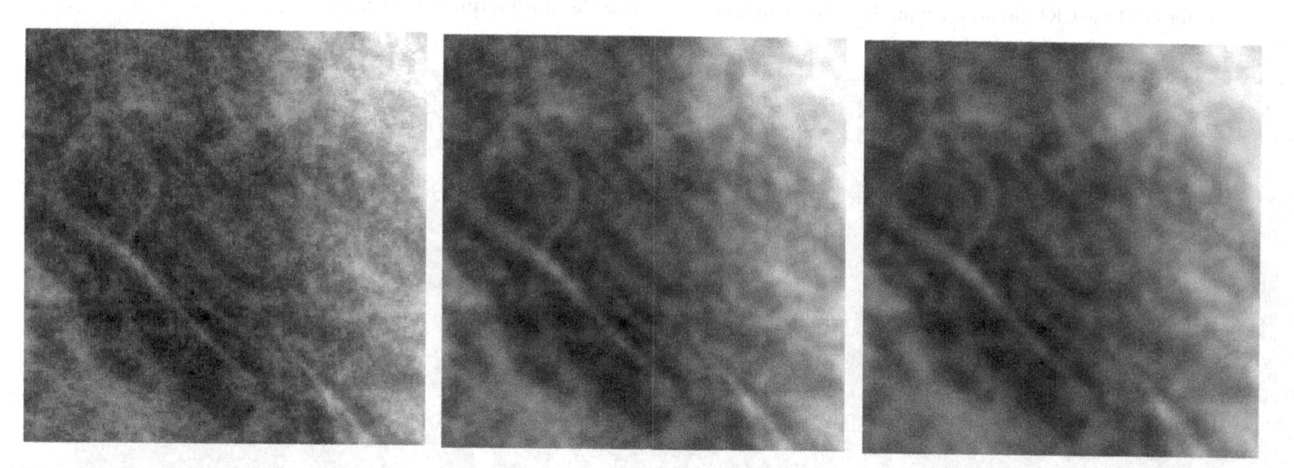

Figure 11.10 A mammographic region displayed with the resolution of an LCD display (left), normal CRT display (center), and degraded CRT display (right).

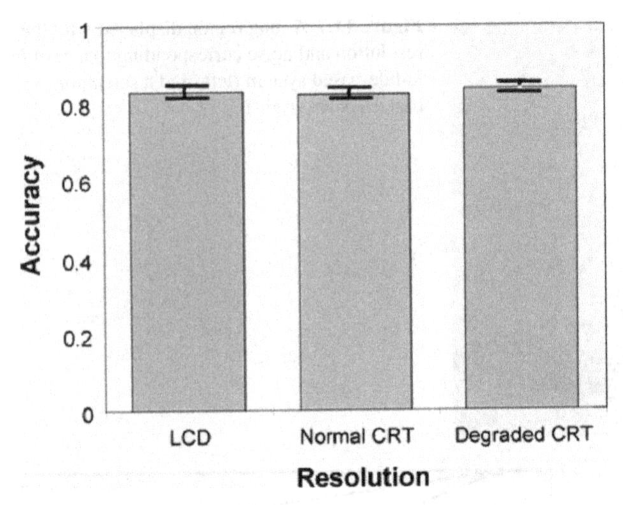

Figure 11.11 Overall accuracy as a function of display.

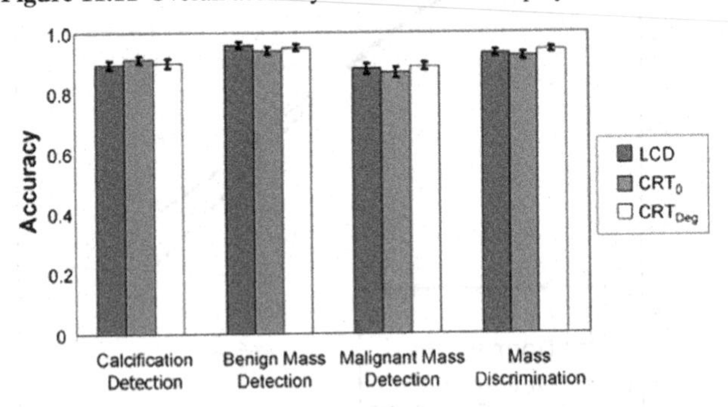

Figure 11.12 Accuracy at four different clinical tasks compared across displays.

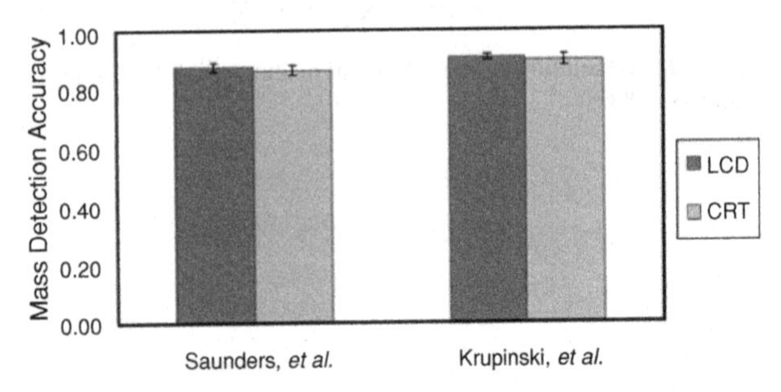

Figure 11.13 Mass detection accuracy as measured by two different investigators for LCD and CRT displays (Krupinski, 2005; Saunders, 2006).

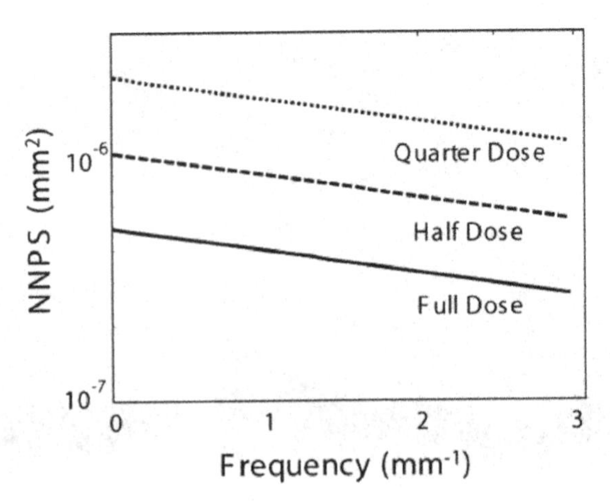

Figure 11.14 Noise, represented by the NNPS, at three different dose levels.

conclusion is supported by prior studies using phantom and human subject images (Obenauer, 2003; Hemdal, 2005), which also found the potential for substantial dose reduction in digital mammography.

Finally, resolution and noise were combined to examine overall accuracy as a function of dose and display (Saunders, 2007). Figure 11.18 shows the results. It was found that noise was the dominant factor and affected accuracy much more strongly than display resolution. This also was similar to a prior study that examined two early digital mammography systems; it found that detection performance was higher on the system with better noise performance even if it offered lower resolution (Roehrig, 1995).

11.3 IMPACT OF IMAGE QUALITY ON RADIOLOGIST EFFICIENCY

While accuracy is an important measure of system performance, it is not the only metric. Another means of evaluating systems is to examine how long it takes radiologists to interpret images on that system. If radiologists have similar accuracy on two systems, the better system would be the one where radiologists have shorter interpretation times.

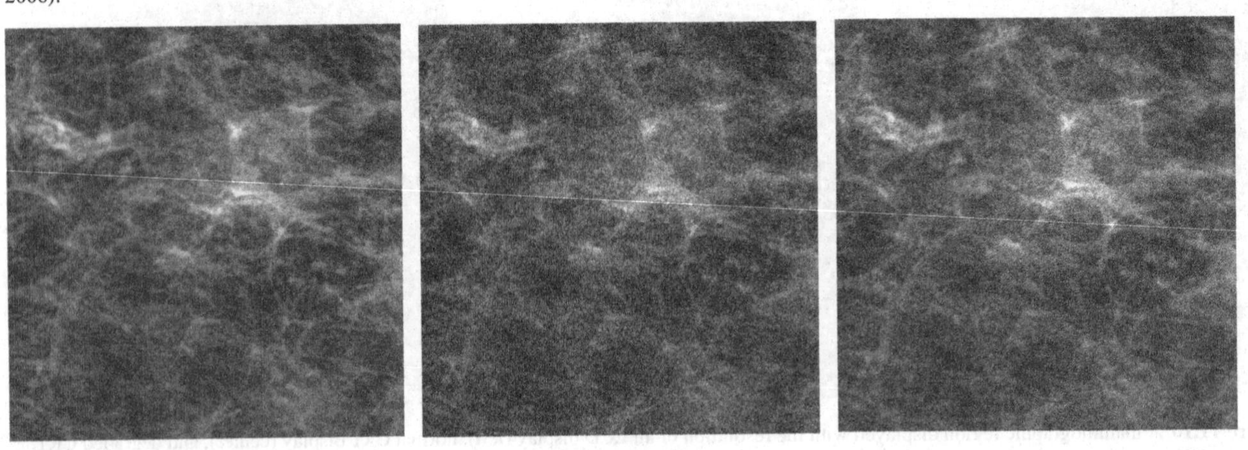

Figure 11.15 A mammographic region displayed with the noise corresponding to a full dose (left), half dose (center), and quarter dose (right).

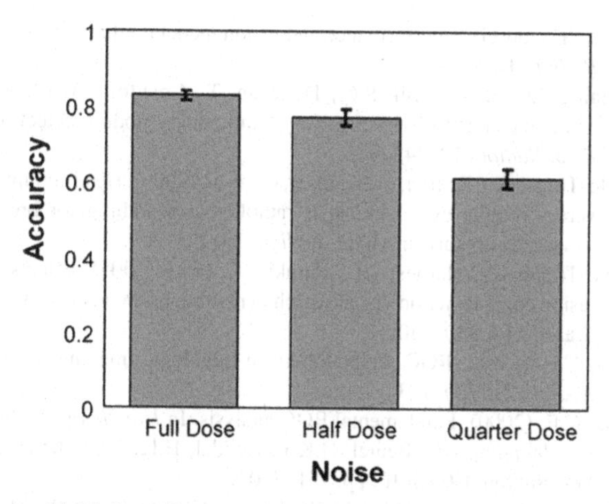

Figure 11.16 Overall accuracy as a function of dose level.

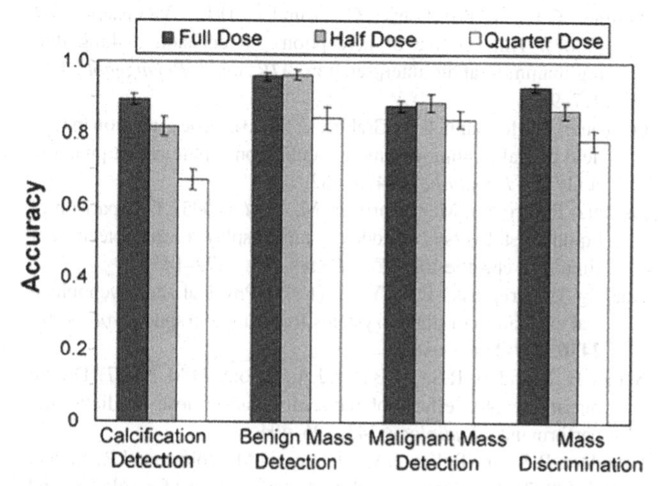

Figure 11.17 Accuracy at four different clinical tasks as a function of dose condition.

To evaluate this factor, radiologists were timed during the previous observer experiment that examined the effect of display and dose in mammography (Saunders, 2006a). It was found that radiologists took the same time to examine images on each of the three displays (Figure 11.19), but that interpretation times increased strongly as noise increased (Figure 11.20).

Specifically, this study found that longer interpretation time correlated with incorrect decisions. This finding was similar to that found by several other studies, which found that incorrect decisions in chest radiography and mammography often occurred after longer interpretation times (Nodine, 1999; Nodine, 2002; Mello-Thoms, 2005; Manning, 2006; Mello-Thoms, 2006). This finding also would explain why decision time was inversely correlated with reduced dose, as reduced dose lowered accuracy, but was largely uncorrelated with display resolution, which had a minimal effect on accuracy.

The result that interpretation time appears to be inversely correlated to accuracy could be exploited in the clinical arena. Clinical images that are analyzed for too long could be flagged. Those flagged images are subject to higher rates of decision errors and may benefit from further review.

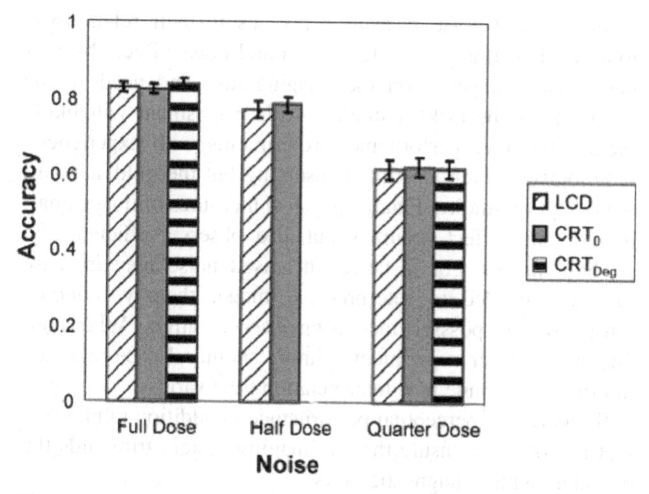

Figure 11.18 Overall accuracy as a function of dose and display.

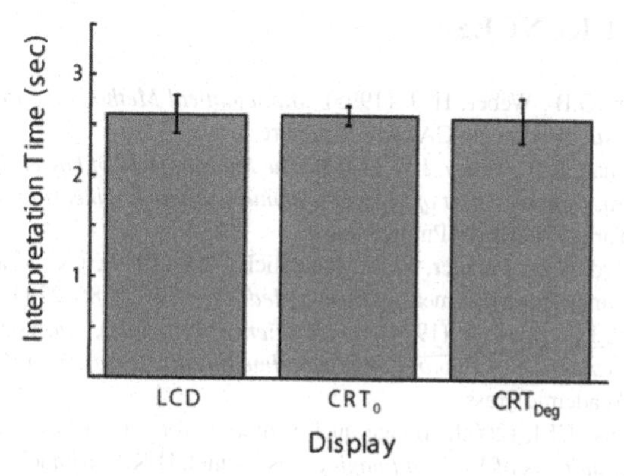

Figure 11.19 Time to interpret a mammographic region as a function of display.

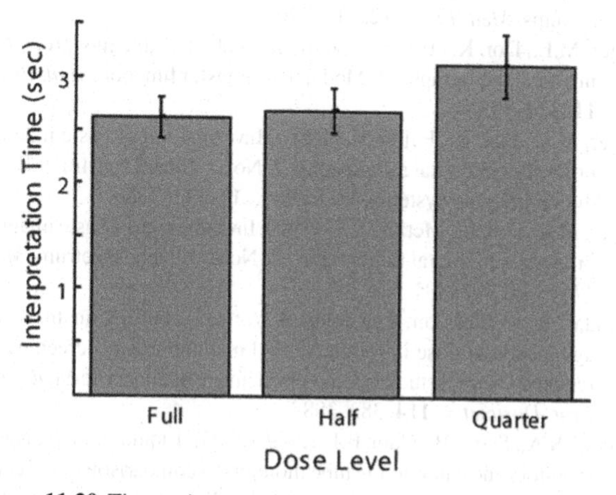

Figure 11.20 Time to interpret a mammogram as a function of dose.

11.4 CONCLUSIONS

Three conclusions can be drawn from this chapter. The first is that physical measures of image quality present an easy means of measuring system performance. The MTF, normalized noise power spectrum (NNPS), and detective quantum efficiency are straightforward to measure and allow for fast comparisons between systems. Second, the value of these resolution and noise measurements lies in their relationship to clinical accuracy. How resolution and noise affect observer performance depends on the imaging task and needs to be assessed for the task of interest. This assessment will likely require observer performance experiments with experienced radiologists, which are time-consuming but the gold standard of perception studies. Finally, physical measures of image quality not only affect accuracy but also observer efficiency in reading images. For instance, increased noise may increase the time needed to interpret an image, thereby reducing throughput and possibly increasing observer fatigue. When conducting observer experiments, timing should also be recorded in order to examine how interpretation time varies.

By assessing perceptual performance in addition to physical metrics, one can ensure that an imaging system truly aids the clinician in their diagnostic tasks.

REFERENCES

Arfken, G.B., Weber, H.-J. (1995). *Mathematical Methods for Physicists.* San Diego, CA: Academic Press.

Blackman, R.B., Tukey, J.W. (1959). *The Measurement of Power Spectra, from the Point of View of Communications Engineering.* New York, NY: Dover Publications.

Bradford, C.D., Peppler, W.W., Waidelich, J.M. (1999). Use of a slit camera for MTF measurements. *Med. Phys.*, **26**, 2286–2294.

Dainty, J.C., Shaw, R. (1974). *Image Science: Principles, Analysis and Evaluation of Photographic-type Imaging Processes.* New York: Academic Press.

Dobbins, J.T.I. (2000). Image quality metrics for digital systems. In *Handbook of Medical Imaging,* eds. Beutel, H.K.J., Kundel, H.L., Van Metter, R. Washington, DC: SPIE, pp. 163–222.

Fujita, H., Doi, K., Giger, M.L. (1985). Investigation of basic imaging properties in digital radiography. 6. MTFs of II-TV digital imaging systems. *Med. Phys.*, **12**, 713–720.

Giger, M.L., Doi, K. (1984a). Investigation of basic imaging properties in digital radiography. 1. Modulation transfer function. *Med. Phys.*, **11**, 287–295.

Giger, M.L., Doi, K., Fujita, H. (1986). Investigation of basic imaging properties in digital radiography. 7. Noise Wiener spectra of II-TV digital imaging systems. *Med. Phys.*, **13**, 131–138.

Giger, M.L., Doi, K., Metz, C.E. (1984b). Investigation of basic imaging properties in digital radiography. 2. Noise Wiener spectrum. *Med. Phys.*, **11**, 797–805.

Hemdal, B., Andersson, I., Grahn, A., *et al.* (2005). Can the average glandular dose in routine digital mammography screening be reduced? A pilot study using revised image quality criteria. *Radiat. Prot. Dosimetry*, **114**, 383–388.

Hwang, S.A., Seo, J.B., Choi, B.K., *et al.* (2003). Liquid-crystal display monitors and cathode-ray tube monitors: a comparison of observer performance in the detection of small solitary pulmonary nodules. *Korean J. Radiol.*, **4**, 153–156.

Krupinski, E.A., Johnson, J., Roehrig, H., Nafziger, J., Lubin, J. (2005). On-axis and off-axis viewing of images on CRT displays and LCDs: observer performance and vision model predictions. *Acad. Radiol.*, **12**, 957–964.

Manning, D., Barker-Mill, S.C., Donovan, T., Crawford, T. (2006). Time-dependent observer errors in pulmonary nodule detection. *Br. J. Radiol.*, **79**, 342–346.

Mello-Thoms, C., Britton, C., Abrams, G., *et al.* (2006). Head-mounted versus remote eye tracking of radiologists searching for breast cancer: a comparison. *Acad. Radiol.*, **13**, 203–209.

Mello-Thoms, C., Hardesty, L., Sumkin, J., *et al.* (2005). Effects of lesion conspicuity on visual search in mammogram reading. *Acad. Radiol.*, **12**, 830–840.

Metz, C.E. (1986). ROC methodology in radiologic imaging. *Invest. Radiol.*, **21**, 720–733.

Metz, C.E. (2000). Fundamental ROC analysis. In *Handbook of Medical Imaging,* eds. Beutel, H.K.J., Kundel, H.L., Van Metter, R. Washington, DC: SPIE, pp. 751–770.

Nodine, C.F., Kundel, H.L., Mello-Thoms, C., *et al.* (1999). How experience and training influence mammography expertise. *Acad. Radiol.*, **6**, 575–585.

Nodine, C.F., Mello-Thoms, C., Kundel, H.L., Weinstein, S.P. (2002). Time course of perception and decision making during mammographic interpretation. *AJR Am. J. Roentgenol.*, **179**, 917–923.

Obenauer, S., Hermann, K.P., Grabbe, E. (2003). Dose reduction in full-field digital mammography: an anthropomorphic breast phantom study. *Br. J. Radiol.*, **76**, 478–482.

Oschatz, E., Prokop, M., Scharitzer, M., *et al.* (2005). Comparison of liquid crystal versus cathode ray tube display for the detection of simulated chest lesions. *Eur. Radiol.*, **15**, 1472–1476.

Roehrig, H., Krupinski, E.A., Yu, T. (1995). Physical and psychophysical evaluation of digital systems for mammography. *Proc. SPIE*, **2436**, 124–134.

Samei, E., Saunders, R.S., Jr., Baker, J.A., Delong, D.M. (2007). Digital mammography: effects of reduced radiation dose on diagnostic performance. *Radiology*, **243**, 396–404.

Saunders, R.S., Jr., Baker, J.A., Delong, D.M., Johnson, J.P., Samei, E. (2007). Does image quality matter? Impact of resolution and noise on mammographic task performance. *Med. Phys.*, **34**, 3971–3981.

Saunders, R.S., Samei, E. (2006a). Improving mammographic decision accuracy by incorporating observer ratings with interpretation time. *Br. J. Radiol.*, **79 Spec No 2**, S117–122.

Saunders, R.S., Samei, E. (2006b). Resolution and noise measurements of five CRT and LCD medical displays. *Med. Phys.*, **33**, 308–319.

Saunders, R.S., Samei, E., Baker, J., *et al.* (2006c). Comparison of LCD and CRT displays based on efficacy for digital mammography. *Acad. Radiol.*, **13**, 1317.

Saunders, R., Samei, E., Hoeschen, C. (2003). Impact of resolution and noise characteristics of digital radiographic detectors on the detectability of lung nodules. *Proc. SPIE*, **5030**, 16–25.

Saunders, R.S., Jr., Samei, E., Hoeschen, C. (2004). Impact of resolution and noise characteristics of digital radiographic detectors on the detectability of lung nodules. *Med. Phys.*, **31**, 1603–1613.

Workman, A., Brettle, D.S. (1997). Physical performance measures of radiographic imaging systems. *Dentomaxillofacial Radiology*, **26**, 139–146.

12

Beyond the limitations of the human visual system

MARIA PETROU

This chapter discusses the limitations of the human visual system in perceiving volume data and differences in high-order statistics, as well as measuring accurately color and shape characteristics. It presents some image-processing techniques that may overcome these limitations and discusses some examples in relation to medical applications.

12.1 INTRODUCTION

A lot of great discoveries have happened because some people dared to question the obvious or demand the outrageous. Think how "nutty" Newton might have been considered by his contemporaries for posing the question of why the apple falls from the tree, and what treatment one might have received if 200 years ago one had demanded to speak here and be heard at the other side of the world. So, if we really want to advance science, we should dare to ask trivial questions and make outrageous demands. Something we all take for granted, never questioning its validity, is the human visual system. My grandmother used to say: "How far is the truth from the lie?" She would then proceed to answer her own question: "Just four fingers; as far as the eye is from the ear: 'I heard it' means it is a lie! 'I saw it' means it is true!" Wise as my grandmother might have been, she was wrong about this point and this chapter is about showing that our own eyes, which we seem to trust so much, have their limitations.

Let us consider a few of these limitations.

- People cannot see in volumes! Indeed, we do not see inside objects; we only see surfaces.
- People cannot perceive variation in volumes (effectively a fourth dimension). If we cannot see inside volumes, we can hardly expect to be able to perceive rates of change inside volumes!
- People cannot measure accurately. The only reason we survive crossing the street is because we can estimate that the coming car goes slowly enough to give us enough time to reach the other side; however, we cannot tell whether it goes at 39 or 41 miles per hour.
- People cannot see variations in high-order statistics. This point will become clearer later in this chapter.

The role of image processing is to allow us to:

- Go beyond the above limitations.
- Measure objectively what people only estimate subjectively.
- Visualize things that are not perceivable or visible. By this point we do not mean developing tomographic sensors that allow us to create 3D images of the human body, for example. Even with the availability of such sensors, the limitation of our vision is obvious: the 3D tomographic data produced by magnetic resonance imagers are printed as a series of 2D plates for the clinician to look at.
- Discover new knowledge.

In the sections that follow I will endeavor to discuss point by point the above limitations and demonstrate some examples of medical applications where image processing has allowed us to go beyond them, and thus enhance the vision of the clinician.

12.2 3D TEXTURE CHARACTERIZATION AND VISUALIZATION OF VOLUME DATA

Let us consider 3D tomographic data. At every voxel, we can define the direction along which the data change most rapidly. This direction is indicated by a vector of unit length, the so-called gradient vector. Imagine now collecting all these vectors from all voxels and making them all start from the same point. We may consider cones of directions starting from this point and count how many vectors fall inside each cone. What we construct this way is the so-called 3D orientation histogram of the data.

Let us try to work out what such a construct tells us about the data, in relation to a 3D set of data from geophysics, depicted in Figure 12.1. This image has been produced by ultrasound technology applied to the sub-sea crust of the Earth. The different types of texture observed in this image correspond to different geological structures. For example, the stratified texture at the top corresponds to layers of sediment created when the bottom of the sea was at that level at some time during the last 200 million years of the history of the Earth. Note that in this case the direction of maximum variation of the data is the vertical direction, so most voxels have gradient vectors pointing either upwards or downwards. There is very little variation along the horizontal direction, and, therefore, very few gradient vectors in the horizontal plane. Thus, when we consider a sub-volume in that image region and count how many vectors we find inside conic bins of orientations, and plot this number as a point away from a center, we create the 3D structure (the so-called indicatrix) at the top of the figure: it is elongated upwards, with a

The Handbook of Medical Image Perception and Techniques, ed. Ehsan Samei and Elizabeth Krupinski. Published by Cambridge University Press.

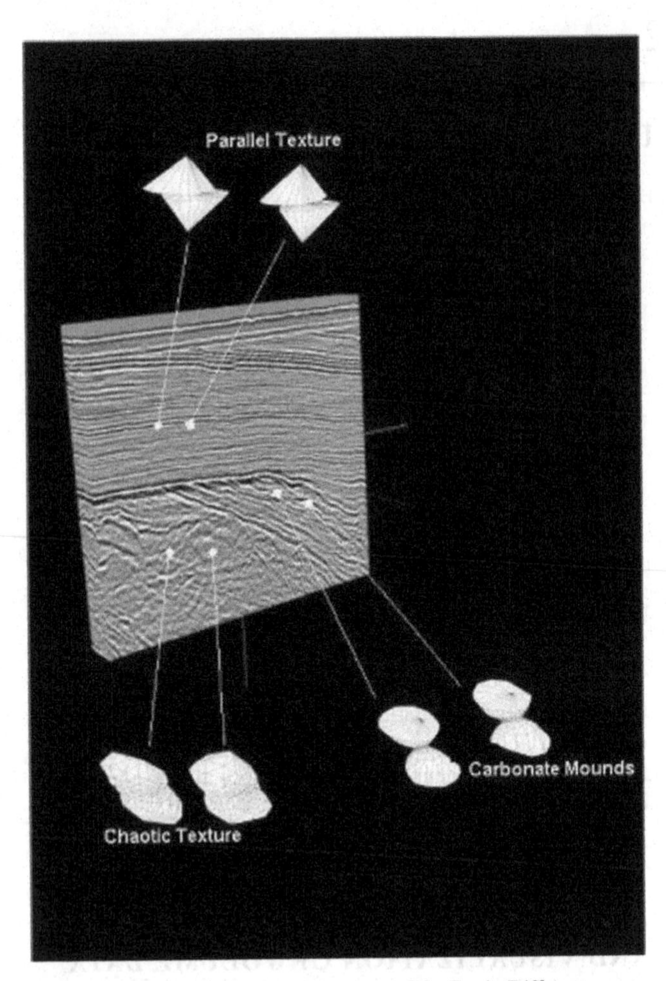

Figure 12.1 3D seismic data of the crust of the Earth. Different textures indicate different geological structures and different potentials to hold gas or oil.

very narrow "waist" in the horizontal direction, indicating the lack of gradient vectors in the horizontal plane, and thus the minimal variation of the data in this plane. On the contrary, the texture at the bottom left of the image is rather chaotic, indicating porous material capable of holding water or oil. This texture is more isotropic, so if we consider a sub-volume of it, we may find gradient vectors pointing more or less in equal numbers in all directions. The 3D indicatrix we may create to visualize such a texture looks like those at the bottom left of the image: much more roundish than those at the top of the image, reflecting the isotropy of the data. An intermediate type of texture is observed on the right of the image, where chaotic sub-parts appear to bulge upwards: these are places that indicate the presence of rising natural gas. The corresponding indicatrices, constructed by considering only the local gradient vectors, have shapes intermediate to those indicated by the two extreme types of texture we encounter in the two other parts of the image.

Constructing the indicatrix of a sub-volume of 3D tomographic data helps us visualize the structure of the data in terms of isotropy/anisotropy. However, it would be helpful also to be able to characterize the level of data isotropy/anisotropy. This is relatively easy: all we have to do is to define some measure that characterizes the shape of the indicatrix we construct. For example, extreme anisotropy, like the one observed at the top of Figure 12.1, is indicated by an indicatrix with a very narrow "waist" while total isotropy will produce a "fat" indicatrix, more or less spherical. We may then define as a measure of data anisotropy the ratio of the maximum value of the indicatrix over the minimum value. Note that these values are actually numbers of gradient vectors found with orientation inside cones of quantized orientations. Due to noise, it is very unlikely that any cone will be entirely empty, and so it is very unlikely that the minimum number will be exactly zero, causing the problem of division by zero. Noise tends to create gradient vectors in all directions, so even small amounts of noise will be enough to populate all bins of the indicatrix. For extreme anisotropy, this number will be very large, tending to infinity if the texture is perfectly stratified and there is no noise at all. For a perfectly isotropic texture, this number will tend to 1. For all intermediate levels of anisotropy, the value of this ratio will be a number more than 1. The higher its value, the more anisotropic the data are. Alternative, more robust 3D shape measures may be defined, like, for example, the standard deviation of the values of the bins we use to construct the indicatrix. More details on how to construct these indicatrices may be found in Kovalev (1999; 2000).

This method of characterizing data anisotropy was applied to MRI data of schizophrenic patients, collected at Modsley Hospital in London (Suckling, 1999). Figure 12.2 shows that statistically significant differences were found between the 3D structures of the brains of schizophrenics and normal controls, as they manifest themselves at scales of the order of 1 mm, i.e. the resolution of these data in MRI-T2 scans. The indicatrices used to compute this indicator were constructed from the bottom quarter of the brain of each subject. The bottom quarter was defined by visually identifying the slices of the data that corresponded to the bottom quarter of the slices in the Talairach and Tournoux atlas (Talairach, 1988). No significant differences were found when the full data sets were used, the top quarter, or the top half of the sets. No significant differences were found either when proton density data were used.

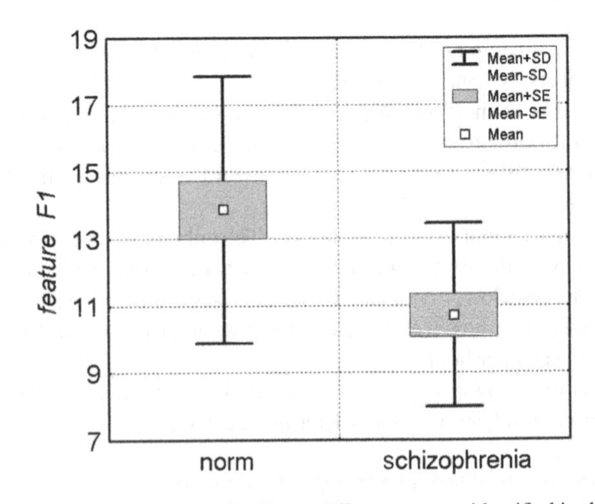

Figure 12.2 Statistically significant differences were identified in the 3D structure between the bottom quarter of the MRI-T2 scans of the brains of schizophrenic patients and normal controls.

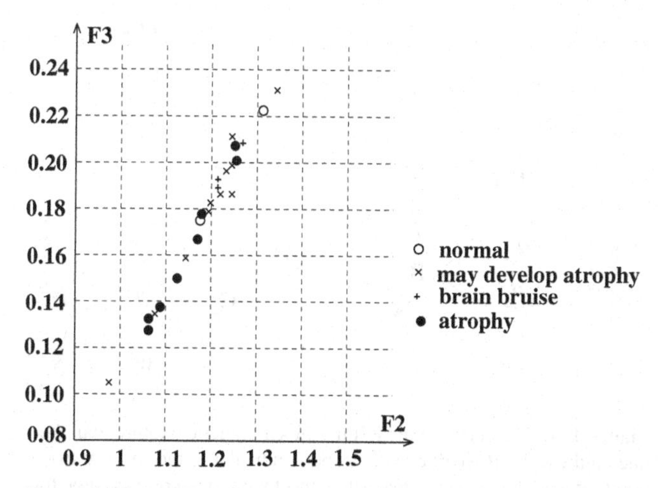

Figure 12.3 Two measures of anisotropy in 3D CT scans of brains of patients with various degenerative brain conditions, plotted against each other.

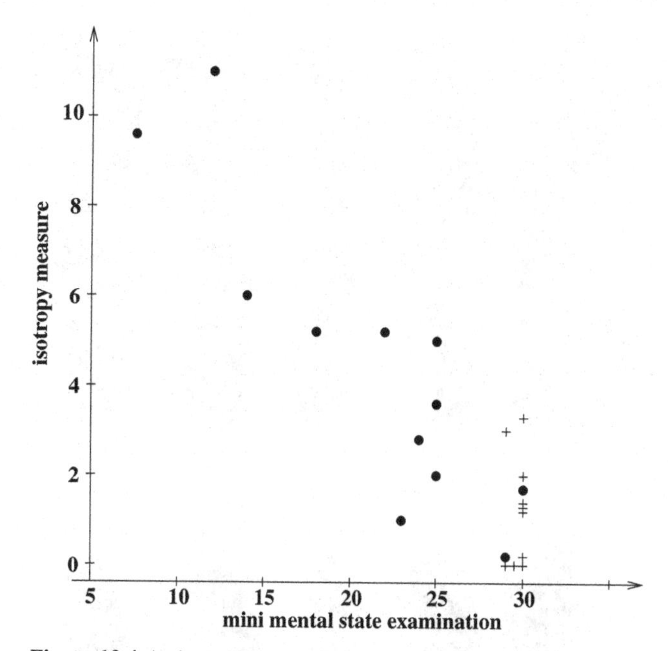

Figure 12.4 An isotropy measure computed from 3D scans anti-correlates very well with the score of a psychological test of patients suffering from dementia. The crosses correspond to normal controls.

The indicator plotted along the vertical axis of Figure 12.2 is the ratio of the largest over the smallest value of the constructed indicatrix. More details of this study may be found in Kovalev (2003).

As mentioned above, alternative measures of the shape of the indicatrix may be used. Figure 12.3 shows the results of applying this method to some CT scans of patients with various levels of brain degenerative conditions. Along the two axes the values of the standard deviation of the bins of the indicatrix are plotted against the value of "peakiness" (an indicator expressing how spiky the constructed indicatrix is). Of course, as both these numbers measure how non-spherical the constructed indicatrix is, one expects that these values are strongly correlated with each other. This strong correlation manifests itself by the fact that all points are along the diagonal. This characteristic of the graph is not particularly interesting. The interesting part is how the points corresponding to normal brains correspond to high values of the anisotropy measures, while brains with advanced states of degenerative conditions correspond to lower values of anisotropy. These studies show that, in general, globally degenerating conditions of the brain tend to manifest themselves as decreased anisotropy of tomographic data at scales of the order of 1 mm. Normal, healthy brains tend to be more anisotropic at these scales. More details of this study may be found in Segovia-Martinez (1999).

Even more striking are the results shown in Figure 12.4. Here, along the horizontal axis, we plot the score of a psychological test that patients referred to the dementia clinic are routinely subjected to. The test is called mini mental state examination (MMSE) and normal people easily score 30 out of 30. Dementia patients usually score less, depending on the severity of their condition. Along the vertical axis we plot an isotropy measure computed from the 3D scans of the patients, i.e. a purely image-processing result. The anti-correlation with the score of the test is very clear: a purely image-processing measurement correlating with a purely psychological measurement. More details about these results may be found in Segovia-Martinez (2001).

12.3 3D SHAPE CHARACTERIZATION

Figure 12.5 shows the 3D shapes of the first section of the colon of people suffering from ulcerous colitis at the top and at the bottom the same section from non-sufferers. These shapes have been extracted from 3D X-ray images. It is very hard for anybody to identify any difference in the shapes of the two classes of data. However, we may objectively try to quantify the shape of each surface by realizing that each surface element (surface patch) may be represented by a vector perpendicular to it, the so-called normal vector. This is demonstrated in Figure 12.6. Let us consider pairs of points on two surfaces, at a certain fixed distance apart, say distance d. For the surface in Figure 12.6(a) two such points are points A and B, while for the surface in Figure 12.6(b) two such points are points C and D. The arrows show the unit vectors that are orthogonal to the surface at these points. Note that if we extrapolate these vectors, they will meet at point O_1 for surface (a) and at point O_2 for surface (b). In the first case, where the surface is highly curved, they will form an angle ϕ_1, which is much larger than the angle ϕ_2 they will form in the second case, where the surface is almost flat. If the surface is perfectly flat, the two extrapolated lines will never meet; they will be parallel and we may say that they form an angle of 0 degrees, with its vertex at infinity. If we consider many pairs of points on each surface that are at the same distance d apart, and in each case we measure the angle ϕ that the two vectors normal to the surface form, we shall produce a distribution of values of ϕ. This distribution will be different for different surfaces. If we consider different values of d, we shall also be able to characterize the surface at various scales. Considering then these distributions of values of the angles normal vectors form at pairs of points certain distances apart, we may compute various features (numbers) from them

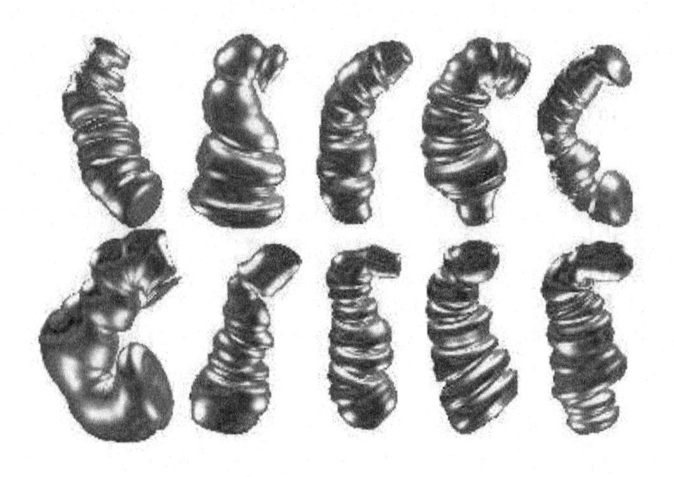

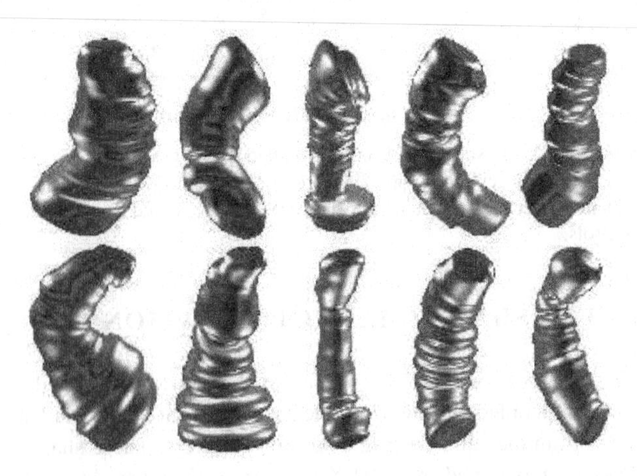

Figure 12.5 3D colon images extracted from X-ray tomography. The images at the top correspond to sufferers of ulcerous colitis, and those at the bottom to non-sufferers.

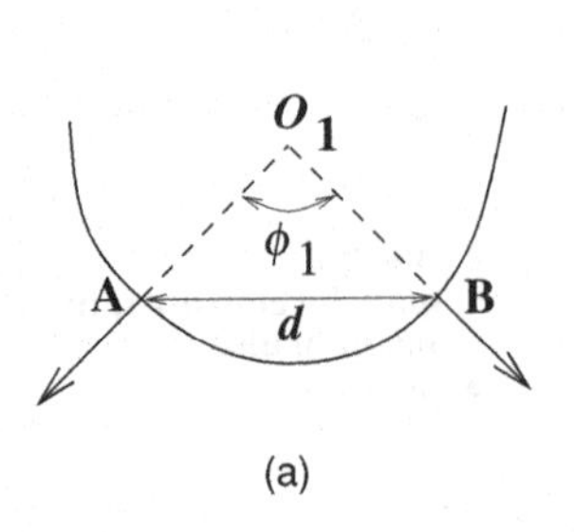

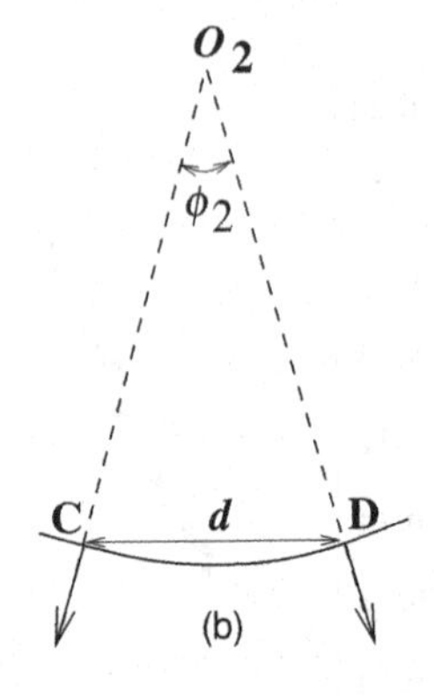

Figure 12.6 The curve on the left has much higher curvature than the one on the right. If we draw vectors perpendicular to the curve at two points at a certain distance apart, they meet with much bigger angle for the curve on the left than for the curve on the right. Computing statistics on the values of this angle over many pairs of points on a surface may help characterize the surface shape.

that characterize the surface. Some of these numbers are shown in the plots of Figure 12.7. Along the horizontal axis we plot an index identifying the patient. The second half of the indices refer to patients that suffer from ulcerous colitis. The first half are patients who are either healthy or suffer from a different condition. In all cases the measurements performed on the surfaces of the colon of the colitis patients are consistent across patients, while the same measurements for the non-sufferers are not consistent. More details on this may be found in Kovalev (1997; 1998).

The same methodology of considering pairs of gradient vectors and the angle they form may be used to characterize the structure of 3D volume data. One may perform statistics on the frequency of appearance of certain angles formed by the gradient vectors of voxels at a certain distance apart, and use them to characterize each volume data set. Once it has been established which distances and which relative angles characterize each class, we may work backwards and check which particular points of the volume data gave rise to these different statistics. This technique was used to identify the specific points in the MRI-T2 data we discussed in the previous section that helped distinguish schizophrenics from normal controls. Figure 12.8 shows how well this technique allows the distinction of the two classes, while Figure 12.9 indicates the parts of the brain that contributed to the different statistics of the particular schizophrenic patient. These parts turn out to be the 3D structure

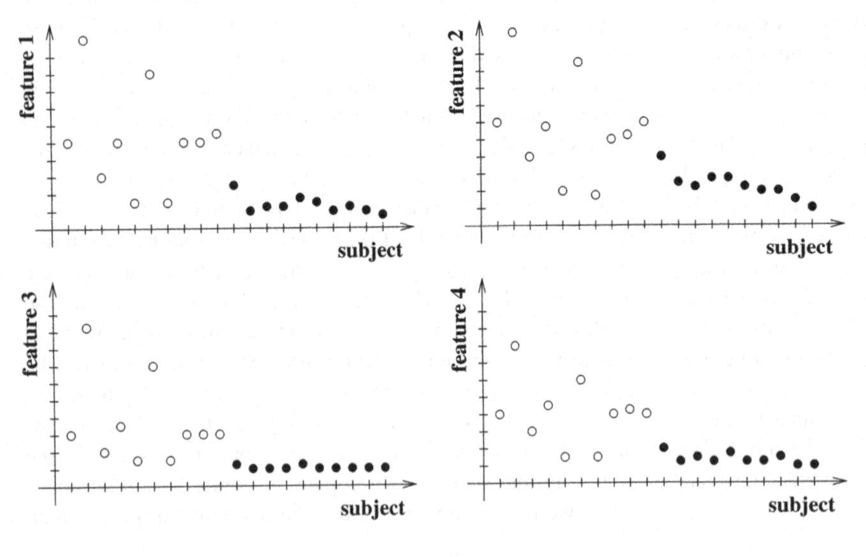

Figure 12.7 Some features computed from the statistics of angle ϕ defined in Figure 12.6 may be used to characterize the two classes of images of Figure 12.5: the filled dots correspond to sufferers of ulcerous colitis, while the empty circles correspond to non-sufferers.

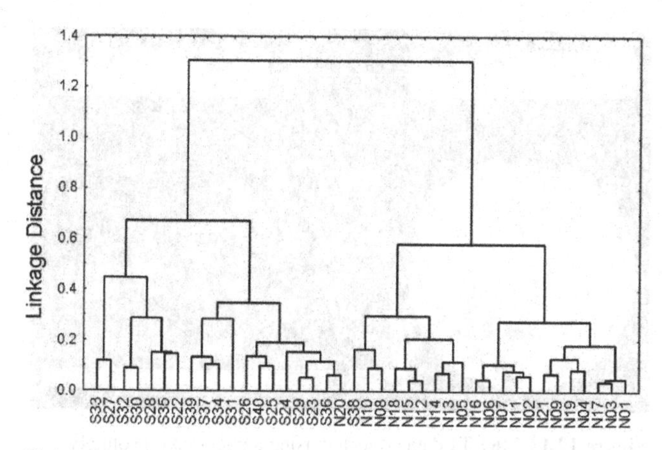

Figure 12.8 Some features computed from the statistics of angle ϕ, defined by considering the gradient vectors of pairs of voxels at a certain distance apart, may be used to separate the schizophrenics from normal controls. Codes starting with S along the horizontal axis refer to "schizophrenia" and codes starting with N refer to "normal."

of the gray matter that makes up the sulci of the patient. This is not the same as the surface shape of the sulci, as sulci surface shape information was deliberately excluded from this study. More information may be found in Kovalev (1999).

12.4 ESTIMATION VERSUS MEASUREMENT

Pleomorphism is one of the indicators used to assess the severity of breast cancer and manage the patient after the diagnosis. Pleomorphism involves the assessment of the degree of deviation of the shape of the cells from the elliptical shape. Figure 12.10 shows two pathology slides, one from a healthy breast and one from a breast with cancer. Image processing may be used to measure accurately the shape of the cells and estimate the fraction of those that deviate from the elliptical shape. The important aspect here is objectivity and repeatability of the measurement. Another indicator used for assessing the severity of the condition and subsequent management of the patient is oestrogen positivity, measured by the degree of color darkness observed in the cells when a particular dye has been used. Figure 12.11 shows such a slide. The more brown the cells are, the better their oestrogen uptake, and the better the prognosis for the patient as oestrogen may be used to deliver the right chemical treatment to the pathogenic cells. However, the human eye cannot really measure color in a repeatable and objective way. The triangle on the right in Figure 12.11 shows the so-called chromaticity diagram of color measurement. In it the various pixels that make up the slide on the left are plotted with their corresponding color. The saturation of a particular color is measured as the distance from the center of the triangle, while the hue of a color is measured as an angle formed by the line connecting the pixel with the center of the triangle with some reference direction. The center of the triangle is supposed to correspond to the ideal white, because at that point all three basic colors, red, green, and blue, that correspond to the three vertices of the triangle meet in equal proportions. However, the white pixels of the slide may appear away from that ideal center from which all objective color measurements have to be made. This is because each viewer perceives white according to their own system of sensors (their specific cone responses in their

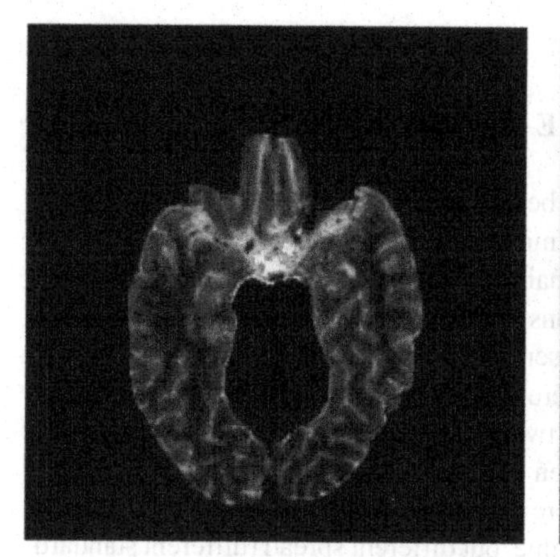
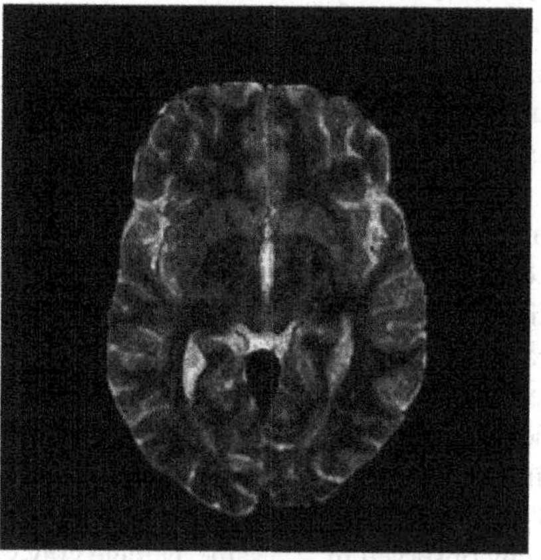

Figure 12.9 By performing statistics on the relative angles of gradient vectors in 3D volume data, we can identify features that separate the schizophrenics from normal controls, as shown in Figure 12.8, and at the same time allow us to trace back the exact locations where the differences come from. In this set of data, the differences are in the 3D texture of the tissue that makes up the sulci in the inferior quarter of the brain.

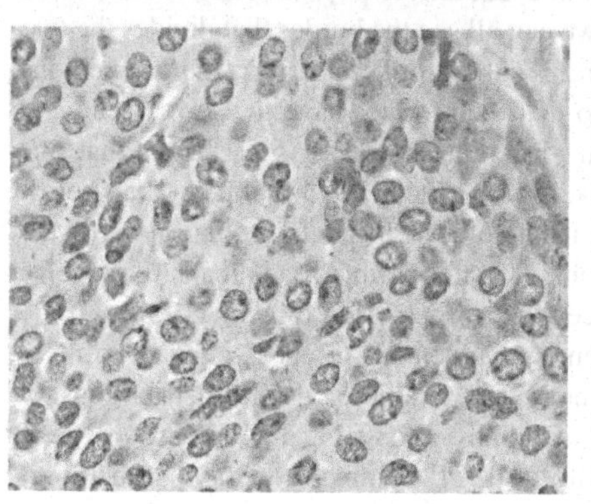
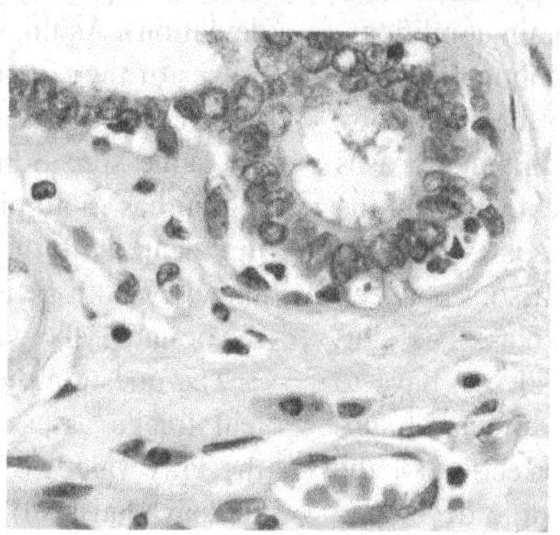

Figure 12.10 The image on the right is from a healthy breast. The image on the left depicts breast cancer. The depicted cells on the left image deviate from perfect ellipses (pleomorphism). Quantifying this shape change may be highly subjective.

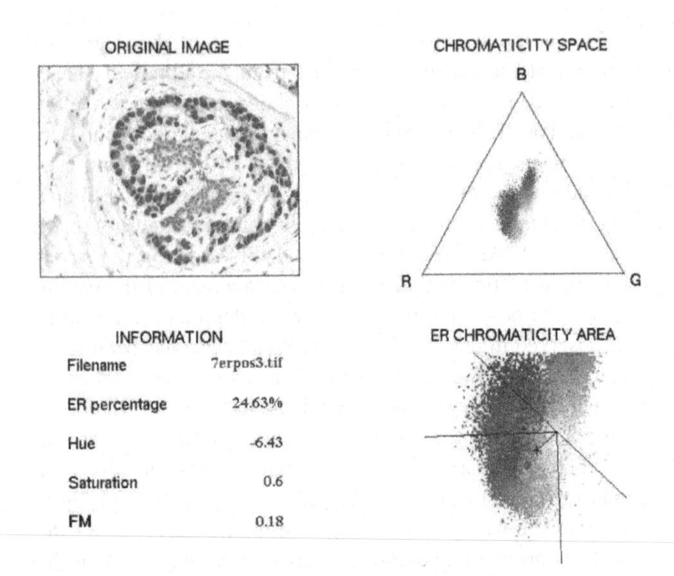

Figure 12.11 Oestrogen positive cells may be identified by appropriate staining. The degree of uptake of the stain by the tissue is manifested by the darkness of the stained cells and their numbers. The level of color saturation (color intensity) may be measured on the chromaticity diagram (top right) as the distance of a point from the center of the triangle, while the actual color shade (hue) may be measured as an angle from some reference direction, with the vertex of the angle also at the center of the triangle. Under ideal conditions, white coincides with the center of the triangle. However, white is different for different people and often the case is that the reference white is away from the center of the triangle. The three principal colors (red, green, blue) of a color camera are fully saturated at the three vertices of the triangle.

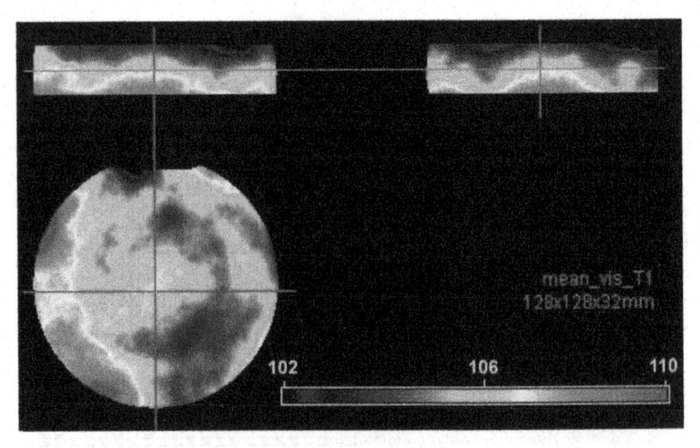

Figure 12.12 MRI-T1 data collected using a phantom (absolutely uniform 3D object), with a Siemens Vision MRI scanner. At each voxel a spherical window of diameter 13 pixels was placed and the mean value of the voxels inside the window was calculated and assigned to the central voxel. The anisotropy and inhomogeneity of the image is obvious.

retinas). This indicates that the ideal (objective) white is very different from the subjective white of each viewer, and this is an important factor that has to be taken into consideration when objective color measurements are made: computer systems that do not take that into consideration are as subjective as the individual human experts!

In general, issues of *objective* measurements from images are much more complicated than the problems of *relative* measurements used to characterize different conditions or different groups of subjects. In the latter case, *relative* calibration of the computed features suffices, while in the former *absolute* calibration is necessary. Figure 12.12 shows the inhomogeneity and anisotropy of the field produced by an MRI instrument. This immediately puts under question the validity of the conclusions we drew in the previous sections on the anisotropy observed in normal controls and sufferers of degenerative brain conditions. How can we be sure that what we observe is attributable to the genuine difference between subjects and not to the anisotropy of the imaging instrument itself? The answer is that in all cases we compare data captured by the *same* instrument and with the *same* protocol. So, any systematic instrument bias is expected to manifest itself equally in all data and thus comparisons between different subjects are expected to remove this bias. The situation for absolute measurements is more crucial: we have to calibrate the processes and the data so that the effects of subjective color balance, or automatic gamma correction of the camera, are removed: no image-processing module developed to work with microscopy images captured under one magnification is expected to operate under another magnification setting; no image-processing color measurement module designed for a particular staining procedure is expected to work correctly under another staining procedure. Image-processing objectivity is only real objectivity as long as the data presented to the system have been captured under the same imaging conditions, and by the same instrument, and with the same protocol followed. Image-processing objective measurement modules may be made adaptable to different staining protocols and imaging procedures, but this will have to be done manually, off the line, and by trained users or developers of image-processing systems.

12.5 INVISIBLE BOUNDARIES

In spite of our eyes being often considered the ultimate proof of truth, there are limitations on what they can recognize. In Figure 12.13 three pairs of adjacent regions are shown: each pair consists of regions that have been made from random numbers chosen to represent pixel brightness. For the first pair, the random numbers were chosen to differ only in the mean, i.e. the averages of the two regions are different. We can all see the boundary between the two regions. In the second pair, the random numbers were chosen so that the two regions have the same average gray value, but different spread (different standard deviation). Again, we can all see the border that divides the two halves of the picture. Finally, in the third pair, the two regions were constructed so that their values have the same mean and the same spread but different skewness. This means that one region has more pixels with value above the mean than below the mean, with the mean and the standard deviation being the same for both regions. It is impossible for humans to perceive the boundary between the two halves of this image.

The question then is whether this effect is relevant to medical image assessment. For example, if the distributions of the gray values of medically interesting regions are symmetric (say Gaussian), the issue of asymmetry in the distribution is

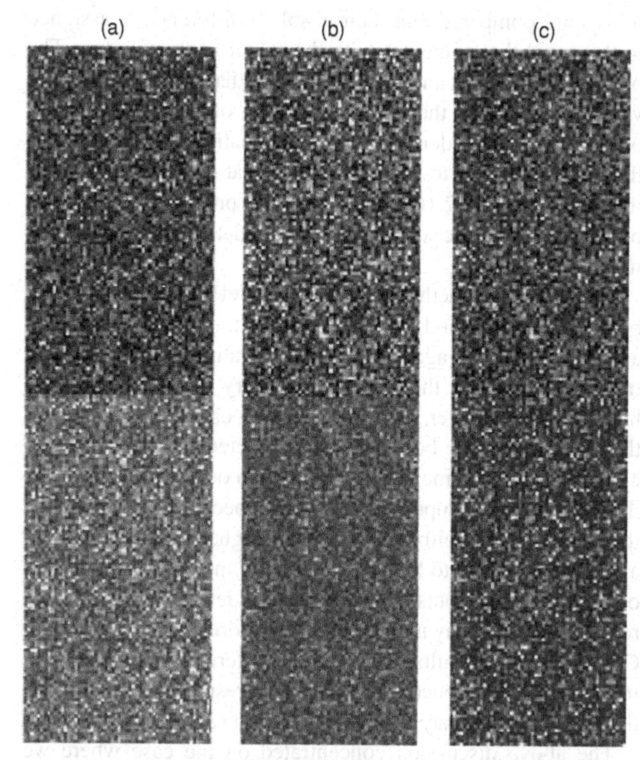

Figure 12.13 In the absence of any spatial pattern, and when the gray values of two regions differ only in the mean or only in the spread, the human brain can perceive the virtual line that divides them, as shown in panels (a) and (b). When the gray values of two regions differ only in the asymmetry of the distribution from which they have been drawn, the boundary between them cannot be perceived, as shown in panel (c) .

irrelevant. Equally, if the gray values of the regions of interest are spatially correlated, i.e. form spatial patterns, again the issue of perceiving differences in high-order statistics (i.e. statistical measurements beyond the well-known mean and standard deviation) is irrelevant. However, there are cases of medical image analysis where the gray values of the data neither form spatial patterns nor come from symmetric distributions. An example is the case of the gray values in the periphery of a glioblastoma. The true boundary of such a lesion is not visible. At best the radiologist can specify an outer annulus inside which the real boundary of the lesion lies. The boundary the surgeon will use for the excision of the tumoar will lie somewhere inside that annulus. Figure 12.14 shows the distribution of the gray values of MRI-T2 scans from the region around the necrotic region of a glioblastoma, and inside an annulus specified by a radiologist. The distribution is obviously asymmetric, indicating that it is possible that useful information concerning the true nature of the depicted tissue lies in the tail of this distribution. It is not unusual for distributions of medical data to be asymmetric: in fact the opposite is unusual. Apart from genuine biochemical reasons that may cause asymmetries in the distributions of the data, the mere fact that gray values are positive numbers and often near zero excludes the existence of tails of the distributions beyond the zero point, creating asymmetries.

It has been advocated, therefore, that hidden boundaries may be revealed if one searches for localities in the data where the asymmetry of the gray value distribution changes. Ordinary edge detectors, often used by image-processing people, search for places where the average gray value of the image changes sharply (Petrou, 1994). Very rarely, non-linear edge detectors have been employed to identify edges where some statistic (usually the variance) changes (Graham, 1988). However, edge detectors have been proposed, at least at the theoretical level (Pitas, 1986) where other statistics may be explored. In practice, once one moves to higher than second-order statistics, one encounters the problem of reliable calculation of their value. This has some implications for the application of this methodology: it cannot be used locally for 2D data, as pixels of 1 mm in size imply the need to use windows much larger than 1 cm × 1 cm in order to have enough pixels to compute reliable statistics. From 100 samples, the skewness of a distribution may be estimated with an error of about 20%. Big windows will produce unacceptable ambiguities in the location of the detected boundary. On the other hand, 3D data with the same resolution, even within windows of 0.5 cm × 0.5 cm × 0.5 cm, contain 625 points, enough for the calculation of third-order statistics with roughly 10% error. Fourth- and higher-order statistics require many more samples. For example, with 1000 samples the fourth-order moment (otherwise known as kurtosis) may at best be estimated with 25% accuracy (Petrou, 2006).

With the above provision then, one may use the following scheme to detect variations in the skewness of the local distributions: a bipolar scanning window is constructed, consisting of two halves. Each half should be big enough to contain a sufficient number of voxels for reliable statistics to be computed. For example, a window consisting of two halves, each made

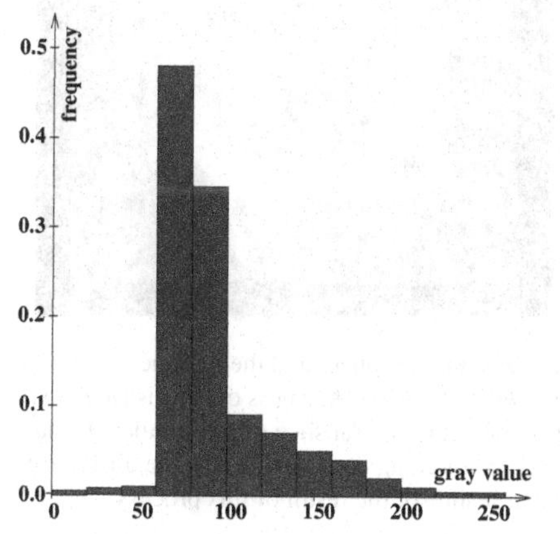

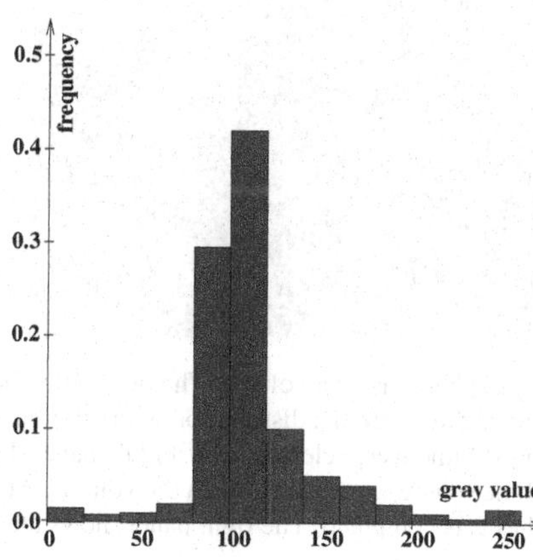

Figure 12.14 The histograms of the gray values of MRI-T2 data around the visible glioblastoma regions of two patients. Their asymmetric shapes indicate the possibility that useful information may be extracted by analyzing this asymmetry locally and detecting places where the degree of asymmetry changes.

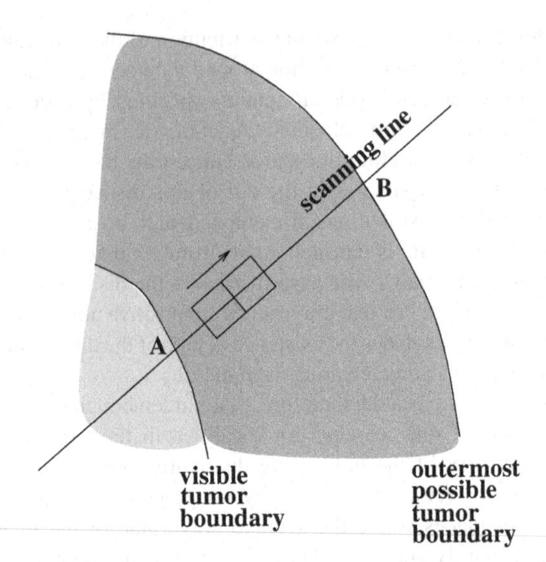

Figure 12.15 Sliding a bipolar window along line AB and computing a statistic in each half of the window allows one to identify the point along segment AB where the value of the statistic between the two window halves is maximally different. That point is designated as the most probably true boundary of the tumor.

up of 5 × 5 × 5 voxels, will use 625 values to compute the skewness of the distribution in each half. Such a number of voxels is expected to produce reasonably reliable statistics. If the pixels correspond to 1 mm, the resolution with which the boundary will be detected will be 5 mm. The process of boundary detection is schematically shown in Figure 12.15. A mask has been manually created to delineate the inner, visible boundary of the tumor, and the outermost possible boundary. Scanning rays are considered emanating from inside the tumor outwards. Along each scanning ray the bipolar window slides, starting at the inner boundary (point A) and finishing at the outer boundary (point B). At each point the skewness (or some other statistic) is computed from the gray values of the voxels covered by the window, for each half of the window. The values from the two halves are compared and their absolute difference is assigned to the voxel that coincides with the center of the window. The point/voxel along scanning segment AB that has the maximum value is marked as the possible location of the true boundary of the tumor. The identified possible location boundaries may then be connected to form a surface that possibly delineates the tumor. Figure 12.16 demonstrates the process of revealing a hidden sphere inside some simulated 3D data, using exactly this method.

Note that this method will always yield a boundary point between points A and B. Note also that there is no such thing as a "ground truth" against which the location of this boundary may be checked, as the detected boundary is not visible to the human eye. However, there are ways to check the validity of the detection: if the boundary surface detected corresponds to a location where some physical transition occurs, the change in the value of the computed statistic is expected to be always of the same polarity; either always from higher to lower value or always from lower to higher value as we move from the inside of the tumor to the outside. Further, if the detected boundary is a real boundary, it may manifest itself in more than one modality. Correlating the results obtained for different modalities may increase the confidence in the produced result. More details on this methodology may be found in Petrou (2006a; 2006b).

The above discussion concentrated on the case where we know that there must be a boundary and we are trying to find its most likely location, according to some criterion. The criterion we used was not deduced by forward thinking, but rather by examining one of the possible options. Forward thinking would have to work out, for example, a model of how the cells of the tumor propagate, with what speed they advance in relation to the speed of the apoptosis, and by taking into consideration other such factors finally deduce that such a boundary is expected to affect the third-order statistic of the observed gray values. No such deductive process took place. Instead, the logic we used was that of the "lost keys at night" paradigm: if you lose your keys in the street at night, and walk back looking around to find them, where do you search? You search under the lamp

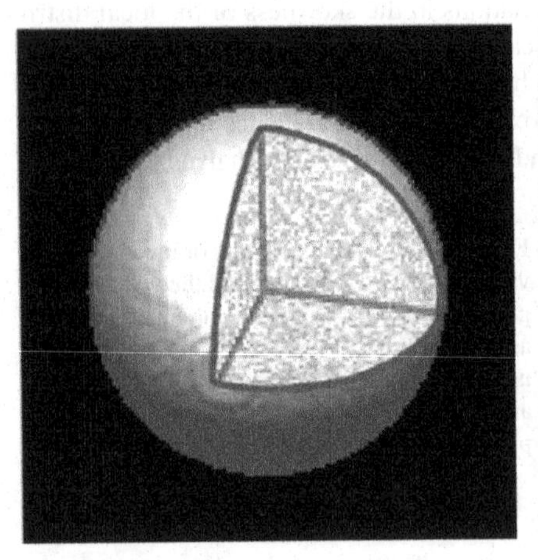

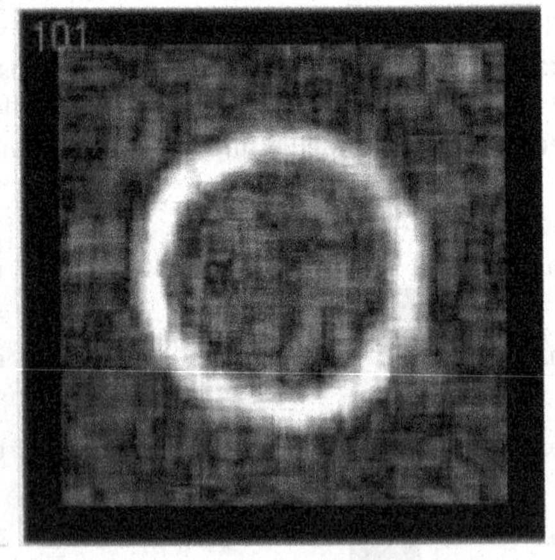

Figure 12.16 Simulated data were created embedding a sphere inside a volume. The only difference between the sphere and the volume surrounding it was that the gray values of the sphere were drawn from a distribution with different skewness than the skewness of the distribution from which the values of the voxels in the surrounding volume were selected. The middle panel shows what the central slice of the simulated data looks like. Tracing lines emanating from the center of the sphere were used to scan the volume. At each point along each tracing line the difference in the skewness between the voxels of two local patches was computed. The right panel shows the central slice of the result of this process.

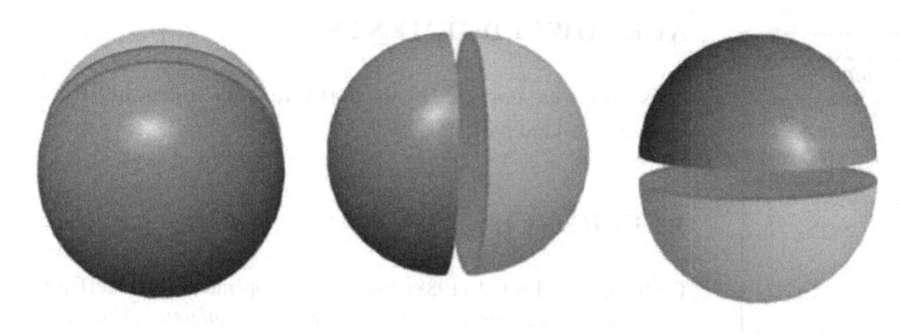

Figure 12.17 A spherical window may be divided into two halves in three different ways, by planes orthogonal to the three coordinate axes. A statistic may be computed from the gray values of the voxels that are covered by each window half, and the difference between these two values may be considered as the difference of the statistic along the corresponding axis. By squaring and adding these differences, the gradient vector of the statistic may be defined and associated with the position where the center of the window is.

posts! Why? Because it is only there that you can see. This does not mean that your keys were lost under the lamp posts, but those places are the only places where you can search. Our logic here is the same: we have no real reason to believe that a tumor boundary should correspond to a change in the third-order statistic, but since we know that the human eye cannot see such changes and since we have the tool of revealing such changes, we might as well use that tool to check. We have, however, some arguments in support of our effort: the distributions of the gray values are indeed asymmetric, so maybe what we are looking for is hidden in those asymmetric tails. Besides, as cancerous cells diffuse into the surrounding healthy tissue, it is not implausible to expect that their rarity among their local surroundings will be manifested in the tails of the distributions, since the tails are formed by the least frequently observed values. If the results produced that way are self-consistent, in the way discussed above, then our confidence in our reverse thinking being fruitful increases.

Moving away from situations where we know *a priori* that a boundary is present and we are trying to reveal it, we may like to search for hidden boundaries even in data where we have no reason to believe that such boundaries are present. Image processing people routinely run edge detectors on images to segment them. Of course, the edge detectors used are designed to detect the visible edges. It would be interesting to see whether scanning data with the type of edge detector we discussed here will reveal any edges that we did not know were there. For such a purpose, a spherical window may be used. At each position of the volume data, the window is divided into two halves in three different ways, as shown in Figure 12.17.

As in this case we do not have any prior knowledge about the orientation of the possible boundary, the difference in the computed statistic will have to be determined along all three orientations, i.e. for all three divisions of the window. If ΔM_x is the difference in the mean, ΔV_x is the difference in the standard deviation, and ΔS_x is the difference in the skewness along the x axis, with similar notation for the differences along the other axes, we may define for each position of the window a mean gradient, a standard deviation gradient, and a skewness gradient, as follows:

$$G_M \equiv \sqrt{\Delta M_x^2 + \Delta M_y^2 + \Delta M_z^2} \tag{12.1}$$

$$G_V \equiv \sqrt{\Delta V_x^2 + \Delta V_y^2 + \Delta V_z^2} \tag{12.2}$$

$$G_S \equiv \sqrt{\Delta S_x^2 + \Delta S_y^2 + \Delta S_z^2} \tag{12.3}$$

We know that significant values of either G_M or G_V will be detected by the naked eye. However, points for which $G_M \simeq 0$ and $G_V \simeq 0$ will not be detected, even if $G_S \gg 0$. It is those points we wish to identify. Figure 12.18 shows the extent of such regions identified in normal controls and in schizophrenic patients. Figure 12.19 shows that these points constitute statistically significant differences between the normal controls and

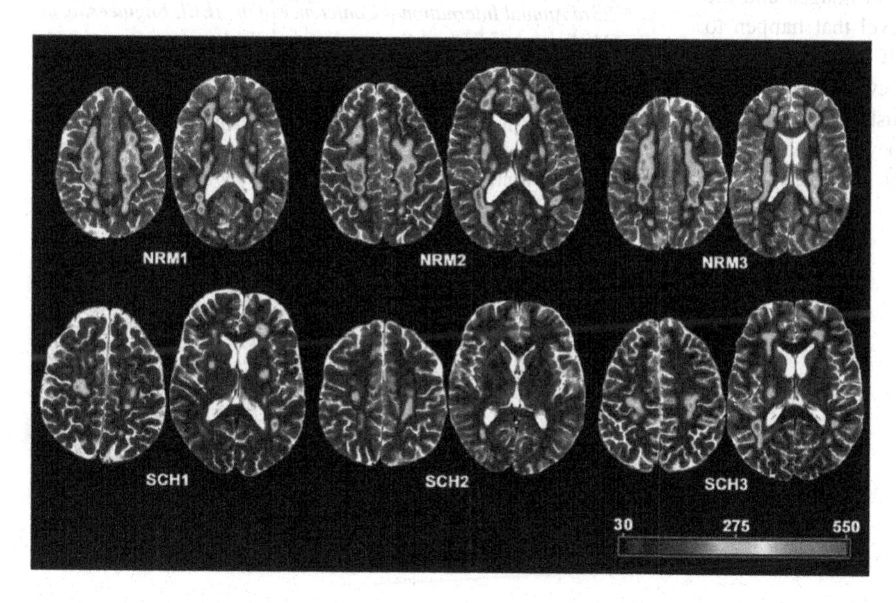

Figure 12.18 The color coding represents regions in the scans of normal controls (top) and schizophrenic patients (bottom) where significant values of skewness gradients were detected, without visible gradients being present.

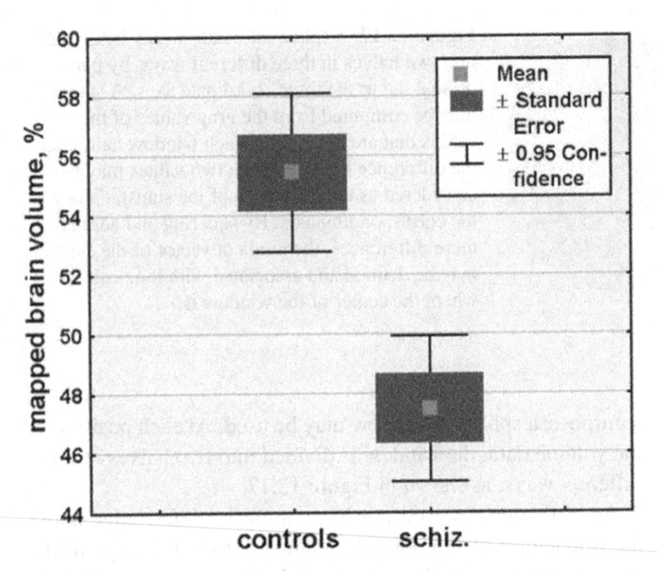

Figure 12.19 Statistically significant differences between the structure of the white matter of schizophrenic patients and normal controls were identified when considering the extent of the regions where skewness gradients were detected.

the patients, computed over data consisting of 19 patients and 21 controls. More details may be found in Petrou (2006).

12.6 CONCLUSIONS

Image processing most of the time is used as a tool that allows the clinician to perform measurements on the data that otherwise have to be done manually or simply estimated without high accuracy. However, image processing may play another role in medical image research: it may allow the researcher to visualize and detect properties in the data that were not even suspected to exist. It is possible then to trigger new directions of research that may lead to the discovery of new knowledge. One of the most challenging aspects of current medical imaging and medical image processing is to find the connection between the macroscopic measurements performed on images and the microscopic changes at the biochemical level that happen to the tissue. This gap can only be bridged if the data we collect at the imaging scales are treated with new methodologies, beyond those clinical practitioners are accustomed to. So, the role of image-processing researchers goes beyond that of technical support for the clinician, and becomes a possible route for the discovery of new knowledge.

ACKNOWLEDGEMENTS

This work was supported by the portfolio grant "Integrated Electronics," funded by EPSRC.

REFERENCES

Graham, J., Taylor, C.J. (1988). Boundary cue operators for model based image processing. *Proc of 4th Alvey Vision Conference*, University of Manchester, UK, pp. 59–64.

Kovalev, V.A., Petrou, M., Bondar, Y.S. (1997). 3D surface roughness quantification. *British Machine Vision Conference*, Colchester, UK, pp. 450–458.

Kovalev, V.A., Petrou, M., Bondar, Y.S. (1998). Using orientation tokens for object recognition. *Pattern Recognition Letters*, **19**, 1125–1132.

Kovalev, V.A., Petrou, M., Bondar, Y.S. (1999). Texture anisotropy in 3D images. *IEEE Transactions on Image Processing*, **8**, 346–360.

Kovalev, V.A., Petrou, M. (2000). Texture analysis in three dimensions as a cue to medical diagnosis. In *Handbook of Medical Imaging, Processing and Analysis*, I Bankman, editor. New York, NY: Academic Press, pp. 231–247.

Kovalev, V.A., Petrou, M., Suckling, J. (2003). Detection of structural differences between the brains of schizophrenic patients and controls. *Psychiatry Research: Neuroimaging*, **124**, 177–189.

Petrou, M. (1994). The differentiating filter approach to edge detection. *Advances in Electronics and Electron Physics*, **88**, 297–345.

Petrou, M., Kovalev, V.A. (2006a). Statistical differences in the grey level statistics of Tl and T2 MRI data of glioma patients. *International Journal of Scientific Research*, **16**, 119–123.

Petrou, M., Kovalev, V.A., Reichenbach, J.R. (2006b). Three dimensional nonlinear invisible boundary detection. *IEEE Transactions on Image Processing*, **15**, 3020–3032.

Pitas, I., Venetsanopoulos, A.N. (1986). Edge detectors based on nonlinear filters. *IEEE Transactions on Pattern Analysis and Machine Intelligence*, **8**, 538–550.

Segovia-Martinez, M., Petrou, M., Kovalev, V.A., Perner, P. (1999). Quantifying level of brain atrophy using texture anisotropy in CT data. *Proceedings of Medical Image Understanding and Analysis*, Oxford, UK, pp. 173–176.

Segovia-Martinez, M., Petrou, M., Crum, W. (2001). Texture features that correlate with the MMSE score. *Proceedings of the 23rd Annual International Conference of the IEEE Engineering in Medicine and Biology Society*, Istanbul, Turkey.

Suckling, J., Brammer, M.J., Lingford-Hughes, A., Bullmore, E.T., (1999). Removal of extracerebral tissues in dual-echo magnetic resonance images via linear scale-space features. *Magnetic Resonance Imaging*, **17**, 247–256.

Talairach, J., Tournoux, P. (1988). *Co-planar Stereotaxic Atlas of the Human Brain*. New York, NY: Thieme.

PART III

PERCEPTION METROLOGY

13

Logistical issues in designing perception experiments

EHSAN SAMEI AND XIANG LI

13.1 INTRODUCTION

Perception experiments are a vital part of medical imaging research. Such experiments involve a group of observers, often clinicians, who provide interpretations of a set of medical images with known characteristics. The results are then analyzed to infer information about the performance of the observers, the utility of the clinical system, and/or the accuracy by which a clinical outcome is rendered. Designing such experiments involves multitudes of planning and logistical decisions about how to conduct the experiment. These include decisions involving defining the objective of the experiment, choosing appropriate methodology, deciding on target samples, and planning for experimental implementation, data collection, and data processing. Proper design would ensure that the desired statistical power and precision are achieved, potential biases and confounding factors are avoided, cost is minimized, and the experimental results are accurate, reproducible, definitive, and generalizable.

Devoted effort at the outset for proper design of a perception experiment cannot be underestimated; even powerful statistical analysis cannot salvage a poorly designed study. For specific design issues, there are numerous resources in the literature that one may turn to. However, these resources are scattered throughout the literature. The purpose of this chapter is to provide a glossary of major design considerations for planning a perception experiment and to discuss basic issues associated with such designs.

13.2 OBJECTIVE

13.2.1 Objective of the experiment

Most perception experiments aim to assess and compare the performance of imaging systems (e.g. imaging modalities, machines, algorithms, or displays). Some perception experiments may also be designed to compare different types of observers in performing imaging tasks on single or multiple imaging systems. Thus, in the context of perception experiments, it is best to incorporate the observer as part of the imaging system, recognizing that the observer is an indispensable element of the "system" as a whole.

In terms of the nature of the hypothesis, the objectives can be broadly divided into three types:

(1) Evaluation of a single imaging system (e.g. baseline evaluations on whether a particular imaging system provides adequate diagnostic value).
(2) Comparing the performance of two or more imaging systems (e.g. comparing the detection of lung nodules with and without CAD input; comparing the efficiency of radiology residents against experienced radiologists for a given diagnostic task).
(3) Assessing the equivalence of one imaging system to another (e.g. comparing digital and analog mammography in terms of diagnostic accuracy; determining whether the dose for a particular imaging exam can be reduced without affecting diagnosis).

In terms of the endpoint of the assessments and comparisons, five additional categories may also be defined:

(1) Studying imaging systems in terms of detection accuracy (e.g. the detection of lung nodules in chest CT exams).
(2) Studying imaging systems in terms of classification accuracy (e.g. distinguishing benign masses from malignant masses or their likelihood of malignancy in mammography).
(3) Assessing the subjective preference of observers for imaging systems (e.g. comparing the aesthetic acceptability of the appearance of images processed by two different image post-processing algorithms).
(4) Studying imaging systems in terms of the efficiency or time required to perform an observer task (e.g. comparing the speed of reading images on two different display devices).
(5) Assessing observer behaviors for a given imaging system (e.g. assessing variability for and among observers, satisfaction of search, memory effect, visual search patterns, and observer fatigue in reading a particular type of image).

Studies in the first and the second of the five categories above provide objective results pertaining to diagnostic accuracy. Therefore the majority of perception experiments fall into those categories. Studies in the third category fall short of providing objective results; however, they are easier to perform. Studies in the fourth and fifth categories focus more on the observer side of the imaging system and have traditionally been less frequently pursued. Interested readers are referred to relevant chapters in Parts I and II of this book for in-depth discussions on observer behaviors. This chapter aims to focus on the more commonly performed studies of categories 1 and 2 above.

Though it seems naive to assume otherwise, it is paramount that at the outset of a perception experiment the objective of the study is explicitly defined and documented. That step alone provides the most important guideline for answering study design questions.

The Handbook of Medical Image Perception and Techniques, ed. Ehsan Samei and Elizabeth Krupinski. Published by Cambridge University Press.

13.2.2 Phased approach

For general diagnostic procedures, including medical imaging, it is often most efficient to adopt a phased approach in that the overall objective of the study is achieved in multiple sequential phases (Zhou *et al.*, 2002). In that way, it is possible to refine the study design along the way to avoid "costly" mistakes in conducting large studies with premature designs.

A study may start with a pilot or exploratory phase in which a limited number of images and observers are employed. The study, while void of significant statistical power, may provide data needed to assess the magnitude of influencing effects and to "test-drive" the study design. The study may then proceed to a comprehensive "laboratory" phase in which a larger number of images and observers are used but the conditions are still kept fairly tractable to minimize variability. In the third and most comprehensive or clinical phase, the imaging systems are tested under a more realistic clinical scenario with added images and observers to fully test the hypothesis in the most clinically representative setting. Different phases of a perception experiment may have slightly different objectives, all aiming towards a broader overall objective.

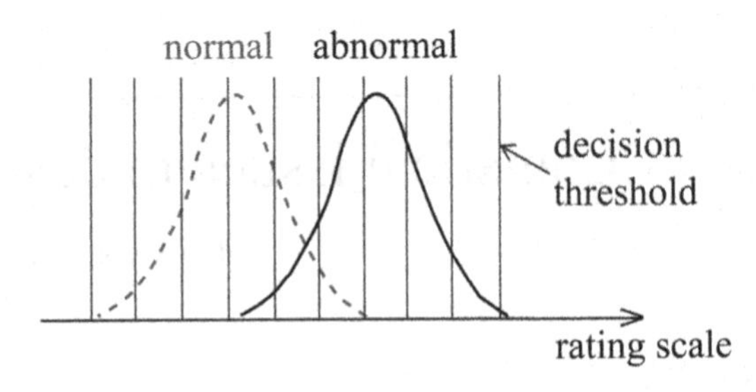

Figure 13.1 In an ROC experiment, the rating data are assumed to come from two underlying distributions: one for normal cases and the other for abnormal cases.

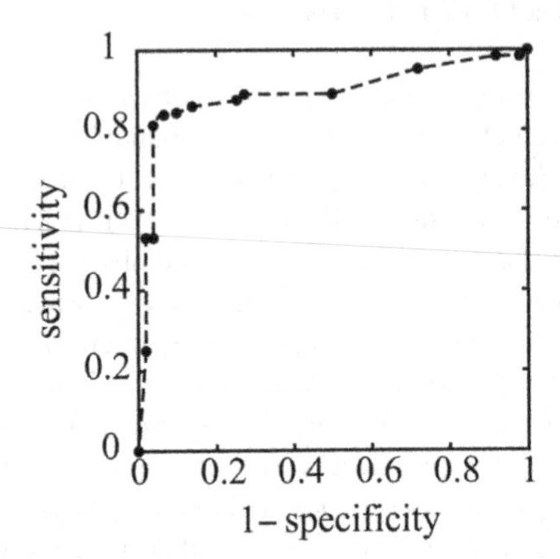

Figure 13.2 An ROC curve is a plot of sensitivity versus one minus specificity.

13.3 CHOICE OF METHODOLOGY

The methodology is an important choice to be made at the outset of an experiment. Generally speaking, for studies focused on assessing diagnostic accuracy (category 1 and 2 studies defined above), two types of methods are used: receiver operating characteristic (ROC) and alternative forced choice (AFC). It is possible for a study design not to conform to these two alternatives, but such approaches are rare and would require "customized" statistical analyses. Both types of methods require a sample of normal (negative) cases and a sample of abnormal (positive) cases. A case can be composed of a single image of a patient (e.g. a patient's mammogram) or a set of images of a patient (e.g. a patient's chest CT examination). "Normal" or "abnormal" condition is defined relative to the clinical task. For example, if the clinical task is the detection of lung nodules, a case with nodule absent is a normal case, whereas a case with nodule present is an abnormal case. Both ROC and AFC methods aim to measure the ability of an imaging system to discriminate between normal and abnormal conditions.

In an ROC experiment, an observer reviews each case and provides a confidence rating regarding "normal" or "abnormal" condition of the case for a specific clinical task. The confidence rating is given on either a categorical scale (e.g. 1: definitely normal, 2: probably normal, 3: unsure, 4: probably abnormal, 5: definitely abnormal) or a continuous probability scale (e.g. between 0% and 100% with 0% being definitely normal and 100% being definitely abnormal). The rating data are assumed to come from two underlying distributions: one for normal cases and the other for abnormal cases (Figure 13.1).

A series of decision thresholds are then placed on the rating scale (Figure 13.1). Each decision threshold represents a possible critical level at which the observer (e.g. a radiologist) operates in his/her daily practice: if the rating assigned to a case is above the critical level, the case is diagnosed as an abnormal case, whereas a case with a rating below the critical level is diagnosed as a normal case. In other words, the decision threshold reflects how strict or lax the observer is when deciding whether a case is normal or abnormal. The application of multiple decision thresholds in ROC analysis recognizes the fact that the strictness of an observer may vary from case to case and from day to day. Changing the decision threshold changes both an imaging system's sensitivity (i.e. the probability that an actually-abnormal case will be correctly classified as "abnormal" with regard to the given clinical task) and the imaging system's specificity (i.e. the probability that an actually-normal case will be correctly classified as "normal"). By measuring sensitivity and specificity at all possible decision thresholds of an observer, one can obtain an ROC curve, which is a plot of sensitivity versus one minus specificity (Figure 13.2). The diagnostic accuracy of the imaging system is then characterized by the area under the ROC curve (AUC), a summation of sensitivity at all specificities. As such, the AUC describes the *intrinsic* accuracy of the imaging system, free from the influence of the decision threshold.

In addition to the total area under the ROC curve, partial area under the ROC curve (pAUC), defined as the area between two specificities or two sensitivities, has also been used to describe diagnostic accuracy over a range of specificities (sensitivities) relevant to a particular clinical application. Furthermore, the ROC method has been extended to a three-way paradigm, in which the observer's rating reflects his/her confidence regarding not the usual two-way classification of "normal" and "abnormal," but a more sophisticated three-way classification such as "normal," "benign," and "malignant." For a comprehensive review of the ROC method, the statistical tests for evaluating difference/equivalence between ROC results, and software available for ROC analysis, the reader is referred to the next chapter. As we will discuss below, most perception experiments require more than one observer/reader. Special methods have been developed for multi-reader multi-case (MRMC) ROC data and they are the focus of Chapter 15.

In conventional ROC, the location of the abnormality within the image is not of primary relevance, an assumption often not reflecting clinical reality. To address this limitation, there have been a number of ROC variants devised including region-of-interest-based ROC (ROIROC), localization ROC (LROC), free-response ROC (FROC), and alternative free-response ROC (AFROC). ROIROC divides each image into multiple zones, each of which is rated as a separate image. In LROC, one rating-location pair is given to each image, while FROC and AFROC allow multiple rating-location pairs per image. Each ROC variant has its own particularities and limitations; some require statistical analyses which are more sophisticated and less standardized. To determine which methodology is most suitable for a particular study, we refer the reader to the detailed discussion of the unconventional ROC methods in Chapter 16.

As an alternative to ROC analysis, some studies have used the AFC methodology, most commonly in the form of the two-alternative forced choice (2AFC). In an AFC experiment, the observer reviews two or more independent images simultaneously (Figure 13.3). Exactly one of the images is actually abnormal with respect to a given clinical task. The observer is asked to indicate which image is abnormal. The data are analyzed to assess the percentage of correct decisions. It has been shown that the percentage correct in 2AFC experiments equals the AUC in the ROC method (Green, 1966; Bamber, 1975; Hanley, 1982). AFC experiments are generally much faster to conduct (i.e. more cases can be read per unit time). However, contrary to ROC, the AFC results do not provide information about the underlying distribution functions, and the tradeoff between sensitivity and specificity is not determined. Furthermore, the setting of an AFC experiment does not resemble an actual clinical paradigm; clinicians usually do not evaluate multiple unrelated images together. For in-depth discussions on AFC methods, the reader is referred to excellent articles on that subject (Burgess, 1995; Metz, 2000).

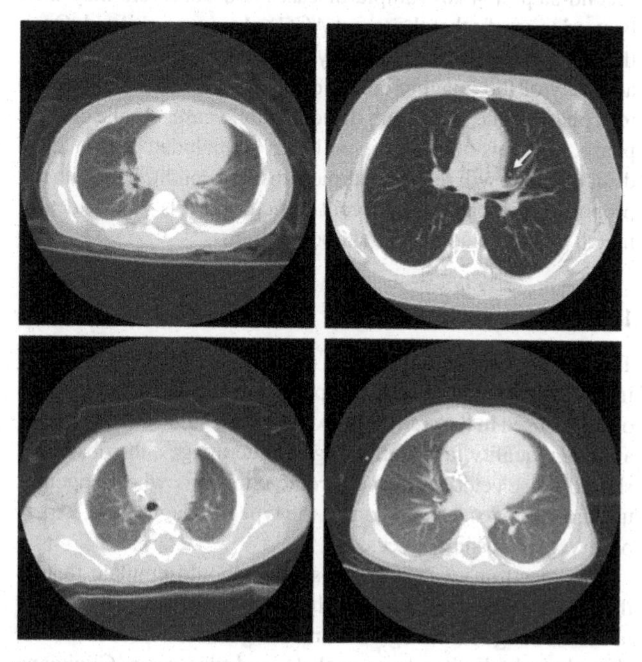

Figure 13.3 An example image panel used in a four-alternative forced choice experiment. Exactly one of the images contains a lung nodule. The observer is asked to identify which image has a nodule.

Both ROC and AFC differ from the clinical paradigm in one way or another. The choice of which one to use for a particular study depends on time/resource constraints, the granularity by which the study aims to assess accuracy (i.e. getting the area under the curve in AFC versus getting the full ROC curve in ROC), and whether the implication of the deviation of the method (as implemented) from clinical practice is adequately considered. Once a choice of methodology is made, that generally informs the other aspects of study design considerations discussed below.

13.4 SAMPLE SELECTION

13.4.1 Case and image selection

A perception experiment may employ three types of images: real images from actual patients, simulated images from computer or physical models of patient anatomy, and hybrid images which contain computer-simulated abnormalities added to real image backgrounds (Figure 13.4).

Among the three, real images are usually the preferred choice as they fully represent the clinical paradigm. However, they have major drawbacks. Collection of clinical cases is a time- and cost-consuming process. The abnormal features within the image often need to be subtle enough to be appropriate for the investigation at hand. Finally, clinical cases frequently contain more than one type and location of abnormality. Adherence to regulatory standards and Health Insurance Portability and Accountability Act (HIPPA) regulations further adds to the complications associated with this particular option.

An important consideration in using actual patient cases: the cases need to have associated "truth" in that either the normalcy or the abnormality of the case needs to be established in reference to the true disease status or the gold standards. Ideally, the gold standard should be based on the results of surgery or pathology (biopsy). However, patients who undergo biopsy represent only a subgroup of the patient population. When biopsy results are unavailable, results from clinical workup have been used to serve as gold standard (Thornbury, 1993; Dobbins, 2008). In the absence of that option, the truth can be established by expert opinion or expert consensus. Reviews of the potential biases associated with using suboptimal gold standards and the approaches to minimize these biases can be found in the next chapter and in Zhou *et al.* (Zhou, 2002).

In any phase of a study, the selected image sample for the perception experiment must reasonably represent the patient population of interest; a lack of important subgroups of patients can bias study results. As an example, bias is introduced into the selection process when the image sample contains only images of patients who are likely to participate in the study or elect a diagnostic test. Proper study design requires careful sampling of the patient population.

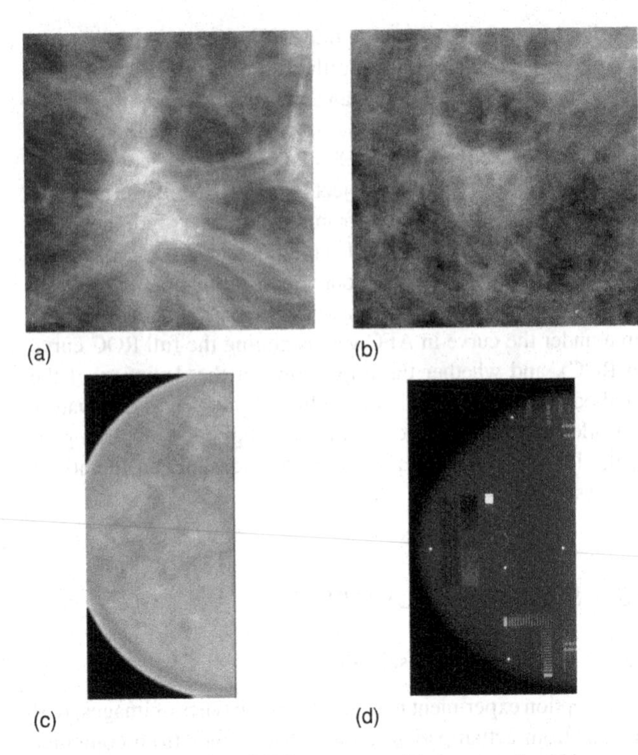

Figure 13.4 Examples of different types of images employed in perception experiments: (a) real image (real mammographic background and lesion), (b) hybrid image (real mammographic background with a simulated lesion), (c) simulated image from computer model of human anatomy (a simulated mammogram from a computerized breast model), (d) simulated image from physical model of human anatomy (an image of a breast phantom (Model 011A, CIRS, Norfolk, VA); image courtesy of Joseph Y. Lo, PhD).

Disease prevalence (i.e. the ratio of abnormal to normal cases) is another important aspect of case selection. In general, the observer results are assumed to be free from the influence of disease prevalence as long as the sample sizes of abnormal and normal cases are large enough. However, studies have shown that the disease prevalence can influence the observer's psychology, and that observer accuracy may increase with disease prevalence (Egglin, 1996). This effect should be taken into consideration in the design of perception experiments.

A second alternative to the use of actual patient data is to rely on fully simulated images. Those include simulations using computational models (Segars, 2008) or physical phantom models (Ko, 2003). The simulated images can represent simple objects (e.g. a uniform cylinder representing the abdomen for CT imaging) (Schindera, 2008), or an anthropomorphic model (Segars, 2008). Realistic simulation of clinical cases is extremely challenging, and no study, up to now, has provided simulated images that are equivalent to real images. Nonetheless, these images can be effective in observer-based assessment of the impact of certain physical aspects of imaging systems (e.g. quantum noise), and can be invaluable in pilot phases of subsequent studies using more clinically relevant images.

An effective alternative to the two choices above, hybrid images use actual clinical cases with added realistic simulation of abnormalities. Simulation of abnormalities alone can be easier to achieve, as showcased by a number of studies modeling mammographic, pulmonary, and hepatic lesions (Saunders, 2006; Samei, 1997; Hoe, 2006). Computer-simulated abnormalities offer several advantages over real abnormalities. First, the characteristics of the abnormalities, such as size, shape, contrast, and location, can be well controlled, permitting the study of abnormalities with desired characteristics. Second, the prevalence of the abnormalities can be controlled. Images containing isolated abnormalities, which may be rare in actual patients, can be readily created. Moreover, the truth is readily known: simulated abnormalities are by definition "real" since they are intentionally inserted within the image. Lastly, compared with the difficult and time-consuming process of collecting patient data with abnormalities of desired characteristics, a large database of cases can be quickly generated to enable a large-scale perception experiment.

As alluded to earlier, the level of difficulty or subtlety of the abnormalities within the image set is an important consideration for a perception experiment. For example, in comparing two different modalities in detecting certain lesions, if the lesions are too apparent, they could easily be identifiable by either modality, thus masking possible differences in performance. Similarly, if the abnormalities are too subtle, they will be missed by both modalities. In ROC experiments, it is generally recommended that a target AUC of 0.7–0.8 is sought, so that the differences would be most discernable. When comparing different imaging systems, AUC values associated with all systems should also ideally fall proximate to this range. The appropriate level of subtlety of abnormalities (or range of subtleties for all imaging systems under investigation) can be determined and confirmed in the pilot/exploratory phase of a study in two steps. In the first step, a representative observer may be presented with cases having a wide range of subtlety levels to select from. Alternatively, software may be used to allow the representative observer to iteratively adjust the conspicuity (e.g. the contrast of a simulated lesion) until the desired subtle range is reached. In the second step, a small sample of cases and observers may then be used to verify that the target AUC values are achieved. Once the appropriate level of subtlety is ascertained, the diagnostic accuracy of the imaging condition(s) is assessed in the laboratory phase of the study with difficult/subtle cases. In the clinical phase, all levels of difficulty will then be included. The level of difficulty here will represent typical disease conditions while the proportions of cases at different difficulty levels would ideally match that of an actual patient population.

13.4.2 Observer selection

The observer is a quintessential element of a perception experiment. The objective of the experiment is tied to the kind of observer used in the study. For example, if a study aims to assess an image quality factor affecting cardiac images, the observers should be selected from the cardiac clinicians who frequently use those images. Depending on the type of images used, expert observers are usually necessary.

The spectrum of selected observers should emulate that in the actual clinical situation. If the users of the imaging system under investigation are general radiologists, observer selection should not be limited to specialists, and vice versa. Clinicians can also come from a broad range of experience, in terms of both years of practice as well as the volume of images read. Again the objective of the experiment would determine that choice. The source of observers is another important consideration. The broader the source of observers, the more generalizable the conclusions can be; if only observers from the same institution/hospital are accessible, the conclusions can only be generalized to comparable-level care units.

In certain types of perception experiments, the focus is on a psychophysical aspect of the human visual system. The objectives of such studies do not involve a clinical condition. As such, the experiment can use non-clinician lay observers such as imaging scientists and students (Pollard, 2008). In certain other types of studies, the imaging system under investigation might involve other members of allied health personnel such as radiological technologists or nurses. Needless to say, such perception experiments need to employ those individuals as observers.

In addition to human observers, model (mathematical) observers have been widely used in perception experiments. Model observers are mathematical formulations that combine the characteristics of the human visual system with the statistical properties of an image. The need for model observers stems from the fact that the diagnostic accuracy of any modern imaging system is often determined by a myriad of parameters. Optimization of these parameters requires the testing of a large number of conditions (i.e. combinations of parameters), which is expensive and time-consuming to perform using human observers. Model observers are designed to emulate the performance of human observers. They play an essential role in the preliminary evaluation of new imaging systems (Chawla, 2008) and are important in subsequent optimizations of imaging protocols with respect to image quality and radiation dose (Boedeker, 2007). Chapter 17 provides a thorough review on model observers. Applications of model observers in perception experiments are discussed in Chapter 18.

13.4.3 Case and observer crossing/matching

When comparing two or more imaging systems, it is often desirable to obtain and present images from each sample case/patient in all image systems, a study design known as case-crossing. This approach rules out the potential contribution of case variation to the difference in diagnostic accuracy and improves statistical power. However, as much as it is statistically desirable, case-crossing has two major problems that might make it infeasible for many experiments. First, it is often not possible to image a patient more than once because of radiation dose or logistical reasons. But even if that can be overcome, second, an observer might remember a patient from his/her first reading so that the second reading of the case in a different imaging system will be affected. This phenomenon is known as the "memory effect." It is possible to design studies with randomization and time lags such that the memory effect is minimized or averaged across observers (see the sections on "Image and viewing order randomizations" and "Session design" in this chapter), but it cannot be fully eliminated.

When case-crossing is not feasible, case-matching can be used, provided that the cases/patients representing different imaging systems are drawn from similar populations to avoid introduction of bias. For example, in a study aimed to assess the effectiveness of two imaging systems for lung cancer screening, the two imaging systems should not represent populations with differing smoking habits.

As with image selection, it is also desirable to have the same sample of observers read images representing all imaging systems, a study design known as observer-crossing. This design eliminates the contribution of observer variation to the measured difference in diagnostic accuracy and improves statistical power. But again, observer-crossing might not always be possible. For example, observers from one institution may have access to only one imaging system. In those situations, observer-matching may be employed in that observers reading different imaging systems have somewhat similar attributes (e.g. similar experience or expertise). That does not guarantee the absence of bias, but at least minimizes the likelihood.

A study design that employs both case-crossing and observer-crossing is often referred to as a fully-crossed design. Because such a design is the most statistically powerful (Dorfman, 1992), it is adopted by many multi-observer ROC studies and is the target of many ROC software tools.

13.4.4 Numbers

The numbers (sample size) are of prime importance in the design of a perception experiment. In that regard, the investigator needs to decide on the number of cases or images, the number of observers, and the number of repeated readings per observer. Integral to these decisions are the methodology that a study will employ. For example, the computation of the numbers depends on whether the study will employ a ROC or an AFC method. In general, the needed numbers depend on the expected magnitude of the difference that exists between different imaging systems in relationship to the expected magnitudes of variability. If there were no sources of variability in an experiment, a single case and a single observer could provide sufficient information regarding whether a particular system is better or worse than another one. However, in any perception experiment, there are three major sources of variability:

(1) The first source of variability is the variability associated with the case sample brought about by the fact that no two cases presenting the same abnormality would invoke the same response from the observer. The case sample variability necessitates that a number of cases larger than one be used in the experiment. Typically numbers range between tens of cases, used for pilot and feasibility experiments, to hundreds or even thousands for laboratory and clinical studies.

(2) The second source of variability is the inter-observer variability. It is brought about by the fact that the same case presented to different observers will not generate the same response from the observers. The inter-observer variability similarly would require the participation of more than one observer in an experiment. In the majority of perception studies, a minimum of three (typically four to six) observers is deployed, with larger-scale clinical trials and observer-focused studies using tens or even hundreds of observers (Obuchowski, 2004).

(3) The third source of variability is the intra-observer variability brought about by the fact that the same case presented to an observer twice will not bring about the same response from the observer. Intra-observer variability is

commonly not isolated out; it is assumed to be part of the all-encompassing "observer" variability. However, if it is measured specifically, it can provide added statistical power to the study. To do this, the observers would need to provide redundant readings of some or all the images. Integral to that design, however, the compounding influence of the memory effect should be taken into consideration.

Here we provide a simple example[1] to illustrate the dependence of sample size on the difference between imaging systems in relationship to the magnitudes of the three sources of variability.

Consider a perception experiment that compares the diagnostic accuracy (AUC) of two imaging systems X and Y for a specific clinical task. For each imaging system, the researcher collects n cases; the cases in X are independent of (unmatched to) the cases in Y. The researcher plans to recruit a group of l observers, each of which will review all cases in both X and Y (i.e. observers crossing). Provided that the number of cases cannot be further increased, the question for the researcher is then how many observers he/she will need in order to demonstrate a statistically significant difference in AUC between the two imaging systems if such a difference does exist.

Let us start by analyzing the different variance terms of AUC. We denote σ_c^2 as the variance in AUC due to case sample variability, σ_{br}^2 as the variance in AUC due to inter-observer variability, and σ_{wr}^2 as the variance in AUC due to intra-observer variability. Because σ_c^2 and σ_{br}^2 cannot be measured directly without including σ_{wr}^2, case sample variability is measured indirectly as $\sigma_{c+wr}^2 (= \sigma_c^2 + \sigma_{wr}^2)$, the variance in AUC that would be obtained by having one observer read once a set of different case samples, and inter-observer variability is measured indirectly as $\sigma_{br+wr}^2 (= \sigma_{br}^2 + \sigma_{wr}^2)$, the variance in AUC that would be obtained by having one case sample read once by all the observers. Assuming σ_{c+wr}^2, σ_{br+wr}^2, and σ_{wr}^2 are the same for the two imaging systems, the standard error for the difference between X and Y can be calculated as (Swets, 1982):

$$\text{S.E.}_{(\text{diff})} = 2^{1/2}\left[\sigma_{c+wr}^2 + \frac{\sigma_{br+wr}^2}{l}(1 - r_{br-wr}) - \sigma_{wr}^2\right]^{1/2} \quad (13.1)$$

where r_{br-wr} is the correlation between the AUC values in imaging system X ($\text{AUC}^{\text{X}}_{observer\ 1}, \text{AUC}^{\text{X}}_{observer\ 2}, \ldots \text{AUC}^{\text{X}}_{observer\ l}$) and those in imaging system Y ($\text{AUC}^{\text{Y}}_{observer\ 1}, \text{AUC}^{\text{Y}}_{observer\ 2}, \ldots \text{AUC}^{\text{Y}}_{observer\ l}$) and is a result of observer crossing in this experiment. In an ideal situation of $r_{br-wr} = 1$, the contribution to the standard error from inter-observer variability vanishes, demonstrating the advantage of crossing the observers. Assuming the case sample is large, the test statistic follows a normal distribution, i.e.

$$z = \frac{\text{AUC}^{\text{X}} - \text{AUC}^{\text{Y}}}{\text{S.E.}_{(\text{diff})}}, \quad \text{or}$$

$$z = \frac{\text{AUC}^{\text{X}} - \text{AUC}^{\text{Y}}}{2^{1/2}\left[\sigma_{c+wr}^2 + \frac{\sigma_{br+wr}^2}{l}(1 - r_{br-wr}) - \sigma_{wr}^2\right]^{1/2}} \quad (13.2)$$

[1] The example is modified from Swets (Swets, 1982). Similar notation and values are used.

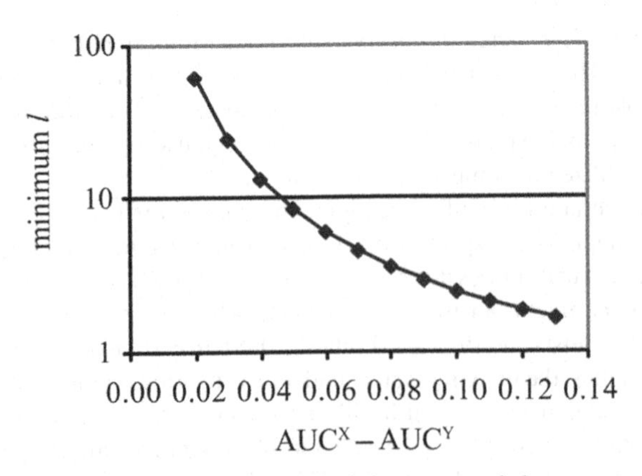

Figure 13.5 The minimum number of observers needed to demonstrate statistically significant difference between X and Y as a function of the AUC difference between the two imaging systems, assuming $\sigma_{c+wr}^2 = 0.00121$, $\sigma_{br+wr}^2 = 0.0050$, $\sigma_{wr}^2 = 0.0012$, and $r_{br-wr} = 0.5$.

where $\text{AUC}^{\text{X/Y}}$ is the observer-averaged AUC value in imaging systems X and Y. Using a significance level of 0.05, the difference between X and Y is significant if z is larger than the 97.5% percentile of the z-distribution, i.e. if $z > 1.96$.

Let us assume $\sigma_{c+wr}^2 = 0.00121$, $\sigma_{br+wr}^2 = 0.0050$, $\sigma_{wr}^2 = 0.0012$, and $r_{br-wr} = 0.5$. Figure 13.5 plots the minimum number of observers needed to demonstrate statistically significant difference between X and Y as a function of the AUC difference between the two imaging systems.

Because the required sample size depends on the difference between image systems in relationship to the magnitude of variability, it is difficult to suggest a universal guideline for the numbers of cases and observers needed for a perception experiment. Obuchowski (Obuchowski, 2000) provided tables of sample size for fully-crossed multi-observer ROC experiments that compare the diagnostic accuracies of two imaging systems. Statistical programs that perform sample size computations for multi-observer ROC studies are also available for download (Hillis, 2006). Readers interested in the theoretical background of sample size calculations are referred to a relevant publication (Zhou *et al.*, 2002).

In designing an observer experiment, operational constraints are an ever-present reality. Generally speaking, those constraints manifest themselves into two forms: case-limited constraints, and observer-limited constraints. In the former, the investigator faces a limited number of cases, which limits the statistical power of any study based on that limited number. In those cases, a larger number of observers and repeated readings may be used to reduce the observer sources of variability. In the latter, the investigator might have access to only a limited number of observers, each with limited time constraints. In that scenario, it might be possible to limit the scope of the experiment, limit the number of questions that the observers need to answer, and streamline the display software to move the observers through the images as efficiently as possible. In many perception experiments, a combination of both constraints exists, necessitating careful planning and execution of the study.

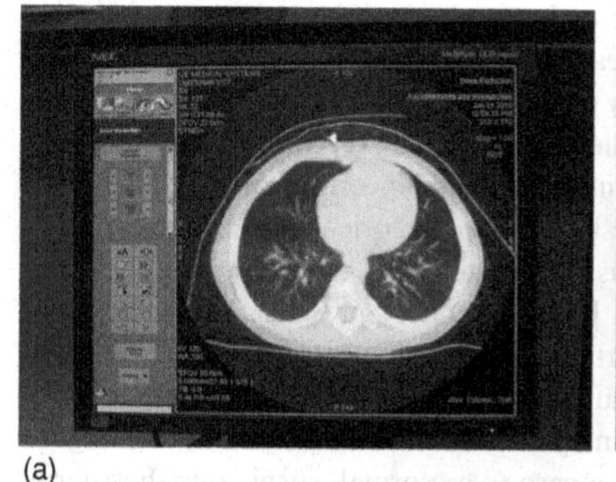
(a)

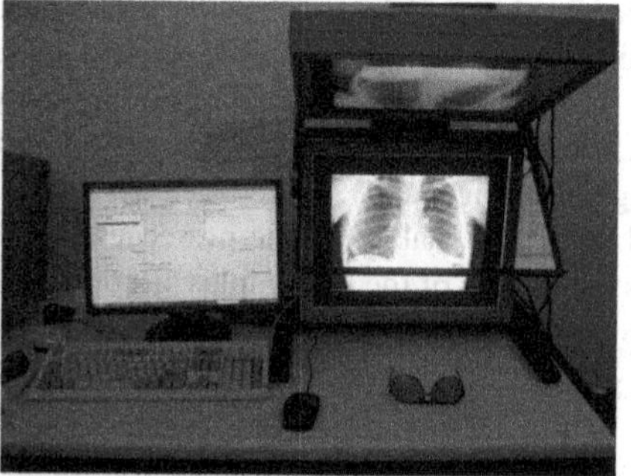
(b)

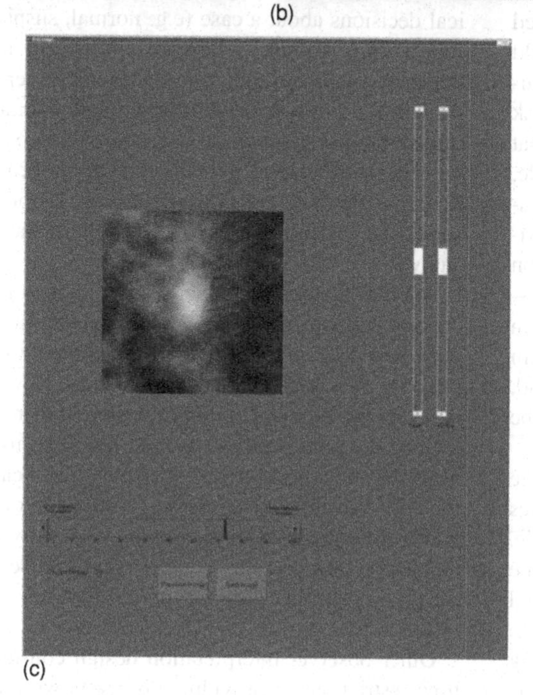
(c)

Figure 13.6 Examples of graphical user interfaces (GUIs) used in perception experiments: (a) the same GUI and display device as used clinically, (b) specialized GUI and display device for stereoscopic viewing of images where image display and controls are located on two different display hardware, (c) specialized GUI that provides rating and recording functionalities.

13.5 DATA COLLECTION

13.5.1 Display hardware and software

A perception experiment would naturally involve presenting images to observers. As such, the display device is an integral part of such experiments. Display devices themselves can be the imaging system of interest. For example, an experiment might aim to assess the impact of display physical attributes (e.g. grayscale rendition, pixel density, etc.) on diagnostic accuracy (Samei, 2008). However, in many studies in which these are not the focus, the choice of the display may have a bearing on the outcome. Care should be exercised so that the characteristics of the display devices used in the perception experiment are not limiting visual presentation and are consistent with the objectives of the study (e.g. experimental displays for a clinical study should have the same grade and calibration as those used clinically).

The display software, generally named the graphical user interface (GUI), is also of utmost importance (Figure 13.6). The software should provide the functionalities that the objective of the study necessitates. For example, if a study aims to compare two CT systems in terms of diagnostic accuracy on a particular task, the observers should be provided with the same functions as those they use in their daily clinical practice to interpret such images (e.g. magnification, cine scrolling, region-of-interest measurements, window/leveling, volume rendering, etc.). To do this, a study might employ the same software as is used clinically (Figure 13.6a). In that approach, paper or electronic scoring sheets will be used.

The use of clinical software for perception experiments usually does not provide certain functionalities that would make perception experiments more robust, namely the ability to record the observer's rating, marking, and timing during the experiment. In order to do that, some studies employ specialized GUI software that provides not only the basic clinical functionality but also additional recording tools. In development of such software, it is important to minimize the observers' burden (in terms of time and attention) in using the recording features. For example, it is important to make sure that only a minimum area of the display is devoted to those features, and the observer input can be provided with a minimum number of key strokes or mouse clicks (Saunders, 2004) (Figure 13.6c).

13.5.2 Reading environment

In addition to the display hardware and software discussed above, the reading environment of a perception experiment should be properly set up. The environment should provide a consistent and optimized ambient illumination condition (Chawla, 2007), unless that attribute alone is the subject of the study (Pollard, 2008). Furthermore, the space should be free of distracting visual clutter and distracting noise. Pager and telephone use should ideally be avoided and limited to enable maximum concentration (again, unless that is the subject of the study).

13.5.3 Observer training

Regardless of the clinical experience of the observers employed in a study, a certain amount of training is essential. That would even be the case if the observers had participated in prior similar experiments. Training is necessary since the perception task that an observer is expected to perform rarely matches what the observer does clinically on a day-to-day basis. For example, most clinical tasks involve a binary decision of "disease presence" or "disease absence," whereas most perception experiment tasks require an observer to render a confidence rating on a categorical or continuous scale. Furthermore, the characteristics, subtlety, and prevalence of abnormalities might differ from those seen clinically. The training provides the observer with an opportunity to fully understand the perception task at hand, and to "calibrate" his/her reading with the images that will be presented.

Furthermore, training familiarizes the observer with the experimental protocol, setup, and display device and gives him/her the opportunity to practice using the full rating scale. Generally speaking, rating on a categorical scale requires more training. During the training session, feedback can be provided to the observer so that observer can adjust his/her response to better comply with the study design. When training is missing or inadequate, the observers have to train themselves during the first part of the actual experiment. As a result, the observer's performance changes during the course of the experiment, leading to a potential bias or reduced statistical power.

13.5.4 Observer interpretation

In a perception experiment, the observer is to view an image (or a series of images) and provide an interpretation. The expected interpretation is a key consideration in designing the perception experiment. The interpretation is usually a response to a specific question or set of questions. For example, an observer might be asked whether a lesion is present within the image that he/she is presented with. That question might be followed with a second one asking the observer to rate the confidence of his/her answer, and then further mark the location of the suspect lesion. Depending on how specifically these questions are asked, the observer might render different responses. For example, the interpretation might be different if the observer is asked about the likelihood of the presence of a lesion versus the likelihood of the presence of a *malignant* lesion. Ambiguous questions lead to ambiguous responses and to ambiguous findings. A proper study design requires that the inquiring questions, and the order in which they are answered, be specific, explicit, and unambiguous, consistent with the specific objective of the study. Furthermore, the investigator should make sure that the observer has a clear understanding of the task at hand by the end of the training session so that he/she can provide consistent readings of the study cases.

As discussed earlier, in an ROC experiment, the observers are often asked to rate their confidence in their judgment (e.g. the presence of a lesion) on a categorical scale. Some studies employ a continuous rating from 0% (a lesion definitely not present) to 100% (a lesion definitely present). While the continuous scale has certain statistical advantages (Metz, 2000), it does not closely emulate the clinical paradigm, where a clinician makes either binary (e.g. normal or abnormal) or categorical decisions about a case (e.g. normal, suspicious/short-term follow-up, long-term follow-up, biopsy). It might be argued that if the interpretation task that the observer is asked to perform is not consistent with the clinical paradigm, that would impact the psychophysical and cognitive processes associated with the interpretation, thus affecting the clinical relevance of the findings. The pros and cons of various interpretational choices should be carefully considered in light of the objective of the study.

Another factor that influences the clinical relevance of perception experiments is observer diligence. In clinical settings, clinicians have an inherent understanding of the significance of possible mistakes in their interpretation. However, in the artificial setting of a perception experiment, there is no penalty in making a wrong decision, at least not in terms of patient care. Thus the study is at the mercy of how much care and diligence the observer will be exerting in the conduct of the experiment. It might be possible to incorporate a penalty mechanism into the design of a perception experiment to assess and to control for the impact that diligence might have on the interpretation results.

Other observer interpretation design considerations include time restrictions in providing observers with unrestricted time or limited time per image. Such a choice will impact the outcome. Furthermore, the observers can be blinded or unblinded to patient history. That choice should also be made in reference to the specific objective of the experiment.

13.5.5 Image and viewing order randomizations

Ideally, the observer's performance should reach a stable level or a plateau at the end of the training session and stay constant during the actual experiment. In reality, the time and number of images available for training are usually limited, and the observer's performance may still change during the experiment. Furthermore, each particular case has unique features that might uniquely impact an observer score, leading to a potential bias. This is particularly problematic when two or more imaging systems are being compared. For example, if each observer first reads images associated with system one, and the observer performance should improve throughout the course of the experiment, system one will be judged to be of inferior quality, a conclusion that would be merely a consequence of the reading order.

In order to overcome such potential biases or otherwise minimize the effect, the order in which the images representing different systems are shown to each observer should be randomized. Furthermore, across observers, a different randomization

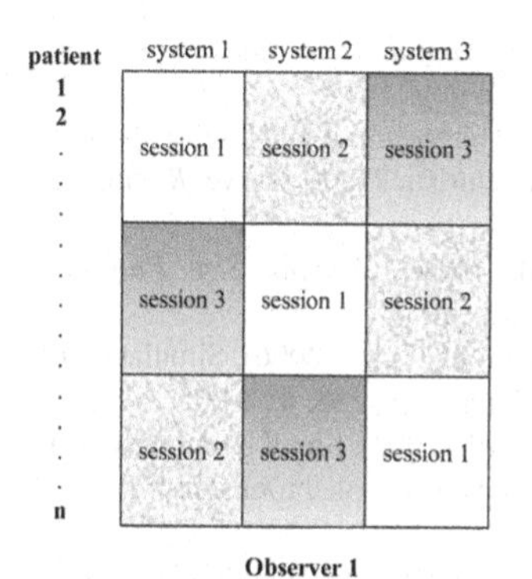

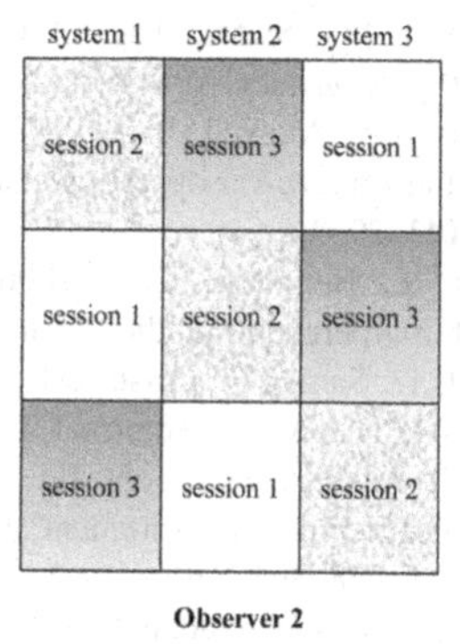

Figure 13.7 An example session design for a fully-crossed perception experiment. Each observer reviews images of all patients in a single session, with one-third of the patients being reviewed in system one, one-third in system two, and another one-third in system three. Images of the same patient will not be presented in different systems in a single session of an observer. The viewing order is varied from observer to observer.

order should be invoked. If this cannot be easily achieved, the study should be designed such that the order is randomized across observers (e.g. the first observer reads images of system one followed by those of system two, while the second observer reads images of these systems in opposite order). That would counterbalance the effect of performance improvement during the experiment if the results from different observers are averaged.

13.5.6 Session design

Unless group consensus is part of the design, in the majority of perception experiments, each observer evaluates the images independently. Small perception experiments with a limited number of cases may be conducted in a single reading session per observer. However, most perception experiments involve a large number of images and cases which cannot be read in a single session. In a multi-session experiment, each observer's reading period is divided into multiple sessions. A study may also deliberately use a multi-session design to reduce observer fatigue, or to allow time gaps between the observer's viewings of the same cases to minimize possible memory effect. In that case, it would be important to have adequately long time gaps in the order of weeks or even months. Figure 13.7 illustrates an example session design for a perception experiment, in which each observer reads each case in each system (i.e. a fully-crossed experimental design).

13.5.7 Inconsistent or missing data

Even in the most carefully designed and executed perception experiments, there can be missing or erroneous data. Depending on the situation, those data may still be "redeemable" to be incorporated in the study. However, any such correction of the data should be explicitly reported as the correction might have unintentionally introduced a bias in the study.

In cases where inconsistent or missing data are isolated and due to recording errors, the erroneous data may be corrected, provided that the study could establish that no systematic bias is introduced by such "corrections." For example, if the timing data from an observer shows erratic values due to pauses in the reading process (e.g. answering a pager), such data can be deleted from the analysis. Outlier data, on the other hand, might require follow-up analysis and reporting. In other cases, a larger portion of the data might be missing. For example, an observer might have left the study unfinished. In those situations, it might be legitimate to include the partial data in the study. Analysis should be performed with and without such inclusion to determine if that changes the trend of the data in any way.

13.6 SUMMARY

A perception experiment can be challenging to design as it involves a multitude of logistic issues. Because many design issues pivot around the objective of the experiment, it is paramount that the objective is clearly defined at the outset of the experiment. This includes the nature of the hypothesis (testing baseline, difference, or equivalence) and its endpoint (detection, classification, preference, efficiency, or observer behavior). It is often most efficient to adopt a phased approach (exploratory, laboratory, and clinical phases) to avoid "costly" mistakes in conducting large studies with premature designs.

As the endpoint of most perception experiments is detection or classification, ROC analysis is the most commonly used methodology. Depending on whether the location of the abnormality is of primary interest, conventional ROC or ROC with localization may be used. If observer time is a major constraint and the full ROC curve is not sought, the AFC method is a good alternative to ROC and may be more appropriate for certain clinical situations.

Both ROC and AFC methods require a sample of (normal and abnormal) cases and a sample of observers. The case sample must represent the patient population of interest. The type of cases/images can be real, simulated, or hybrid. Each case must have associated truth. The ratio of normal to abnormal cases may have a bearing on the observer's psychology and hence performance. The difficulty/subtlety of the cases should be appropriate for the phase of the study. The observer sample must represent the observer population of interest, taking into account specialty and experience. Model (mathematical) observers can substitute human observers in initial evaluation or multi-parameter optimization of a system. Whenever possible, case/observer crossing should be sought to improve statistical power or alleviate sample size requirements. The needed sample size depends on the expected magnitude of the difference between imaging systems in relationship to the expected magnitudes of case sample, inter-observer, and intra-observer variability.

As with sample selection, collection of observer data also deserves careful planning. Generally speaking, the display hardware and software should have similar characteristics and functionalities as used clinically unless they are the subject of investigation. Specialized GUIs may also be devised to streamline the rating and recording process. The reading environment should be controlled and free of distraction. The questions asked to the observer need to be specific, explicit, and consistent with the objective of the study. To minimize the drift of observer performance during the course of the experiment, adequate training should be provided, especially when a categorical rating scale is used. The viewing order should be randomized for different observers to further reduce the effect of the learning period. It might be necessary to conduct the experiment in multiple sessions to minimize memory effect. Lastly, care should be exercised when handling inconsistent or missing data to avoid introducing bias to the study result.

The success of a perception experiment relies on careful consideration of all the issues discussed in this chapter prior to the actual experiment. The researchers are advised to seek the help of clinicians and statisticians at the design stage of the study.

REFERENCES

Bamber, D. (1975). Area above ordinal dominance graph and area below receiver operating characteristic graph. *J Math Psychol*, **12**, 387–415.

Boedeker, K.L., McNitt-Gray, M.F. (2007). Application of the noise power spectrum in modern diagnostic MDCT: part II. Noise power spectra and signal to noise. *Phys Med Biol*, **52**, 4047–4061.

Burgess, A.E. (1995). Comparison of receiver operating characteristic and forced choice observer performance measurement methods. *Med Phys*, **22**, 643–655.

Chawla, A.S., Samei, E. (2007). Ambient illumination revisited: a new adaptation-based approach for optimizing medical imaging reading environments. *Med Phys*, **34**, 81–90.

Chawla, A.S., Samei, E., Saunders, R.S., Lo, J.Y., Baker, J.A. (2008). A mathematical model platform for optimizing a multiprojection breast imaging system. *Med Phys*, **35**, 1337–1345.

Dobbins, J.T., McAdams, H.P., Song, J.W., *et al.* (2008). Digital tomosynthesis of the chest for lung nodule detection: interim sensitivity results from an ongoing NIH-sponsored trial. *Med Phys*, **35**, 2554–2557.

Dorfman, D.D., Berbaum, K.S., Metz, C.E. (1992). Receiver operating characteristic rating analysis. Generalization to the population of readers and patients with the jackknife method. *Invest Radiol*, **27**, 723–731.

Egglin, T.K., Feinstein, A.R. (1996). Context bias. A problem in diagnostic radiology. *JAMA*, **276**, 1752–1755.

Green, D.M., Swets, J.A. (1966). *Signal Detection Theory and Psychophysics*. New York, NY: Wiley.

Hanley, J.A., McNeil, B.J. (1982). The meaning and use of the area under a receiver operating characteristic (ROC) curve. *Radiology*, **143**, 29–36.

Hillis, S., Berbaum, K.S. (2006). *MRMC Sample Size Program*. http://perception.radiology.uiowa.edu

Hoe, C.L., Samei, E., Frush, D.P., Delong, D.M. (2006). Simulation of liver lesions for pediatric CT. *Radiology*, **238**, 699–705.

Ko, J.P., Rusinek, H., Jacobs, E.L., *et al.* (2003). Small pulmonary nodules: volume measurement at chest CT – phantom study. *Radiology*, **228**, 864–870.

Metz, C. (2000). Fundamental ROC analysis. In *Handbook of Medical Imaging, Vol. 1, Physics and Psychophysics*, eds. Beutel, J., Kundel, H.L., Van Metter, R.L. Washington, DC: SPIE Press, pp. 751–769.

Obuchowski, N.A. (2000). Sample size tables for receiver operating characteristic studies. *AJR Am J Roentgenol*, **175**, 603–608.

Obuchowski, N.A. (2004). How many observers are needed in clinical studies of medical imaging? *AJR Am J Roentgenol*, **182**, 867–869.

Pollard, B.J., Chawla, A.S., Delong, D.M., Hashimoto, N., Samei, E. (2008). Object detectability at increased ambient lighting conditions. *Med Phys*, **35**, 2204–2213.

Samei, E., Flynn, M.J., Eyler, W.R. (1997). Simulation of subtle lung nodules in projection chest radiography. *Radiology*, **202**, 117–124.

Samei, E., Ranger, N.T., Delong, D.M. (2008). A comparative contrast-detail study of five medical displays. *Med Phys*, **35**, 1358–1364.

Saunders, R., Samei, E., Baker, J., Delong, D. (2006). Simulation of mammographic lesions. *Acad Radiol*, **13**, 860–870.

Saunders, R.S., Jr., Samei, E., Hoeschen, C. (2004). Impact of resolution and noise characteristics of digital radiographic detectors on the detectability of lung nodules. *Med Phys*, **31**, 1603–1613.

Schindera, S.T., Nelson, R.C., Mukundan, S., Jr., *et al.* (2008). Hypervascular liver tumors: low tube voltage, high tube current multidetector row CT for enhanced detection – phantom study. *Radiology*, **246**, 125–132.

Segars, W.P., Mahesh, M., Beck, T.J., Frey, E.C., Tsui, B.M. (2008). Realistic CT simulation using the 4D XCAT phantom. *Med Phys*, **35**, 3800–3808.

Swets, J.A., Pickett, R.M. (1982). *Evaluation of Diagnostic Systems: Methods from Signal Detection Theory*. New York, NY: Academic Press.

Thornbury, J.R., Fryback, D.G., Turski, P.A., *et al.* (1993). Disk-caused nerve compression in patients with acute low-back pain: diagnosis with MR, CT myelography, and plain CT. *Radiology*, **186**, 731–738.

Zhou, X.-H., Obuchowski, N.A., McClish, D.K. (2002). *Statistical Methods in Diagnostic Medicine*. New York, NY: Wiley-Interscience.

14

Receiver operating characteristic analysis: basic concepts and practical applications

GEORGIA TOURASSI

14.1 INTRODUCTION

Receiver operating characteristic (ROC) analysis is commonly used to assess the performance of prediction and classification models developed to separate two mutually exclusive classes (i.e. normal vs. diseased patients). Although the fundamental principles of ROC analysis were introduced in the 1950s, there have been significant developments over the past 20 years with respect to intricate statistical concepts. Nowadays, ROC analysis enjoys widespread acceptance in the medical imaging community as well as in clinical radiology. There are numerous reviews and tutorials on the topic, focusing mainly on the statistical aspects of ROC and its proper utilization in the medical domain. The purpose of this chapter is to review the basic theoretical concepts of ROC analysis, provide examples of how to construct ROC curves depending on the type of available data, discuss how to conduct ROC analysis with readily available software, and explain how to summarize the results of an ROC study. The ultimate goal is to provide the newcomer with the basic skills needed to recognize when ROC analysis is appropriate and conduct it properly so that valid conclusions are drawn. Advanced topics and current developments will be mentioned with relevant citations for those interested in more detailed reading on the topic.

14.2 DEFINING THE ROC CURVE

14.2.1 When is ROC analysis appropriate?

14.2.1.1 The binary classifier

The fundamental building block of ROC analysis is the binary prediction problem, the simplest scenario encountered in decision-making. Consider a situation where the researcher wants to develop a diagnostic test to discriminate two classes such as two different medical conditions (i.e. "disease present vs. disease absent" or "disease X vs. disease Y"). The desired diagnostic test will yield two possible outcomes ("yes/no" or "X/Y") that will separate the two conditions as well as possible. Such a diagnostic test operates as a binary classifier. To evaluate its classification performance, it is understood that the classifier needs to be tested extensively on a large number of test cases. However, certain assumptions need to be made. First, it is assumed that the two classes are mutually exclusive. Mutual exclusivity is a fundamental assumption to avoid the complicated situation where the diagnostic test is asked to discriminate between two conditions that may co-exist in the same test case. Such a scenario is not covered by ROC analysis. Second, it is assumed that the truth of each test case is established by an undeniable gold standard. Without knowing the truth prior to evaluation, it is impossible to assess how good the diagnostic test truly is. A typical example in medical imaging is the development of a new imaging modality (e.g. digital breast tomosynthesis (DBT)) for non-invasive detection of a disease (e.g. breast cancer). To evaluate the diagnostic performance of the modality, it is expected that a pool of test patients is available for whom the truth has been established prior to applying the new imaging modality (e.g. by means of invasive breast biopsy). Of course, someone may ask, "If I know the truth, why bother to develop a new test?". Well, the purpose of the overall research experiment is to set the stage so that any new binary classifier can be properly tested before deployment in clinical practice. To achieve that, the truth of the test cases needs to be established undeniably using a "perfect" gold standard to ensure that the proposed classifier is indeed acceptable for the specific decision task.

The results of a binary classifier can be summarized with a 2×2 table that completely covers every possible type of decision. There are four possible types of decisions, two of them correct and two of them incorrect. The diagnostic test is correct if its outcome agrees with the truth as established by the gold standard. In contrast, the diagnostic test is incorrect if its outcome disagrees with the gold standard. For example, let's consider the previous example with the DBT imaging modality. If DBT is tested on a patient with breast cancer and it says, "Yes, breast cancer is present in this patient," then this is considered a *true positive* (TP) decision because the diagnostic test came back positive and in agreement with the truth. If, on the other hand, DBT is tested on a cancer-free patient and it says, "No, breast cancer is absent," then this is considered a *true negative* (TN) decision because the diagnostic test came back negative and in agreement with the truth. Both answers are correct. There are however two possible scenarios in which the diagnostic test is incorrect. If DBT is tested on a patient without breast cancer and it says, "Yes, breast cancer is present," then this is considered a *false positive* (FP) decision because the diagnostic test came back positive and in disagreement with the truth. The extension of this terminology to *false negative* (FN) is straightforward and this indicates a breast cancer patient missed by DBT.

The Handbook of Medical Image Perception and Techniques, ed. Ehsan Samei and Elizabeth Krupinski. Published by Cambridge University Press.

14.2.1.2 Classification accuracy, sensitivity, specificity, and beyond

There are several ways to summarize the performance of a binary classifier. From an engineering point of view, the most popular performance measure is classification accuracy (or its opposite, the misclassification error). Basically, out of all test subjects, what is the proportion of correct decisions?

$$\text{Classification Accuracy} = \frac{\text{TP} + \text{TN}}{\text{TP} + \text{TN} + \text{FP} + \text{FN}} \tag{14.1}$$

Classification accuracy is simple to compute and understand but it is suboptimal for medical decision-making problems. This performance measure does not provide any information regarding the *relative* reliability of the diagnostic test for the two classes. Therefore, classification accuracy provides only a partial view. Consider a clinical scenario where two different diagnostic tests are developed for the same classification problem. Both tests are applied on the same population of 150 test subjects (50 with disease and 100 without disease). Let's assume that Test 1 detects correctly all non-diseased patients but none of the diseased ones (TP = 0, TN = 100, FP = 0, FN = 50). Let's also assume that Test 2 detects correctly 50 diseased patients and 50 non-diseased patients (TP = 50, TN = 50, FP = 50, FN = 0). The calculated classification accuracy of both tests is the same (67%) since both tests identify correctly 100 of the 150 test subjects. If the only information reported at the end of the experiment is that both tests are 67% accurate, one would conclude that there is no benefit in using one test over the other. However, the clinical implications are very different. One could easily argue that Test 2 is superior since it has perfect sensitivity (100%) and decent enough specificity (50%) while Test 1 is no better than blindly calling every test patient "non-diseased," clearly a clinically unacceptable scenario.

To address this limitation, sensitivity and specificity are preferred to assess a binary classifier's performance on each class separately. Sensitivity is a classifier's accuracy with patients who have the disease (or in general the condition of main interest). Similarly, specificity is the classifier's accuracy with patients who do not have the disease.

$$\begin{aligned} \text{Sensitivity} &= \frac{\text{TP}}{\text{TP} + \text{FN}} \\ \text{Specificity} &= \frac{\text{TN}}{\text{TN} + \text{FP}} \end{aligned} \tag{14.2}$$

Sensitivity is also known as true positive fraction (TPF) while specificity is also called true negative fraction (TNF). Note that the denominator for sensitivity is the total number of test cases who truly have the disease (some of whom were correctly detected (TP) and the rest were incorrectly misclassified as non-diseased (FN)). The denominator for specificity is the total number of test cases without the disease (some of whom were correctly detected (TN) and the rest were incorrectly misclassified as diseased (FP)). For the example given above, the sensitivity of Test 1 is 0% and its specificity is 100%. For Test 2, the sensitivity is 100% and the specificity is 50%.

Two other measures of performance that are popular among clinicians are the positive predictive value (PPV) and the negative predictive value (NPV). PPV and NPV are defined as follows:

$$\begin{aligned} \text{Positive Predictive Value} &= \frac{\text{TP}}{\text{TP} + \text{FP}} \\ \text{Negative Predictive Value} &= \frac{\text{TN}}{\text{TN} + \text{FN}} \end{aligned} \tag{14.3}$$

Note that the definitions for sensitivity and positive predictive value are very similar with the exception of the second term in the denominator. The denominator for sensitivity (or TPF) is the total number of diseased test cases while the one for PPV is the total number of "diseased" test responses. A similar distinction can be made for the definitions of specificity (or TNF) and NPV. This seemingly minor difference has important implications. The (TPF, TNF) pair is independent of the relative class prevalence in the test cases. In contrast, PPV and NPV are affected by class prevalence. We will illustrate this important property with an example. Let us assume that a diagnostic test with 90% sensitivity and 50% specificity has been developed in a research laboratory and it is deployed for clinical testing. The test is developed to screen populations for disease X. Two different cities have been selected as possible testing sites, with 500 people agreeing to be tested at each city. Let us also assume that the disease prevalence is different between the two test samples (as determined prior to testing by the gold standard). In test site 1, 10% of the test cases have disease X. Thus, 50 (= 500 × 10%) test subjects are diseased while the remaining 450 are healthy. In test site 2, the disease prevalence is 20% among the test sample. Thus, 100 (= 500 × 20%) test subjects are diseased while the remaining 400 are healthy. The observed results are summarized below:

Test site 1:	TP = 45	TN = 225	FN = 5	FP = 225
Test site 2:	TP = 90	TN = 200	FN = 10	FP = 200

Based on the observed results, the calculated accuracy, PPV, and NPV of the same diagnostic test differ between the two sites:

For test site 1:

$$\begin{aligned} \text{Accuracy} &= \frac{45 + 225}{45 + 225 + 5 + 225} = 54\% \\ \text{PPV} &= \frac{45}{45 + 225} = 16.7\% \\ \text{NPV} &= \frac{225}{225 + 5} = 97.8\% \end{aligned}$$

For test site 2:

$$\begin{aligned} \text{Accuracy} &= \frac{90 + 200}{90 + 200 + 10 + 200} = 58\% \\ \text{PPV} &= \frac{90}{90 + 200} = 31\% \\ \text{NPV} &= \frac{200}{200 + 10} = 95.2\% \end{aligned}$$

Thus, if two researchers conduct their evaluation studies in two different cities with otherwise similar populations but different disease prevalence, they will report different performance for the exact same diagnostic test! This example highlights the most important limitation of accuracy, PPV, and NPV. They are all dependent on class prevalence.

Table 14.1 *Hypothetical data for a diagnostic test with rating-type output*

Gold standard diagnosis	1	2	3	4	5
Diseased	4	1	15	30	50
Not diseased	120	140	0	80	60

14.2.1.3 When sensitivity and specificity are not enough

Clearly, sensitivity and specificity are independent of class prevalence. As long as the difficulty level of the test cases remains the same among different test sites, the duplex (TPF, TNF) provides a complete description of performance for a binary classifier. However, in clinical applications, most binary classifiers are not designed to give a yes/no answer but rather an ordinal or continuous output response. For example, a radiologist is asked to determine the malignancy status of a breast lesion based on available imaging and clinical data. The radiologist could express this information in an ordinal manner using a rating scale (1: definitely benign, 2: probably benign, 3: uncertain, 4: probably malignant, 5: definitely malignant). For the same problem, an artificial neural network classifier could be developed producing a numeric value on a continuous scale where a lower value indicates a lower likelihood of malignancy. To assess either the radiologist's or the neural network's diagnostic accuracy in terms of classification accuracy, TPF, TNF, PPV, and NPV, it is necessary to convert their outputs into a binary "yes, lesion is malignant"/"no, lesion is benign" label. To achieve that, a decision threshold must be selected first. Output values that exceed the threshold would be assigned one label (i.e. "yes, lesion is malignant") while the rest would be assigned the "no, lesion is benign" label. The choice of decision threshold is subjective and it alters substantially the reported results.

To illustrate this limitation, consider a hypothetical scenario where a diagnostic test is applied on 500 test subjects. Let us also assume that according to the gold standard, 100 of the test subjects are diseased while the remaining 400 subjects are disease-free. The diagnostic test provides rating-type output values ranging from 1 through 5 where 1 suggests that disease is definitely absent and 5 suggests that disease is present. The observed results are shown in Table 14.1.

To report diagnostic accuracy using the standard measures, a decision threshold needs to be decided. Table 14.2 shows the calculated measures as the decision threshold varies from an extremely conservative value (≥1) to an extremely liberal value (>5). The table clearly highlights the influence of the decision threshold on the reported measures of diagnostic accuracy. For the same diagnostic test and the same test subjects, two researchers would report completely different results depending on where they choose to set the decision threshold. In addition, the table highlights the trade-off between sensitivity and specificity. As the decision threshold is progressively raised, the sensitivity decreases while the reported specificity increases. The relationship between sensitivity and specificity is reciprocal.

Clearly, the ideal measure of diagnostic performance is one that is independent of class prevalence and decision threshold.

14.2.2 What is an ROC curve?

As illustrated before, for ordinal and continuous output value classifiers, there is not a single value of sensitivity and specificity that completely describes such classifiers. For completeness, the entire range of (TPF, TNF) values for every possible decision threshold need to be provided to fully characterize the diagnostic accuracy of the classifier. An ROC curve captures graphically the reciprocal relationship between sensitivity (or TPF) and specificity (or TNF) for the entire range of decision thresholds. Specifically, the ROC curve plots the sensitivity versus (1 – specificity) for every possible decision threshold. Note that based on the definitions provided before, (1 – specificity) is the ratio of false positive test responses, or FPF:

$$1 - \text{Specificity} = 1 - \frac{\text{TN}}{\text{TN}+\text{FP}} = \frac{\text{FP}}{\text{TN}+\text{FP}} = \text{FPF} \quad (14.4)$$

Although plotting sensitivity vs. specificity conveys the exact same information, for historical reasons ROC analysis is focused on positive test responses, both true positive and false positive (Lusted, 1960; Lusted, 1961; Swets, 1979).

Figure 14.1 shows a typical ROC curve. Each ROC data point (FPF_i, TPF_i) represents a particular setting of the decision threshold θ_i. It should be noted however that the threshold value is not provided in the graph. The ROC curve clearly captures the trade-off between sensitivity and specificity. For a low decision threshold (e.g. see Table 14.2), the diagnostic test has the tendency to over-call test cases as positive. Consequently, many true positive test cases are correctly diagnosed as "diseased," thus resulting in high sensitivity. However, many true negative cases are falsely called "diseased," resulting in low specificity. Low specificity translates into a high FPF. The opposite is true for high decision thresholds. High decision thresholds imply a tendency to over-call test cases as negative, thus resulting in low

Table 14.2 *Performance of the hypothetical test for various decision thresholds*

Decision threshold	Accuracy	Sensitivity	Specificity	PPV	NPV
≥1	20% (100/500)	100% (100/100)	0% (0/400)	20% (100/500)	N/A (0/0)
≥2	43.2% (216/500)	96% (96/100)	30% (120/400)	25.5% (96/376)	96.8% (120/124)
≥3	71% (355/500)	95% (95/100)	65% (260/400)	40.4% (95/235)	98.1% (260/265)
≥4	68% (340/500)	80% (80/100)	65% (260/400)	36.4% (80/220)	92.9% (260/280)
≥5	74% (370/500)	50% (50/100)	85% (340/400)	45.5% (50/110)	86.5% (320/370)
>5	80% (400/500)	0% (0/100)	100% (400/400)	N/A (0/0)	80% (400/500)

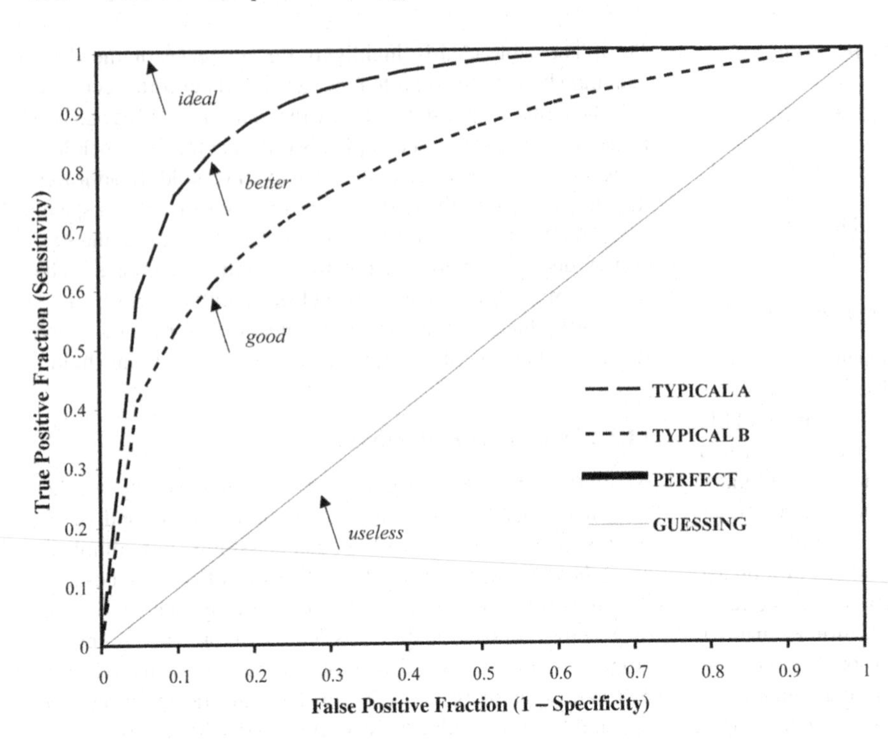

Figure 14.1 Typical ROC curves along with the upper (perfect) and lower (guessing) bounds.

sensitivity but high specificity (or low FPF). By definition, the ROC curve has a fundamental advantage over accuracy, PPV, and NPV. It is independent of disease prevalence since it focuses on sensitivity and specificity.

It is important to emphasize the following points for the newcomer in the field. First, an ROC curve is always plotted in a square since both axes are bounded in [0,1]. Second, the ROC curve always includes the (0,0) point (i.e. the lower left corner of the square) and the (1,1) point (i.e. the upper right corner of the square). The (0,0) point describes the extreme scenario where the diagnostic test blindly calls every test case negative, thus resulting in 100% specificity (or 0% FPF) and 0% sensitivity. Similarly, the (1,1) point describes the extreme scenario where the diagnostic test blindly calls every test case positive, thus resulting in 0% specificity (or 100% FPF) and 100% sensitivity. Third, a perfect diagnostic test implies 100% sensitivity and 100% specificity for every possible decision threshold. Therefore, the perfect ROC curve runs vertically from the (0,0) point to the (0,1) point and then horizontally to the (1,1) point of the square. In contrast, a completely inaccurate diagnostic test performs no better than random guessing. Random guessing should theoretically result in 50% sensitivity and 50% specificity regardless of the decision threshold, since the diagnostic test operates similar to flipping a coin every time it is presented with a test case. Thus, the completely inaccurate ROC curve is the 45-degree diagonal line that connects the (0,0) and (1,1) points of the customary plotting square. A typical ROC curve lies always between these visual lower and upper bounds provided by the two extreme situations. In general, the closer the ROC curve is to the upper left corner of the plotting square, the better the diagnostic accuracy of the test. Figure 14.1 illustrates the corresponding ROC curves for all scenarios described above. It should be noted however that it is possible to derive an ROC curve that lies below the lower bound diagonal. Let us consider the extreme case where the ROC curve runs horizontally from (0,0) to (1,0) and then vertically to (1,1). Such an ROC curve implies that the diagnostic test is consistently incorrect. In other words, by simply reversing all test responses from positive to negative and vice versa, the diagnostic test becomes perfectly accurate.

14.3 PLOTTING THE ROC CURVE

14.3.1 How to derive the ROC curve

ROC analysis is appropriate for diagnostic tests that produce either continuous or ordinal data. For example, measuring the blood glucose level to determine the presence of diabetes is a diagnostic test with continuous output. Using the standardized uptake value (SUV) in PET scans to determine the likelihood of malignancy is another example. In both cases, it is assumed that the gold standard for the test cases is available prior to diagnostic testing. Examples of medical diagnostic tests that produce ordinal-type output data are those where the findings are reported using a rating scale with a fixed number of categories (e.g. ejection fraction <25%, 25–50%, 50–75%, >75%). Often, qualitative-type data can be treated as ordinal, and thus analyzed with ROC analysis. For example, if a radiologist is asked to assess the malignancy status of a lesion using a five-category scale such as "definitely benign," "probably benign," "uncertain," "probably malignant," and "definitely malignant," then such test data can be treated as ordinal.

To construct an ROC curve, it is assumed that a diagnostic test has been applied to a number of test cases, some of them positive (e.g. "diseased") and some negative (e.g. "healthy") as established by the gold standard. Although the relative prevalence of each class (positive and negative) is not important, it is

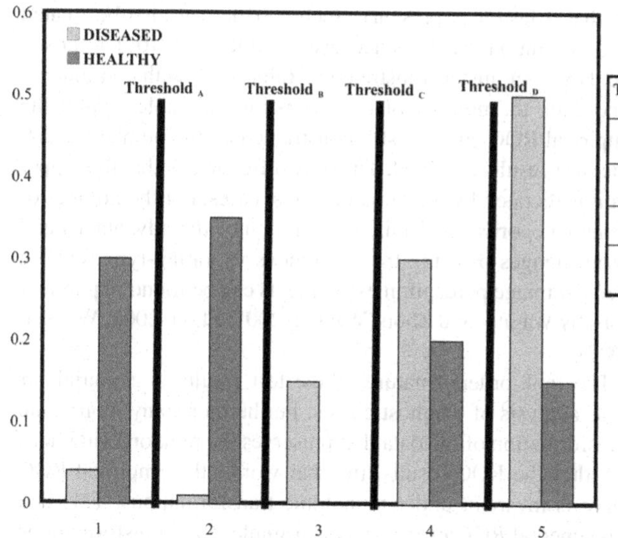

Threshold	FPF	TPF
A	0.70	0.96
B	0.35	0.95
C	0.30	0.80
D	0.15	0.50

Figure 14.2 Histogram of the ordinal output data for a hypothetical test applied to discriminate healthy from diseased patients. The table shows the corresponding (FPF, TPF) pairs for every possible decision threshold.

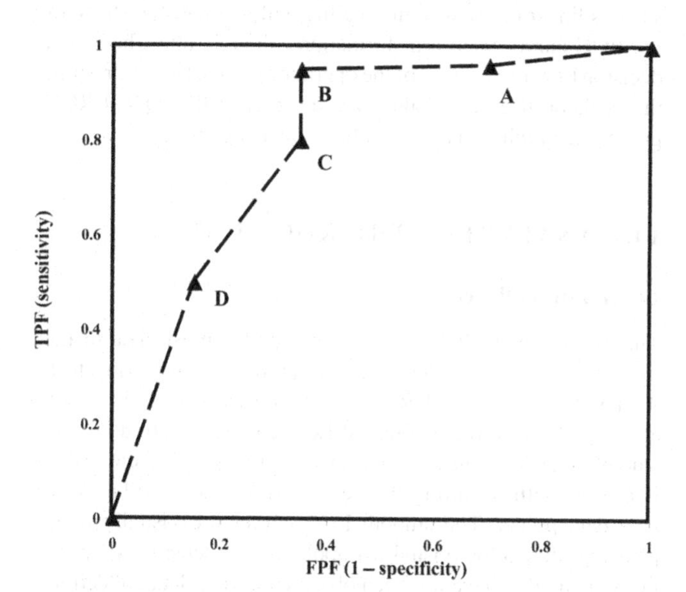

Figure 14.3 ROC curve for the ordinal test output data shown in Figure 14.2.

critical that there is available test data from both classes. It is not possible to perform ROC analysis if there is available test data from only one class. In such a scenario, one of the two coordinates (TPF or FPF) cannot be calculated and thus ROC analysis is not feasible. The following subsections demonstrate two examples of how to construct the empirical ROC curve based on ordinal and continuous test data.

14.3.1.1 Ordinal data

Let us consider the test data shown in Table 14.1. In this example, there are 500 test cases, 100 positive (i.e. diseased) and 400 negative (i.e. healthy). The diagnostic test under evaluation reports the outcome using a five-point ordinal scale. By changing the decision threshold from "below 1," to "between 1 and 2," "between 2 and 3," all the way to "above 5," the corresponding TPF and FPF pairs can be calculated. Figure 14.2 shows the test data histograms for each class and it illustrates the process of sweeping across every possible decision threshold θ_i setting to derive the ROC data points (FPF_i, TPF_i).

Note that the five-category rating scale results in four distinct ROC data points. In general, a diagnostic test with N different possible outputs results in (N – 1) distinct ROC operating points. By default, the points (0,0) and (1,1) are included in the ROC curve. By connecting the available data points with linear segments the empirical ROC curve is derived, as shown in Figure 14.3. By connecting the data points, it is implied that any point on the ROC curve is a feasible operating point. However, with ordinal data this is not true.

14.3.1.2 Continuous data

Plotting the empirical ROC curve for continuous test data is similar to dealing with ordinal-type data, with the only exception that there are typically many more ROC data points available. The underlying methodology is the same. Basically, the decision threshold is progressively swept across the whole range of possible values while the corresponding (FPF, TPF) pairs are calculated.

This process is illustrated in Table 14.3 for a hypothetical diagnostic test that produces continuous results. The test is applied on ten test subjects, five of them healthy and five diseased. Initially, the obtained test results are ranked ordered from lowest to highest. As the decision threshold progressively increases by moving from one row to the next, the number of diseased and healthy test cases that fall below the decision line changes, thus generating a new data point (FPF, TPF).

With ten test subjects and nine distinct test results (notice that one test output value appears for a healthy *and* a diseased subject), the process shown in Table 14.3 results in eight distinct ROC data points. Including the (0,0) and (1,1) default ROC points produces the empirical ROC curve shown in Figure 14.4.

14.3.2 General comments

The examples provided in the previous sections illustrate how empirical ROC curves can be plotted with both ordinal and

Table 14.3 *Hypothetical test data for a diagnostic test with continuous output values. The ROC data points are calculated assuming the following. If the decision threshold is set at the value shown at each row of the first two columns, any test output equal or above this value is considered a positive test response.*

HEALTHY	DISEASED	FPF	TPF
3.15		1.0 (5/5)	1.0 (5/5)
	4.48	0.8 (4/5)	0.8 (4/5)
5.12		0.8 (4/5)	0.8 (4/5)
5.66		0.6 (3/5)	0.8 (4/5)
6.34	6.34	0.4 (2/5)	0.8 (4/5)
	7.45	0.2 (1/5)	0.6 (3/5)
	8.34	0.2 (1/5)	0.4 (2/5)
9.12		0.2 (1/5)	0.2 (1/5)
	10.26	0.0 (0/5)	0.2 (1/5)

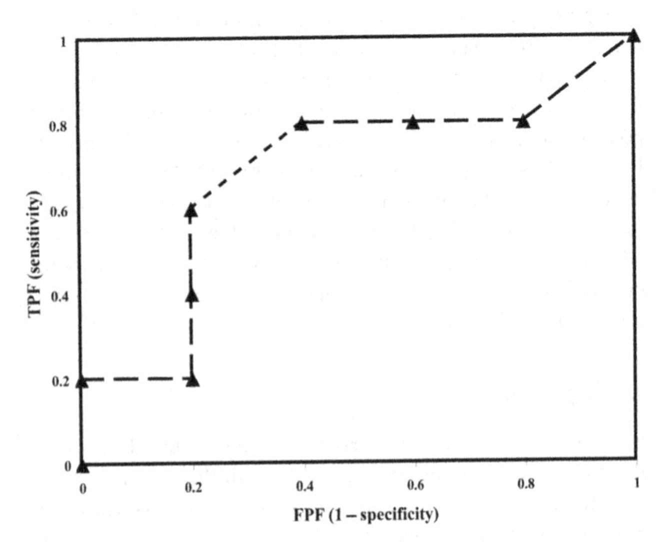

Figure 14.4 ROC curve for the continuous test output data shown in Table 14.3.

continuous test data as long as the data are rank ordered in a meaningful way. The number of distinct ROC data points available for plotting the ROC curve depends on the number of distinct test results collected during an experiment. For ordinal-type data, this number is limited by the number of categories present in the rating scale, regardless of the size of the dataset. For example, a three-point scale results in only two distinct data points. But even a five- or ten-point scale results in four or nine distinct points, respectively. However, if there are no test cases assigned in a particular rating category, then the resulting data points are even fewer. The limited number of ROC data points generated from ordinal data is a common problem with empirical ROC curves. The problem is further exacerbated if the data points are poorly distributed, not adequately sampling the full range of TPF and FPF values. To address this limitation, it is common practice for medical image perception applications (i.e. human observer studies) to use the 0–100% rating scale. This scale makes intuitive sense for human observers who tend to assign a probabilistic interpretation (e.g. 0–100% probability that a lesion is present). Such a scale can also be treated as a continuous scale, since every value in [0,100] is possible. However, just as illustrated in Table 14.3, with continuous-type data, the number of distinct points available to plot the empirical ROC curve is still constrained by the number of distinct test results obtained. The only difference is that this upper limit is dictated by the number of test cases, not the number of rating categories. A detailed discussion of the advantages and disadvantages of using the continuous vs. rating-type scale in medical image perception experiments can be found in publications by Wagner and Zhou (Wagner, 2001; Zhou, 2002; Wagner, 2007).

The rank-ordered nature of the test results is essential for ROC analysis of diagnostic tests. Furthermore, any monotonic transformation of the data that preserves the rank ordering does not alter the ROC results. In other words, the empirical ROC curve is invariant under monotonic transformations. This is a fundamental ROC property. For example, data transformations such as linear or logarithmic scaling will not alter the shape of the ROC curve. Such transformations, though, alter the actual decision threshold values of the operating points. However, since the decision threshold values are not part of the typical ROC graph, the graph remains essentially unchanged.

14.4 ANALYZING THE ROC CURVE

14.4.1 Fitting the curve

The experimental ROC curve is essentially an estimate of the true ROC curve that describes the particular diagnostic test. Although the empirical ROC curves are easy to derive, they are jagged. Since the theoretical ROC curve is continuous and smooth, it is desirable to derive an experimental ROC curve that is also smooth by fitting the best possible curve to the available data points. The smooth ROC curves are visually more pleasing, they allow visual extrapolation of (sensitivity, specificity) pairs that have not been observed during data collection, and they facilitate precise statistical inferences when comparing ROC curves. There is a wide range of algorithms available for the task. The algorithms are mathematically elaborate and their sophistication extends beyond the scope of this chapter. The following is a brief description of some popular fitting algorithms and their underlying assumptions, advantages, and limitations. Appropriate references are provided for readers interested in more details.

14.4.1.1 Non-parametric

The underlying philosophy of the non-parametric fitting techniques is to smooth the histograms of the test output data for each one of the two classes (e.g. healthy and diseased). Smooth histograms result in smooth ROC curves. The kernel-based approach is the best known for the task (Zou, 1997). Basically, a different kernel density function is used to smooth each one of the histograms and derive the "kernel-estimate" of each probability density function. The main advantage of such an approach is that it is free of any structural assumptions regarding the underlying data distributions. Therefore, it is applicable regardless of

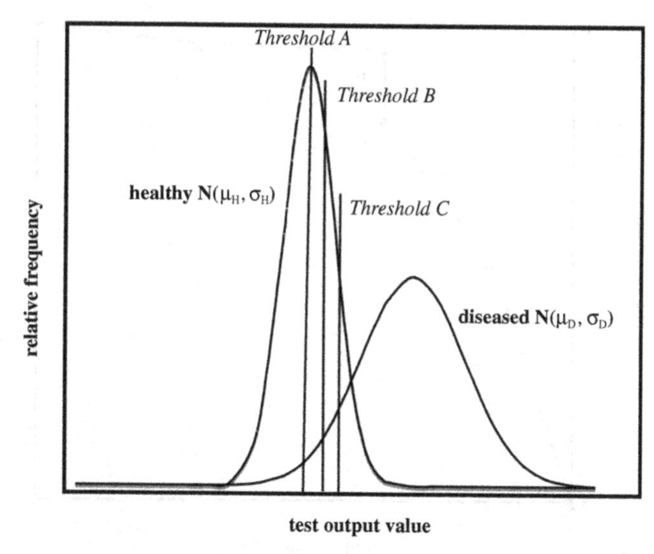

Figure 14.5 The binormal model.

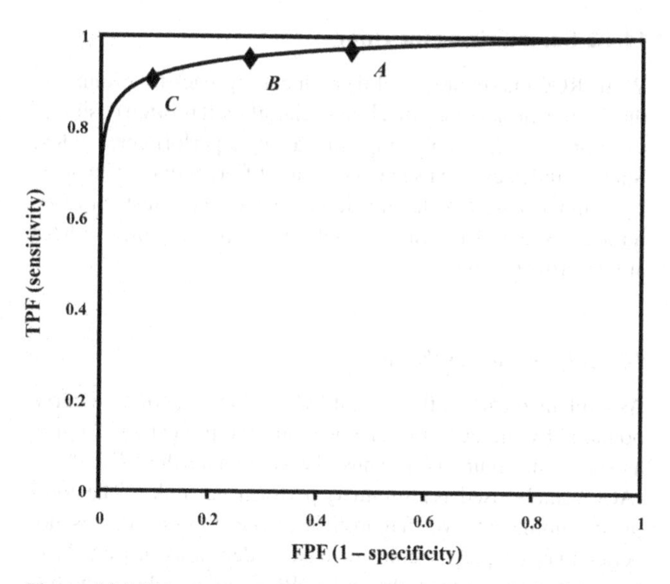

Figure 14.6 The parametric ROC curve that is based on the test data distributions shown in Figure 14.5.

the test data complexity (e.g. bimodal histograms). Setting the kernel parameters for each class is not a trivial issue. Furthermore, the method is not reliable with histograms close to zero and when estimating the extreme portions of the ROC curve. It should be emphasized that the term "non-parametric ROC curve" is most often used to denote the unsmoothed, empirical ROC curve.

14.4.1.2 Parametric

The parametric methods rely on parametric modeling of the true ROC curve. As with any modeling approach, the validity of the underlying assumptions is critical. The most common among all parametric techniques is the one based on the binormal model (Dorfman, 1968; Dorfman, 1969; Hanley, 1988; Hanley, 1996; Metz, 1998). The underlying assumption of this model is that the test output data for each class are, or can be, monotonically transformed into normal distributions (Figure 14.5). Some techniques require that the transformation function be carefully specified while others do not pose such a requirement. The latter are often called semi-parametric techniques and they are used much more often in clinical applications. For simplicity, the term "parametric" appears throughout this chapter to denote both fully parametric and semi-parametric methods.

Let us assume that the test output data for healthy subjects are normally distributed with mean μ_H and standard deviation σ_H. For the test output data collected on diseased subjects, the mean and standard deviation are μ_D and σ_D respectively. Then, for any decision threshold θ, the TPF and FPF are defined using the definition of the normal cumulative distribution function $\Phi(x)$:

$$\text{TPF} = \Phi\left(\frac{\mu_D - \theta}{\sigma_D}\right) \quad \text{FPF} = \Phi\left(\frac{\mu_H - \theta}{\sigma_H}\right) \tag{14.5}$$

Solving the above equations for θ leads to the following:

$$\text{TPF} = \Phi\left(\frac{\mu_D - \mu_H + \sigma_H \Phi^{-1}(\text{FPF})}{\sigma_D}\right) = \Phi(\alpha + \beta\Phi^{-1}(\text{FPF})) \tag{14.6}$$

where $\alpha = \frac{\mu_D - \mu_H}{\sigma_H}$ and $\beta = \frac{\sigma_D}{\sigma_H}$. Notice that a binormal ROC curve (shown in Eq. 14.6) is transformed into a straight line when plotted on double probability paper, with slope α and intercept β. Thus, under the binormal model, the ROC curve can be completely described by two parameters, α and β. The beauty of the binormal model is its simplicity and elegance. It is applicable with a wide range of experimental data as long as there is an increasing monotonic transformation (for both classes) to satisfy the binormal assumption. Studies have also shown that the binormal model is effective even when the data deviate mildly from the underlying assumption (Swets, 1986; Hanley, 1988). Also, with a properly chosen transformation such as a log transformation or the Box-Cox transformation proposed by Zou *et al.* (Zou, 1998), the binormal model may produce a superior fit compared to non-parametric models (Walsh, 1999). However, with bimodal data (for either or both of the class distributions), the binormal model is not recommended (Zou, 2005).

Figure 14.6 shows the corresponding fitted ROC curve (using the binormal model) for the example distributions depicted in Figure 14.5. The operating points that correspond to the three decision thresholds (A, B, and C) are highlighted.

While the parametric fitting approach based on the binormal model is the most popular in the medical field, it is recommended to err on the side of caution and apply a statistical test such as the one proposed by Dorfman and Alf (Dorfman, 1968) to determine whether a particular data set violates the underlying assumptions of the binormal model. Goddard and Hinberg (Goddard, 1990) have addressed the consequences of violating the binormal assumption. Dorfman *et al.* (Dorfman, 1974) have discussed improper ROC curves; namely ROC curves with an unnatural hook-shape ending at the upper right corner of the ROC plot. This is the result of applying the binormal model on small datasets and/or ordinal-type data generated with a limited rating scale. A "proper" binormal model was introduced by Metz *et al.* to address these rare but possible cases (Metz, 1999).

14.4.2 Summarizing the curve

While ROC curves are a useful graphical approach for evaluating the diagnostic accuracy of a binary classifier, it is often desirable to summarize the whole graph into a single performance index. Such an index can be used to compare different diagnostic tests by summarizing the diagnostic content of each test into one number. Some of the most popular ROC performance indices are described below.

14.4.2.1 Area under the curve

As explained earlier, the typical ROC curve is a concave curve bounded by the guessing diagonal and the perfect ROC curve as shown in Figure 14.1. Thus, the area under the ROC curve (AUC) can be used as a summary performance index. The completely uninformative diagnostic test (i.e. the one that is not better than flipping a coin to make a decision) has an AUC equal to 0.5. In contrast, the perfect ROC curve includes the full square and has an AUC of 1.0. Thus, the AUC summary index is constrained between 0.5 and 1.0. In general, the higher the AUC value, the better the diagnostic test is. In 1988, Swets provided a general stratification guideline on how to qualitatively assess the diagnostic accuracy of a diagnostic test based on its AUC index: completely uninformative test (AUC = 0.5), less accurate test ($0.5 < \text{AUC} \leq 0.7$), moderately accurate test ($0.7 < \text{AUC} \leq 0.9$), highly accurate test ($0.9 < \text{AUC} < 1.0$), and perfect test (AUC = 1.0) (Swets, 1988).

The AUC index is the most frequently used (Hanley, 1982). It is often misinterpreted as equivalent to classification accuracy. In other words, it is common for newcomers to perceive an AUC of 0.85 as 85% classification accuracy. However, this is not true. Actually, two diagnostic tests with the same classification accuracy may have substantially different AUCs. This was shown in the study by Cortes and Morhi, who derived a closed expression relating the expecting value and variance of the AUC to a fixed classification error rate (Cortes, 2003). There are several different interpretations of the AUC value: (1) it is the average sensitivity across all possible values of specificity (Metz, 1986b); (2) it is the average specificity across all possible values of sensitivity (Metz, 1986b); (3) it is the probability that a randomly selected diseased test subject has a test result more indicative of disease than that of a randomly chosen healthy test subject (Bamber, 1975; Hanley, 1982).

There are different techniques to estimate the AUC index, depending on whether we are dealing with an empirical or a fitted ROC curve. With an empirical ROC curve, the AUC index can be estimated using the trapezoidal rule (Figure 14.7). This is a straightforward way to summarize the AUC curve, assuming that there are a few well-distributed data points along the empirical curve.

The trapezoidal rule can be particularly burdensome with continuous data and many data points. For continuous data, the non-parametric (trapezoidal) AUC estimate is equivalent to the two-sample Mann-Whitney-Wilcoxon rank-sum U statistic (Harrell, 1982). Within this context, the U statistic is used to determine whether the diagnostic test result is generally more indicative of disease in diseased test subjects than in healthy ones. Basically, AUC is equal to the proportion of all possible pairs of healthy and diseased test subjects for which the diseased test result is more indicative of disease than the healthy test result, plus one half the proportion of ties. The statistical relevance of the non-parametric AUC index to the U statistic is very convenient. It lends itself to additional analysis, such as estimation of the standard error according to the Hanley-McNeil algorithm (Hanley, 1982) or using bootstrap techniques (Hanley, 1983; Efron, 1993). Overall, this is the most convenient method to derive the AUC performance index. It is valid with all types of data although it tends to underestimate the true AUC with few data points on the curve.

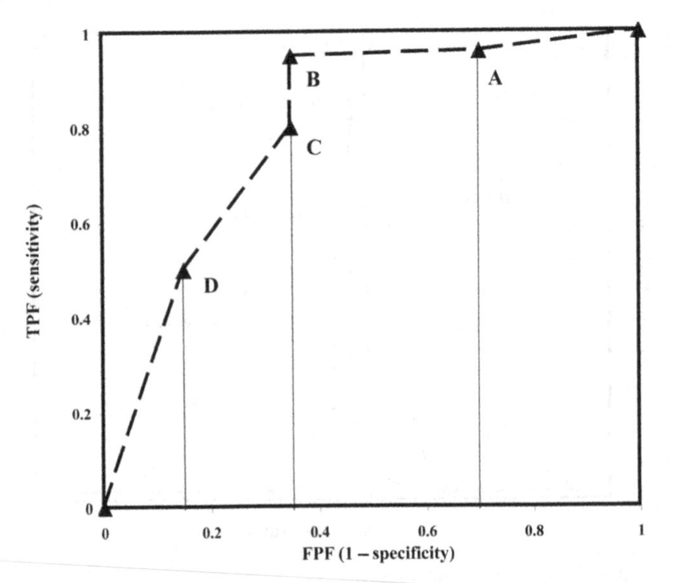

Figure 14.7 The non-parametric estimation of AUC for the data shown in Figure 14.3. The AUC index calculated using the trapezoid rule is 0.82.

With parametric, fitted ROC curves, the AUC index is expressed as a function of the model parameters. For example, for the binormal model, the AUC is a simple function of the slope and the intercept parameters:

$$\text{AUC} = \Phi\left(\frac{a}{\sqrt{1+b^2}}\right) \tag{14.7}$$

The parametric expression facilitates estimation of the standard error. This issue will be addressed in the next section.

There is no general consensus as to which estimation technique is best. The best choice depends on the type of data (ordinal or continuous), the number of distinct ROC data points available, and the complexity of the classification task. According to Faraggi and Reiser, who studied the problem with Monte Carlo simulations for a wide range of conditions, no particular technique emerged as the superior one for all studied scenarios (Faraggi, 2002). Similar findings were reported by Hajian-Tilaki (Hajian-Tilaki, 1997). Overall though, the empirical method behaved robustly, and consistently emerged as one of the top performers.

14.4.2.2 Partial area

The strength of the AUC index is its ability to summarize the full ROC curve into one concise numerical value. The underlying

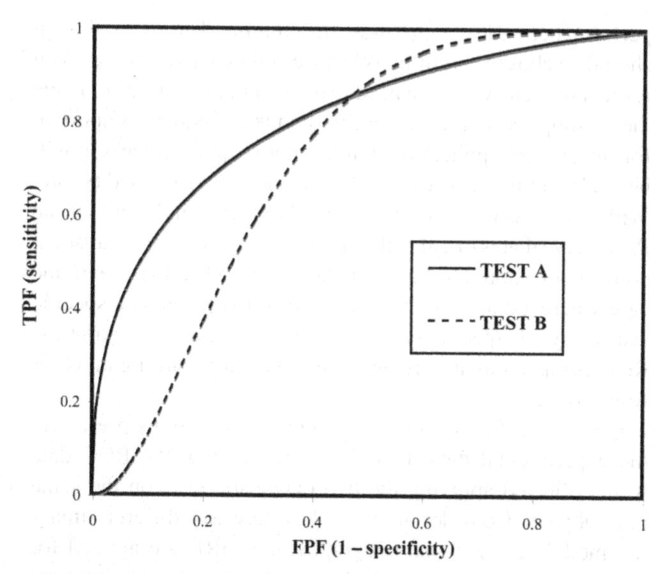

Figure 14.8 Intersecting ROC curves for two diagnostic tests A and B. Although AUC(A) > AUC(B), it is not clear that test A is superior. There is a certain portion of the graph (for FPF > 0.45) where test B is superior. In this case, AUC is a suboptimal performance index.

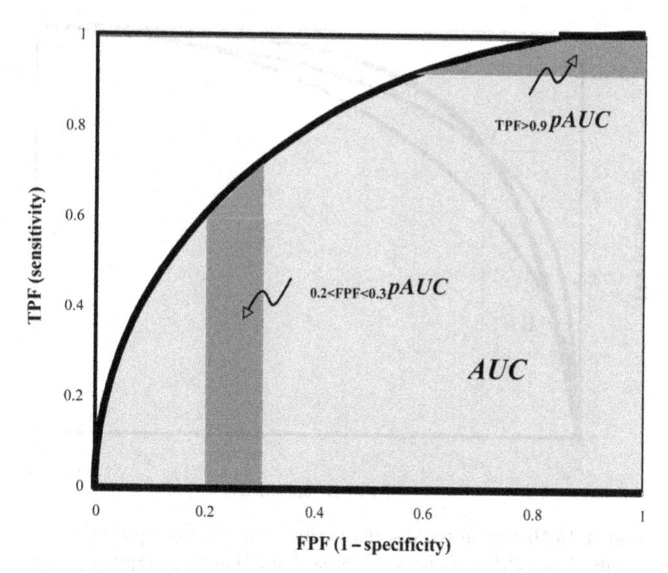

Figure 14.9 Schematic illustration of the AUC area indices. The light gray area under the ROC curve corresponds to the overall area index AUC. The two dark gray portions correspond to partial AUC area indices, one calculated for FPF values [0.2, 0.3] and the other for TPF > 0.9.

assumption of the summarization is that all decision thresholds (i.e. operating points) are equally important. There are, however, clinical situations when this is not the case. For example, in some cases, operating with a low-sensitivity decision threshold may not be clinically acceptable for a diagnostic test. Then, assigning an equal weight to all portions of the ROC curve is not clinically relevant. This scenario is particularly true when comparing ROC curves that intersect (Figure 14.8). The two ROC curves shown in Figure 14.8 have similar AUC values but clearly there are portions of the graph where A is superior to B and vice versa.

To address this limitation, the partial ROC area index was proposed as a more refined measure for regional assessment of the curve (McClish, 1989; Jiang, 1996). The partial ROC area index limits the AUC evaluation to the portion of the curve that is clinically important (e.g. TPF > 90%, FPF < 20%). Specifically, McClish proposed a partial index for a preset range of specificity values. Jiang proposed a partial index for a preset range of sensitivity values. Thus, analogous to the full AUC, the partial AUC (pAUC) can be interpreted as the average sensitivity for the range of specificities examined or as the average specificity for the range of sensitivities examined. In addition, the pAUC can be estimated either non-parametrically (Zhang, 2002) or parametrically (McClish, 1989). Just like the full AUC, pAUC may be normalized so that it is always bounded between 0.5 and 1.0 (McClish, 1989). Jiang proposed a different standardization; dividing the estimated area under the desired portion of the ROC curve by the maximum possible area under the specific ROC curve portion. Such standardization leads to pAUC indices that range from 0.0 to 1.0.

Figure 14.9 is a schematic illustration of the AUC and pAUC area indices.

14.4.2.3 Other indices

The full and partial AUC indices are by far the most common, especially in medical imaging. However, there have been several other ROC performance indices proposed (Pepe, 2003). The simplest example is choosing a specific operating point on the ROC curve. The clinical application dictates the selection of the point (e.g. the point that corresponds to 20% FPF). Another example is the Youden index proposed in 1950 (Youden, 1950) and still frequently used in practice (e.g. Aoki, 1997; Grmec, 2004):

$$Y = \text{Sensitivity} + \text{Specificity} - 1 = \text{TPF} + (1 - \text{FPF}) - 1 = \text{TPF} - \text{FPF} \quad (14.8)$$

The Youden index is bounded in [0,1] where Y = 1.0 indicates perfect performance and Y = 0.0 indicates guessing performance. The main disadvantage of the Youden index is that it depends on the decision threshold.

Other examples include the KS index and the symmetry index. KS is the maximum distance between the ROC curve and the 45-degree diagonal line. This index also ranges between 0 and 1. Similarly, the symmetry index is the ROC point where sensitivity = specificity. The visual interpretation of the symmetry index is the intersection point between the ROC curve and the negative 45-degree diagonal. More complex measures have been proposed (Lee, 1996; Lee, 1999) but none is remotely as popular in medical imaging as the AUC and pAUC indices.

14.5 USING THE ROC CURVE

14.5.1 Comparing ROC curves

ROC analysis is very useful for comparing the diagnostic accuracy of diagnostic tests (e.g. imaging modalities, human observers) because it is comprehensive and independent of the decision criteria. Consider two diagnostic tests A and B, for

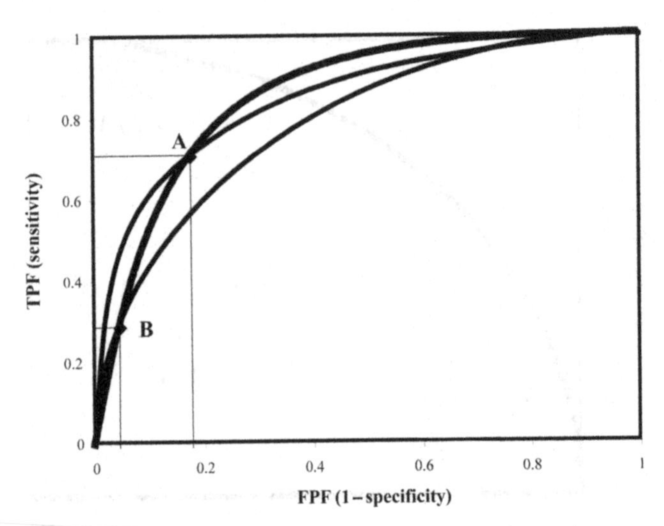

Figure 14.10 Hypothetical scenario where the reported operating points of two different diagnostic tests A and B may correspond to two different ROC curves (solid lines) or the same ROC curve (dotted line).

which the accuracy is reported only in terms of specificity and sensitivity. If both the sensitivity and specificity of one test are higher than the sensitivity and specificity of the other test, then it is clear which test is superior. The situation becomes complex when one test is more sensitive but less specific than the other test. In such cases it is unclear whether A and B represent points on the same ROC curve or points on two different ROC curves (Figure 14.10). The first scenario implies that the only difference between tests A and B is that they were applied with a different decision threshold. Therefore, the tests are equivalent in terms of diagnostic accuracy. The second scenario, though, shows clearly that test A is superior to B since A belongs to an ROC curve that is closer to the upper left corner of the ROC graphing square. However, without full knowledge of the ROC curves it is impossible to distinguish the two different scenarios.

Although visual comparison of ROC curves is the first level of analysis, it is important to apply statistical techniques that take into account the standard error when comparing ROC curves estimated from experimental data. The comparison of ROC curves is approached as a typical hypothesis-testing problem. The null hypothesis H_0 is that there is no difference between the two ROC curves. The next step is to determine whether there is enough evidence to reject the null hypothesis, and thus conclude that the two curves are different. To select the appropriate statistical test certain questions must be answered first: (1) Which ROC characteristics will be used to do the comparison? (2) Is the comparison based on unpaired or paired data? (3) What type of data (ordinal or continuous) is the comparison based upon? It is important to understand the implications of each question.

The clinical application should dictate the answer to question 1. The obvious scenario is to determine whether the two ROC curves are identical. While this is seemingly intuitive, it is not very informative. Even if hypothesis testing shows that there is a difference between the two ROC curves, it does not show which one is better. Most clinical applications want to determine whether the diagnostic accuracy of one diagnostic test is superior to that of the other. Therefore, focusing on an ROC performance index is the most meaningful. The index though should be chosen carefully. When we compare ROC curves with respect to their AUCs we need to understand that we have made the assumption that all operating points are equally important for the clinical application. Furthermore, such comparison will be irrelevant for ROC curves that intersect, as discussed before. With intersecting curves, it is critical to focus on the part of the ROC curve that is of clinical significance. To do so, the statistical comparison should be based on the pAUC index. In the extreme case where a diagnostic test is expected to operate at a specific sensitivity (or specificity) level, it is appropriate to restrict the ROC comparison at only one point, thus disregarding all other ROC points.

Question 2 focuses on any potential correlations present in the experimental data. Paired data means that the ROC data were collected applying the two diagnostic tests on the same test subjects. Consider an example where two different imaging modalities (e.g. mammography and MRI) are applied for breast cancer screening. If each test subject undergoes imaging with both modalities, then the collected data are paired since for every test subject there are two test results (i.e. one from mammography and one from MRI). However, there are many clinical situations where it is dangerous and/or unethical to expose a test subject to both medical tests. In such situations, the test subjects are randomized in two different arms, each undergoing a different test. This is the typical study design for randomized clinical trials. If dealing with paired data, the estimated ROC curves have an implicit correlation that needs to be taken into account when comparing the ROC curves. Such a correlation does not exist with unpaired data. Although paired data lead to conclusions with higher statistical power, it is not always feasible to implement paired study designs.

Finally, the answer to question 3 determines the statistical methods that should be used to derive the standard error of the performance index estimate and to determine the correlations in paired designs. As discussed before, based on the type and amount of data, non-parametric or parametric techniques are used to estimate the ROC characteristic of interest and its standard error. All techniques derive a general test statistic T to be used for hypothesis testing,

$$T = \frac{\hat{Z}_1 - \hat{Z}_2}{\sqrt{\sigma^2(\hat{Z}_1) + \sigma^2(\hat{Z}_2) - 2\mathrm{cov}(\hat{Z}_1, \hat{Z}_2)}} \tag{14.9}$$

where $\hat{Z}_i$ is the estimated ROC characteristic for diagnostic test i (e.g. AUC, pAUC, TPF at a fixed FPF, α and β parameters of the binormal ROC curve), $\sigma^2(\hat{Z}_i)$ is the variance of the estimate, and $\mathrm{cov}(\hat{Z}_1, \hat{Z}_2)$ is the covariance between the two estimates. For paired data, the covariance term is equal to 0.

The final outcome of the statistical comparison is either a confidence interval of the difference between the ROC summary index values of choice or a two-tailed p-value. Confidence intervals that do not include 0 (or p-values < 0.05) are interpreted as enough statistical evidence that the two tests are indeed different. If, however, the above conditions are not met, then it is not possible to draw conclusions of equivalence. In hypothesis testing, the absence of statistical evidence that two tests are different should not be misinterpreted as statistical evidence for absence of difference. In other words, if we fail to conclude that two ROC curves are different, we cannot state that the ROC curves are equivalent. ROC equivalence testing is a different concept.

The following sections will present techniques for comparing ROC curves with respect to the most popular summary indices: AUC and pAUC. The first subsection summarizes techniques that look for enough statistical evidence that two ROC curves are different. The second subsection discusses briefly techniques that look for statistical evidence that two ROC curves are equivalent. Detailed mathematical descriptions can be found in Zhou (Zhou, 2002).

14.5.1.1 Looking for statistical evidence of difference

There are several different methods proposed on how to construct confidence intervals (Hanley, 1982; Hanley, 1983; Wieand, 1989) for differences between ROC curves. Delong *et al.* (Delong, 1988) proposed one of the most popular non-parametric options, the non-parametric asymptotic method. A detailed implementation guide for the technique can be found in a more recent publication (Hanley, 1997). The technique was originally presented based on AUC comparisons but it was later extended to pAUC comparisons (Zhang, 2002). Among the parametric techniques, the most popular is the one based on the binormal model (Metz, 1980; Metz, 1986a). McClish *et al.* (McClish, 1989) described how the binormal assumption could be applied to estimate the variance of the pAUC estimate. Another parametric method was also introduced by Zou *et al.* (Zou, 2001).

There is also a wide range of resampling techniques proposed for comparing ROC curves. These techniques are applicable with both parametric and non-parametric curves. The techniques include the jackknife approach (Dorfman, 1992) and the bootstrap methods (Obuchowski, 1998). The bootstrap methods are preferred over jackknife with small datasets. Overall though there is no single method that is considered superior. In a detailed comparison for both continuous and ordinal data and variable sample sizes, Obuchowski and Liebler (Obuchowski, 1998) provided some general recommendations. One last point that needs to be emphasized is that when comparing ROC curves, it is important to view the statistical results within the context of the clinical application. Even if there is a statistically significant difference between two diagnostic tests, this difference does not necessarily translate into significantly different clinical impact.

14.5.1.2 Looking for statistical evidence of equivalence

There are some situations where proving the superiority of one diagnostic test over another is not necessary. This is certainly the case when a new technology is considered as an alternative to an existing technology. If the new technology offers an alternative diagnostic solution that is less invasive or less expensive than the standard technique, it is not necessary that the new technology is also diagnostically superior. Although replacing the standard way of clinical practice with a new technique that is not only less invasive and less expensive but also more accurate would be ideal, the new technique offers enough advantages beyond diagnostic accuracy. Simply showing non-inferiority in diagnostic accuracy is sufficient to establish the new technology as the improved way of clinical practice. For example, in this digital era, softcopy display of medical images is only natural. If, for a particular diagnostic task, hardcopy display is the standard way of medical practice and softcopy display is considered as an alternative, it is important to establish that softcopy display does not deteriorate the diagnostic performance of radiologists compared to hardcopy display.

As discussed before, not finding statistical evidence of difference between two diagnostic tests should not be viewed as statistical evidence of equivalence. To prove equivalence, a different statistical test needs to be performed (Schuirmann, 1987; Obuchowski, 1997; Zhou, 2002). Actually, the equivalence test is based on the two one-sided tests (TOST) method. According to this method, equivalence is defined as the presence of difference that is bounded between a lower and an upper bound. If $\hat{Z}_i$ is the estimated performance index of diagnostic test i, then two tests are considered equivalent as long as

$$\Delta_L < (\hat{Z}_1 - \hat{Z}_2) < \Delta_U \tag{14.10}$$

where Δ_L and Δ_U are the lower and upper bounds respectively. The two bounds are set according to clinical criteria; what amount of difference would be acceptable for the two techniques to be considered clinically equivalent? (Obuchowski, 1997). Therefore, the null hypothesis is set so that the difference between the diagnostic tests is less than or equal to a predetermined lower bound Δ_L or equal to or greater than a predetermined upper bound Δ_U. In other words, the null hypothesis is set by (1) assuming that the two tests are indeed very different and (2) the difference can be either positive or negative (i.e. test 1 is superior to test 2 and vice versa). For example, it may be acceptable clinically to consider two diagnostic tests as equivalent if their AUC indices are at most 3% different. In that case, the lower and upper bounds are –0.03 and 0.03, respectively. As long as there is statistical evidence that the two diagnostic tests result in AUCs with a difference in the range (–0.03,0.03), the diagnostic tests are considered diagnostically similar.

The equivalence test proceeds by performing two separate one-sided tests with the following test statistics:

$$T_L = \frac{(\hat{Z}_1 - \hat{Z}_2) - \Delta_L}{\sqrt{\sigma^2(\hat{Z}_1) + \sigma^2(\hat{Z}_2) - 2\mathrm{cov}(\hat{Z}_1, \hat{Z}_2)}} \quad \text{and}$$

$$T_U = \frac{\Delta_U - (\hat{Z}_1 - \hat{Z}_2)}{\sqrt{\sigma^2(\hat{Z}_1) + \sigma^2(\hat{Z}_2) - 2\mathrm{cov}(\hat{Z}_1, \hat{Z}_2)}} \tag{14.11}$$

If there is enough statistical evidence (i.e. one-sided p-value < 0.05 for both tests) to reject the null hypothesis, then tests 1 and 2 are equivalent. Note that to prove strict equivalence, it is necessary to reject both parts of the null hypothesis. A different twist to equivalence testing is when emphasis is placed on only one of the two one-sided tests. For example, it may be appropriate to test only whether the new diagnostic test is not worse than the standard one. That would involve only one one-sided test and that is considered a non-inferiority (rather than equivalence) test. Proving with statistical significance that test 1 is *not* inferior to test 2 leaves open the possibility that test 1 may actually be significantly better. Regardless, equivalence and non-inferiority tests can be performed with various performance indices (AUC, pAUC, TPF at a fixed FPF) using either non-parametric (Obuchowski, 1997), parametric (Ikeda, 2003), or resampling methods (Liu, 2006; Li, 2008).

14.5.2 Choosing the optimal operating point

For research purposes, ROC analysis is used in a relative sense; namely to compare two or more diagnostic tests. While this is the most common utilization, it is only the first step to clinical translation. After the best diagnostic test has emerged with the help of ROC analysis, clinical application dictates that an operating point needs to be selected before the diagnostic test is applied clinically. In other words, the result of a binary diagnostic test is expected to determine patient management in some way. To get the binary test result, a decision threshold needs to be set. This is equivalent to selecting an appropriate operating point on the ROC curve. By selecting the operating point (i.e. fixing the decision threshold), the expected diagnostic performance of the test (in terms of sensitivity and specificity) is known. Of course, this statement assumes that the general patient population on which the diagnostic test will be applied was adequately represented by the patient sample on which the test was evaluated and the ROC curve was derived. Testing the validity of such assumptions is a completely different issue and beyond the scope of this chapter.

The selection process of the optimal operating point is guided primarily by clinical criteria. For example, it may be critical that a diagnostic test operates at a specific sensitivity level. Or, two different settings may be desired to stratify test cases in those considered safely as healthy, those considered unquestionably diseased, and those that are intermediate cases. Then, presetting the sensitivity (TPF) or specificity (TNF) dictates which operating point(s) is selected on the ROC curve (Shafer, 1989; Greiner, 1995). Clearly, the financial consequences of a test result cannot be disregarded when trying to determine the best operating point. For example, when screening for breast cancer it is important to set the decision threshold low to ensure that false negative decisions are minimized. These low thresholds, which have high specificity, correspond to ROC points close to the upper boundary of the graphing square. Detecting breast cancer early translates into significant gains for the patient (extended lifetime) and the society. The extra cost imposed by unnecessary workup for false positive patients is relatively less than the therapeutic costs incurred for the treatment of advanced cancer. In contrast, if a test result will lead to a costly treatment with questionable benefit for those tested positive who are truly diseased but dangerous implications for those tested positive who are truly non-diseased, then the decision threshold should be raised significantly. In this case, choosing operating points of high specificity (i.e. ROC points closer to the lower boundary of the graphing square) is recommended. In practice, the choice of the optimal ROC operating point should be guided by clinical cost-effectiveness analysis that takes into account disease prevalence and the relative costs of false positive vs. false negative diagnoses (DeNeef, 1993; Halpern, 1996).

From a mathematical point of view, there have been various methods proposed on how to select the optimal ROC operating point. The methods vary mainly in how they set the selection criterion and/or if they are designed for parametric or non-parametric ROC curves. The simplest method picks the operating point that results in a binary classifier as close to the perfect classifier (point (1,1) on the ROC graph) as possible. Mathematically, the optimal operating point is the one that minimizes the following expression:

$$\min\left\{(1-\text{sensitivity})^2+(1-\text{specificity})^2\right\} \qquad (14.12)$$

Alternatively, the best operating point may be the one that results in a binary classifier as far away as possible from the uninformative classifier (the 45-degree diagonal on the ROC graph):

$$\max\{\text{sensitivity}+\text{specificity}-1\} \qquad (14.13)$$

Note that this criterion is equivalent to maximizing the Youden index (Schisterman, 2005). Both criteria make intuitive sense; the first tries to maximize perfection while the second tries to minimize imperfection. However, they do not necessarily lead to the same operating point.

The methods discussed above consider false positive and false negative decisions as equally detrimental. This is rarely the case in a clinical application. When guided by cost-effectiveness analysis, the selection of the operating point includes disease prevalence and the relative utilities (i.e. consequences) of TP, TN, FP, and FN decisions. Specifically, it has been shown that the best operating point is the one for which the slope of the ROC curve is given by the following equation,

$$\text{slope}=\frac{\text{P(healthy)}}{\text{P(diseased)}}\times\frac{C_{FP}-C_{TN}}{C_{FN}-C_{TP}} \qquad (14.14)$$

where P is the disease prevalence and C_{FP}, C_{FN}, C_{TP}, and C_{TN} are the utilities associated with each one of the four possible decisions (Metz, 1978; Swets, 1992). The utility terms express the risk and benefits associated with each decision. Quantifying the four utilities though is not a trivial task. For the special case where $P = 0.5$, $C_{FP} = C_{FN} = 1.0$, and $C_{TP} = C_{TN} = 0.0$, the optimal operating point is the one for which slope = 1. This is the line tangent to the ROC curve at a 45-degree angle. Thus, this is the ROC point that is closest to the upper left corner of the graphing square.

The following point needs to be emphasized. The above discussion is only relevant to ROC curves that are derived based on continuous data. For continuous test results all decision thresholds are possible, even if not clinically meaningful. In contrast, smooth ROC curves derived from ordinal-type test data give a false impression that every ROC point on the curve is a possible operating point. This is not the case though. For diagnostic tests that produce ordinal outcomes, most ROC points do not correspond to clinically realistic decision threshold values (Dwyer, 1997). The same can be said for laboratory, human observer ROC studies that are popular for assessing medical imaging modalities and algorithms. Often these studies are based on rating-type data where each rating category reflects a general impression of the probability of disease. The decision thresholds that correspond to the operating points derived from such experiments do not necessarily reflect the actual decision threshold imposed in clinical practice. Therefore, setting the optimal decision threshold based on laboratory-type ROC data should be approached with extra caution.

Table 14.4 *ROC software packages*

SOFTWARE PACKAGE	ROC CURVE	AUC ESTIMATION	CURVE COMPARISON	PLATFORM	FREE
Analyse-It	Empirical	Empirical with parametric CIs	Non-parametric	Windows	No
GraphROC	Empirical	Empirical with non-parametric CIs	Non-parametric	Windows	No
MedCalc	Empirical	Empirical with parametric CIs	Non-parametric	Windows	No
NCSS	Empirical Parametric	Parametric and non-parametric	Parametric and non-parametric	Windows	No
ROCKIT	Parametric	Parametric	Parametric	Windows/ Macintosh/Unix	Yes
SPSS	Empirical	Empirical with parametric CIs	N/A	Windows/ Macintosh/Linux	No
STAR	Empirical	Empirical with non-parametric CIs	Non-parametric	Web-based	Yes
STATA	Parametric	Parametric and non-parametric	Parametric and non-parametric	Windows/ Macintosh/Unix	No

14.6 ROC SOFTWARE

It must be clear by now that the statistical analysis of ROC curves is not a trivial matter. To be done properly, it requires advanced statistical knowledge and understanding. Since ROC has been established as the standard way of assessing diagnostic accuracy, several software programs have become available to researchers. These programs are available as either stand-alone for ROC analysis or as components of comprehensive statistical packages. Many of them are regularly updated to address previous limitations. In 2003, Stephan *et al.* reported a comparison of eight popular software packages available for IBM-compatible computers (Stephan, 2003). The comparison was done using a clinical dataset and it was based on three criteria, namely correctness of results, completeness in terms of estimating the ROC curve along with its performance indices including the necessary confidence intervals, and difficulty of use. No single software package emerged as the superior according to the evaluation criteria. The take-home message of the comparative study is still relevant today. Basically, before relying on a particular software package, the researcher needs to understand the underlying statistical methods in the software. Only this way can the researcher truly appreciate the limitations of the data analysis performed using the particular package.

The following is a survey of popular ROC software packages that are currently available. The list includes only packages that have the capability to do something more than simply display the ROC curve. Several popular and user-friendly statistical packages such as JMP (http://www.jmp.com) and SIMSTAT (http://www.simstat.com) provide the empirical ROC curve and calculate the corresponding AUC. However, they do not provide anything more such as confidence intervals or the capability for statistical comparisons among several ROC curves. Thus, they are not included in the survey. Each surveyed package is described individually below but Table 14.4 provides a general summary. The packages are listed in alphabetical order.

Analyse-It (http://www.analyse-it.com) is a commercial software package that includes ROC analysis in the methods evaluation edition. This software provides empirical ROC curves. It uses parametric techniques to estimate confidence intervals on AUC and specific operating points. Also, the non-parametric estimation technique proposed by Delong *et al.* is used to compare ROC curves (Delong, 1988). This software package is available with a 30-day free trial period for Windows operating systems.

GraphROC (http://www.netti.fi/~maxiw/) is a commercial ROC software package. It provides empirical ROC curves and empirical estimates of AUC and pAUC with non-parametric confidence intervals. Actually, this is one of very few packages that provides pAUC estimates. Comparison of ROC curves is based on the non-parametric technique by Hanley *et al.* (Hanley, 1983). However, GraphROC does not estimate the standard error of pAUC and as such, ROC comparisons based on the pAUC performance index are not possible. The theoretical foundation of this software package has been described in detail before (Kairisto, 1995). GraphROC is also available with a 30-day free trial period for Windows operating systems.

MedCalc (http://www.medcalc.be) is a commercial statistical package that offers non-parametric ROC analysis (Schoonjans, 1995). The package displays the empirical ROC curve, a non-parametric estimate of AUC, and parametric estimates of the 95% confidence interval. Comparison of ROC curves is done using the non-parametric technique by Hanley *et al.* (Hanley, 1983). The software is available for Windows operating systems.

NCSS (http://www.ncss.com) is a general commercial statistical package that offers both non-parametric and parametric ROC analysis. The package provides the estimated AUC and pAUC indices along with the 95% confidence interval. It is one of the few software packages capable of performing statistical comparisons of ROC curves based on both the AUC and pAUC indices. The package also provides power calculations and sample size analysis for both ordinal data and continuous-type data. The software is available only for Windows operating systems and it offers a seven-day free trial.

ROCKIT (http://xray.bsd.uchicago.edu/krl/roc_soft.htm) is by far the most popular and often used ROC software among researchers in the medical imaging community. This software includes a suite of programs for parametric ROC fitting, plotting, analyzing, and comparing (Metz, 1984; Metz, 1986a; Dorfman, 1992). The software provides smooth ROC curves according to the binormal model and parametric estimates of AUC along with 95% confidence intervals. ROC curves can be compared using parametric modeling based on unpaired, paired, or partially paired data. Comparing ROC curves based on their slope and intercept parameters or on a particular operating point is also possible. This software package is made freely available to the research community by Dr. Charles E. Metz (University of Chicago). The same investigator has also made available an additional software package known as LABMRMC for analysis of multi-reader, multi-case ratings-type data. This software provides a parametric estimate of the AUC index but if the data is degenerate, then it reports the empirical estimate. The jackknife technique is used to derive 95% CIs and for comparisons of ROC curves (Dorfman, 1992).

SPSS (http://www.spss.com) is a popular commercial statistical package that offers ROC analysis. The package provides the empirical ROC curve and non-parametric estimates of the AUC and its 95% CI. Statistical comparison of ROC curves is not possible with SPSS, therefore it is not a competitive choice.

STAR (http://protein.bio.puc.cl/cardex/servers/roc/home.php) is a free, web-based software which follows non-parametric ROC estimation techniques for paired data and comparison of their AUCs (Vergara, 2008).

STATA (http://www.stata.com) is a frequently used general statistical package that includes ROC analysis as part of its tools for epidemiologists. The package provides parametric and non-parametric ROC curves and comparisons of their AUCs. This is a commercial product.

A notable omission from the above list is the AccuROC software package (http://www.accuroc.com). This is a commercial product that rated the best in the comparative study by Stephan *et al.* (Stephan, 2003). However, the site does not distribute new licenses any more; it only supports existing ones. Therefore, it is not a viable option for the newcomers in the field and as such was not included in the above survey.

14.7 ADVANCED TOPICS

Advanced topics in ROC analysis span a wide range of statistical issues, techniques to address well-known caveats of the ROC methodology, as well as recent developments on extending the methodology from two-class to multi-class applications. The following is a limited list of topics with appropriate references for further reading.

14.7.1 Sample size calculations

Although not a standard practice in published studies, sample size calculations are essential to understand the implications of the results of an ROC study. An adequate sample size ensures the reliability of the conclusions drawn by the study. The sample size calculation is performed during the planning phase of an ROC study to determine the number of subjects needed to properly test the underlying study hypothesis. The calculation depends on the performance index selected (e.g. AUC, pAUC) and the study design (i.e. a single diagnostic test compared to a fixed standard or to another diagnostic test studied in the same experiment). There are different methods proposed for sample size calculations in ROC studies based on parametric or non-parametric assumptions (Zhou, 2002). Specifically, for ROC studies related to medical image applications where multiple observers may be involved, there are methods for single-observer (Hanley, 1983; Obuchowski, 1994) and multiple-observer designs (Obuchowski, 2005; Beiden, 2000b). Regardless of the methodology employed, the study designer needs to specify the smallest expected difference in the selected performance index. This specification should be clinically relevant and based on pilot data. However, in the absence of pilot data, designers tend to assume a larger difference than realistically possible so that a smaller sample size requirement is estimated for the study. The drawback of this approach is that ultimately the executed studies can be seriously underpowered. Obuchowski *et al.* have provided some general guidelines on the required sample size based on the underlying assumptions of an ROC study (Obuchowski, 2000). Regardless of whether the actual sample size achieved in the study is less than the desired sample size determined according to the sample size calculations, it is still important to report the statistical power of the ROC study based on its actual sample size.

14.7.2 Suboptimal gold standard

The material presented in the previous sections relies on the strong assumption that a perfect binary gold standard exists for all test cases. In clinical applications, the gold standard is established with nearly infallible methods such as surgery, autopsy, or long-term clinical follow-up. However, it is not always possible to obtain a perfect gold standard. First, there are several situations where a gold standard is practically unattainable. For example, when developing a new diagnostic test for screening patients for lung cancer, biopsy results may not be possible to obtain for all patients due to the invasiveness of the procedure. Lung biopsy cannot be justified for patients who have a negative diagnostic test so that the proper gold standard is established. For such situations, it is often common to settle for a less-than-perfect alternative such as the opinion of a committee of experts who review the available data for each test subject to provide a consensus opinion regarding truth ("expert panel"). In this scenario, the gold standard suffers from verification bias (Begg, 1983; Toledano, 1999; Zhou, 2000). Second, the gold standard technique itself may be imperfect. For example, the biopsy gold standard can be trusted completely only if the biopsy specimen is taken exactly from the right location. This is also known as a fuzzy gold standard because it suffers from measurement error (Phelps, 1995; Johnson, 2001). A special case of measurement error is the complete absence of a gold standard, first introduced by Henkelman *et al.* in 1990 (Henkelman, 1990). More recently, Beiden *et al.* (Beiden, 2000a) offered a maximum likelihood solution to the problem based on the expectation-maximization algorithm. The details of the algorithm are beyond the scope of this chapter. It should be noted however that the study determined that a large multiple of cases (to the order of 25 times more) are required to achieve the same level of precision in an ROC study with unknown gold standard than when solving the same problem with a perfect gold standard. In general, researchers should not neglect the bias or uncertainty introduced due to a suboptimal gold standard. Although the existing ROC software packages are not capable of dealing with this challenge, there are several methods available for analysis of ROC data burdened by a suboptimal gold standard (Zhou, 2002; Pepe, 2003). Particularly, in medical imaging applications where an expert panel is often used to establish the truth, resampling techniques have been proposed to incorporate this source of variability in the reported results (Miller, 2004; Petrick, 2005).

14.7.3 ROC deviations

Although conventional ROC analysis is well studied and there is sophisticated ROC software available, it is not equipped to handle the variety of classification problems encountered in medical imaging. ROC analysis is designed to handle binary decision problems. However, disease diagnosis in medical images rarely conforms to the binary problem assumption. For example, the detection of abnormalities in a medical image involves more than simply deciding if there is disease present or not. Typically, diagnostic image interpretation requires that the physician identifies and correctly localizes all abnormalities present in the image. Correct identification and localization of all abnormalities are necessary steps for successful patient outcome through surgery or targeted treatment planning, for example. Although diagnostic image interpretation could be approached as a binary problem by ignoring the exact number of abnormalities present or their precise location, ROC analysis under such circumstances could easily lead to overestimation of performance. For example, if a physician calls a case abnormal, it does not necessarily mean that all possible abnormalities have been detected and correctly localized. Variants of ROC analysis such as localization ROC (LROC) and free-response ROC (FROC) have been introduced to deal with these situations and they are covered in other chapters of this book.

One of the latest developments in ROC analysis is its extension from the two-class to the multi-class classification problem. The computational complexity of the problem increases as more classes are considered. Conceptually, multi-class ROC analysis is equivalent to extending the traditional ROC methodology to problems where the gold standard is not binary anymore but rather it is expressed on a nominal or ordinal scale. Several investigators have attempted formalization of the problem within the typical ROC framework (Mossman, 1999; Kijewski, 1989; Obuchowski, 2001). Concepts such as the AUC index have been extended to the volume under the ROC surface (VUS) index. An interesting discussion on the topic can be found in the article by Hand and Till (Hand, 2001). Recently, Obuchowski explored extension of the ROC framework to problems where the gold standard is expressed on a continuous scale (Obuchowski, 2006). Regardless of the challenges it poses, multi-class ROC is an active field of investigation with exciting progress expected in the upcoming years.

14.8 SUMMARY

ROC analysis is widely accepted in the medical imaging community as the standard evaluation methodology for imaging modalities and diagnostic tests as long as they can be viewed as binary predictive models. ROC curves provide a comprehensive measure of diagnostic accuracy, independent of disease prevalence and decision thresholds. They are suitable for quantifying the diagnostic accuracy of a predictive model and for selecting its optimal decision threshold depending on the clinical circumstances. ROC analysis is also valuable for comparing the accuracy of different, competing predictive models. The chapter provided an introduction to the basic mathematical concepts, statistical issues, hands-on implementation, and limitations associated with conventional ROC methodology. Understanding the theoretical foundation of ROC analysis is critical to help avoid common errors of misuse and misinterpretation sometimes found in the literature. For the more interested reader, advanced ROC topics were briefly introduced with appropriate references. Some of these topics are covered more extensively in other chapters.

REFERENCES

Aoki, K., Misumi, J., Kimura, T., Zhao, W., Xie, T. (1997). Evaluation of cutoff levels for screening of gastric cancer using serum pepsinogens and distributions of levels of serum pepsinogen I, Ii and of Pg I/Pg Ii ratios in a gastric cancer case-control study. *Journal of Epidemiology*, **7**, 143–151.

Bamber, D. (1975). The area above the ordinal dominance graph and the area below the receiver operating characteristic graph. *Journal of Mathematical Psychology*, **12**, 387–415.

Begg, C.B., Greenes, R.A. (1983). Assessment of diagnostic tests when disease verification is subject to selection bias. *Biometrics*, **39**, 207–215.

Beiden, S.V., Campbell, G., Meier, K.L., Wagner, R.F. (2000a). The problem of ROC analysis without truth: the EM algorithm and the information matrix. *Proceedings of SPIE*, **3981**, 126–134.

Beiden, S.V., Wagner, R.F., Campbell, G. (2000b). Components-of-variance models and multiple-bootstrap experiments: an alternative method for random effects, receiver operating characteristic analysis. *Academic Radiology*, **7**, 341–349.

Cortes, C., Mohri, M. (2003). AUC optimization vs. error rate. *Advances in Neural Information Processing Systems 16: Proceedings of the 2003 Conference*. Cambridge, MA: The MIT Press.

Delong, E.R., Delong, D.M., Clarke-Pearson, D.L. (1988). Comparing the areas under two or more correlated receiver operating characteristics curves: a non-parametric approach. *Biometrics*, **44**, 837–845.

Deneef, P., Kent, D.L. (1993). Using treatment-tradeoff preferences to select diagnostic strategies. *Medical Decision Making*, **13**, 126–132.

Dorfman, D.D., Alf, E. (1968). Maximum likelihood estimation of parameters of signal detection theory: a direct solution. *Psychometrika*, **33**, 117–124.

Dorfman, D.D., Alf, E. (1969). Maximum likelihood estimation of parameters of signal detection theory and determination of confidence intervals – rating method data. *Journal of Mathematical Psychology*, **6**, 487–496.

Dorfman, D.D., Berbaum, K.S., Metz, C.E. (1992). Receiver operating characteristic rating analysis: generalization to the population of readers and patients with the jackknife method. *Investigative Radiology*, **27**, 723–731.

Dwyer, A.J. (1997). In pursuit of a piece of the ROC. *Radiology*, **202**, 621–625.

Efron, B., Tibshirani, R.J. (1993). *An Introduction to the Bootstrap*. New York, NY: Chapman & Hall.

Faraggi, D., Reiser, B. (2002). Estimation of the area under the ROC curve. *Statistics in Medicine*, **21**, 3093–3106.

Goddard, M.J., Hinberg, I. (1990). Receiver operating characteristic (ROC) curves and non-normal data: an empirical study. *Statistics in Medicine*, **9**, 325–337.

Greiner, M., Sohr, D., Gobel, P. (1995). A modified ROC analysis for the selection of cut-off values and the definition of intermediate results for serodiagnostic tests. *Journal of Immunological Methods*, **185**, 123–132.

Grmec, I., Kupnik, D. (2004). Does the Mainz emergency evaluation scoring (MEES) in combination with capnometry (MEESC) help in the prognosis of outcome from cardiopulmonary resuscitation in a prehospital setting? *Resuscitation*, **58**, 89–96.

Hajian-Tilaki, K.O., Hanley, J.A., Joseph, L., Collet, J.P. (1997). A comparison of parametric and nonparametric approaches to ROC analysis of quantitative diagnostic tests. *Medical Decision Making*, **17**, 94–102.

Halpern, E.J., Albert, M., Krieger, A.M., Metz, C.E., Maidment, A.D. (1996). Comparison of receiver operating characteristic curves on the basis of optimal operating points. *Academic Radiology*, **3**, 245–253.

Hand, D.J., Till, R.J. (2001). A simple generalization of the area under the ROC curve to multiple class classification problems. *Machine Learning*, **45**, 171–186.

Hanley, J.A., McNeil, B.J. (1982). The meaning and use of the area under a receiver operating characteristic curve. *Radiology*, **143**, 29–36.

Hanley, J.A., McNeil, B.J. (1983). A method for comparing the areas under receiver operating characteristic curves derived from the same cases. *Radiology*, **148**, 839–843.

Hanley, J.A. (1988). The robustness of the "binormal" assumptions used in fitting ROC curves. *Medical Decision Making*, **8**, 197–203.

Hanley, J.A. (1996). The use of the ''binormal'' model for parametric ROC analysis of quantitative diagnostic tests. *Statistics in Medicine*, **15**, 1575–1585.

Hanley, J.A., Hajian-Tilaki, K.O. (1997). Sampling variability of nonparametric estimates of the areas under receiver operating characteristic curves: an update. *Academic Radiology*, **4**, 49–58.

Harrell, F.E., Jr., Califf, R.M., Pryor, D.B., Lee, K.L., Rosati, R.A. (1982). Evaluating the yield of medical tests. *Journal of the American Medical Association*, **247**, 2543–2546.

Henkelman, R.M., Kay, I., Bronskill, M.J. (1990). Receiver operator characteristic (ROC) analysis without truth. *Medical Decision Making*, **10**, 24–29.

Ikeda, M., Ishigaki, T., Yamauch, K. (2003). How to establish equivalence between two treatments in ROC analysis. *Proceedings of SPIE*, **5034**, 383–392.

Jiang, Y., Metz, C.E., Nishikawa, R.M. (1996). A receiver operating characteristic partial area index for highly sensitive diagnostic tests. *Radiology*, **201**, 745–750.

Johnson, W.O., Gastwirth, J.L., Pearson, L.M. (2001). Screening without a "gold standard": the Hui-Walter paradigm revisited. *American Journal of Epidemiology*, **153**, 921–924.

Kairisto, V., Poola, A. (1995). Software for illustrative presentation of basic clinical characteristics of laboratory tests – Graphroc for Windows. *Scandinavian Journal of Clinical Laboratory Investigations*, **55**, 43–60.

Kijewski, M.F., Swennson, R.G., Judy, P.F. (1989). Analysis of rating data from multiple-alternative tasks. *Journal of Mathematical Psychology*, **33**, 1–23.

Lee, W.-C., Hsiao, C.K. (1996). Alternative summary indices for the receiver operating characteristic curve. *Epidemiology*, **7**, 605–611.

Lee, W.-C. (1999). Probabilistic analysis of global performances of diagnostic tests: interpreting the Lorenz curve-based summary measures. *Statistics in Medicine*, **18**, 455–471.

Li, C.-R., Liao, C.-T., Liu, J.-P. (2008). A non-inferiority test for diagnostic accuracy based on the paired partial areas under ROC curves. *Statistics in Medicine*, **27**, 1762–1776.

Liu, J.-P., Ma, M.-C., Wu, C.-Y., Tai, J.-Y. (2006). Tests of equivalence and non-inferiority for diagnostic accuracy based on the paired areas under ROC curves. *Statistics in Medicine*, **25**, 1219–1238.

Lusted, L.B. (1960). Logical analysis in roentgen diagnosis. *Radiology*, **74**, 78–93.

Lusted, L.B. (1961). Signal detectability and medical decision making. *Science*, **171**, 1217–1219.

McClish, D.K. (1989). Analyzing a portion of the ROC curve. *Medical Decision Making*, **9**, 190–195.

Metz, C.E. (1978). Basic principles of ROC analysis. *Seminars in Nuclear Medicine*, **8**, 283–298.

Metz, C.E., Kronman, H.B. (1980). Statistical significance tests for binormal ROC curves. *Journal of Mathematical Psychology*, **22**, 218–243.

Metz, C.E., Wang, P.-L., Kronman, H.B. (1984). A new approach for testing the significance of differences between ROC curves measured from correlated data. In: Deconinck, F., ed. *Information Processing In Medical Imaging*. The Hague: Nijhoff, 432–445.

Metz, C.E. (1986a). Statistical analysis of ROC data in evaluating diagnostic performance. In: Herbert, D., Myers, R., eds. *Multiple Regression Analysis: Applications in the Health Sciences*. New York, NY: American Institute of Physics, 365–384.

Metz, C.E. (1986b). ROC methodology in radiologic imaging. *Investigative Radiology*, **21**, 720–733.

Metz, C.E., Herman, B.A., Shen, J.-H. (1998). Maximum-likelihood estimation of ROC curves from continuously-distributed data. *Statistics in Medicine*, **17**, 1033–1053.

Metz, C.E., Pan, X. (1999). "Proper" binormal roc curves: theory and maximum-likelihood estimation. *Journal of Mathematical Psychology*, **43**, 1–33.

Miller, D.P., O'Shaughnessy, K.F., Wood, S.A., Castellino, R.A. (2004). Gold standards and expert panels: a pulmonary nodule case study with challenges and solutions. *Proceedings of SPIE*, **5372**, 173.

Mossman, D. (1999). Three-way ROCs. *Medical Decision Making*, **19**, 78–89.

Obuchowski, N.A. (1994). Sample size for receiver operating characteristic studies. *Investigative Radiology*, **29**, 238–243.

Obuchowski, N.A. (1997). Testing for equivalence of diagnostic tests. *American Journal of Radiology*, **168**, 13–17.

Obuchowski, N.A., Liebler, M.L. (1998). Confidence intervals for the receiver operating characteristic area in studies with small samples. *Academic Radiology*, **5**, 561–571.

Obuchowski, N.A. (2000). Sample size tables for receiver operating characteristic studies. *American Journal of Roentgenology*, **175**, 603–608.

Obuchowski, N.A., Goske, M.J., Applegate, K.E. (2001). Assessing physicians' accuracy in diagnosing pediatric patients with acute abdominal pain: measuring accuracy for multiple diseases. *Statistics in Medicine*, **20**, 3261–3278.

Obuchowski, N.A. (2005). Multi-reader multi-modality ROC studies: hypothesis testing and sample size estimation using an ANOVA approach with dependent observations. *Academic Radiology*, **2**, 522–529.

Obuchowski, N.A. (2006). An ROC-type measure of diagnostic accuracy when the gold standard is continuous-scale. *Statistics in Medicine*, **25**, 481–493.

Pepe, M.S. (2003). *The Statistical Evaluation of Medical Tests for Classification and Prediction*. Oxford, UK: Oxford University Press.

Petrick, N., Gallas, B.D., Samuelson, F.W., Wagner, R.F., Myers, K.J. (2005). Influence of panel size and expert skill on truth panel performance when combining expert ratings. *Proceedings of SPIE*, **5749**, 49.

Phelps, C.E., Hutson, A. (1995). Estimating diagnostic test accuracy using a "fuzzy gold standard." *Medical Decision Making*, **15**, 44–57.

Schafer, H. (1989). Constructing a cut-off point for a quantitative diagnostic test. *Statistics in Medicine*, **8**, 1381–1391.

Schoonjans, F., Zalata, A., Depuydt, C.E., Comhaire, F.H. (1995). Medcalc: a new computer program for medical statistics. *Computer Methods and Programs in Biomedicine*, **48**, 257–262.

Schisterman, E.F., Perkins, N.J., Aiyi, L., Bondell, H. (2005). Optimal cut-point and its corresponding Youden index to discriminate individuals using pooled blood samples. *Epidemiology*, **16**, 73–81.

Schuirmann, D.U.I. (1987). A comparison of the two 1-sided tests procedure and the power approach for assessing the equivalence of average bioavailability. *Journal of Pharmacokinetics and Pharmacodynamics*, **15**, 657–680.

Stephan, C., Wesseling, S., Schink, T., Jung, K. (2003). Comparison of eight computer programs for receiver-operating characteristic analysis. *Clinical Chemistry*, **49**, 433–439.

Swets, J.A. (1979). ROC analysis applied to the evaluation of medical imaging techniques. *Investigative Radiology*, **14**, 109–121.

Swets, J.A. (1986). Empirical ROCs in discrimination and diagnostic tasks: implications for theory and measurement of performance. *Psychology Bulletin*, **99**, 181–198.

Swets, J.A. (1988). Measuring the accuracy of diagnostic systems. *Science*, **240**, 1285–1293.

Swets, J.A. (1992). The science of choosing the right decision threshold in high-stakes diagnostics. *American Psychology*, **47**, 522–532.

Toledano, A.Y., Gatsonis, C. (1999). Generalized estimating equations for ordinal categorical data: arbitrary patterns of missing responses and missingness in a key covariate. *Biometrics*, **55**, 488–496.

Vergara, I.A., Norambuena, T., Ferrada, E., Slater, A.W., Melo, F. (2008) StAR: a simple tool for the statistical comparison of ROC curves. *BMC Bioinformatics*, **9**, 265.

Wagner, R.F., Beiden, C.V., Metz, C.E., Campbell, G. (2001). Continuous versus categorical data for ROC analysis: some quantitative considerations. *Academic Radiology*, **8**, 328–34.

Wagner, R.F., Metz, C.E., Campbell, G. (2007). Assessment of medical imaging systems and computer aids: a tutorial review. *Academic Radiology*, **14**, 723–748.

Walsh, S.J. (1999). Goodness-of-fit issues in ROC curve estimation. *Medical Decision Making*, **19**, 193–201.

Wieand, S., Gail, M.H., James, B.R., James, K.L. (1989). A family of nonparametric statistics for comparing diagnostic markers with paired or unpaired data. *Biometrika*, **76**, 585–592.

Youden, W.J. (1950). Index for rating diagnostic tests. *Cancer*, **3**, 32–35.

Zhang, D.D., Zhou, X.-H., Freeman, D.H., Jr., Freeman, J.L. (2002). A non-parametric method for the comparison of partial areas under ROC curves and its application to large health care data sets. *Statistics in Medicine*, **21**, 701–15.

Zhou, X.-H., Higgs, R.E. (2000). Assessing the relative accuracies of two screening tests in the presence of verification bias. *Statistics in Medicine*, **19**, 1697–1705.

Zhou, X.-H., Obuchowski, N.A., McClish, D.K. (2002). *Statistical Methods in Diagnostic Medicine*. New York, NY: Wiley.

Zou, K.H., Hall, W.J., Shapiro, D.E. (1997). Smooth nonparametric receiver operating characteristic (ROC) curves for continuous diagnostic tests. *Statistics in Medicine*, **16**, 2143–2156.

Zou, K.H., Tempany, C.M., Fielding, J.R., Silverman, S.G. (1998). Original smooth receiver operating characteristic curve estimation from continuous data: statistical methods for analyzing the predictive value of spiral CT of ureteral stones. *Academic Radiology*, **5**, 680–687.

Zou, K.H. (2001). Comparison of correlated receiver operating characteristic curves derived from repeated diagnostic test data. *Academic Radiology*, **8**, 225–233.

Zou, K.H., Resnic, F.S., Talos, I.F., *et al.* (2005). A global goodness-of-fit test for receiver operating characteristic curve analysis via the bootstrap method. *Journal of Biomedical Informatics*, **38**, 395–403.

15

Multireader ROC analysis

STEPHEN HILLIS

15.1 INTRODUCTION TO MULTIREADER ROC ANALYSIS

Receiver operating characteristic (ROC) curve analysis is a well-established method for evaluating and comparing the performance of diagnostic tests, as discussed in Chapter 14. ROC curves allow us to quantify the ability of a reader to discriminate between diseased and normal images in a way that does not depend on the thresholds used by the reader. Throughout this chapter we assume that we are considering diagnostic imaging studies where there are multiple readers (e.g. radiologists) who assign disease-severity or disease-likelihood ratings, using one or more modalities, to the same images using either a discrete (e.g. 1, 2, 3, 4, or 5) or a continuous (e.g. 0–100%) ordinal scale. From these ratings, ROC curves and corresponding accuracy estimates are computed for each reader and each modality (if there are multiple modalities), in order to assess how well a modality performs or to compare the performance of modalities.

In imaging studies there often is variation in ratings due to readers. For example, for a disease such as breast cancer, the rating assigned to an image can be different for two radiologists, even if they have similar experience and training. In such situations we say there is *reader variability*. When human readers are involved in evaluating images, either unaided or using computer-aided diagnosis (CAD), then the assumption of no reader variability is typically not realistic.

The methods discussed in Chapter 14 only take into account variation in ratings due to subject variability when making inferences (e.g. computing *p*-values and confidence intervals); hence these methods are limited either to studies where there is no reader variability, or where the researcher is only interested in the performance of the modalities when used by the particular readers in the study. An example where there is no reader variability is when the scoring of an image is completely done by a computer and it is reasonable to assume that another computer would give the same score; for this situation there is no reader (i.e. computer) variability, and thus the methods from Chapter 14 would be suitable for such a study. An example where the researcher is only interested in the performance of the particular readers in the study is when the modality will be used only by the study readers; in this situation there is no need to generalize results to other readers, and hence the Chapter 14 methods would be suitable.

For studies where there is reader variability, the researcher typically will prefer to have conclusions apply to both the case and reader populations. (Note that *case* is used synonymously with *subject* throughout.) For example, in comparing modality A with modality B, the conclusion, "modality A performs better than modality B, taking into account both reader and case variability," is typically preferable to the conclusion, "modality A performs better than modality B when images are read by the five radiologists used in the study." The second conclusion is restricted to the readers used in the study; in contrast, the first conclusion applies to the population of readers from which the study readers can be treated as a random sample. Although an extension of the methods in Chapter 14 can yield a conclusion such as the second one, they cannot yield a conclusion like the first.

The purpose of this chapter is to provide an introduction to the analysis of multireader ROC data that takes into account reader variability. The remainder of the chapter is organized as follows. In Section 15.2 I discuss an example of a multireader study where the researcher wants to account for reader variability. I describe the major approach for multireader ROC analysis that accounts for reader variability in Section 15.3, and illustrate its use for the Section 15.2 example in Section 15.4. Concluding remarks are made in Section 15.5.

15.2 EXAMPLE: SPIN-ECHO VERSUS CINE MRI

In this section I consider a motivating example. The data for the example consist of rating data for a study where several radiologists read the same set of images. The researcher wants to compare two modalities and wants conclusions about the modalities to apply to both the reader and case populations. Methods that accomplish this goal are commonly referred to as *multireader multicase (MRMC) methods*, and data for which these methods are designed are referred to as *MRMC data*. In Section 15.3 I discuss the most popular MRMC analysis methods.

The data are provided by Carolyn Van Dyke, MD. The study (Van Dyke, 1993) compared the relative performance of single spin-echo (SE) magnetic resonance imaging (MRI) to cine MRI for the detection of thoracic aortic dissection. There were 114 patients: 45 patients with an aortic dissection and 69 patients without a dissection, imaged with both SE and cine MRI. Five radiologists independently interpreted all of the images using a five-point ordinal scale: 1 = definitely no aortic dissection, 2 = probably no aortic dissection, 3 = unsure about aortic dissection, 4 = probably aortic dissection, and 5 = definitely aortic dissection. Each radiologist read each image using each

The Handbook of Medical Image Perception and Techniques, ed. Ehsan Samei and Elizabeth Krupinski. Published by Cambridge University Press.

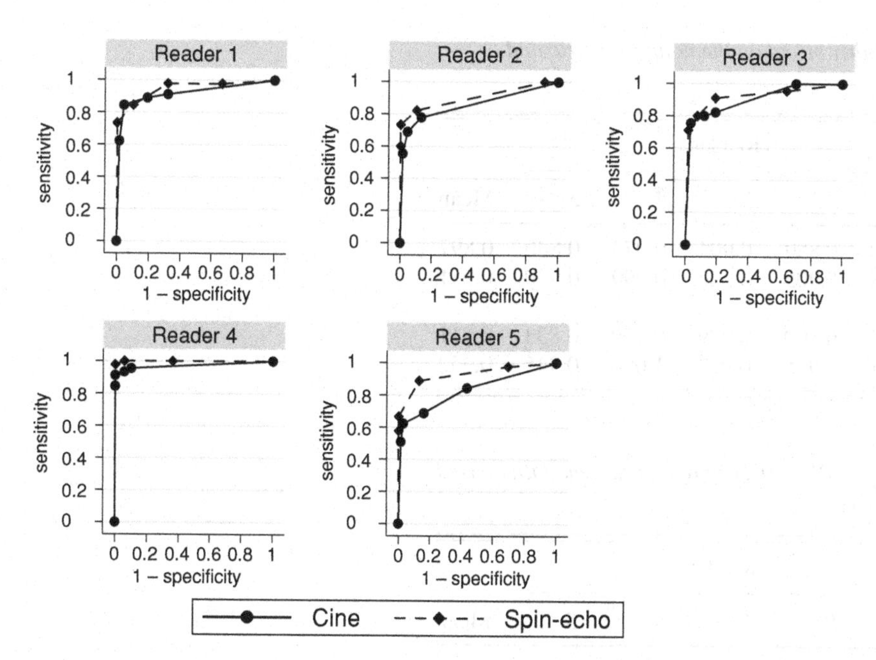

Figure 15.1 Observed (1 – specificity, sensitivity) points and nonparametric ROC curves for spin-echo and cine MRI in the detection of aortic dissection.

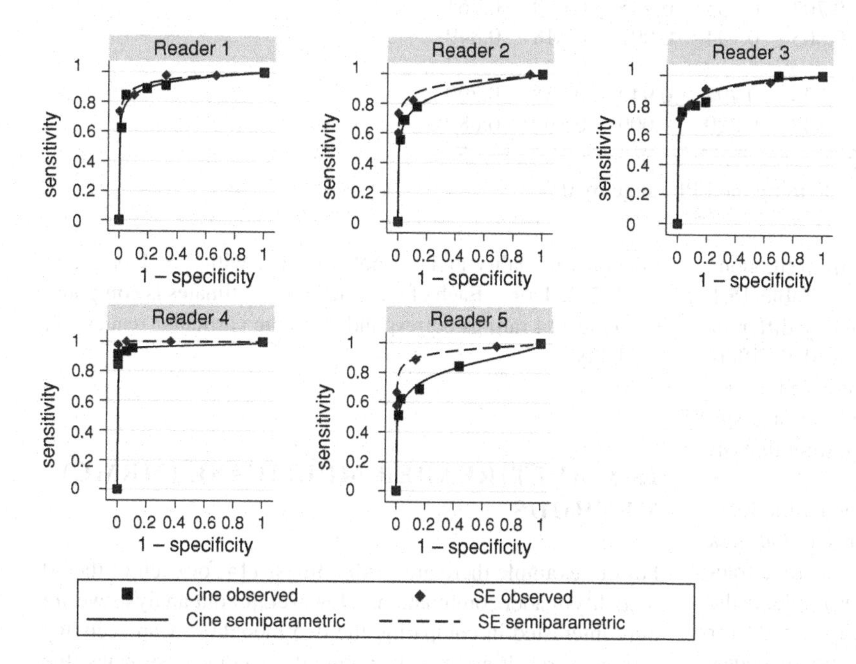

Figure 15.2 Observed (1 – specificity, sensitivity) points and semiparametric ROC curves for spin-echo and cine MRI in the detection of aortic dissection.

of the two modalities, SE and cine MRI. Because each reader reads each image under each modality, this experiment has a factorial design with three factors: modality, reader, and case.

The performance of the two modalities can be compared descriptively by visually comparing plots of the ROC curves that show the performance of the modalities for each reader. Two common methods for computing the ROC curve are (1) the nonparametric or empirical method, where the ROC curve is constructed by connecting the observed (1 – specificity, sensitivity) points with straight lines; and (2) the semiparametric method that assumes a latent binormal model and estimates the ROC curve using maximum likelihood estimation, resulting in a smooth ROC line. These methods have been previously discussed in Chapter 14.

The nonparametric ROC curves are presented in Figure 15.1 and the semiparametric ROC curves in Figure 15.2. In both figures we see that the ROC curve for SE MRI tends to be above that for cine MRI for each reader (except for the semiparametric curves for reader 3, which look very similar), suggesting that SE MRI performs better.

In addition to visually comparing ROC curves, we can compare accuracy measure estimates, computed from the ROC curves, that describe the discriminatory ability of each modality and reader pair. A commonly used accuracy measure is the area under the ROC curve (AUC), which was previously discussed in Chapter 14. This measure has the interpretation of being the probability that the reader will assign a higher rating to a diseased case image compared to a normal case image. Table 15.1

Table 15.1 *AUC estimates by reader and modality based on the nonparametric and semiparametric ROC curves*

Estimation method	Modality	Reader 1	2	3	4	5	Mean
nonparametric	Cine	0.920	0.859	0.904	0.973	0.830	0.897
	Spin-echo (SE)	0.948	0.905	0.922	0.999	0.930	0.941
semiparametric	Cine	0.933	0.890	0.929	0.970	0.833	0.911
	Spin-echo (SE)	0.951	0.935	0.928	1.000	0.945	0.952

Table 15.2 *Partial AUC (pAUC) estimates ($0 \leq FPF \leq 0.2$) by reader and modality based on the nonparametric and parametric ROC curves*

Estimation method	Modality	Reader 1	2	3	4	5	Mean
nonparametric	Cine	0.808	0.703	0.735	0.945	0.629	0.764
	Spin-echo (SE)	0.831	0.805	0.794	0.997	0.818	0.849
semiparametric	Cine	0.822	0.726	0.794	0.949	0.658	0.790
	Spin-echo (SE)	0.868	0.843	0.820	1.000	0.869	0.880

Note: The pAUC values have been normalized by dividing the FPF range by 0.2.

presents the AUC accuracy estimates computed from the nonparametric and semiparametric ROC curves. From Table 15.1 we see that the two methods give similar but slightly different results, with the nonparametric method showing an AUC difference (SE − cine) of $0.941 - 0.897 = 0.044$, averaged across readers, and the semiparametric method showing an average difference of $0.952 - 0.911 = 0.041$, again suggesting that SE MRI performs better.

Another accuracy measure is the partial area under the ROC curve (pAUC), as discussed in Chapter 14. This is the area under the curve for a specified range of the false positive fraction (FPF), or equivalently, 1 − specificity. For example, if the researcher wants each modality to have a specificity of ≥ 0.8 (or equivalently, $FPF \leq 0.2$), then it makes sense to compare modalities using the pAUC for $0 \leq FPF \leq 0.2$. Table 15.2 presents the pAUC ($0 \leq FPF \leq 0.2$) for each reader-modality pair, computed from the nonparametric and semiparametric ROC curves. In Table 15.2 the pAUCs have been normalized by dividing by the FPF range of 0.2; the normalized pAUC can be interpreted as the average sensitivity over the restricted FPF interval (McClish, 1989; Jiang, 1996). From Table 15.2 we see that the two methods give similar results, with the nonparametric method showing an average pAUC difference (SE − cine) of $0.849 - 0.764 = 0.085$ and the semiparametric method showing an average difference of $0.880 - 0.790 = 0.090$, again suggesting that SE MRI performs better.

In this example the outcomes of interest are the ten accuracy estimates corresponding to the five readers and two modalities. That is, the outcomes are either the ten AUC estimates or the ten pAUC estimates, which we denote by AUC_{ij} or $pAUC_{ij}$, where i denotes modality and j denotes reader, with $i = 1$ or 2 and $j = 1, 2, 3, 4$, or 5. Each of these accuracy estimates is computed from the 114 ratings corresponding to the particular reader and modality.

15.3 MULTIREADER MULTICASE (MRMC) METHODS

For our example there are 1140 ratings, 114 for each of the 10 modality-reader combinations. However, for our analysis we are only interested in comparing the two modality mean accuracy estimates, resulting from averaging the accuracy estimates (the AUCs or pAUCs) across readers for each modality. Because each reader reads each image under each modality, the 10 accuracy estimates are correlated with each other. This correlation must be taken into account when making inferences, e.g. when computing p-values for hypotheses tests or constructing confidence intervals.

The two conventional ways to account for correlated data are random effects models and generalized estimating equations (GEEs) (Zeger, 1986). For our example neither of these approaches works: the conventional random effects analysis of variance (ANOVA) model is not appropriate because the accuracy measure is not the mean of the case-level ratings, and GEE is not applicable since there are no independent clusters of outcomes.

In this section I give an introduction to the main MRMC methods. While some of the details require basic knowledge of

ANOVA methods, it is not necessary to understand the details of this section in order to use the results. Thus I recommend that you skim over those statistical details that you may not have the necessary statistical background to understand.

15.3.1 The Dorfman-Berbaum-Metz (DBM) method

The first practical way to analyze MRMC data was the procedure proposed by Dorfman, Berbaum, and Metz (DBM) (Dorfman, 1992); this method is commonly referred to as the *DBM method.* This method involves computing jackknife *pseudovalues* (Shao, 1995) and then applying a conventional random effects ANOVA to the pseudovalues.

Jackknife pseudovalues are computed by leaving out one case image at a time, computing the resulting accuracy estimate, then computing a weighted difference between the accuracy estimate based on all of the data and the accuracy estimate computed with the case omitted. Specifically, for modality i and reader j, the AUC pseudovalue for case k, denoted by Y_{ijk}, is defined by

$$Y_{ijk} = c\widehat{\text{AUC}}_{ij} - (c-1)\,\widehat{\text{AUC}}_{ij(k)} \qquad (15.1)$$

where $\widehat{\text{AUC}}_{ij}$ is the AUC estimate for modality i and reader j, computed from all of the data, $\widehat{\text{AUC}}_{ij(k)}$ is the AUC estimate computed from the same data but with data for case k removed, and c is the number of cases. Pseudovalues for pAUC are similarly defined. For our example there will be 1140 computed pseudovalues, one for each modality, reader, and case combination.

A random effects ANOVA is then applied to the pseudovalues, treating pseudovalue as the dependent variable. For this ANOVA there are three factors: modality, reader, and case. Two-way and three-way interactions are included in the model. Modality is treated as a *fixed factor* because we are only interested in the modalities in the experiment. Reader and case are treated as *random factors* because we are interested in comparing the modalities for the *populations* of readers and cases. That is, we are interested to know what the expected difference in the modality accuracy measures is for a *randomly selected reader* reading images from *randomly selected cases.* A typical shorthand notation for this model that shows the effects in the model, with "*" used to indicate an interaction, is given by

$$\begin{aligned}\text{pseudovalue} &= \text{treatment} + \text{reader} + \text{case} + \text{treatment*reader}\\ &\quad + \text{treatment*case} + \text{reader*case}\\ &\quad + \text{treatment*reader*case} + \text{error}\end{aligned}$$

Here and elsewhere I use *treatment* synonymously with *modality* for consistency with the computer output that will be presented and discussed in Section 15.4.

Although the original DBM method performed satisfactorily in simulations, there were some problems with it: (1) it was limited to jackknife accuracy estimates (e.g. the jackknife AUC estimate); (2) it was substantially conservative (i.e. it accepted the null hypothesis too often and confidence intervals were too wide); and (3) it lacked a firm statistical foundation, since pseudovalues are not observed outcomes. Problem (1) was remedied by using normalized pseudovalues (Hillis, 2005a). Normalized pseudovalues are the same as raw pseudovalues, defined by Equation 15.1, except that a normalizing constant is added to the pseudovalues for each modality-reader pair so that the mean of the pseudovalues will be equal to the accuracy estimate. Problem (2) was remedied by using less data-based model reduction (Hillis, 2005b) and a new denominator degrees of freedom (Hillis, 2007a). The remedy for problem (3) will be discussed in Section 15.3.3.

15.3.2 The Obuchowski-Rockette (OR) method

Recall that the DBM method performs a conventional three-factor ANOVA of the pseudovalues. The three factors are modality (fixed), reader (random), and case (random). For our example there would be 1140 pseudovalues, one for each modality.

A seemingly different approach was suggested by Obuchowski and Rockette (OR) (Obuchowski, 1995), which we refer to as the *OR method.* Their approach applies an unconventional two-factor ANOVA to the outcomes of interest, the accuracy measure estimates; for our example these would be the ten accuracy estimates (AUCs or pAUCs), one for each modality-reader combination. The factors for the OR ANOVA are modality (fixed) and reader (random). Using shorthand notation, the OR model, with AUC as the accuracy estimate, is given by

$$\text{AUC} = \text{treatment} + \text{reader} + \text{treatment*reader} + \text{error}$$

What is unconventional about the OR model is that the error terms are allowed to be correlated to account for the correlation due to each reader reading the same cases. The OR procedure defines the following three error covariances that are allowed to be different: (1) Cov_1 is the covariance between error terms for the same reader using different modalities; (2) Cov_2 is the covariance between error terms for two different readers using the same modality; and (3) Cov_3 is the covariance between error terms for two different readers using different modalities. If this was a conventional ANOVA model, the errors would be assumed to be uncorrelated and thus there would be zero covariance between the error terms.

In the OR procedure these error term covariances are estimated using methods such as jackknifing, bootstrapping (Shao, 1995), and the method proposed by DeLong *et al.* (DeLong, 1988). The OR test statistic is similar to that obtained from a conventional ANOVA that treats the errors as uncorrelated, but it contains a correction factor in the denominator that is a function of the error covariance estimates. For this reason this method is sometimes referred to as the *corrected F* method. Specifically, if this was a conventional ANOVA model with independent errors, the F statistic for testing for a modality difference would be

$$F = \frac{\text{MS(T)}}{\text{MS(T*R)}} \qquad (15.2)$$

where MS(T) is the mean square due to treatment (i.e. modality) and MS(T*R) is the mean square due to the treatment-by-reader interaction. In contrast, the OR procedure uses the following statistic:

$$F = \frac{\text{MS(T)}}{\text{MS(T*R)} + r(\widehat{\text{Cov}}_2 - \widehat{\text{Cov}}_3)} \qquad (15.3)$$

Here $\widehat{\text{Cov}}_2$ and $\widehat{\text{Cov}}_3$ are estimates of the corresponding covariances. Note that this F statistic differs from the conventional

ANOVA F statistic only in the addition of the "correction" factor, $r(\widehat{\text{Cov}}_2 - \widehat{\text{Cov}}_3)$, in the denominator.

An advantage of the OR procedure, compared to the DBM procedure, is that it is based on a firm statistical model. However, a disadvantage of the OR procedure, as originally presented, was that it did not perform as well as the DBM procedure. Specifically, it tended to be even more conservative than the DBM procedure, especially when the number of readers was small, as is often the case for imaging studies.

15.3.3 Relationship between the DBM and OR procedures

For more than ten years the DBM and OR procedures were considered to be different procedures, which is understandable since they appear to be quite different in their original formulations. However, recently it was shown that the two procedures are closely related, when Hillis *et al.* (Hillis, 2005a) showed that the modified DBM method that uses normalized pseudovalues and less data-based model reduction yields the same F statistic as the OR method, if OR uses jackknife covariance estimates. Furthermore, they showed that if OR does not use jackknife covariance estimates (e.g. it uses bootstrapping or DeLong's method), the same F statistic can still be obtained using DBM with *quasi-pseudovalues* (which I will not try to define in this chapter). However, the two methods still gave different results because they used different denominator degrees of freedom (DDF) formulas. This discrepancy was resolved by a new DDF formula (Hillis, 2007a) that could be used with both procedures, improved the performance of both procedures, and made the procedures equivalent (since they have the same F statistic and the same DDF).

Thus the updated DBM procedure can be viewed as equivalent to the updated OR procedure. (The updated DBM procedure uses normalized or quasi-pseudovalues, less data-based model reduction, and the new DDF; the updated OR procedure uses the new DDF). This equivalency establishes the theoretical foundation for the DBM method by allowing it to be viewed as an implementation of the OR model, which has a firm statistical foundation. Since one can obtain the same results using either method, the choice of which procedure to use is mainly dependent on availability of software. Historically the DBM procedure has been more popular, due to the availability of free stand-alone software and due to its improved performance in its original formulation compared to the original OR formulation. However, an analysis performed by either of the updated DBM or OR methods that incorporates the recent improvements is probably best referred to as a *DBM/OR* analysis, since the updated approaches give equivalent results. A summary of the recent developments in the DBM and OR procedures and a discussion of their relationship are provided by Hillis *et al.* (Hillis, 2008).

15.4 ANALYSIS OF VAN DYKE DATA

15.4.1 Software

DBM MRMC 2.1 (Berbaum, 2006a; Berbaum, 2006b), available for download at http://perception.radiology.uiowa.edu, and *LABMRMC*, written by Charles Metz and colleagues and available at http://xray.bsd.uchicago.edu/krl/index.htm, are, to my knowledge, the only free stand-alone software packages for implementing the DBM/OR procedure. *DBM MRMC 2.1* resulted from a collaboration between researchers at the University of Chicago, led by Dr. Charles Metz, and researchers at the University of Iowa, led by Dr. Kevin Berbaum. In the near future *DBM MRMC 2.1* will completely replace *LABMRMC*, since *DBM MRMC 2.1* has an improved algorithm and additional features. In this section we use *DBM MRMC 2.1* to analyze the data from the example in Section 15.2. This software is easy to use and requires no other statistical software. While this program uses the DBM three-factor ANOVA of pseudovalues, the same results can be obtained using the OR two-factor ANOVA of accuracy estimates with jackknife covariance estimates.

For readers familiar with the statistical software SAS, *DBM MRMC Procedure for SAS* (Hillis, 2007b) does the same analyses as *DBM MRMC 2.1* and is available for free from the same website. For readers familiar with the FORTRAN programming language, another option for MRMC analysis is the program *OBUMRM*, written by Dr. Nancy Obuchowski and available at http://www.bio.ri.ccf.org/html/rocanalysis.html.

The SAS programs *MRMC sample size input1* and *MRMC sample size input2* (Hillis, 2007c) perform sample size computations for MRMC studies and are also available for download from the same website as *DBM MRMC 2.1*. These programs compute the sample size needed to detect a specified difference in the AUC (or other ROC accuracy estimate such as the pAUC or sensitivity for a fixed specificity) between two modalities, using an updated version of the sample size method discussed by Hillis and Berbaum (Hillis, 2004).

15.4.2 Analysis options

Figure 15.3 shows the menu of analysis options in *DBM MRMC 2.1*. I have indicated *trapezoidal/Wilcoxon* as the method for estimating the ROC curve, *area* as the accuracy estimate, and have requested all three ANOVA analyses, which I refer to as *Analysis 1*, *Analysis 2*, and *Analysis 3*. I now discuss each of these options separately.

The *trapezoidal/Wilcoxon* curve-fitting option provides the nonparametric empirical ROC curve, discussed in Section 15.2, which results from connecting the observed (1 – specificity, sensitivity) points with straight lines. In addition to the trapezoidal/Wilcoxon curve-fitting option, there are other curve-fitting options: *PROPROC, Contaminated Binormal*, and *RSCORE*, as well as two grayed-out options, meaning they were not available when this chapter was written. The RSCORE option uses the semiparametric method, discussed in Chapter 14, which assumes a latent binormal model. The PROPROC (Pan, 1997; Metz, 1999) and contaminated binormal (Dorfman, 2000a; Dorfman, 2000b; Dorfman, 2000c) methods are alternative semiparametric methods that can be used when a smooth ROC curve is desired, but the RSCORE method produces a curve that is improper (i.e. crosses the chance line) or does not appear to fit the data well.

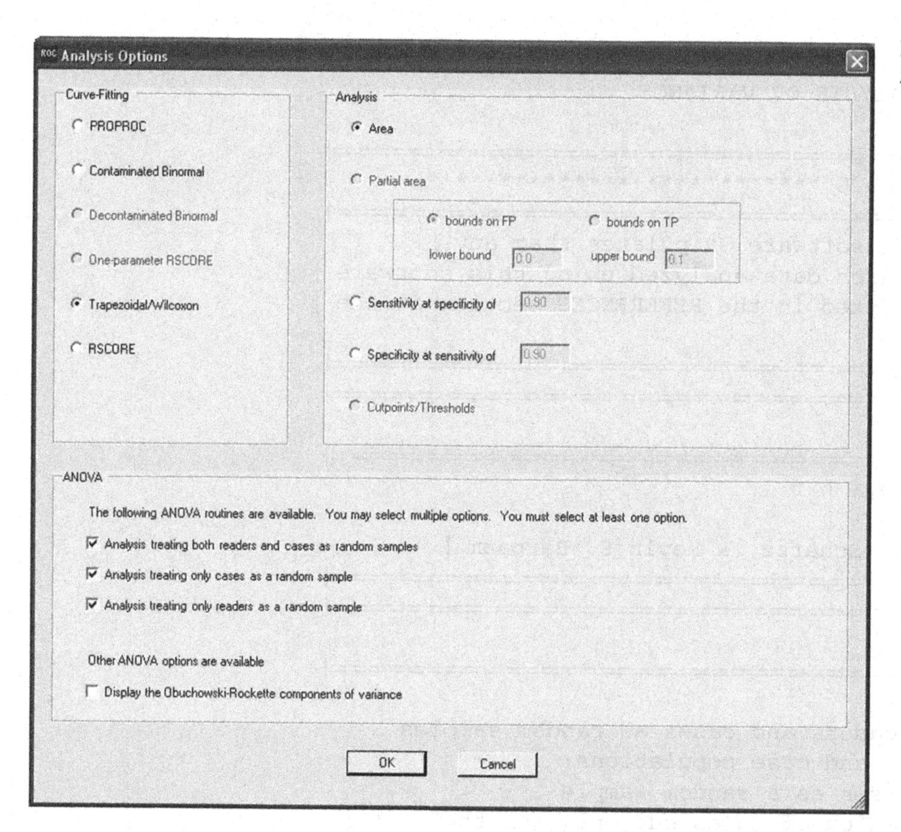

Figure 15.3 Analysis options menu for *DBM MRMC 2.1* software.

The *area* option specifies that AUC is the accuracy measure estimate to be analyzed. Other options include the partial AUC, sensitivity for a given specificity, and specificity for a given sensitivity. The AUC and partial AUC were discussed in Section 15.2, as well as in Chapter 14.

Analysis 1 treats both readers and cases as random. This is the DBM/OR analysis. This is the main analysis and is the only one that needs to be performed if the researcher wants to make inferences about both the reader and case populations. The other two analyses should be considered exploratory – they are primarily useful for the situation when Analysis 1 does not result in a significant finding, to determine if inference to only one population might result in a significant finding.

Analysis 2 treats cases as random but readers as fixed. Thus conclusions apply to the population of cases but only for the readers in the study. Although this analysis is performed within the DBM framework of a three-factor ANOVA of pseudovalues, this should technically not be referred to as a DBM/OR analysis. This analysis is equivalent to a treatment-by-reader ANOVA of the accuracy estimates that treats both treatment and reader as fixed and uses jackknifing of cases to estimate the error covariance matrix; see Hillis (Hillis, 2005a, p. 1593) for a discussion of this test.

Analysis 2 is included for comparison with Analysis 1. Typically, but not always, Analysis 2 will give more significant results than Analysis 1 because it only makes inferences about one population instead of two. Thus if Analysis 1 does not yield a significant finding but Analysis 2 does yield a significant finding, the Analysis 2 finding would be of interest to report since it suggests that the insignificant finding using Analysis 1 can be attributed to insufficient sample size, i.e. not enough readers, cases, or both. In this situation one could use the sample size programs mentioned in Section 15.4.1 to determine the necessary number of readers and cases for achieving adequate power using Analysis 1, based on estimates obtained from the present study. Analysis 2 can also be used for comparing modalities for individual readers.

Analysis 3 treats readers as random but cases as fixed. The comments about Analysis 2 also apply to Analysis 3: this is technically not DBM/OR, it is included for comparison with Analysis 1, and a nonsignificant Analysis 1 coupled with a significant Analysis 3 suggests insufficient sample size for Analysis 1 (in particular, not enough cases). This analysis is equivalent to a conventional repeated measures treatment-by-reader ANOVA of the accuracy estimates, with reader treated as a random factor and modality treated as a fixed repeated-measures factor. Jackknifing plays no role in this analysis. For two treatments, this analysis is the same as a paired t test, which performs a t test on the modality differences, one for each reader.

15.4.3 Annotated output for AUC accuracy estimate, nonparametric ROC curve

Below is the output from the program with my comments interspersed.

```
DBMMRMC 2.1 BETA 3
MULTIREADER-MULTICASE ROC ANALYSIS OF VARIANCE
TRAPEZOIDAL AREA ANALYSIS
|===============================================================|
|************************** NOTE ***************************|
|===============================================================|
| The user agreement for this software stipulates that any |
| publications based on research data analyzed using this software |
| must cite the references listed in the REFERENCES section at the |
| end of the output. |
|===============================================================|
|===============================================================|
| ***** Credits *****|
|===============================================================|
| ANOVA Computations & Display: |
| -------------------------- - |
| Stephen L. Hillis, Kevin M. Schartz, & Kevin S. Berbaum |
|===============================================================|
|===============================================================|
***** Overview *****
|===============================================================|
Three analyses are presented:
(1) Analysis 1 treats both readers and cases as random samples
--results apply to the reader and case populations;
(2) Analysis 2 treats only cases as a random sample
--results apply to the population of cases but only for the
readers used in this study; and
(3) Analysis 3 treats only readers as a random sample
--results apply to the population of readers but only for the
cases used in this study.

For all three analyses, the null hypothesis of equal treatments is
tested in part (a), treatment difference 95% confidence intervals
are given in part (b), and treatment 95% confidence intervals are
given in part (c). Parts (a) and (b) are based on the treatment x
reader x case ANOVA while part (c) is based on the reader x case
ANOVA for the specified treatment; these ANOVA tables are displayed
before the analyses. Different error terms are used as indicated
for parts (a), (b), and (c) according to whether readers and cases
are treated as fixed or random factors. Note that the treatment
confidence intervals in part (c) are based only on the data for the
specified treatment, rather than the pooled data. Treatment
difference 95% confidence intervals for each reader are presented
in part (d) of Analysis 2; each interval is based on the treatment
x case ANOVA table (not included) for the specified reader.

Data file: C:\Dorfman\data sets -- important\VanDyke.in
2 treatments, 5 readers, 114 cases (69 normal, 45 abnormal)
Curve fitting methodology is TRAPEZOIDAL/WILCOXON
Dependent variable is AREA
Normalized pseudovalues used in analyses.
```

We see above that the program specifies the following: the data set is VanDyke.in; there are 2 modalities (recall that *treatment* is used synonymously with *modality*), 5 readers, and 114 cases with 69 normal and 45 abnormal cases; the ROC curve is computed using the nonparametric method; the accuracy estimate analyzed is AUC; and normalized pseudovalues are used. This last statement about normalized pseudovalues is included because the DBM procedure had been based on raw pseudovalues until Hillis *et al.* (Hillis, 2005a) noted the advantage of using normalized pseudovalues, as discussed in Section 15.3.1.

```
=====================================
***** Estimates *****
=====================================
```

	TREATMENT	
READER	1	2
1	0.91964573	0.94782609
2	0.85877617	0.90531401
3	0.90386473	0.92173913
4	0.97310789	0.99935588
5	0.82979066	0.92995169
TREATMENT	MEANS (averaged across readers)	
1	0.89703704	
2	0.94083736	
TREATMENT	MEAN DIFFERENCES	
1 -- 2	-0.04380032	

Above are the AUC estimates for each modality-reader pair, the modality AUC means, and the difference of the modality AUCs. Treatment 1 is cine MRI and treatment 2 is SE MRI. The nonparametric AUC estimates in Table 15.1 are taken from this output.

```
=====================================
***** ANOVA Tables *****
=====================================
```

TREATMENT X READER X CASE ANOVA			
Source	SS	DF	MS
T	0.54676344	1	0.54676344
R	1.74930720	4	0.43732680
C	44.84629692	113	0.39686988
TR	0.25126996	4	0.06281749
TC	11.28283352	113	0.09984808
RC	29.15447929	452	0.06450106
TRC	18.06716464	452	0.03997160
TOTAL	105.89811497	1139	

Above is the three-factor ANOVA table based on the pseudovalues. This is used to test the null hypothesis of equal modality AUCs and to give confidence intervals for differences between modality AUCs.

READER X CASE ANOVAs for each treatment			
		Mean Squares	
Source	Df	Treatment 1	Treatment 2
R	4	0.35141967	0.14872462
C	113	0.33629430	0.16042367
RC	452	0.06043607	0.04403659

Above are two-factor (treatment-by-reader) ANOVA mean squares computed for the data for each modality separately. These are used for constructing single modality confidence intervals.

```
=====================================
***** Variance components estimates *****
=====================================
DBM Variance Component Estimates
(for sample size estimation for future
studies)
Note: These are unbiased ANOVA estimates
which can be negative
```

DBM Component	Estimate
Var(R)	0.00153500
Var(C)	0.02724923
Var(T*R)	0.00020040
Var(T*C)	0.01197530
Var(R*C)	0.01226473
Var(T*R*C) + Var(Error)	0.03997160

The variance component estimates given above are useful for determining reader and case sample sizes needed as inputs for the sample size programs mentioned in Section 15.4.1.

```
=====================================
***** Analysis 1: Random Readers and
Random Cases *****
=====================================
(Results apply to the population of
readers and cases)
a) Test for H0: Treatments have the same
AUC
```

Source	DF	Mean Square	F value	Pr > F
Treatment	1	0.54676344	4.46	0.0517
Error	15.26	0.12269397		
Error term: MS(TR) + max[MS(TC)-MS(TRC),0]				

```
Conclusion: The treatment AUCs are not
significantly different, F(1,15) = 4.46,
p =.0517.
```

Above we see that the two modalities show a close-to-significant difference.

```
b) 95% confidence intervals for treatment
differences
```

Treatment	Estimate	StdErr	DF	t	Pr > t	95% CI
1 − 2	−0.04380	0.02075	15.26	−2.11	0.0517	−0.08796, 0.00036

```
H0: the two treatments are equal.
Error term: MS(TR) + max[MS(TC)-
MS(TRC),0]
```

Above we see that a 95% confidence interval for the modality difference in AUCs is given by (−0.08796, 0.00036). That is, we have 95% confidence that the AUC for cine (treatment 1) is between 0.08796 less and 0.00036 more than the AUC for SE (treatment 2). A more clinically meaningful interpretation is the following: we have 95% confidence that the average reader accuracy for discriminating between diseased and normal images for cine is between 0.08796 less and 0.00036 more than the average reader accuracy for SE. Since this is Analysis 1, these results apply to the reader and image *populations*, and hence are not limited to the reader or case samples used in the study. If there were more than two modalities used in the study, then there would be a confidence interval for each possible modality pair, as well as a *p*-value (given under *Pr* > F in part (a)) for testing for equal modality AUCs.

```
c) 95% treatment confidence intervals
based on reader x case ANOVAs for
each treatment (each analysis is
based only on data for the specified
treatment
```

Treatment	Area	Std Error	DF	95% Confidence Interval
1	0.89703704	0.03317360	12.74	(0.82522360, 0.96885048)
2	0.94083736	0.02156637	12.71	(0.89413783, 0.98753689)
Error term: MS(R) + max[MS(C)-MS(RC),0]				

Above are confidence intervals for the AUC for each modality. From these we have 95% confidence that the average reader accuracy for cine is between 0.825 and 0.969; similarly, we have 95% confidence that the average reader accuracy for SE is between 0.894 and 0.988.

Since Analysis 1 did not result in a significant finding (throughout we use the standard significance level of alpha = 0.05), it is of interest to investigate if inference to one population instead of two results in a significant finding. Thus we now consider the results from Analysis 2 and Analysis 3.

```
=====================================
***** Analysis 2: Fixed Readers and
Random Cases *****
=====================================
(Results apply to the population
of cases but only for the readers
used in this study)
a) Test for H0: Treatments have the
same AUC
```

Source	DF	Mean Square	F value	Pr > F
Treatment	1	0.54676344	5.48	0.0210
Error	113	0.09984808		
Error term: MS(TC)				

```
Conclusion: The treatment AUCs are not
equal, F(1,113) = 5.48, p =.0210.
```

Above we see that the test for the null hypothesis of equal modalities attains significance (p = 0.0210) when only cases are treated as random, in contrast to Analysis 1 (p = 0.0517) where both cases and readers were treated as random.

```
b) 95% confidence intervals for treatment
differences
```

Treatment	Estimate	StdErr	DF	t	Pr > t	95% CI
1 − 2	−0.04380	0.01872	113	−2.34	0.0210	−0.08088, −0.00672

```
* H0: the two treatments are equal.
Error term: MS(TC)
```

Above we see that a 95% confidence interval for the modality difference in AUCs is given by (−0.08088, −0.00672). That is, we have 95% confidence that the AUC for cine is between 0.08088 and 0.00672 less than the AUC for SE. However, this conclusion holds only for the *readers in this study*, since Analysis 2 treats readers as fixed rather than random. This conclusion is of interest since it implies that SE is superior, on average, when used by one of the study readers reading randomly selected images; this conclusion suggests that the insignificant finding in Analysis 1 can be attributed to insufficient sample size.

```
c) 95% treatment confidence intervals
based on reader x case ANOVAs
for each treatment (each analysis
is based only on data for the
specified treatment
```

Treatment	Area	Std Error	DF	95% Confidence Interval
1	0.89703704	0.02428971	113	(0.84891474, 0.94515933)
2	0.94083736	0.01677632	113	(0.90760044, 0.97407428)
Error term: MS(C)				

Note that the above confidence intervals are narrower than the corresponding ones in part (c) of Analysis 1. This is typical since Analysis 2 conclusions generalize only to the case population.

d) Treatment-by-case ANOVA CIs for each reader (each analysis is based only on data for the specified reader)

Reader	Treatment	Estimate	StdErr	DF	t	Pr > t	95% CI
1	1--2	−0.02818	0.02551	113	−1.10	0.2717	−0.07872, 0.02236
2	1--2	−0.04654	0.02630	113	−1.77	0.0795	−0.09865, 0.00557
3	1--2	−0.01787	0.03121	113	−0.57	0.5680	−0.07971, 0.04396
4	1--2	−0.02625	0.01729	113	−1.52	0.1318	−0.06051, 0.00801
5	1--2	−0.10016	0.04406	113	−2.27	0.0249	−0.18745, −0.01288

Part (d) above gives the modality-difference confidence intervals for each reader, based only on the data for that reader. For example, we can conclude that there is a significant difference between cine and SE for reader 5 ($p = 0.0249$), with SE performing better, and we have 95% confidence that the cine AUC is between 0.18745 and 0.01288 less than that for SE. It may at first seem strange that only one reader would show a significant difference between modalities after concluding in part (a) of Analysis 2 that there is a significant population difference between modalities; however, this is easily explained by the fact that the part (a) analysis is a global test based on the data from all five readers.

===================================

***** Analysis 3: Random Readers and Fixed Cases *****

===================================

(Results apply to the population of readers but only for the cases used in this study)

a) Test for H0: Treatments have the same AUC

Source	DF	Mean Square	F value	Pr > F
Treatment	1	0.54676344	8.70	0.0420
Error	4	0.06281749		
Error term: MS(TR)				

Conclusion: The treatment AUCs are not equal, F(1,4) = 8.70, p =.0420.

Again we see that when we generalize to just one population (now we treat only readers as random), we obtain a slightly more significant result ($p = 0.0420$) for testing the null hypothesis of equal modalities than we did in Analysis 1 ($p = 0.0517$). For this data set where there are two modalities, this Analysis 3 test is the same as a paired t test performed on the modality differences, one for each reader.

b) 95% confidence intervals for treatment differences

Treatment	Estimate	StdErr	DF	t	Pr > t	95% CI
1 − 2	−0.04380	0.01485	4.00	−2.95	0.0420	−0.08502, −0.00258

H0: the two treatments are equal.

Above we see that a 95% confidence interval for the modality difference in AUCs is given by (−0.08502, −0.00258). That is, we have 95% confidence that the AUC for cine is between 0.08502 and 0.00258 less than the AUC for SE. However, this conclusion only holds for the *cases in this study*, since Analysis 3 treats cases as fixed rather than random. This conclusion is of interest since it implies that SE is superior, on average, when used by a randomly selected reader reading the images used in this study, suggesting that a study with a larger sample size would likely show a significant result for Analysis 1.

c) Reader-by-case ANOVAs for each treatment (each analysis is based only on data for the specified treatment)

		Mean Squares	
Source	df	Treatment 1	Treatment 2
R	4	0.35141967	0.14872462
C	113	0.33629430	0.16042367
RC	452	0.06043607	0.04403659

Estimates and 95% Confidence Intervals

Treatment	Area	Std Error	DF	95% Confidence Interval
1	0.89703704	0.02482994	4	(0.82809808, 0.96597599)
2	0.94083736	0.01615303	4	(0.89598936, 0.98568536)

Similar to Analysis 2, the above confidence intervals are narrower than the corresponding ones in Analysis 1. This is typical since Analysis 3 conclusions generalize only to the reader population.

15.4.4 Annotated output for other accuracy estimates and ROC curve-fitting methods

For the sake of comparison I show in this section the results for Analysis 1 using (1) AUC accuracy estimate and semiparametric (RSCORE) curve fitting; (2) pAUC accuracy estimate and nonparametric (trapezoidal/Wilcoxon) curve fitting; and (3) pAUC accuracy estimate and semiparametric (RSCORE) curve fitting. For the pAUC I use a range of $0 \leq FPF \leq 0.2$, as discussed in Section 15.2. The partial output shown below is limited to the test for equality of modalities in Analysis 1.

Below are the AUC accuracy estimate and semiparametric (RSCORE) curve fitting results. See Table 15.1 for the corresponding AUC estimates.

a) Test for H0: Treatments have the same AUC

Source	DF	Mean Square	F value	Pr > F
Treatment	1	0.46899647	2.62	0.1339
Error	10.99	0.17908808		
Error term: MS(TR) + max[MS(TC)-MS(TRC),0]				

Conclusion: The treatment AUCs are not significantly different, F(1,11) = 2.62, p = 0.1339.

Below are the partial AUC accuracy estimate and nonparametric (trapezoid/Wilcoxon) curve fitting results. See Table 15.2 for the corresponding pAUC estimates.

a) Test for H0: Treatments have the same pAUC

Source	DF	Mean Square	F value	Pr > F
Treatment	1	2.0608	5.05	0.0448
Error	11.64	0.4078		
Error term: MS(TR) + max[MS(TC)-MS(TRC),0]				

Conclusion: The treatment pAUCs are significantly different, F(1,11.6) = 5.05, p = 0.0448.

Below are the partial AUC accuracy estimate and semiparametric (RSCORE) curve fitting results. See Table 15.2 for the corresponding AUC estimates.

a) Test for H0: Treatments have the same pAUC

Source	DF	Mean Square	F value	Pr > F
Treatment	1	2.3177	4.49	0.0600
Error	10.01	0.5157		
Error term: MS(TR) + max[MS(TC)-MS(TRC),0]				

Conclusion: The treatment pAUCs are not significantly different, F(1,10) = 4.49, p = 0.0600.

Recall that we obtained $p = 0.0517$ for the nonparametric AUC analysis in Section 15.4.3 in Analysis 1. Taking into account the results from the preceding output, we see that the nonparametric AUC ($p = 0.0517$), nonparametric pAUC ($p = 0.0448$) and semiparametric pAUC ($p = 0.0600$) analyses all yield similar p-values, while the semiparametric AUC analysis ($p = 0.1339$) differs somewhat.

It should be noted that one should not try numerous accuracy estimates and ROC curve methods and then pick the one that gives the most significant results, since this will increase the probability of a type I error (rejecting H_0 when it is true); rather, the researcher should decide in advance what accuracy estimate and curve method are most appropriate. A problem with the nonparametric method is that it underestimates the true AUC or pAUC, especially for discrete rating data with only a few categories (Bamber, 1975; Hanley, 1982). Since the AUC is the average sensitivity, averaged uniformly across the whole range of specificity, a problem with the AUC is that it typically reflects the sensitivity of the test for a range of specificity values that are not clinically meaningful, e.g. too low to use in practice. For these reasons I would recommend the parametric pAUC analysis, provided $0 \leq FPF \leq 0.2$ represents a clinically meaningful FPF range for these modalities.

Finally, it is important to visually check the fit of ROC curves and to try a different method if they appear not to fit the data suitably. From Figure 15.2 it appears that the semiparametric ROC curves provide an acceptable fit to the data.

15.5 CONCLUDING REMARKS

In diagnostic studies involving human readers reading images, the researcher typically will want conclusions to generalize to both the reader and case populations, rather than be limited to only the readers used in the study. I have discussed the main methods for analyzing such MRMC studies, the DBM and OR methods. Although it had been thought that the DBM and OR methods were different for over ten years, recently it has been shown that the updated DBM and OR procedures are two different ways to reach the same conclusions, and hence it makes sense to refer to an analysis using either of the updated methods as a *DBM/OR* analysis.

The analysis of an MRMC data set was illustrated using free stand-alone software. Although some statistical details concerning the DBM/OR method were given, the main intent of this chapter has been to illustrate the need for MRMC analysis and to demonstrate how it can be easily performed using readily available software. Although the example used had only two modalities, analysis for more than two modalities can be easily performed using *DBM MRMC 2.1*.

MRMC methodology is presently an active area of research. My colleagues and I are presently developing MRMC methodology and software for other sampling designs. These include split-plot designs, where different readers for each modality evaluate the same images or where different images for each modality are evaluated by the same readers; and ANCOVA designs, where estimates are adjusted for the effect of reader-level covariates. I note that although MRMC analysis for split-plot designs has been discussed by Dorfman *et al.* (Dorfman, 1999), my colleagues and I are presently updating this earlier work to reflect recent methodological developments.

ACKNOWLEDGEMENTS

I thank Carolyn Van Dyke, MD, for sharing her data set. This research was supported by Grant R01EB000863 from the National Institutes of Health, Bethesda, MD. The views expressed in this chapter are those of the author and do not necessarily represent the views of the Department of Veterans Affairs.

REFERENCES

Bamber, D. (1975). Area above ordinal dominance graph and area below receiver operating characteristic graph. *Journal of Mathematical Psychology*, **12**(4), 387–415.

Berbaum, K.S., Schartz, K.M., Pesce, L.L., Hillis, S.L. (2006a). *DBM MRMC 2.1* (computer software). Available at: http://perception.radiology.uiowa.edu

Berbaum, K.S., Metz, C.E., Pesce, L.L., Schartz, K.M. (2006b). *DBM MRMC 2.1 User's Guide* (software manual). Available at: http://perception.radiology.uiowa.edu

DeLong, E.R., DeLong, D.M., Clarke-Pearson, D.L. (1988). Comparing the areas under two or more correlated receiver operating characteristic curves: a nonparametric approach. *Biometrics*, **44**(3), 837–845.

Dorfman, D.D., Berbaum, K.S., Metz, C.E. (1992). Receiver operating characteristic rating analysis: generalization to the population of readers and patients with the jackknife method. *Investigative Radiology*, **27**(9), 723–731.

Dorfman, D.D., Berbaum, K.S., Lenth, R.V., Chen, Y.-F. (1999). Monte Carlo validation of a multireader method for receiver operating characteristic discrete rating data: split plot experimental design. *SPIE Proceedings Series*, **3663**, 91–99.

Dorfman, D.D., Berbaum, K.S. (2000a). A contaminated binormal model for ROC data – Part III. Initial evaluation with detection ROC data. *Academic Radiology*, **7**(6), 438–447.

Dorfman, D.D., Berbaum, K.S. (2000b). A contaminated binormal model for ROC data – Part II. A formal model. *Academic Radiology*, **7**(6), 427–437.

Dorfman, D.D., Berbaum, K.S., Brandser, E.A. (2000c). A contaminated binormal model for ROC data – Part I. Some interesting examples of binormal degeneracy. *Academic Radiology*, **7**(6), 420–426.

Hanley, J.A., McNeil, B.J. (1982). The meaning and use of the area under a receiver operating characteristic (ROC) curve. *Radiology*, **143**(1), 29–36.

Hillis, S.L., Berbaum, K.S. (2004). Power estimation for the Dorfman-Berbaum-Metz method. *Academic Radiology*, **11**(11), 1260–1273.

Hillis, S.L., Obuchowski, N.A., Schartz, K.M., Berbaum, K.S. (2005a). A comparison of the Dorfman-Berbaum-Metz and Obuchowski-Rockette methods for receiver operating characteristic (ROC) data. *Statistics in Medicine*, **24**, 1579–1607.

Hillis, S.L., Berbaum, K.S. (2005b). Monte Carlo validation of the Dorfman-Berbaum-Metz method using normalized pseudovalues and less data-based model simplification. *Academic Radiology*, **12**, 1534–1542.

Hillis, S.L. (2007a). A comparison of denominator degrees of freedom methods for multiple observer ROC analysis. *Statistics in Medicine*, **26**(3), 596–619.

Hillis, S.L., Schartz, K.M., Pesce, L.L., Berbaum, K.S., Metz, C.E. (2007b). *DBM MRMC Procedure for SAS* (computer software). Available at: http://perception.radiology.uiowa.edu

Hillis, S.L., Berbaum, K.S. (2007c). *MRMC Sample Size Program User's Guide* (software manual). Available at: http://perception.radiology.uiowa.edu

Hillis, S.L., Berbaum, K.S., Metz, C.E. (2008). Recent developments in the Dorfman-Berbaum-Metz procedure for multireader ROC study analysis. *Academic Radiology*, **15**, 647–661.

Jiang, Y.L., Metz, C.E., Nishikawa, R.M. (1996). A receiver operating characteristic partial area index for highly sensitive diagnostic tests. *Radiology*, **201**(3), 745–750.

McClish, D.K. (1989). Analyzing a portion of the ROC curve. *Medical Decision Making*, **9**(3), 190–195.

Metz, C.E., Pan, X.C. (1999). "Proper" binormal ROC curves: theory and maximum-likelihood estimation. *Journal of Mathematical Psychology*, **43**(1), 1–33.

Obuchowski, N.A., Rockette, H.E. (1995). Hypothesis testing of diagnostic accuracy for multiple readers and multiple tests: an ANOVA approach with dependent observations. *Communications in Statistics – Simulation and Computation*, **24**(2), 285–308.

Pan, X.C., Metz, C.E. (1997). The "proper" binormal model: parametric receiver operating characteristic curve estimation with degenerate data. *Academic Radiology*, **4**(5), 380–389.

Shao, J., Dongshen, T. (1995). *The Jackknife and Bootstrap*. New York, NY: Springer-Verlag.

Van Dyke, C.W., White, R.D., Obuchowski, N.A., *et al.* (1993). Cine MRI in the diagnosis of thoracic aortic dissection. 79th RSNA Meetings, Chicago, IL.

Zeger, S.L., Liang, K.Y. (1986). Longitudinal data analysis for discrete and continuous outcomes. *Biometrics*, **42**(1), 121–130.

16

Recent developments in FROC methodology

DEV CHAKRABORTY

16.1 INTRODUCTION

Receiver operating characteristic (ROC) analysis, which is widely used for imaging system assessment, is reviewed in Chapter 14 of this book. The most common application of observer performance studies is comparing imaging systems, i.e. determining which has the higher figure-of-merit and whether the difference is statistically significant (i.e. sufficiently large so as to be unlikely to be attributable to chance). In ROC analysis a commonly used figure-of-merit is the area under the ROC curve (AUC). A recent review (Wagner, 2002) describes the advantages of Dorfman-Berbaum-Metz (DBM) analysis (Dorfman, 1992) of ROC ratings obtained when a group of readers read the same cases in two or more modalities, commonly termed the multiple-reader multiple-case (MRMC) paradigm. This design yields the maximum statistical power, i.e. the ability to detect small differences in AUC between two or more modalities, and the methods described in this chapter assume this design.

It is not widely appreciated that the ROC method does not use all available information when the decision task is non-binary, i.e. involves more than a determination of whether the patient is diseased or normal. In addition to *detecting* an abnormal condition, the radiologist often *locates* specific image regions that are suspicious for disease. The location information cannot be used by ROC analysis and its neglect leads to a loss of statistical power and differences between modalities may go undetected. The statistical power is the probability of declaring modalities different when in fact they are different. In planning an observer performance study it is desirable to have high statistical power (a typical goal is 80%). One way of compensating for reduced statistical power is to increase the number of radiologists and cases, which of course increases the cost of the study. For these reasons location-specific approaches that do not ignore the location information are being pursued by some investigators.

There are three location specific approaches: the free-response operating characteristic (FROC; Bunch, 1978), location receiver operating characteristic (LROC; Starr, 1975; Starr, 1977; Swensson, 1981; Swensson, 1996), and region of interest (ROI; Obuchowski, 2000; Rutter, 2000) paradigms. Figure 16.1 shows a mammogram as it might be interpreted by the same mammographer according to current observer performance paradigms. Two real lesions are indicated by the arrows and the three light crosses indicate suspicious regions found by the mammographer. Evidently the mammographer saw one of the lesions, missed the other lesion, and mistook two normal structures for lesions. In the ROC paradigm in Figure 16.1 (top-left), the mammographer assigns a single rating, say 1 through 5, indicating the confidence level that there is at least one lesion somewhere in the image. If the left-most cross is a very suspicious region then the ROC rating might be 5 (highest confidence). In Figure 16.1 (top-right), illustrating the free-response paradigm, the dark crosses indicate suspicious regions *that are marked*, and the numerals are the corresponding ratings. Let us assume the allowed ratings are 1 through 4. Two marks are shown, one rated 4 close to a lesion, and the other rated 1 not close to any lesion. The third suspicious region was not marked as its confidence level did not exceed the reporting threshold, i.e. that corresponding to the 1-rating. The marked region rated 4 (highest confidence) is likely what caused the mammographer to assign the ROC rating of 5 to this image (in the ROC paradigm an image without any marks would be rated 1, the lowest available bin, and a 4-rating free-response scale corresponds to a 5-rating). Figure 16.1 (bottom-left) illustrates the LROC paradigm. Here the mammographer provides a rating that there is at least one lesion somewhere in the image *and* marks the most suspicious region (LROC analysis is limited to one lesion per image). In this example the rating might be 5, as in the ROC paradigm, and the mark may be the one rated 4 in the free-response task, and in LROC terminology it would be recorded as a correct localization (if the mark is not near a lesion it would be recorded as an incorrect localization). In the ROI paradigm illustrated in Figure 16.1 (bottom-right) the investigator segments the image into a number of ROIs and the mammographer rates each ROI for presence of at least one lesion somewhere within the ROI. In this example there are four regions and the region at ~9 o'clock might be rated 5 as it contains the most suspicious region, the one at ~11 o'clock might be rated 1 as it does not contain any suspicious regions, the one at ~3 o'clock might be rated 2 or 3 (the unmarked suspicious region would tend to increase the confidence level), and the one at ~7 o'clock might be rated 1.

With the exception of the free-response paradigm, the other paradigms yield data structures that do not vary between images. In ROC a single rating, in LROC a single rating plus a single mark, and in ROI a fixed number of ratings per image, are obtained. In contrast, the free-response data structure varies between images. The number of marks is *a priori* unknown; indeed it must be regarded as a case (patient)-dependent *random* integer ≥ 0. As one may suspect, analysis of this type of data is challenging. According to Egan (Egan, 1961), "the method of free-response is particularly difficult to analyze simply because a trial is not defined." Here "trial" refers to a defined

The Handbook of Medical Image Perception and Techniques, ed. Ehsan Samei and Elizabeth Krupinski. Published by Cambridge University Press.

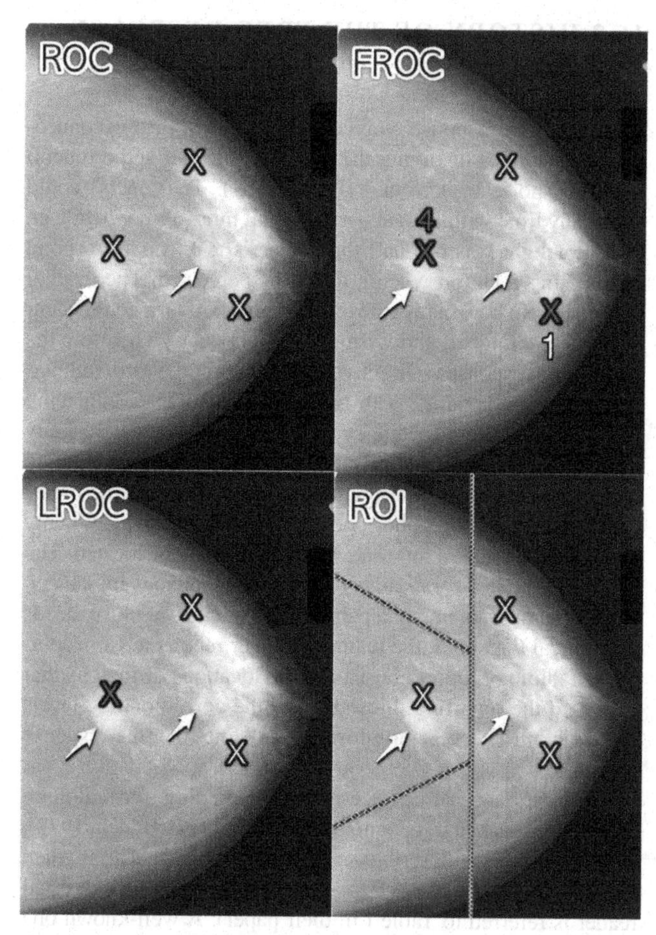

Figure 16.1 A mammogram as it might be interpreted by the same observer according to current observer performance paradigms. Two real lesions are indicated by the arrows and the three light crosses indicate suspicious regions found as a result of the search. Evidently the mammographer saw one of the lesions, missed the other lesion, and mistook two normal structures for lesions. ROC: the mammographer assigns a single confidence level that *somewhere* in the image there is a lesion(s). FROC: the dark crosses indicate suspicious regions that are *marked* (i.e. the confidence level exceeded the observer's lowest threshold) and the accompanying numerals are the FROC ratings. In LROC the mammographer provides a rating that somewhere in the image there is a lesion(s) and marks the most suspicious region. In ROI the image is divided into a number of regions-of-interest by the investigator and the mammographer rates each ROI for presence of lesion(s) somewhere within the ROI. The image to ROI segmentation shown is intended as a schematic. In practice it is based on clinical considerations.

observational unit (e.g. an image) yielding a single response. Analysis of free-response data has been one of the major challenges in imaging systems assessment. Work in this area has spanned four decades and only recently have some of us begun to come to grips with the inherent randomness of the free-response data structure and its close relationship to visual search.

Free-response data consists of mark-rating pairs. A "mark" is the location of a suspicious region, and the "rating" is the corresponding confidence level. For each mark the investigator decides whether it is "close enough" to a lesion to be credited as having localized the lesion. The definition of a closeness or proximity criterion is presently arbitrary, but some progress has been made that may enable objective selection of a proximity criterion (Chakraborty, 2007b). In my terminology a mark close to a lesion is termed *lesion localization* (LL) rather than true positive. Marks that do not qualify as lesion localizations are termed *non-lesion localizations* (NLs) instead of false positives. True positive and false positive terminology risks serious confusion with the ROC paradigm where the same terms are used for the whole image, whereas in the free-response context the data units correspond to suspicious regions. While this may take some getting used to, I believe in the long run it is desirable to clearly separate the terminology used to describe two intrinsically different paradigms. Likewise the term "lesion" is reserved for a true (e.g. biopsy-proven) localized abnormality and "suspicious region" for a region in the image that the radiologist examines for possible marking, which may or may not correspond to a lesion. Also, I refer to the general paradigm itself as free-response, and FROC (see definition below) refers to the curve that visually summarizes free-response performance (Bunch, 1978).

In free-response studies the concept of "true negatives" is undefined and un-measurable. A free-response "true negative" would be a region that was identified as suspicious but not marked (e.g. the light cross in the upper-right panel of Figure 16.1). Eye-tracking equipment could be used to attempt to define the "true negatives," but such information is not collected in a free-response study, and moreover, the identified "true negatives" are not definitive, since they depend on investigator-dependent choices of several eye-tracking parameters. Asking the radiologist to show the unmarked regions is not definitive either, since some of the regions may have been examined and dismissed at the subconscious level. Attempting to define the "true negatives" by, for example, assuming the radiologist segments the image into lesion-sized regions and inspects each of them for possible marking is an assumption, and moreover changes the paradigm from free-response to ROI. *The only observed events in free-response studies are the mark-rating pairs. Any proposed analysis of free-response data should use only the observed data and not make assumptions that cannot be tested. I believe the term "true negative" should never be used in the free-response context.*

The FROC curve is defined (Bunch, 1978) as the plot, as the threshold for reporting a suspicious region is varied, of lesion localization fraction (LLF) along the y-axis vs. non-lesion localization fraction (NLF) along the x-axis, where the denominators for the fractions are the total number of lesions and the total number of images, respectively. FROC curve operating points are generated by cumulating LL and NL counts in a manner completely analogous to that used to plot ROC operating points. A typical FROC curve such as might be observed for a computer-aided detection (CAD) algorithm is shown in Figure 16.2. An FROC curve starts at the origin with a steep slope and, as the confidence level is reduced, the operating point moves up the curve, the slope decreases monotonically, and the curve terminates at the *end-point* where the slope is zero. (The origin corresponds to infinite confidence and the *end-point* corresponds to negative infinity. Like the ROC curve, the FROC curve is unaffected by a strictly monotonically increasing transformation of the confidence scale, and any finite-range confidence

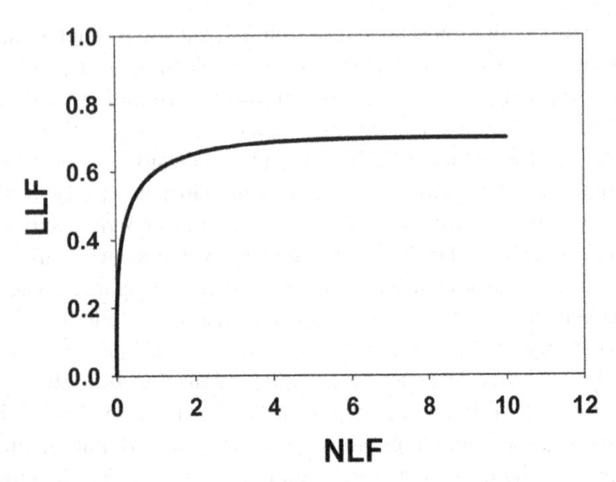

Figure 16.2 A typical FROC curve such as might be observed for a CAD algorithm. It is characterized by a steep high confidence region starting at (0, 0), a shoulder corresponding to the intermediate confidence region, and a plateau in the lowest confidence region, terminating at the *end-point* (10, 0.7) where the confidence is at its lowest value. FROC curves do not extend to (∞, 1), a common misconception. (The curve was generated with search model parameters $\mu = 2.34$, $\lambda = 10$, and $\nu = 0.7$.)

scale, e.g. 1 to 100, can be transformed to one ranging from negative infinity to positive infinity via a strictly monotonically increasing transformation. The curve shown was generated with search-model parameters, defined later, $\mu = 2.34$, $\lambda = 10$, and $\nu = 0.7$.)

If the ROC quantities true positive fraction (TPF) and false positive fraction (FPF) are calculated from a free-response dataset by using the rating of the highest-rated mark on an image as the *ROC-equivalent* rating for that image, free-response data can be used to plot an ROC curve. A four-rating free-response study will yield four operating points on the derived ROC, in other words the same number of operating points as a five-rating ROC study. In Bunch *et al.* (Bunch, 1978), a curve was also described, which I later termed (Chakraborty, 1990) the alternative free-response receiver operating characteristic (AFROC) which is a plot of LLF vs. FPF. The AFROC is a hybrid plot whose ordinate is the same as the ordinate of the FROC curve and whose abscissa is the same as the abscissa of the ROC curve. The LROC curve is a plot of probability of correct localization (PCL) vs. FPF, where PCL is the joint probability that a case is reported positive and the lesion is correctly localized.

The following is an outline of this chapter. Section 16.2 is a historical account of the free-response paradigm and briefly covers the early methods and recent advances. The latter are developed in greater detail later in the chapter. The practical aspects of conducting free-response studies and collecting optimal data are described in Section 16.3. Section 16.4 describes two models for fitting FROC curves. Section 16.5 describes current methods for analyzing free-response data. Section 16.6 describes a simulation study assessing the validity of these methods and comparing their statistical powers. The assumptions of the models described in Section 16.4 are examined in Section 16.7. The chapter concludes with a discussion of the main findings and directions for future research.

16.2 HISTORY OF THE FREE-RESPONSE PARADIGM

The term "free-response" was coined by Egan in 1961 in connection with studies involving the detection of brief audio tone(s) against a noise background (Egan, 1961). The tone(s) could occur any time within an active listening interval (e.g. while an indicator light was on) and the listener's task was to respond by pressing a button when a tone was perceived. The listener was uncertain how many true tones (signals), if any, could occur in the active interval and when they might occur. Therefore the number of responses per active interval could be ≥ 0 and was *a priori* unpredictable. With two-dimensional space replacing time, and excepting for the causality effect, the acoustic study mirrors a common task in medical imaging, namely finding lesions in images. In interpreting an image for possible breast cancer the mammographer does not know *a priori* how many lesions are present, if any, and where they could be located. The image is searched for regions that appear suspicious for cancer. If the level of suspicion of a particular suspicious region exceeds the clinical threshold, the mammographer reports it. Conceptually a radiology report consists of the locations of regions that exceeded the reporting threshold and the corresponding levels of suspicion. This type of information defines the free-response paradigm.

The importance of the free-response paradigm for radiology applications was first recognized by Bunch *et al.* (Bunch, 1978). Their classic paper describes several ambiguities that arise when the ROC method is applied to a localization task (the interested reader is referred to Table I in their paper). A well-known one is the ambiguity when a false positive and a false negative occur on the same image. The two mistakes effectively "cancel" each other in ROC analysis and the image is scored as a "perfect" true positive. In other words the radiologist was right – he detected that the image was abnormal – but for the wrong reason – he reported the incorrect location; this is termed the "right for wrong reason" scenario. The clinical consequences of the two canceling mistakes can be serious; the missed lesion may grow and the marked region could lead to a biopsy at the wrong location. Bunch *et al.* introduced the concept of the FROC curve and performed a free-response study using X-ray images of small beads superimposed on a uniform phantom. They also developed theoretical relations between the FROC curve and the ROC curve implied by the ROC-equivalent rating of the image. However, they did not describe how an FROC curve could be fitted to the mark-rating data.

In 1986 my colleagues and I at the University of Alabama at Birmingham reported the first clinical FROC study which compared conventional chest radiography to a prototype digital chest imaging system by Picker International (Chakraborty, 1986). The method was applied in a second study (Niklason, 1986) to evaluate a dual-energy chest imaging system by the same manufacturer. Since no method was then available for fitting the FROC curve, the analysis involved linear interpolation between neighboring operating points to compare LLFs of the two modalities at a common NLF. Around 1989 two parametric methods for fitting FROC curves, termed FROCFIT and AFROC, were described. (Chakraborty, 1989; Chakraborty, 1990). They gave good fits to human observer data. In 1996 Swensson described a parametric model for the LROC paradigm that also predicted FROC and AFROC curves (Swensson, 1996). Recently I showed that the FROCFIT, AFROC, and Swensson models are identical and that the approaches differ only in the parameter estimation methods (Chakraborty, 2008b). While all of these methods were previously found to give good fits to human observer data, they failed to fit CAD data in this study, especially in the low-confidence region. CAD algorithms yield many marks and the ratings are finely spaced floating-point numbers, which means a detailed FROC curve with closely spaced operating points is available. With human observers it is generally not possible to get more than five operating points. Therefore lack of fit is easier to detect with CAD than with human observer data.

In addition to the usual parametric assumptions similar to those made in ROC analysis (e.g. the binormal model), the fitting methods described so far assumed that ratings of multiple marks on the same image were independent. The independence assumption drew justifiable criticism (Metz, 1996; Swensson, 1996) that has discouraged usage of these methods. Several intrinsically free-response studies were analyzed by the ROC method (Chan, 1990; Kobayashi, 1996; Li, 2005; Petrick, 2008). To compare modalities using any observer performance paradigm it is only necessary to define a figure-of-merit and a method for testing the significance of observed differences in figures-of-merit between two or more modalities. Case-based resampling techniques (Efron 1982) make no assumptions regarding the correlations of ratings on the same image. These have been available all along and could have been used to perform valid significance testing with the prior "faulty" models such as FROCFIT, AFROC, or the Swensson model, whose only role would have been to calculate parametric figures-of-merit from the data. However, this possibility went unrecognized and resampling methods were not applied to free-response data until quite recently (Chakraborty, 2004; Samuelson, 2006; Samuelson, 2007). All of the methods described in this chapter use case-based resampling and are quite robust to data correlations and therefore the original objection to free-response studies is no longer valid.

CAD evaluation has been a notable exception to the general reluctance to use the free-response paradigm. In CAD evaluation the FROC curve has been the primary method of summarizing performance. The CAD developer generally has access to the mark-rating pairs of many suspicious regions found by the algorithm. Most of them have low ratings and would not be shown to the radiologist. This type of detailed information is termed *designer-level* data. *All references to CAD in this chapter are to designer-level data.* In mammography CAD systems the average number of marks per image can be about ten and ROC methodology is clearly inadequate as it would entail ignoring about 90% of the data. However, due to the lack of suitable analytical tools, most developers have had to resort to reporting a particular operating point on the FROC curve. This makes comparison between different CAD systems difficult, just as reporting an operating point on an ROC curve makes it difficult to separate an improvement in LLF ("sensitivity") from a change in threshold effect.

The importance of assessing CAD algorithms for mammography and low-dose CT screening for lung cancer, and other applications involving localization (e.g. breast tomosynthesis or breast CT), has spurred recent research in free-response methodology. In 2002 the initial detection and candidate analysis (IDCA) method was proposed (Edwards, 2002) for fitting CAD-generated FROC curves, thereby formalizing and putting on a firm statistical footing an ad hoc procedure that was being used by some algorithm designers. The IDCA method predicted that the FROC curve does not extend to infinitely large values of NLF, and not all lesions are necessarily localized, even at the lowest confidence level. The significance of these predictions was not recognized until recently, for they are consistent with a model of image interpretation that is very non-intuitive to most imaging physicists, but is quite familiar to vision psychologists, namely the Kundel-Nodine (KN) perceptual model (Kundel *et al.* 1983; Kundel *et al.* 2004; Kundel *et al.* 2007) of how radiologists interpret images. In 2004 the jackknife alternative free-response operating characteristic (JAFROC) method was proposed for human observer data (Chakraborty, 2004). This was the first application of a resampling method to free-response data. JAFROC applies to an MRMC free-response study and allows generalization to the population of readers and cases. In other words, JAFROC did for free-response what the DBM method (Dorfman, 1992) did for ROC. Software for this method was made available on the web starting April 16, 2004 (www.devchakraborty.com). In 2006 I described a search-model (Chakraborty, 2006b; Chakraborty, 2006c) for the free-response paradigm that resembled IDCA, but was in fact a more general parameterization of the KN model. The differences are discussed in Section 16.4.3. Other approaches have been proposed for analyzing free-response data (Anastasio, 1998; Irvine, 2004; Bornefalk, 2005a; Bornefalk, 2005b; Popescu, 2006; Hutchinson, 2007) but have not found wide acceptance and software implementations are not available. Another group at the University of Pittsburgh is currently active in free-response research (Bandos, 2009; Song, 2008). This chapter will focus on JAFROC, a related method termed JAFROC1, IDCA, and a non-parametric approach (Samuelson, 2006; Samuelson, 2007).

16.3 CONDUCTING A FREE-RESPONSE STUDY

Since the free-response method is less familiar, it is appropriate to describe the practical aspects of conducting such studies especially since this area has been much neglected. An exception is the paper (Bunch, 1978) that described training issues in some depth. No doubt this neglect was partly because good methods for analyzing the data were not available. That is no longer the case and careful attention to data collection, particularly in human observer studies, is essential if the new methods are to be used effectively. It is important to realize that while the free-response method is closer to the clinical paradigm, it is not the "real thing." Like ROC, LROC, and ROI, it is a level-2 laboratory study in the Fryback *et al.* (Fryback, 1991) hierarchy, according to which assessment methods form a six-level hierarchy of efficacies: technical, diagnostic, diagnostic-thinking, therapeutic, patient outcome, and societal. The "clinical relevance" of a measurement is its hierarchy level. The difficulty/cost of measurement increases rapidly as one moves up the hierarchy.

Therefore one is usually limited to lower-level measurements and hopes that conclusions reached using them (e.g. modality A is better than modality B) will be confirmed at higher levels. To have the best chance of correlating with higher levels it is important that any level 2 laboratory study be conducted conscientiously. Data quality has a strong effect on statistical power. If an observer does not provide operating points that adequately sample the FROC curve, not only is statistical power compromised but the analysis may be unreliable.

16.3.1 Truth (gold standard)

This follows standard procedures developed for ROC except here one needs to also record the locations of the lesions. For normal images truth can be retrospectively established by follow-up mammograms interpreted as normal. Malignant lesions are proven by biopsy and/or other imaging. A truth panel consisting of expert mammographers, using all available clinical information, should locate the known malignancies on the mammograms, preferably by outlining them. Outlining is preferable to simply indicating the center, as the latter is prone to variability due to subjectivity, especially for larger lesions. For three dimensional (3D) modalities a lesion should be outlined in the slice in which it is most visible. The truth panel should not participate in the actual study.

The degree of visibility of the lesions (i.e. difficulty level of the study) needs to be controlled. If the lesions are too easily visible the radiologists may not generate appreciable numbers of NLs and the operating points may be on the initial near-vertical portion of the FROC curve, and the dataset cannot be analyzed optimally. One way to increase the level of difficulty is to use images from a previous screening where the lesion may be retrospectively visible. Also one could include difficult normal images in the case set. On the other hand an unduly difficult task may discourage active participation in the study. As a rule of thumb, radiologists should be able to detect between 60% and 80% of the lesions at their most lax criterion and the corresponding FPF, using the ROC-equivalent rating, should not be less than about 30%. In other words at least 30% of images should contain at least one NL mark. These guidelines are based on my experience analyzing this type of data but the general issue of quality control metrics for free-response studies needs further research.

The radiologist should be told the general characteristics of the lesions in the dataset, e.g. the range of lesion sizes, contrasts, average number of lesions per image, etc. They should not be forced to make a minimum or maximum number of marks. Large numbers of marks per image and/or multiple marks in the same vicinity are unusual and probably indicate misunderstanding of the task. The last problem is more likely to occur in CAD where the algorithm may generate multiple marks in the same vicinity. These are usually resolved by the algorithm designer using a clustering step.

Phantom or simulation studies are a convenient way of controlling truth and level of difficulty. The number of lesions per image should not be too large, especially for large lesions. A rule of thumb is one large lesion per abnormal image, or one to three small masses per image. While statistical power does increase with more lesions per image, having too many lesions detracts from the realism of the simulation. Attention should be paid to the quality of the simulated images (do the simulated masses resemble real masses?) and the locations where the lesions are superimposed (not all locations are equally likely to have lesions). If a normal image with a superimposed lesion is used as a simulated abnormal image, the same image without the superimposed lesion *must* not be used as a normal image, as all methods of analysis assume that different cases are independent. Likewise, images from other views of an abnormal breast, or the contralateral breast, must not be used as normal images. If the backgrounds are also simulated, e.g. power law noise (Burgess, 2001; Chakraborty, 2006a), there is no justification, in my opinion, for using radiologists to interpret such images. There is little in their training that specially qualifies them to read such images and using them in this mode is wasteful of resources. Anybody with good eye-sight and the ability to follow directions can, with sufficient training, be used to read simulated-background simulated-lesion images.

16.3.2 Localization accuracy

An important aspect of a free-response study is localization accuracy. A certain amount of marking inaccuracy is unavoidable with human observers but if the spread is too large it may be difficult to score the marks unambiguously. For digital images the location should be indicated *directly* on the image with a mouse click. Anatomically guided location statements, such as "lesion is in the upper outer quadrant," or "lesion is in the periareolar region," etc., are too ambiguous for a laboratory free-response study. Having the radiologist mark a hard-copy schematic (as they sometimes do in the clinic) and subsequently transferring the location to the digital image can lead to high localization inaccuracy. The reason why the free-response method gives larger statistical power than ROC is because it gives credit for true detections and penalizes for marks far from lesions; with increasing marking inaccuracy this distinction gets blurred. With sufficiently large spread the free-response dataset will effectively degenerate into an ROC dataset, where any mark on an abnormal image would count as having "detected" the lesion, and the statistical power advantage of conducting a free-response study will have been lost. For small lesions the radiologist can generally indicate the center fairly accurately, but for larger ones they could be encouraged to outline the suspected lesion. The centroid of the outlined region can be regarded as the location of the mark. Radiologists have different perceptions about what constitutes the lesion boundary, and the centroid method should result in higher inter- and intra-reader agreement.

16.3.3 Workstation issues

For digital images the interface should be kept as simple as possible; in my experience this issue is often overlooked. Ideally the workstation should resemble the clinical workstation as much as possible. Unless the purpose of the study dictates otherwise, common display functions such as window/level, zoom, pan, and other available tools should be enabled. When the radiologist clicks on a region a cross should be overlaid at that location and the rating recorded via a "pop-up" window displayed adjacent to the mark (to minimize mouse movements).

Table 16.1 *Malignancy rating scale.*

Meaning	Mark?	Probability of malignancy (%)	Discrete BIRADS-like rating
Assessment incomplete OR normal image OR definitely benign finding	No	N/A	
Possibly suspicious for malignancy	Yes	<2	3
Probably suspicious for malignancy	Yes	3–33	3½
Suspicious for malignancy	Yes	34–64	4
Quite suspicious for malignancy	Yes	65–94	4½
Highly suspicious for malignancy	Yes	≥95	5

The numerical value of the rating should be overlaid next to the mark or "pop-out" when the cursor hovers near the mark. This is to help the radiologist keep track of the marks and ratings and not inadvertently mark a region twice. All overlay information should be low-contrast and capable of being toggled off and on as otherwise it will interfere with the interpretation. The location information (x and y for projection images and x, y, and slice number for tomographic images) must be recorded. When the radiologist has concluded interpreting they should be allowed to review their marks before moving on to the next case. A lesion tracing function should be provided. It is my hope that standardized display software for free-response studies will be developed soon and made freely available.

16.3.4 Scoring the marks

Scoring is the act of classifying each mark as LL or NL. A mark is scored as an LL if it is close to a lesion and otherwise it is scored as an NL. Unfortunately at this time there is no consensus on what is a suitable closeness or proximity criterion. This makes it difficult to compare different free-response studies. For example, one does not know if a particular CAD manufacturer's FROC curve performance numbers are good because the algorithm is good or a lax proximity criterion was adopted. The proximity criterion issue should be addressed in consultation with the radiologists and the truth panel prior to commencement of the study and it must be reported in publications. The proximity criterion could be an acceptance radius surrounding each lesion (a mark inside is counted as an LL) or a percentage overlap of the outlined region and the true lesion. One approach to minimizing the subjectivity in choice of acceptance radius is to plot a histogram of the number of marks summed over all abnormal cases as a function of radial distance from the nearest lesion. The histogram applies only to abnormal cases since any mark on a normal case is obviously an NL. The histogram will typically show a narrow high peak, corresponding to marks made when the observer actually saw the lesion and the small spread is due to marking inaccuracy, followed by a minimum, and a subsequent broad low peak (Chakraborty, 2007b), corresponding to suspicious looking normal regions, which typically bear no fixed spatial relationship to the lesions. The radial distance defined by the minimum is a reasonable objective choice (in my opinion) for acceptance radius. It is good practice to re-analyze the data with different choices of proximity criteria (e.g. 20% higher or smaller than the baseline value) to determine if the conclusions are sensitive to the choice.

16.3.5 Rating scale

For human observers particular attention needs to be paid to the rating scale to obtain operating points that adequately sample (or "straddle") the FROC curve. In screening mammography the radiologist assigns a Breast Imaging Reporting and Data System (BIRADS) rating 0 through 6 with the following meanings (Eberl, 2006): 0 = assessment incomplete – need additional imaging evaluation; 1 = negative; 2 = benign finding; 3 = probably benign finding – short interval follow-up; 4 = suspicious abnormality – biopsy should be considered; 5 = highly suggestive of malignancy – appropriate action should be taken; 6 = known biopsy-proven malignancy – treatment pending.

It is difficult to convert this to a monotonically increasing rating scale. In particular it is not clear what to do with the 0 rating. An image rated 0 has suspicious findings but more information is needed before a final rating can be assigned and the patient is recalled. So a 0 is actually more suspicious than a 1. A finding of 1, 2, and possibly 3 would result in a "return the screening" action, while the others would result in a "recall." To construct a unidirectional scale one instructs the mammographer to assign ratings to suspicious regions even if they would have assigned the case a BIRADS 0 rating. A simple unidirectional scale with ratings linked to probabilities of malignancy ranges is shown in Table 16.1. A similar scale can be designed describing visibility of the lesion (visibility refers to presence of a benign or malignant lesion). In my experience radiologists prefer to use a discrete scale with at most four or five ratings.

16.3.6 Preliminary FROC curves and feedback

It is important to ascertain that the radiologists are familiar with the task and the user interface. Training sessions should be conducted using images from a separate training set to familiarize the radiologists with the task, and to provide feedback to both the radiologists and the investigator so that good FROC data can be collected in the actual study. Preliminary FROC curves should be constructed from the training set data. (See, for example, Figure 6 in Bunch *et al.* (Bunch, 1978).) If all points fall on the y-axis the radiologist needs to be told that their NLF rate is zero, which is good, but they are missing a significant fraction of lesions that were visible to the expert panel. This knowledge, and the fact that this is a laboratory study with no clinical consequences, may induce the radiologist to be more aggressive in reporting lesions.

16.3.7 Sample protocol

It is essential to have a written protocol, duly acknowledged by all participants, prior to commencement of the study. An example follows.

Purpose of the study

The purpose of the study is to compare digital mammography (DM) and breast tomosynthesis (BT) for the detection of malignant lesions.

The task

You will interpret 25 DM cases (craniocaudal – CC – and medi-olateral oblique – MLO) and 25 BT volumes per session, in a total of eight sessions. For each case you will indicate, with mouse clicks, regions that are suspicious for malignancy. Upon a mouse click a "cross" will be overlaid at the marked location and a pop-up window will allow you to rate the probability of malignancy on a rating scale (Table 16.1).

Notes

- You may find nothing to mark on a case, or one mark, two marks, etc.
- If the same lesion is visible in both CC and MLO views, mark it in both views and rate each mark.
- Try to mark the center of the suspicious region accurately, or use the workstation's TRACE function; additionally for BT images choose the slice where the lesion appears in best focus.
- There is a HIDE/SHOW toggle button which does not erase the data, but only suppresses them for your convenience.
- Details of the user interface are in the "*Free-Response Study User Guide*" document which we will discuss during the training sessions.

16.4 MODELS OF FREE-RESPONSE DATA

As noted above, early models for free-response data, namely FROCFIT (Chakraborty 1989), AFROC (Chakraborty, 1990), and LROC (Swensson, 1996), do not fit the low-confidence region of CAD-generated FROC curves. This is because they predict the FROC curve extends to $(\infty, 1)$ which is inconsistent with data (Chakraborty, 2008b). Two other models (Irvine, 2004; Hutchinson, 2007) that predict the curve extends to $(\lambda, 1)$, where λ is a finite number, will give somewhat better fits in the low-confidence region. IDCA (Edwards, 2002) and the search-model (Chakraborty, 2006c; Chakraborty, 2006b; Chakraborty, 2007a; Yoon, 2007; Chakraborty, 2008b) are more flexible in that the end-point is not constrained to $(\infty, 1)$. They yield good fits to CAD and human observer data (Yoon, 2007; Chakraborty, 2008b) and in my opinion they are the only viable models at this time. Moreover they are consistent with the KN perceptual model described in Section 16.4.2.1. The early methods are not discussed in this chapter and the interested reader is referred to a recent publication (Chakraborty, 2008b) that summarizes them.

16.4.1 IDCA

The IDCA sampling model assumes a binormal model for the z-samples (z-sample = decision variable or confidence level for a suspicious region). The z-samples for noise-sites are sampled from N(0, 1) and z-samples for signal-sites are sampled from $N(\mu, \sigma^2)$, where $N(\mu, \sigma^2)$ is the normal distribution with mean μ and variance σ^2. Define F the total number of NLs, T the total number of LLs, N_I the total number of images, and N_L the total number of lesions. Then the coordinates of the *observed* end-point (λ', ν') are given by (the primes are needed to distinguish from similar quantities in the search-model described later):

$$\left.\begin{aligned} \lambda' &= \frac{F}{N_I} \\ \nu' &= \frac{T}{N_L} \end{aligned}\right\} \tag{16.1}$$

Defining $\Phi(z)$ the cumulative normal distribution function for N(0, 1), the fitted FROC curve (Edwards, 2002) is given by (note that $\Phi(-z) = 1 - \Phi(z)$):

$$\left.\begin{aligned} NLF(z) &= \lambda'\Phi(-z) \\ LLF(z) &= \nu'\Phi\left(\frac{\mu' - z}{\sigma'}\right) \end{aligned}\right\} \tag{16.2}$$

These expressions will become clearer in Section 16.4.2.3, where the corresponding search-model expressions are derived. According to Eqn. 16.2, as z decreases from $+\infty$ to $-\infty$, the IDCA FROC curve starts at (0,0) and terminates at (λ', ν'). The parameters μ' and σ' can be estimated by regarding the NL and LL marks as normal and abnormal "images," and fitting their ratings to the binormal model (Dorfman, 1969), whose parameters, a and b, are related to μ' and σ' by $\mu' = a/b$ and $\sigma' = a/b$. Alternatively, "proper" ROC models may be used to fit the pseudo-ROC (Dorfman, 1997; Pan, 1997; Metz, 1999; Dorfman, 2000b; Pesce, 2007). The fitted curve is referred to as a *pseudo-ROC* since the marks do not refer to images; rather they refer to suspicious regions in the images, i.e. parts of images.

The coordinates of a general point on the pseudo-ROC are $[\Phi(-z), \Phi((\mu' - z)/\sigma')]$. It follows from Eqn. 16.2 that the IDCA-predicted FROC curve has a simple geometric interpretation, namely it is a "two-dimensionally stretched" version of the pseudo-ROC where the scaling factors along the x and y axes are λ' and ν', respectively. Figure 16.3 illustrates the mapping of the pseudo-ROC to the FROC curve when $\lambda' = 2.5$ and $\nu' = 0.7$. The origin of the pseudo-ROC corresponds to cumulating *none* of the marks and the corresponding cutoff z is positive infinity. Likewise the (1, 1) point corresponds to cumulating *all* of the marks and the cutoff is negative infinity. Since (1, 1) on the pseudo-ROC is mapped to (λ', ν') on the FROC curve, the cutoff corresponding to the observed end-point of the IDCA-predicted FROC must be negative infinity. This can be confirmed by

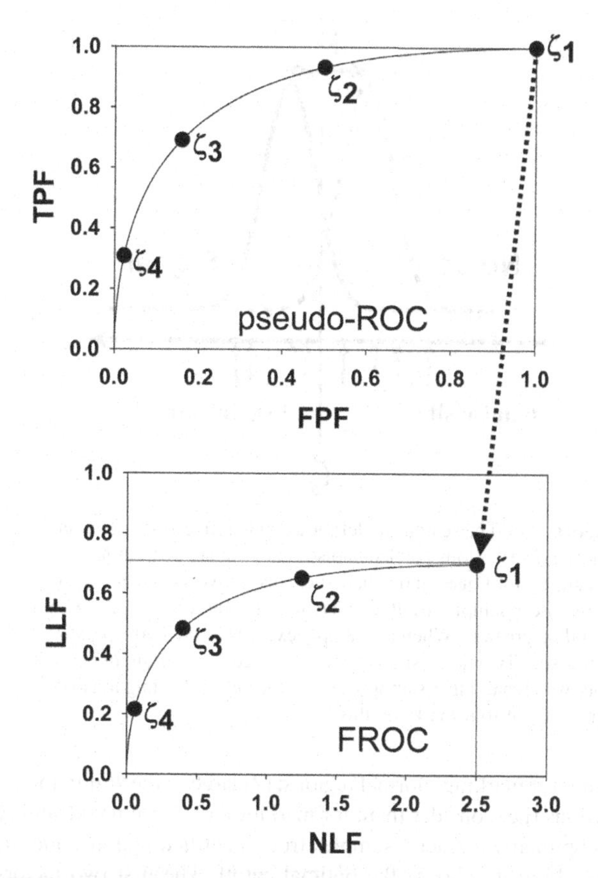

Figure 16.3 This figure illustrates the IDCA approach to fitting FROC operating points. Upper panel: the NL and LL marks are regarded as normal and abnormal "images" and used to plot pseudo-ROC operating points, which are fitted by ROC curve-fitting software. Lower panel: the FROC curve is obtained by a mapping or "stretching" operation indicated by the arrow, consisting of a point-by-point multiplication of the pseudo-ROC curve along the y-axis by ν' and along the x-axis by λ', where (λ', ν') are the coordinates of the observed end-point of the FROC curve, in this case (2.5, 0.7). Four pseudo-ROC and four FROC operating points and the corresponding cutoffs ς_i ($i = 1, 2, 3, 4$) are shown. Since ς_1 corresponds to the upper right-hand corner of an ROC curve, it must be equal to $-\infty$.

setting $z = -\infty$ in Eqn. 16.2 which yields NLF $= \lambda'$ and LLF $= \nu'$.

16.4.2 The search-model

The search-model is a statistical parameterization of a perceptual model (Kundel, 1983; Nodine, 1987; Kundel, 2004; Kundel, 2007) of how experts interpret images. The KN model has similarities to models developed in non-medical imaging contexts (Gregory, 1970; Rock, 1983; Rock, 1984; Wolfe, 1994) and is consistent with eye-tracking recordings of the search patterns of radiologists.

16.4.2.1 Perceptual basis

According to KN, image perception involves (i) a global-response stage that identifies, using peripheral vision and in about 300 msec, perturbations or "suspicious regions" that figuratively speaking "pop out" of the image, and (ii) a decision-making stage during which the gaze is directed towards each perturbation in turn, feature analysis is performed using foveal inspection of the region and the surround, and a decision is made whether or not to mark (i.e. report) the region. Swensson referred to these two stages as pre-attentive and cognitive, respectively (Swensson, 1980). In the CAD literature the suspicious regions are termed "initial detections" and the decision-making stage as "candidate analysis" (Edwards, 2002). It is interesting but perhaps not surprising that CAD developers apparently independently arrived at a two-stage process for finding lesions that is remarkably similar to the KN model.

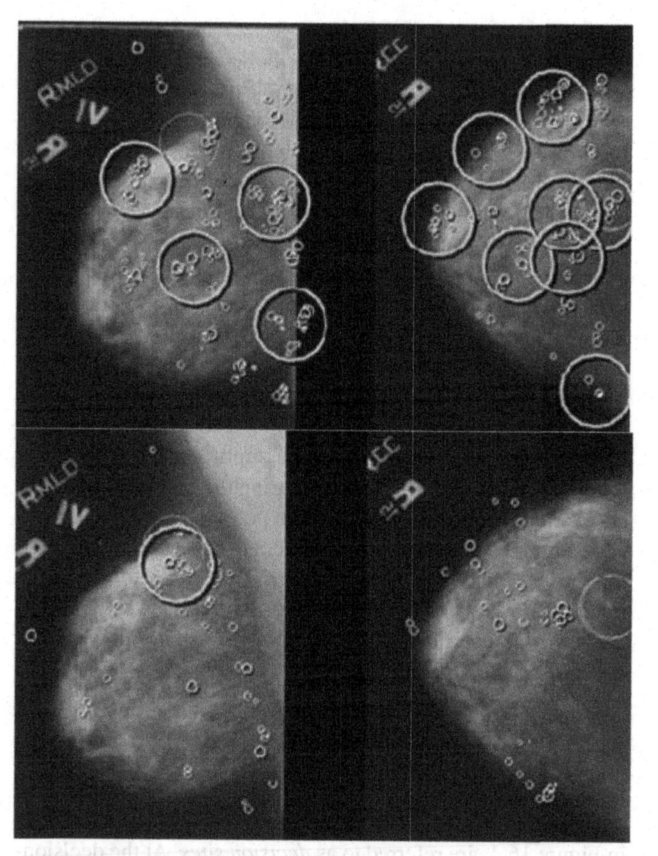

Figure 16.4 This figure shows eye-tracking recordings for a two-view soft-copy mammogram display for two observers, an inexperienced observer (upper two panels) and an expert mammographer (lower two panels). Individual fixations (dwell time ~100 msec) are indicated by the small circles and clustered fixations (cumulative dwell time ~1 sec) are indicated by the larger high-contrast circles. The latter correspond to the decision sites in the search-model. A mass visible on both views is indicated by the larger low-contrast circles. Note that the inexperienced observer finds more suspicious regions than does the expert mammographer but misses the lesion in the MLO view. The mammographer finds the lesion in the MLO view without finding suspicious regions in the normal parenchyma.

By monitoring corneal reflections from an infrared light source one can measure the line-of-gaze of an observer (Duchowski, 2002) in real-time. Shown in Figure 16.4 are eye-tracking recordings for a two-view mammogram soft-copy display for two observers, an inexperienced observer (upper two panels) and an expert mammographer (lower two panels). Individual fixations (dwell time ~100 msec, insufficient time to make decisions) are indicated by the small circles. Clustered fixations (cumulative dwell time $\sim$1 sec) are indicated by the larger high-contrast circles and it is believed (Hillstrom, 2000) the observer makes conscious decisions to report or not to report only at these locations. The clustered fixations are the suspicious "pop-out" regions identified by the global response, the first stage of the KN model. A malignant mass visible in both views is indicated by the larger low-contrast circles. Note that the inexperienced observer makes decisions at several more locations than does the experienced mammographer. On the MLO view the inexperienced observer identified four normal regions as possible lesion candidates and missed the lesion, whereas the mammographer identified the lesion without having to make decisions in any normal regions (making a decision – to mark or not mark – in a normal region always runs the risk of reporting it). The perceptual model does not imply that the radiologist ignores large areas of the image. During the first stage all areas are examined by the lower-resolution peripheral vision but more attention units are applied at the second stage to regions identified as suspicious by the first stage. The degree to which the radiologist is successful at screening out normal regions from the need for decision-making, without screening out lesions, depends on expertise.

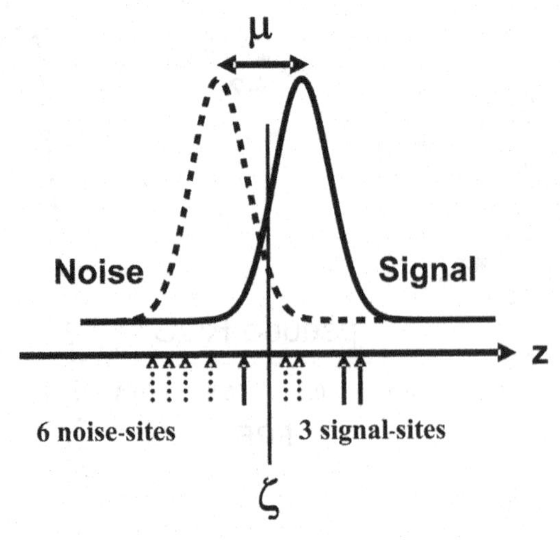

Figure 16.5 The search-model for a single rating study. The unit normal distributions labeled "noise" and "signal" determine the z-samples for noise- and signal-sites, respectively. Shown are six noise-site z-samples (dotted up arrows) and three signal-site z-samples (solid up arrows). When a z-sample exceeds ζ, the corresponding site is marked. Two noise-site z-samples exceed ζ, resulting in two NLs, and two signal-site z-samples exceed the cutoff, leading to two LLs, for a total of four marks on this image.

16.4.2.2 The statistical parameterization

In the search-model parameterization the suspicious regions identified during the global response, i.e. the clustered fixations in Figure 16.4, are referred to as *decision sites*. At the decision-making stage a z-sample is calculated for each decision site and if it exceeds the observer's lowest reporting threshold the site is marked. Decision sites representing normal anatomy are termed *noise-sites* and decision sites that represent actual lesions are termed *signal-sites*. The signal-to-noise-ratio between noise-sites and signal-sites is the μ parameter of the model, i.e. it is the separation of two unit-variance normal density functions (Figure 16.5). The z-samples for noise-sites are sampled from $N(0, 1)$ and z-samples for signal-sites are sampled from $N(\mu, 1)$. The number of noise-sites on an image is modeled as a random integer sampled from the Poisson distribution with mean λ ($\lambda > 0$) and the number of signal-sites on an image is modeled as sampled from the binomial distribution with success probability ν ($0 \leq \nu \leq 1$) and trial size s (s is the number of lesions in the image). The random sampling allows the number of NL and LL marks to vary from image to image and represents an important advance in the understanding of the free-response paradigm. Note that the search-model does not have a σ-parameter. An R-rating free-response study requires R cutoffs $\zeta_1, \zeta_2, \ldots, \zeta_R$. If a z-sample exceeds the observer's lowest cutoff ζ_1, the observer marks (or "reports") the corresponding region. The *rating* assigned to a mark is the value of the z-sample binned according to the set of ordered cutoffs, or the actual value could be reported as with a CAD algorithm. Marks made because of noise-site z-samples exceeding ζ_1 are assumed to fall far from lesions and are classified as NLs and those made because of signal-site z-samples exceeding ζ_1 are assumed to fall close to lesions and are classified as LLs.

Experts are characterized by small λ, large ν (subject to the constraint that $\nu \leq 1$), and large μ. Expertise consists of four factors: (1) small λ: the ability to avoid identifying (i.e. considering for marking) normal regions; (2) large ν: the ability to find lesions (i.e. consider them for marking); (3) large μ: the ability to optimally extract z-samples from identified regions; and (4) the ability to choose the optimal cutoff. The first two factors represent search expertise. The third and fourth factors represent decision-making: is the observer using the right template, is the observer properly accounting for the background, is the observer employing a proper threshold – not too strict, not too lax? A "found" lesion is only marked if its z-sample is larger than the lowest cutoff. For example, the expert mammographer who found the lesion (lower panel of Figure 16.4) may not mark it if its z-sample is low and/or the reporting criterion is very conservative.

The search-model for a single rating study is illustrated in Figure 16.5, which is for a hypothetical image with 6 noise-sites (dotted up arrows), and 3 signal-sites (solid up arrows). Two noise-site z-samples exceed the cutoff ζ, leading to 2 NLs, and 2 signal-site z-samples exceed the cutoff ζ, leading to 2 LLs, for a total of 4 marks on this image. Assuming $s = 5$ (it must be at least 3, the local value of the number of signal-sites), the estimates of λ and ν based on this one image sample are 6 and 0.6, respectively, and the corresponding FROC operating point is (2, 0.4). Averaging these numbers over a large number of images yields better estimates.

16.4.2.3 The FROC curve

The probability that a z-sample from $N(0, 1)$ exceeds z is $\Phi(-z)$. Since λ is the expected number of noise-sites per image and each noise-site yields a z-sample from $N(0, 1)$, the expected number of noise-sites with z-samples exceeding z is $\lambda\Phi(-z)$, which is the x-coordinate of the FROC curve, i.e. $NLF(z) = \lambda\Phi(-z)$. *Note that this result does not depend on the validity of the Poisson assumption.* As long as λ is interpreted as the mean number of noise-sites per image, the sampling distribution of the number of noise-sites per image is irrelevant.

The probability that a z-sample from $N(\mu, 1)$ exceeds z is $\Phi(\mu - z)$. Since the expected number of signal-sites in an image is $s\nu$ (s is the number of lesions in the image), the expected number of signal-sites with decision variable exceeding z is $s\nu\Phi(\mu - z)$. Assuming s is constant, LLF(z) is obtained by summing $s\nu\Phi(\mu - z)$ over all abnormal images and dividing by the total number of lesions, leading to $LLF(z) = \nu\Phi(\mu - z)$. The result also applies to the case where s is variable (Chakraborty, 2007a). *The expression for LLF does not depend on the validity of the binomial assumption.* As long as ν is interpreted as the probability that a lesion is identified by the global response, the sampling distribution of the number of signal-sites per image is irrelevant. Summarizing, the coordinates [NLF(z), LLF(z)] of the operating point on the FROC curve defined by threshold z are

$$\begin{aligned} NLF(z) &= \lambda\,\Phi(-z) \\ LLF(z) &= \nu\,\Phi(\mu - z) \end{aligned} \tag{16.3}$$

Since $0 \leq \Phi(z) \leq 1$, the FROC is contained within the rectangle with corner points (0, 0) and (λ, ν), where the origin corresponds to $z = \infty$ and the end-point (λ, ν) corresponds to $z = -\infty$. The FROC curve predicted by Eqn. 16.3 is concave down and the slope decreases monotonically and asymptotically approaches zero at the end-point ("proper" FROC curve). Note that Eqn. 16.3 defines the *population* FROC. For finite numbers of images the observed operating point will generally not lie on the asymptotic curve.

16.4.2.4 Estimation

The log-likelihood function for the search-model is (this expression corrects an error in (Yoon, 2007)):

$$\begin{aligned} LL_{SM} = \sum_{i=1}^{R} \{F_i \ln p_i + T_i \ln q_i\} + [F \ln(\lambda) - N_I\lambda] \\ + T \ln(\nu) + N_I\lambda p_0 + (N_L - T)\ln(1 - \nu + \nu q_0) \end{aligned} \tag{16.4}$$

Here F_i is the observed number of NLs in bin i over the entire image set, T_i is the corresponding number of LLs, and p_i and q_i are defined by $(\zeta_0 = -\infty)$

$$\begin{aligned} p_i &= \Phi(\zeta_{i+1}) - \Phi(\zeta_i) \\ q_i &= \Phi(\zeta_{i+1} - \mu) - \Phi(\zeta_i - \mu) \end{aligned} \tag{16.5}$$

This function can be maximized to determine the parameters. The procedure is to pick a value of λ, minimize negative log-likelihood with respect to all parameters except λ, and vary λ until a global minimum is found. Figure 16.6 is a typical plot of $-LL_{SM}(\lambda)$ vs. λ for a CAD dataset where the ordinate is the value of negative log-likelihood after it has been minimized with respect to all parameters except λ (Yoon, 2007). It is seen that $-LL_{SM}(\lambda)$ has a minimum as a function of λ. Based on our experience, Newton-Raphson and similar iterative methods do not work well for this function as they tend to find false maxima, but the simulated annealing procedure (Press, 1988) implemented in the GNU library (Galassi, 2005) works rather well. But there is a problem. The term $\lambda\, p_0$, which accounts for the unmarked noise-sites, can be held constant by decreasing

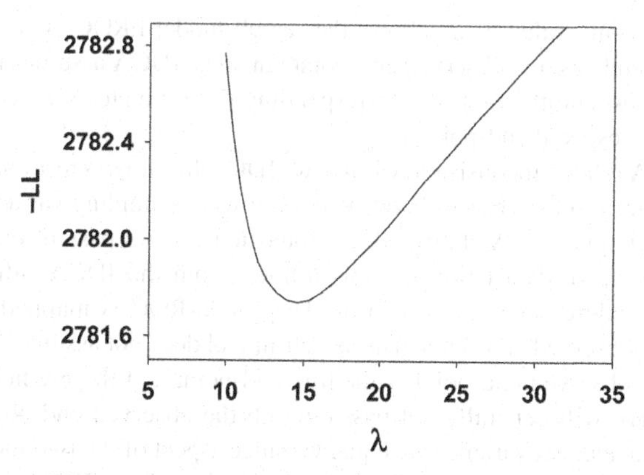

Figure 16.6 A typical plot of $-LL_{SM}(\lambda)$ vs. λ, where the ordinate is the value of the negative of the log likelihood after it has been minimized with respect to all parameters except λ. It is seen that $-LL_{SM}(\lambda)$ has a minimum at $\lambda = \lambda_M = 14.6$. Used with permission from Yoon *et al.* (Yoon, 2007).

ζ_1 and increasing λ appropriately, and likewise νq_0, which accounts for the unmarked signal-sites, can be held constant by decreasing ζ_1 and increasing ν. If there is only one cutoff there is complete degeneracy and one cannot uniquely estimate λ and ν. The presence of multiple cutoffs breaks the degeneracy but nevertheless estimation is difficult, especially for human observer data.

16.4.3 Comparison of IDCA and the search-model

If one sets $\sigma' = 1$ the IDCA (Eqn. 16.2) and search-model (Eqn. 16.3) expressions for the FROC curve look similar, excepting for the primes, and one may ask if the two models are not in fact the same. The difference is that (λ', ν') is the *observed* end-point (that generated by the least confident rating) whereas (λ, ν) is the *true* end-point if all decision sites are marked (because of the finite reporting threshold, not all decision sites are marked). The observed and true end-points are related by

$$\begin{aligned} \lambda' &= \lambda\,\Phi(-\zeta_1) \\ \nu' &= \nu\,\Phi(\mu - \zeta_1) \end{aligned} \tag{16.6}$$

This result follows from Eqn. 16.3 if one sets $z = \zeta_1$, because ζ_1, the lowest cutoff, defines the observed end-point. For IDCA, see Figure 16.3, the pseudo-ROC to FROC mapping is from (1, 1) to the *observed* end-point (λ', ν') whereas for the search-model the mapping is from (1, 1) to the *true* end-point (λ, ν). According to Eqn. 16.6 the two will coincide only if $\zeta_1 = -\infty$. For finite ζ_1 the observed end-point will generally be left and downward shifted with respect to the true end-point, but sampling variability may lead to a different shift. Therefore, for finite ζ_1 the FROC curve will terminate with finite slope at the observed end-point in the sense that the next lower operating point is below and to the left of the uppermost point – i.e. the points are not leveling out. This of course implies that more lesions could be found by relaxing the criterion and marking more LLs at the expense of more NLs. This leads to an inconsistency since according to the IDCA model one cannot relax the criterion since it is already at its lowest possible value, namely $-\infty$. (A subtle point: the z-scales in the two models are different: the finite cutoff ζ_1 in the search-model corresponds to $-\infty$ in the IDCA model;

the finite value of ζ_1 allows the search-model FROC curve to extend past the observed end-point but since IDCA assumes the lowest cutoff is $-\infty$ the corresponding fit terminates *exactly* at the observed end-point.)

A related inconsistency is that all IDCA fits *always* terminate exactly at the observed end-point even when sampling variability is present. A statistical fit should not *always* pass through any given data point, but this follows from the IDCA fitting procedure, since the (1, 1) on the pseudo-ROC is mapped to the observed end-point. The search-model does not assume that the observed end-point is the true end-point and the predicted curve will generally not pass through the observed end-point. The unequal variance vs. equal variance aspect of the two models is a relatively minor difference. In our experience IDCA with the σ' parameter yields better fits to CAD and human observer data than the same model with $\sigma' = 1$; the search-model without the σ' parameter achieves equally good fits since it does not constrain the curve to terminate at the observed end-point. (While IDCA was intended to analyze designer-level CAD data, I have found that it also yields excellent fits to human observer operating points.)

Both models predict operating characteristics with the *finite end-point property* (Chakraborty, 2008b). The true end-point is reached when the observer adopts an infinitely small threshold ($-\infty$) so that all decision sites are marked. Since the number of noise-sites is finite, the limiting search-model NLF is finite. Likewise, unless all lesions are identified, the limiting search-model LLF is less than unity. The IDCA end-point is further limited by the finite lowest cutoff value (not all decision sites are marked). Earlier free-response models (Bunch, 1978; Chakraborty, 1989; Chakraborty, 1990; Swensson, 1996) predicted that the end-point is at (∞, 1). These models imply an infinite number of decision sites, i.e. effectively the observer performs an infinite point-by-point inspection of the image which should be contrasted to the KN model and Figure 16.4. The search-model and IDCA predict ROC, LROC, and AFROC curves, all of which share the finite end-point characteristic of the FROC curve (Chakraborty, 2008a; Chakraborty, 2008b). The finite end-point property of the ROC is due to images with no decision sites. In an ROC study such images are binned in the lowest available bin (e.g. 1 in a quasi-continuous 100-rating ROC study, with allowed ratings 1 through 100) and only when this bin is included in the counting is the (1, 1) point trivially – and discontinuously – reached. Bin 2 corresponds to images with at least one decision site, and no matter how small the threshold, the point generated by this bin cannot be infinitely close to (1, 1) because of the finite number of images with no decision sites in bin 1 (Chakraborty, 2006b). The discontinuity is particularly pronounced for observers characterized by small λ. In such cases the operating points are clustered near the initial near-vertical section of the curve. There is evidence (Dorfman, 2000c) that radiologists sometimes provide data like these and this type of data clustering presents degeneracy problems for the binormal model analysis. Instructions to the observers to "spread their ratings" or use "continuous" ratings (Metz, 1998) and to be more aggressive in their reporting style do not always seem to work. To the experimenter such observers will appear to be not heeding the advice to "spread their ratings." To our knowledge no one has shown that radiologists can move the operating point infinitely close to (1, 1) in ROC space, or to (∞, 1) in FROC space, and the same is true for clinical CAD algorithms where marks at very low confidence levels are available. For example, confidence levels for a CAD algorithm might range from 0 to 1. Some CAD images may have no marks, even at the 0.001 level. About 12% of the images in our CAD datasets have no marks (Yoon, 2007; Chakraborty, 2008a).

Figure 16.7, (a) through (d), illustrates the finite end-point characteristics of the search-model fits, the IDCA fits, and the CAD data. Figure 16.7(a) shows Swensson model fits to the observed ROC, AFROC, and LROC curves (Chakraborty, 2008b). Figure 16.7(b) shows corresponding search-model fits, Figure 16.7(c) shows IDCA fits, and Figure 16.7(d) shows Swensson and search-model fits to FROC curves (the IDCA fit, not shown, is visually identical to the search-model fit but it stops exactly at the observed end-point). Note that the observed ROC and AFROC data points do not extend to (1, 1), the observed LROC data point abscissa does not extend to 1, and the observed FROC curve does not extend to (∞, 1). The Swensson model and earlier free-response models deviate from the data, especially at low confidence levels, and especially for the FROC curve.

16.5 ANALYSIS OF FREE-RESPONSE DATA

This section describes methods for analyzing free-response data and summarizes results from a recent study. There are currently several options for analyzing free-response data that differ essentially in the definition of the figure-of-merit. Free-response figures-of-merit generally do a better job of appropriately rewarding/penalizing observer decisions than ROC figures-of-merit. An example of "right for wrong reason" was given in section 16.2. In the ROC paradigm, a less experienced radiologist who misses a lesion but reports a normal region is scored identically as an expert radiologist who sees and reports the lesion at the same confidence level but ignores the normal region, i.e. both are credited with true-positives at identical confidence levels. In free-response the non-expert makes two mistakes (missed lesion and the NL) and the figure-of-merit is smaller relative to the expert. All methods in this chapter are applicable to the MRMC study design. Following principles outlined by Obuchowski (Obuchowski, 2009) the methods could be extended to *partially-paired* designs, which are often more practical.

16.5.1 Hypothesis testing

As with any statistical measurement, one cannot state with 100% certainty that a modality has greater figure-of-merit than another, but one can place probabilistic bounds, e.g. 95% confidence intervals, on the difference. If the 95% confidence interval does not include zero, then one can state that the difference is significant at the 5% level. In the language of statistical hypothesis testing there are two possibilities: the difference is zero and the null hypothesis (NH) is true, or the difference is non-zero and the alternative hypothesis (AH) is true (we only consider two-tailed tests here). One selects a *significance level* α (e.g. $\alpha = 5\%$) for the test, the nominal probability of a Type 1 error, i.e. declaring modalities different when in fact they have the same figure-of-merit. Statistical validity of an analysis is tested

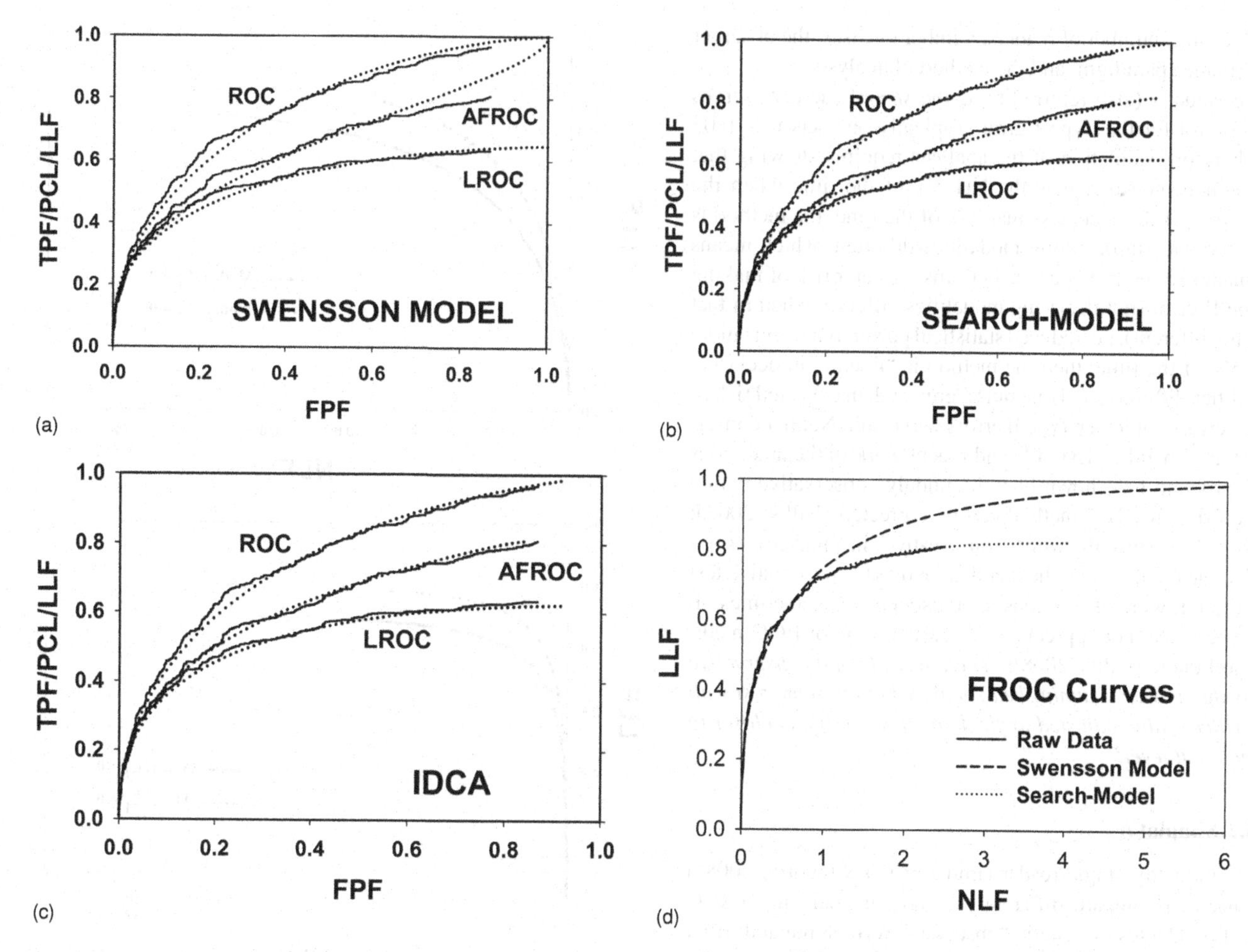

Figure 16.7 This figure shows Swensson model, search-model, and IDCA fits to different operating characteristics for CAD data. It illustrates the finite end-point characteristic of the search-model and IDCA fits *and* the CAD data, a property not shared by the Swensson model and earlier free-response models. (a) Swensson LROC-model, ROC, AFROC, and LROC curves; (b) search-model fits; (c) IDCA model fits; and (d) FROC curves predicted by the Swensson and search-models (the IDCA fit, not shown, is visually identical to the search-model fit but it stops exactly at the observed end-point). Note that the observed ROC and AFROC curves do not extend to (1, 1), the observed LROC abscissa does not extend to 1, and the observed FROC curve does not extend to (∞, 1). The Swensson model deviates from the data at low confidence levels, especially for the FROC curve. Used with permission from Chakraborty *et al.* (Chakraborty, 2008b).

with simulations – hence the need for a simulator (see Section 16.4.2). One simulates a dataset under NH conditions, ensured by equating the modality 1 and 2 simulator parameters so they have identical FROC curves and figures-of-merit. For each simulated MRMC dataset the analysis yields a test statistic (e.g. the t-statistic for a single reader and two modalities or an F-statistic for MRMC) and if the test statistic falls in the acceptance region for the NH (determined by α and the statistical distribution of the test statistic in question), the NH is not rejected and otherwise it is rejected. Equivalently the test statistic can be converted to a p-value and if $p < \alpha$, the NH is rejected. The p-value is defined as the probability, under the NH condition, that chance could account for a test statistic at least as large in magnitude as the observed value. It is essentially a statement of the *unlikelihood* of the observed test statistic if the two modalities actually have identical performance. For example, a p-value of 0.15% means that the observed test-statistic is quite unlikely (1 in 667) to be attributable to chance under the NH condition. Making α smaller reduces the probability of a Type I error but increases the probability of a Type II error, i.e. accepting the NH when it is false, which decreases statistical power.

The test statistic is determined by the magnitude of the observed difference in the figures-of-merit relative to the variability of the difference. Large values of the test statistic imply the difference is unlikely to be due to chance. The variability of the difference can be determined by resampling methods such as the jackknife or the bootstrap, described below, and this yields the p-value. This step is termed *significance testing*. The simulations are repeated with different seeds and the average rejection fraction (often termed rejection rate in the statistical literature) is determined over many simulations. Typically one performs 2000 NH simulations and a valid method should yield about 100 rejections for $\alpha = 5\%$. For 2000 simulations a 95% confidence interval for the rejection rate is ± 0.01, so an observed rejection rate between 0.04 and 0.06 means the method passes the NH test.

Statistical power is defined as the probability of rejecting the NH when it is false. To determine statistical power, simulations are conducted under AH conditions and the observed NH rejection rate is the statistical power. Modality 1 and 2 parameters are chosen different, which implies a non-zero effect size, i.e. a non-zero difference in figures-of-merit. Statistical power depends on α, effect size, number of readers, number of cases, the observer performance paradigm, and the method of analysis.

The value α (also referred to as the *size* of the test) serves as a control for the Type I error. Typically one sets it to 0.05 and therefore validation of the analysis requires showing that the method indeed rejects the NH 5% of the time when the NH is true. If it rejects less than 5% of the time, the method is "conservative" in declaring modalities different, which means one makes fewer Type I errors but runs a greater risk of making a Type II error (not declaring modalities different when in fact they are different), i.e. reduced statistical power. If it rejects more than 5% of the time, then the method is "liberal" in declaring modalities different, and one runs a greater than expected risk of Type I errors but fewer Type II errors will result. Neither conservative nor liberal is desirable and recent work in the analogous ROC problem has shown how the unduly conservative nature of the original DBM method can be corrected (Hillis, 2005a; Hillis, 2008). Since the word "conservative" in some circles has a desirable connotation, the fact that an overly conservative test is not good, which is obvious to statisticians, is sometimes, in my experience, not appreciated by casual users of ROC methods, so I quote (Hillis, 2008): "*The downside of a conservative test is that power is diminished compared to the same test with the critical value adjusted to yield significance levels closer to the nominal level.*"

16.5.2 Simulator

A two-modality single-reader simulator (Chakraborty, 2008c) was used to compare different methods of analyzing FROC data. Two classes of generic "observers" were simulated: (a) a human observer who finds, on the average, about 1.3 actually normal regions (noise-sites) per image, and (b) a CAD algorithmic "observer" that finds, on the average, about 10 noise-sites per image. For the human observer the term "modality" has the conventional meaning, e.g. CT or digital chest radiography. The testing also applies to two observers (human or different CAD algorithms) interpreting the same set of images. The two modalities are labeled 1 and 2, where modality 1 had lower performance. Modality 1 parameters ($\mu = 1.5$, $\lambda = 1.3$, $\nu = 0.8$ for the human observer and $\mu = 2.34$, $\lambda = 10$, $\nu = 0.9$ for the CAD observer) were chosen so that for both classes of observers the area under the search-model-predicted ROC curve (Chakraborty, 2006b; Chakraborty, 2006c), was 80%, i.e. $AUC_1 = 0.80$. Parameters for the modality-2 observers were chosen to yield $AUC_2 = 0.85$ (λ was 20% smaller, ν was 10% greater, and μ was adjusted to get the desired AUC value (Chakraborty, 2008c); this yielded $\mu = 1.55$ for the human observer and $\mu = 2.37$ for the CAD observer). In most instances 100 normal and 100 abnormal images, the latter with one lesion per image, were simulated, but other conditions were also investigated. The lowest threshold ζ_1 determines the fraction of noise-sites that are not marked, i.e. the *truncation fraction*, $\Phi(\zeta_1)$. This follows from Section 16.4.2.3, since the probability that a z-sample from N(0, 1) not exceeding ζ_1 is $\Phi(\zeta_1)$. The truncation fraction is the fraction of the maximum horizontal extent of the FROC curve that is *not* observed. Values chosen for ζ_1, namely $-\infty$, -0.674, 0.0, and 0.674, correspond to truncation fractions 0%, 25%, 50%, and 75%, respectively. A large ζ_1 characterizes a more conservative observer who is reluctant to generate NLs. For such an observer the end-point is on the ascending portion of the FROC curve, i.e. the observer provides an *incomplete* curve. The *complete* curve corresponds to $\zeta_1 = -\infty$. Search-model-predicted FROC curves for both modalities are shown in Figure 16.8 where the top panel is for the human observers and the bottom panel is for the CAD algorithms. The sections labeled $\zeta_1 = 0$ represent the incomplete curves for 50% truncation fraction. For modality 1 the complete curve extends to NLF = 1.3 for the human observer and to NLF = 10 for CAD. The corresponding incomplete curves extend to 0.65 and 5, respectively. Modality 2 has smaller λ and larger ν and therefore both complete and incomplete curves end to the upper left of the corresponding modality 1 curves. The dotted lines represent γ, the upper limit

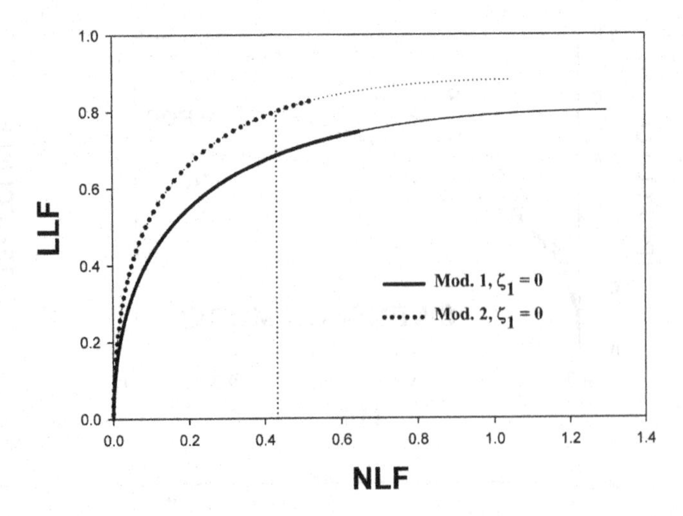

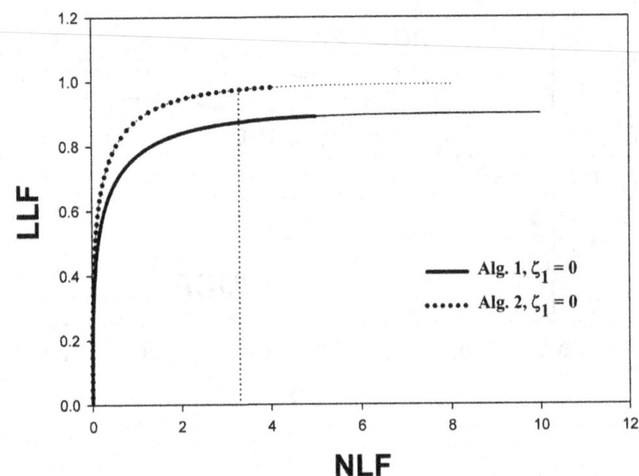

Figure 16.8 Search-model-predicted FROC curves for the human observer (top panel) and CAD (bottom panel) used in the simulation testing reported in Tables 16.2 through 16.5. The parameters were chosen so that the areas under the search-model-predicted ROC curves were 0.80 for modality 1 and 0.85 for modality 2. The lowest cutoff ($\zeta_1 = 0$ in this figure) corresponds to the observed end-point of the incomplete curve (darker lines and dots) and the complete curve (light lines) corresponds to $\zeta_1 = -\infty$. Large ζ_1 corresponds to a more conservative observer and lower observed end-point. The lowest cutoff was a variable in the simulations. The vertical lines denote γ, the upper limits of integration for the $AUFC_\gamma$ figure of merit (for $\zeta_1 = 0$, $\gamma = 0.433$ and 3.33 for the human and CAD observer, respectively).

of integration of a figure-of-merit discussed later (see Section 16.5.5).

The simulator includes parameters to account for inter-modality (ρ_{inter}) and intra-image correlations (ρ_{intra}) of the z-samples. The correlations arise because (a) the reader interprets the same image in the two modalities – this is referred to as inter-modality correlation, ρ_{inter}; and (b) because the ratings of marks on the same image could be correlated – this is referred to as intra-image correlation, ρ_{intra}.

16.5.3 ROC analysis

ROC analysis based on the ROC-equivalent ratings inferred from the free-response data (i.e. the rating of the highest-rated mark) was included for comparison since it is the current gold standard. The ROC figure-of-merit was the area under the ROC curve (AUC). For the simulated human observers the data were binned to ≤ 6 bins and ROCFIT was used to estimate AUC. For the simulated CAD algorithms, the data were not binned and the trapezoidal area under the ROC curve was used as the figure-of-merit. The implementation of ROCFIT was based on the FORTRAN source code that used to be posted on the University of Chicago ROC website, but is no longer available (checked April 21, 2009). Our implementation was validated by comparing to the online ROC calculator available at http://www.rad.jhmi.edu/jeng/javarad/roc/JROCFITi.html, which is based on ROCFIT. The significance testing procedure uses a resampling technique known as the jackknife (Efron, 1982) and is conceptually identical to the DBM-MRMC method (Dorfman, 1992). When a case is "jackknifed" or removed from the analysis, the figure-of-merit is recomputed and converted to a pseudovalue (see Eqn. 16.8 below). The matrix of pseudovalues is analyzed by a mixed-model analysis of variance (ANOVA) procedure which yields the p-value for rejecting the NH and other useful statistics. Our ANOVA code is functionally equivalent to that in DBM-MRMC Version 2.2 and includes recent enhancements (Hillis, 2007; Hillis, 2008). (The DBM-MRMC method has been shown (Hillis, 2005b) to be identical to an alternative method for analyzing MRMC data (Obuchowski, 1995b).)

16.5.4 JAFROC

The JAFROC (Chakraborty, 2004) figure-of-merit is the non-parametric (Mann-Whitney-Wilcoxon U-statistic) area θ^{JAFROC} under the AFROC curve, defined as the plot of LLF vs. FPF. Here FPF is computed in the usual (ROC) manner over *normal* cases, i.e. it is the number of normal cases with highest-rated NLs rated higher than a threshold rating, divided by the total number of normal cases. An additional figure-of-merit was introduced (Chakraborty, 2004) termed JAFROC1, defined as the non-parametric area $\theta^{JAFROC\,1}$ under the AFROC′ curve, defined as the plot of LLF vs. FPF′, where FPF′ is the FPF computed over normal *and* abnormal cases, i.e. it is the number of normal and abnormal cases with highest-rated NLs rated higher than a threshold divided by the total number of cases (FPF′ can only be defined in the free-response context since in ROC abnormal images cannot yield false positives). The two JAFROC figures-of-merit are defined by

$$\left.\begin{aligned} \theta^{JAFROC} &= \frac{1}{N_N N_L}\sum_{i=1}^{N_N}\sum_{j=1}^{N_L}\psi(X_i, Y_j) \\ \theta^{JAFROC1} &= \frac{1}{(N_N + N_A)N_L}\sum_{i=1}^{N_N+N_A}\sum_{j=1}^{N_L}\psi(X_i, Y_j) \\ \psi(X, Y) &= \begin{bmatrix} 1.0 \; \textit{if}\; Y > X \\ 0.5 \; \textit{if}\; Y = X \\ 0.0 \; \textit{if}\; Y < X \end{bmatrix} \end{aligned}\right\} \tag{16.7}$$

Here X_i is the highest noise rating for case i, Y_j is the rating for the jth lesion, N_N is the number of normal cases, N_A is the number of abnormal cases, and N_L is the number of lesions. The ψ function involves comparisons between lesion ratings and the highest noise ratings. Unmarked lesions and images with no NL marks are formally assigned the "negative infinity" lesion and highest noise rating, respectively (in the software "negative infinity" is -2000).

If θ_{ij} is the figure-of-merit for modality-i and reader-j when all cases are included, and $\theta_{ij(k)}$ is the corresponding figure-of-merit when case-k is deleted, and N_T is the total number of cases, the pseudovalue (PV_{ijk}) for modality-i, reader-j, and case-k is defined by

$$PV_{ijk} = N_T\theta_{ij} - (N_T - 1)\theta_{ij(k)} \tag{16.7}$$

The significance testing procedure is identical to DBM-MRMC. The jackknife method makes no assumptions regarding the correlations of the ratings on the jackknifed image. Each jackknifed image is represented by a single pseudovalue, not a multiplicity of mark-rating pairs. This is in the same spirit as an approach (Rutter, 2000) suggested for analyzing ROI data where one has a similar issue with correlations between multiple ratings on an image. The free-response pseudovalue matrix has the same structure as the ROC pseudovalue matrix used in DBM analysis (Dorfman, 1992) where each image is also represented by a single pseudovalue. Therefore the free-response pseudovalue matrix can be analyzed by DBM-MRMC-ANOVA yielding the p-value for rejecting the NH that the reader-averaged θ are identical for all modalities.

JAFROC has been extensively validated using a different MRMC simulator that included intra- and inter-image correlations (Chakraborty, 2004). *The simulation testing confirmed that the method had the correct NH behavior even in the presence of strong intra-image correlations.* That simulator assumed a constant total number of suspicious regions per image and hence did not account for the observed randomness of these numbers. It also assumed that all lesions were decision sites (implying $\nu = 1$) which is generally not the case. Recently JAFROC has been re-validated using a search-model-based simulator which accounts for these factors (Chakraborty, 2008c). JAFROC has been acknowledged (Dodd, 2004; Wagner, 2007) to be a rigorous method for analyzing free-response data. JAFROC software reports the p-value and degrees of freedom for rejecting the NH, the individual reader figures-of-merit θ, the difference $\Delta\theta$ averaged over all readers between pairs of modalities, and a 95% confidence interval for $\Delta\theta$. If for a particular modality pairing the 95% confidence interval does not include zero, the NH that the modalities are identical is rejected at the 5% level. The method has been used in a number of studies (Penedo, 2005; Zheng, 2005; Volokh, 2006; Brennan, 2007; Ruschin, 2007; Svahn, 2007; Vikgren, 2008).

Users of JAFROC are sometimes surprised that a modality (for example, A) on which the observer marks *some* of the lesions without marking any normal image does not yield a perfect figure-of-merit ($\theta = 1$) and conversely a modality (for example, B) on which the observer marks *some* of the normal images and does not mark any of the lesions does not yield a zero figure-of-merit ($\theta = 0$). In fact it is observed that $0 < \theta_B < \theta_A < 1$. The observer who marks only lesions is obviously better than the observer who marks only normal images. If the observer marked *every* lesion and did not mark any normal images, the figure-of-merit would be unity and if the observer marked *every* normal image and did not mark any lesions, the figure-of-merit would be zero.

If the observer does not provide operating points adequately straddling the AFROC curve, the non-parametric area under the AFROC may be unreliable. The current software (JAFROC1_V1.1) detects this condition and issues a warning. Until the method gains broad acceptance, it is good practice to also present results of ROC analysis and comment on whether the study conclusions change appreciably. The software produces an ROC-equivalent (highest rating) data file in the correct LABMRMC format (*.lrc) for analysis by DBM-MRMC V2.2.

We previously concluded (Chakraborty, 2004) that JAFROC1 had incorrect NH behavior. However, recent (Chakraborty, 2008c) retesting of JAFROC1 showed that the earlier conclusion was incorrect. Since the new results became available at the galley stage of writing this chapter, we summarize them in this chapter and refer the interested reader to (Chakraborty, 2008c) for more details.

16.5.5 IDCA analysis

The original IDCA paper described how to *fit* free-response data (Edwards, 2002), not how to *compare* modalities using the method. The IDCA fitting method can be used to compute a figure-of-merit and this together with a method for significance testing defines an IDCA analysis. The maximum likelihood procedure used in IDCA to fit the pseudo-ROC curve requires the data to be binned. For human observers the number of bins was set to be ≤ 6 and the pseudo-ROC curve was fitted by ROCFIT. For CAD the number of bins was < 20 and the pseudo-ROC curve was fitted by ROCFIT.

As regards choice of the figure-of-merit, the area under the pseudo-ROC curve is a very poor choice because it totally ignores search performance. It is a measure of *classifier performance* on regions that have already been identified. For example, the area under the pseudo-ROC curve could be unity, corresponding to perfect classifier performance, i.e. all initial detections are correctly classified as lesions or non-lesions, but if ν is small, few lesions are being found and if λ is large, most of the identified regions are NLs, and free-response performance will be poor as the FROC curve will tend to "hug" the x-axis. Both classifier (the second stage of the KN model) and search (the first stage) performance need to be accounted for in a good figure-of-merit.

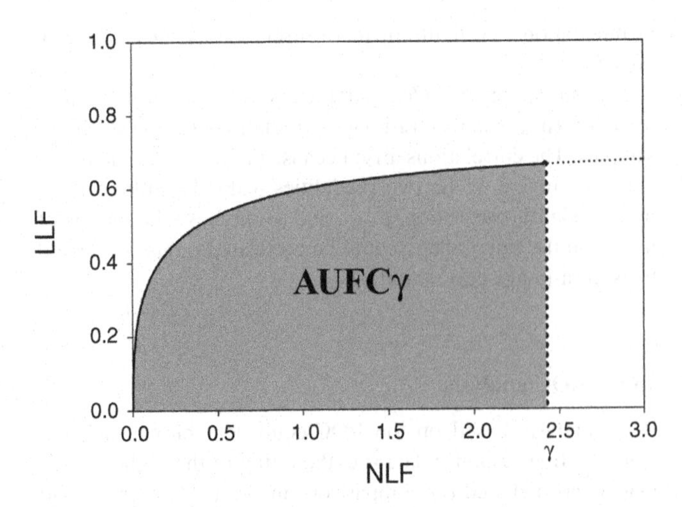

Figure 16.9 The partial area $AUFC_\gamma$ under the FROC curve to the left of a specified value of γ is an FROC curve-based figure of merit that was used for the IDCA and NP methods.

A better figure-of-merit (Samuelson, 2006) is the partial area $AUFC_\gamma$ under the FROC curve to the left of $NLF = \gamma$, illustrated in Figure 16.9. It can be calculated by numerical integration of the IDCA fitted curve. Significance testing was performed using the jackknife-pseudovalue DBM-MRMC-ANOVA method.

16.5.6 Non-parametric analysis

In non-parametric (NP) analysis, $AUFC_\gamma$ is estimated by connecting adjacent points with straight lines and calculating the trapezoidal area under the curve. This is illustrated in Figure 16.10, which shows four operating points, as might be the case with a human observer, and the dotted curve is the true curve. The upper limit of integration $NLF = \gamma$ must be to the left of the last operating point and a linearly interpolated part of the area between the last two operating points needs to be included in the integral. For CAD data the FROC curve operating points are closely spaced and the trapezoidal method estimate introduces minimal error. However, for human observers, as is evident in Figure 16.10, the trapezoidal area underestimates the true area. Sampling-related variability in the placement of the operating points will induce further variability.

Significance testing was performed based on a resampling technique known as the bootstrap. In the bootstrap method, for each simulated dataset, 2000 bootstrap samples were generated (i.e. resampled with replacement) and the distribution of the figures-of-merit (not pseudovalues) was percentiled to determine two cut-points c1 and c2 such that 2.5% of the differences are below c1 and 2.5% of the differences are above c2. If the 95% confidence interval for the difference, namely (c_1, c_2), did not include zero, the NH was rejected. The procedure was repeated for 2000 simulated datasets to obtain the average NH rejection rate.

16.5.7 Choice of γ

It will be shown in the next section that *for maximum statistical power it is desirable to choose γ as large as possible*. However,

Table 16.2 *This table shows null hypothesis rejection rates for the human observer for different choices of the lowest cutoff ζ_1 and correlations $\rho = \rho_{inter} = \rho_{intra}$; γ is the upper limit of integration for the $AUFC_\gamma$ figure-of-merit, which applies to IDCA and NP methods only. The search-model predicted AUCs for both modalities was 0.80. The human observer generated about 1.3 noise-sites per image. The last rate is the average over all conditions. Note that all methods had good NH behavior provided the appropriate method is used for significance testing. The jackknife significance testing method was used for all methods except NP, where the bootstrap method was used.*

ζ_1 and γ	ρ	ROC	JAFROC	IDCA@γ	NP@γ
$\zeta_1 = -\infty$	0.1	0.0605	0.0475	0.0500	0.0625
$\gamma = 0.867$	0.5	0.0565	0.0500	0.0565	0.0660
	0.9	0.0560	0.0505	0.0640	0.0555
$\zeta_1 = -0.674$	0.1	0.0435	0.0490	0.0455	0.0465
$\gamma = 0.650$	0.5	0.0470	0.0380	0.0505	0.0555
	0.9	0.0495	0.0505	0.0540	0.0485
$\zeta_1 = 0.0$	0.1	0.0525	0.0555	0.0480	0.0455
$\gamma = 0.433$	0.5	0.0560	0.0435	0.0415	0.0620
	0.9	0.0495	0.0620	0.0505	0.0490
$\zeta_1 = 0.674$	0.1	0.0520	0.0525	0.0605	0.0480
$\gamma = 0.217$	0.5	0.0435	0.0540	0.0555	0.0575
	0.9	0.0495	0.0440	0.0535	0.0470
AVG		0.0513	0.0498	0.0525	0.0536

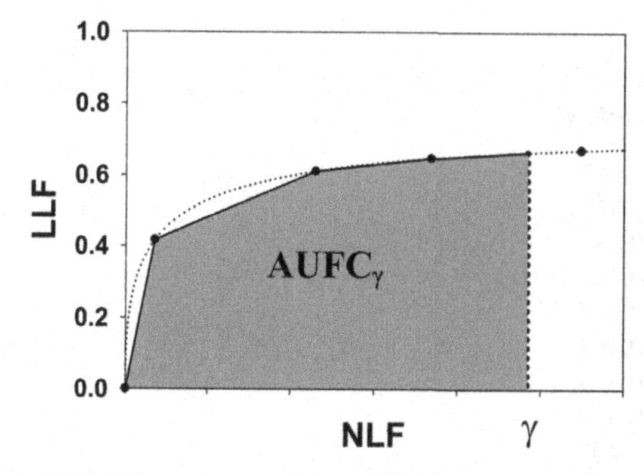

Figure 16.10 The non-parametric area $AUFC_\gamma$ under a human observer FROC curve with four operating points. The "jerkiness" of the figures-of-merit when individual cases are removed causes problems for the jackknife significance testing method (NH behavior is unacceptable) that are resolved by using the bootstrap.

if γ is chosen larger than the observed end-point then $AUFC_\gamma$ cannot be calculated. For a fair comparison the same γ must be chosen for both modalities, as otherwise one will introduce bias in favor of the modality with the larger γ. Choosing a large common value for γ is relatively easy with CAD algorithms since they generate many NLs but is more difficult with human observers since conservative observers may not provide appreciable numbers of NLs. In that case the maximum common value of γ would be limited by the NLF of the observer who generates the least number of NLs, and the data to the right of this value provided by other observers has to be excluded from the analysis. Since the true curve ends at NLF $= \lambda$ the maximum value of γ is determined by the modality with the smaller λ, i.e. λ_2 in the present case. Including the effect of the cutoff ζ_1, the maximum possible choice for γ is $\lambda_2\Phi(-\zeta_1)$ (see Eqn. 16.3). Due to sampling considerations γ has to be somewhat smaller as otherwise sometimes the observed end-point NLF may be less than $\lambda_2\Phi(-\zeta_1)$ in which case integration to $\lambda_2\Phi(-\zeta_1)$ would fail. It was found empirically that $\gamma = \lambda_2\Phi(-\zeta_1)/1.2$ was sufficiently small (but not too small) to ensure that the end-point abscissa almost always was larger than this value.

16.6 SIMULATION STUDY

16.6.1 NH behavior

This section summarizes results from a recently conducted study (Chakraborty, 2008c) of NH validity and statistical power for the different methods for analyzing human observer and CAD free-response data. The study was for a single observer interpreting the same images in two modalities and the corresponding data simulator is described in the cited paper. For the human observer, NH rejection rates of ROC, JAFROC, IDCA, and NP are summarized in Table 16.2 for different choices of lowest cutoff ζ_1 and correlations ($\rho = \rho_{inter} = \rho_{intra}$). The simulation conditions were 100 normal images, 100 abnormal images, and 1 lesion per abnormal image, henceforth abbreviated 100/100/1. The rejection rates, averaged over all conditions, were 0.0513, 0.0498, 0.0525, and 0.0536 for the four methods. *For the human observer data, use of the jackknife for the NP method led to unacceptable NH rejection rates.* The rejection rate range was (0.0025, 0.0920). Table 16.3 shows NH rejection rates for the CAD observer. The rejection rates, averaged over all conditions, were 0.0515, 0.0476, 0.0505, and 0.0539 for the four methods. *For the CAD observer, use of the jackknife for the NP method also led to unacceptable NH rejection rates.* The rejection rate range was (0.0310, 0.1035). JAFROC1, not shown in the tables,

Table 16.3 *This table shows null hypothesis rejection rates for the CAD observer. The CAD observer generated about ten noise-sites per image. Other conditions are as in Table 16.2. Note that all methods had good NH behavior.*

ζ_1 and γ	ρ	ROC	JAFROC	IDCA@γ	NP@γ
$\zeta_1 = -\infty$	0.1	0.0545	0.0455	0.0495	0.0590
$\gamma = 6.67$	0.5	0.0590	0.0495	0.0505	0.0630
	0.9	0.0355	0.0455	0.0460	0.0525
$\zeta_1 = -0.674$	0.1	0.0475	0.0485	0.0450	0.0500
$\gamma = 5$	0.5	0.0590	0.0550	0.0550	0.0450
	0.9	0.0550	0.0400	0.0560	0.0600
$\zeta_1 = 0.0$	0.1	0.0600	0.0525	0.0520	0.0580
$\gamma = 3.33$	0.5	0.0470	0.0485	0.0425	0.0595
	0.9	0.0485	0.0485	0.0550	0.0595
$\zeta_1 = 0.674$	0.1	0.0480	0.0430	0.0470	0.0515
$\gamma = 1.67$	0.5	0.0560	0.0445	0.0590	0.0440
	0.9	0.0480	0.0500	0.0485	0.0445
AVG		0.0515	0.0476	0.0505	0.0539

Table 16.4 *Listed are the statistical powers for the human observer for an effect size $\Delta(AUC) = 0.05$. Other conditions are as in Table 16.2. The bottom half of the table is expected to be most relevant to human observers. Note that the ROC method has the least power, while the others had comparable powers when averaged over all conditions but in the relevant region the JAFROC method had the highest power.*

ζ_1 and γ	ρ	ROC	JAFROC	IDCA@γ	NP@γ
$\zeta_1 = -\infty$	0.1	0.231	0.456	0.452	0.417
$\gamma = 0.867$	0.5	0.252	0.443	0.490	0.441
	0.9	0.272	0.475	0.523	0.482
$\zeta_1 = -0.674$	0.1	0.234	0.414	0.415	0.403
$\gamma = 0.650$	0.5	0.248	0.421	0.460	0.417
	0.9	0.283	0.461	0.516	0.467
$\zeta_1 = 0.0$	0.1	0.224	0.384	0.349	0.319
$\gamma = 0.433$	0.5	0.242	0.376	0.393	0.390
	0.9	0.282	0.409	0.417	0.402
$\zeta_1 = 0.674$	0.1	0.183	0.287	0.205	0.210
$\gamma = 0.217$	0.5	0.161	0.299	0.249	0.251
	0.9	0.161	0.361	0.284	0.299
Average (ALL)		0.231	0.399	0.396	0.375
Average (relevant)		0.209	0.353	0.316	0.312

was found to have satisfactory NH behavior for both human and CAD observers.

16.6.2 AH behavior

Under the AH condition, $AUC_2 = 0.85$ for both classes of observers, i.e. the effect size was 0.05 AUC units. Tables 16.4 and 16.5 list the statistical powers of the four methods for the human observer and the CAD observer, respectively. For the human observer $\zeta_1 = -\infty$ is not relevant as radiologists do not mark extremely low-confidence-level suspicious regions. Based on FROC datasets available to us they mark fewer than 0.5 NL locations per image so $\zeta_1 > 0$ is a reasonable choice for the relevant region. For CAD the opposite is true. In each table the next-to-last row is the statistical power averaged over all conditions of ζ_1 and ρ and the last row is the average over the *relevant* portion of the table, i.e. the top half of the table ($\zeta_1 = -\infty$ and $\zeta_1 = -0.674$) for the CAD observer and the bottom half ($\zeta_1 = 0.433$ and $\zeta_1 = 0.674$) for the human observer.

For the human observer, Table 16.4, the statistical powers, averaged over the relevant portion of the table, for the ROC, JAFROC, IDCA, and NP methods were 0.209, 0.353, 0.316, and 0.312. The ordering was JAFROC > (IDCA ~ NP) > ROC. The corresponding values averaged over the entire table were 0.231, 0.399, 0.396, and 0.375, respectively, and the ordering was (JAFROC ~ IDCA ~ NP) > ROC. For the CAD observer,

Table 16.5 *Listed are the statistical powers for the CAD observer for an effect size $\Delta(AUC) = 0.05$. Other conditions are as in Table 16.3. The bottom half of the table is expected to be most relevant to CAD. Note that IDCA and the NP methods had similar powers, followed by JAFROC, and ROC had the least power.*

ζ_1 and γ	ρ	ROC	JAFROC	IDCA@	NP@
$\zeta_1 = -\infty$	0.1	0.251	0.536	0.815	0.847
$\gamma = 6.67$	0.5	0.410	0.721	0.867	0.902
	0.9	0.570	0.886	0.934	0.954
$\zeta_1 = -0.674$	0.1	0.244	0.524	0.801	0.799
$\gamma = 5$	0.5	0.366	0.737	0.858	0.859
	0.9	0.584	0.882	0.924	0.945
$\zeta_1 = 0.0$	0.1	0.239	0.566	0.700	0.712
$\gamma = 3.33$	0.5	0.375	0.695	0.787	0.807
	0.9	0.616	0.854	0.919	0.952
$\zeta_1 = 0.674$	0.1	0.233	0.491	0.554	0.530
$\gamma = 1.67$	0.5	0.380	0.651	0.666	0.629
	0.9	0.574	0.795	0.835	0.853
Average (ALL)		0.403	0.695	0.805	0.816
Average (relevant)		0.404	0.714	0.866	0.884

Table 16.5, the values for the ROC, JAFROC, IDCA, and NP methods averaged over the relevant portion were 0.404, 0.714, 0.866, and 0.884, respectively, with ordering (NP $\sim$ IDCA) > JAFROC > ROC. When averaged over the entire table the values were 0.403, 0.695, 0.805, and 0.816, respectively, with the same ordering. Including JAFROC1, not shown in the tables, for the human observer the ordering was JAFROC1 > JAFROC > (IDCA $\sim$ NP) > ROC in the relevant region, and the corresponding ordering for the CAD observer was (NP $\sim$ IDCA) > (JAFROC1 $\sim$ JAFROC) > ROC. JAFROC1 has higher power than JAFROC because it uses all highest rated NL marks while JAFROC only uses the highest rated NL mark on normal cases. The power loss is large when the relative number of normal cases is small.

Figure 16.11 shows that for the CAD observer *as γ increases the statistical power of IDCA increases by about a factor of three over the indicated range.* The reason for this behavior should be fairly obvious: with larger γ more data, i.e. mark-rating pairs, are included in the analysis, resulting in a more stable estimate of the figure-of-merit and more sensitivity at detecting differences in figures-of-merit. Therefore, as noted earlier, for maximum statistical power it is desirable to include as much of the FROC curve as possible in the range of integration.

For either human observer or CAD, the ROC method had the least statistical power. Table 16.4 and Table 16.5 show the substantial power penalty of using the ROC method to analyze a localization task. For more details of this study, including the results for JAFROC1, we refer the reader to (Chakraborty, 2008c).

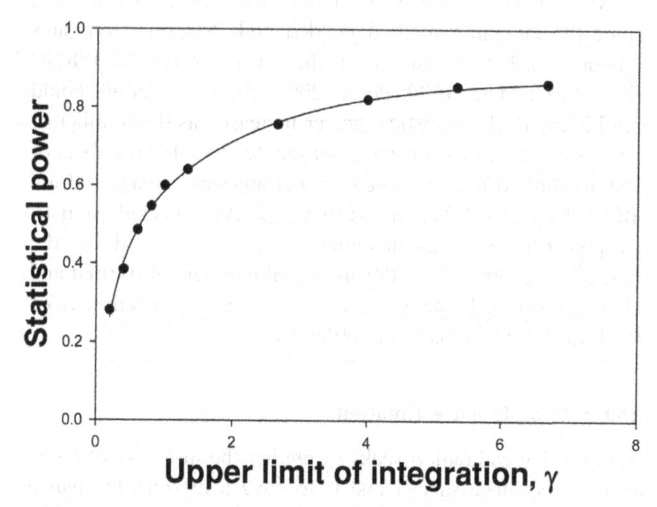

Figure 16.11 This figure shows that statistical power of IDCA increases with γ, the upper limit of integration in the $AUFC_\gamma$ figure-of-merit. A similar effect is observed for the non-parametric method. For maximum statistical power it is desirable to include as much of the FROC curve as possible in the range of integration.

16.6.3 JAFROC vs. IDCA and NP

For the human observer (see Table 16.4), JAFROC, IDCA, and NP had similar statistical powers, except at the highest cutoff value where JAFROC had more power. It was initially surprising that JAFROC was even competitive with IDCA/NP, let alone surpassing them in some cases. Since JAFROC ignores the highest rated NL mark on abnormal images and uses only the highest-rated NL mark on normal images, whereas IDCA/NP use all the marks contributing to the FROC curve below γ, one expects IDCA/NP to have greater power. However, in JAFROC *the mark that is used is a special one*, as it is the highest-rated mark, and contains more information about the distribution than a randomly selected mark. Another reason is that unlike JAFROC, $AUFC_\gamma$ is insensitive to the *distribution* of NL marks between the images. For example, all NL marks on normal images could occur on a single normal image without affecting the FROC curve or $AUFC_\gamma$, but θ^{JAFROC} would increase sharply since the observer is perfect on all but one normal image (a normal image with no marks is a perfect response) and is terrible on the one normal image with all NL marks. A third reason is that JAFROC uses the entire area under the AFROC as the figure-of-merit whereas IDCA/NP uses a partial area figure-of-merit. Partial area measures sacrifice statistical power and the effect is expected to be particularly pronounced at $\zeta_1 = 0.674$ where only 25% of the noise-sites are marked.

For the CAD observer (see Table 16.5) there are much fewer unmarked normal images and almost all marks are used in the relevant range; IDCA and NP win over JAFROC. Since the operating points on the FROC curve are closely spaced, there is no need to perform a parametric fit as the NP area is a good approximation to the true area and avoids the need for any assumptions. This is why, for designer-level CAD data, the NP method may be desirable over IDCA.

16.6.4 Other simulation conditions

(a) In preliminary studies the contaminated binormal model (CBM) fitting method (Dorfman, 2000a; Dorfman, 2000b; Dorfman, 2000c) did not yield appreciably higher statistical power than ROCFIT for ROC analysis. It was also found that with CBM, one had to use bootstrapping for significance testing, since the jackknife method yielded high NH rejection rates (about 8%). For reasons of a technical nature the PROPROC (Pan, 1997; Metz, 1999; Pesce, 2007) fitting procedure could not be tested. (b) Statistical power increased as the number of lesions per abnormal image increased. (c) Variable numbers of lesions (one to five; average two) per abnormal image did not affect the NH validity of JAFROC. (d) As expected, statistical power increased as the number of cases increased. For the human observer at $\zeta_1 = 0.0$ and equal numbers of normal and abnormal cases, the statistical power of JAFROC increased from 0.14 for 25/25/1 to 0.68 for 200/200/1.

16.6.5 Sample size estimation

Sample size estimation seeks to predict the numbers of cases and readers necessary if one is to have a reasonable chance (i.e. statistical power) of detecting (i.e. rejecting the NH) a specified difference (i.e. effect size) in performance between two modalities. With the high cost of conducting observer performance studies, sample size estimation plays an important role at the planning stage as it can be used to ensure that available resources are used optimally. Hillis and Berbaum (HB) have described a sample size estimation procedure for the ROC paradigm (Hillis, 2004). The procedure uses the DBM-MRMC-ANOVA calculated pseudovalue variance-components which are input to SAS software that can be downloaded from the University of Iowa website (http://perception.radiology.uiowa.edu/Software/ReceiverOperatingCharacteristicROC/DBMMRMC/tabid/116/ctl/Login/Default.aspx). Since the underlying figure-of-merit variance-components models are the same, the HB sample-size estimation procedures should work with FROC data (although the z-sampling models are very different, the FOM models are the same; i.e. in spite of the complication introduced by the multiple marks and ratings, it remains true that each dataset yields a single FOM). This is another example of a recurring theme of this chapter, namely FROC and ROC analyses are very similar: the only essential difference is the figure-of-merit, and therefore tools developed for ROC analysis are often applicable to FROC analysis. With appropriate substitution of the figure-of-merit, the HB method is also expected to work with the LROC and ROI paradigms. Additionally, an alternate method developed for sample size estimation for ROC studies (Obuchowski, 1994; Obuchowski, 1995a) is also applicable to FROC/LROC/ROI studies.

16.7 THE VALIDITY OF THE MODEL ASSUMPTIONS

Since the resampling-based significance testing methods are quite general, the model assumptions are only relevant to the design of the simulator used for testing the analyses and for calculating parametric figures-of-merit. The search-model simulator assumes independent and normally distributed z-samples, and Poisson and binomial sampling for the numbers of noise- and signal-sites, respectively. These assumptions can be relaxed and alternate distributions can be used, e.g. instead of normal distributions one could use Laplace of logistic distributions, and instead of Poisson and binomial one could use any non-negative integer value distribution. Simulators with these changes can be useful to establish robustness of the analysis. Regarding the parametric figures-of-merit, note that the assumptions are only relevant for the analysis of human observer data. For CAD data the NP method is preferable as it has similar power to IDCA (see Table 16.5) and none of the IDCA/SM parametric assumptions are needed.

16.7.1 Poisson and binomial assumptions

The shape of the IDCA (or search-model) predicted FROC curve is independent of the Poisson or binomial assumptions (see Section 16.4.2.3). The assumptions are convenient for *fitting* a curve to observed data points, since closed-form expressions for the likelihood function are relatively easy to calculate using symbolic mathematical software. Given the sparse nature of human observer data, the parametric fits could be regarded as convenient interpolation tools that allow parametric figures-of-merit to be readily calculated. Earlier models that allow the end-point to go to $(\infty, 1)$ gave good fits to human observer FROC data and it was only when they were applied to CAD data that their shortcomings became apparent (Chakraborty, 2008b).

16.7.2 Gaussian assumption

The shape of the FROC curve depends on the Gaussian assumptions (see Eqn. 16.3). For sparse human observer data, failure of the normality assumptions may not be as bad as may appear at first sight. In the ROC context, it has been shown (Hanley, 1988) that for the relatively small datasets such as are usually encountered in medical imaging, the ROC binormal model fits are quite robust to deviations from normality. Since interest is in determining if two modalities are different, systematic bias due to lack of fit may be expected to cancel out to first order,

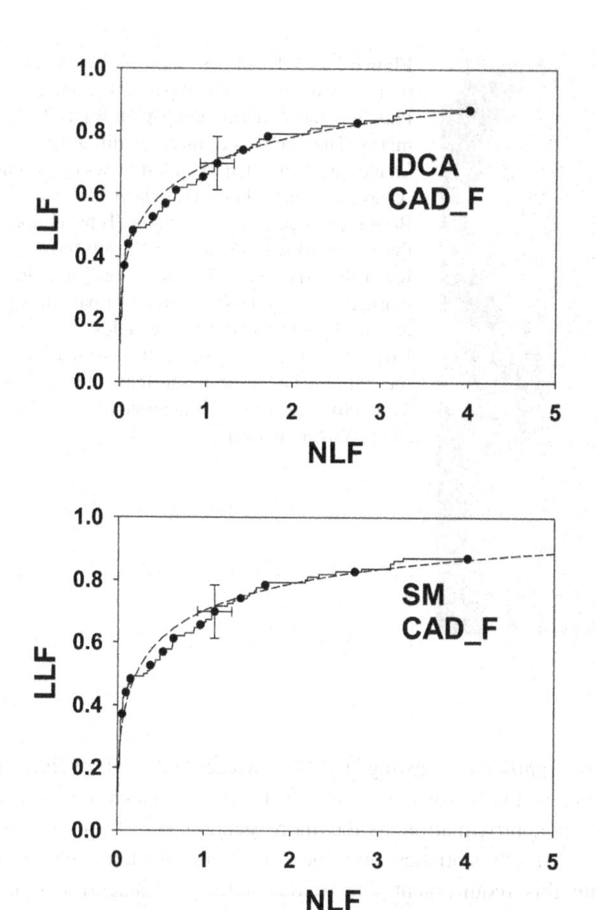

Figure 16.12 This figure illustrates the effect of failure of the normality assumption. IDCA (upper panel) and search-model (lower panel) fits to a CAD dataset. Both model fits show significant deviation from the raw data, due to failure of the normality assumption, and will lead to errors in estimating $AUFC_\gamma$. For this reason the parametric methods may not be suitable for the analysis of designer-level CAD data. Used with permission from Yoon *et al.* (Yoon, 2007).

which could further dilute the effect of failure of the normality assumptions. In preliminary testing with Laplace and logistic distributions instead of Gaussian, unacceptable NH behavior has not been observed.

For CAD, the raw data gives a detailed empirical curve and effects of deviations from normality are more readily visible. Figure 16.12 is an example of apparent failure of the normality assumption (Yoon, 2007). The upper and lower panels show IDCA and search-model fits to a CAD dataset. Both model fits deviate from the raw data, most likely due to failure of the normality assumption which cannot account for the quasi-linear part of the data, and this will lead to errors in estimating the area under the FROC curve. Although NH behavior may not be compromised, the NP method is a safer choice for CAD data.

16.7.3 Correlation of the ratings

The results reported in this chapter were obtained with deliberately introduced correlations ranging from weak (0.1) to strong (0.9). Since data are resampled at the case level, one expects and observes correct NH behavior even in the presence of strong correlations. Therefore, as suggested earlier (Yoon, 2007), for a human observer a parametric figure-of-merit, for example $AUFC_\gamma$, can be used provided a resampling method is used for significance testing.

16.7.4 Multiple views per case

So far it was implicitly assumed that each image corresponds to a different case (patient) so that the number of images and cases are equal. In mammography there are four views per case: two views per breast and two breasts per case. With multiple views per case it is possible to regard all the images associated with the case as one "big" image and analyze the mark-rating pairs from such "big" images using any of the methods described previously. Provided resampling is done at the case level, this is expected to have the correct NH behavior.

For two-dimensional projection images of the same anatomy, e.g. the two views of the breast, it is not clear how to score the marks. Should LL marks on the two views near the same physical lesion be counted as two LLs or as one LL? Is the effective number of lesions two or one? This problem was encountered by CAD developers, which led them to *case-based* and *view-based* methods for plotting FROC curves (Yoon, 2007). For either method the calculation of NLF is the same – the denominator is always the total number of views. The two methods essentially employ different definitions of the total number of lesions and what constitutes lesion localization. When multiple LLs occur on a case the highest rated one is used. This is expected to decrease statistical power since the other LLs are not used to calculate the figure-of-merit. In the case-based method the denominator for LLF is the number of abnormal cases. In the view-based method the denominator for LLF is the total number of lesions visible to the truth panel, i.e. a lesion visible in two views counts as two lesions. In scoring the marks each LL is counted individually. Both IDCA and the SM can fit FROC curves for either case-based or view-based methods provided appropriate values for NL and LL counts and total number of lesions are supplied (Yoon, 2007). With patient-level resampling for significance testing, NH behavior should be satisfactory with either method.

16.7.5 Correlation of the numbers of decision sites

While intra-case (between-view) correlations of the *ratings* are considered in the simulator, correlation of the *numbers* of decision sites between views, e.g. correlation of the number of noise-sites on a CC view with the number of noise-sites on the corresponding MLO view, was not considered in the current simulator. Figure 16.13 shows CAD marks on four-view displays for two patients A and B. Patient A has 6 marks, and patient B has 28 marks. The numbers of marks in different images are correlated: both left views of patient A have one mark and both right views have two marks. Likewise, both right breast views of patient B have more than six marks and both left breast views have fewer than six marks. The ratings given to the marks on the left breast of patient A (not shown) were similar because they represent the same lesion. Likewise the ratings given to the two marks on the right breast were similar as they refer to the same physical regions. Because of the nature of the case-based

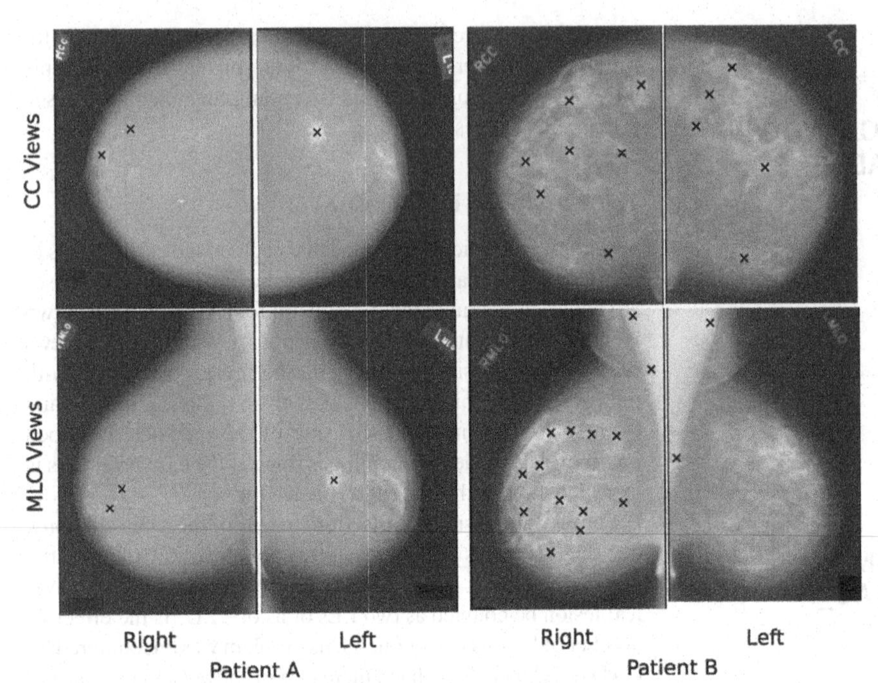

Figure 16.13 This figure shows CAD marks on four-view displays for two patients A and B. Patient A has 6 marks, and patient B has 28 marks. The numbers of marks in different images are correlated: both left views of patient A have one mark, both right views have two. Both right breast views of patient B have more than six marks. Both views of the left have fewer than six marks. The ratings given to the marked lesions on the left breast of patient A are similar because it is the same lesion. Likewise the ratings given to the two marks on the right breast are similar as they refer to the same physical region. (Images courtesy of Dr. Frank W. Samuelson.)

resampling, NH behavior is unlikely to be adversely affected by these correlations, but statistical power is expected to be affected. A more realistic simulator that includes these correlations is needed.

16.8 DISCUSSION

In this chapter recent advances in free-response modeling and analyses have been reviewed. While the free-response paradigm is more than four decades old, practical methods of analyzing the data, namely JAFROC, JAFROC1, IDCA, and NP, have only recently become available. The methods differ essentially in the figure-of-merit and how significance testing is done. For JAFROC the figure-of-merit is the NP estimate of the area under the AFROC curve using only normal cases to calculate FPF, for JAFROC1 non-lesion localizations on all cases are used to calculate FPF, while IDCA and NP use parametric and non-parametric estimates, respectively, of $AUFC_\gamma$, the area under the FROC curve to the left of $NLF = \gamma$. Except for NP, where one must use bootstrapping, the rest can use the DBM-MRMC jackknife method for significance testing. In all cases the resampling is done at the patient (case) level, so that when a case is jackknifed or bootstrapped, all marks related to the case are removed from the analysis. The methods make no assumptions regarding the intra-case correlations of the numbers and ratings of marks on the same case.

Results are presented of a simulation study comparing the statistical validity and statistical powers of ROC, JAFROC, JAFROC1, IDCA, and NP for a typical human observer and a typical CAD algorithm. The bootstrap method for significance testing worked (i.e. yielded acceptable NH behavior) in all cases but the computationally less intensive jackknife method was adequate for ROC, JAFROC, JAFROC1, and IDCA. For NP analysis use of the jackknife method for significance testing led to unacceptable NH behavior. This is likely related to the fact that the jackknife, which is an approximation to the more general bootstrap, is valid for smooth statistics and the NP figure-of-merit may violate this requirement due to the staircase ("jagged") nature of the FROC curve, particularly for human observers. For human observer datasets with few marks (less than 0.65 marks per image) the ordering of the methods was JAFROC1 $>$ JAFROC $>$ (IDCA $\sim$ NP) $>$ ROC. For CAD in the relevant range (more than five marks per image), the ordering of the methods was (NP $\sim$ IDCA) $>$ (JAFROC1 $\sim$ JAFROC) $>$ ROC. Although IDCA had comparable power to NP for CAD, the NP method may be preferable because it does not make any distributional assumptions. In all cases ROC had the least statistical power. For diagnostic tasks that do not involve localization, e.g. classification tasks, ROC methods are appropriate, but for localization tasks usage of ROC entails a substantial power penalty and is not recommended. Most simulations were conducted with 100 normal, 100 abnormal cases, and 1 lesion per abnormal case. Given the number of parameters that can be changed (e.g. AUC, ratio of normal to abnormal cases, total number of cases, average number of lesions per case, etc.), a comprehensive report is outside the scope of this chapter.

Free-response performance combines search expertise and decision-making expertise. Search expertise is the ability to label lesions as suspicious, while not labeling normal regions as suspicious. Decision-making refers to making the right decision (to report or not to report) at each labeled region. Much progress has been made in modeling decision-making expertise, as evidenced by the extensive literature on model-observer methodologies (Barrett, 2003). These generally presuppose a location-known-exactly scenario, which ignores search expertise. Search expertise is an important component of radiologist expertise (see Figure 16.4). Radiologists hone their search expertise by reading thousands of images, and this aspect of expertise should not be ignored in the modeling. It is highly desirable, in my opinion, to extend model-observer research to include search.

The search-model described in this chapter is relatively simple. As noted earlier it is desirable to extend it to multiple views per case. A more sophisticated model could help us understand effects such as satisfaction-of-search (Berbaum, 1990), where finding a lesion can affect the probability of finding other lesions in the image. For example, in Figure 16.4 the expert mammographer did not fixate the lesion in the CC view, presumably because the same lesion was "convincingly discovered" in the MLO view, or perhaps because the CC view lesion was seen using peripheral vision, which is not measured by eye-trackers. The effect of changing the proximity criterion needs to be studied. These are just a sample of issues that make free-response research interesting.

As expected with any new paradigm, free-response methodology has generated healthy debate. ROC is the established gold standard and challenges to a gold standard need to be examined carefully. We refer the interested reader to two recent papers, one (Gur, 2008) citing issues with the free-response paradigm and the other (Chakraborty, 2009a) a counterpoint responding to these issues. One of the issues raised is instructive to examine here. It is an argument I have heard several times that for the screening mammography, task localization "does not make a difference to the final clinical outcome" and therefore the ROC method is appropriate. Here is the counter-argument. Suppose that during screening mammography the radiologist reports a false positive region on an image that actually has a lesion but does not mark the lesion. The patient would be referred for a diagnostic workup during which the true lesion may be incidentally identified, perhaps with high probability, in which case appropriate clinical action would be taken. For this patient the two incorrect decisions did not make a difference to the final clinical outcome. The fallacy is that the radiologist (or imaging equipment) that is prone to this kind of error (rating normal regions higher than lesions on abnormal images) will also tend to generate high-rated NLs on *normal* images. NLs on normal images result in unnecessary recalls and cost to the healthcare system and emotional suffering for the patient. To pose a rhetorical question, given a choice between two modalities, on one of which the radiologist is "right for the right reason," and on the other where the radiologist is "right for the wrong reason," which modality would regulatory agencies, insurance providers, hospital administrators, radiologists, and their patients trust? I believe the free-response paradigm is highly relevant to the screening task, and indeed for any task involving lesion localization. This is not to suggest that free-response is the panacea to diagnostic imaging system assessment. Indeed Gur *et al.* (Gur, 2008) raise a number of provocative issues that need to be addressed by improvements to free-response methodology, and some recent progress has been reported (Chakraborty, 2009b).

ACKNOWLEDGEMENTS

The author is grateful to Hong-Jun Yoon, MS, for developing the software necessary for this work; to Dr. Claudia Mello-Thoms for discussions regarding the Kundel-Nodine model and for providing Figure 16.4; to Dr. Frank W. Samuelson for providing Figure 16.13, and to Dr. Bin Zheng and Dr. Berkman Sahiner for providing CAD datasets. The author is grateful to Dr. Kevin Berbaum and Dr. Steve Hillis for discussions on the analysis, to Dr. Berkman Sahiner and Dr. Harold Kundel for comments on the manuscript, and to Mr. Tony Svahn and Ms. Federica Zanca for providing FROC datasets. I thank the Editor of *Medical Physics*, Dr. William R. Hendee, for permission to reproduce some of the figures shown in this chapter. This work was supported by a grant from the Department of Health and Human Services, National Institutes of Health, R01-EB005243 and R01-EB008688.

REFERENCES

Anastasio, M.A., Kupinski, M.A., Nishikawa, R.M. (1998). Optimization and FROC analysis of rule-based detection schemes using a multiobjective approach. *IEEE Trans Med Imaging* **17** (6), 1089–1093.

Bandos, A.I., Rockette, H.E., Song, T., Gur, D. (2009). Area under the free-response ROC curve (FROC) and a related summary index. *Biometrics* **65** (1), 247–256.

Barrett, H.H., Myers, K. (2003). *Foundations of Image Science*. Hoboken, NJ: John Wiley and Sons.

Berbaum, K.S., Franken, E.A., Dorfman, D.D. *et al.* (1990). Satisfaction of search in diagnostic radiology. *Invest Radiol* **25** (2), 133–140.

Bornefalk, H. (2005a). Estimation and comparison of CAD system performance in clinical settings. *Acad Radiol* **12**, 687–694.

Bornefalk, H., Hermansson, A.B. (2005b). On the comparison of FROC curves in mammography CAD systems. *Med Phys* **32** (2), 412–417.

Brennan, P.C., McEntee, M., Evanoff, M. *et al.* (2007). Ambient lighting: effect of illumination on soft-copy viewing of radiographs of the wrist. *Am J Roentgenol* **188** (2), W177–W180.

Bunch, P.C., Hamilton, J.F., Sanderson, G.K., Simmons, A.H. (1978). A free-response approach to the measurement and characterization of radiographic-observer performance. *J Appl Photogr Eng* **4** (4), 166–171.

Burgess, A.E., Jacobson, F.L., Judy, P.F. (2001). Human observer detection experiments with mammograms and power-law noise. *Med Phys* **28** (4), 419–437.

Chakraborty, D.P. (1989). Maximum likelihood analysis of free-response receiver operating characteristic (FROC) data. *Med Phys* **16** (4), 561–568.

Chakraborty, D.P. (2006a). An alternate method for using a visual discrimination model (VDM) to optimize softcopy display image quality. *Journal of the Society for Information Display* **14** (10), 921–926.

Chakraborty, D.P. (2006b). ROC curves predicted by a model of visual search. *Phys Med Biol* **51**, 3463–3482.

Chakraborty, D.P. (2006c). A search model and figure-of-merit for observer data acquired according to the free-response paradigm. *Phys Med Biol* **51**, 3449–3462.

Chakraborty, D.P. (2007a). FROC curves using a model of visual search. *Proc. SPIE Medical Imaging: Image Perception, Observer Performance, and Technology Assessment*. San Diego, CA: SPIE.

Chakraborty, D.P. (2008c). Validation and statistical power comparison of methods for analyzing free-response observer performance studies. *Acad Radiol* **15** (12), 1554–1566.

Chakraborty, D.P. (2009a). Counterpoint to "Performance assessment of diagnostic systems under the FROC paradigm" by Gur and Rockette. *Acad Radiol* **16**, 507–510.

Chakraborty, D.P., Berbaum, K.S. (2004). Observer studies involving detection and localization: modeling, analysis and validation. *Med Phys* **31** (8), 2313–2330.

Chakraborty, D.P., Breatnach, E.S., Yester, M.V. *et al.* (1986). Digital and conventional chest imaging: a modified ROC study of observer performance using simulated nodules. *Radiology* **158**, 35–39.

Chakraborty, D.P., Winter, L.H.L. (1990). Free-response methodology: alternate analysis and a new observer-performance experiment. *Radiology* **174**, 873–881.

Chakraborty, D.P., Yoon, H.J. (2008a). Investigation of methods for analyzing location specific observer performance data. *Proc. SPIE: Biomedical Optics and Imaging*. San Diego, CA: SPIE.

Chakraborty, D.P., Yoon, H.J. (2008b). Operating characteristics predicted by models for diagnostic tasks involving lesion localization. *Med Phys* **35** (2), 435–445.

Chakraborty, D.P., Yoon, H.J. (2009b). JAFROC analysis revisited: figure-of-merit considerations for human observer studies. *Proc SPIE Med Imag* **7263**.

Chakraborty, D.P., Yoon, H.J., Mello-Thoms, C. (2007b). Spatial localization accuracy of radiologists in free-response studies: inferring perceptual FROC curves from mark-rating data. *Acad Radiol* **14**, 4–18.

Chan, H., Doi, K., Vyborny, C.J. *et al.* (1990). Improvement in radiologists' detection of clustered microcalcifications on mammograms. *Invest Radiol* **25** (10), 1102–1110.

Dodd, L.E., Wagner, R.F., Armato, S.G. *et al.* (2004). Assessment methodologies and statistical issues for computer-aided diagnosis of lung nodules in computed tomography: contemporary research topics relevant to the lung image database consortium. *Acad Radiol* **11** (4), 462–475.

Dorfman, D.D., Alf, E. (1969). Maximum-likelihood estimation of parameters of signal-detection theory and determination of confidence intervals – rating-method data. *J Math Psychol* **6**, 487–496.

Dorfman, D.D., Berbaum, K.S. (2000a). A contaminated binormal model for ROC data: part II. A formal model. *Acad Radiol* **7** (6), 427–437.

Dorfman, D.D., Berbaum, K.S. (2000b). A contaminated binormal model for ROC data: part III. Initial evaluation with detection. *Acad Radiol* **7** (6), 438–447.

Dorfman, D.D., Berbaum, K.S., Brandser, E.A. (2000c). A contaminated binormal model for ROC data: part I. Some interesting examples of binormal degeneracy. *Acad Radiol* **7** (6), 420–426.

Dorfman, D.D., Berbaum, K.S., Metz, C.E. (1992). ROC characteristic rating analysis: generalization to the population of readers and patients with the jackknife method. *Invest Radiol* **27** (9), 723–731.

Dorfman, D.D., Berbaum, K.S., Metz, C.E. *et al.* (1997). Proper receiving operating characteristic analysis: the bigamma model. *Acad Radiol* **4** (2), 138–149.

Duchowski, A.T. (2002). *Eye Tracking Methodology: Theory and Practice*. Clemson, SC: Clemson University.

Eberl, M.M., Fox, C.H., Edge, S.B., Carter, C.A., Mahoney, M.C. (2006). BI-RADS classification for management of abnormal mammograms. *J Am Board Fam Med* **19**, 161–164.

Edwards, D.C., Kupinski, M.A., Metz, C.E., Nishikawa, R.M. (2002). Maximum likelihood fitting of FROC curves under an initial-detection-and-candidate-analysis model. *Med Phys* **29** (12), 2861–2870.

Efron, B. (1982). *The Jackknife, the Bootstrap and Other Resampling Plans*. Montpelier, VT: Capital City Press.

Egan, J.P., Greenburg, G.Z., Schulman, A.I. (1961). Operating characteristics, signal detectability and the method of free response. *J Acoust Soc Am* **33**, 993–1007.

Fryback, D.G., Thornbury, J.R. (1991). The efficacy of diagnostic imaging. *Med Decis Making* **11** (2), 88–94.

Galassi, M., Davies, J., Theiler, J. *et al.* (2005). *GNU Scientific Library Reference Manual 1.6*. Bristol, UK: Network Theory Limited.

Gregory, R.L. (1970). *The Intelligent Eye*. New York, NY: McGraw-Hill.

Gur, D., Rockette, H.E. (2008). Performance assessment of diagnostic systems under the FROC paradigm: experimental, analytical, and results interpretation issues. *Acad Radiol* **15**, 1312–1315.

Hanley, J.A. (1988). The robustness of the "binormal" assumptions used in fitting ROC curves. *Med Decis Making* **8** (3), 197–203.

Hillis, S.L. (2007). A comparison of denominator degrees of freedom methods for multiple observer ROC studies. *Stat Med* **26**, 596–619.

Hillis, S.L., Berbaum, K.S. (2004). Power estimation for the Dorfman-Berbaum-Metz Method. *Acad Radiol* **11** (11), 1260–1273.

Hillis, S.L., Berbaum, K.S. (2005a). Monte Carlo validation of the Dorfman-Berbaum-Metz method using normalized pseudovalues and less data-based model simplification. *Acad Radiol* **12**, 1534–1541.

Hillis, S.L., Berbaum, K.S., Metz, C.E. (2008). Recent developments in the Dorfman-Berbaum-Metz procedure for multireader ROC study analysis. *Acad Radiol* **15** (5), 647–661.

Hillis, S.L., Obuchowski, N.A., Schartz, K.M., Berbaum, K.S. (2005b). A comparison of the Dorfman-Berbaum-Metz and Obuchowski-Rockette methods for receiver operating characteristic (ROC) data. *Stat Med* **24**, 1579–1607.

Hillstrom, A. (2000). Repetition effects in visual search. *Percept Psychophys* **2**, 800–817.

Hutchinson, T.P. (2007). Free-response operator characteristic models for visual search. *Phys Med Biol* **52**, L1–L3.

Irvine, J.M. (2004). Assessing target search performance: the free-response operator characteristic model. *Opt Eng* **43** (12), 2926–2934.

Kobayashi, T., Xu, X.W., MacMahon, H., Metz, C.E., Doi, K. (1996). Effect of a computer-aided diagnosis scheme on radiologists' performance in detection of lung nodules on radiographs. *Radiology* **199**, 843–848.

Kundel, H.L., Nodine, C.F. (1983). A visual concept shapes image perception. *Radiology* **146**, 363–368.

Kundel, H.L., Nodine, C.F. (2004). Modeling visual search during mammogram viewing. *Proc SPIE* **5372**, 110–115.

Kundel, H.L., Nodine, C.F., Conant, E.F., Weinstein, S.P. (2007). Holistic component of image perception in mammogram interpretation: gaze-tracking study. *Radiology* **242** (2), 396–402.

Li, F., Arimura, H., Suzuki, K. *et al.* (2005). Computer-aided detection of peripheral lung cancers missed at CT: ROC analyses without and with localization. *Radiology* **237**, 684–690.

Metz, C.E. (1996). Evaluation of digital mammography by ROC analysis. In *Digital Mammography '96*. Eds. Doi, K., Giger, M.L., Nishikawa, R.M., Schmidt, R.A. Amsterdam, the Netherlands: Elsevier Science: 61–68.

Metz, C.E., Herman, B., Shen, J. (1998). Maximum-likelihood estimation of receiver operating characteristic (ROC) curves from continuously-distributed data. *Statistics in Medicine* **17**, 1033–1053.

Metz, C.E., Pan, X. (1999). Proper binormal ROC curves: theory and maximum-likelihood estimation. *J Math Psychol* **43** (1), 1–33.

Niklason, L.T., Hickey, N.M., Chakraborty, D.P. *et al.* (1986). Simulated pulmonary nodules: detection with dual-energy digital versus conventional radiography. *Radiology* **160**, 589–593.

Nodine, C.F., Kundel, H.L. (1987). Using eye movements to study visual search and to improve tumor detection. *RadioGraphics* **7** (2), 1241–1250.

Obuchowski, N.A. (1994). Computing sample size for receiver operating characteristic studies. *Invest Radiol* **29** (2), 238–243.

Obuchowski, N.A. (1995a). Multireader, multimodality receiver operating characteristic curve studies: hypothesis testing and sample size estimation using an analysis of variance approach with dependent observations. *Acad Radiol* **2**, S22–S29.

Obuchowski, N.A. (2009). Reducing the number of reader interpretations in MRMC studies. *Acad Radiol* **16**, 209–217.

Obuchowski, N.A., Lieber, M.L., Powell, K.A. (2000). Data analysis for detection and localization of multiple abnormalities with application to mammography. *Acad Radiol* **7** (7), 516–525.

Obuchowski, N.A., Rockette, H.E. (1995b). Hypothesis testing of diagnostic accuracy for multiple readers and multiple tests: an ANOVA approach with dependent observations. *Comm Stat* **24**, 285–308.

Pan, X., Metz, C.E. (1997). The proper binormal model: parametric receiver operating characteristic curve estimation with degenerate data. *Acad Radiol* **4** (5), 380–389.

Penedo, M., Souto, M., Tahoces, P.G. *et al.* (2005). Free-response receiver operating characteristic evaluation of lossy JPEG2000 and object-based set partitioning in hierarchical trees compression of digitized mammograms. *Radiology* **237** (2), 450–457.

Pesce, L.L., Metz, C.E. (2007). Reliable and computationally efficient maximum-likelihood estimation of proper binormal ROC curves. *Acad Radiol* **14** (7), 814–829.

Petrick, N., Haider, M., Summers, R.M. *et al.* (2008). CT colonography with computer-aided detection as a second reader: observer performance study. *Radiology* **246** (1), 148–156.

Pisano, E.D., Gatsonis, C., Hendrick, E. *et al.* (2005). Diagnostic performance of digital versus film mammography for breast-cancer screening. *N Engl J Med* **353** (17), 1–11.

Popescu, L.M., Lewitt, R.M. (2006). Small nodule detectability evaluation using a generalized scan statistic model. *Phys Med Biol* **51** (23), 6225–6244.

Press, W.H., Flannery, B.P., Teukolsky, S.A., Vetterling, W.T. (1988). Cambridge, UK: *Numerical Recipes in C: The Art of Scientific Computing* Cambridge University Press.

Rock, I. (1983). *The Logic of Perception*. Cambridge, MA: The MIT Press.

Rock, I. (1984). *Perception*. New York, NY: WH Freeman and Company.

Ruschin, M., Timberg, P., Bath, M. *et al.* (2007). Dose dependence of mass and microcalcification detection in digital mammography: free response human observer studies. *Med Phys* **34**, 400–407.

Rutter, C.M. (2000). Bootstrap estimation of diagnostic accuracy with patient-clustered data. *Acad Radiol* **7** (6), 413–419.

Samuelson, F.W., Petrick, N. (2006). Comparing image detection algorithms using resampling. *3rd IEEE International Symposium on Biomedical Imaging: From Nano to Micro*, 1312–1315.

Samuelson, F.W., Petrick, N., Paquerault, S. (2007). Advantages and examples of resampling for CAD evaluation. *4th IEEE International Symposium on Biomedical Imaging: From Nano to Macro*, 492–495.

Song, T., Bandos, A.I., Rockette, H.E., Gur, D. (2008). On comparing methods for discriminating between actually negative and actually positive subjects with FROC type data. *Med Phys* **35** (4), 1547–1558.

Starr, S.J., Metz, C.E., Lusted, L.B. (1977). Comments on generalization of receiver operating characteristic analysis to detection and localization tasks. *Phys Med Biol* **22**, 376–379.

Starr, S.J., Metz, C.E., Lusted, L.B., Goodenough, D.J. (1975). Visual detection and localization of radiographic images. *Radiology* **116**, 533–538.

Svahn, T., Hemdal, B., Ruschin, M. *et al.* (2007). Dose reduction and its influence on diagnostic accuracy and radiation risk in digital mammography: an observer performance study using an anthropomorphic breast phantom. *Br J Radiol* **80**, 557–562.

Swensson, R.G. (1980). A two-stage detection model applied to skilled visual search radiologists. *Perception and Psychophysics* **27** (1), 11–16.

Swensson, R.G. (1996). Unified measurement of observer performance in detecting and localizing target objects on images. *Med Phys* **23** (10), 1709–1725.

Swensson, R.G., Judy, P.F. (1981). Detection of noisy visual targets: models for the effects of spatial uncertainty and signal-to-noise ratio. *Perception and Psychophysics* **29** (6), 521–534.

Vikgren, J., Zachrisson, S., Svalkvist, A. *et al.* (2008). Comparison of chest tomosynthesis and chest radiography for detection of pulmonary nodules: human observer study of clinical cases. *Radiology* **249** (3), 1034–1041.

Volokh, L., Liu, C., Tsui, B.M.W. (2006). Exploring FROC paradigm – initial experience with clinical applications. *Proc. SPIE Medical Imaging: Image Perception, Observer Performance, and Technology Assessment*. San Diego, CA: SPIE.

Wagner, R.F., Beiden, S.V., Campbell, G., Metz, C.E., Sacks, W.M. (2002). Assessment of medical imaging and computer-assist systems: lessons from recent experience. *Acad Radiol* **9** (11), 1264–1277.

Wagner, R.F., Metz, C.E., Campbell, G. (2007). Assessment of medical imaging systems and computer aids: a tutorial review. *Acad Radiol* **14** (6), 723–748.

Wolfe, J.M. (1994). Guided search 2.0: a revised model of visual search. *Psychonomic Bull Rev* **1** (2), 202–238.

Yoon, H.J., Zheng, B., Sahiner, B., Chakraborty, D.P. (2007). Evaluating computer-aided detection algorithms. *Med Phys* **34** (6), 2024–2038.

Zheng, B., Chakraborty, D.P., Rockette, H.E., Maitz, G.S., Gur, D. (2005). A comparison of two data analyses from two observer performance studies using jackknife ROC and JAFROC. *Med Phys* **32** (4), 1031–1034.

17

Observer models as a surrogate to perception experiments

CRAIG K. ABBEY AND MIGUEL P. ECKSTEIN

17.1 INTRODUCTION

Modern medical imaging systems deliver an enormous quantity of information for the purpose of diagnosing disease. These systems are also increasingly complex from an engineering standpoint, with numerous free parameters to optimize during image acquisition, processing, and display. Ultimately, imaging systems are established based on their performance in clinical trials, where the patients and image readers are drawn from clinical populations (for examples in mammography see Pisano, 2005, or Gur, 2004). However, it is unfeasible from both ethical and economic standpoints to run such large studies on systems that have not already undergone extensive preliminary validation (Obuchowski, 2007). Receiver operating characteristic (ROC) studies using sub-selected patient populations serve an important role in these preliminary validations (see for example Swets, 1982, Metz, 1986, or Obuchowski, 2007). However, even these studies are difficult and time-consuming to conduct, and typically they involve at most five to ten different experimental conditions. This allows for optimization of a single parameter or perhaps a coarse optimization of two parameters, but makes them of limited use for answering the many system design trade-offs that arise in virtually every medical imaging modality. More general observer performance studies are somewhat easier to conduct (Burgess, 1995), since they typically use non-clinical (but trained) observers performing a simple visual task that is an abstraction of clinical image reading. While observer studies of this sort do allow for more extensive evaluations, they are still unwieldy for what would ideally be a multi-parameter optimization problem, which would typically involve the testing of several tens to hundreds of conditions.

Model observers were conceived and developed with the goal of predicting human observer performance for these early-stage optimizations (Wagner, 1985; Judy, 1987; Barrett, 1993). Since they generally involve simple detection and discrimination tasks – in which a known target profile embedded in noise is discriminated from a known alternative profile – the development of models has typically considered diagnostic features in terms of signal and noise characteristics. Thus models serve as a way to relate statistical properties of the images to diagnostic performance at this early stage. However, they are not intended – nor should they be interpreted – as a replacement for subsequent clinical validation.

The purpose of this chapter is to give a unified description of model observers, and how they fit into the more general context of signal detection theory. We begin with a review of signal detection theory that closely follows Green and Swets' (Green, 1966) classic treatment for human observers. Here we emphasize the one additional piece of information that a model observer gives us – an explicit formula for determining a decision variable from an image, with different models consisting of different ways to formulate this variable. We then describe a number of linear model observers that have been used for tasks relevant to medical imaging. These are specified in terms of the linear template used by the model, and we discuss how the template incorporates statistical properties of the images.

17.2 MODELING VISUAL PERFORMANCE IN SIMPLE TASKS

17.2.1 Imaging preliminaries

For the purposes of theoretical analyses, images have been modeled both as continuously defined functions of space (and possibly time), as well as finite arrays of pixel intensities. For the most part we will adopt the latter approach since we are mostly concerned with digital imaging modalities in this treatment. We make an exception to this in Section 17.2.6, where we relate the discrete matrix-vector approach to continuously defined stochastic processes.

We begin by specifying a generic image as the column vector $\mathbf{g}$ (boldface symbols are used to indicate vectors and matrices), with elements identified by g_m, $m = 1, \ldots, M$. While we may be used to thinking of images as multidimensional arrays with up to three spatial dimensions and possibly additional dimensions for time, photon energy, etc., reformatting them as a one-dimensional list of pixel intensities simplifies many expressions in the development of model observers. For example, the covariance structure of a three-dimensional image is a six-dimensional object listing the covariance between any two three-dimensional indices. This high-dimensional formulation becomes more complicated when an inverse or decorrelating transform is needed. By contrast, when the image is reformatted as a one-dimensional column vector, the covariance structure is a two-dimensional matrix familiar from linear algebra.

Image statistics play an important role in both the functional form and the performance of model observers. The most fundamental and complete description of the statistical variability in a class (or ensemble) of images is a probability density function, $p(\mathbf{g})$. As an example, the often-used multivariate Gaussian distribution has a density function of

$$p(\mathbf{g}) = \frac{1}{(2\pi)^{M/2}\,|\boldsymbol{\Sigma}|^{1/2}} e^{-\frac{1}{2}(\mathbf{g}-\boldsymbol{\mu})^t \boldsymbol{\Sigma}^{-1}(\mathbf{g}-\boldsymbol{\mu})} \tag{17.1}$$

The Handbook of Medical Image Perception and Techniques, ed. Ehsan Samei and Elizabeth Krupinski. Published by Cambridge University Press.

where $\boldsymbol{\mu}$ and $\boldsymbol{\Sigma}$ are the multivariate mean vector and covariance matrix respectively, the superscript t indicates the transpose of a multivariate quantity, and $|\boldsymbol{\Sigma}|$ indicates the determinant of the covariance matrix. The mean vector and covariance matrix are often used as summary statistics for a distribution of images since they are considered to be measures of distribution "location" (mean) and "spread" (covariance). This is often the case when the full probability distribution of an ensemble of images is unknown or intractable.

An advantage of working with only the mean and covariance of images is that these quantities can be estimated from samples, allowing them to be determined with some degree of accuracy depending on the number of samples available. Let $\mathbf{g_n}$, $n = 1, \ldots, N$, be a sample of images from the distribution of interest. Then the mean vector may be estimated from the average

$$\hat{\boldsymbol{\mu}} = \frac{1}{N}\sum_{n=1}^{N} \mathbf{g}_n \tag{17.2}$$

and the covariance matrix is estimated by

$$\hat{\boldsymbol{\Sigma}} = \frac{1}{N-1}\sum_{n=1}^{N} (\mathbf{g}_n - \hat{\boldsymbol{\mu}})(\mathbf{g}_n - \hat{\boldsymbol{\mu}})^t \tag{17.3}$$

Note that each component in the sum of Equation 17.3 is an outer product and therefore an M by M matrix.

Covariance matrices for classes of images with even modest size can be prohibitively large, and special consideration must be given to how they will be evaluated. For example, a relatively modest 512 by 512 pixel image (i.e. 0.25 megapixels in digital-camera terminology) has a covariance matrix that is 262,144 elements by 262,144 elements or 68,719,476,736 elements in total (with 34,359,869,440 elements in total after removing symmetric redundancies). This makes working with and storing such matrices difficult and has led to various tricks for accomplishing the necessary computations with them. Most of these are based on the fact that we generally only use the covariance matrix in some sort of a matrix-vector product.

For example, consider the estimated covariance matrix in Equation 17.3. Say we wished to compute the product $\hat{\boldsymbol{\Sigma}}\mathbf{x}$ for some M-element vector $\mathbf{x}$. Creating $\hat{\boldsymbol{\Sigma}}$ first and then multiplying by $\mathbf{x}$ would be extremely memory intensive because of the large size of the matrix. However, we can easily compute an intermediate product, an N-element vector $\mathbf{y}$ defined by

$$y_n = (\mathbf{g}_n - \hat{\boldsymbol{\mu}})^t\, \mathbf{x} \tag{17.4}$$

The desired matrix-vector product is then obtained by

$$\hat{\boldsymbol{\Sigma}}\mathbf{x} = \frac{1}{N-1}\sum_{n=1}^{N} (\mathbf{g}_n - \hat{\boldsymbol{\mu}})\, y_n \tag{17.5}$$

This computation never involved storage of the matrix $\hat{\boldsymbol{\Sigma}}$.

This simple example serves to illustrate an important point in much of what follows. Mathematical equations are used to specify the quantities that need to be evaluated, but the actual computations used to evaluate them may be much involved.

17.2.2 Decision variables from model observers

At the heart of the model-observer approach, and the distinguishing feature of different models, is the formulation of a decision variable from an image. In fact, we would define this as the working definition of a model observer in contrast to other sorts of models that predict performance from image statistics, etc. A model observer is defined as an explicit formula for determining a decision variable. The decision variable is thus a transformation of the image into a quantity (typically scalar-valued) that can be used directly to perform the task. In this section we regard the decision variable as something of a "black box," with specifics to be given in Section 17.3. This allows us to describe how model observers perform a task somewhat more generically, and then focus on the differences in models more specifically.

Simple tasks are generally modeled with a scalar decision variable, which we designate r. Thus the role of the model is to compress all of the image data into a single number for the purpose of performing the task. This transformation is denoted by the scalar-valued function w, which takes the image vector, $\mathbf{g}$, as its argument. Model-observer analyses often attempt to incorporate sources of variability within human observers (i.e. internal noise). Under an additive model of internal noise, made explicit in the variable ε, the decision variable is related to the image, the model, and the internal noise by

$$r = w(\mathbf{g}) + \varepsilon \tag{17.6}$$

We note that this additive model is general if we allow that the distribution of ε may not be independent of $\mathbf{g}$ (although this is a common assumption).

An important special case of Equation 17.6 is linear model observers which are defined by a linear transformation of the image. In this case, the model may be described by a vector of weights with the same dimension as $\mathbf{g}$. This vector is often called the observer template, and is denoted by $\mathbf{w}$ here. The linear model of the decision variable is given by

$$r = \mathbf{w}^t\mathbf{g} + \varepsilon \tag{17.7}$$

where $\mathbf{w}^t\mathbf{g}$ is an inner product between the column vectors $\mathbf{w}$ and $\mathbf{g}$, and thus implements the transform as a weighted sum.

17.2.3 Observer decisions

Once the observer decision variable has been determined, the task can be performed directly from the response. For example, in a simple "yes-no" detection task where the possible responses are "target present" and "target absent," the observer decides by comparison to a threshold, r_{thresh}, according to

$$\text{Decision} = \begin{cases} \text{Target Present} & \text{if } r > r_{\text{thresh}} \\ \text{Target Absent} & \text{if } r \leq r_{\text{thresh}} \end{cases} \tag{17.8}$$

For a model observer, the threshold is considered part of the model and must be specified for decisions to be rendered.

In forced-choice experiments, the observer is shown two or more different images, and is asked to identify the one image that contains the target. When M images are shown in a trial, this is referred to as M-alternative forced choice (MAFC). In this case, decision variables from the M images are compared

Table 17.1 **Components of simple detection and discrimination experiments.** *Note that detectability-index formulae assume equal variance of the decision variable (i.e. b = 1) under target-present and target-absent conditions.*

Experiment	Decision	Performance measure(s)	Detectability computation (assuming $b = 1$)
"Yes-no"	Comparison to threshold	True- and false-positive rates (P_{TP} and P_{FP})	$d' = \Phi^{-1}(P_{TP}) - \Phi^{-1}(P_{FP})$
Forced-choice	Maximum decision variable	Proportion correct (P_C)	Solve: $P_C = \int dr\, \phi(r - d')\, \Phi(r)^{M-1}$
Rating scale	Categorized decision variable	ROC curve and area under curve (AUC)	$d' = \sqrt{2}\Phi^{-1}(\text{AUC})$

to each other rather than to a threshold. The observer chooses the image corresponding to the largest decision variable. Let $r_1, r_2, \ldots, r_M$ be decision variables for the M images in a single trial. The observer decision is given by

$$\text{Decision} = \begin{cases} 1 & \text{if } r_1 > \max(r_2, r_3, \ldots, r_M) \\ 2 & \text{if } r_2 > \max(r_1, r_3, \ldots, r_M) \\ \vdots & \vdots \\ M & \text{if } r_M > \max(r_1, r_2, \ldots, r_{M-1}) \end{cases} \tag{17.9}$$

Note that forced-choice experiments do not require specification of thresholds.

Rating-scale experiments are a generalization of "yes-no" experiments. Here the observer rates their confidence on an ordinal scale. The yes-no experiment is when the scale consists of 1 (signal absent) and 2 (signal present), a two-point categorical scale. For an M-point categorical scale, the observer decision is given by

$$\text{Decision} = \begin{cases} 1 & \text{if } r \le r_{\text{thresh},1} \\ 2 & \text{if } r_{\text{thresh},1} < r \le r_{\text{thresh},2} \\ \vdots & \vdots \\ M & \text{if } r_{\text{thresh},M-1} < r \end{cases} \tag{17.10}$$

Here again, the $M - 1$ thresholds are considered part of the model.

Once the decision has been made, the observer has effectively "performed" the task. Table 17.1 lists some simple tasks that are commonly used in psychophysical experiments with medical images, along with the way a scalar decision variable is used to perform the task. These include yes-no, forced-choice, and rating-scale detection tasks. These are equally applicable to simple discrimination tasks as well.

17.2.4 Task performance

One of the strengths of the model-observer approach is that by actually doing the task, the natural measures of task performance can be used, as shown in the last column of Table 17.1. Performance measures are defined by converting decisions into outcomes by comparing them to ground truth.

The detectability index, d', is often used as a general measure of task performance since it can be readily formulated for all three tasks in Table 17.1. It is based on the mean and variance of the decision variable under the target-present and target-absent classes. Let μ_{TP} and μ_{TA} be the mean of the decision variable r in Equation 17.6 (or 17.7) under the target-present and target-absent cases, and let σ^2_{TP} and σ^2_{TA} be the variances. We define d' as the ratio of the difference in means to the square-root of the average variance,

$$d' = \frac{\mu_{TP} - \mu_{TA}}{\sqrt{0.5\left(\sigma^2_{TP} + \sigma^2_{TA}\right)}} \tag{17.11}$$

We note that historically d' has often been defined as the special case of Equation 17.11 in which $\sigma^2_{TP} = \sigma^2_{TA} = \sigma^2$. We use the more general form (denoted by $D(\Delta m, s)$ in Green, 1966) because it appears to be more generally applicable in medical tasks where unequal variances have often been observed. We now turn to how d' is related to measures of performance in yes-no, forced-choice, and rating-scale experiments.

For the yes-no task, performance is typically reported in terms of true-positive and false-positive fractions or equivalently, sensitivity and specificity. The true-positive fraction, P_{TP}, is the probability of detecting the target when it is present in the image. The false-positive fraction, P_{FP}, is the probability of erroneously detecting a target when none is present. Let us assume that the decision variable is Gaussian distributed in both classes, with parameters defined as in Equation 17.11. Let b denote the ratio of standard deviations ($b = \sigma_{TP}/\sigma_{TA}$), and let Φ denote the cumulative normal distribution function. Then for a known value of b, the detectability index is related to P_{TP} and P_{FP} by

$$d' = \frac{1}{\sqrt{0.5\left(b^{-2} + 1\right)}}\Phi^{-1}(P_{TP}) - \frac{1}{\sqrt{0.5\left(b^{2} + 1\right)}}\Phi^{-1}(P_{FP}) \tag{17.12}$$

Note that with the assumption of $b = 1$, both constants multiplying Φ^{-1} simplify to 1.

In forced-choice tasks, the natural performance measure is the proportion of correct responses, P_C. A correct outcome in a trial is achieved when the decision variable corresponding to the target-present image is greater than the maximum decision variable from the target-absent images. For MAFC experiments (with $M \ge 2$), proportion correct is related to the distributions of decision variables by

$$P_C = \int_{-\infty}^{\infty} dr\, p_{TP}(r)\, P_{TA}(r)^{M-1} \tag{17.13}$$

where $p_{TP}(r)$ is the probability density function for the target-present decision variables and $P_{TA}(r)$ is the cumulative distribution function for the signal-absent function. This expression implicitly assumes that the decision variables are independent and identically distributed across trials as well as within a trial for the target-absent decision variables. With these assumptions, $P_{TA}(r)^{M-1}$ is the cumulative distribution function for the maximum of the target-absent decision variables. Under Gaussian assumptions on the decision variables, Equation 17.13 can be rewritten as

$$P_C = \int_{-\infty}^{\infty} dr \frac{1}{\sqrt{2\pi b^2}} e^{-\frac{1}{2b^2}\left(r-\sqrt{0.5(b^2+1)}d'\right)^2} \Phi(r)^{M-1} \tag{17.14}$$

This equation can be solved implicitly for known b using numerical integration techniques. For 2AFC, Equation 17.14 may be solved explicitly, yielding

$$d' = \sqrt{2}\Phi^{-1}(P_C) \tag{17.15}$$

Notably in this case, the ratio of variances, b, need not be known.

Analysis of rating scale data is the basis of ROC analysis, and an entire field unto itself (for more information see Chapter 14). Essentially this is an application of categorical regression. Gaussian assumptions on the decision variable are referred to as the binormal model. Traditionally, analysis of the binormal model parameterizes the ROC curve using two parameters, $a = (\mu_{TP} - \mu_{TA})/\sigma_{TA}$ and b which is the ratio of standard deviations as defined above. In this case, the detectability index is defined

$$d' = \frac{a}{\sqrt{0.5\,(b^2+1)}} \tag{17.16}$$

Note that if $b = 1$, then $d' = a$.

17.2.5 Observer efficiency

The notion of observer efficiency in visual tasks is closely tied to the ideal observer, and is covered in detail elsewhere in this volume. However, it is important to mention it here because efficiency measures are often used to evaluate model observers. The term "efficiency" is generally reserved to indicate performance relative to the ideal observer, while "relative efficiency" is used when the reference is not the ideal observer. The observer efficiency can be thought of generally as representing the fraction of diagnostic information available in the image that is being used to perform the task. The value of the measure ranges between 0 and 1, with 0 implying chance performance and 1 implying equivalent performance to the ideal observer. Efficiency scores less than 1 can be thought of in terms of Equation 17.6 as the consequence of a decision variable arising from a suboptimal transform of the image data or from dilution due to internal noise.

A common way to compute efficiency is to estimate d' for the observer of interest, d'_{obs}, and for the ideal observer, d'_{IO}. Efficiency is then defined as the ratio

$$\eta = \left(\frac{d'_{obs}}{d'_{IO}}\right)^2 \tag{17.17}$$

A second approach is to use a ratio of contrast thresholds. Here, target contrast is adjusted to achieve equal performance between the ideal observer and the observer of interest. Let C_{obs} be the contrast of the observer of interest, and let C_{IO} be the contrast needed to match ideal observer performance. Then observer efficiency is given by the ratio of squared contrast thresholds

$$\eta = \left(\frac{C_{IO}}{C_{obs}}\right)^2 \tag{17.18}$$

Since the ideal observer requires the minimal contrast needed to achieve a set level of performance, this ratio is guaranteed to be less than or equal to 1. The two definitions of efficiency are equivalent when d' is linear with target contrast.

17.2.6 Evaluating model observer performance

We have described above how model observers can be used to generate a decision variable, and that decision variable can be used to perform a task leading naturally to various relevant measures of performance which can be used for testing and early validation of new methods in medical imaging. This process forms the basis for model-observer evaluation from sample images. When a sample of images is available (both target-present and target-absent), Equation 17.6 (or Equation 17.7) can be used to generate model-observer decision variables and then a decision depending on the nature of the task. The decisions can be converted to outcomes by comparison with ground truth, which allows various performance measures to be computed. This approach essentially applies psychophysical methodology to the model observer in order to measure task performance.

However, because the functional form of models is often known explicitly, there are commonly used short-cuts to determine model-observer performance. These virtually always require additional assumptions about the perceptual process to be considered if they are to be used. Below we describe some simple alternatives that are often used for computing detectability.

The statistical properties of the response variable can often be determined directly from the statistical properties of the images. For example, consider a linear observer as in Equation 17.7, and an image with Gaussian statistics in each class, as in Equation 17.1. Let class 1 define the target-absent images, and class 2 define target presence, and let them be parameterized by mean vectors $\boldsymbol{\mu}_1$ and $\boldsymbol{\mu}_2$, and covariance matrices $\boldsymbol{\Sigma}_1$ and $\boldsymbol{\Sigma}_2$, respectively. Then the detectability index in Equation 17.11 is defined as

$$d' = \frac{\mathbf{w}^t\left(\boldsymbol{\mu}_2 - \boldsymbol{\mu}_1\right)}{\sqrt{0.5\mathbf{w}^t\left(\boldsymbol{\Sigma}_2 + \boldsymbol{\Sigma}_1\right)\mathbf{w} + \sigma_\varepsilon^2}} \tag{17.19}$$

If we define $\mathbf{s}$ as the difference in mean vectors and make the further assumption that image noise is not class-dependent (i.e. $\boldsymbol{\Sigma}_1 = \boldsymbol{\Sigma}_2 = \boldsymbol{\Sigma}_{\mathbf{g}}$), then we get the simpler form

$$d' = \frac{\mathbf{w}^t\mathbf{s}}{\sqrt{\mathbf{w}^t\boldsymbol{\Sigma}_{\mathbf{g}}\mathbf{w} + \sigma_\varepsilon^2}} \tag{17.20}$$

We note that performance in both Equations 17.19 and 17.20 can be estimated using sample statistics as in Equations 17.2 and 17.3.

As a simple example, let us consider Equation 17.20 when the noise is white (i.e. $\boldsymbol{\Sigma}_{\mathbf{g}} = \sigma_n^2\mathbf{I}$, where $\mathbf{I}$ is the identity matrix), and there is no internal noise ($\sigma_\varepsilon = 0$). Furthermore, let us assume that the observer template is proportional to the signal ($\mathbf{w} = k\mathbf{s}$, where k is the constant of proportionality), which makes it a matched filter. The resulting detectability is

$$d' = \frac{\sqrt{\mathbf{s}^t\mathbf{s}}}{\sigma_n} \quad (17.21)$$

which can be seen to be the root-mean-squared (RMS) signal energy divided by the standard deviation of the noise. This example shows the clear connection between detectability and signal-to-noise ratio that goes back to the early work of Rose (Rose, 1953; Burgess, 1999). If we relax the assumption of white noise, Equation 17.21 generalizes to

$$d' = \sqrt{\mathbf{s}^t\boldsymbol{\Sigma}_{\mathbf{g}}^{-1}\mathbf{s}} \quad (17.22)$$

which is the well-known Mahalanobis distance between the class probability distributions (Mahalanobis, 1936; Fukunaga, 1972).

The detectability index is often formulated in terms of continuous Fourier integrals (for example, see Wagner, 1985, and Burgess, 1997). This is done by analogy to Equation 17.20, with the additional assumption of stationary random noise. In this case, letting u and v be the two-dimensional Fourier variables, we define $s(u, v)$ and $w(u, v)$ as the Fourier-domain continuous analogs of $\mathbf{s}$ and $\mathbf{w}$, and $N(u, v)$ is the noise power spectrum (NPS) characterizing the variability in the image. The detectability index can then be written in terms of integrals over the two-dimensional Fourier plane

$$d' = \frac{\int dudv\, \overline{w(u,v)}s(u,v)}{\sqrt{\int dudv N(u,v)|w(u,v)|^2 + \sigma_\varepsilon^2}} \quad (17.23)$$

where the over line indicates the conjugate of this complex quantity.

Equation 17.23 has an important connection to commonly used physical measures of medical imaging systems. Let us again assume that there is no internal noise, and that the difference signal can be decomposed into the product of the difference in the mean profiles of the objects being images, $s_{\mathrm{obj}}(u, v)$, and the modulation transfer function (MTF) of the system (i.e. $s(u, v) = M(u, v)s_{\mathrm{obj}}(u, v)$). Let $w(u, v) = s(u, v)/N(u, v)$, which is referred to as a prewhitened matched filter. In this case Equation 17.23 simplifies to

$$d' = \left(\int dudv \frac{|M(u,v)|^2 \left|s_{\mathrm{obj}}(u,v)\right|^2}{N(u,v)}\right)^{1/2} \quad (17.24)$$

Recognizing the ratio of MTF-squared to NPS as the noise-equivalent quanta (NEQ) we obtain

$$d' = \left(\int dudv\, \mathrm{NEQ}(u,v) \left|s_{\mathrm{obj}}(u,v)\right|^2\right)^{1/2} \quad (17.25)$$

which nicely decomposes the detectability into components representing the object being imaged and the imaging system used to produce the image, assuming that system noise is the only source of variability in the image. When anatomical variability is included in $N(u, v)$, the generalized NEQ is used in Equation 17.25 (Barrett, 1995; Richard, 2007), and the decomposition imaging system and object being imaged are less clear.

The relationships between the detectability index and many other useful characterizations of imaging systems such as image statistics, Fourier-domain representations, and NEQ provide insights to how image quality impacts task performance. However, it is important to be aware of the strong assumptions of Gaussian distributions, stationarity of noise, and continuum representation that are necessary to make these associations. Investigators using these approaches should always be asking how appropriate the assumptions are for their data, and how the failure of the assumptions impacts their conclusions.

17.2.7 More complex tasks

So far we have considered simple detection of a target with a fixed profile, size, and location within the image. This is referred to as the signal-known-exactly (SKE) paradigm. There are many reasons to consider relaxing this. None of these components are actually fixed in patient images, and therefore every fixed parameter represents an additional level of abstraction from the clinical setting. Many medical imaging modalities are nonstationary, and hence image statistics can vary substantially across different locations in the image. In this case it would make sense to evaluate performance at a representative set of locations. The issue of image size is somewhat more subtle. Certainly disease rarely comes in a fixed size, but it is often the case that larger targets are easier to detect. Hence when "stress testing" two or more imaging methods, a small and difficult target is usually chosen, since that is where limitations will most readily be apparent. Similar considerations apply to signal profiles. A sharper-edged target will typically place a higher demand on system resolution, and is therefore useful for testing different methods in this context. Nonetheless, target uncertainty may still be desirable for size and profile parameters as a way to see the effect of uncertainty on the observer or when an aggregate measure is more relevant.

Here we consider two approaches to adding signal complexity. When the target profile is known, but allowed to vary from trial to trial, the paradigm is called signal-known-exactly but variable (SKEV). One example of an SKEV task occurs when the location is variable from trial to trial, but the observer is told what location the target will be in, if it is present. This can be done with cross-hairs for location cues, or by a precue before the image comes up. In the second approach to signal variability, the observer is intrinsically uncertain about properties of the target. This is referred to as signal-known-statistically (SKS). If the location cues are removed in the example SKEV task just described (and presuming the observer knows what the possible target locations are), then the task becomes SKS. Typically SKS tasks are substantially more difficult than SKE or SKEV tasks because of the additional uncertainty in the target.

17.2.7.1 Signal-known-exactly but variable

The SKEV experiment is essentially an average of many SKE tasks, one for each possible parameter value or combination of parameter values. Let u be an index for parameter uncertainty. For example, if there were five possible locations, then $u = 1, \ldots, 5$. The decision variable is then formed with knowledge of this index. Assuming a linear observer, the decision variable is then given by

$$r_u = \mathbf{w}_u^t \mathbf{g} + \varepsilon_u \tag{17.26}$$

where ε_u allows for internal noise that is dependent on the target parameters. Decision variables associated with a given u are used to obtain a measure of performance for that target parameter setting. For example, in a forced-choice experiment, a proportion correct would be computed for each value of u, $P_{C,u}$. These parameter-dependent performance measures are then averaged to give the SKEV performance. In the forced-choice experiment just described,

$$P_{C,SKEV} = \sum_u \pi_u P_{C,u} \tag{17.27}$$

where π_u is the prior probability on index u. From the averaged performance measure, a representative detectability can be computed using the methods described in Section 17.2.4.

17.2.7.2 Signal-known-statistically tasks

Whereas the SKEV task is essentially the average of many SKE tasks, SKS tasks are fundamentally different since the inherent target uncertainty requires consideration of all possible target parameters. This is reflected in a fundamentally different computation for the observer's decision variable. It also makes very questionable the Gaussian assumptions underpinning any of the transforms to a detectability index.

We will describe one implementation of the SKS observer that generalizes the linear SKE model in Equation 17.7. In this case let us define r_u in Equation 17.26 for all possible target parameter settings in an image. So unlike the SKEV task, where a particular parameter setting – indexed by u – is evaluated, in this case all possible parameter settings are considered in a given trial. These individual decision variables must then be combined into a single decision variable for the trial. This is done using an independent Gaussian likelihood ratio (Zhang, 2004) for each of the parameter-specific decision variables. Let $\bar{r}_u^-$ be the mean decision variable under the signal-absent hypothesis at parameter setting u, and let $\bar{r}_u^+$ be the mean decision variable under the signal-present hypothesis. Furthermore, let σ_u^2 be the variance of the response, which is assumed to be the same whether the target is present or absent. Then the trial decision variable is defined by the sum

$$r = \sum_u \pi_u \exp\left(-\frac{1}{2}\left(\left(\frac{r_u - \bar{r}_u^+}{\sigma_u^2}\right)^2 - \left(\frac{r_u - \bar{r}_u^-}{\sigma_u^2}\right)^2\right)\right) \tag{17.28}$$

which integrates over the uncertainty in target parameter settings for each trial, rather than in the performance metric as was the case for the SKEV approach in Equation 17.27.

17.3 MODEL OBSERVERS

The key component of the process for evaluating performance in image-based tasks spelled out in the previous section is the functional form of $w(\mathbf{g})$ in Equation 17.6. Here we examine several proposed models for a linear observer template, $\mathbf{w}$ in

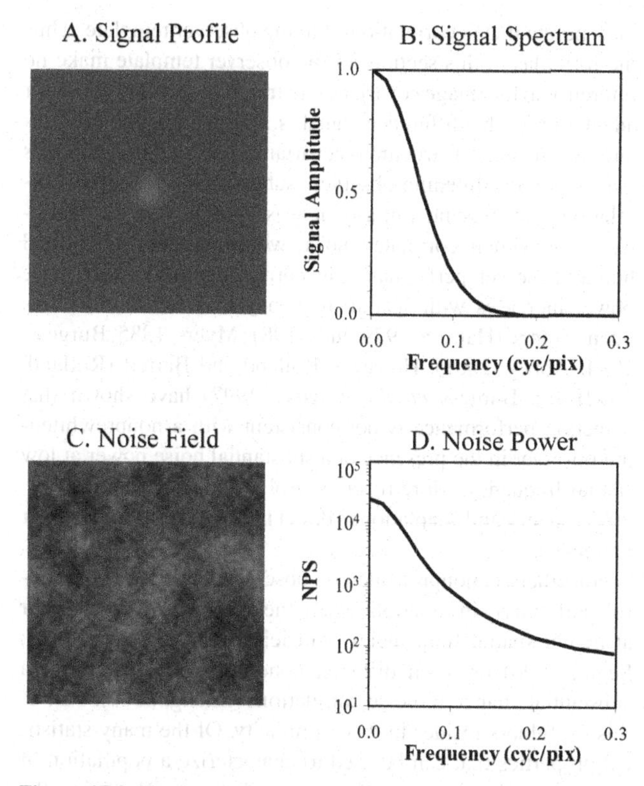

Figure 17.1 Test-case signal and noise. The target profile (A) and profile spectrum (B) are shown along with a sample noise field (C) and noise-power spectrum (D). Note that a logarithmic scale is used for noise power.

Equation 17.7. We begin with nonprewhitening observers, which do not attempt to explicitly accommodate the correlation structure of noise in the images. These models are tuned only to properties of $\mathbf{s}$, the difference signal in Equation 17.20. Nonprewhitening observers are followed by prewhitening observers that do account for noise correlations. We then discuss channelized observers, which occupy a middle ground between prewhitening and nonprewhitening observers. These models posit a limited number of fixed (i.e. non-adaptive) features that constitute the initial step of the model. Any accommodations of noise correlation and/or signal profile occur in the channel weights after the channel features have been extracted. The channels therefore represent a way to implement a limited form of prewhitening on the image data. They are also an attempt to bring known features of the human visual system into models of medical image perception.

For the purposes of illustration we consider a detection task in correlated Gaussian noise as a way to show how the different models result in different templates. The SKE target is a Gaussian profile, as shown in Figure 17.1, and the noise is described by a low-pass NPS. Since both target and noise are isotropic, we will display a radial plot of the two-dimensional Fourier transform in addition to showing the templates rendered as an image.

17.3.1 Nonprewhitening observers

Nonprewhitening observer models are essentially defined by what they don't do. They make no attempt to explicitly incorporate image correlations into the observer template. Thus the formulas in this section for the observer template make no reference to the image covariance matrix, $\mathbf{\Sigma}_g$, and only allow for adaptation to the difference signal, **s**. The inability to decorrelate (or prewhiten) structured covariance between image pixels makes nonprewhitening observers suboptimal when these correlations are present. Initially, it was thought that the inability to prewhiten correlated noise would explain suboptimal human-observer performance in correlated noise when noise power increases with spatial frequency over the relevant frequency range (Hanson, 1979; Judy, 1981; Myers, 1985; Burgess, 1994; Abbey, 2001). However, Rolland and Barrett (Rolland, 1992) and Burgess *et al.* (Burgess, 1997) have shown that detection performance is not consistent with a nonprewhitening observer in the presence of a substantial noise power at low spatial frequency. More recently, Abbey and Eckstein (Abbey, 2007) have found adaptation in direct measurements of observer templates.

Nonetheless, nonprewhitening observers have remained useful, particularly in situations where the differences mainly occur at higher spatial frequencies (Aufrichtig, 2000; Reiser, 2006; Segui, 2006) or when different conditions do not lead to a substantial change in noise correlations (Zhang, 2004). Part of this usefulness resides in their simplicity. Of the many statistical properties that can be used to characterize a population of images, nonprewhitening observers only require knowing the mean signal profile.

The nonprewhitening matched filter (NPW) observer was previewed in the previous section near Equation 17.21. This model uses an observer template that is proportional to the difference signal

$$\mathbf{w}_{\mathrm{NPW}} = \mathbf{s} \qquad (17.29)$$

where we have set the proportionality constant to 1 for simplicity. The NPW observer is optimal in Gaussian white noise (assuming no internal noise), but can be highly suboptimal when strong low-pass correlations are present. This is because most simple "lesion" type signals have a significant low-frequency component. As a result, the NPW template is sensitive to noise in these frequencies, which can substantially degrade performance in the presence of low-pass object variability.

The problem of low-frequency variability in the NPW observer led Burgess (Burgess, 1994) to develop the eye-filtered nonprewhitening observer model, abbreviated NPWE. The idea was to build the contrast sensitivity of the human visual system into the model observer template. Burgess viewed this as convolving the incoming image with a filter matched to the frequency response of measured human contrast sensitivity functions (CSFs). Based on the work of Barten (Kelly, 1975; Barten, 1987; Barten, 1999), he used a CSF profile with frequency response $E(f) = f^a e^{-bf}$, with a set to 1.3 and b set in a viewing distance-dependent manner so that peak sensitivity would occur at a frequency of 4 cycles per degree visual angle. Note that this form for the CSF will suppress input from low spatial frequencies.

If we generically refer to convolution with the CSF by the matrix **E** (for **E**ye filter), then the filtered input is given by **Eg**. Applying the NPW strategy to this input leads to a difference signal of **Es**. Therefore the NPWE template uses the convolution twice, once on **g** and once on **s**, and thus the form of the observer template is

$$\mathbf{w}_{\mathrm{NPWE}} = \mathbf{E}^t\mathbf{E}\mathbf{s} \qquad (17.30)$$

In terms of Barten's CSF, the NPWE observer template is the convolution of the signal with a filter having an isotropic filter response of $f^{2a}\, e^{-2bf}$.

Recently, Abbey and Eckstein (Abbey, 2007) described an alternative approach to deriving the NPWE observer. Drawing on previous work by Ahumada (Ahumada, 1987), they proposed an internal noise component defined in the image domain and effectively added to the image. If we define the image-domain internal noise by the vector, $\boldsymbol{\varepsilon}$, with elements indicating the additional positive or negative intensity in each pixel, then we can modify the observer response in Equation 17.7 to be

$$r = \mathbf{w}^t(\mathbf{g} + \boldsymbol{\varepsilon}) \qquad (17.31)$$

This is equivalent to Equation 17.7 if we define the scalar internal noise component, $\varepsilon = \mathbf{w}^t\boldsymbol{\varepsilon}$. We assume that $\boldsymbol{\varepsilon}$ is zero mean with a covariance matrix, $\mathbf{\Sigma}_{\boldsymbol{\varepsilon}}$. If we now assume that the observer prewhitens the internal noise, but does not incorporate the external noise covariance into its observer template, then the observer template is

$$\mathbf{w}_{\mathrm{NPWE}} = \mathbf{\Sigma}_{\boldsymbol{\varepsilon}}^{-1}\mathbf{s} \qquad (17.32)$$

Positing an internal noise power spectrum that is isotropic with frequency profile $N_\varepsilon(f) = f^{-2a}e^{2bf}$ results in $\mathbf{\Sigma}_{\boldsymbol{\varepsilon}}^{-1}$ being equivalent to $\mathbf{E}^t\mathbf{E}$, and the identical observer template is obtained. While this approach produces the same observer template in nonprewhitening models, it is more straightforward to apply to prewhitening and channel models where the eye-filtering approach is more problematic. For example, Barrett points out how any effects of an eye filter applied through filtering as in Equation 17.30 would be negated by a prewhitened matched filter (Barrett, 1993).

The NPW and NPWE models for the example detection task are shown in Figure 17.2. The main effect of the eye filter in this case is to suppress low spatial frequencies.

17.3.2 Prewhitening observers

By contrast with the nonprewhitening observer models described previously, prewhitening observers do explicitly incorporate the correlation structure of the noise into the observer template (Wagner, 1985; Rolland, 1991; Tapiovaara, 1993). The basic idea here is that the image first undergoes a whitening transformation, effectively multiplying the image vector, **g**, by $\mathbf{\Sigma}_g^{-1/2}$ and then applying the matched filter, $\mathbf{\Sigma}_g^{-1/2}\mathbf{s}$. This two-step process is implemented in the prewhitening matched-filter observer template

$$\mathbf{w}_{\mathrm{PW}} = \mathbf{\Sigma}_g^{-1}\mathbf{s} \qquad (17.33)$$

The prewhitening matched filter is equivalent to an ideal observer when random variability in **g** is governed by a multivariate Gaussian with the same covariance structure in both classes. The observer template is also considered optimal because it can be shown to maximize detectability in Equation 17.20 in the absence of internal noise. Finally, we note that the prewhitening observer as we have defined it has also been

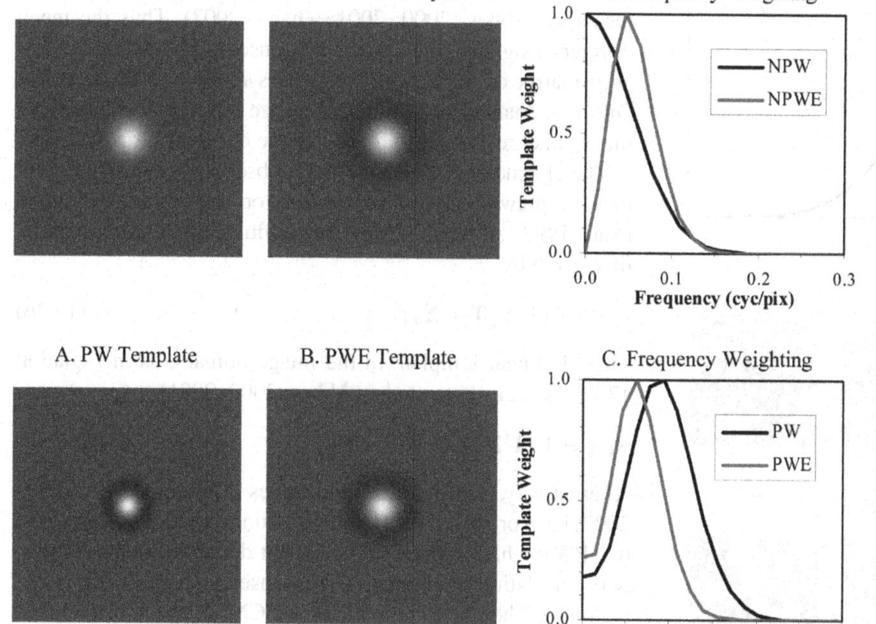

Figure 17.2 Nonprewhitening observer templates. The NPW observer template (A) is equivalent to the target profile, while the eye-filtered NPW template (B) has a visible dark surrounding region. This is seen in the frequency weights of the templates (C) as suppression of low spatial frequencies. Note that the 4 cyc/deg peak in the CSF has been set at 0.07 cyc/pixel.

Figure 17.3 Prewhitening observer templates. The PW observer template (A) and PWE observer template (B) both have inhibitory surrounding regions, with the PWE having a broader spatial extent. Both profiles exhibit low-frequency suppression (C) with the PWE model peaking at somewhat lower frequency due to the CSF peak at 0.07 cyc/pixel.

referred to as the Hotelling observer, Fisher-Hotelling observer, and Fisher linear discriminant, with distinctions being drawn on the basis of whether the signal and/or covariance matrix are estimated from samples or derived from a model. We will generically refer to all such variants simply as prewhitening models.

The rationale for using the prewhitening observer as a model of a human observer rests on the idea that humans can change and adapt the template they use to perform a task based on the structure of the noise present. While the early work of Myers and Barrett (Myers, 1985) showed that this was not the case for strong high-pass noise correlations, the prewhitening model has been successful at predicting observer performance in the presence of low-pass correlations that arise from the presence of normal anatomy in the images (Rolland, 1992; Burgess, 2007).

As mentioned above in the development of the NPWE model in Equation 17.32, an internal noise spectrum (Zhang, 2007) can be used to implement a form of CSF eye-filtering on the prewhitening observer model. The approach (Abbey, 2007) specifies internal noise in the image domain, which allows for an internal noise covariance matrix, $\mathbf{\Sigma}_{\varepsilon}$. The internal noise is presumed to be independent of the (external) image noise, and therefore the total noise covariance the observer must adapt to is $\mathbf{\Sigma}_{\mathbf{g}} + \mathbf{\Sigma}_{\varepsilon}$, and the resulting eye-modeled prewhitening (PWE) observer template is given by

$$\mathbf{w}_{\text{PWE}} = \left(\mathbf{\Sigma}_{\mathbf{g}} + \mathbf{\Sigma}_{\varepsilon}\right)^{-1}\mathbf{s} \tag{17.34}$$

This observer template has shown relatively good agreement with human observers in direct evaluations of observer templates. However, the approach is still relatively untested as a practical model of performance.

The PW and PWE models for the example detection task are shown in Figure 17.3. Here the internal noise spectrum (i.e. eye filter) serves to shift the spatial-frequency weighting of the observer template downward. This is due to the trough of the internal noise spectrum at approximately 0.07 cyc/pixel corresponding to a presumed minimum at 4 cycles per degree visual angle (Abbey, 2007).

17.3.3 Channelized observers

Channelized observers (Myers, 1987) utilize the notion of visual channels – derived from the literature on human and animal vision – as an initial step in the detection process. The channels define a set of linear features that have certain spatial frequency characteristics (octave bandwidth, center frequency, etc.), and the responses of these features are used to formulate the decision variable. In principle, the channel features are intended to model the relevant components of the early visual system. However, the relatively limited number of channels used in most medical imaging applications (3–50 total) makes this something of a stretch by comparison with the many thousands of receptive fields that are postulated to operate in early visual areas such as V1 (Watson, 1987). Nonetheless, channels allow for limited forms of partial prewhitening to be used in formulating a decision variable. Typically channels are used that have constant bandwidth of about an octave or so, and therefore on an absolute scale the low-frequency channels have a narrower spatial frequency band than high-frequency channels. This allows for more precise prewhitening at low spatial frequencies, consistent with experimental findings.

Examples of linear features in a channel model are shown in Figure 17.4, where we have the frequency response of each channel shown in Figure 17.4A, and then two sets of channels that are consistent with this frequency response. The first is an isotropic set of channel features with one feature per channel and $N_C = 4$ shown in Figure 17.4B. The second set of channel features in Figure 17.4C has multiple orientation selective

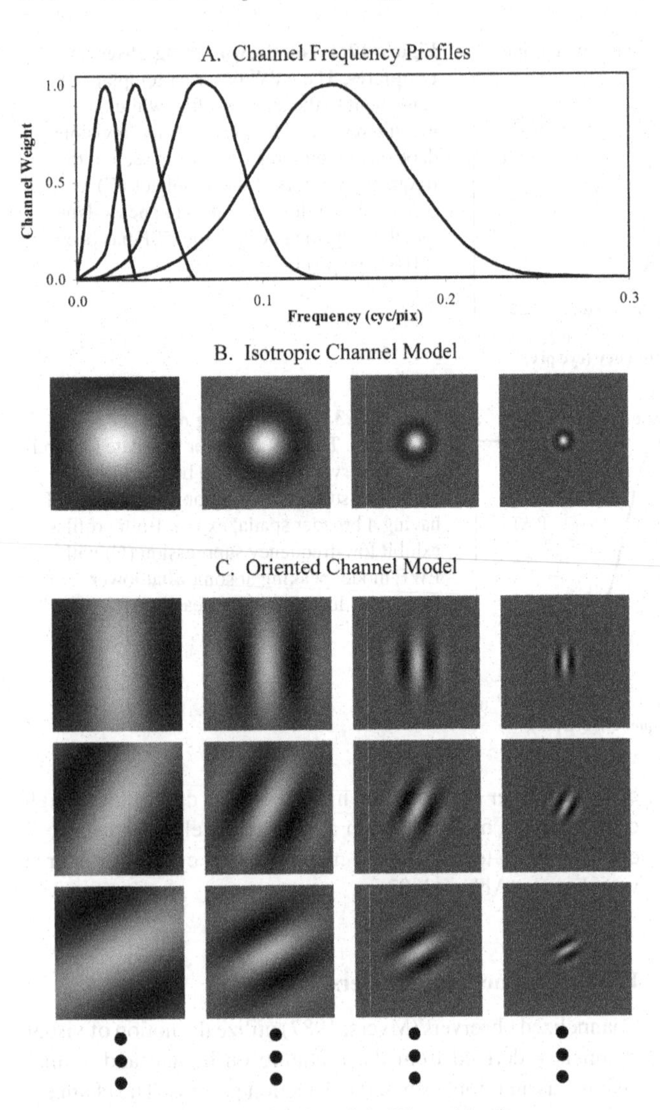

Figure 17.4 Example channel models. Plots of the frequency weights (A) for four spatial frequency channels. Channel feature weights are shown for two models consistent with these channels. A set of four isotropic channel features (B) have a single radially symmetric feature for each channel. A larger set of non-isotropic channel features (C) has multiple features for each channel, each tuned to a different orientation.

features per frequency. Isotropic channels have been used often in simple studies related to medical imaging, where a rotationally symmetric target is masked by noise with an isotropic NPS. In this case there is no need for orientation selective features, and the basic assumption is that responses from orientation selective features within a channel are averaged. For cosine phase features like those used in Figure 17.4C, this is roughly equivalent to having a single isotropic feature. Larger feature sets are needed in tasks with asymmetric targets or noise correlations (Eckstein, 1995).

Let $\mathbf{t}_c$ with $c = 1, \ldots, N_C$ be channel templates with elements representing the weightings of N_C channel features, and let each of these be a column of the channel matrix $\mathbf{T}$. Then we can write the channel responses, $\mathbf{v}$, as the noisy matrix-vector product

$$\mathbf{v} = \mathbf{T}^t\mathbf{g} + \mathbf{z} \tag{17.35}$$

where $\mathbf{z}$ represents zero-mean internal noise in each channel response (Abbey, 2000; 2001; Zhang, 2007). Thus the mean difference signal in the channel responses is $\mathbf{T}^t\mathbf{s}$, and the covariance matrix of the channel responses is $\mathbf{\Sigma_v} = \mathbf{T}^t\mathbf{\Sigma_g}\mathbf{T} + \mathbf{\Sigma_z}$. Different channelized observers utilize different weightings of the channel responses to formulate the decision variable.

The channelized Hotelling (CH) observer essentially implements a prewhitening matched filter on the channel responses (Yao, 1992; Abbey, 2001). The resulting decision variable is then given by

$$r = \mathbf{s}^t\mathbf{T}\left(\mathbf{T}^t\mathbf{\Sigma_g}\mathbf{T} + \mathbf{\Sigma_z}\right)^{-1}\mathbf{v} \tag{17.36}$$

The CH linear template in the image domain, as in Equation 17.7, has been shown to be (Abbey, 2000; 2001)

$$\mathbf{w}_{\mathrm{CH}} = \mathbf{T}\left(\mathbf{T}^t\mathbf{\Sigma_g}\mathbf{T} + \mathbf{\Sigma_z}\right)^{-1}\mathbf{T}^t\mathbf{s} \tag{17.37}$$

Alternatively, Gifford and colleagues (Gifford, 2003; 2005; 2007) have proposed the channelized nonprewhitening observer (CNPW), which utilizes channels but does not attempt to deal with correlations in the channel responses arising from the external noise. The linear template of the CNPW observer template is given by

$$\mathbf{w}_{\mathrm{CNPW}} = \mathbf{T}\left(\sigma_n^2\mathbf{T}^t\mathbf{T} + \mathbf{\Sigma_z}\right)^{-1}\mathbf{T}^t\mathbf{s} \tag{17.38}$$

which is equivalent to replacing $\mathbf{\Sigma_g}$ in Equation 17.37 with σ_n^2. In effect, the CNPW observer uses the channels to make an eye filter for interpreting the signal.

The internal noise covariance matrix in Equation 17.37 can have important consequences on the model (Zhang, 2007). For example, in the absence of internal noise, and using a dense channel model, Abbey and Barrett (Abbey, 2001) found that channel covariance matrices could be poorly conditioned, requiring more than five digits of accuracy to effectively invert. It would seem unlikely that the human eye-brain system would be able to implement such a complex computation with sufficiently high accuracy. However, in the presence of internal noise in each channel, the condition of the matrix improved to the extent that only two digits were required.

The image-domain internal noise model described by Abbey and Eckstein (Abbey, 2007) can be used to implement an effective CSF eye-filter in the CH observer model. Internal noise is assumed to be independent in each channel, forcing $\mathbf{\Sigma_z}$ to be diagonal, and the variance of each element is

$$\sigma_{z_c}^2 = \mathbf{t}_c^t\mathbf{\Sigma}_\varepsilon\mathbf{t}_c \tag{17.39}$$

where $\mathbf{\Sigma}_\varepsilon$ is the covariance of the internal noise in Equation 17.31.

The CH, CH with an eye filter (CHE), and CNPW models for the example detection task are shown in Figure 17.5. Note that the somewhat rough appearance of the frequency plots in Figure 17.5D is a result of the coarse channel model shown in Figure 17.4.

17.4 SUMMARY

In this chapter we have reviewed model observers for simple detection and discrimination experiments masked by noise. The

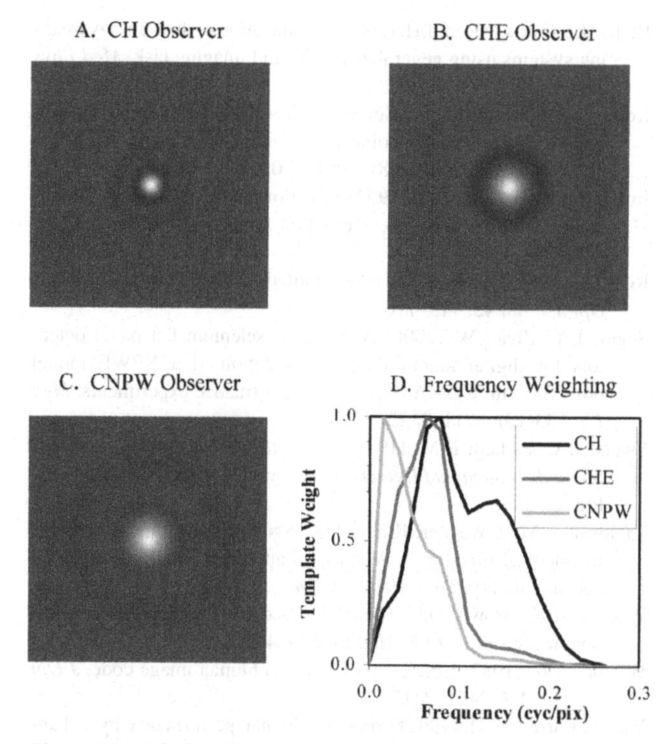

Figure 17.5 Channelized observer templates. All observer templates use the channel model in Figure 17.4B. The CH observer template (A), CHE observer template (B), and CNPW observer template (C) all have some degree of inhibitory surrounding region. The somewhat jagged appearance of the frequency weights (C) is due to the coarse frequency sampling of the channel model.

methodology for evaluating model observers essentially parallels the way signal detection theory has taught us to analyze human-observer decisions. The difference is that the model observers are associated with an explicit formulation of the decision variable.

We have described a number of commonly used linear model observers. These are distinguished most critically by the way they incorporate image correlations into the model-observer template. The models span the gamut from total incorporation to non-incorporation going from the prewhitening matched filter to the nonprewhitening matched filter and with partial incorporation via channels in the middle. We have deliberately avoided trying to argue for one particular model or approach as best. Instead we have tried to give a concise development of each, and we hope that readers will be able to use their insight of the problem at hand to guide them along with a more focused evaluation of the relevant literature.

While the purpose of the chapter is to act as a tutorial and reference for those considering the use of such models for evaluating imaging methodology, we believe strongly that there are still many important questions to answer in this field. These include a better fundamental understanding of the mechanisms at play in medical image perception as well as more practical knowledge and experience of fine-tuning models to specific imaging modalities and relevant tasks. The methods covered in this chapter represent more than 20 years of investigation into this topic by many investigators. But given the combined complexity of human biology to be imaged, medical imaging systems themselves, and visual perception of the images they produce, it is not surprising to us that there are still many issues to be resolved.

REFERENCES

Abbey, C.K., Bochud, F.O. (2000) Modeling visual detection tasks in correlated image noise with linear model observers. In *The Handbook of Medical Imaging: Volume 1, Progress in Medical Physics and Psychophysics*, ed. Kundel, H., pp. 629–654.

Abbey C.K., Barrett, H.H. (2001) Human and model-observer performance in ramp-spectrum noise: effects of regularization and object variability. *J Opt Soc Am A* **18**(3): 473–488.

Abbey, C.K., Eckstein, M.P. (2007) Classification images for simple detection and discrimination tasks in correlated noise. *J Opt Soc Am A* **24**(12): B110–B124.

Ahumada, A.J., Jr. (1987) Putting the visual system noise back in the picture. *J Opt Soc Am A* **4**(12): 2372–2378.

Aufrichtig, R., Xue, P. (2000) Dose efficiency and low-contrast detectability of an amorphous silicon X-ray detector for digital radiography. *Phys Med Biol* **45**(9): 2653–2669.

Barrett, H.H., Yao, J., Rolland, J.P., Myers, K.J. (1993) Model observers for assessment of image quality. *Proc Natl Acad Sci USA* **90**(21): 9758–9765.

Barrett, H.H., Denny, J.L., Wagner, R.F., Myers, K.J. (1995) Objective assessment of image quality. II. Fisher information, Fourier crosstalk, and figures of merit for task performance. *J Opt Soc Am A* **12**(5): 834–852.

Barten, P.G.J. (1987) The SQRI method: a new method for the evaluation of visible resolution on a display. *Proc Soc Inf Disp* **28**: 253–262.

Barten, P.G.J. (1999) *Contrast Sensitivity of the Human Eye and its Effect on Image Quality*. Bellingham, WA: SPIE Press.

Burgess, A.E. (1994) Statistically defined backgrounds: performance of a modified nonprewhitening observer model. *J Opt Soc Am A* **11**(4): 1237–1242.

Burgess, A.E. (1995) Comparison of receiver operating characteristic and forced choice observer performance measurement methods. *Med Phys* **22**(5): 643–655.

Burgess, A.E., Li, X., Abbey, C.K. (1997) Visual signal detectability with two noise components: anomalous masking effects. *J Opt Soc Am A* **14**(9): 2420–2442.

Burgess, A.E. (1999) The Rose model, revisited. *J Opt Soc Am A* **16**(3): 633–646.

Burgess, A.E., Judy, P.F. (2007) Signal detection in power-law noise: effect of spectrum exponents. *J Opt Soc Am A* **24**(12): B52–B60.

Eckstein, M.P., Whiting, J.S. (1995) Lesion detection in structured noise. *Acad Radiol* **2**(3): 249–253.

Fukunaga, K. (1972) *Introduction to Statistical Pattern Recognition*. New York, NY: Academic Press.

Gifford, H.C., Pretorius, P.H., King, M.A. (2003) Comparison of human- and model-observer LROC studies. *Proc SPIE* **5034**: 112–122.

Gifford, H.C., King, M.A., Pretorius, P.H., Wells, R.G. (2005) A comparison of human and model observers in multislice LROC studies. *IEEE Trans Med Imaging* **24**(2): 160–169.

Gifford, H.C., Kinahan, P.E., Lartizien, C., King, M.A. (2007) Evaluation of multiclass model observers in PET LROC studies. *IEEE Trans Nucl Sci* **54**: 116–123.

Green, D.M., Swets, J.A. (1966) *Signal Detection Theory and Psychophysics*. New York, NY: Wiley.

Gur, D., Sumkin, J.H., Rockette, H.E. *et al.* (2004) Changes in breast cancer detection and mammography recall rates after the introduction of a computer-aided detection system. *J Natl Cancer Inst* **96**(3): 185–190.

Hanson, K.M. (1979) Detectability in computed tomographic images. *Med Phys* **6**: 441–451.

Judy, P.F., Swensson, R.G. (1981) Lesion detection and signal-to-noise ratio in CT images. *Med Phys* **8**: 13–23.

Judy, P.F., Swensson, R.G. (1987) Display thresholding of images and observer detection performance. *J Opt Soc Am A* **4**(5): 954–965.

Kelly, D.H. (1975) Spatial frequency selectivity in the retina. *Vision Res* **15**: 665–672.

Mahalanobis, P. (1936) On the generalized distance in statistics. *Proc Nat Inst Sci India (Calcutta)* **2**: 49–55.

Metz, C.E. (1986) ROC methodology in radiologic imaging. *Invest Radiol* **21**: 720–733.

Myers, K.J., Barrett, H.H., Borgstrom, M.C., Patton, D.D., Seeley, G.W. (1985) Effect of noise correlation on detectability of disk signals in medical imaging. *J Opt Soc Am A* **2**(10): 1752–1759.

Myers, K.J., Barrett, H.H. (1987) Addition of a channel mechanism to the ideal-observer model. *J Opt Soc Am A* **4**(12): 2447–2457.

Obuchowski, N.A., Schoenhagen, P., Modic, M.T., Meziane, M., Budd, G.T. (2007) Incidence of advanced symptomatic disease as primary endpoint in screening and prevention trials. *Am J Roentgenol* **189**(1): 19–23.

Pisano, E.D., Gatsonis, C., Hendrick, E. *et al.* (2005) Diagnostic performance of digital versus film mammography for breast-cancer screening. *N Engl J Med* **353**(17): 1773–1783.

Reiser, I., Nishikawa, R.M. (2006) Identification of simulated microcalcifications in white noise and mammographic backgrounds. *Med Phys* **33**(8): 2905–2911.

Richard, S., Siewerdsen, J.H. (2007) Optimization of dual-energy imaging systems using generalized NEQ and imaging task. *Med Phys* **34**(1): 127–139.

Rolland, J.P., Barrett, H.H., Seeley, G.W. (1991) Ideal versus human observer for long-tailed point spread functions: does deconvolution help? *Phys Med Biol* **36**(8): 1091–1109.

Rolland, J.P., Barrett, H.H. (1992) Effect of random background inhomogeneity on observer detection performance. *J Opt Soc Am A* **9**: 649–658.

Rose, A. (1953) Quantum and noise limitations of the visual process. *J Opt Soc Am* **43**: 715–716.

Segui, J.A., Zhao, W. (2006) Amorphous selenium flat panel detectors for digital mammography: validation of a NPWE model observer with CDMAM observer performance experiments. *Med Phys* **33**(10): 3711–3722.

Swets, J.A., Pickett, R.M. (1982) *Evaluation of Diagnostic Systems: Methods from Signal Detection Theory*. New York, NY: Academic Press.

Tapiovaara, M.J., Wagner, R.F. (1993). SNR and noise measurements for medical imaging. I. A practical approach based on statistical decision theory. *Phys Med Biol*, **38**(1): 71–92.

Wagner, R.F., Brown, D.G. (1985) Unified SNR analysis of medical imaging systems. *Phys Med Biol* **30**: 489–518.

Watson, A.B. (1987) Efficiency of a model human image code. *J Opt Soc Am A* **4**: 2401–2417.

Yao, J., Barrett, H.H. (1992) Predicting human performance by a channelized Hotelling observer model. *Proc SPIE* **1768**: 161–168.

Zhang, Y., Pham, B.T., Eckstein, M.P. (2004) Automated optimization of JPEG 2000 encoder options based on model observer performance for detecting variable signals in X-ray coronary angiograms. *IEEE Trans Med Imaging* **23**(4): 459–474.

Zhang, Y., Pham, B.T., Eckstein, M.P. (2007) Evaluation of internal noise methods for Hotelling observer models. *Med Phys* **34**(8): 3312–3322.

18

Implementation of observer models

MATTHEW KUPINSKI

18.1 INTRODUCTION

A guiding principle of this book is that image quality should be defined using measures that account for the task to be performed, the observer performing this task, and the patient population of interest. Task-based measures are those that correlate with, or directly measure, observer performance and can thus be used to evaluate image quality for the purpose of rank-ordering imaging systems, optimizing free parameters of post-processing algorithms, or evaluating the effectiveness of computer-aided diagnosis, among other applications. Measures such as resolution, contrast, and pixel mean-squared error are quantitative but are not directly related to task performance and, thus, have a less meaningful interpretation. Task-based measures of image quality are, in general, more difficult to compute due to the statistical nature of image data. Prior knowledge, or reliable models, of image statistics are required to calculate task-based measures of image quality. In this chapter, we will review a selection of observer models and describe practical methods for their implementation.

18.1.1 The imaging equation

The relationship between an object and its image is fundamental to image analysis. Digital imaging can be represented mathematically as

$$\mathbf{g} = Hf(\mathbf{r}) + \mathbf{n} \tag{18.1}$$

Here, $f(\mathbf{r})$ represents the object being imaged as a function of a spatial coordinate $\mathbf{r}$. This function could, for example, represent the distribution of the radiotracer in nuclear-medicine imaging or the distribution of X-ray attenuation coefficients in computed tomography (CT). For notational convenience, the object $f(\mathbf{r})$ is represented as an infinite-dimensional vector $\mathbf{f}$. The operator H represents the imaging system; it describes how an object is mapped to a noise-free, discrete set of measurements. The vector $\mathbf{n}$ is the noise due to the measurement process. We have represented the noise here as additive, which is always valid since the definition of $\mathbf{n}$ is simply $\mathbf{g} - H\mathbf{f}$. Finally, $\mathbf{g}$ is the image data that are returned from the system. This M-dimensional vector may be a collection of data from many cameras or projections in tomographic imaging or it may be an image that is ready to be displayed to a human observer as in planar X-ray imaging. We will treat the vectors $\mathbf{g}$ and $\mathbf{n}$ as $M \times 1$ column vectors. Each measurement from the imaging system is one element in the vector $\mathbf{g}$. Thus, for two-dimensional image data, there is an ordering of the two-dimensional data into the vector $\mathbf{g}$ that must be consistent from acquisition to acquisition. The same is true for tomographic image data which also includes angle or projection number as well as the two-dimensional measurement data.

For tomographic imaging systems, a subsequent reconstruction operator O is required to produce a reconstructed image for viewing, i.e.

$$\hat{\mathbf{f}} = O\mathbf{g} \tag{18.2}$$

With this notation it is important to note that the reconstruction $\hat{\mathbf{f}}$ is finite-dimensional whereas the patient being imaged $\mathbf{f}$ is infinite-dimensional. Thus, $\mathbf{f}$ and $\hat{\mathbf{f}}$ cannot be directly compared using measures such as mean-squared error.

Some observer models act on image data $\mathbf{g}$ whereas others use reconstructed images $\hat{\mathbf{f}}$. Indeed, some imaging systems, such as planar X-ray imaging, do not require an image reconstruction step. We will develop our object models using only image data $\mathbf{g}$. In many situations, the observer model can just as easily be applied to reconstructed imaged $\hat{\mathbf{f}}$. The imaging application determines whether $\mathbf{g}$ or $\hat{\mathbf{f}}$ should be used by the observer.

Image quality is a statistical concept and thus, a characterization of the random components in the imaging chain is necessary. The noise vector $\mathbf{n}$ accounts for randomness due to the measurement process. If the same object is imaged multiple times, the image data returned are not exactly the same due to measurement noise. The object $\mathbf{f}$ being imaged is also a random quantity because the same patient is not being imaged every time. Thus, the image data $\mathbf{g}$ has random contributions from at least two different sources: the object being imaged and the measurement noise. The measurement noise can depend on the object being imaged. For example, in nuclear-medicine imaging, the image data $\mathbf{g}$ is conditionally Poisson. That is, the probability density function $pr(\mathbf{g} \mid \mathbf{f})$ is Poisson distributed with mean $H\mathbf{f}$. In MRI, for example, the measurement noise is often assumed to be an uncorrelated Gaussian distribution that is independent of the object. Thus, $pr(\mathbf{g} \mid \mathbf{f})$ is modeled as an uncorrelated Gaussian with mean $H\mathbf{f}$. In both cases, we represent the distribution on the measurement noise using $pr(\mathbf{g} \mid \mathbf{f})$.

We can represent the full randomness in the image data $\mathbf{g}$ as

$$pr(\mathbf{g}) = \langle pr(\mathbf{g} \mid \mathbf{f}) \rangle_{\mathbf{f}} \tag{18.3}$$

We use angle brackets to denote expectations in this chapter. For example, the average of the function $\beta(\mathbf{g})$ over the randomness in the image data $\mathbf{g}$ is

$$\langle \beta(\mathbf{g}) \rangle_{\mathbf{g}} = \int \beta(\mathbf{g}) pr(\mathbf{g}) d^M g \tag{18.4}$$

The Handbook of Medical Image Perception and Techniques, ed. Ehsan Samei and Elizabeth Krupinski. Published by Cambridge University Press.

The average of the image data when the object **f** is fixed is given by

$$\bar{\mathbf{g}}(\mathbf{f}) = \langle \mathbf{g} \rangle_{\mathbf{g}|\mathbf{f}} \tag{18.5}$$

Here, the mean is a function of **f** because it depends on the fixed object **f**. The overall average is given by

$$\bar{\bar{\mathbf{g}}} = \langle \bar{\mathbf{g}}(\mathbf{f}) \rangle_{\mathbf{f}} \tag{18.6}$$

Using this notation, $\bar{\mathbf{g}}(\mathbf{f})$ is an average over just the measurement noise $pr(\mathbf{g}\,|\,\mathbf{f})$, and $\bar{\bar{\mathbf{g}}}$ is an average over $pr(\mathbf{g})$ which includes both the measurement noise and the object variability.

18.1.2 Tasks in medical imaging

Tasks in medical imaging are generally classification tasks, estimation tasks, or a combination of both. Classification tasks include the detection of a signal such as a tumor or the classification of a tumor as malignant or benign. The observer's decision is to classify the image into one of L classes, where L is a finite number. If only two classes are present, i.e. $L = 2$, then the classification is called a signal-detection task. Estimation tasks are broad and can include the estimation of quantities related to a patient's health status, such as cardiac ejection fraction. Estimation tasks also include the estimation of signal parameters such as tumor size or location. Estimation tasks can be thought of as a generalization of classification tasks where the number of possible classes L approaches infinity. Thus, there is a strong connection between the mathematics that describes the observers for classification and estimation tasks (Barrett, 1990). Combined tasks are very common in medical imaging and include the detection of an unknown signal and the estimation of its location and size. The task in screening mammography is a combined task; the radiologist must determine if a lesion is present, and, if it is present, he or she must determine the type of cancer, the location, and the extent of the possible lesion.

For a signal-detection task, there are two possible classes for the image data: the signal may be present (H_2 hypothesis), or the signal is absent in the image (H_1 hypothesis). The image data are given by

$$H_2 : \mathbf{g} = H\mathbf{f} + \mathbf{n} = H(\mathbf{f}_b + \mathbf{f}_s) + \mathbf{n} \tag{18.7}$$

$$H_1 : \mathbf{g} = H\mathbf{f} + \mathbf{n} = H(\mathbf{f}_b) + \mathbf{n} \tag{18.8}$$

where $\mathbf{f}_b$ and $\mathbf{f}_s$ are the background and signal portions of the object **f**. Both $\mathbf{f}_b$ and $\mathbf{f}_s$ may be random. The image data are sampled from one of two probability density functions. If the signal is present (H_2 hypothesis), then the image data are sampled from $pr(\mathbf{g} \mid H_2)$. If the signal is not present (H_1 hypothesis), then the image data are sampled from $pr(\mathbf{g} \mid H_1)$. In general,

$$pr(\mathbf{g}\,|\,H_j) = \langle pr(\mathbf{g}\,|\,\mathbf{f}) \rangle_{\mathbf{f}|H_j} \tag{18.9}$$

In the above equation, **f** can be from either the H_2 or H_1 class.

Any observer performing a signal-detection task can be thought of as mapping the image data **g** to a scalar decision variable (or test statistic) t, i.e.

$$t = T(\mathbf{g}) \tag{18.10}$$

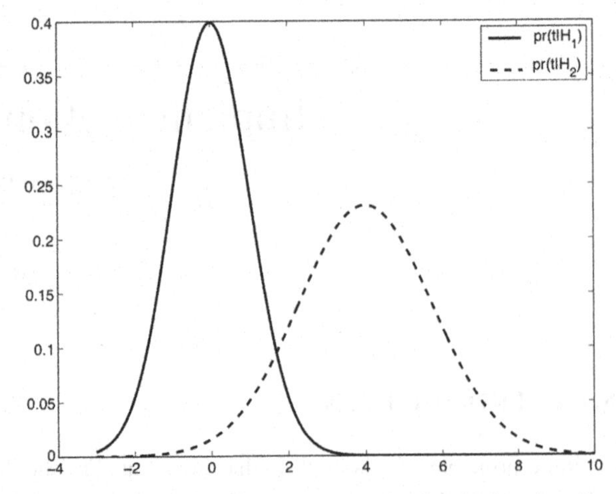

Figure 18.1 Example test statistic probability density functions for an observer performing a signal-detection task.

Here the function $T(\mathbf{g})$ is the observer's mapping function. If the test statistic is above a threshold t_c, then the observer decides that the signal is present. If the test statistic is below the threshold, then the observer decides that the signal is absent. All observers performing binary classification tasks can be mathematically represented in this manner. There are two probability density functions which characterize the observer's decision variable: $pr(t \mid H_2)$ and $pr(t \mid H_1)$. These are the density functions for the test statistics when the signal is present (i.e. H_2) and when the signal is absent (i.e. H_1). Example densities are depicted in Figure 18.1. In general, the more overlap between these two densities, the poorer the observer performance.

For estimation tasks, the observer is attempting to estimate a set of parameters θ which could, for example, represent the location of a tumor in an image. The imaging equation becomes

$$\mathbf{g} = H\mathbf{f} + \mathbf{n} = H\mathbf{f}(\theta) + \mathbf{n} \tag{18.11}$$

where $\mathbf{f}(\theta)$ represents the random object as a function of the parameters θ to be estimated. These parameters θ are not the same from patient to patient and, thus, there is a probability density function which characterizes the randomness in this vector, i.e. $pr(\theta)$. The image data are characterized by a likelihood given by

$$pr(\mathbf{g}\,|\,\theta) = \langle pr(\mathbf{g}\,|\,\mathbf{f}, \theta) \rangle_{\mathbf{f}|\theta} \tag{18.12}$$

where $pr(\mathbf{g}|\mathbf{f}, \theta)$ is the measurement noise. An observer performing an estimation task takes the image data and produces a set of estimates $\hat{\theta}$ via

$$\hat{\theta} = \Theta(\mathbf{g}) \tag{18.13}$$

where the vector-valued function $\Theta(\mathbf{g})$ maps the image data to a vector of estimates. The vector $\hat{\theta}$ is an estimate of the true parameters θ.

Observers performing combined detection/estimation tasks first produce a test statistic, as do observers performing just a signal-detection task. However, if the test statistic is above the threshold, then the observer also produces a vector of estimates $\hat{\theta}$.

18.1.3 Figures of merit

A figure of merit characterizes the average performance of the observer on the task. For signal-detection tasks, there are two common figures of merit. The first is the signal-to-noise ratio (SNR) of the observer's test statistics, defined by

$$\mathrm{SNR}^2 = \frac{(\bar{t}_2 - \bar{t}_1)^2}{\frac{1}{2}(\sigma_2^2 + \sigma_1^2)} \tag{18.14}$$

where $\bar{t}_i = \langle t \rangle_{t|\mathrm{H_i}}$ is the mean test statistic under the $\mathrm{H_i}$ hypothesis and $\sigma_{\mathrm{i}}^2 = \langle (t - \bar{t}_i)^2 \rangle_{t|\mathrm{H_i}}$ is the variance of the test statistic under the $\mathrm{H_i}$ hypothesis. Given a set of test statistics from images with and without signals, the SNR in Eqn. 18.14 can be estimated using sample means and sample variances. The area under the receiver operating characteristic (ROC) curve, or AUC, is another common figure of merit. The AUC can be estimated using a two-alternative forced-choice (2AFC) experiment. In this experiment, the test statistic for an image with a signal is compared to that of an image without a signal present. The observer makes a correct decision if the test statistic for the image with the signal is greater than that of the image without the signal. Performing this computation for all possible pairings of images with a signal to images without a signal and computing the fraction of correct decisions produces an unbiased estimate of the observer's AUC.

For estimation tasks, the ensemble mean-square error (EMSE) is

$$\mathrm{EMSE} = \left\langle \left\langle \|\hat{\theta} - \theta\|^2 \right\rangle_{\mathbf{g}|\theta} \right\rangle_\theta \tag{18.15}$$

This quantity is called *ensemble* because the figure of merit is averaged over the ensemble of the parameters being estimated θ. For a testing dataset of images $\mathbf{g}$ with known θ, the EMSE can be estimated using sample averages.

Finally, for combined detection/estimation tasks, there is the newly developed estimation ROC (EROC) analysis (Clarkson, 2007). The x-axis of an EROC curve is the false-positive fraction as in the standard ROC curve. The y-axis is the average value of a utility function on the parameters being estimated for the true-positive cases. The area under the EROC curve (AEROC) is a figure of merit that characterizes the ability of the observer to perform the combined task. Location ROC analysis (LROC) is a specific form of EROC analysis where the estimation task is to locate the signal.

18.1.4 Training and testing

Most observers, both mathematical and human, require training to make decisions. Once trained, observers are tested to assess their performance. For model observers, training takes the form of a training dataset that is used to set the parameters of the model. It is generally considered "unfair" to test an observer on the same set of images used for training since the observer is expected to perform well on these images. Thus, two sets of data are required to assess model observers: the training set used to determine the parameters of the observer model and the testing set used to measure the performance of this observer by the computation of a figure of merit.

Let us consider a detection task and we will characterize observer performance using AUC. Let us also assume that we have an infinite amount of training data. The performance of the model observer if such an infinite dataset existed is the best possible performance (i.e. highest AUC) that the model observer can achieve. With finite training datasets, the performance of the model observer will be, on average, lower than this best possible performance. This downward bias is caused by the finite amount of training data. Training and testing on the same dataset causes an upward bias. While the goal is to train the observer model as best as possible, bias is often introduced with finite training and testing data.

For signal-detection tasks, the training images are indicated using $\mathbf{g}_i^{(1)}$ and $\mathbf{g}_i^{(2)}$ where the subscript i indicates the image number and the superscript indicates the truth state ($\mathrm{H_1}$ signal absent or $\mathrm{H_2}$ signal present). For estimation tasks, the training images will be indicated by $\mathbf{g}_i$; the corresponding parameters to be estimated for this ith image are given by θ_i . Knowledge of the truth in the training dataset, either the class for signal detection or the parameters to be estimated for estimation tasks, will be assumed for the observer models discussed below. Observer performance can be assessed without knowledge of the truth in certain situations (Henkelman, 1990; Hoppin, 2002; Kupinski, 2002). We will limit our discussion to training datasets that have known truth.

18.2 LINEAR OBSERVER MODELS

18.2.1 Linear detection tasks

A linear observer is one that uses only linear manipulations on the image data to make decisions. For example, a linear observer performing a detection task produces test statistics using

$$t = T(\mathbf{g}) = \mathbf{w}^t \mathbf{g} \tag{18.16}$$

where $\mathbf{w}$ is a vector with the same dimension as $\mathbf{g}$ and $\mathbf{w}^t$ denotes the transpose of $\mathbf{w}$. We will assume that the image data $\mathbf{g}$ are always real-valued (i.e. never complex) so we will not worry about any complex conjugates in this chapter. The linear observer that maximizes SNR (Eqn. 18.14) is known as the Hotelling observer (Barrett, 2004). The Hotelling observer produces test statistics using

$$t = \mathbf{w}^t \mathbf{g} = \mathbf{s}^t K^{-1} \mathbf{g} \tag{18.17}$$

where

$$\mathbf{s} = \langle \mathbf{g} \rangle_{\mathbf{g}|\mathrm{H_2}} - \langle \mathbf{g} \rangle_{\mathbf{g}|\mathrm{H_1}} = H\bar{\mathbf{f}}_s \tag{18.18}$$

The signal $\mathbf{s}$ is the difference between the mean data under the two hypotheses, i.e. it is the average signal to be detected. The covariance matrix K is an $M \times M$ matrix that measures the variability in the image data. The ith diagonal element of K is the variance in g_i. The off-diagonal element K_{ij} denotes the covariance between the ith and jth elements of $\mathbf{g}$. A large covariance indicates that the ith and jth components of $\mathbf{g}$ trend with one another. A near-zero covariance indicates that these components do not vary with one another.

Equation 18.17 describes how the optimal linear observer makes decisions. It does not, however, describe how to compute this quantity if $\mathbf{s}$ and K are unknown. For this computation, we need sample statistics computed using a training dataset. An unbiased estimate of $\mathbf{s}$ is given by

$$\hat{\mathbf{s}} = \frac{1}{N_2}\sum_{i=1}^{N_2}\mathbf{g}_i^{(2)} - \frac{1}{N_1}\sum_{i=1}^{N_1}\mathbf{g}_i^{(1)} = \hat{\bar{\mathbf{g}}}_2 - \hat{\bar{\mathbf{g}}}_1 \quad (18.19)$$

where N_2 and N_1 are the numbers of signal-present and signal-absent training images, respectively, and $\hat{\bar{\mathbf{g}}}_j$ is the sample mean under the H_j hypothesis. An estimate of the covariance matrix K can be computed by

$$K = \frac{1}{2}\left[\frac{1}{N_2 - 1}\sum_{i=1}^{N_2}(\mathbf{g}_i^{(2)} - \hat{\bar{\mathbf{g}}}_2)(\mathbf{g}_i^{(2)} - \hat{\bar{\mathbf{g}}}_2)^t + \frac{1}{N_1 - 1}\sum_{i=1}^{N_1}(\mathbf{g}_i^{(1)} - \hat{\bar{\mathbf{g}}}_1)(\mathbf{g}_i^{(1)} - \hat{\bar{\mathbf{g}}}_1)^t\right] \quad (18.20)$$

This estimate of the covariance matrix is an average over the covariance matrices for the two classes of data. Note that the image data $\mathbf{g}$ is a column vector and hence the terms in the sums in Eqn. 18.20 are outer products that produce $M \times M$ matrices. Once the training dataset has been used to determine these quantities, then the Hotelling observer model acts on an image $\mathbf{g}$ via

$$\hat{t} = \hat{\mathbf{s}}\hat{K}^{-1}\mathbf{g} \quad (18.21)$$

It is important to note that even though $\hat{\mathbf{s}}$ is an unbiased estimate of $\mathbf{s}$ and $\hat{K}$ is an unbiased estimate of K, it is not true that the Hotelling observer in Eqn. 18.21 will be an unbiased estimate of t. In fact, there is a bias introduced by this method caused by the finite amount of training data (Fukunaga, 1990).

For most modern medical imaging systems, a direct application of Eqns. 18.19–18.21 is difficult because the dimension M of the image data is very large. For example, modern CT systems collect hundreds of thousands of measurements, which implies that $\mathbf{s}$ is a very long vector and K is a huge matrix. Furthermore, using Eqn. 18.20 to directly estimate the covariance matrix causes problems if the number of images in the training dataset is less than the number of measurements M in the image-data vector. When $N_1 + N_2 < M$, the sample covariance matrix is not full rank and cannot be inverted using conventional means. An alternative method for estimating and inverting this large covariance matrix K is to use a covariance matrix decomposition and the Woodbury matrix inversion lemma (Woodbury, 1950). We start by decomposing the covariance matrix K into two components using techniques described fully by Barrett and Myers (Barrett, 2004), to achieve

$$K = \bar{K}_n + K_{\bar{g}} \quad (18.22)$$

Here $\bar{K}_n$ is the measurement noise covariance matrix averaged over all patients and the term $K_{\bar{g}}$ is the covariance of noise-free images. This decomposition is completely general and does not require any assumptions about linear imaging systems or even statistical independence of the objects $\mathbf{f}$ and the measurement noise $\mathbf{n}$. The key advantage of the decomposition shown in Eqn. 18.22 is that the term $\bar{K}_n$ is often known from the physical and statistical model or can be easily estimated. For example, in nuclear-medicine imaging, the measurement noise is known to be Poisson and, hence, the matrix $\bar{K}_n$ is a diagonal matrix whose diagonal elements are given by the mean of the image data (which is also the variance for Poisson statistics). If the noise in the measurement process is modeled as uncorrelated Gaussian, then $\bar{K}_n$ is again diagonal with diagonal components given by the variances of the Gaussian distributions. Because $\bar{K}_n$ is known and is, most often, diagonal, then any estimate of K produced using Eqn. 18.22 is necessarily full rank. So once we have determined what $\bar{K}_n$ is from the physics of the imaging process, we can use any number of methods to estimate $K_{\bar{g}}$ and the resulting estimate of the full covariance matrix will be full rank and invertible.

The term $K_{\bar{g}}$ is the covariance of noise-free images of objects. This covariance can be determined using simulations of the imaging system and objects or it can be measured for some modalities. For example in nuclear medicine, long-exposure images of objects will result in projections that are essentially noise-free. These noise-free projections can then be used to produce a sample estimate of $K_{\bar{g}}$. Thus the sample covariance matrix to be used in Eqn. 18.21 is

$$\hat{K} = \hat{\bar{K}}_n + \hat{K}_{\bar{g}} \quad (18.23)$$

where $\hat{\bar{K}}_n$ is an estimate of the diagonal $\bar{K}_n$ matrix and $\hat{K}_{\bar{g}}$ is estimated from noise-free sample images.

Another benefit of this matrix decomposition is that the burden of the inversion of a large covariance matrix is mitigated by using the matrix inversion lemma (Barrett, 2001). We can rewrite Eqn. 18.23 as

$$\hat{K} = \hat{\bar{K}}_n + XX^t \quad (18.24)$$

where X is a matrix whose dimensions are the number of detector elements M by the number of samples $N_1 + N_2$. The matrix X is a compact means of representing sample covariance matrices as in Eqn. 18.20. The matrix inversion lemma states that

$$\left[\hat{\bar{K}}_n + XX^t\right]^{-1} = \left[\hat{\bar{K}}_n\right]^{-1} - \left[\hat{\bar{K}}_n\right]^{-1} X\left[I + X^t\left[\hat{\bar{K}}_n\right]^{-1}X\right]^{-1} X^t\left[\hat{\bar{K}}_n\right]^{-1} \quad (18.25)$$

This equation is composed of two separate inverses: $\left[\hat{\bar{K}}_n\right]^{-1}$, which is the inverse of a diagonal matrix, and the inverse of a matrix whose dimension is the number of samples by the number of samples. Thus, the matrix inversion lemma greatly simplifies the computational effort required to invert large covariance matrices.

The matrix decompositions discussed are far easier to implement on image data $\mathbf{g}$ as opposed to reconstructed images $\hat{\mathbf{f}}$. This is because the noise properties of $\mathbf{g}$ are easily separated into measurement noise and object variability. For reconstructions $\hat{\mathbf{f}}$, this separation is not as straightforward but still may be accomplished using techniques described by Barrett *et al.* and Wilson *et al.* (Barrett, 1994; Wilson, 1994).

18.2.2 Linear estimation tasks

A linear observer performing an estimation task computes

$$\hat{\theta} = W^t\mathbf{g} + \mathbf{c} \quad (18.26)$$

where W is a matrix whose dimension is the number of detector elements M by the number of parameters to be estimated P. The vector constant $\mathbf{c}$ is an $M \times 1$ vector. The linear estimator that minimizes the EMSE is called the Wiener estimator and is given by

$$\hat{\theta} = K_{\theta,g} K^{-1} (\mathbf{g} - \bar{\mathbf{g}}) + \bar{\theta} \quad (18.27)$$

where $K_{\theta,g}$ is the cross-covariance between the image data $\mathbf{g}$ and the parameters to be estimated θ, and $\bar{\theta}$ is the mean of the parameters to be estimated.

Using the training dataset, the mean parameters can be estimated via

$$\hat{\bar{\theta}} = \frac{1}{N} \sum_{i=1}^{N} \theta_i \quad (18.28)$$

where N is the total number of images in the training dataset. Recall that truth for the ith image in the training dataset was given by θ_i. It is important to distinguish the sample mean of the parameters to estimate $\hat{\bar{\theta}}$ from the estimates themselves given by $\hat{\theta}$. An unbiased estimate of the cross-covariance $K_{\theta,g}$ can be determined using

$$\hat{K}_{\theta,g} = \frac{1}{N-1} \sum_{i=1}^{N} \left(\theta_i - \hat{\bar{\theta}}\right) \left(\mathbf{g}_i - \hat{\bar{\mathbf{g}}}\right)^t \quad (18.29)$$

The covariance matrix K can be estimated directly from samples using an equation similar to Eqn. 18.20. A matrix decomposition can also be employed to ease the burden of estimating and inverting K. As before,

$$K = \bar{K}_n + K_{\bar{g}} \quad (18.30)$$

However, $\bar{K}_n$ is the noise covariance averaged over object variability and parameter θ variability. The second term $K_{\bar{g}}$ is the covariance matrix of the noise-free image data but including parameter variations. As before, the first term in Eqn. 18.30 is often diagonal and can be estimated using image data. The matrix inversion lemma can again be employed to invert estimates of the covariance matrix computed using the above expansion.

18.2.3 Bias and the matrix inversion lemma

As mentioned earlier, any observer that is trained using finite data will exhibit some bias because sample statistics instead of population-based statistics are used to determine the parameters of the observer model. It is important to note that the bias introduced by using the matrix inversion lemma is not well characterized and can, in certain circumstances, be quite large (Kupinski, 2007). While the matrix inversion lemma is a powerful technique and makes an intractable problem possible, it should not be blindly applied. There is a great deal of literature on the effect of training dataset size on observer performance (Fukunaga, 1990; Chan, 1999).

18.3 CHANNELIZED OBSERVER MODELS

The principal difficulty in implementing the ideal linear observer models discussed in the previous section is the estimation and the inversion of the image-data covariance matrix K. The matrix inversion lemma eases this computational burden but still may require very large training datasets to achieve reasonable results. Another alternative is to use channels to compress the image data to a smaller data vector. Channelized observers act on the image data using a linear operator

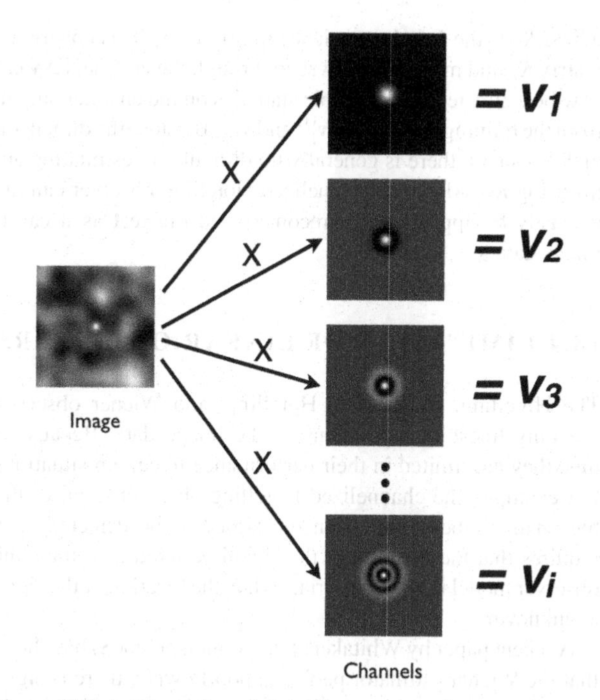

Figure 18.2 An illustration of channel operators acting on an image.

$$\mathbf{v} = U\mathbf{g} \quad (18.31)$$

where U is an $L \times M$ matrix and $\mathbf{v}$ are the channel outputs. Here, L is generally much smaller than M. This channel operator can be thought of as a series of image templates that are applied to images. This process is illustrated in Figure 18.2. We then apply the Hotelling observer to these channel outputs instead of the image data itself. Because the dimension of $\mathbf{v}$ is much less than that of $\mathbf{g}$, the computational burden associated with estimating means and covariance matrices is mitigated.

The choice of channels depends on the type of observer model desired. One common choice is to use frequency and orientation selective channels designed to mimic the human visual system (Abbey, 2001). These channels are centered on the known signal location. A second choice for the channel operator are the efficient channels (Barrett, 1998). These channels are designed to optimally represent the signal to be detected using just a few channel outputs. Because the signal is well represented in the basis of the efficient channels, the performance of the Hotelling observer on these channels will be close to that of the Hotelling observer acting on the entire image. For a radially symmetric signal at a known location, the Laguerre-Gaussian channels are one example of efficient channels. Images of the Laguerre-Gaussian channels are shown in Figure 18.2.

The channelized Hotelling observer is computed using the channel outputs. Thus, the test statistic for a channelized Hotelling observer is given by

$$[U\mathbf{s}]^t K_v^{-1} U\mathbf{g} = \Delta\bar{\mathbf{v}} K_v^{-1} \mathbf{v} \quad (18.32)$$

where K_v is the $L \times L$ covariance matrix for $\mathbf{v}$. This covariance matrix K_v and mean signal as seen through the channels $\Delta \bar{\mathbf{v}}$ can now be estimated using sample statistics on the channel outputs from the training dataset, i.e. $\mathbf{v}_i^{(1)}$ and $\mathbf{v}_i^{(2)}$. Because the dimension of K_v is small, there is generally no difficulty in estimating and inverting K_v. Also, the channelized Hotelling observer can just as easily be applied to the reconstructed image $\hat{\mathbf{f}}$ as it can to image data $\mathbf{g}$.

18.4 LIMITATION OF LINEAR OBSERVERS

The Hotelling, channelized Hotelling, and Wiener observers use only linear manipulations of the image data. Because of this, they are limited in their performance in certain situations. For example, the channelized Hotelling observer requires that the channels be centered on the signal to be detected. This requires that the location of the signal be fixed and, thus, this observer model is not appropriate when the location of the signal is unknown.

A recent paper by Whitaker *et al.* (Whitaker, 2008) has shown that the Wiener estimator performs poorly when there is signal location and/or shape variability. The authors showed that the Wiener estimator was unable to use the image data to simultaneously estimate signal size, location, and amplitude. It is also expected that the Hotelling observer will perform very poorly if the location of the signal is random. While the computational efforts required for linear estimation are not great, their application is generally limited to specific tasks such as the detection of a known signal at a known location or the estimation of signal amplitude when the signal location and size are known.

18.5 IDEAL OBSERVERS

The observers described thus far have been limited to performing linear manipulations of the image data to make decisions. Ideal observers are not restricted to performing linear manipulations of the image data. For signal-detection tasks, the observer that maximizes AUC is known as the Bayesian ideal observer and uses the likelihood ratio to make decisions. That is,

$$t = \Lambda(\mathbf{g}) = \frac{pr(\mathbf{g}\,|\,\mathrm{H}_2)}{pr(\mathbf{g}\,|\,\mathrm{H}_1)} \tag{18.33}$$

The Bayesian ideal observer requires extensive knowledge of the statistical distributions of the image data to make decisions. In the absence of object variability, the Bayesian ideal observer is relatively easy to compute since $pr(\mathbf{g}\mid \mathrm{H}_j)$ is the measurement noise. However, object variability is almost always present in medical imaging, which makes computation of the ideal observer difficult. Using techniques described by Kupinski *et al.* (Kupinski, 2003), we can rewrite Eqn. 18.33 as

$$\begin{aligned}\Lambda(\mathbf{g}) &= \int \frac{pr(\mathbf{g}\,|\,\mathbf{b},\mathrm{H}_2)}{pr(\mathbf{g}\,|\,\mathbf{b},\mathrm{H}_1)} pr(\mathbf{b}\,|\,\mathbf{g},\mathrm{H}_1) d^M b \\ &= \langle \Lambda_{\mathrm{BKE}}(\mathbf{g}\,|\,\mathbf{b})\rangle_{\mathbf{b}|\mathbf{g},\mathrm{H}_1}\end{aligned} \tag{18.34}$$

where $\mathbf{b} = H\mathbf{f}$ is a noise-free image of an object $\mathbf{f}$, and $\Lambda_{\mathrm{BKE}}(\mathbf{g} \mid \mathbf{b})$ is the likelihood ratio when the background is known to be $\mathbf{b}$. If we draw N samples $\mathbf{b}_n$ from the distribution $pr(\mathbf{b} \mid \mathbf{g}, \mathrm{H}_1)$, then we can estimate the likelihood ratio for a given image $\mathbf{g}$ as

$$\hat{\Lambda}(\mathbf{g}) = \frac{1}{N}\sum_{n=1}^{N} \Lambda_{\mathrm{BKE}}(\mathbf{g}\,|\,\mathbf{b}_n) \tag{18.35}$$

Markov chain Monte Carlo (MCMC) techniques have been successfully used to sample from the distribution $pr(\mathbf{b} \mid \mathbf{g}, \mathrm{H}_1)$ (Kupinski, 2003; He, 2008). Ideal observers have been computed for a number of different nuclear medicine imaging systems using lumpy background object models (Gross, 2003; Kupinski, 2003) and realistic digital phantoms (He, 2008). In addition, these techniques have been extended to include random signals as well as the background variability (Park, 2003; 2005).

To further explain how the MCMC technique is implemented to compute ideal observer performance, the steps of the process will be described. More information on these techniques can be obtained in Gilks *et al.* (Gilks, 1996). To implement MCMC it is necessary to parameterize the object model with a finite number of parameters γ. Then, starting at an initial γ_0 which may be random, we propose a new background by sampling from a user-specified distribution $pr(\tilde{\gamma}|\gamma_i)$. This distribution should be designed to move from one object realization to another slightly different object realization. For lumpy backgrounds, this procedure involves choosing a particular lump center and shifting by a random amount. Once this new proposed background is sampled, then we must compute the acceptance probability, which is given by

$$\alpha = \min\left(1, \frac{pr(\tilde{\gamma}\,|\,\mathbf{g},\mathrm{H}_1)pr(\tilde{\gamma}\,|\,\gamma_i)}{pr(\gamma_i\,|\,\mathbf{g},\mathrm{H}_1)pr(\gamma_i\,|\,\tilde{\gamma})}\right) \tag{18.36}$$

As explained in Kupinski *et al.* (Kupinski, 2003), the ratio of probabilities is straightforward to compute as long as the distribution of the measurement noise is known. The proposed background $\tilde{\gamma}$ is accepted with probability α. If accepted, then $\gamma_{i+1} = \tilde{\gamma}$; otherwise $\gamma_{i+1} = \gamma_i$. At each iteration i, $\Lambda_{\mathrm{BKE}}(\mathbf{g} \mid \gamma_i)$ is computed and stored. After many iterations, sometimes millions, the $\Lambda_{\mathrm{BKE}}(\mathbf{g} \mid \gamma_i)$ values are averaged to produce an estimate of the likelihood ratio for that particular image. This process is then repeated for many images so that a figure of merit can be computed.

When $pr(\mathbf{g} \mid \mathrm{H}_2)$ and $pr(\mathbf{g} \mid \mathrm{H}_1)$ are both Gaussian distributed with the same covariance matrix K, then the Bayesian ideal observer performs the same operations as the Hotelling observer. Thus, when the data are Gaussian, the Bayesian ideal observer is linear and reduces to the Hotelling observer.

For estimation tasks, the posterior mean estimator is an ideal estimator that is known to minimize the EMSE among all possible estimators. The posterior mean estimate of θ for a given image $\mathbf{g}$ is given by

$$\hat{\theta} = \int \theta\, pr(\theta \mid \mathbf{g}) d^P \theta \tag{18.37}$$

Equation 18.37 can be rewritten using the Bayes rule as

$$\hat{\theta} = \int \theta \frac{pr(\mathbf{g}\,|\,\theta)pr(\theta)}{pr(\mathbf{g})} d^P \theta \tag{18.38}$$

Again, the posterior mean estimator requires complete knowledge of all random components in the imaging chain including

$pr(\theta)$, $pr(\mathbf{g} \mid \theta)$, and $pr(\mathbf{g})$. As before, we can use MCMC techniques to produce estimates of the posterior mean estimator. However, precise knowledge of $pr(\theta)$ is also necessary. When the joint statistics governing the image data and the parameters to be estimated are Gaussian, the posterior mean reduces to the Wiener estimator.

18.5.1 Limitations of the ideal observers

The ideal observers discussed above are optimal observers and cannot be outperformed by any other observer models. However, the computation of these observers is difficult. MCMC techniques may require many iterations (sometimes millions) to make a decision for a single image. To measure observer performance, this process must be repeated for many images (i.e. the testing dataset). In addition, the MCMC techniques can only be applied in simulations where complete knowledge of the object model is known. Examples include the lumpy object model and the NCAT phantom. With real patient data, there is little hope of computing the ideal-observer decision strategy. The uses of these techniques are mainly in imaging system design, which is performed in simulation.

18.6 COMBINED DETECTION/ESTIMATION TASKS AND OTHER OBSERVER MODELS

The observer models we have discussed thus far have been for detection tasks or estimation tasks only. Combined observer models do, however, exist and these models have led to a new class of observer model that can be applied to any type of detection task, estimation task, or combination detection/estimation task. The pioneering work of Khurd and Gindi (Khurd, 2005) developed an observer model that maximized that area under the LROC curve. This observer model required a search over potential signal locations and the application of an observer that computes a test statistic as a function of signal location. That is, the observer's test statistic is given by

$$t = \max_{r'} \left\{ \langle \Lambda(\mathbf{g} \mid \mathbf{r}) u(\mathbf{r}, \mathbf{r}') \rangle_r \right\} \tag{18.39}$$

where $\Lambda(\mathbf{g} \mid \mathbf{r})$ is the ideal observer when the signal is located at position $\mathbf{r}$, and $u(\mathbf{r}, \mathbf{r}')$ is the utility function of locating the signal at position $\mathbf{r}'$ when the true location is $\mathbf{r}$ (Khurd, 2005). If t is above a detection threshold, then the estimate of the signal location is given by

$$\mathbf{r} = \arg\max_{r'} \left\{ \langle \Lambda(\mathbf{g} \mid \mathbf{r}) u(\mathbf{r}, \mathbf{r}') \rangle_r \right\} \tag{18.40}$$

Eric Clarkson generalized this analysis to allow for the estimation of other parameters instead of just signal location (Clarkson, 2007) and he generated a new form of ROC analysis called estimation ROC or EROC analysis.

The work of Meredith Whitaker (Whitaker, 2008) went a step further and generated an estimation observer model that uses only the second step shown in Eqn. 18.40. In addition, Whitaker *et al.* also used a simplifying assumption to replace the expectation in Eqn. 18.40 with a Gaussian function. Using Gaussian assumptions, scanning Hotelling observers have also been developed that implement Eqn. 18.39. Both the scanning linear estimator and scanning Hotelling observer are nonlinear because of the maximization step. However, the computation of the objective function depends on only first- and second-order statistics of the image data $\mathbf{g}$. While the objective function may be a straightforward computation, the maximization of this objective function can be difficult. For example, in Whitaker *et al.* (Whitaker, 2008), a combination of simulated annealing with a Nelder-Mead algorithm was used.

18.7 FINAL THOUGHTS

Methods for implementing linear-observer models that perform well when the location of the signal is known *a priori* were presented. The principal difficulty in implementing these methods is the estimation and inversion of a large data covariance matrix. The use of channels and covariance matrix decomposition may ease the burden of implementing linear-observer models but they are still limited in the tasks for which they are useful. A method for implementing the ideal observer was described. However, this technique is computationally burdensome and requires precise knowledge of an object model. In addition, these techniques can only be applied in simulations and not using real images. Finally, we ended with a discussion of new nonlinear observer models that require many of the same computations as were needed for the linear observers but are known to perform well in tasks where the signal location is unknown. While these new observer models have already been used to assess task performance, they have not yet proven themselves as models for human-observer performance although there is work ongoing in this area (Gifford, 2000; Lehovich, 2008).

Image quality is a statistical concept and assessing image quality using task-based measures requires knowledge of the sources of variability in the imaging equation. We have considered object variability, measurement noise, and task variability (i.e. randomness in θ and randomness in whether an image has a signal or not). A source of variability not considered in this chapter but discussed elsewhere is imaging-system variability (Barrett, 2006). In addition, knowledge of the forward problem H is beneficial for implementing observer models. Models of H may be needed to determine noise-free images of fixed objects, or to implement the MCMC technique. While the computational burden of implementing task-based measures of image quality is greater than most other measures such as contrast and resolution, the extra effort results in a measure of performance that directly accounts for the task to be performed with the images.

REFERENCES

Abbey, C.K., Barrett, H.H. (2001). Human- and model-observer performance in ramp-spectrum noise: effects of regularization and object variability. *Journal of the Optical Society of America A* **18**, 473–488.

Barrett, H.H. (1990). Objective assessment of image quality: effects of quantum noise and object variability. *Journal of the Optical Society of America A* **7**(7), 1266–1278.

Barrett, H.H., Wilson, D.W., Tsui, B.M.W. (1994). Noise properties of the EM algorithm I: theory. *Physics in Medicine and Biology* **39**, 833–846.

Barrett, H.H., Abbey, C., Gallas, B., Eckstein, M. (1998). Stabilized estimates of Hotelling-observer detection performance in patient-structured noise. *Proceedings of SPIE Medical Imaging*, **3340**, 27–43.

Barrett, H.H., Myers, K.J., Gallas, B.D., Clarkson, E., Zhang, H. (2001). Megalopinakophobia: its symptoms and cures. In Antonuk, L.E., Yaffe, M.J., eds., *Medical Imaging 2001: Physics of Medical Imaging*, 299–307. San Diego, CA: SPIE.

Barrett, H.H., Myers, K.J. (2004). *Foundations of Image Science*. Hoboken, NJ: John Wiley.

Barrett, H.H., Myers, K.J., Devaney, N., Dainty, C. (2006). Objective assessment of image quality. IV. Application to adaptive optics. *Journal of the Optical Society of America A* **23**(12), 3080–3105.

Chan, H.-P., Sahiner, B., Wagner, R.F., Petrick, N. (1999). Classifier design for computer-aided diagnosis: effects of finite sample size on the measured performance of classical and neural-network classifiers. *Medical Physics* **26**(12), 2654–2669.

Clarkson, E. (2007). The estimation receiver operating characteristic curve and ideal observers for combined detection (estimation tasks. *Journal of the Optical Society of America A* **24**(12), B91–B98.

Fukunaga, K. (1990). *Statistical Pattern Recognition*. San Diego, CA: Academic Press.

Gifford, H.C., King, M.A., Wells, R.G. (2000). Single-photon emission computed tomography: LROC analysis of detector-response compensation in SPECT. *IEEE Transactions on Medical Imaging* **19**, 463–473.

Gilks, W.R., Richardson, S., Spiegelhalter, D.J. (1996). *Markov Chain Monte Carlo in Practice*. Boca Raton, FL: Chapman and Hall/CRC.

Gross, K., Kupinski, M.A., Peterson, T., Clarkson, E. (2003). Optimizing a multiple-pinhole SPECT system using the ideal observer. In Charkraborty, D.P., Krupinski, E.A., eds., *Medical Imaging 2003: Image Perception, Observer Performance, and Technology Assessment*, 314–322. San Diego, CA: SPIE.

He, X., Caffo, B.S., Frey, E.C. (2008). Toward realistic and practical ideal observer estimation for the optimization of medical imaging systems. *IEEE Transactions on Medical Imaging* **27**(10), 1535–1543.

Henkelman, R.M., Kay, I., Bronskill, M.J. (1990). Receiver operator characteristic (ROC) analysis without truth. *Medical Decision Making* **10**, 24–29.

Hoppin, J.W., Kupinski, M.A., Kastis, G., Clarkson, E., Barrett, H.H. (2002). Objective comparison of quantitative imaging modalities without the use of a gold standard. *IEEE Transactions on Medical Imaging* **21**, 441–449.

Khurd, P., Gindi, G. (2005). Decision strategies maximizing the area under the LROC curve. *Proceedings of SPIE Medical Imaging*, **5749**, 150–161.

Kupinski, M.A., Hoppin, J.W., Clarkson, E., Barrett, H.H. (2002). Estimation in medical imaging without a gold standard. *Acad Radiol* **9**, 290–297.

Kupinski, M.A., Hoppin, J.W., Clarkson, E., Barrett, H.H. (2003). Ideal-observer computation in medical imaging with use of Markov-chain Monte Carlo techniques. *Journal of the Optical Society of America A* **20**(3), 430–438.

Kupinski, M.A., Clarkson, E., Hesterman, J.Y. (2007). Bias in Hotelling observer performance computed from finite data. *Proceedings of SPIE Medical Imaging*, **6515**, 65150S-1–65150S-7.

Lehovich, A., Gifford, H.C., King, M.A. (2008). Model observer to predict human performance in LROC studies of SPECT reconstruction using anatomical priors. *Proceedings of SPIE Medical Imaging*, **6917**, 67170R-1–69170R-7.

Park, S., Kupinski, M.A., Clarkson, E. (2003). Ideal-observer performance under signal and background uncertainty. *Information Processing in Medical Imaging* **18**, 342–353.

Park, S., Clarkson, E., Kupinski, M.A., Barrett, H.H. (2005). Efficiency of the human observer detecting random signals in random backgrounds. *Journal of the Optical Society of America A* **22**(1), 3–16.

Whitaker, M.K., Clarkson, E., Barrett, H.H. (2008). Estimating random signal parameters from noisy images with nuisance parameters: linear and scanning-linear methods. *Optics Express* **16**(11), 8150–8173.

Wilson, D.W., Tsui, B.M., Barrett, H.H. (1994). Noise properties of the EM algorithm: II. Monte Carlo simulations. *Physics in Medicine and Biology* **39**(5), 847–871.

Woodbury, M.A. (1950). *Inverting modified matrices*. Statistical research group. Princeton, NJ: Princeton University.

PART IV

DECISION SUPPORT AND COMPUTER AIDED DETECTION

19

CAD: an image perception perspective

MARYELLEN GIGER AND WEIJIE CHEN

19.1 INTRODUCTION

Computer-aided diagnosis (CAD) in medical imaging is generally defined as a diagnosis made by a radiologist who uses the output from a computerized analysis of medical images as "another opinion" in detecting abnormalities, assessing extent of disease, characterizing lesions, and making diagnostic decisions. The major goal of CAD is to help the radiologist improve diagnostic accuracy, efficiency, and consistency in medical image interpretation tasks where perception plays a substantial role. In this chapter, we discuss CAD from an image perception perspective with emphasis on impacts that CAD and CAD output can have on image perception and how CAD can be best utilized.

The benefit of a medical imaging exam depends on both the physical quality of the medical images and the ability of the radiologist interpreting them. Studies indicate that radiologists do not detect all abnormalities on images that are visible on retrospective review, and they do not always correctly characterize abnormalities that are found (Renfrew, 1992). Causes of detection and interpretation errors in the clinical interpretation of medical images include limitations in the human eye-brain visual system, reader fatigue and distraction, the presence of overlapping structures that camouflage disease in images, and the vast number of normal cases seen in screening programs, to name a few.

With the goal of reducing image perception errors and improving image interpretation accuracy, CAD has seen tremendous growth over the past 20 years. CAD has evolved rapidly from the early days of time-consuming film digitization and computations on a limited number of cases to its current uses in a broadening range of medical imaging applications and clinical workstations. Basic CAD research involves a variety of activities – collecting relevant normal and abnormal images of clinical patients and the associated pathological exam results serving as ground truth for diagnosis; developing computer algorithms appropriate to the medical interpretation task; validating the algorithms using appropriate cases (distribution and sizes) and study designs to measure performance and robustness; evaluating radiologists' performance in the relevant diagnostic task without and with the use of the computer aid; and then ultimately assessing performance with a clinical trial. At present, CAD research includes analysis of images of a number of disease types – breast cancer, lung cancer, colon cancer, interstitial disease, osteoporosis, osteolysis, vascular plaque, aneurysms, and others – and from various modalities, including analog and digital X-ray radiography, ultrasonography, CT, MRI, PET, and others.

In this chapter, we first present a broad overview of CAD techniques and methodology. Then we discuss factors that affect radiologists' performance in image perception and interpretation, and how CAD can impact radiologists' perception and performance. Finally, we discuss how CAD can be best utilized from an image perception perspective.

19.2 CAD: AN OVERVIEW OF TECHNIQUES

CAD can broadly be categorized into two types – computer-aided detection (CADe) and computer-aided diagnosis (CADx). The word "diagnosis" is used in a broad sense in "CAD" – including both disease detection and diagnostic decision-making – and is used in a narrow sense in "CADx," referring specifically to the diagnostic characterization and clinical decision-making of a lesion already detected. Nonetheless, this categorization is useful from both a clinical point of view and a CAD developer's point of view: "detection" and "diagnosis" are two clinical tasks; the design and development of CADe and CADx systems are different, although some image analysis and pattern recognition techniques are commonly used in both types of systems. In this section, we give a brief introduction to the computerized techniques for the two types of CAD systems, i.e. we aim to shed some light on the "computer part" or the commonly called "CAD black box" and defer our discussions on the "computer-aid" part or the "human-computer interactions" to the next section. While computerized techniques are of interest to CAD developers, they also are important to CAD users: if the user knows what the computer is calculating, he/she will be better able to understand the output, especially if the computer indicates a region as suspicious that appears to the user to be normal.

19.2.1 CADe as an aid for detection of disease

CADe implies the use of an output from a computerized image analysis as an aid in the localization of an image region that is suspected of being abnormal. CADe is basically a detection or localization task, i.e. the computer analyzes medical images to find and output locations of suspect regions to "alert" the radiologist, leaving the final localization of lesions and patient management to the radiologist (Figure 19.1a). CADe systems

The Handbook of Medical Image Perception and Techniques, ed. Ehsan Samei and Elizabeth Krupinski. Published by Cambridge University Press.

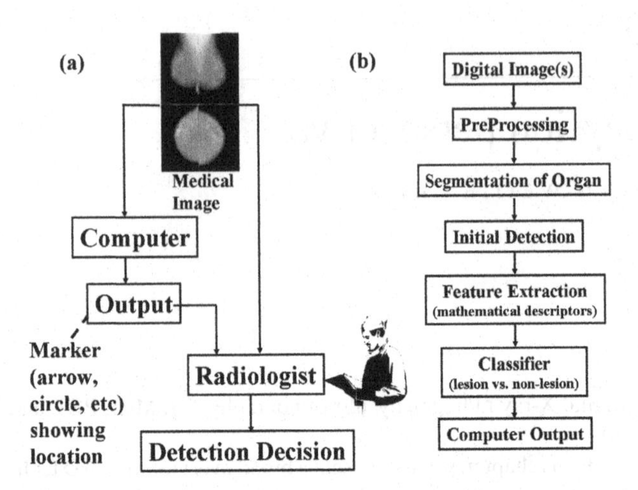

Figure 19.1 Schematic diagrams illustrating (a) the use of a CADe system and (b) the components of a CADe system.

are most useful in imaging examinations in which most cases are normal – such as in screening programs, examples of which include screening mammography, low-dose thoracic CT for smokers, and colon cancer screening. Figure 19.2a shows the first prototype CADe system, which was developed for screening mammography at the University of Chicago. The system took as input a screen/film mammogram, which was subsequently digitized and automatically passed to the computer for analysis. The output annotation from the system would indicate suspect locations on a thermal paper printout or monitor, as demonstrated in Figure 19.2b for a mammogram with a mass lesion. While the output may be quite simple, complex mathematical calculations are performed within the CADe "black box."

Computerized techniques for CADe include the detection of abnormalities that may be signs of potential cancer. This involves several steps of image analyses and computations. Figure 19.1b schematically illustrates the components of a CADe system. Once image data are input to the computer, by either film digitization or directly from an image acquisition device, various image processing and analysis steps are performed. Depending upon the quality of the input images, the initial step may involve *image recovery* techniques for preprocessing, such as noise reduction and/or artifact correction. Following the preprocessing, *image segmentation* techniques are applied to delineate the organ of interest (e.g. the breast in mammography CADe) from the image background and other body parts in the image, allowing for subsequent analyses and computations to focus only on the organ of interest. The "initial detection" step involves using *lesion enhancement* techniques to enhance the signal of the lesion relative to the camouflaging normal anatomy background, and this step results in a number of candidate lesions. Next, numerical features are derived from the image data of candidate lesions, and these features are mathematical descriptors of characteristic features that potentially help distinguish whether a candidate lesion is truly a lesion or normal tissue (Giger, 2000). *Feature extraction* involves computer vision techniques, clinical observations, and/or other prior domain knowledge. These numerical features are "interpreted" by the computer using *pattern recognition* techniques such as discriminant analysis, rule-based methods, and/or artificial neural networks to merge the extracted features into a decision on the likelihood that the region being analyzed is a potential lesion (Sonka, 2000). Such a classifier is trained, using a data set of images from normal subjects as well as patients with histologically-verified lesions, to distinguish between actual lesions and normal regions. In these later stages, the processing methods are mainly to reduce the number of false-positive (FP) candidates, i.e. to improve specificity.

19.2.2 CADx as an aid for differential diagnosis

Once a lesion is detected, the radiologist must extract its characteristic features and then merge these into a decision on patient

(a)

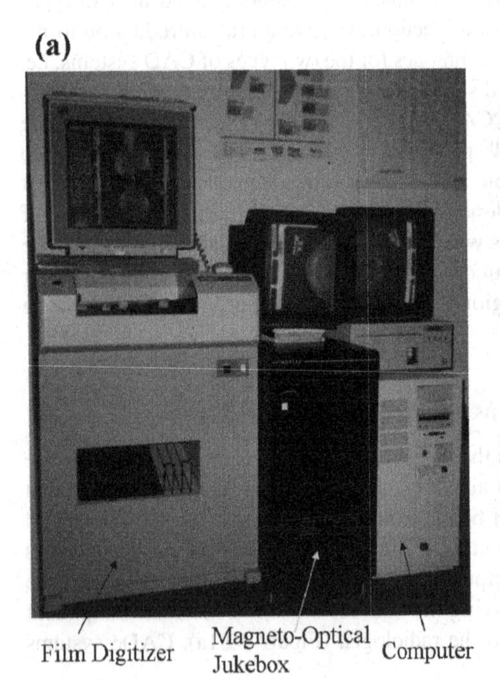

(b)

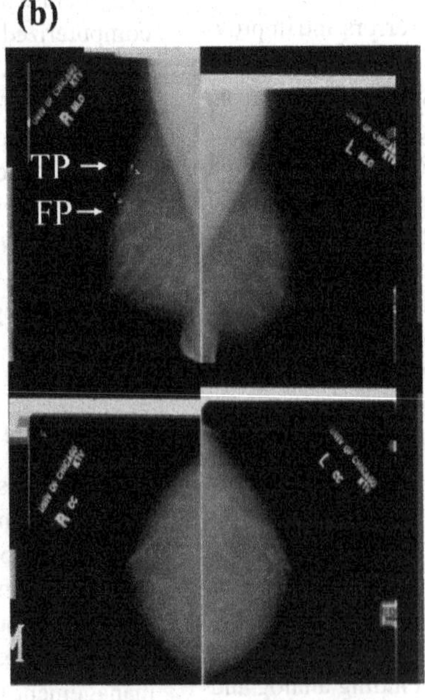

Figure 19.2 (a) shows the first prototype CADe system, which was developed for screening mammography at the University of Chicago, in which output annotation from the system would indicate suspect locations on a thermal paper printout or monitor, as demonstrated in (b) for a mammogram with a mass lesion.

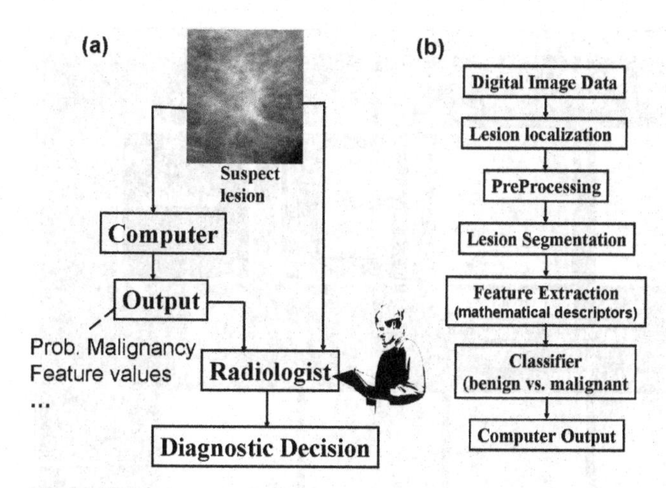

Figure 19.3 Schematic diagrams illustrating (a) the use of a CADx system and (b) the components of a CADx system.

management, e.g. biopsy or no biopsy. CADx is a classification or differential diagnosis task, i.e. the computer analyzes a suspicious region or lesion to calculate and output mathematical descriptors (tumor features) and/or estimated probability of malignancy that are potentially helpful to classify a suspicious lesion as benign or malignant, also leaving the final diagnosis and patient management to the physician (Figure 19.3a). Thus, the role of a CADx system is to characterize an already-found lesion or other abnormality in terms of its morphological or functional features, and to potentially estimate its probability of malignancy or other disease. Such a system would aid a radiologist in their differential diagnosis of malignancy (or disease state) and recommendation on patient management.

Computerized techniques for CADx include the classification of detected lesions as subgroups (e.g. benign and malignant) that warrant different clinical actions (e.g. biopsy or no biopsy, different therapeutic interventions). This involves several steps of image analyses and computations. Figure 19.3b schematically illustrates the components of a CADx system. Unlike CADe, which takes the whole image as input, CADx takes as input either a radiologist-detected or a computer-detected lesion. Note that for such a system, manual indication of the lesion location may be required, since with CADx, this is a classification task, not a detection task. Typical input may include a seed point or a region of interest indicating the location of the lesion. Preprocessing of the image data may be performed to enhance the contrast of the lesion relative to the camouflaging normal anatomy background. Next, the lesion is segmented from the surrounding background (Huo, 1995; Kupinski, 1998; Chen, 2006a; Yuan, 2007). This is generally regarded as a critical step in computerized lesion characterization because analysis of a poorly segmented lesion would result in an erroneous characterization of the lesion features and an incorrect assessment of disease severity (e.g. malignancy). Following lesion segmentation, numerical features are derived from the segmented lesions and these features are mathematical descriptors of characteristic features that potentially help classify a lesion into different subgroups (e.g. benign or malignant). Like in CADe, feature extraction in CADx is often motivated or guided by clinical observations and/or other prior domain knowledge, which are represented by mathematical descriptors derived with computer vision techniques. Finally, these numerical features are "interpreted" by the computer using *pattern recognition* techniques such as discriminant analysis, rule-based methods, and/or artificial neural networks to combine the extracted features into a decision on the likelihood that the lesion being analyzed is, for example, benign or malignant.

Unlike CADe, the output of which is typically a marker (e.g. arrow, circle, etc.) indicating the location of a suspicious lesion, the output from a CADx system can contain diverse information. For example, it can be presented in terms of a numerical estimate of the probability of malignancy, a retrieval of similar images from an online database, a graphical presentation that relates the lesion in question to the computer analyses of cases in a known database with certain cancer prevalence, and/or other useful information from the computer analyses that is potentially helpful to the radiologist for diagnostic decision-making. An example of such an interface appears in Figure 19.4, which displays the computer outputs for both mammography and sonography CADx.

19.2.3 Assessment of the stand-alone performance of CAD systems

Assessment of the ultimate benefit of CAD requires evaluation of the CAD system with observer studies, i.e. studies that involve a comparison of the detection/diagnosis performance of radiologists without the computer aid and with the aid in a clinical realm or a setting that mimics the clinical arena. This is a major topic of the next section in which the impact of CAD on image perception is presented. However, to demonstrate the effectiveness of a CAD system, it is also essential to appropriately assess the stand-alone performance of the computer algorithms, i.e. the detection performance or diagnosis performance of the computer without a physician. Stand-alone performance assessment is useful in comparing the current system with an existing one upon which the current system claims to improve. Additionally, stand-alone performance assessment is needed to demonstrate an adequate level of performance prior to recruiting radiologists for observer studies, which are generally more time-consuming and expensive. General principles and considerations in the assessment of stand-alone computer performance are presented, noting that these aspects also apply to the evaluation of radiologist performance without and with the computer aid.

There are three basic elements in performance assessment of CAD systems: (1) gold standard or truth; (2) performance figure of merit; (3) study designs to appropriately collect and utilize a data set to estimate the performance and its uncertainty.

The gold standard or truth in CADe is the true location of abnormalities which must be determined by well-established processes such as clinical follow-up exams and/or biopsy to determine the presence and location of a lesion, or a consensus panel of experts for determining the extent of the lesion. The truth in CADx is typically obtained from the histo-pathological analysis of biopsied tissue samples or evaluations from a panel of radiology experts. This truth information is usually summarized as binary clinical endpoints, e.g. normal or abnormal, benign or malignant.

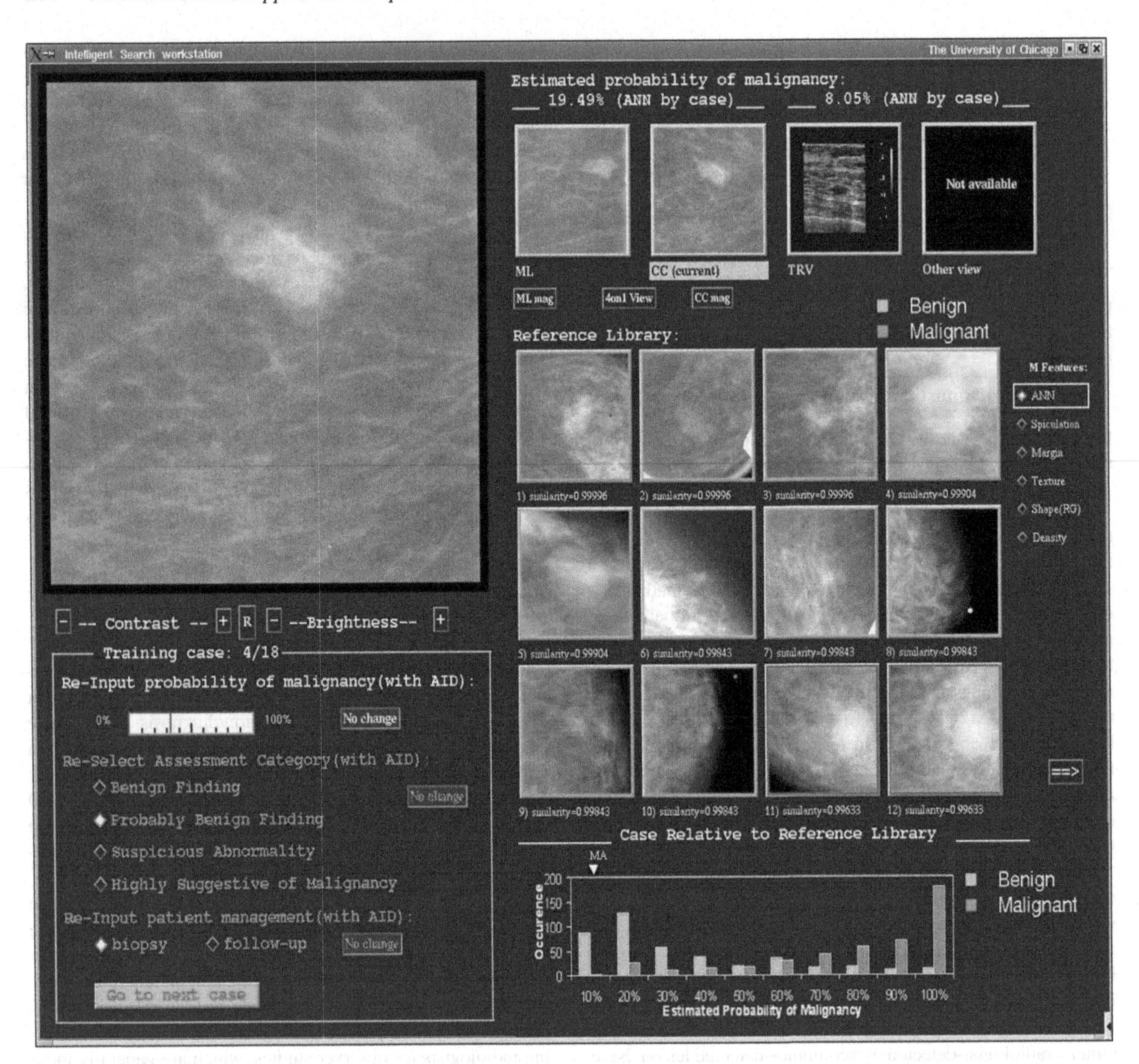

Figure 19.4 Human/computer interface for a CADx system in which the output is presented in terms of numerical values related to the likelihood of malignancy, through a display of similar images of known diagnoses, and/or with a graphical representation of the unknown lesion relative to all lesions in an online database atlas (Giger, 2000; Giger, 2003; Horsch, 2006). The display uses color coding to indicate whether the similar images are malignant or benign.

Performance of a CADe system can be given in terms of sensitivity and FP detections per image. Sensitivity is calculated as the number of lesions detected divided by the total number of real lesions in the data set. The number of FP detections is the average number of FPs per image over the data set. A *free-response receiver operating characteristic (FROC)* curve shows the sensitivity as a function of mean FPs/image. Such a curve can be obtained for a data set by varying one of the algorithm's parameters. Various performance indices exist for use in the evaluation of computerized methods, such as receiver operating characteristic (ROC) (Metz, 1978; Metz, 1986; Metz, 2000; Wagner, 2007) and FROC analyses (Bunch, 1977; Chakraborty, 2000; Samuelson, 2006). Performance of a CADx system can be given in terms of sensitivity and specificity or ROC curve which, by convention, plots sensitivity versus (1–specificity) at varying cut-off thresholds. The area under the ROC curve (AUC) is a standard summary performance figure of merit (Wagner, 2007). A partial area index from an ROC curve is useful when evaluating a system that requires a high level of sensitivity, such as in the task of determining the likelihood of malignancy of lesions seen on mammograms (Jiang, 1996).

In the assessment of CAD systems, it is critically important to apply appropriate study designs to collect and utilize a data set to estimate the performance and its uncertainty. It is essential to collect a sample data set that is representative of the general population to which the CAD system is designed to apply. Study designs for evaluation of clinical benefits of CADe are discussed in detail in the next chapter, on "Common designs of computer-aided detection studies," by Jiang. Here we emphasize

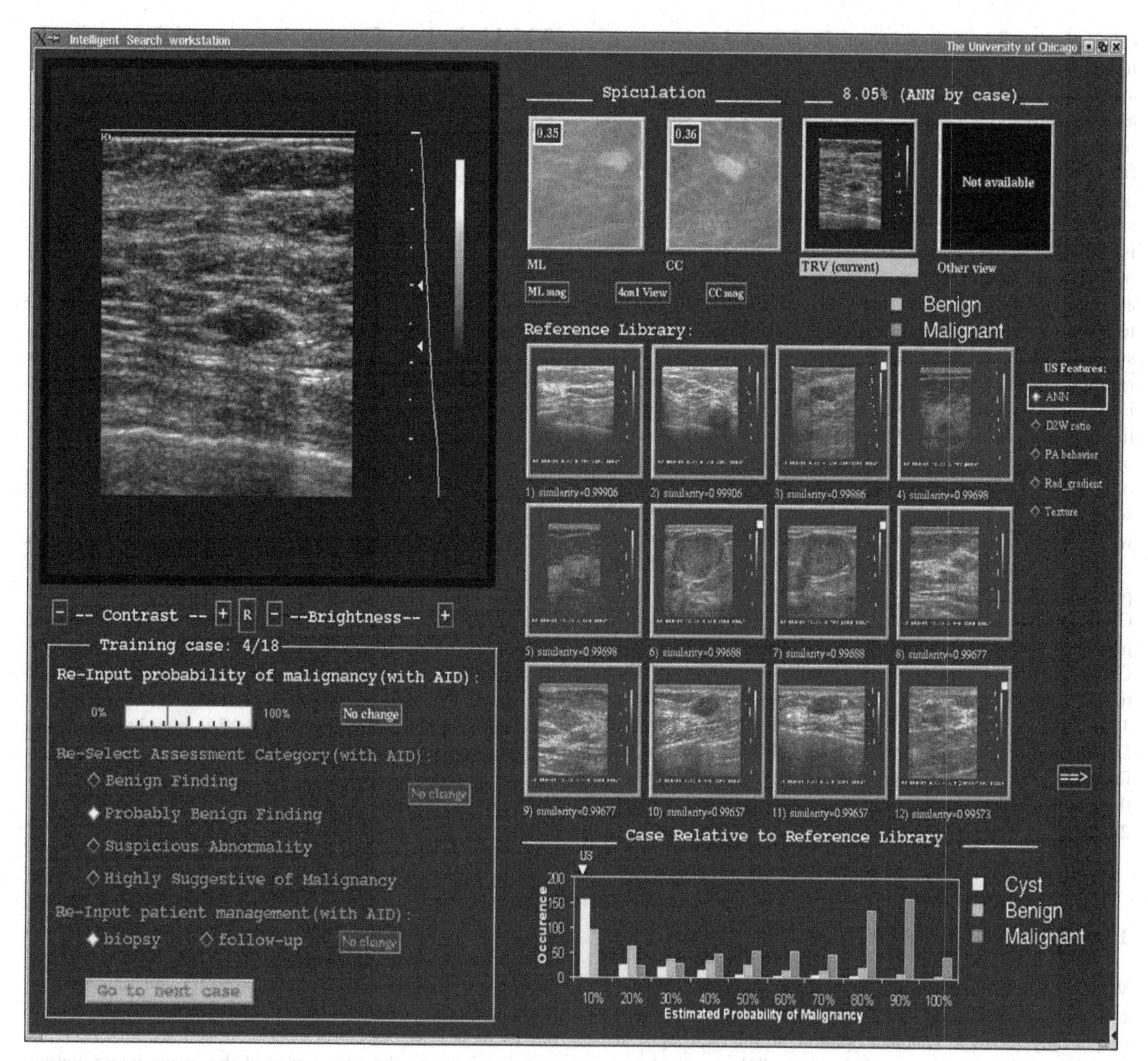

Figure 19.4 *(cont.)*

some general considerations in evaluation of stand-alone performance of CAD algorithms with typically a finite data set of images.

One consideration involves the proper use of a finite data set for the assessment task. There are multiple tasks that need to be assessed in developing a CAD system, e.g. feature selection, classifier model selection, training, and testing the model. Ideally, each task needs to be assessed by an independent data set. When only a limited data set is available, however, one may use resampling techniques such as cross-validation or bootstrapping. It is critically important to be clear about which task is being assessed when utilizing a data set and be cautious of the bias of the results introduced by re-using a data set to assess multiple tasks. For example, it is well known that training and testing a classifier with the same data set would introduce large optimistic bias to the performance results. Also, if a data set is initially used for feature selection, and then partitioned to a training set and a testing set (or if cross-validation is used to train/test the model) to assess the performance, the results would also be optimistically biased (Sahiner, 2000). Another less noticeable example is when a data set is used in the examination of multiple models and/or parameter tunings; then the performance of the best model, as assessed with the very same data set, is subject to model selection bias. In other words, if a data set has been used for the model selection task, another data set is needed to assess the performance of the selected model.

Another consideration is related to the uncertainty estimation problem. Appropriate assessment of the variability of the stand-alone performance of CAD is important as the performance is usually estimated from a data set of limited size. Variability estimation is particularly useful in, for example, calculating confidence intervals on the estimated performance, model selection, model comparison, etc. In the conventional approach, the variability of the estimated performance is ascribed only to the random choice of testing cases. One argument for this approach is the claim that the developers have frozen their training algorithm – so that it is a *fixed effect*, contributing no randomness to the analysis. However, a classifier is rarely frozen until a sufficiently large training sample is employed in the training. This means that the random choice of a training sample also contributes to the variability of the estimated performance and hence the training variability has to be assessed. Assessment of training variability can demonstrate the stability of a CAD algorithm – a more stable algorithm would be less sensitive to the varying training sample, i.e. would have smaller training variability. Methods are emerging in the literature for the assessment of classifiers in terms of uncertainty due to both finite size of trainers and finite size of testers, e.g. a bootstrap-based approach when only one data set is available (Yousef, 2005) or a U-statistics-based approach when two data sets are available for training and testing the classifier (Yousef, 2006). Additional random effects are also present in these evaluation situations. For example, for some CAD applications, the truth status is determined by an expert panel for which the variability should also be accounted. In CADe, additional random effects include the number and location of multiple CAD cues and/or multiple lesions in images. There will be further uncertainty associated with specifying a test cut-off value corresponding to a desired level of performance, e.g. sensitivity or true-positive rate, and estimating the corresponding number of FP cues per image (Bornefalk, 2005).

19.3 CAD AND IMAGE PERCEPTION

We have so far discussed the development and assessment of computerized techniques of a CAD system. Since the computer output only serves as "another opinion" to the radiologist and since the final diagnosis is made by the radiologist, the human observer (i.e. the radiologist) is inevitably an integral part of a CAD system. CAD is motivated by limitations of human observers in interpreting medical images and is designed in the hope of overcoming these limitations, and thus, as mentioned earlier, the ultimate benefit of CAD requires evaluation of the CAD system with observer studies, i.e. compare the image interpretation performances of the radiologist without the computer aid and with the aid. Many types of CAD, especially those related to detection, are designed to improve the detection/diagnosis accuracy of disease in medical image interpretation where perception plays a significant role, and hence one of the major goals of CAD is to enhance the perception of abnormalities. Although the exact mechanism of how CAD output may impact human perception of medical images is not completely understood, a large variety of studies, in laboratory settings or in clinical trials, have been conducted to examine the effect of CAD on image perception and interpretation in terms of the overall diagnostic performance, and are reviewed in this section.

19.3.1 Limitations of human observers and the promise of computer analyses

Misses of lesions on radiographic images may be because of the presence of noise, i.e. quantum mottle or overlapping normal structures, or because of the radiologist's inadequate search patterns or lapses in perception (Kundel, 1975). It has been reported that at least half of the errors made in clinical image interpretation practice are perceptual (Bird, 1992; Renfrew, 1992). Factors affecting radiologists' performance levels may be summarized as the following:

(1) Physical quality of the image such as poor spatial resolution, noise, low contrast, and image artifacts. This represents one of the most fundamental limitations of the medical image itself.
(2) Attributes of the disease state such as lesion size, lesion contrast, and complexity of the anatomical background ("structure noise").
(3) Amount of data: medical imaging is evolving from traditional 2D imaging to 3D and 4D (functional imaging) with multiple views. The amount of information presented to the radiologist can be overwhelming.
(4) Interpretation conditions such as presence of distractions, fatigue, mental distractions, and/or interpretation time limitations (workload) which occur in screening programs or when a huge amount of data is available for review.
(5) Ability of the radiologist: radiologists may have different levels of prior training and knowledge.

Overcoming the image quality limitations may eventually rely on the improvement of imaging hardware and optimal acquisition protocols. However, with advances in computer vision, artificial intelligence (AI), and computer technology, along with the availability of large databases of cases, CAD could potentially impact all aspects of the image interpretation performance. As a second reading tool, CAD could potentially reduce human perceptual search errors and interpretation errors, and improve diagnostic *accuracy*. As an automatic technique, CAD has potential to reduce the variability between and within observers, and thus improve the *consistency* of the image interpretation. With advanced AI and computer graphics techniques, CAD yields quantitative measures such as morphology and physiological parameters and also improved visualizations, which could potentially reduce image interpretation time and improve *efficiency*.

Lung cancer ranks as the leading cause of cancer deaths in the USA, and early detection is essential for effective therapeutic intervention (Flehinger, 1992; Shah, 2003). Because of its low cost, simplicity, and low radiation dose, chest radiography is used for the clinical assessment of many thoracic diseases including the detection of lung nodules as indicators of primary lung cancer or metastases. Radiologists may miss nodules on radiographic chest images due to the presence of overlying ribs, bronchi, blood vessels, and other anatomic lung structures. A CADe system for the detection of lung nodules aims to direct a radiologist's attention to potential nodule sites.

Due to the elimination of overlapping structures in thoracic CT, CT is more sensitive than chest radiography in the detection of lung cancer at an early stage (Henschke, 2001). However, even with the reduction of overlapping structures, nodule detection is still limited due to the presence of blood vessels. The advent of low-dose CT has led to lung cancer screening programs (Henschke, 2001; Sone, 2001; Swensen, 2003). Also, the high number of CT slices in a single case, which need to be reviewed by a radiologist, makes lesion detection a burdensome task that is complicated by fatigue and distractions. A CADe system for the detection of nodules on CT images is expected to help radiologists focus their attention on regions suspected of being cancerous, and avoid missing nodules. This seems especially advantageous in low-dose CT lung cancer screening procedures, in which most cases will be normal – similar to the large number of normal mammograms requiring review in a breast cancer screening program.

Figure 19.5(a), for example, reproduces a chest radiograph with a subtle nodule camouflaged by a rib in the upper right lung making it difficult to detect. While one of the benefits of thoracic CT is the removal of overlapping tissues in the image, structure noise can still cause problems in detection as in Figure 19.5(b), in which a nodule is located near a neighboring vessel. If a computer could indicate the location of the suspect lesion, the radiologist's attention would be drawn to the region, potentially avoiding the miss.

Another cause for misses in *detection tasks* is when the radiologist must review a large amount of image data. Such a tedious task can occur in a screening program where most cases are normal or when searching a large amount of image data such as in thoracic CT (which can reach hundreds of slices). Detecting abnormalities in screening cases, in which most are normal, has been compared to the task of finding the character "Waldo" (also known as "Wally") in a *Where's Waldo* book in which he appears only on 5 of 1000 pages! The vast amount of data for human interpretation in a cancer screening program makes lesion detection a burdensome task, impacts radiologists' workloads, and causes oversight errors.

For example, Figure 19.6 shows mammograms from a screening program. They had been input to the University of Chicago prototype CADe system for analysis of performance in the prospective computer analysis of cancers missed on screening mammography (Nishikawa, 2001). After the automatic analysis, the computer correctly detected the lesion in the right mediolateral oblique (MLO) view, but it also placed one false mark in the left MLO view and two false marks in the left craniocaudal (CC) view. Note that in this study, it was found that the lesion had been clinically missed, and also missed by the three readers in the study (Nishikawa, 2001).

Medical imaging in radiology practice is evolving to include multimodalities, multidimensions, and multiviews. For example, a clinical breast imaging reading room, which had traditionally included hardcopy screen/film mammograms, nowadays is equipped with state-of-the-art LCD monitors displaying full-field digital mammograms, computer workstations showing 2D and 3D breast ultrasound images, and scanner workstations displaying 4D dynamic MRI. The radiologists are overwhelmed with a huge amount of information, and computer aids definitely have great potential to extract, combine, and display the relevant information. An example in breast MRI shows that a computer may help extract useful diagnostic information from a large amount of data. A typical dynamic breast MRI exam consists

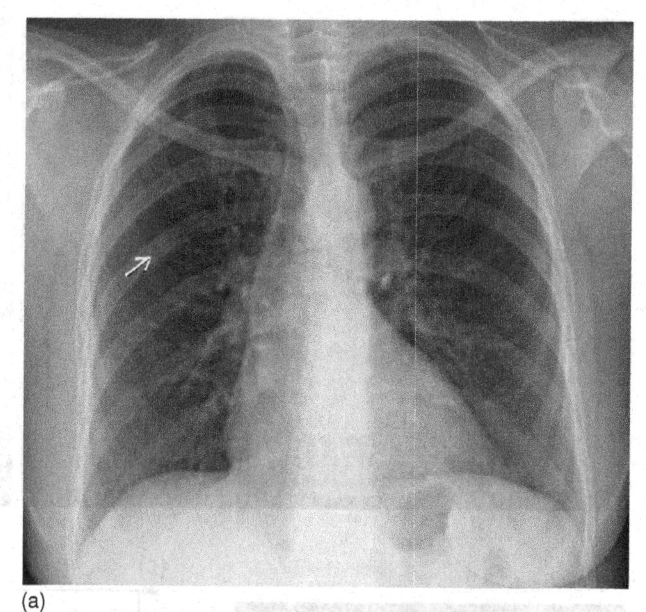

(a)

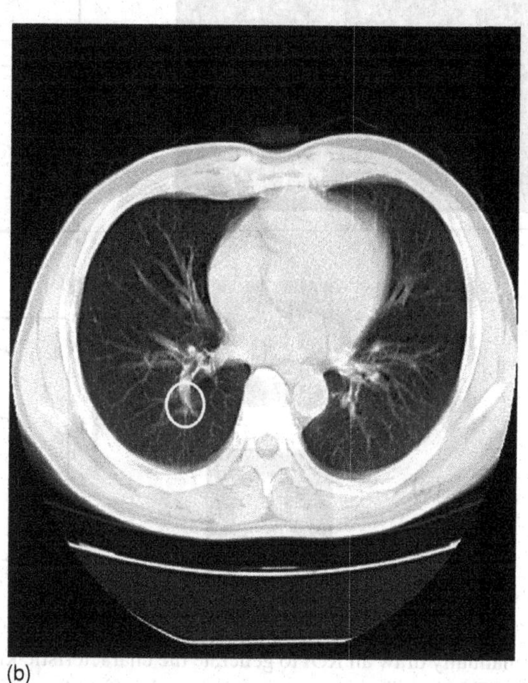

(b)

Figure 19.5 (a) A chest radiograph with a subtle nodule camouflaged by a rib in the upper right lung making it difficult to detect. (b) A thoracic CT image, with a nodule located near a neighboring vessel, illustrating that even with the removal of overlapping tissues in the image, structure noise can still cause problems in detection.

of 3D images acquired before and repeatedly after the injection of a contrast agent with each voxel possessing a MR-signal-intensity-time curve. One of the commonly used features for diagnosing a breast lesion as benign or malignant is the shape of the kinetic curve (Kuhl, 1999; ACR, 2003): a persistent type of curve typically is associated with a benign diagnosis, a washout type of curve is an indicator of malignancy, and a plateau type of curve is intermediate. Due to uptake inhomogeneity, however, averaging over the entire lesion to obtain a kinetic curve is suboptimal. In practice, radiologists search over the lesion to find the most enhancing region ("hot" area) within the lesion

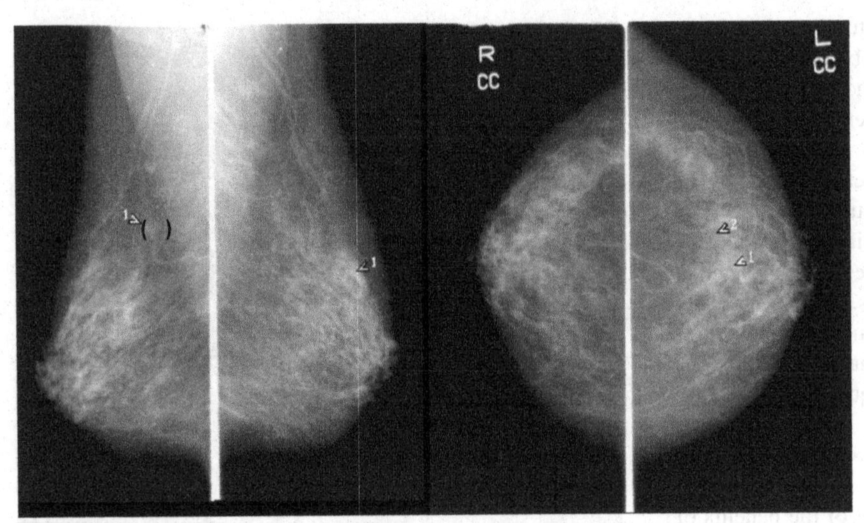

Figure 19.6 Mammograms from a screening program that had been input to the University of Chicago prototype CADe system for analysis of performance in the prospective computer analysis of cancers missed on screening mammography. After the automatic analysis, the computer correctly detected the lesion in the right mediolateral oblique (MLO) view, but it also placed one false mark in the left MLO view and two false marks in the left craniocaudal (CC) view. Note that in this study, it was found that the lesion had been clinically missed, and also missed by the three readers in the study (Nishikawa, 2001).

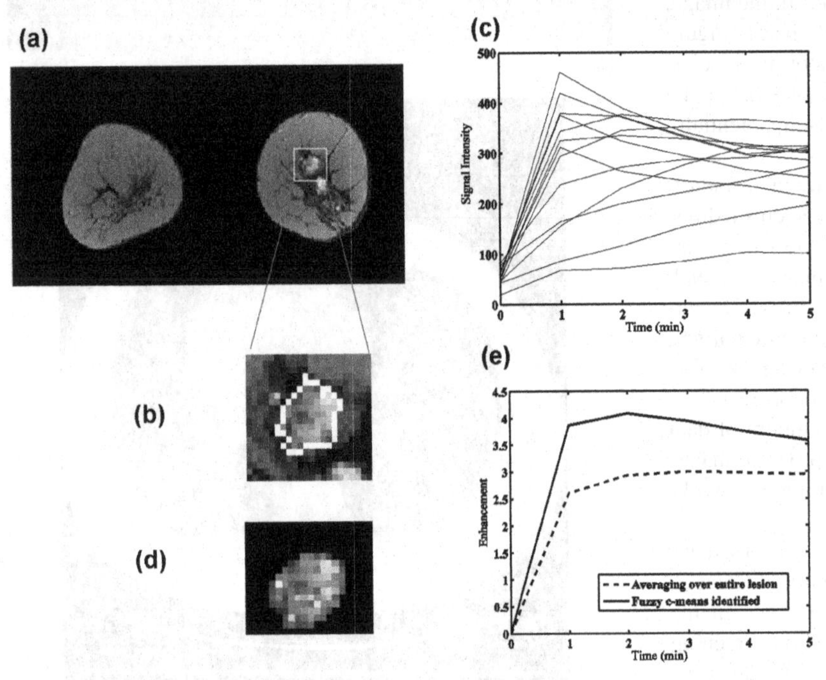

Figure 19.7 (a) shows a slice containing a lesion highlighted in (b). (c) shows the signal-time curves of randomly selected voxels in the lesion. (d) and (e) [solid line] show the most-enhancing region (color-coded) and the corresponding characteristic kinetic curve, respectively, as identified by a computerized approach (Chen, 2006b). The computer-identified washout curve leads to a correct diagnosis of malignancy (for comparison, the curve averaging over the entire lesion is also shown in (e) [dash line]).

and manually draw an ROI to generate the characteristic kinetic curve. This is a time-consuming task and suffers from significant inter- and intra-observer variability, and thus a computer may play a role in identifying the characteristic kinetic curve from a given lesion automatically and efficiently. For example, Figure 19.7(a) shows a slice containing a lesion highlighted in Figure 19.7(b) (note that there are 60 slices for a 3D image at a particular time point and this just shows one slice at one time point). Figure 19.7(c) shows the signal-time curves of randomly selected voxels in the lesion. It can be easily seen that all three types of curves are found in one lesion. Figures 19.7(d) and (e) [solid line] show the most-enhancing region (color-coded) and the corresponding characteristic kinetic curve, respectively, as identified by a computerized approach (Chen, 2006b). The computer-identified washout curve leads to a correct diagnosis of malignancy (for comparison, the curve averaging over the entire lesion is also shown in Figure 19.7(e) [dash line]). This research has shown that computerized approaches are promising in extracting relevant and manageable information from a large amount of image data.

CAD can also play a critical role in visualization. Various investigators are developing CADe systems for CT colonography. It is important to note that these systems require both an effective visual interface for the 3D CT colonography images (Royster, 1997; McFarland, 2001) and reliable CADe output (Summers, 2001; Yoshida, 2002; Nappi, 2003; Suzuki, 2006) to aid in the detection of colonic polyps. With advanced computer graphics techniques, the interface provides the fly-through of the colon in the virtual colonoscopy – and the computer is additionally tasked to label regions suspicious for colonic polyps in these CT-based images (Yoshida, 2004; Summers, 2005). In the interpretation process, the combination of an efficient interface and accurate CADe output has the potential to improve both the interpretation accuracy and interpretation time. Figure 19.8 shows the interface of a colon CADe system developed in the University of Chicago.

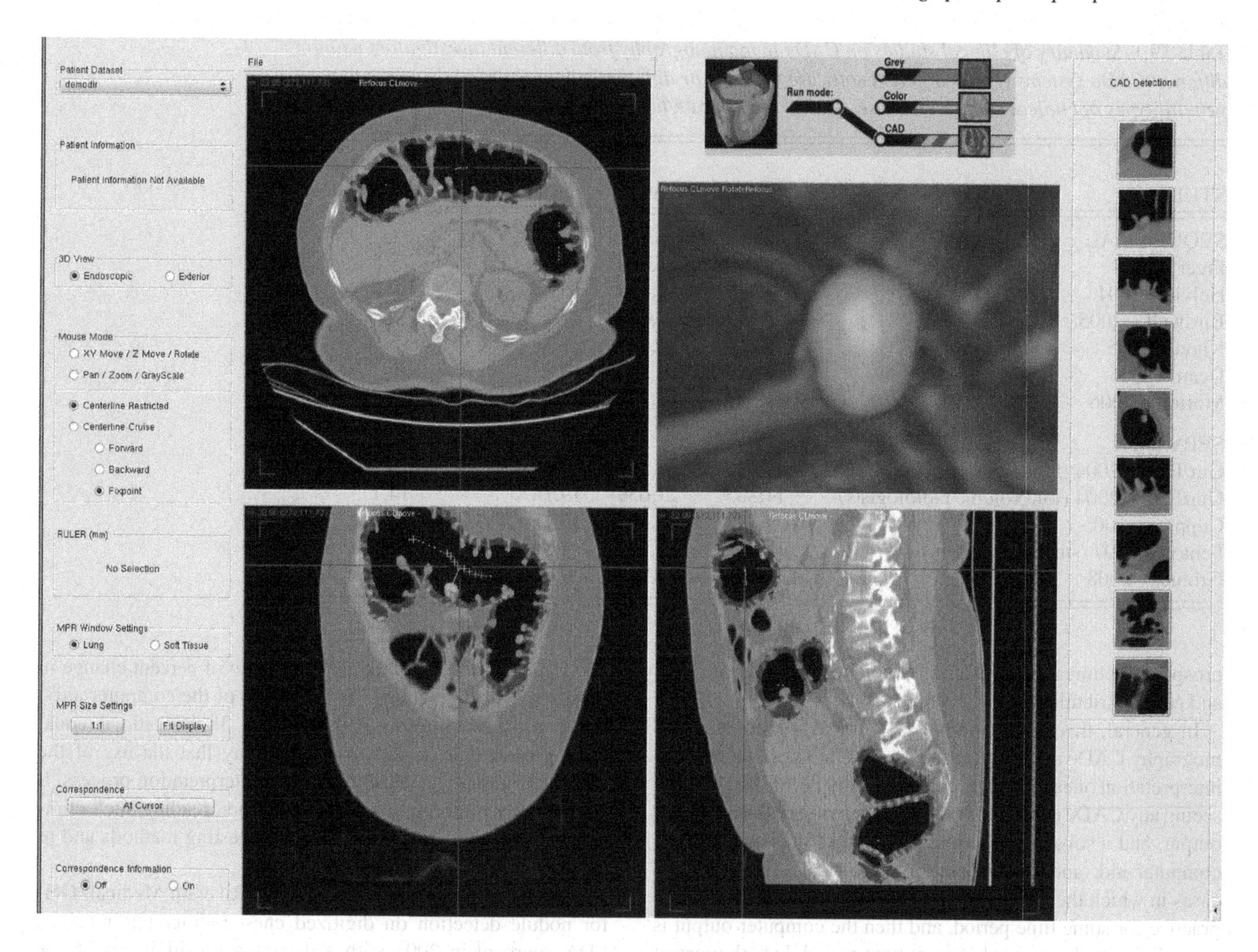

Figure 19.8 The interface of the prototype CAD system developed at the University of Chicago to assist radiologists in finding polyps on CT colonography. The upper left, lower left, and lower right images show the multiplanar reformatted views of the colon. The upper right image shows a 3D endoluminal view of the colon. The CAD system finds the highly suspicious polyps and colors them green. (Courtesy of Kenji Suzuki, Ph.D., University of Chicago.)

19.3.2 Observer studies: an overview of methodology

An observer study (or "reader study") in CAD of medical images refers to a study of observer performances in medical image perception, interpretation, and/or diagnosis without the use of CAD compared with user performance with CAD – either in a real clinical realm ("clinical study") or in a simulated clinical setting ("laboratory study"). A well-designed observer study provides direct evidence of the impact of CAD on image perception and interpretation.

Modern experience in medical imaging has indicated that at least two sources of variability – reader variability and patient case variability – need to be accounted for when comparing two imaging modalities/conditions so that any conclusion on the impact of a new modality (e.g. CAD) can be generalized to a population of readers and a population of patients. The experimental paradigm within this context has come to be referred to as the *multiple-reader, multiple-case (MRMC) ROC* paradigm. In this study design for assessing CAD, a sample of readers reads a sample of patient cases under both conditions: without the computer aid and with the computer aid. The most powerful study design – for a given number of cases with verified truth-status – is the *fully-crossed* design, in which the same readers read the same cases in both situations. A reading consists of a radiologist providing a level of suspicion or rating of the probability of the abnormal condition of interest for each patient case. Methods for designing and analyzing MRMC experiments include the "DBM" method (Dorfman, Berbaum, and Metz, 1992), the "BWC" method (Beiden, Wagner, and Campbell, 2000), and the "BCK" method (Barrett, 2005; Clarkson, 2006; Kupinski, 2006) with the associated "one-shot" estimate of MRMC variance of AUC (Gallas, 2006). Alternative study designs are possible, and sometimes necessary, because it may be impractical for the same readers to read the same cases. Recent progress includes methods on a "doctor-patient" design in which every doctor reads his/her own patients and arbitrary designs (Gallas, 2007; Gallas, 2008). For an introduction to multireader ROC analysis, the reader should consult other chapters in this book and reviews by Wagner *et al.* (Wagner, 2007).

As mentioned earlier in this section, an observer study in CAD can be either a laboratory study or a clinical study. The advantages of a laboratory study include having well-controlled experimental conditions and protocols that are also less costly. However, a laboratory study can be less realistic than a real

Table 19.1 *Summary of clinical studies on CADe in mammography from different investigators using different CADe systems (either different software versions or different manufacturers). Note the two general types of clinical study – either sequential or separate/historical study designs.*

STUDY	Unaided	Aided	% change in cancer detected	% change in recall rates
SEQUENTIAL				
Freer – 2001	12,860	12,860	19.5	18.5
Helvie – 2004	2,389	2,389	10	9.9
Birdwell – 2005	8,692	8,692	7.4	8
Khoo – 2005	6,111	6,111	1.3	5.8
Dean – 2006	9,520	9,520	13.3	26
Morton – 2006	21,349	21,349	7.6	10.8
SEPARATE				
Gur/Feig – 2004 (high volume radiologists)	44,629	37,500	−3.3	−4.9
Gur/Feig – 2004 (low volume radiologists)	11,803	21,639	19.7	14.1
Cupples – 2005	7,872	19,402	16.1	8.1
Fenton – 2007 (survey study)	313,259	31,186	4.5	31
Gromet – 2008	231,221	231,221	1.9	3.9

prospective clinical study in terms of both reading environments and case distributions (Gur, 2008).

In general, there are two types of clinical studies of mammography CADe systems: (a) "sequential" studies in which the interpretation of each clinical case is initially performed without seeing any CADe output, followed by a viewing of the computer output, and a subsequent re-interpretation of the case with the computer aid; and (b) "separate or historical" assessment of cases in which the CADe system is not used within a radiology practice for some time period, and then the computer output is used clinically during a subsequent time period. In both types of clinical studies, the performance of the radiologists interpreting without the CADe output is compared to the performance of the radiologists interpreting with the CADe output. Note that in the second type of study, radiologists may be interpreting different mammograms from different patients.

19.3.3 Impact of CAD: evidence from observer studies

19.3.3.1 Impact of CADe

Since the first CADe prototype, there have been multiple CADe systems approved by the FDA, with the first approval being of an R2 Technology CADe system (now Hologic) in 1998. These various systems (R2/Hologic; ISSI; Kodak) have been applied on numerous systems for screen-film mammography and full-field mammography units. Websites of the systems describe performance, physical space requirements, workflow aspects, and options. The stand-alone performances of some of these systems tend to achieve a high detection sensitivity for clustered microcalcifications with a lower detection sensitivity for mass.

To date, multiple clinical studies/surveys have been performed and are listed in Table 19.1 (Freer, 2001; Feig, 2004; Gur, 2004; Helvie, 2004; Birdwell, 2005; Cupples, 2005; Khoo, 2005; Dean, 2006; Morton, 2006; Fenton, 2007; Gromet, 2008), which also gives the percent change in sensitivity and call-back rate for each study. Of interest is the ratio of percent change in sensitivity (from "without" to "with use of the computer aid") to the percent change in call-back rate. If this ratio is equal to or greater than 1, then one might say that the use of the computer output was beneficial to the interpretation process. It is extremely important to be careful when reading such clinical studies to understand the statistical testing methods and to recognize limitations.

The first commercial CADe system (Riverain Medical, OH) for nodule detection on digitized chest radiographs received FDA approval in 2001 with a detection sensitivity of 65.0% with 5.3 FP marks per image (Freedman, 2001). Radiologist performance was shown to improve even at moderate levels and large FP rates.

Colon cancer is another leading cause of cancer deaths. The early detection of precursor colonic polyps and their subsequent removal can drastically improve survival. CT colonography (virtual colonoscopy) is being investigated as a screening alternative to conventional colonoscopy for the early detection of such colonic polyps (Hara, 1997). Radiologist interpretation of a CT colonography exam can be quite time consuming, and, because of the potentially large number of axial CT images (greater than 500 slices), can result in oversight errors.

A number of CADe algorithms have been developed in colon CADe and the impact of these aids on radiologist performance is currently being evaluated in the clinical setting. Recently, Petrick *et al.* (Petrick, 2008) conducted an observer performance study to evaluate the effect of CADe as a second reader on radiologists' diagnostic performance in interpreting CT colonographic examinations. With a modest number of patients (i.e. 60 CT exams) and a modest number of observers (i.e. four board-certified radiologists), the authors demonstrated that use of CADe led to a significant increase in sensitivity for detecting polyps in the 6 mm or larger (sensitivity increased 15%, $p<0.01$) and 6–9 mm (sensitivity increased 16%, $p<0.02$) groups at the expense of a similar significant reduction in specificity. Larger, prospective clinical trials are needed to further demonstrate the effect of CADe on CT colonography.

19.3.3.2 Impact of CADx

Through various laboratory observer studies, CADx has been shown to aid radiologists in the task of distinguishing between malignant and benign breast lesions (Chan, 1999; Jiang, 1999; Huo, 2002; Horsch, 2006). Use of a computer diagnostic aid has the potential to increase sensitivity, specificity, or both in the work-up of breast lesions. Investigators have demonstrated that radiologists showed an increase in both sensitivity and specificity in the characterization of clustered microcalcifications and in the associated recommendation for biopsy (Jiang, 1999). In addition, it was shown that improvement in performance can be obtained both by expert mammographers and by community-based radiologists who used CADx information, with the increase greater for the non-experts (Huo, 2002).

Also, use of computer output is expected to reduce the variability among radiologists' interpretations (Jiang, 2001). Various CADx observer studies have been performed for individual breast imaging modalities as well as for multimodality CADx workstations. These have demonstrated that the use of computer-estimated probabilities of malignancy led to a statistically significant improvement in radiologists' performances in the task of interpreting single-modality breast images and multimodality breast images (Giger, 2003; Horsch, 2006).

When evaluating a potential cancer case (or other disease state), radiologists consider all available information including the entire case of images (single or multimodality) and not just individual images. Therefore, CADx systems are being developed to analyze multiple views within a single modality as well as across multimodality images. Thus, the computer/human interface of a CADx system needs to communicate both the multiple images and the CADx output to the radiologist. Figure 19.4 shows such a CADx interface. Here, the output of the CADx system can be presented in terms of numerical values related to the likelihood of malignancy, through a display of similar images of known diagnoses, or with a graphical representation of the unknown lesion relative to all lesions in an online database atlas (Giger, 2000; Giger, 2003; Horsch, 2006). This interface displays similar images and uses color coding to indicate whether the similar images are malignant or benign. In addition, with the graphical option, the probability of malignancy of the unknown case can be shown relative to the probability distributions of all the malignant and benign cases in the database or relative to the distributions of a specific feature (i.e. lesion characteristic).

Radiologists learn to interpret cases during their radiology residencies through the review of hundreds of cases, and thus the access to online cases of known pathology during a radiologist's daily practice may be helpful for continuous learning. A search of an online image atlas can be based on individual features, on likelihoods of malignancy, or on psychophysical measures of similarity. The system in Figure 19.4 searches either via the computer-determined estimate of the probability of malignancy or by way of computer-extracted lesion features (characteristics) (Giger, 2003; Horsch, 2006). Others have combined the computer-extracted lesion characteristics with subjective similarity measures obtained from observers reviewing pairs of images (Muramatsu, 2005) or from observers giving subjective perceived ratings of lesion features (Zheng, 2007).

19.3.4 Summary of the impact of CAD

We have discussed the limitations of human observers in the perception and interpretation of medical images, and the promise of computerized methods for overcoming these limitations. Observer studies have provided evidence of the contribution of CAD to the diagnostic performance of human observers. The positive impact of CAD is typically that the computer's true positive findings contain abnormalities initially missed by the human observer and hence, appropriate use of CAD increases sensitivity. However, this positive impact is often associated with an expense of more FP findings, e.g. increased recall rate in screening mammography. Whether the positive impact can counteract the negative impact depends on both the CAD system and the appropriate use of the system by the human observer. How to best utilize CAD is the topic of the next section.

19.4 HOW TO BEST UTILIZE CAD?

CAD in medical imaging is designed to assist radiologists to better interpret medical images, improve diagnostic accuracy, and ultimately, benefit patient care. The most advantageous utilization of CAD in medical image interpretation relies on both the detection/diagnosis ability (i.e. stand-alone performance) of the computer system and proper use of the CAD output by the clinicians. Studies in mammography CADe have shown how the CAD performance level could affect radiologists' performance in detecting subtle masses and microcalcification clusters. Results reported by Zheng *et al.* (Zheng, 2001; Zheng, 2004) suggest that highly performing CAD schemes have the potential to significantly improve detection performance of radiologists, whereas poorly performing schemes had little or negative effect on radiologists' performance in identifying abnormalities depicted on mammograms.

CAD can be used in two reading paradigms (or modes): the so-called "second reader" paradigm and the "concurrent" paradigm. In the "second reader" paradigm, radiologists view the CAD output *after* they have made their initial interpretation without computer aid. This is the classic paradigm that most CAD systems are currently designed, evaluated, and labeled to use. In this paradigm, CAD acts as a checker of the abnormalities missed by an unassisted reader. One may argue that such a paradigm increases interpretation time as compared to a single reading without computer aid since it requires further assessing the computer output in addition to a regular single reading. A potentially more time-efficient paradigm is emerging as "concurrent reading," which displays the CAD output at the start of image perception and interpretation. Although it is an intuitively attractive proposition, concurrent application of CAD may reduce observer vigilance and therefore reduce sensitivity. For example, Zheng *et al.* (Zheng, 2004) reported that, in CADe of mammographic masses, viewing CADe cues during the initial display consistently resulted in fewer abnormalities being identified in noncued regions. In a recent investigation of the optimal CAD reader paradigm on CT colonography, Taylor

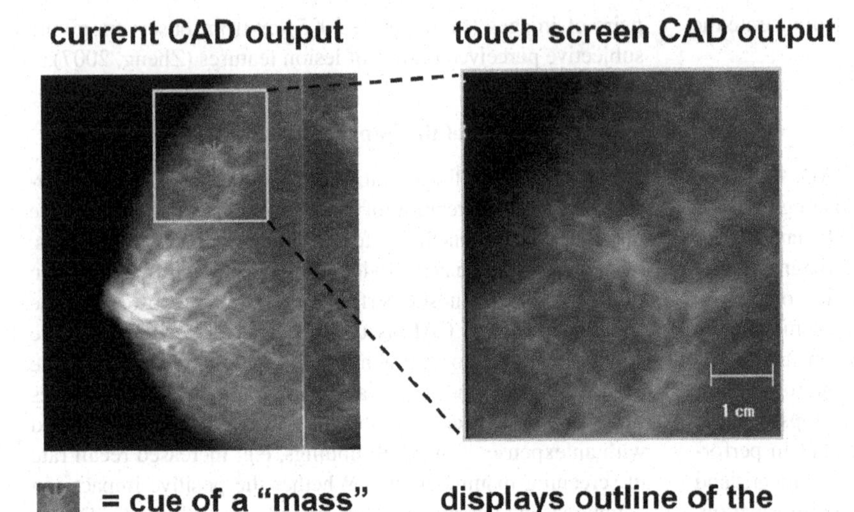

Figure 19.9 Output from a commercial CADe system in which the user can touch the lesion on the screen and the system will show the computer segmentation. (Courtesy of R2 Technology, a Hologic Company, Sunnyvale, CA.)

et al. (Taylor, 2007) reported that CAD is more time-efficient in its "concurrent" mode than "second reader" mode, with similar sensitivity for polyps 6 mm or larger. However, the "second reader" paradigm maximized sensitivity, particularly for small lesions. The relative pros and cons of different reading paradigms may need further investigation in the real clinical settings, especially in the potentially different uses of CADe and CADx. However, it is crucially important to use CAD in the mode for which it is designed and evaluated.

Proper use of a CADe output in the classic "second reader" paradigm may require that any initial human-detected region/lesion remains as a detection even if the CADe system does not indicate it. Such requirements ensure that the radiologists' detection sensitivity remains either constant or improves with computer assistance. In the mean time, such requirements may not be necessary if the performance level of a CADe system is adequately high or the system is designed to be used in the "concurrent" mode.

Sufficient training of the radiologist is important for the best utilization of CAD since, as with any new modality, there will be a learning curve. Such training would include the use of the CAD interface so that the radiologist could extract and interpret the CAD output information appropriately. Getting the radiologist familiar with the typical correct and incorrect findings from the CAD may help the radiologist better combine the CAD output into his/her final diagnostic decisions.

Also, radiologists may better use CAD, and with more confidence, if intermediate CAD output is shown to help radiologists understand the image analyses in the "black box." For example, Figure 19.9 shows the output from a commercial CADe system in which the radiologist can touch the lesion on the screen and the system will show the computer segmentation. Different algorithms may yield different segmentations (i.e. delineations) of lesions, as demonstrated in Figure 19.10, which shows a breast mass in mammography as outlined ("perceived") by a radiologist and three different computer algorithms (region-growing method, radial-gradient-index-based method, and snake-based method, respectively; Huo, 1995; Kupinski, 1998; Yuan, 2007). If the segmentation is "poor" from the radiologist's perspective, then the radiologist can assume that the subsequent computer analysis of the "lesion" will be erroneous, and thus he/she should proceed with caution in using the computer output.

19.5 SUMMARY

The purpose of this chapter has been to present a broad overview of CAD in medical imaging from an image perception perspective. As a tool to assist humans in overcoming perceptual errors and other human limitations in the interpretation of medical images, CAD has seen tremendous growth in technical development, assessment methodology, and clinical applications. Laboratory and clinical observer studies have shown evidence that CAD can help detect cancers missed by human perception. Nonetheless, the increased sensitivity is often associated with a cost of decreased specificity. Thus research is needed for further development of computerized techniques in CAD systems, further development of human-computer interactions in image perception tasks, and further assessment of the acceptable levels for sensitivity and specificity. While we have focused on the overall impact of CAD on image perception, it should be noted that image perception research can have fundamental impacts on CAD – since CAD is designed to catch human perceptual errors, an overall understanding of the perceptual process may fundamentally impact the way CAD is designed and used (Krupinski, 1998).

ACKNOWLEDGEMENT AND DISCLAIMER

The authors are grateful for the many fruitful discussions with the faculty and research staff in the Department of Radiology and Committee on Medical Physics at the University of Chicago. Certain parts of the chapter are the result of research supported in parts by USPHS grants from NCI, NIBIB, and NIAMS, as well as from the U.S. Army Breast Cancer Research Program,

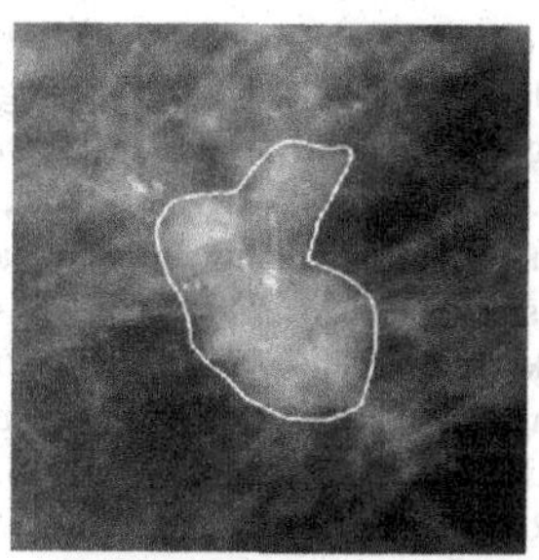

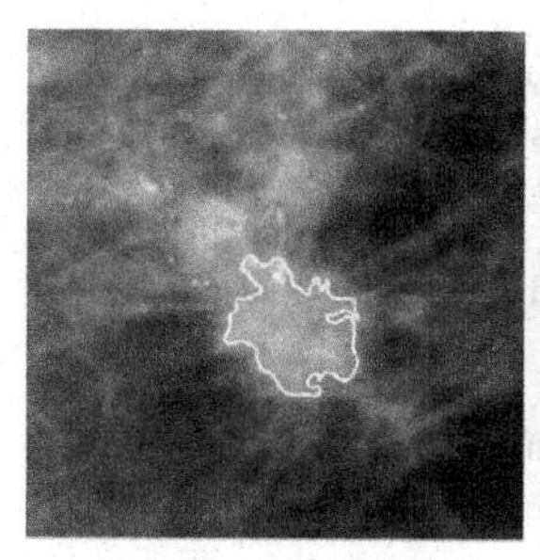

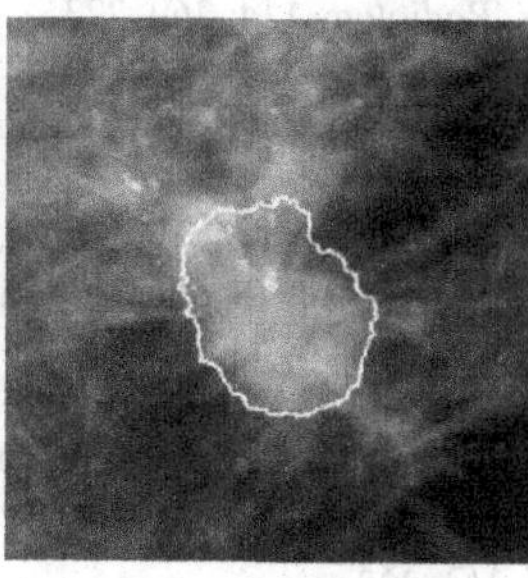

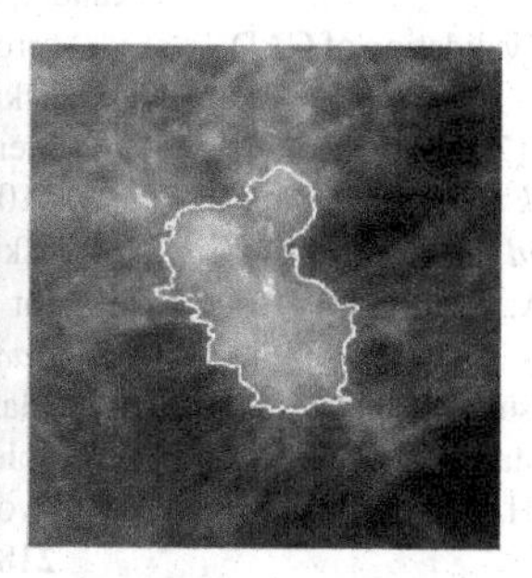

Figure 19.10 Image of a breast mass in mammography as outlined ("perceived") by a radiologist and three different computer algorithms (region-growing method, radial-gradient-index-based method, and snake-based method, respectively; Huo, 1995; Kupinski, 1998; Yuan, 2007).

the American Cancer Society, the Whitaker Foundation, and the University of Chicago Cancer Research Center. M. Giger is a stockholder in R2 Technology, a Hologic Company (Sunnyvale, CA). It is the University of Chicago conflict-of-interest policy that investigators disclose publicly actual or potential significant financial interests that may appear to be affected by the research activities.

This chapter represents the professional views of the authors. It is not an official document, guidance, or policy of the US government, Department of Health and Human Services, or the Food and Drug Administration, nor should any official endorsement be inferred. No FDA endorsement of any product or company mentioned in this manuscript should be inferred.

REFERENCES

American College of Radiology. (2003). ACR BI-RADS: MRI. In *ACR BI-RADS: Breast Imaging Reporting and Data System: Breast Imaging Atlas*. Reston, VA: ACR.

Barrett, H.H., Kupinski, M.A., Clarkson, E. (2005). Probabilistic foundations of the MRMC method. *Proc SPIE* **5749**: 21–31.

Beiden, S.V., Wagner, R.F., Campbell, G. (2000). Components-of-variance models and multiple-bootstrap experiments: an alternative method for random-effects, receiver operating characteristic analysis. *Acad Radiol* **7**: 341–349.

Bird, R.E., Wallace, T.W., Yankaskas, B.C. (1992). Analysis of cancers missed at screening mammography. *Radiology* **184**: 613–617.

Birdwell, R.L., Bandodkar, P., Ikeda, D.M. (2005). Computer-aided detection with screening mammography in a university hospital setting. *Radiology* **236**: 451–457.

Bornefalk, H., Hermansson, A.B. (2005). On the comparison of FROC curves in mammography CAD systems. *Med Phys* **32**: 412–417.

Bunch, P.C., Hamilton, J.F., Sanderson, G.K., Simmons, A.H. (1977). A free response approach to the measurement and characterization of radiographic observer performance. *Proc SPIE* **127**: 124–135.

Chakraborty, D.P. (2000). The FROC, AFROC and DROC variants of the ROC analysis. In *Handbook of Medical Imaging, Volume 1. Physics and Psychophysics*, Beutel, J., Kundel, H., Van Metter, R. (eds.), Bellingham, WA: SPIE, pp. 771–798.

Chan, H.P., Sahiner, B., Helvie, M.A. *et al.* (1999). Improvement of radiologists' characterization of mammographic masses by using computer-aided diagnosis: an ROC study. *Radiology* **212**: 817–827.

Chen, W., Giger, M.L., Bick, U. (2006a) A fuzzy c-means (FCM) based approach for computerized segmentation of breast lesions in dynamic contrast-enhanced MR images. *Acad Radiol* **16**: 63–72.

Chen, W., Giger, M.L., Bick, U., Newstead, G. (2006b). Automatic identification and classification of characteristic kinetic curves of breast lesions on DCE-MRI. *Med Phys* **33**: 2878–2887.

Clarkson E., Kupinski, M.A., Barrett, H.H. (2006). A probabilistic model for the MRMC method. Part I. Theoretical development. *Acad Radiol* **13**: 1410–1421.

Cupples, T., Cunningham, J.E., Reynolds, J.C. (2005). Impact of computer-aided detection in a regional screening mammography program. *AJR* **185**: 944–950.

Dean, J.C., Ilvento, C.C. (2006). Improved cancer detection using computer-aided detection with diagnostic and screening mammography: prospective study of 104 cancers. *AJR* **187**: 20–28.

Dorfman, D.D., Berbaum, K.S., Metz, C.E. (1992). Receiver operating characteristic rating analysis: generalization to the population of readers and patients with the jackknife method. *Invest Radiol* **27**: 723–731.

Feig, S.A., Sickes, E.A., Evans, W.P., Linver, M.N. (2004). Re: Changes in breast cancer detection and mammography recall rates after the introduction of a computer-aided detection system. *J Natl Cancer Inst* **96**: 1260–1261.

Fenton, J.J., Taplin, S.H., Carney, P.A. *et al.* (2007). Influence of computer-aided detection on performance of screening mammography. *N Engl J Med* **356**: 1399–1409.

Flehinger, B.J., Kimmel, M., Melamed, M.R. (1992). The effect of surgical treatment on survival from early lung cancer. Implications for screening. *Chest* **101**: 1013–1018.

Freedman, M., Lo, S., Lure, F. *et al.* (2001). Computer-aided detection of lung cancer on chest radiographs: algorithm performance vs. radiologists' performance by size of cancer. *Proc SPIE* **4319**: 150–159.

Freer, T.W., Ulissey, M.J. (2001). Screening mammography with computer-aided detection. Prospective study of 12,860 patients in a community breast center. *Radiology* **222**: 781–786.

Gallas, B.D. (2006). One-shot estimate of MRMC variance: AUC. *Acad Radiol* **13**: 353–362.

Gallas, B.D., Pennello, G.A., Myers, K.J. (2007). Multi-reader multi-case variance analysis for binary data. *Journal of the Optical Society of America A*, **24**(12): B70–B80.

Gallas, B.D., Brown, D.G. (2008). Reader studies for validation of CAD systems. *Neural Networks* **21**: 387–397.

Giger, M.L., Huo, Z., Kupinski, M.A., Vyborny, C.J. (2000). Computer-aided diagnosis in mammography. In *Handbook of Medical Imaging, Volume 2. Medical Imaging Processing and Analysis*, Sonka, M., Fitzpatrick, M.J. (eds.), Bellingham, WA: SPIE, pp. 915–1004.

Giger, M.L., Huo, Z., Vyborny, C.J. *et al.* (2003). Results of an observer study with an intelligent mammographic workstation for CAD. In *Digital Mammography, IWDM 2002*, Peitgen, H.-O. (ed.), Berlin: Springer, pp. 297–303.

Gromet, M. (2008). Comparison of computer-aided detection to double reading of screening mammograms: review of 231,221 mammograms. *Am J Roentgenol* **190**(4): 854–859.

Gur, D., Sumkin, J.H., Rockette, H.E. *et al.* (2004). Changes in breast cancer detection and mammography recall rates after the introduction of a computer-aided detection system. *J Natl Cancer Inst* **96**: 185–190.

Gur, D., Bandos, A.I., Cohen, C.S. *et al.* (2008). The "laboratory" effect: comparing radiologists' performance and variability during prospective clinical and laboratory mammography interpretations. *Radiology* **249**: 47–53.

Hara, A.K., Johnson, C.D., Reed, J.E. *et al.* (1997). Detection of colorectal polyps with CT colography: initial assessment of sensitivity and specificity. *Radiology* **205**: 59–65.

Helvie, M., Hadjiiski, L., Makariou, E. *et al.* (2004). Sensitivity of noncommercial computer-aided detection system for mammographic breast cancer detection: pilot clinical trial. *Radiology* **231**: 208–214.

Henschke, C.I., Naidich, D.P., Yankelevitz, D.F. *et al.* (2001). Early lung cancer action project: initial findings on repeat screenings. *Cancer* **92**: 153–159.

Horsch, K., Giger, M.L., Vyborny, C.J. *et al.* (2006). Multi-modality computer-aided diagnosis for the classification of breast lesions: observer study results on an independent clinical dataset. *Radiology* **240**: 357–368.

Huo, Z., Giger, M.L., Vyborny, C.J. *et al.* (1995). Analysis of spiculation in the computerized classification of mammographic masses. *Med Phys* **22**: 1569–1579.

Huo, Z., Giger, M.L., Vyborny, C.J. *et al.* (2002). Effectiveness of CAD in the diagnosis of breast cancer: an observer study on an independent database of mammograms. *Radiology* **224**: 560–568.

Jiang, Y., Metz, C.E., Nishikawa, R.M. (1996). A receiver operating characteristic partial area index for highly sensitive diagnostic tests. *Radiology* **201**: 745–750.

Jiang, Y., Nishikawa, R.M., Schmidt, R.A. *et al.* (1999). Improving breast cancer diagnosis with computer-aided diagnosis. *Acad Radiol* **6**: 22.

Jiang, Y., Nishikawa, R.M., Schmidt, R.A. *et al.* (2001). Potential of computer-aided diagnosis to reduce variability in radiologists' interpretations of mammograms depicting microcalcifications. *Radiology* **220**: 787–794.

Kakeda, S., Moriya, J., Sato, H. *et al.* (2004). Improved detection of lung nodules on chest radiographs using a commercial computer-aided diagnosis system. *Am J Roentgenol* **182**: 505–510.

Khoo, L.A.L., Taylor, P., Given-Wilson, R.M. (2005). Computer detection in the United Kingdom national breast screening programme: prospective study. *Radiology* **237**: 444–449.

Krupinski, E.A., Kundel, H.L., Judy, P.F. *et al.* (1998). Key issues for image perception research. *Radiology* **209**: 611–612.

Kuhl, C.K., Mielcareck, P., Klaschik, S. *et al.* (1999). Dynamic breast MR imaging: are signal intensity time course data useful for differential diagnosis of enhancing lesions? *Radiology* **211**: 101–110.

Kundel, H. (1975). Peripheral vision, structured noise and film reader error. *Radiology* **114**: 269–273.

Kupinski, M.A., Giger, M.L. (1998). Automated seeded lesion segmentation on digital mammograms. *IEEE Trans Med Imaging* **17**: 510–517.

Kupinski, M., Clarkson, E., Barrett, H. (2006). A probabilistic model for the MRMC method, part 2: validation and applications. *Acad Radiol* **13**: 1422–1430.

McFarland, E.G., Brink, J.A., Pilgram, T.K. *et al.* (2001). Spiral CT colonography: reader agreement and diagnostic performance with two- and three-dimensional image-display techniques. *Radiology* **218**: 375–383.

Metz, C.E. (1978). Basic principles of ROC analysis. *Semin Nucl Med* **8**: 283–298.

Metz, C.E. (1986). ROC methodology in radiologic imaging. *Invest Radiol* **21**: 720–733.

Metz, C.E. (2000). Fundamental ROC analysis. In *Handbook of Medical Imaging, Volume 1. Physics and Psychophysics*, Beutel, J., Kundel, H., Van Metter, R. (eds.), Bellingham, WA: SPIE, pp. 751–769.

Morton, M.J., Whaley, D.H., Brandt, K.R., Amrami, K.K. (2006). Screening mammograms: interpretation with computer-aided detection – prospective evaluation. *Radiology* **239**: 375–383.

Muramatsu C., Li, Q., Suzuki, K. *et al.* (2005). Investigation of psychophysical measures for evaluation of similar images for mammographic masses: preliminary results. *Med Phys* **32**: 2295–2304.

Nappi, J., Yoshida, H. (2003). Feature-guided analysis for reduction of false positives in CAD of polyps for computed tomographic colonography. *Med Phys* **30**: 1592–1601.

Nishikawa, R.M., Giger, M.L., Vyborny, C.J. *et al.* (2001). Prospective computer analysis of cancers missed on screening mammography. In *Digital Mammography 2000, Proc 5th International Workshop on Digital Mammography*. Madison, WI: Medical Physics Publishing, pp. 493–498.

Petrick, N., Haider, M., Summers, R.M. *et al.* (2008). CT colonography with computer-aided detection as a second reader: an observer performance study. *Radiology* **246**(1): 148–156.

Renfrew, D.L., Franken, E.A., Jr., Berbaum, K.S., Weigelt, F.H., Abu-Yousef, M.M. (1992). Error in radiology: classification and lessons in 182 cases presented at a problem case conference. *Radiology*, **183**: 145–150.

Royster, A.P., Fenlon, H.M., Clarke, P.D., Nunes, D.P., Ferrucci, J.T. (1997). CT colonoscopy of colorectal neoplasms: two-dimensional and three-dimensional virtual-reality techniques with colonoscopic correlation. *Am J Roentgenol*, **169**: 1237–1242.

Sahiner, B., Chan, H.-P., Petrick, N., Wagner, R.F., Hadjiiski, L. (2000). Feature selection and classifier performance in computer-aided diagnosis: the effect of finite sample size. *Med Phys* **27**: 1509–1522.

Samuelson, F.W., Petrick, N. (2006). Comparing image detection algorithms using resampling. *Proceedings of the 2006 IEEE International Symposium on Biomedical Imaging*, pp. 1312–1315.

Shah, P.K., Austin, J.H., White, C.S. *et al.* (2003). Missed non-small cell lung cancer: radiographic findings of potentially resectable lesions evident only in retrospect. *Radiology* **226**: 235–241.

Sone, S., Li, F., Yang, Z.G. *et al.* (2001). Results of three-year mass screening programme for lung cancer using mobile low-dose spiral computed tomography scanner. *Br J Cancer* **84**: 25–32.

Sonka, M., Fitzpatrick, M.J. (eds.) (2000). *Handbook of Medical Imaging, Volume 2. Medical Imaging Processing and Analysis*. Bellingham, WA: SPIE.

Summers, R.M., Johnson, C.D., Pusanik, L.M. *et al.* (2001). Automated polyp detection at CT colonography: feasibility assessment in a human population. *Radiology* **219**: 51– 59.

Summers, R.M., Yao, J., Pickhardt, P.J. *et al.* (2005). Computed tomographic virtual colonoscopy computer-aided polyp detection in a screening population. *Gastroenterology* **129**: 1832–1844.

Suzuki, K., Yoshida, H., Nappi, J., Dachman, A.H. (2006). Massive-training artificial neural network (MTANN) for reduction of false positives in computer-aided detection of polyps: suppression of rectal tubes. *Med Phys* **33**: 3814–3824.

Swensen, S.J., Jett, J.R., Hartman, T.E. *et al.* (2003). Lung cancer screening with CT: Mayo Clinic experience. *Radiology* **226**: 756–761.

Taylor, S.A., Charman, S.C., Lefere, P. *et al.* (2007). CT colonography: investigation of the optimum reader paradigm by using computer-aided detection software. *Radiology* **246**(2): 463–471.

Wagner, R.F., Metz, C.E., Campbell, G. (2007). Assessment of medical imaging systems and computer aids: a tutorial review. *Acad Radiol* **14**: 723–748.

Yoshida, H., Masutani, Y., MacEneaney, P., Rubin, D.T., Dachman, A.H. (2002). Computerized detection of colonic polyps at CT colonography on the basis of volumetric features: pilot study. *Radiology*, **222**: 327–336.

Yoshida, H., Dachman, A. (2004). Computer-aided diagnosis for CT colonography. *Semin Ultrasound CT MR* **25**: 419–431.

Yousef, W.A., Wagner, R.F., Loew, M.H. (2005). Estimating the uncertainty in the estimated mean area under the ROC curve of a classifier. *Patt Recog Lett* **26**, 2600–2610.

Yousef, W.A., Wagner, R.F., Loew, M.H. (2006). Assessing classifiers from two independent data sets using ROC analysis: a nonparametric approach. *IEEE Trans Pattern Anal Mach Intell* **28**, 1809–1817.

Yuan, Y., Giger, M.L., Li, H., Suzuki, K., Sennett, C. (2007). A dual-stage method for lesion segmentation on digital mammograms. *Med Phys* **34**: 4180–4193.

Zheng, B., Ganott, M.A., Britton, C.A. *et al.* (2001). Soft-copy mammographic reading with different computer-assisted detection cuing environments: preliminary findings. *Radiology* **221**: 633–640.

Zheng, B., Swensson, R.G., Golla, S. *et al.* (2004). Detection and classification performance levels of mammographic masses under different computer-aided detection cueing environments. *Acad Radiol* **11**: 396–406.

Zheng, B., Mello-Thoms, C.C., Wang, X.-H. *et al.* (2007). Interactive computer-aided diagnosis of breast masses: computerized selection of visually similar image sets from a reference library. *Acad Radiol* **14**: 917–927.

20

Common designs of CAD studies

YULEI JIANG

20.1 INTRODUCTION

Computer-aided detection (CADe) is a form of radiology practice in which the radiologist makes a clinical diagnosis by interpreting the images with the assistance of a CADe computer, which analyzes the images independently, and detects and marks potential abnormalities in the image. The premise of CADe is that the CADe computer, and this form of radiology practice, can help radiologists improve diagnostic performance by reducing the number of missed cancers that the radiologist could have detected, which the CADe computer would help the radiologist detect. Several of the preceding chapters discuss CADe in various aspects; in this chapter, we focus on the common evaluation studies that assess the clinical effect of CADe.

It is important to evaluate the clinical effect of CADe. This evaluation is needed to verify the premise of CADe that it benefits clinical practice and it benefits the performance of radiologists. More importantly, this evaluation is needed to clearly indicate that the benefit of CADe outweighs any "cost" associated with its use. There is a financial cost from CADe because it represents an advanced technology. But there are also other costs that may be more important. At the present time, CADe computers invariably mark false positives as they mark cancers. Radiologists can easily dismiss most of the CADe computer false positives, but it takes time for radiologists to review the false positive marks. If radiologists do not dismiss all of the CADe computer false positives, then additional diagnostic work-up studies will become necessary, without yielding cancer diagnosis in the end. Such CADe-prompted false positive diagnostic studies are an important clinical concern. Furthermore, the assurance of improved diagnostic performance that radiologists and patients may feel from the use of CADe is justified only if CADe lives up to its promise of helping radiologists detect cancers. A false assurance, if the improvement in diagnostic performance did not materialize, would be dangerous.

To evaluate the clinical effect of CADe is a challenging endeavor. This is true for most new imaging technologies but perhaps more so for CADe because it is not simply a new imaging modality which can be compared straightforwardly with the conventional imaging modality. CADe is a technology that enhances the conventional imaging modality (e.g. conventional screening mammography) and, because of that, one must compare the conventional imaging modality against the conventional imaging modality plus CADe. That the conventional imaging modality appears in both sides of the comparison complicates the evaluation studies. In addition, to evaluate CADe, one must consider several distinct aspects involved in the practice of interpreting images with the assistance of CADe: the CADe computer, the radiologist, and the interaction of the radiologist with the CADe computer. There are several types of common evaluation studies, which we will discuss in this chapter. These are: (1) studies of the clinical potential of CADe, (2) laboratory observer performance studies, (3) clinical "head-to-head" comparisons, (4) clinical "historical-control" studies, and (5) randomized controlled clinical trials.

20.2 SOME COMMON TYPES OF CADe STUDY

20.2.1 Studies of clinical potential

For CADe to help radiologists detect cancer, the CADe computer must be able to detect cancer on its own. Further, the CADe computer must be able to detect cancer that radiologists miss. Without these capabilities, the CADe computer cannot possibly help the radiologist. However, these capabilities are only an indication that the CADe computer could help – it does not necessarily mean that the CADe computer will help. For the CADe computer to help a radiologist detect cancer, the radiologist must first miss the cancer, then the CADe computer must detect the cancer, and finally the radiologist must recognize the cancer once the CADe computer points it out. Therefore, the ability of a CADe computer to detect cancers that radiologists tend to miss is a necessary – but not sufficient – condition for CADe to help radiologists detect cancer.

It is nevertheless important to test a CADe computer to see whether it can detect cancers that radiologists tend to miss. This is the first necessary test of the CADe computer before it advances to the clinical arena. Historically, this was also an important test for the field of CADe research; this test demonstrated the potential of CADe at a time when it was uncertain in the scientific and medical community whether it was at all possible for a computer to be good enough to help the well-trained and sophisticated specialist observer – the radiologist – in a complex image-perception and interpretation task. This test provided a reason for CADe to be later evaluated in clinical studies.

To demonstrate this potential benefit, researchers concentrated on cancers that have been missed by radiologists in routine clinical practice. A cancer is missed if it is present in a mammogram but the interpreting radiologist did not recognize the lesion as abnormal or did not consider the lesion suspicious for cancer, and the cancer is detected subsequently after the screening

The Handbook of Medical Image Perception and Techniques, ed. Ehsan Samei and Elizabeth Krupinski. Published by Cambridge University Press.

examination. Because it is easier for one to see the cancer once it is pointed out by someone else, it is also common to require that not only the cancer is visible retrospectively in the mammogram but also a panel of experienced radiologists consider, retrospectively, that the lesion is suspicious enough that the original interpreting radiologist should have acted on it and recalled the patient for additional imaging study. Using mammograms containing missed cancers, researchers showed that CADe computers were indeed able to detect many of those cancers in the mammograms where the original interpreting radiologist failed to recall the patient for additional imaging study. These studies provided tangible evidence of the potential for the CADe computer to help radiologists detect cancer.

The design of this type of study is straightforward. It involves the assembly of a missed-cancer image database and the analysis of the images with a CADe computer. The most demanding task is to assemble the image database because it is not easy to identify a large number of cancers that are clinically missed. After such cases are identified, a panel of expert radiologists usually reviews and assesses the appropriateness of the cases for the study to the best of their ability. The main outcome of this type of study is how many missed cancers the CADe computer can detect. It is not uncommon for a CADe computer to detect about half of missed cancers. A second important outcome is how many false positives the CADe computer also detects as it detects the cancers. If the CADe computer makes several detection marks in each image, then it is possible for missed cancers to be marked by chance rather than by actual computer detection – the greater the number of computer marks per image, the greater the possibility of a cancer being marked by chance. Large numbers of false-positive computer marks are undesirable also because they indicate that the computer detection results are nonspecific and radiologists must review many computer false-positive marks before finding a cancer. Researchers must decide before the study the scoring criteria to consider a CADe computer mark as a "hit" on a cancer. For obvious reasons, the computer performance is inversely related to the scoring criteria: the computer will appear to detect more cancers with a set of lax scoring criteria and detect fewer cancers with a set of stringent scoring criteria, all without actually changing the computer technique or its performance. If the criterion is that the center of a computer-detected object be within a specified distance from the center of a lesion, then obviously the greater the specified distance the more likely computer detections will be scored as true positives, and vice versa. It is difficult to come up with a rule that evaluates the computer-detection performance entirely objectively and prevents chance detections from being counted as true positives. It is common to require that a computer detection overlaps with a lesion for the computer mark to be counted as a true positive. It is also helpful if a radiologist reviews the computer marks to decide whether each computer mark is a true positive.

An example of a study of the clinical potential of CADe is the study of Warren Burhenne *et al.* (Warren Burhenne, 2000). In this study, the researchers analyzed 1083 biopsy-proven breast cancers and found 427 cases had a prior mammogram. A panel of radiologists evaluated these prior mammograms and determined that 115 mammograms could be considered as missed cancer cases. Of these "missed" cancers, the CADe computer marked 89 cancers, or 77%. Therefore, this study demonstrated a potential for the CADe computer to help radiologists detect breast cancer in mammograms earlier.

20.2.2 Laboratory observer performance studies

To test whether a CADe computer can help radiologists detect cancers, radiologists must be involved in the test so that the scenario in which the CADe computer helps a radiologist detect cancer can be actually observed in the test. One can conduct such a test in a controlled, laboratory environment, known as a laboratory observer performance study.

The purpose of a laboratory observer study is to measure the performance of observers in the interpretation of a specific kind of image or, alternatively, to compare observer performance in the interpretation of two (or more) kinds of images. In the case of CADe, an observer study compares the observer performance in interpreting a specific kind of image (such as mammograms) alone versus interpreting the same images with the assistance of a CADe computer. Both the observer performance in interpreting the images without and with the CADe computer assistance are measured in the experiment, and a comparison is made between the two. Laboratory observer performance studies are commonly analyzed with receiver operating characteristic (ROC) analysis; although not necessarily required, an observer study almost invariably implies an ROC study. ROC analysis provides a fundamentally meaningful depiction of the observer performance by simultaneously examining sensitivity, specificity, and the tradeoffs that are available between them. In the last 20 years, ROC analysis has been developed successfully to address many complex issues in the measurement and the comparison of observer performance with statistical validity, great attention to experimental details, and improved efficiency. For an introduction to ROC analysis, the reader should consult other chapters in this volume and the reviews by Wagner *et al.* (Wagner, 2007) and Metz (Metz, 1978; 1989).

In a laboratory observer performance study, one tries to replicate as closely as possible the clinical conditions in which images are interpreted. One also tries to control the experiment to eliminate as much as possible distractions and uncertainties that are part of clinical practice. To facilitate ROC analysis, the experimenter must know in every case the diagnostic truth with respect to the diagnostic task being studied, but this knowledge is withheld from the observers to simulate the condition of clinical image interpretation. The general design of a laboratory observer study includes the collection of a set of images and the selection of observers, having the observers interpret the images both with and without the CADe computer assistance in a carefully designed, controlled environment during which data required for ROC analysis are collected, and subsequently performing ROC analysis on the collected data. In the following, we describe these components in more detail.

For a laboratory observer study to replicate the conditions of clinical image interpretation, the images should be representative of those encountered clinically. However, in many situations this is neither practical nor desirable. Take breast cancer for example. Because the cancer prevalence is extremely small (about 5 cancers per 1000 women screened in an average-risk screening population), to ensure that a study includes a sufficient number of cancers, the study must also include a much larger number of non-cancer cases, causing the study to become extremely large. This is a particularly difficult problem for studies that address screening (asymptomatic) patient populations. For diagnostic examination (i.e. the follow-up or work-up imaging studies after an abnormality is identified in screening) patient populations, although cancer prevalence is greater, this is still a problem because the number of non-cancer cases is still several times that of cancer cases. In a study with small cancer prevalence, much of the observers' time will be spent on reading cases that are not cancer. Further, although the total number of cases is large, the statistical power of the study is determined primarily by the smaller number of cancer cases, rather than the larger number of non-cancer cases. For these reasons, laboratory observer performance studies usually do not randomly sample clinical cases. Instead, it is common to enrich the cancer prevalence in an observer study such that there are disproportionately more cancer cases compared with routine clinical practice. A common practice is to balance the numbers of cancer and non-cancer cases about equally because in this way they both contribute equally to the statistical power and, therefore, improve the efficiency of the experiment. However, a fundamental question arises when one artificially increases the number of cancer cases: does the change in cancer prevalence affect observer performance and decision-making? It is important to realize that the ROC curve does not explicitly depend on the disease prevalence, i.e. the ROC curve is invariant as one changes the disease prevalence. An intuitive explanation for this is that the innate diagnostic ability of the observers is not affected by the disease prevalence. To prove this assertion experimentally is extremely difficult because small cancer prevalence requires extremely large studies. Nevertheless, studies have been done with progressively smaller disease prevalence that showed little effect of the disease prevalence on the ROC curve (Gur, 2007). However, questions still remain of whether the clinical decisions that radiologists make are affected by the change in cancer prevalence. For example, radiologists strive in screening mammography to detect as many breast cancers as possible while simultaneously maintaining a reasonably low percentage of patients recalled for additional imaging study, knowing that a great majority of the recalled patients do not have breast cancer. If the cancer prevalence is artificially changed substantially, then a radiologist will not be able to maintain both identical sensitivity and identical recall rate in the observer study as in routine clinical practice. Under those conditions, how radiologists change their decision-making in the observer study and how those changes affect their sensitivity, or recall rate, or both, is not clear.

Similar to case selection, the observers who take part in an observer study also should be representative of those who interpret the images clinically. In the USA, radiologists must meet the requirements of the Mammography Quality Standard Act (MQSA) to interpret mammograms clinically. It is therefore common to use that requirement as a threshold to select observers in observer studies. A key component of the MQSA is that a radiologist must interpret a minimum of 960 mammograms in a two-year period, or 480 mammograms per year. During the early development of CADe, it was common for highly experienced radiologists to serve as observers. However, because the number of highly experienced radiologists is small and they interpret only a small fraction of the total clinical cases, it is also important that observer studies include radiologists from a broader pool of practice styles and interpretation-skill levels. Subsequent observer studies often include community radiologists as observers.

Many study-design details of how observers interpret images can potentially bias the results when comparing two imaging modalities in a laboratory observer performance study. For example, it is conceivable that radiologists perform better in interpreting an image simply by spending more time on it. It is also possible that if a radiologist always interprets one imaging modality first and then, immediately after, interprets the second imaging modality, then the performance measured from interpreting the second image modality is likely the result of interpreting both imaging modalities, because the information that the radiologist has attained from the first imaging modality remains present as the radiologist interprets the second imaging modality. One can design the experiment in such a way that the effects of these potential biases tend to cancel each other rather than accumulate during the course of the study. A popular study design, sometimes referred to as the counter-balanced study design, can be summarized as follows. Divide the cases into two comparable halves and divide the observers into two comparable groups. The first group of observers will read the first half of cases in imaging modality A and read the second half of cases in imaging modality B. The second group of observers will read the first half of cases in imaging modality B and read the second half of cases in imaging modality A. After all images are read, wait several weeks to discourage the observers' memory of reading the images for the first time from influencing their reading of the images for a second time. After this memory "washout" interval, all observers read the images for a second time in the imaging modality that they have not read. This study design minimizes potential biases from reading-order effects because such effects tend to cancel between the two halves of the study.

In general, the interval between the two readings is important to ensure that the observers read the two imaging modalities independently. However, observer studies of CADe are special in that the reading of the second modality (images plus CADe assistance) can immediately follow the reading of the first modality (images alone). The reason for this is that that is how CADe may be used clinically: a radiologist interprets the images and then, immediately after, consults the CADe-computer results for concurrence or for potential abnormalities that he or she has not noticed. In the spirit of duplicating how cases are interpreted clinically in the laboratory observer performance study, it is reasonable to modify the standard study design specifically for CADe studies, such that each observer interprets all cases twice in a single sitting, and interprets each case first with the images alone and, immediately after rendering a diagnostic opinion, interprets the images again with the CADe-computer results present, followed by the rendering of a second, potentially modified, diagnostic opinion. This study design is known sometimes as the "sequential study," whereas the standard, counter-balanced study design is referred to as the "independent study." Observers may behave differently in an independent study compared with in a sequential study. The independent study is designed specifically to discourage the observer from remembering, when he or she interprets an image for the second time, what his/her impression was when he/she interpreted the image for the first time. On the other hand, in a sequential study the observer knows exactly what his/her impression was from reading the images alone when the CADe computer results are made available to him/her. Therefore, observers tend to "change their mind" less frequently in a sequential study compared with in an independent study in which they would appear to "change their mind" simply because they could not duplicate their previous impression exactly. This may seem to be an advantage of the independent study given that CADe only helps radiologists if it can persuade a radiologist to change his/her mind toward the correct diagnosis. However, the sequential study can yield greater statistical power in comparing reading images alone against reading images with the assistance of CADe because of improved observer consistency and reduced random statistical noise between the observers' diagnostic opinions attributed to interpreting the images with and without the assistance of CADe (Beiden, 2002). It has been shown that, reassuringly, a study conducted twice with both the independent and the sequential designs produced similar results (Kobayashi, 1996).

One of the early laboratory observer performance studies was done by Chan *et al.* (Chan, 1990). They selected 60 single-view mammograms, of which 30 mammograms each contained a subtle cluster of microcalcifications and the other 30 mammograms did not contain clustered microcalcifications. Seven attending radiologists and eight radiology residents participated in the study as observers. The results of the observers' "composite" ROC curves (Figure 20.1) show that CADe statistically significantly ($p < 0.001$) improved the radiologists' performance interpreting these mammograms in the detection of clustered microcalcifications.

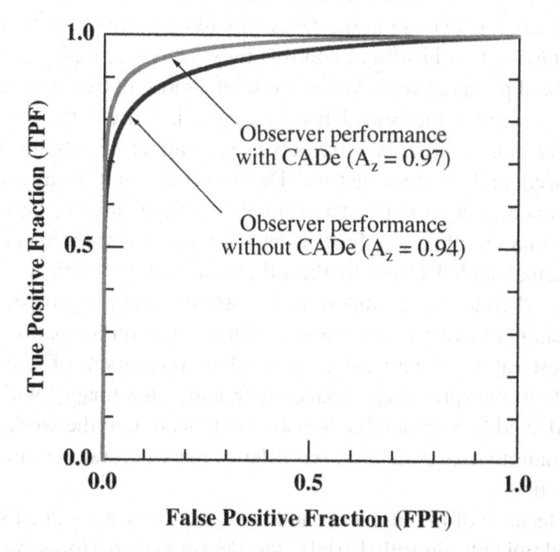

Figure 20.1 ROC curves adapted from the observer performance study of Chan *et al.* (Chan, 1990) showing statistically significant ($p < 0.001$) improvement in 15 radiologists' performance in detecting clustered microcalcifications from interpreting mammograms with the assistance of CADe.

20.2.3 Clinical head-to-head comparisons

One way to test in the clinical setting whether CADe helps a radiologist detect cancers is to compare on a case-by-case basis the radiologist's interpretation of images with and without CADe assistance. We call this the "head-to-head" comparison.

The goal of the head-to-head comparison is to determine the frequency and the specific cases in which CADe helps a radiologist detect cancer. To do this, a radiologist interprets each case first without the CADe computer assistance and formally records his/her findings in the case. Then, the radiologist reviews the CADe computer results and interprets the case again, potentially modifying his/her findings prompted by the marks of the CADe computer, and formally records a second set of findings (or modifies the initial set of findings). The radiologist or radiologists do so in every case over a period of time, during which cancer cases will be encountered and, hopefully, cases in which CADe helps the radiologist or radiologists detect the cancer also will be encountered. Because of low cancer prevalence in screening and because CADe is expected to help radiologists only when they initially miss the cancer, which presumably is infrequent, the head-to-head comparison must go through a large number of cases to accumulate the encounters of cancer cases in which CADe helps a radiologist detect the cancer. This translates into long periods of study. Because of the additional work required to record two sets of findings in each case, this type of study usually accumulates only a limited number of cancer cases.

The advantages of head-to-head comparisons are that CADe is used in the daily clinical practice as it is intended and that one can identify the specific cases in which CADe helps radiologists detect the cancer. However, one could raise the question of whether the performance of the radiologist in the study is indeed the same as his/her ordinary performance outside of the study. The presumption of the head-to-head study is that when the radiologist first interprets a case without the assistance of the CADe computer, he/she performs in exactly the same way as he/she does ordinarily outside of the study. However, there are plausible arguments that claim the radiologist might be more – or less – vigilant than usual. Because it is apparent to the radiologist that his/her performance interpreting the cases without the assistance of the CADe computer will be compared with the performance of the CADe computer, the radiologist may feel pressured to compete with the CADe computer and, consciously or subconsciously, be more vigilant than usual in detecting abnormalities. On the other hand, because the radiologist knows that any abnormality missed in his or her initial interpretation has a chance to be caught by the CADe computer and by his/her second read of the images, the radiologist may be less vigilant in his/her initial interpretation than usual. It is commonly accepted that CADe computers can detect clustered microcalcifications in mammograms with high sensitivity (in excess of 90%). Therefore, it may seem reasonable to some radiologists to do only a cursory search for calcifications on their own and rely on the CADe computer to find other subtle calcifications, which the radiologist would then interpret after the CADe computer marks them. If a radiologist operates in this way in a head-to-head comparison study, then his/her performance reading the images alone will not be the same as his/her performance without the assistance of CADe.

An example of the clinical head-to-head comparison was done by Freer and Ulissey (Freer, 2001). Their study lasted 12 months and included 12,860 screening mammograms and 2 radiologists. Each mammogram was initially interpreted without the assistance of CADe, followed immediately by a re-evaluation of areas marked by the CADe computer. They reported an increase in the number of cancers detected of 19.5% (from 41 cancers detected without the assistance of CADe to 49 cancers detected with the assistance of CADe) and a corresponding increase in the recall rate from 6.5% to 7.7%.

20.2.4 Clinical historical-control studies

Another way to test in the clinical setting whether CADe helps radiologists detect cancer is to compare radiologists' performance interpreting some cases without the assistance of CADe against their performance interpreting some other cases with the assistance of CADe. In this way, radiologists avoid having to interpret and record the findings twice in every case. One way to implement such a study efficiently is to compare the performance of a group of radiologists after they have begun to use CADe against their historical performance before they began to use CADe. In this way, the radiologists act as their own controls (hence the term "historical control"). However, such a study does not isolate the effect of CADe alone; any change in the radiologists' performance is the summary result of whatever has changed between the two historical time periods, including the use of CADe, but also including any other changes such as radiologist staffing changes, improvements in their interpretation skills, and any underlying changes in the patient population. Therefore, such a study is sensible only for well-established practices with a stable patient population and a stable radiologist staff, for which the assumption would appear plausible that little else has changed except for the addition of CADe during the study.

In addition to the requirement of stable practice, detailed and high-quality audit of practice data is also critical to a historical-control study. At a minimum, the audit data should show the total number of patients undergoing an imaging study, the number of cancers diagnosed as a result of the imaging study, the number of false-positive imaging findings (e.g. a recall for diagnostic breast imaging study in a patient who does not have breast cancer), and the frequency of unintended downtime of CADe in the study. More detailed audit data at the level of individual radiologist or individual cases are desirable. Aside from maintaining a high-quality audit, the study is not different from routine clinical practice in other ways. Every patient, every radiologist, and every case are entered into the study as the clinical imaging study is performed in its routine fashion. Therefore, the study provides a good snapshot of the clinical performance.

An example historical-control study is the Gur *et al.* study (Gur, 2004). This study included the screening mammography studies during 2000, 2001, and 2002 interpreted by 24 radiologists at an academic medical center and its satellite breast imaging clinics. The control arm included 56,432 screening mammograms interpreted without the assistance of CADe before the installation of the CADe system and the study arm included 59,139 screening mammograms interpreted with the assistance of CADe after the CADe system was used consistently in their clinical practice. They found a similar cancer detection rate with and without the assistance of CADe and also similar recall rates. The cancer detection rate was 3.49 without the assistance of CADe and 3.55 with the assistance of CADe per 1000 screening mammograms. The recall rate was 11.39% without the assistance of CADe and 11.40% with the assistance of CADe.

20.2.5 Randomized controlled clinical trial

The randomized controlled clinical trial offers perhaps the most rigorous design of clinical studies. The historical-control study achieves high efficiency in taking an accurate snapshot of clinical performance with the tradeoff of not being able to isolate the effect of one particular change – the use of CADe – from other longitudinal changes during the study. In comparison, the randomized controlled clinical trial overcomes this shortcoming by taking snapshots of clinical performance with and without the assistance of CADe at the same time. To do this, a scheduled imaging study is assigned randomly with equal probability to either the study arm or the control arm. Radiologists interpret cases assigned to the study arm with the assistance of CADe and interpret cases assigned to the control arm without the assistance of CADe. Each radiologist should interpret an equal number of cases with and without the assistance of CADe. In this way, the clinical performance with and without the assistance of CADe can be compared directly because the patient populations are comparable, the radiologists are comparable, and, except for the assistance of CADe, the imaging studies are interpreted in a comparable way.

The randomized controlled clinical trial is the gold standard in drug and interventional studies. However, whether the randomized controlled trial is appropriate for CADe studies also depends on cost-effectiveness. A randomized controlled trial is highly demanding on resources. Because every case must be randomly assigned to either the study arm or the control arm, it is not adequate to audit the cases only retrospectively as is commonplace in a historical-control study, but every case must be tracked prospectively. And because in almost all cancer screening situations the prevalence of cancer is extremely low, the study must accrue an extremely large number of patients and, consequently, last a long time. The historical-control study must also accrue a similarly large number of patients, but because the data are obtained through retrospective audit, there is only minimal added effort to the otherwise routine clinical practice, whereas in the randomized controlled trial the prospective tracking of every case demands a large effort on the part of the investigators. In particular, the random assignment of radiologists to interpret some images with and some images without CADe adds considerable logistic complication to the workflow of radiologists compared with their otherwise routine clinical practice.

Because of the high cost and the complexity associated with randomized controlled trials, the decision to perform such a trial is not made lightly. A number of randomized controlled trials have been done to determine the efficacy of screening mammography for the detection of breast cancer in asymptomatic women. A similar trial (rather than randomized, in this

Table 20.1 *Studies of clinical potential of mammography CADe*

Lead author, year	Journal	Location
te Brake, 1998	*Radiology*	Netherlands
Warren Burhenne, 2000	*Radiology*	Canada, CA
Birdwell, 2001	*Radiology*	CA
Zheng, 2002	*Academic Radiology*	PA
Brem, 2003	*AJR*	PA
Karssemeijer, 2003	*Radiology*	Netherlands
Destounis, 2004	*Radiology*	NY
Ikeda, 2004	*Radiology*	CA
Ciatto, 2006	*Breast*	Italy
Skaane, 2007	*AJR*	Norway

trial every patient was imaged with both imaging modalities) has been done to determine the efficacy of full-field digital mammography as compared with conventional screening-film mammography (Pisano, 2005). However, no randomized controlled trial has been done to study CADe.

20.3 WHAT WE HAVE LEARNED FROM CADe STUDIES

20.3.1 Summary of each type of CADe study

A large number of studies have been done to demonstrate the clinical potential of mammography CADe. Many of these studies were done in the early phase as CADe was being introduced into clinical practice and were done to collect evidence in support of the clinical use of CADe. Table 20.1 lists ten of this type of study. In total, these studies include about 1000 "missed" cancers – cancers not diagnosed in the mammograms studied but diagnosed subsequently. The CADe computer flagged about 50% of these "missed" cancers, with a range of 13% to 77% reported in the individual studies. Therefore, these studies indicate that the CADe computer can flag about 50% of breast cancers in mammograms before the cancers are diagnosed in routine clinical practice without the assistance of CADe. This finding suggests that there is a great clinical potential for the CADe computer to help radiologists detect breast cancers earlier. However, these studies indicate only the clinical potential – they do not indicate what fraction of the cancers that the CADe computer flags the radiologist recognizes as cancer.

Several observer studies have been done to study the effect of mammography CADe on radiologists' performance in detecting breast cancer in screening mammograms. Table 20.2 lists eight of these studies. All of these studies reported improved performance of the radiologists when they used the assistance of CADe compared with their unaided performance. Although there is the possibility that the literature is biased toward publishing studies with positive findings and, therefore, there could have been studies that did not find a statistically significant difference in performance that were not published, we are unaware of any such null-result studies. Given the large effort required to conduct these observer performance studies, such null-result studies would have been rare if they existed.

Table 20.2 *Laboratory observer performance studies of mammography CADe*

Lead author, year	Journal	Location
Chan, 1990	*Investigative Radiology*	IL
Kegelmeyer, 1994	*Radiology*	CA
Moberg, 2001	*EJR*	UK
Marx, 2004	*EJR*	Germany
Alberdi, 2005	*BJR*	UK
Taylor, 2005	*BJR*	UK
Gilbert, 2006	*Radiology*	UK
Taplin, 2006	*AJR*	WA

Table 20.3 *Clinical head-to-head comparisons of mammography CADe*

Lead author, year	Journal	CADe	Location	Practice
Freer, 2001	*Radiology*	R2	TX	Community
Helvie, 2004	*Radiology*	In-house	MI	Academic
Birdwell, 2005	*Radiology*	R2	CA	Academic
Khoo, 2005	*Radiology*	R2	UK	Program
Dean, 2006	*AJR*	iCAD	CA	Private
Ko, 2006	*AJR*	iCAD	MA	Academic
Morton, 2006	*Radiology*	R2	MN	Academic

Table 20.4 *Clinical historical-control studies of mammography CADe*

Lead author, year	Journal	CADe	Location	Practice
Gur, 2004*	*JNCI*	R2	PA	Academic
Cupples, 2005	*AJR*	R2	SC	Community
Fenton, 2007	*NEJM*	R2	USA	Community
Gromet, 2008	*AJR*	R2	NC	Community

*See also Feig, 2004.

Several clinical "head-to-head" comparisons of radiologists' performance in detecting breast cancers in screening mammograms with and without the assistance of CADe have been done. Table 20.3 lists seven of these studies. In all, these studies include about 60,000 screening mammography cases, from which the radiologists detected 319 cancers without the assistance of CADe, and detected 31 additional cancers with the assistance of CADe. These results amount to an increase in sensitivity of 9.7% with a concurrent increase in recall rate of 12.5%. The individual studies reported a range of increase in sensitivity from 4.7% to 19.5%, and a range of increase in recall rate from 8.2% to 25.8%. Note that because of the extremely low breast cancer prevalence in the average-risk screening population, the total number of cancers and the number of cancers detected because of the CADe computer assistance are small even with a total number of 60,000 screening mammograms.

Four historical-control studies have been conducted (Table 20.4). These are large studies. In total, these studies include about 228,000 screening mammogram cases read with the assistance of CADe and 687,000 screening mammogram cases read without the assistance of CADe – the large imbalance is due, mostly, to the Fenton study, which included disproportionally more cases read without the assistance of CADe than the cases read with the assistance of CADe. The results of these studies do not agree. The Gur study reported an increase in the cancer detection rate of 1.7% and an increase in the recall rate of 0.1%. A subsequent re-analysis of that study by Feig *et al.* of a subset of radiologists excluding those who were high-volume readers found a 19.7% increase in the cancer detection rate and 14.1% increase in the recall rate. The Cupples study reported a 16.3% increase in the cancer detection rate and an 8.1% increase in the recall rate. The Fenton study reported a statistically non-significant 1.2% increase in the cancer detection rate and a 30.7% increase in the recall rate. The Gromet study reported a 1.9% increase in the cancer detection rate and a 3.9% increase in the recall rate. These disparate results raise questions and have prompted confusion and debate over the clinical benefit of CADe. The Fenton study offers the most extreme position in concluding[1] that "the use of computer-aided detection is associated with reduced accuracy of interpretation of screening mammograms."

20.3.2 The relations between laboratory observer studies and historical-control clinical trials and why they may find different results

The disparate findings in the historical-control trials contradict the consistency in the positive findings from laboratory observer performance studies. Why does CADe, a new technology that has been tested extensively in laboratory observer studies and has consistently produced strong evidence of decision-making benefits, fail to produce the same consistent results in large clinical trials? Are these inconsistent trial results valid grounds to question whether CADe is clinically beneficial? If CADe is clinically beneficial, what are its clinical effects and how can we measure those beneficial effects consistently and unambiguously? These are some important questions that are not easy to answer. Of the several types of study that we have described, we focus on the differences between the laboratory observer studies and historical-control clinical trials because a laboratory observer study is probably the most rigorous study that can be done in the laboratory environment outside of the realm of clinical practice and the historical-control clinical trials are by far the largest clinical studies (it is difficult for head-to-head comparison studies to achieve a similarly large size).

There are important differences between laboratory observer studies and clinical trials and there are important differences between the data analyses of these studies. Laboratory observer studies often use ROC analysis, sensitivity, and specificity as the fundamental performance metrics. The sensitivity and specificity can be calculated in an observer study because the study uses only cases of known diagnosis. For all practical purposes, sensitivity and specificity cannot be calculated in a clinical trial because the diagnostic "truth" in each and every case – required to calculate sensitivity and specificity – is not known. In clinical trials the performance metrics most often calculated are the cancer detection rate and the recall rate. The cancer detection rate is the number of cancers detected in a particular cohort of patients. The cancer detection rate in breast cancer screening is approximately 4 to 5 cancers detected per 1000 average-risk women screened. The recall rate refers to the fraction of patients recalled for additional diagnostic studies after the initial screening study resulted in an abnormal finding. The recall rate for screening mammography ranges from a few percent to over 10%, even approaching 20%. The cancer detection rate and the recall rate are related to the sensitivity and specificity through the cancer prevalence, which is an unknown quantity. Therefore, one can neither calculate sensitivity and specificity from cancer detection rate and recall rate, nor vice versa.

20.3.2.1 The cancer detection rate is not expected to increase

The premise of using the cancer detection rate as one of the primary endpoints in a clinical trial is based on the expectation that if CADe helps radiologists detect cancers then the cancer detection rate will increase. However, that expectation may be flawed. Cancer is relatively rare in the average-risk asymptomatic population and the cancer prevalence is generally stable over time. The prevalence of breast cancer in the average-risk, asymptomatic population is approximately 5 per 1000 women and is believed to be stable over time (Jemal, 2008). If we ignore, for a moment, the plausible assertion that cancer prevalence is approximately constant over time, then if CADe helps radiologists detect more cancers, it would be plausible to expect the cancer detection rate to increase. However, if we take into account that cancer prevalence remains approximately constant over time, then the effect of CADe helping radiologists detect more cancers on the cancer detection rate becomes more complicated, and the effect varies depending on at which time point we look at the cancer detection rate.

Initially, when CADe is introduced into clinical practice, if CADe helps radiologists detect more cancers, then the cancer detection rate will be greater compared with the baseline cancer detection rate without the assistance of CADe. However, in the next screening round, because CADe has already helped radiologists detect some cancers that normally would not have been detected already (in the last screening round) had CADe not been used, the number of cancers that normally could have been detected in the current screening round would be smaller compared with the conventional practice without the assistance of CADe. If CADe again helps radiologists detect some cancers that they normally would not have detected, then the total number of cancers detected – i.e. the sum of the smaller number of cancers that radiologists normally would have detected without the assistance of CADe and the number of cancers that CADe helps radiologists to detect – might be similar to the number of cancers that radiologists normally would detect without the assistance of CADe.

[1] This conclusion was based primarily on an ROC analysis of the BI-RADS final assessment ratings that radiologists provide in clinical practice, which is probably a scientifically flawed use of the ROC analysis, as we will discuss later. The authors reported reduced overall accuracy in the ROC curve associated with CADe compared with the ROC curve associated with interpreting mammograms without the assistance of CADe.

Over time, the cancer detection rate probably does not change substantially compared with the baseline cancer detection rate without the assistance of CADe, even as CADe helps radiologists detect cancers that they are normally not able to detect without the assistance of CADe. But an important difference is that the cancers that radiologists detect because of the assistance of CADe are detected earlier than they normally would have been without the assistance of CADe. This is analogous to the benefit of screening mammograms, which primarily help to detect breast cancers early when they can still be treated effectively, rather than detecting more of them. Nishikawa has studied these dynamics in detail with stochastic modeling and has found quantitative evidence that the cancer detection rate might not change substantially as CADe helps radiologists detect more cancers (Nishikawa, 2006).

20.3.2.2 The cancer detection rate is difficult to measure

When the cancer detection rate does increase, it is entirely another challenge to measure that increase accurately. In an average-risk asymptomatic population, the prevalence of breast cancer is approximately 5 per 1000 women, and the cancer detection rate with mammography is somewhat smaller than that, perhaps 4 per 1000. It should be clear that the statistical uncertainty in measuring these small rates is quite large, especially when the total number of cancers is small. The binomial statistics indicate that if the expected cancer detection rate is 4 per 1000 women, then the standard deviation in the observed cancer detection rate is expected to be 2 per 1000 women for a cohort study of 1000 women, 0.9 per 1000 women for a cohort study of 5000 women, and 0.6 per 1000 women for a cohort study of 10,000 women. Often, this problem is recognized and a common remedy is to combine the cases interpreted by several radiologists to calculate the aggregate cancer detection rate of a group of radiologists instead of calculating the cancer detection rate of each individual radiologist, because the uncertainty in the calculated cancer detection rate decreases as the total number of cases becomes larger. However, what may not be obvious is that when comparing two cancer detection rates, e.g. when comparing the cancer detection rates with and without the assistance of CADe, the uncertainty is greater than the uncertainty of the individual cancer detection rates. This is illustrated schematically in Figure 20.2.

In addition to this statistical uncertainty, measurement of the cancer detection rate is influenced also by variability in the performance of radiologists – whereas some radiologists are able to operate at a relatively high cancer detection rate and maintain the recall rate at a reasonably low level, other radiologists operate at a smaller cancer detection rate and are often compelled to operate at a relatively high recall rate. This variability is well known and it is generally agreed that large variability in radiologists' performance in the interpretation of screening mammograms exists, although accurate quantification of the extent of it is often difficult (Elmore, 1994; Beam, 1996; Schmidt, 1998). Inter-radiologist variability compounded with statistical uncertainty increase the uncertainty in the measurement of the cancer detection rate and the measurement of the difference between two cancer detection rates (with and without the assistance of CADe).

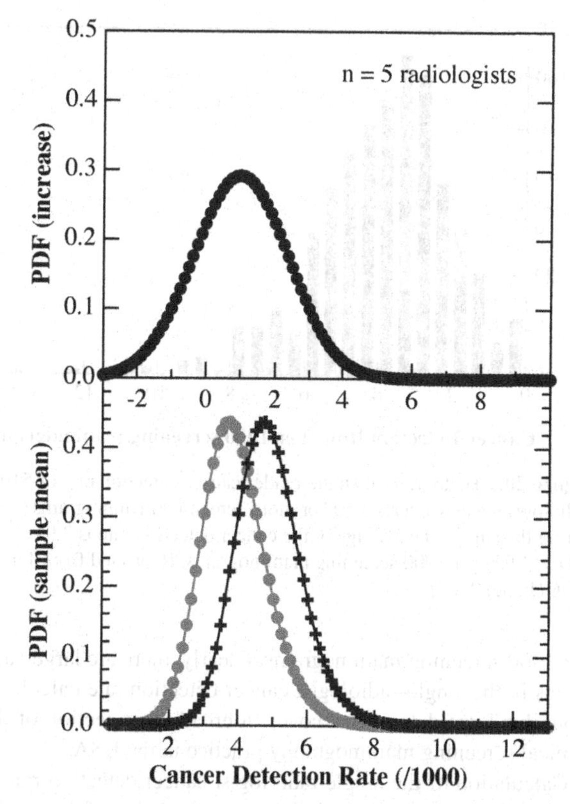

Figure 20.2 Illustration that the aggregate cancer detection rate of five radiologists reduces the uncertainty in the cancer detection, but calculating the difference between two cancer detection rates results in greater uncertainty. **Bottom**: aggregate cancer detection rate calculated from five radiologists based on data in Jiang *et al.* (Jiang, 2007; also shown in Figure 20.3). Data shown as circles are calculated directly from Jiang *et al.* (Jiang, 2007) and data shown as crosses are postulated with an increase of one additional cancer detected per 1000 screening mammograms by every radiologist. **Top**: the difference in the two cancer detection rates. The mean of the difference is 1/1000 as is postulated. Note the greater uncertainty associated with the difference of the cancer detection rates.

How large is the uncertainty in the measurement of the cancer detection rate? Jiang *et al.* (Jiang, 2007) calculated the single-radiologist cancer detection rate based on the clinical practice data collected by the Breast Cancer Surveillance Consortium (Ballard-Barbash, 1997) in seven US regional registries: the Carolina Mammography Registry, Chapel Hill, NC; the Colorado Mammography Project, Denver, CO; the New Hampshire Mammography Network, Lebanon, NH; the New Mexico Mammography Project, Albuquerque, NM; the San Francisco Mammography Registry, San Francisco, CA; the Vermont Breast Cancer Surveillance System, Burlington, VT; and Group Health Cooperative, Seattle, WA. The data cover the period between January 1, 1996 and December 31, 2002, and include 510 radiologists, each of whom read at least 500 mammograms within the study during the study period. A total of 2,289,132 screening mammograms, and 9,030 screen-detected breast cancer cases, are included in the study. Analysis of these data showed that the average single-radiologist cancer detection rate was 3.91 cancers per 1000 screening mammograms, with a standard deviation of 1.93 cancers per 1000 screening mammograms (Figure 20.3). The range of the cancer detection rate was 0.25 to 13.75 cancers

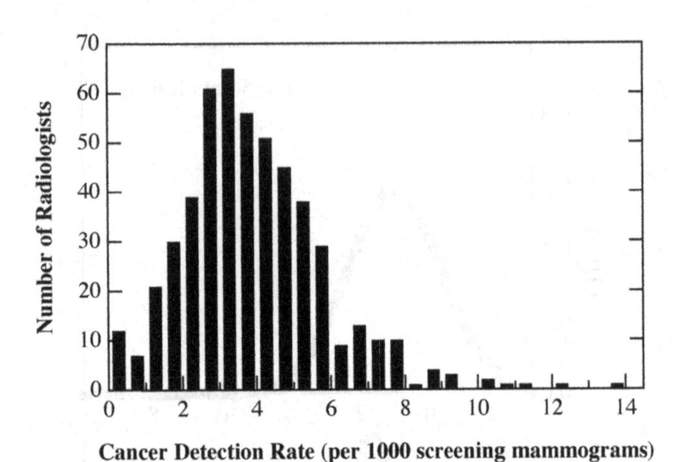

Figure 20.3 Histogram of single-reader cancer detection rate of 510 radiologists who each read 500 or more screening mammograms during the study. The average of the cancer detection rate is 3.91 (SD = 1.93) per 1000 screening mammograms. Reprinted from Jiang *et al.* (Jiang, 2007).

per 1000 screening mammograms. Clearly, there are large variations in the single-radiologist cancer detection rate calculated from this large dataset that covers a broad cross-section of the clinical screening mammography practice in the USA.

Calculation of the single-radiologist cancer detection rate is not typically done in clinical trials, but calculation of the single-radiologist cancer detection rate from a large number of radiologists allows one to model the effect in large clinical trials of the uncertainty in the measurement of the cancer detection rate, and the effect of that uncertainty in the comparison of cancer detection rates. By treating the single-radiologist cancer detection rate calculated from a large number of radiologists as empirical population data, one can simulate large clinical trials by randomly sampling individual radiologists with particular known clinical performance characteristics from the empirical population data and, based on those performance characteristics, predict the aggregate performance of the radiologists in large clinical trials. In this way, one can estimate the size necessary for a clinical trial to uncover a certain amount of change in the cancer detection rate.

An important finding from this analysis is that clinical trials need to be extremely large to ascertain even large changes in the cancer detection rate, requiring both large numbers of patients and large numbers of radiologists (Figure 20.4). To achieve 80% statistical power to detect a postulated increase in the cancer detection rate of one per 1000 screening mammograms for every radiologist requires at least 25 radiologists each reading 8000 or more screening mammograms (a total of at least 200,000 screening mammograms), or at least 91 radiologists each reading 1000 to 2000 screening mammograms (a total of at least 91,000 or 182,000 screening mammograms). The postulated increase of one cancer detected per radiologist per 1000 screening mammograms is an extremely large one in comparison with the average single-radiologist cancer detection rate of 3.91 cancers per 1000 screening mammograms, and in comparison with the overall cancer rate – including those detected by screening mammography and those not detected by screening mammography – of 5.09 cancers per 1000 screening mammograms (Jiang, 2007). The reported increases in the cancer detection rate from CADe are typically not as large as what is postulated here – therefore, even larger trials likely will be needed for CADe. If the trials are not large enough, then there is considerable risk of not being able to detect a true underlying increase in the cancer detection rate, and there is even the risk of observing a decrease in the cancer detection rate when actually

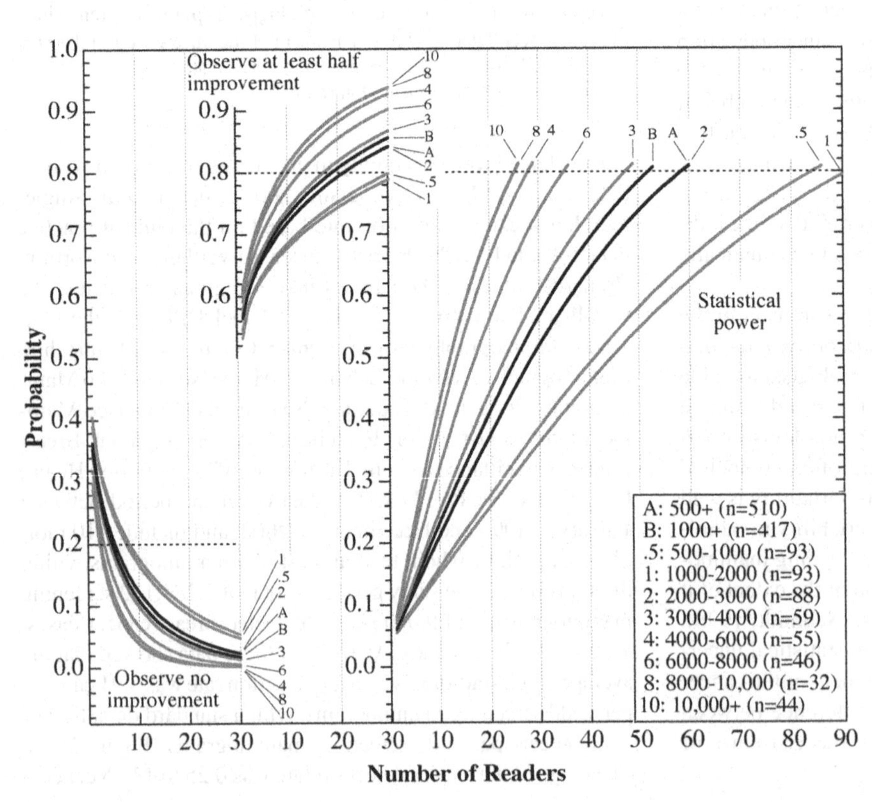

Figure 20.4 Probabilities as a function of the number of readers in a trial for observing, given a postulated increase of one additional cancer detected per 1000 screening mammograms, (**left**) no additional cancer detected, (**middle**) one half or more additional cancers detected per 1000 screening mammograms, and (**right**) statistically significant increase in the cancer detection rate (i.e. statistical power). Correlation of individual reader's cancer detection rates between the two arms of a trial is not included. Variability in the observed single-reader cancer detection rates causes the discrepancies between the postulated and observed changes in the cancer detection rates. Curves are grouped by single-reader total screening mammogram volume; n is the number of readers on whom the calculation was based. Sample sizes refer to half of a two-arm trial. Reprinted from Jiang *et al.* (Jiang, 2007).

there is an underlying increase in the cancer detection rate. A trial of three to eight radiologists (depending on the number of mammograms each radiologist reads) can have a 20% risk of not observing any increase in the cancer detection rate (including observing a decrease) when there is an underlying increase in the cancer detection rate of one additional cancer detected per radiologist per 1000 screening mammograms. An analysis of several published clinical trials of CADe and full-field digital mammography showed that all but the one largest trial are likely not large enough to observe even the large postulated increase in the cancer detection rate of one additional cancer detected per radiologist per 1000 screening mammograms.

20.3.2.3 The cancer detection rate is probably not an ideal performance metric

The cancer detection rate is one of the most commonly used end-points in clinical trials in the evaluation of CADe. However, the arguments that we present here suggest that the cancer detection rate is probably not an ideal clinical performance metric for the evaluation of CADe and other new imaging technologies that aim to improve the clinical performance of cancer screening. The cancer detection rate is a favorite metric in part because it can be calculated readily from data collected in a clinical trial. However, a new imaging technique that results in improved cancer detection in screening does not necessarily cause the overall cancer detection rate to increase. Further, statistical uncertainty and inter-radiologist variability in the clinical performance make it a difficult task to measure the cancer detection rate with sufficient precision. Therefore, other diagnostic performance metrics should also be considered in designing large clinical trials.

20.3.2.4 Other subtle issues need to be investigated: definition of sensitivity and use of BI-RADS assessment ratings for ROC analysis

Here we discuss two other clinical performance metrics that also have been used in published clinical studies. We will identify some potential problems with how these metrics have been used. To assess the clinical effect of CADe properly will likely require more than simply choosing a different clinical performance metric.

Sensitivity is a fundamental measure of one's ability to detect cancers in a screening population. Sensitivity is difficult to measure in a screening setting because it is difficult to ascertain the total number of cancers in a screening population – the sum of the number of cancers detected during screening and the number of cancers not detected during screening. Nevertheless, sensitivity has been calculated in clinical studies (e.g. Fenton, 2007). Sensitivity can be calculated if a comprehensive registry is available that can reliably track the total number of cancers in a screening population during a particular time period. In the Fenton study, the number of cancers not detected during screening is estimated from the number of cancers diagnosed by other means in the interim between two successive screening rounds (also known as interval cancers). However, the calculation of sensitivity based on interval cancers may be biased in favor of the conventional imaging technologies against which a new imaging technology (in this case CADe) is being compared.

The key issue here is that the criteria used to identify cancers that are not detected in screening must be identical between the calculations of sensitivity associated with image interpretation with and without the assistance of CADe. While this may sound trivial, to satisfy this requirement is not easy – if CADe does help radiologists detect cancers early. If CADe does help radiologists detect cancers early, then treating only interval cancers (cancers not detected at a screening study but detected in the interim before the next screening study) as the not-detected cancers does not satisfy this requirement. Had CADe not been used, the cancers that are detected early because of the assistance of CADe would either be detected in the interim before the next screening study (interval cancers) or be detected at or after the next screening study. Those cancers that are detected early because of the assistance of CADe and would have been detected at or after the next screening study had CADe not been used are treated differently in the calculations of sensitivity associated with image interpretation with and without the assistance of CADe. For the calculation of sensitivity associated with image interpretation with the assistance of CADe, these cancers are included in both the number of detected cancers and the total number of cancers. However, for the calculation of sensitivity associated with image interpretation without the assistance of CADe, these cancers are not included in either the number of detected cancers or the total number of cancers. The effect of this is that the sensitivity associated with image interpretation with the assistance of CADe is biased lower compared with the sensitivity associated with image interpretation without the assistance of CADe. The correct calculation is to include these cancers in the total number of cancers for the calculations of sensitivity associated with image interpretation both with and without the assistance of CADe. However, in a clinical trial, the exact number of this kind of cancer – cancers that would have been detected at, or after, the next screening study without the assistance of CADe but are detected early because of the assistance of CADe – is difficult to ascertain and, therefore, it is not readily clear how the calculation of sensitivity can be modified in practice to eliminate this bias.

Another fundamental performance metric is the ROC curve. ROC analysis has been used extensively in laboratory observer studies. Several recent large clinical studies also have used ROC analysis (Pisano, 2005; Fenton, 2007). While it is encouraging and commendable that ROC analysis has entered the realm of large clinical studies, care must be given to ensure that the ROC analysis is conducted in a meaningful way. One fundamental requisite of ROC analysis is that observers – radiologists – report their diagnostic confidence on an ordinal scale. This is done carefully and deliberately in laboratory observer studies. But this report of the diagnostic confidence is not part of routine clinical practice and to conduct ROC analysis in clinical practice one must obtain these data in some way. Some researchers have used the BI-RADS final assessment ratings (ACR, 2003) in lieu of the diagnostic confidence for the purpose of ROC analysis (Barlow, 2004; Fenton, 2007). Unfortunately, this practice is flawed for the following reasons. The BI-RADS final assessment categories are designed to be a uniform language to help improve communications between radiologists, clinicians, and patients – they are not designed for ROC analysis. The BI-RADS final assessment categories are not designed to be an

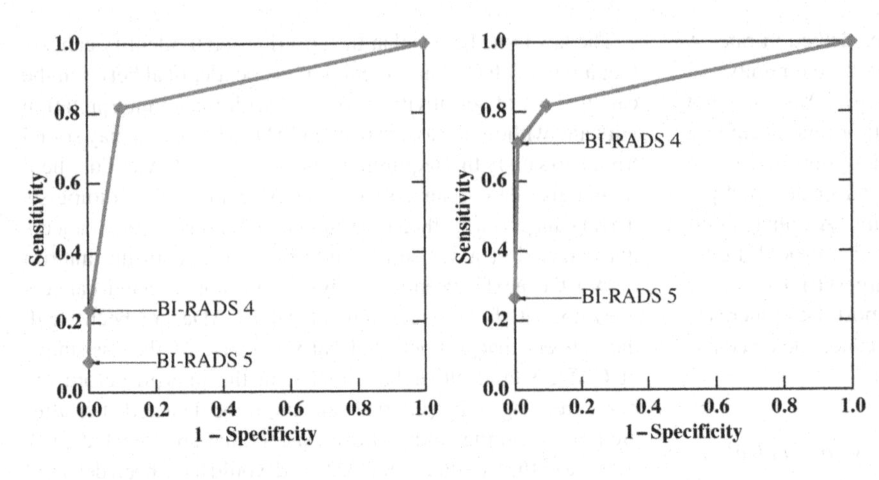

Figure 20.5 Illustration of the dependence of the ROC curve on the frequency with which radiologists use BI-RADS assessment ratings 4 and 5 in screening mammography examinations. The data in both panels that generated the ROC operating points and the ROC curves are identical except that **(left)** 5% of the cases that should have been given the BI-RADS rating 0 are given BI-RADS ratings 4 or 5, and **(right)** 15% of the cases that should have been given the BI-RADS rating 0 are given BI-RADS ratings 4 or 5. The differences in the frequency with which radiologists use BI-RADS ratings 4 and 5 causes the ROC curves to differ even though the radiologists' diagnostic performance depicted by the two ROC curves does not differ.

ordinal scale. The definition of BI-RADS 2 (benign abnormality) does not imply greater suspicion for cancer than BI-RADS 1 (no abnormality), and BI-RADS 0 (incomplete study, additional imaging required) does not imply less suspicion for cancer than any other BI-RADS assessment rating. Further, BI-RADS 3, 4, and 5 are not designed for use in screening examinations.

Some have argued that because the cancer detection rates associated with the BI-RADS assessment ratings are monotonic, it is evidence that radiologists use the BI-RADS assessment ratings as an ordinal scale even though the ratings are not designed for that purpose (Barlow, 2004). However, that cancers are found at all in cases given BI-RADS assessment ratings of 1 and 2 does not necessarily mean that the radiologist who assigned these ratings used the BI-RADS ratings 1 and 2 to indicate their level of diagnostic concern. Radiologists may have considered the mammogram to contain no evidence of cancer, but cancer still can be found in those patients if the cancer is occult in the mammogram, or if the radiologist missed the abnormality in the mammogram corresponding to the cancer, or if the radiologist misinterpreted the abnormality as benign. It is of course also possible that some radiologists do use the BI-RADS ratings 1 and 2 to indicate low – but greater than zero – suspicion for cancer, but that is not the most plausible explanation, and certainly not the only explanation, for why cancers are found in the cases that are given BI-RADS assessment ratings 1 and 2. And there is no indication of a tendency for radiologists to use BI-RADS ratings 1 and 2 to indicate low – but greater than zero – suspicion for cancer, whereas there are indications that sometimes radiologists do use the BI-RADS assessment ratings in ways that the ratings are not intended to be used (e.g. use of BI-RADS ratings 4 and 5 in screening cases before a diagnostic work-up).

Can BI-RADS assessment ratings be used meaningfully for ROC analysis even though the ratings may not be an ordinal scale? Figure 20.5 illustrates a potential problem with doing that. BI-RADS assessment ratings 3, 4, and 5 are not intended to be used for screening examinations. However, some radiologists sometimes do use the BI-RADS ratings 4 and 5 in cases of conspicuous cancer. In published studies that have calculated ROC curves from BI-RADS assessment ratings, BI-RADS ratings 4 and 5 are treated as distinct entries in the ROC data and correspond to distinct operating points (Barlow, 2004). (If one combines the data of BI-RADS ratings 0, 3, 4, and 5 into a single operating point, and combines the data of BI-RADS ratings 1 and 2 into another single operating point, so that the BI-RADS ratings are arranged into an ordinal scale based on the intended use of the BI-RADS ratings, then there will be only one operating point within the ROC-curve space, making it difficult mathematically to estimate the ROC curve.) Figure 20.5 illustrates that the operating points derived from BI-RADS ratings 4 and 5 are affected by how frequently radiologists use BI-RADS ratings 4 and 5 – the location of these operating points can change as the frequency changes in the use of BI-RADS ratings 4 and 5, even though the performance of the radiologists, or their accuracy, has not changed. This raises questions of the meaningfulness of the ROC curves fitted to the operating points calculated from BI-RADS assessment ratings 4 and 5, and the meaningfulness of the calculated area under the ROC curve as a summary index of diagnostic performance.

20.3.3 How do we evaluate CADe more effectively?

20.3.3.1 Develop new techniques to make laboratory observer study methodologies available to clinical trials

Laboratory observer studies often use ROC analysis to show that the assistance of CADe improves radiologists' diagnostic performance. Large historical-control clinical trials tend to use the cancer detection rate as one of the most important primary endpoints. The findings from laboratory observer studies and from large historical-control clinical trials have not been consistent. The different choices in the primary performance metrics between the ROC curve and the cancer detection rate are probably one reason for such inconsistencies. Therefore, a possible solution to resolve these inconsistencies is to use a single set of performance metrics in both laboratory observer performance studies and large clinical trials. The ROC analysis is a good candidate for that purpose. ROC analysis is a fundamentally meaningful way to characterize diagnostic performance because it provides an accurate and complete description of diagnostic performance and because it depicts all tradeoffs that are available between the sensitivity and specificity.

Currently, ROC analysis is not amenable to be applied to analyze routine clinical practice data without some specifically designed clinical trial (such as the Digital Mammographic Imaging Screening Trial; Pisano, 2005) to facilitate that analysis. However, with a few specific and rather straightforward changes in data collection, it is possible to change that to make ROC analysis of routine clinical practice possible. One of the changes that are necessary is to collect data on radiologists' diagnostic confidence. It should be clear that diagnostic confidence does not equal the BI-RADS assessment ratings and radiologists must understand the differences to ensure that the data collected represent diagnostic confidence meaningfully. If the collection of the diagnostic confidence data is integrated seamlessly into the electronic reporting (or voice dictation) of clinical case findings, then this data collection could be done without significant added effort on the part of radiologists and without interruption or significant change to their workflow.

The second change that is necessary is to track patient cases through time so that one can determine, through follow-up imaging studies and biopsies (when applicable) at some time point after a screening mammography study, whether cancer was present at the time of the screening study. With the migration of patient clinical information from paper records to electronic records, this seemingly labor-intensive and daunting task will become increasingly approachable. Once the necessary data are collected of radiologists' diagnostic confidence and of the truth status of whether cancer was present at the time of a screening mammogram, ROC analysis can be applied readily and straightforwardly. Unlike in a laboratory observer study, there will be a time lag between a screening mammogram study and the calculation of the ROC curve that includes the screening study because ascertaining the diagnostic truth status of a screening case requires a year or longer of follow-up time. Nevertheless, calculation of an ROC curve that depicts the routine clinical practice of screening mammography will enable us to compare laboratory observer performance studies and large clinical trials. More research is needed to make this comparison possible, but this may be just what is needed to resolve the inconsistencies that we have seen between laboratory observer performance studies and large clinical trials (Krupinski, 2008).

20.3.3.2 Separating the CADe computer training from its evaluation

Like many imaging technologies, the CADe computer systems are being improved continuously. Such improvement of the computer software happens probably more frequently than similar improvement of imaging hardware components. How do we evaluate a new version of the CADe computer software? One of the important aspects of this evaluation is that the images that one uses to evaluate the CADe computer must not be the same as those that are used to train the CADe computer. Obviously, if the same images are used both to train and evaluate the CADe computer, then the observed CADe computer performance might not represent the computer performance when it encounters new cases in routine clinical practice. Care must be taken to ensure that as more and more cases are used to improve the training of the CADe computer, the evaluation of the CADe computer is still done objectively using new cases that are representative of routine clinical practice.

How do we evaluate the interaction of radiologists with the CADe computer as the CADe computer is being improved? At one extreme, one could argue that improvement in the CADe computer automatically improves the performance of the radiologist who interprets images with the assistance of CADe. Therefore, if the CADe computer algorithm is improved, then there is no need to evaluate how radiologists interact with the new version of the CADe computer marks. At the other extreme, one could also argue that because how radiologists interact with the CADe computer marks depends on the characteristics of the computer marks, any change in the CADe computer algorithm must be validated with some form of observer performance study. Neither of these extreme positions is completely valid and satisfactory. There is no question that how radiologists interact with the CADe computer marks depends on the characteristics of the computer marks. Therefore, automatically accepting all improvements in the CADe computer algorithm is risky. However, conducting observer performance studies for every CADe computer algorithm upgrade is neither practical nor likely to be necessary. While how radiologists interact with the CADe computer marks cannot be completely predicted, those interactions are not completely random either. As we become more sophisticated in our understanding of the image perception process, we need to find a valid and satisfactory balance between the two extreme positions in the evaluation of CADe computer algorithm improvements.

20.4 SUMMARY AND CONCLUSIONS

Studies that evaluate the clinical effect of CADe are important because we must determine whether the premise of CADe – that it helps radiologists detect cancers – materializes in routine clinical practice and whether the clinical benefit of CADe outweighs its associated "costs," such as increased recall rate in screening mammography because of CADe computer false positives. There are a number of common types of CADe study. Some, such as those that demonstrate the clinical potential of CADe, are limited because they evaluate only some, but not all, aspects of interpreting images with the assistance of CADe. Other study designs that evaluate the entire process of interpreting images with the assistance of CADe in the laboratory environment or in the clinical settings each have advantages and disadvantages. Laboratory observer performance studies have consistently shown that CADe can help radiologists improve diagnostic performance, but large historical-control clinical trials do not consistently show the same benefit. Such inconsistencies may be a consequence of how these studies are designed rather than necessarily an indication of the clinical effect of CADe. For example, the cancer detection rate, which tends to be used in large clinical trials as a primary endpoint, may not be an appropriate fundamental clinical performance metric for the evaluation of the clinical effect of CADe. Future research on the clinical effects of CADe should try to reconcile the fundamental differences between laboratory observer performance studies and large clinical trials, e.g. to develop new techniques to extend ROC analysis from being used exclusively in the laboratory environment to broader applications in routine clinical practice. Such effort might help us understand the current inconsistencies between the large number of laboratory observer performance studies and large clinical trials. Clinical evaluation of the effect of CADe remains important until studies can show definitively and unequivocally whether CADe helps radiologists detect cancers early in routine clinical practice.

REFERENCES

Alberdi, E., Povyakalo, A.A., *et al.* (2005). Use of computer-aided detection (CAD) tools in screening mammography: a multidisciplinary investigation. *Br J Radiol*, **78 Spec No 1**, S31–S40.

American College of Radiology (ACR) (2003). *The Breast Imaging Reporting and Data System Atlas*. Reston, VA: American College of Radiology.

Ballard-Barbash, R., Taplin, S.H., *et al.* (1997). Breast Cancer Surveillance Consortium: a national mammography screening and outcomes database. *AJR Am J Roentgenol*, **169**, 1001–1008.

Barlow, W.E., Chi, C., *et al.* (2004). Accuracy of screening mammography interpretation by characteristics of radiologists. *J Natl Cancer Inst*, **96**, 1840–1850.

Beam, C.A., Layde, P.M., *et al.* (1996). Variability in the interpretation of screening mammograms by US radiologists. Findings from a national sample. *Arch Intern Med*, **156**, 209–213.

Beiden, S.V., Wagner, R.F., *et al.* (2002). Independent versus sequential reading in ROC studies of computer-assist modalities: analysis of components of variance. *Acad Radiol*, **9**, 1036–1043.

Birdwell, R.L., Bandodkar, P., *et al.* (2005). Computer-aided detection with screening mammography in a university hospital setting. *Radiology*, **236**, 451–457.

Birdwell, R.L., Ikeda, D.M., *et al.* (2001). Mammographic characteristics of 115 missed cancers later detected with screening mammography and the potential utility of computer-aided detection. *Radiology*, **219**, 192–202.

Brem, R.F., Baum, J., *et al.* (2003). Improvement in sensitivity of screening mammography with computer-aided detection: a multiinstitutional trial. *AJR Am J Roentgenol*, **181**, 687–693.

Chan, H.P., Doi, K., *et al.* (1990). Improvement in radiologists' detection of clustered microcalcifications on mammograms. The potential of computer-aided diagnosis. *Invest Radiol*, **25**, 1102–1110.

Ciatto, S., Ambrogetti, D., *et al.* (2006). Computer-aided detection (CAD) of cancers detected on double reading by one reader only. *Breast*, **15**, 528–532.

Cupples, T.E., Cunningham, J.E., *et al.* (2005). Impact of computer-aided detection in a regional screening mammography program. *AJR Am J Roentgenol*, **185**, 944–950.

Dean, J.C., Ilvento, C.C. (2006). Improved cancer detection using computer-aided detection with diagnostic and screening mammography: prospective study of 104 cancers. *AJR Am J Roentgenol*, **187**, 20–28.

Destounis, S.V., DiNitto, P., *et al.* (2004). Can computer-aided detection with double reading of screening mammograms help decrease the false-negative rate? Initial experience. *Radiology*, **232**, 578–584.

Elmore, J.G., Wells, C.K., *et al.* (1994). Variability in radiologists' interpretations of mammograms. *N Engl J Med*, **331**, 1493–1499.

Feig, S.A., Sickles, E.A., *et al.* (2004). Re: Changes in breast cancer detection and mammography recall rates after the introduction of a computer-aided detection system. *J Natl Cancer Inst*, **96**, 1260–1261; author reply 1261.

Fenton, J.J., Taplin, S.H., *et al.* (2007). Influence of computer-aided detection on performance of screening mammography. *N Engl J Med*, **356**, 1399–1409.

Freer, T.W., Ulissey, M.J. (2001). Screening mammography with computer-aided detection: prospective study of 12,860 patients in a community breast center. *Radiology*, **220**, 781–786.

Gilbert, F.J., Astley, S.M., *et al.* (2006). Single reading with computer-aided detection and double reading of screening mammograms in the United Kingdom National Breast Screening Program. *Radiology*, **241**, 47–53.

Gromet, M. (2008). Comparison of computer-aided detection to double reading of screening mammograms: review of 231,221 mammograms. *AJR Am J Roentgenol*, **190**, 854–859.

Gur, D., Bandos, A.I., *et al.* (2007). The prevalence effect in a laboratory environment: Changing the confidence ratings. *Acad Radiol*, **14**, 49–53.

Gur, D., Sumkin, J.H., *et al.* (2004). Changes in breast cancer detection and mammography recall rates after the introduction of a computer-aided detection system. *J Natl Cancer Inst*, **96**, 185–190.

Helvie, M.A., Hadjiiski, L., *et al.* (2004). Sensitivity of noncommercial computer-aided detection system for mammographic breast cancer detection: pilot clinical trial. *Radiology*, **231**, 208–214.

Ikeda, D.M., Birdwell, R.L., *et al.* (2004). Computer-aided detection output on 172 subtle findings on normal mammograms previously obtained in women with breast cancer detected at follow-up screening mammography. *Radiology*, **230**, 811–819.

Jemal, A., Siegel, R., *et al.* (2008). Cancer statistics, 2008. *CA Cancer J Clin*, **58**, 71–96.

Jiang, Y., Miglioretti, D.L., *et al.* (2007). Breast cancer detection rate: designing imaging trials to demonstrate improvements. *Radiology*, **243**, 360–367.

Karssemeijer, N., Otten, J.D., *et al.* (2003). Computer-aided detection versus independent double reading of masses on mammograms. *Radiology*, **227**, 192–200.

Kegelmeyer, W.P., Jr., Pruneda, J.M., *et al.* (1994). Computer-aided mammographic screening for spiculated lesions. *Radiology*, **191**, 331–337.

Khoo, L.A., Taylor, P., *et al.* (2005). Computer-aided detection in the United Kingdom National Breast Screening Programme: prospective study. *Radiology*, **237**, 444–449.

Ko, J.M., Nicholas, M.J., *et al.* (2006). Prospective assessment of computer-aided detection in interpretation of screening mammography. *AJR Am J Roentgenol*, **187**, 1483–1491.

Kobayashi, T., Xu, X.W., *et al.* (1996). Effect of a computer-aided diagnosis scheme on radiologists' performance in detection of lung nodules on radiographs. *Radiology*, **199**, 843–848.

Krupinski, E.A., Jiang, Y. (2008). Anniversary paper: evaluation of medical imaging systems. *Med Phys*, **35**, 645–659.

Marx, C., Malich, A., *et al.* (2004). Are unnecessary follow-up procedures induced by computer-aided diagnosis (CAD) in mammography? Comparison of mammographic diagnosis with and without use of CAD. *Eur J Radiol*, **51**, 66–72.

Metz, C.E. (1978). Basic principles of ROC analysis. *Semin Nucl Med*, **8**, 283–298.

Metz, C.E. (1989). Some practical issues of experimental design and data analysis in radiological ROC studies. *Invest Radiol*, **24**, 234–245.

Moberg, K., Bjurstam, N., *et al.* (2001). Computed assisted detection of interval breast cancers. *Eur J Radiol*, **39**, 104–110.

Morton, M.J., Whaley, D.H., *et al.* (2006). Screening mammograms: interpretation with computer-aided detection – prospective evaluation. *Radiology*, **239**, 375–383.

Nishikawa, R.M. (2006). Modeling the effect of computer-aided detection on the sensitivity of screening mammography. In *Digital Mammography*, ed. Astley, S.M., Brady, M., Rose, C., Zwiggelaar, R. London: Springer, pp. 46–53.

Pisano, E.D., Gatsonis, C., *et al.* (2005). Diagnostic performance of digital versus film mammography for breast-cancer screening. *N Engl J Med*, **353**, 1773–1783.

Schmidt, R.A., Newstead, G.M., *et al.* (1998). Mammographic screening sensitivity of general radiologists. In *Digital Mammography*, ed. Karssemeijer, N., Thijssen, M., Hendriks, J., van Erning, L. Dordrecht: Kluwer Academic Publishers, pp. 383–388.

Skaane, P., Kshirsagar, A., *et al.* (2007). Effect of computer-aided detection on independent double reading of paired screen-film and full-field digital screening mammograms. *AJR Am J Roentgenol*, **188**, 377–384.

Taplin, S.H., Rutter, C.M., *et al.* (2006). Testing the effect of computer-assisted detection on interpretive performance in screening mammography. *AJR Am J Roentgenol*, **187**, 1475–1482.

Taylor, P., Given-Wilson, R.M. (2005). Evaluation of computer-aided detection (CAD) devices. *Br J Radiol*, **78 Spec No 1**, S26–S30.

te Brake, G.M., Karssemeijer, N., *et al.* (1998). Automated detection of breast carcinomas not detected in a screening program. *Radiology*, **207**, 465–471.

Wagner, R.F., Metz, C.E., *et al.* (2007). Assessment of medical imaging systems and computer aids: a tutorial review. *Acad Radiol*, **14**, 723–748.

Warren Burhenne, L.J., Wood, S.A., *et al.* (2000). Potential contribution of computer-aided detection to the sensitivity of screening mammography. *Radiology*, **215**, 554–562.

Zheng, B., Shah, R., *et al.* (2002). Computer-aided detection in mammography: an assessment of performance on current and prior images. *Acad Radiol*, **9**, 1245–1250.

21

Perceptual effect of CAD in reading chest radiographs

MATTHEW FREEDMAN AND TERESA OSICKA

21.1 INTRODUCTION

The detection of small lung cancers on chest radiographs is a difficult task for radiologists. In prospective clinical trials for lung cancer screening with chest radiographs, radiologists, on average, detected two-thirds or less of those cancers that could be identified in retrospect (Stitik, 1978, 1985; Muhm, 1983; Heelan, 1984). Because of the inherent difficulties that radiologists have in perceiving the small lung nodules that could represent cancers, a major effort has been made for approximately the last 20 years to develop computer systems to assist radiologists in the detection and diagnosis of small lung cancers. These computer programs have emphasized the detection of small pulmonary nodules, usually in the range of 10–30 mm in diameter.

These programs are of two types, computer aided detection (CADe) and computer aided diagnosis (CADx). CADe is designed to help radiologists detect small lung nodules and CADx is designed to help radiologists determine whether or not these small lung nodules are cancers; these CADx systems, for use with chest radiographs, are still under development. There has been a series of studies showing the effectiveness of CADe systems as machines and showing that radiologists using these systems do improve their detection of lung nodules (Giger, 1988, 1990; Lo, 1995; Kobayashi, 1996; Lin, 1996; Xu, 1997; Shiraishi, 2000, 2006; Abe, 2003, 2004; Li, 2008). A summary of several of these reports on the effectiveness of CADe, both tests of the machines and tests showing the improvement of radiologists when they use these machines, is available (Freedman, 2004).

With CADe for chest radiographs, the chest radiograph is converted into digital form. The computer then analyzes the image using specially designed algorithms. It then identifies locations where the image matches the mathematical characteristics that are similar to those seen in the disease in study. In Figure 21.1, we can see a circle placed at the location of a small primary lung cancer. In Figure 21.2, we can see a circle placed at the site of a small area of pneumonia. Most of the research in CADe related to chest radiographs has focused on efforts to help radiologists identify small cancers, but these systems can be designed to analyze other lung diseases (Ishida, 1997; Katsuragawa, 1997).

There are common terms used in this chapter. These are listed in Table 21.1. It is important to separately consider the three types of effects to be discussed: (1) the machine's performance, (2) the radiologist's performance, and (3) the combined performance of the radiologist when using the machine. In addition, there are different types of tests of reader performance (independent, sequential, and concurrent).

21.2 IMAGE PERCEPTION AND COMPUTER AIDED DETECTION: KEY CLINICAL STUDIES

This chapter discusses the perceptual effects of chest CADe; it does not review the multiple publications showing that radiologists can detect more cancers using CADe machines. The information on perceptual effects is based largely on two studies. Both studies showed that, in experimental settings, radiologists did improve their detection of lung nodules caused by cancer when they used the CADe prompts. In addition, both studies addressed certain perceptual issues and therefore form the basis of the discussion in this chapter.

The design of these two studies and their findings will be briefly reviewed. Following this, specific issues in image perception tasks will be discussed, drawing on these two studies to illustrate the findings.

21.2.1 The first study

This study from the University of Chicago was reported in 1996 (Kobayashi, 1996). It provides information on factors of image perception related to:

1. The effect of three different levels of reader skill on the benefits of CADe.
2. The effect of different reading patterns – independent, sequential, and concurrent – on the benefits of CADe.
3. The effect of true positive, false positive, and false negative CADe prompts.
4. Inter-observer variability, where two or more observers viewed the same images.
5. The effect of CADe on the time required to interpret cases.

In this study, 120 chest radiographs were prepared, of which 60 contained a single pulmonary nodule and 60 were nodule-free. The nodules ranged from 9 to 25 mm and were assigned a subtlety index value. Sixteen radiologists participated. These were of three skill levels – two thoracic radiology specialists, six general radiologists, and eight radiology residents. The radiologists interpreted the chest radiographs and used the CADe output

The Handbook of Medical Image Perception and Techniques, ed. Ehsan Samei and Elizabeth Krupinski. Published by Cambridge University Press.

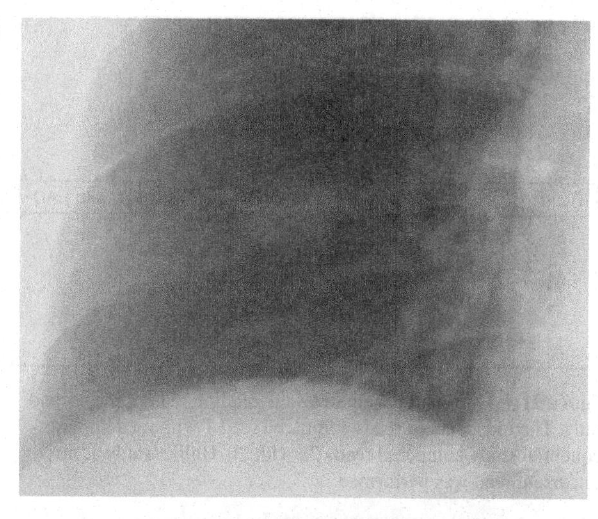

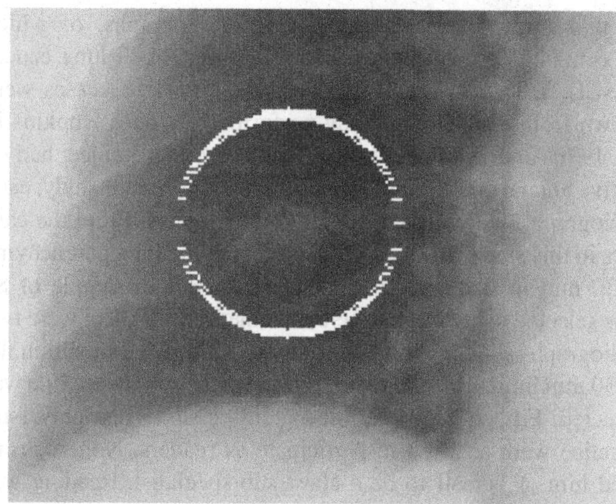

Figure 21.1 An example of CADe. The image at the top shows a portion of the lungs containing a small nodule. The image below shows a CADe circle placed around it. This was a small primary lung carcinoma. (Courtesy Riverain Medical Group, Miamisburg, OH, USA.)

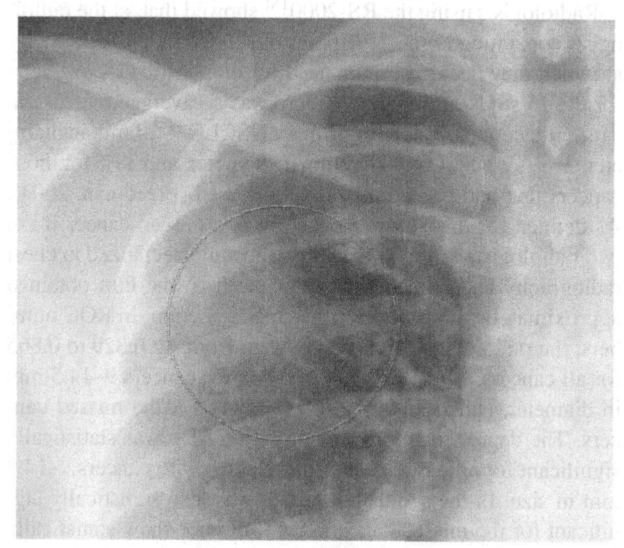

Figure 21.2 An example of CADe for pneumonia. A CADe program for detecting small areas of pneumonia has placed a circle around a small area of SARS pneumonia. (Courtesy Riverain Medical Group, Miamisburg, OH, USA.) (Freedman, 2003.)

Table 21.1 *This table provides definitions of special terms used in this chapter*

Machine true positive	The machine has correctly identified the location of the cancer
Machine false negative	The machine has not correctly identified the location of the cancer and has marked no location
Machine false positive	The machine has marked a location that is not a cancer
Radiologist true positive	The radiologist has correctly identified the location of the cancer
Radiologist false negative	The radiologist has identified no location on a radiograph that contains cancer
Radiologist false positive	The radiologist has marked a location that is not cancer
CADe prompt	A mark placed on an image by a CADe system to locate the location of interest
Independent test	A reader study done either without or with CADe, but not both
Sequential test	A reader study where the radiologist first interprets the image without CADe and then, immediately after, with CADe
Concurrent test	A reader study done when both the image without CADe and the image with CADe are presented at the same time
Intra-observer variability	The variability of interpretation that occurs when the same radiologist is shown the same image(s) twice
Inter-observer variability	The variability of interpretation that occurs when two or more radiologists are shown the same image(s)
Sensitivity	The percentage of correct diagnoses in people who have a disease
Specificity	The percentage of correct diagnoses that the disease is not present in people who do not have the disease

in several ways. The approach to the reader study is shown in Figure 21.3. Each radiograph was interpreted once independent of CADe (read 1), once when the radiograph and the CADe were presented concurrently (read 3), and once using a sequential method where the radiograph was interpreted first without CADe, then, immediately after, with CADe (reads 2a and 2b).

The ROC result of the independent reading without CADe (read 1) was area under the ROC curve (Az) 0.894. For the independent reading with CADe (read 3), Az was 0.940; this is the value for concurrent interpretation. In the sequential study,

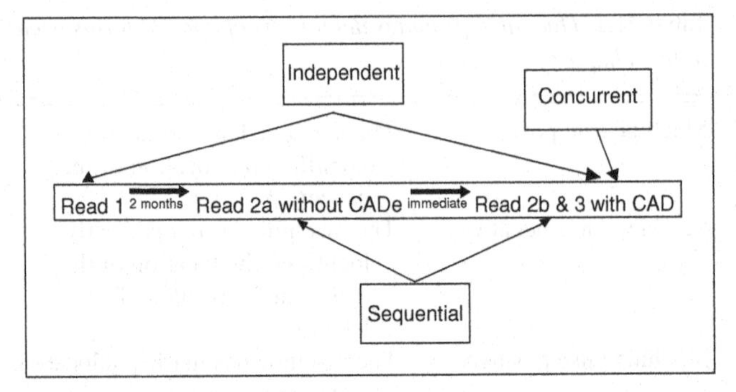

Figure 21.3 This chart schematizes the reading pattern for this study. The radiologists reviewed the cases three times. The independent study compares read 1 with read 3 where the chest radiograph and the chest radiograph with CADe were read at different times. The sequential study compares read 2a with read 2b where the chest radiographs were read in rapid sequence. The concurrent read was a separate read done of the chest radiograph with CADe provided concurrently; this is the same as the second independent read (with CADe), read 3.

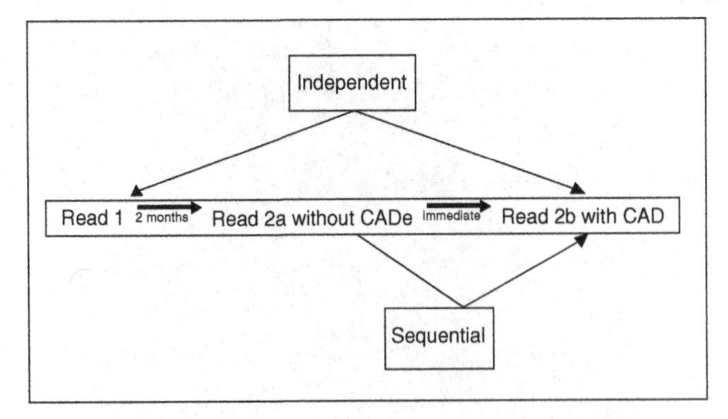

Figure 21.4 This chart schematizes the reading plan of the second study. The independent reads compared read 1 with read 2b. The sequential study compared reads 2a with 2b. Unlike study 1, no concurrent read was performed.

without CADe (read 2a), Az was 0.906 and improved with the sequential CADe (read 2b) to 0.948. With both the independent and sequential methods, the improvements with CADe were statistically significant and were greater for those radiologists with less training and experience. The Az seen with the independent with CADe and sequential with CADe results were not significantly different ($p = 0.205$).

CADe prompts can be correct or incorrect. The effect of CADe true positive, false positive, and false negative prompts varied with the reading test performed. Overall, for the independent tests, the CADe output affected 6.3 cases positively and 2.5 negatively; for the sequential reading test, it was 2.9 positively and 0.1 negatively. Thus, for all CADe prompts, CADe was, on balance, beneficial. The false negative effect, where the CADe missed the cancer, was, in the independent test, 1.0 case positively and 1.8 cases negatively; for the sequential test, 0.4 cases were affected positively and 0.6 cases negatively. This slight negative effect is not statistically significant.

Reader variability and the separate effects of CADe true positives, false positives, and false negatives are shown graphically in the original article. The amount of time the radiologists used to read the cases was not affected.

21.2.2 The second study

This study was performed at Georgetown University using a pre-commercial version of the Deus (now Riverain Medical Group, Miamisburg, OH, USA) R1.0 system. This study provides information on the factors of image perception related to:

1. The effect of CADe results on the radiologists' confidence of the presence of cancer. The effect of true positive, false positive, and false negative CADe prompts.
2. Intra-observer variability where the same radiologist interprets the same images twice.
3. Inter-observer variability where two or more radiologists interpret the same images.
4. The effect of nodule size on nodule detection without and with CADe.
5. The effect of CADe on the time required to interpret cases.

In this study, there were 240 chest radiographs, of which 80 contained a confirmed primary non-small cell lung cancer (NSCLC). There were 160 cancer-free cases. The cases were drawn from a screening study performed at Johns Hopkins in the 1970s and 1980s. In the original study, each subject had up to five annual chest radiographs and, in the original study, each radiograph was interpreted by two radiologists. From the cancers in this study, all cases with cancers in the lung parenchyma 9–30 mm in size were included and a random sample of 80 was selected using random selection methods. If there were two radiographs showing the nodule at different size, but within the 9–30 mm limitation, the radiograph with the smaller nodule was selected. Fifteen board-certified radiologists in non-university practice were selected to participate as readers. None considered him or herself to be a chest sub-specialist. Reading was done as an independent test without CADe, followed one to four months later with a rapid sequential test where the radiologist first interpreted the radiographs without CADe and then, immediately after, with CADe. Figure 21.4 displays the reading schema.

Radiologists using the RS-2000™ showed that, at the radiologists' operating points (the point of sensitivity and specificity at which they decided on the need for diagnostic CT, averaged for the 15 study radiologists), there was an average increase in cancer detection of 11% for primary NSCLC 9–27 mm in diameter, 21% for NSCLC 9–14.5 mm in diameter, and 38% for those cancers that had been prospectively missed (Freedman, 2004). As defined for this study, a cancer was a missed cancer if the two radiologists at Johns Hopkins who sub-specialized in chest radiography had both missed the cancer on the film obtained approximately one year prior to actual detection. In ROC numbers, the radiologists, on average, went from Az 0.829 to 0.865 for all cancers, from Az 0.798 to 0.848 for cancers 9–14.5 mm in diameter, and from Az 0.702 to 0.744 for the missed cancers. The degree of improvement with CADe was statistically significant for all cases and for those cases with cancers 9–14.5 mm in size. In the sequential test, it was also statistically significant for the missed cancers, but it did not show statistically significant improvement for the independent tests, likely due to greater reader variability in this experimental approach.

Studies of the effect of CADe on reader confidence levels, the effect on these confidence levels of true positive, false positive, and false negative CADe prompts, intra-observer and inter-observer variability, and the effect of nodule size on nodule detection are described below.

21.3 THE EFFECT OF CADe ON READER CONFIDENCE LEVELS

The goal of CADe, in these experiments, is to help radiologists recognize small lung cancers on chest radiographs. This can be measured in two ways: (1) by asking the radiologists to record their confidence that a film contains cancer, and/or (2a) by asking them to identify the location of suspicion and (2b) what action they would take (e.g. CT or no CT, biopsy or no biopsy). The first, the confidence rating method, provides a multistep or continuous response curve; the second approach is usually binary. In each case, the interpretations made by the radiologists should be compared to the ground truth – what is the actual or operationally defined diagnosis.

When one looks at confidence levels, each radiologist is likely to have a somewhat different degree of confidence in each case. Therefore, the measure that is of interest can be based on summary statistical methods (such as receiver operating characteristic (ROC)) or recordings of the changes in confidence that occur when the CADe prompts are provided. Changes in a radiologist's individual confidence level can be small or large. When small, they may represent a small amount of minor variation – not reflecting purposeful changes, but, rather, what can be considered jitter from the design of the system used to record the confidence level; in addition, the radiologist might have intended the small changes to represent purposefully small change rather than jitter. These small purposeful changes could be a way the radiologist indicates that the CADe system has confirmed their initial impression, making them feel more confident, or, conversely, disagreed, making them slightly less confident, but in both cases, the CADe results were not sufficient to lead to a change in the final decision to obtain a follow-up CT or not. Large changes are likely to be purposeful. These are unlikely to result from minor problems or uncertainties in the recording system; rather they likely represent a change that is reflected in a change in the binary decision – workup for cancer, or not – and, in study two, they were associated with changes in the binary decisions.

An analysis of the changes in confidence rating can also be applied to the effect on confidence levels when cases are reinterpreted without CADe, or, separately, when the CADe prompts are true positive, false positive, or false negative prompts.

Examples to illustrate these are shown in Figures 21.5, 21.6, and 21.7. In these three charts, we have displayed the changes in confidence ratings under different states. On the y-axis, the zero line is in the vertical center of the chart. Columns above that center line represent an increase in confidence ratings; columns below indicate a decrease in confidence ratings. On the x-axis, the different cases are arrayed by case number. The nature of the CADe prompts varies with the chart: Figure 21.5 shows the lack of effect on one radiologist of CADe false negative prompts. Figure 21.6 shows the effect of true positive CADe prompts. Figure 21.7 shows the effect where the radiologist

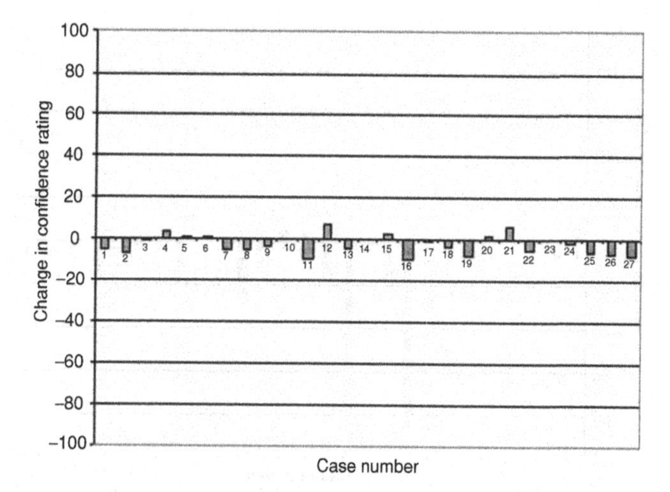

Figure 21.5 In this chart, the changes in confidence rating from the interpretation without CADe and with CADe for one radiologist are displayed. The zero change line is horizontal in the middle of the chart. One can see that only small changes – all less than 15 points on the −100 to +100 point scale – are shown. This is one effect of machine false negative cases.

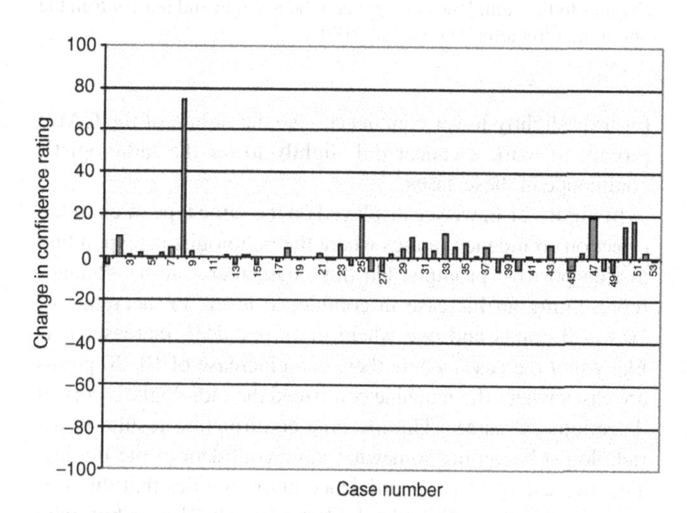

Figure 21.6 In this chart, changes in confidence rating are seen as the radiologist responds to true positive prompts in cancer cases. This radiologist shows one case where the confidence rating was greatly increased after viewing the CADe mark. This implies that the mark pointed to a lesion that was not previously of much concern for cancer and that the radiologist became concerned that this was a cancer and assigned it a much higher confidence rating. For the other CADe true positive cases, in general the correct marks increased this radiologist's confidence rating by a small amount; in only a few cases did the confidence ratings decrease. This is the expected effect where the CADe mark confirms the radiologist's original interpretation (Freedman, 2001).

responded to CADe false negative prompts by changing from correct decisions that cancer was present to an opinion that cancer was not present.

In Figure 21.5, the changes in confidence ratings of one radiologist are shown for cases where the machine did not detect the cancers. One can see that the change in confidence ratings on these cases shows only minor change – up to about ten points. This likely represents jitter; but it is not solely jitter, because we see that more of the changes in the confidence ratings are

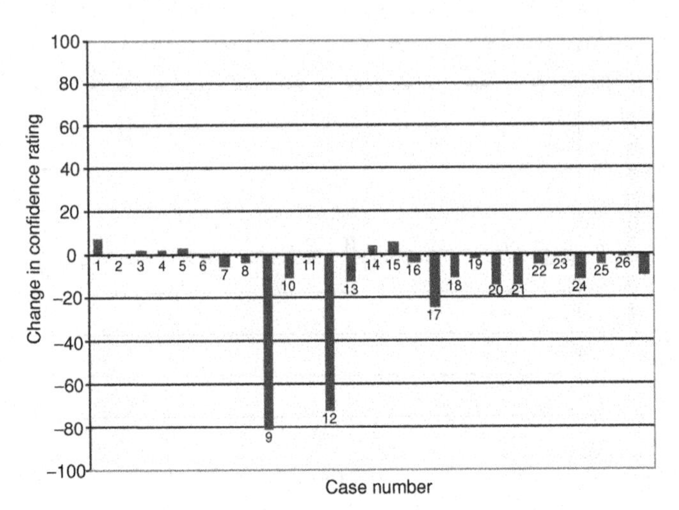

Figure 21.7 In this chart, we see a different response to false negative computer marks than was shown in Figure 21.5. This radiologist, when his initial decisions were not confirmed by the CADe algorithm, greatly decreased his rating for two of the cases. This is shown by the few marks that extend far below the zero change line. Contrast this with Figure 21.5 where a different radiologist made only minor changes to the confidence rating when the machine did not confirm the initial interpretations (Freedman, 2001).

towards slightly lower confidence – so the failure of the CADe prompt to mark a cancer did slightly lower the radiologist's confidence in these cases.

In Figure 21.6, we see displayed on the same type of chart the effect on confidence ratings where the radiologist is responding to true positive prompts. On this chart, there are 41 changes representing an increase in confidence levels in the range of 10 to 20 points and one where there is a 75% increase. It is likely that the cases where there is an increase of 10–20 points are cases where the machine confirmed the radiologist's correct detections of cancer. The machine confirmation results in the radiologist becoming somewhat more confident of the finding. The increase of 75% in confidence rating implies that this was a cancer that the radiologist did not originally detect, but, with CADe, did detect it and was highly confident that this detection was correct.

In Figure 21.7, we see a different effect of false negative CADe prompts than shown in Figure 21.5. In these cases, the machine failed to mark the location of the cancer. In this radiologist's responses, we see two markedly negative columns, with decreases of 70–80 points, and a series of decreases that are substantially smaller. Only a few columns demonstrate a minor increase in confidence. This means that there were two cases where the radiologist was convinced by the lack of a CADe prompt that the original level of suspicion was much too high, and that cancer was likely not present. For the other cases, where the decrease is less, it shows that the radiologist became less confident, but, given the small amount of change, probably did not change the binary decision to either workup the case further for cancer, or not.

These three charts also demonstrate that radiologists are responding to the computer output – it is not being ignored, but is considered as part of the decision process. When CADe confirms the radiologist's decision, only a small change in confidence is seen. When it does not confirm it, the change in confidence rating can be small or large. Because the radiologists are experts, most of the time an incorrect CADe prompt will not cause a change in the original decision on the presence or absence of cancer; but in a few cases it will.

21.3.1 Are minor changes in confidence rating important?

It does not appear that minor changes affect the outcome of observer performance studies. One might think that a small change might switch an observer past some threshold that causes a change in decision from a case not needing a workup to needing one. We did not observe this. When the radiologist made a change to recommend a CT instead of nothing or from recommending a CT without CADe to no CT with CADe, there was always a major change in confidence rating. This probably reflects an underlying pattern in radiology interpretation where radiologists, once they've made a decision, do not want to provide an uncertain conclusion.

21.3.2 Do false negative CADe prompts represent a risk to patients?

In both the University of Chicago and Georgetown studies, there is evidence of a minor risk because a few cases that were initially considered to represent cancer were changed to a non-cancer decision based on the failure of the CADe prompt to mark them. In both studies, the benefit from correct prompts greatly outweighed the negative effect of incorrect prompts.

This bidirectional possibility of change does point to an interesting quandary: if a CADe system is effective, the benefit outweighs the risk; if a CADe system were ineffective in detecting cancers, the effect on the radiologist's decisions might be detrimental. This is because the negative decisions from the lack of correct prompts would not be offset by the benefit from the detection of additional cancers from the correct prompts. It also points to the tradeoff that exists in many medical technologies between risk and benefit. With any new device or medicine, it is essential that experiments demonstrate that the risks, which are almost always present, are offset by a greater benefit resulting from use of the system.

21.3.3 Are the radiologists' opinions stable in the presence of CADe prompts?

One of the interesting perceptual features of tests of CADe is that the radiologists are expert rather than novice observers. Because they are experts, they are less likely to be misled by errors made by CADe and CADx machines. This question was directly addressed within the Georgetown experiment. While newer systems provide many fewer false positive marks, the CADe pre-market system we tested was tuned to produce an average of 5.3 marks on each film (Freedman, 2002); thus there were many extra marks present that might mislead.

When the radiologist and the CADe system both correctly identified the location of the cancer, it should not be surprising that radiologists, by and large, did not change their opinions. Since radiologists improve their detection of cancer when they use CADe, it follows that the radiologists do accept some of the correct CADe prompts. What is of interest is what happens when both the radiologist and the CADe are wrong, given that this CADe system almost always provided location marks on each image, and what happens when the radiologist is correct and the CADe missed the lesion.

21.3.3.1 Agreement of radiologist and machine: effect was lack of shift of the locations marked by the radiologist

The radiologist and the CADe can agree when both are correct or both are incorrect. When there was agreement between the radiologist and the machine, only minimal shift in the marks of "cancer" locations occurred. Of the 37.9 cases that, on average, were true positive for both the radiologists and machine, 0.4 cases moved to a false negative location. Of the 12.7 cases that, on average, were false negative, the radiologists moved 0.3 cases to a true positive location. Thus, when the computer and the radiologist agreed, the CADe prompts had little effect on the radiologists' decisions.

21.3.3.2 Lack of agreement of radiologist and machine: when the radiologist is correct and the machine wrong, little change occurs; when the machine is correct and the radiologist is initially wrong, the radiologist does improve

Of the 14.3 cases, on average, where the radiologist's initial location was true positive and the machine was false negative, 0.4 cases were moved to the false negative location. Of the 15.1 cases, on average, where the machine marked the correct location and the radiologist had marked the wrong or no location, 3 cases were moved to the true positive location. Thus, when the CADe prompt might have indicated that the radiologist was wrong, the radiologist's expert opinion without CADe was dominant. When the radiologist was wrong and the computer was correct, the machine prompt did persuade the radiologist to change to the correct location in some cases.

Thus, radiologists, as experts in the field, were able to use CADe information when it was correct, but largely avoid errors when the CADe information was incorrect. This was particularly true when the radiologist had initially correctly detected the location of the cancer.

21.4 CADe AND SATISFACTION OF SEARCH

These results on the effect of CADe true positives can be interpreted in a different way. When radiologists interpret radiographs, they always will have some point at which they consider themselves to be satisfied and, therefore, stop searching. When errors occur, one of the types of error is a "satisfaction of search" error – that is, they stopped too soon and missed either the disease they were looking for or some other important finding on the radiograph. One of the ways to interpret the positive effect of true positive CADe prompts is that they lead the radiologist to look again at certain portions of the image. If there is something there and they then find it, one could consider this as overcoming the error of satisfaction of search, at least for the lung nodules.

Even when the CADe prompts are incorrect, the radiologists do detect additional cancers. This occurred, on average, in 2.1% of the 28 cases where the CADe prompts did not mark the cancer (Freedman, 2002). This would suggest that CADe helps to correct the errors that can occur when the radiologist is satisfied that search has been sufficient when it was not.

CADe is focused on a single disease, lung nodules that could be cancer. While CADe can help prevent errors from satisfaction of search regarding lung nodules, this does not mean that it helps overcome this problem with other diseases that can affect the chest. A study by Berbaum (Berbaum, 2007) indicates that the problem of satisfaction of search is increased by CADe for diseases other than lung nodules. While radiologists detected more lung nodules with CADe in his experiment, other chest diseases were recognized less often. This would likely represent a disadvantage of concurrent reading as tested in the University of Chicago study. While Chicago found that concurrent read was essentially as good as and not significantly different from the independent readings and the sequential reading method, they did not include diseases other than lung nodules in their dataset. Berbaum's study suggests that concurrent reading could result in radiologists overlooking other diseases present in the chest. While further tests should be performed to confirm Berbaum's findings, in the mean time, concurrent reading should be discouraged.

21.5 INTRA-OBSERVER AND INTER-OBSERVER VARIATIONS IN RESPONSE

Radiologists, when they interpret images, do not always agree with themselves or with other radiologists. In normal clinical practice, the rate of disagreement between radiologists is, in one study, 4.4% (range 0.8%–9.2%) (Siegle, 1998). There is greater reader disagreement when the disease pattern is subtle. If you show the same subtle cases to the same radiologist at two different times, a different interpretation may result in up to 20% of the cases. This is called "intra-observer variability." If you show two or more different radiologists the same subtle cases, they may disagree with each other in up to 30% of cases (Yerushalmy, 1969). This is called "inter-observer variability." This observer variation occurs because the findings on chest radiographs can be ambiguous and because of perceptual errors. These forms of interpretation variability interact with the observers' use of CADe in interesting ways that will be described below.

21.5.1 Certainty and uncertainty of chest radiographic findings

Findings on chest radiographs vary in their certainty, both of detection and diagnosis. The findings that suggest cancer, for example, can vary from a well defined nodule, to a poorly defined nodule, to a focal area of increased radiodensity without an identifiable boundary. On chest radiographs, the smallest nodules have a degree of uncertainty to them caused by the presence

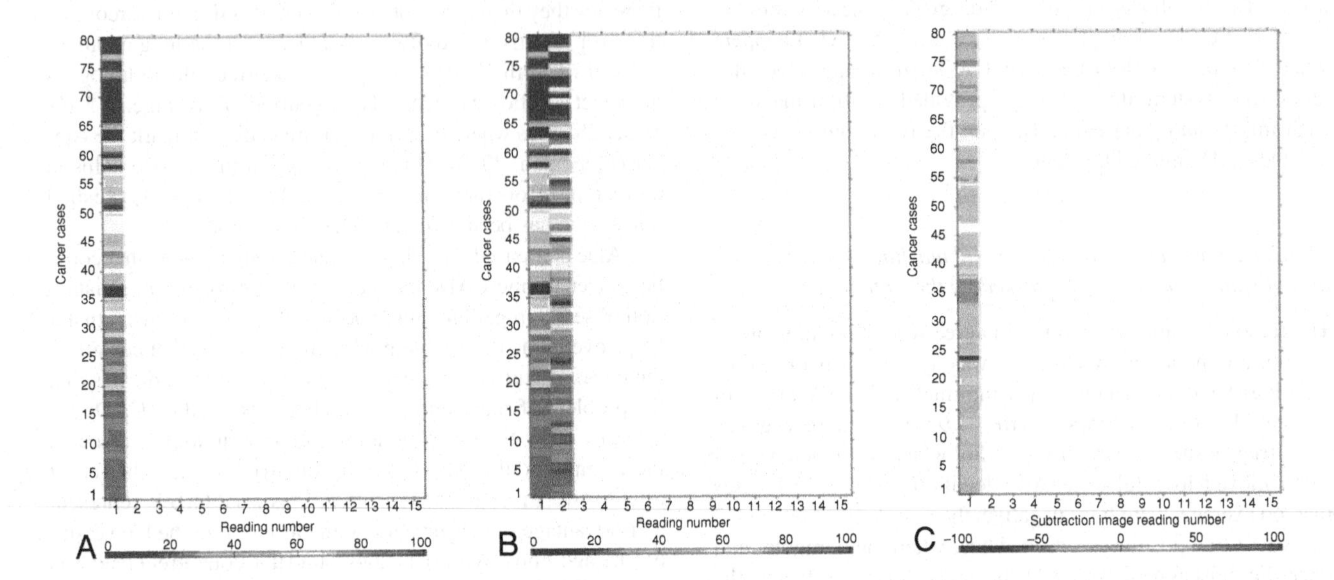

Figure 21.8 These charts show the confidence levels for one radiologist interpreting the chest radiographs of 80 cases with cancer at two different times, without CADe. The left chart shows the confidence levels for the first reading; the center chart shows the confidence ratings for the two readings side by side; the right chart shows the subtraction from the two readings. This shows a moderate amount of change in confidence levels between the two reads 1 and 2a (Freedman, 2006). These charts are available in color on the web (Freedman, 2008a).

of an indistinct or no boundary and blurring of the image that occurs during image acquisition. These structural features of the images of small lung cancers increase the perceptual difficulty in detecting them.

21.5.2 Intra-observer variability

In the Georgetown experiment, the radiologists interpreted the chest radiographs twice without CADe (from Figure 21.4, reads 1 and 2a). The first time, it was an independent without CADe interpretation; the second time, it was as the first part of a sequential reading test, first interpreted without CADe and then with CADe. It is the two readings without CADe that interest us here. Figure 21.8 shows the changes shown by a single reader interpreting the chest radiographs with cancer twice, both without CADe. The left chart shows the confidence levels for the first read; the center chart shows side by side the confidence levels for the first and second reads. The right chart shows the subtraction image. These three images show that if you test on subtle cases, the same radiologist will give different confidence ratings on the same cases when they are presented a second time.

21.5.3 Inter-observer variability

Figures 21.9 and 21.10 demonstrate the comparison of confidence rating changes contrasting the effect of intra-observer and inter-observer variation of 15 radiologists to the effect of CADe for these 15 radiologists. The top set of charts (Figure 21.9) show the variation seen from two separate interpretations of the 80 cancer cases. Referring to Figure 21.4, these are the results comparing read 1 with read 2a. The subtraction image shows that these changes are bidirectional – some cases were considered more suspicious and others less suspicious. This is easier to visualize in the color images available at the website (Freedman, 2008a). Overall, 9% of cases moved in one direction and 9.9% of cases in the other direction. This is similar to the 20% changes reported by Yerushalmy (Yerushalmy, 1969).

The lower set of charts (Figure 21.10) show the changes in the confidence ratings of radiologists that occurred with CADe. Referring to Figure 21.4, these charts are comparing reads 2a and 2b. In this case, the subtraction images show that the changes induced by the use of CADe were mainly unidirectional towards greater suspicion. This is easier to visualize in the color images available at the website (Freedman, 2008a). Overall, 4.1% of cases were considered more suspicious for cancer and 1.1% less suspicious. This is similar to the changes seen by Kobayashi (Kobayashi, 1996) where he found for the independent test, the CADe output affected 6.3 cases positively and 2.5 negatively; for the sequential reading test, it was 2.9 positively and 0.1 negatively.

To confirm that these changes from CADe are related to the CADe prompts, one can compare the changes that occur with machine false negatives and machine true positives. Figure 21.11 demonstrates the subtraction images showing the changes in confidence ratings that occurred with CADe, but in this case, we have separated the cases into the effects of machine false negatives and machine true positives: the left chart shows those changes that occurred when the CADe failed to detect the cancers. In this case, the pattern shown likely represents mainly the pattern of intra-observer variability, but there may be some decrease in confidence levels occurring from the failure of the machine to confirm the location of a cancer already detected by the radiologist without CADe. The right chart demonstrates the effect on confidence ratings when the machine correctly identified the location of the cancers. In this case, there are many more changes towards greater suspicion than towards less suspicion. This is easier to visualize in the color images available at the website (Freedman, 2008a).

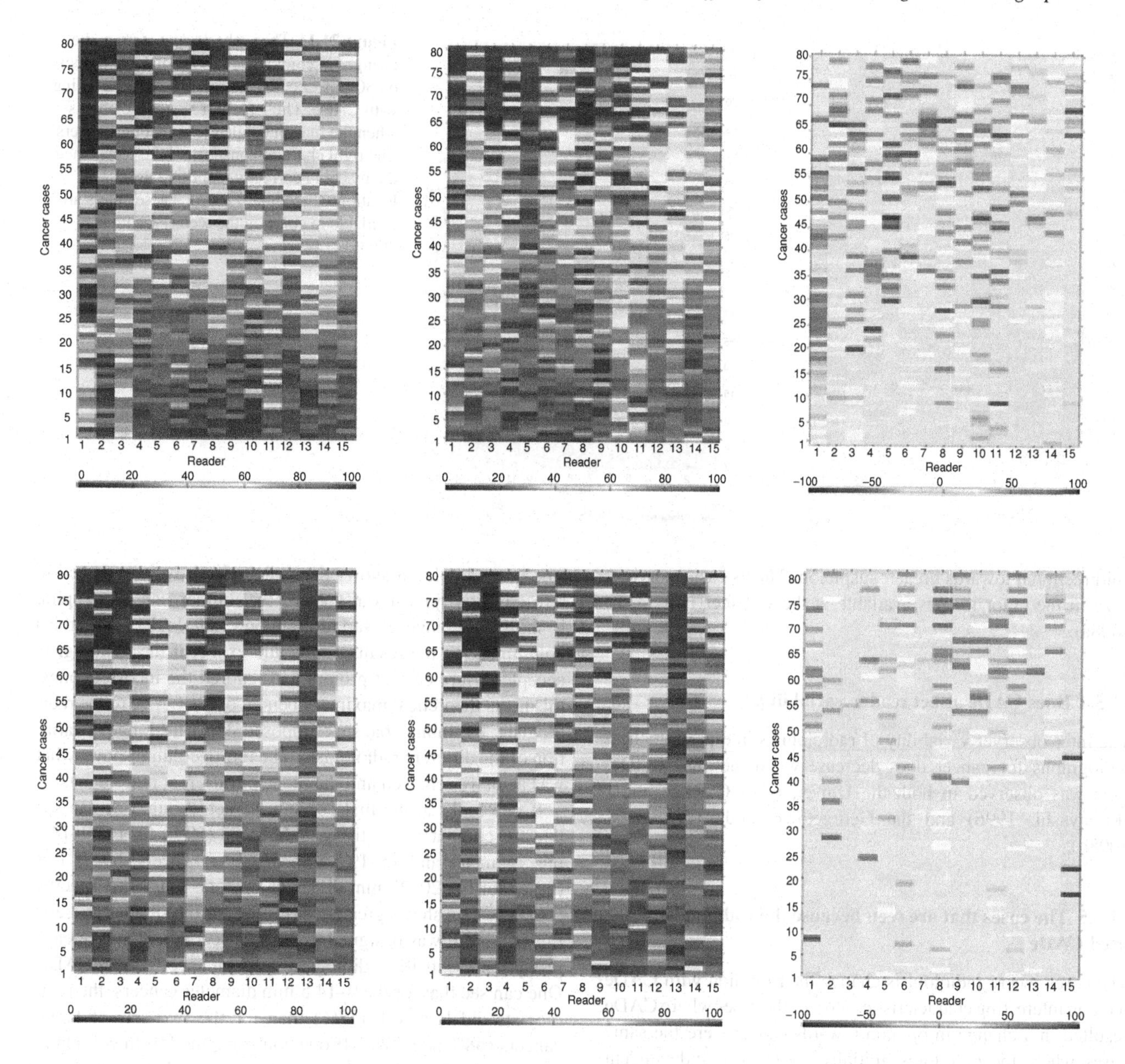

Figure 21.9 (top) and **Figure 21.10** (bottom) These charts contrast the effect of intra-observer variability for 15 radiologists compared to the effect of CADe for these 15 radiologists. The top charts show the confidence ratings for the two interpretations done without CADe. The y-axis represents the 80 cases of cancer. The x-axis arrays the 15 radiologists. The top left chart is the first read without CADe (read 1); the top center chart is the second read without CADe (read 2a). The top right chart is the subtraction showing the changes in confidence ratings between the two readings. The bottom charts demonstrate the effect of CADe. The bottom left chart is of the confidence ratings without CADe (read 2a). The bottom center chart shows the confidence ratings with CADe (read 2b). The bottom right chart shows the subtraction to show the change in confidence ratings when the radiologists used CADe (Freedman, 2006). These charts are available in color on the web (Freedman, 2008a).

Most of the cases radiologists interpret do not have cancer, so it is also important to determine the effect of CADe on cancer-free cases. Many older individuals, those at greater risk for lung cancer, have developed over the years small scars, often from infections. These scars can resemble cancer. For this reason, a moderate number of chest radiographs in smokers, and fewer in non-smokers, show findings that can be confused with cancer (Oken, 2005). In one recent study, findings suspicious for cancer were seen in 7.4% of those aged 55–59. This progressively increased to 11.6% of those aged 70–74. Never-smokers showed suspicious findings in 8.0%. Current and former smokers showed suspicious findings in up to 11.6%, depending on smoking status and smoking history (Oken, 2005).

When radiologists interpret chest radiographs, therefore, it is not surprising that they find some areas that are of concern for cancer. Figures 21.12 and 21.13 display the confidence ratings of 15 radiologists in 160 cancer-free cases. The top set of charts shows the effect of intra-observer variability of 15 radiologists; referring to Figure 21.4, comparing reads 1 and 2a. The lower set of charts shows the effect of CADe; referring to Figure 21.4, comparing reads 2a and 2b. As was seen in the cancer cases, the changes occurring from two separate interpretations without CADe are bidirectional; those with CADe are more

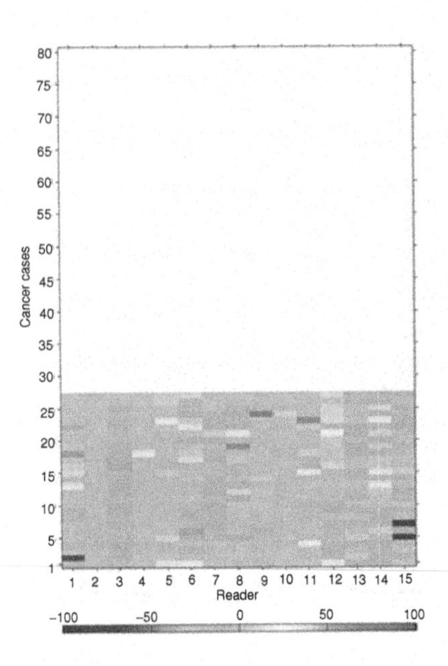

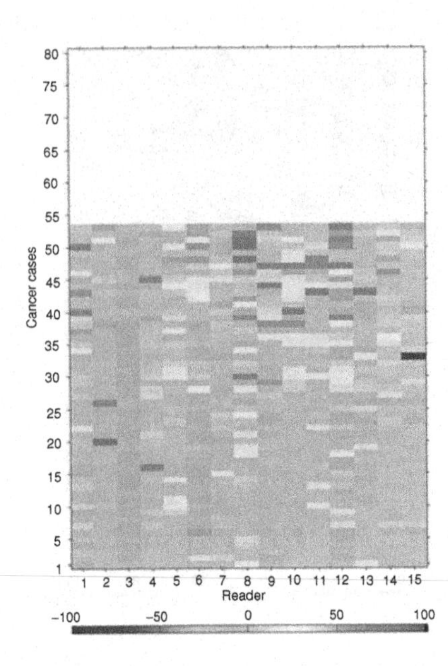

Figure 21.11 These charts demonstrate the changes in confidence ratings of 15 radiologists on 80 cancer cases that occurred from reading with CADe. The left chart shows the effects when the machine failed to detect the cancers. The right chart demonstrates the effects when the machine correctly identified the cancer locations (Freedman, 2006). These charts are available in color on the web (Freedman, 2008a).

unidirectional towards greater suspicion. This is easier to visualize in the color images available at the website (Freedman, 2008a).

21.5.4 Does CADe affect reader variability?

The inter-observer variability of radiologists interpreting chest radiographs for cancer does decrease when they use CADe. This was observed in both the University of Chicago study (Kobayashi, 1996) and the Georgetown study (Freedman, 2008b).

21.5.5 The cases that are seen because the radiologist used CADe

The cases where radiologists change their opinion with CADe have an interesting characteristic. Most of the cases where CADe resulted in a change in opinion towards cancer were the same cases where the radiologist initially, on the first independent reading (Figure 21.4, read 1), called the case positive, but then switched on the second read without CADe (Figure 21.4, read 2a) to call it negative. In effect, CADe helped the radiologists to see again the lesions that they had seen once, but then overlooked or dismissed as not suspicious on the second interpretation done without CADe.

21.5.6 CADe provides greater aid to radiologists on cases that are smaller and harder to detect

CADe is designed to help the radiologist find cancers they might otherwise overlook. Thus, its benefit will be greatest in those cases where it is more likely that cancer will be missed. While small size is only one factor in this, the Georgetown study showed that radiologists showed greater improvement in detection rate with the smaller cancers. Figure 21.14 shows the results of the CADe system divided for cancers of different sizes as shown in four groups of columns. In each group of columns, the left column indicates the results of the computer program. The second column shows the radiologists' average detection rate (sensitivity) without CADe; the thinner bar overlapping it shows the range of variation among radiologists. The third column shows the results one would expect if the radiologists accepted every correct prompt and ignored the incorrect ones; this is the theoretical maximum benefit of the CADe to the radiologists. The thinner bar superimposed shows the range of variation of the different radiologists. The fourth column shows what was actually achieved and the range of different radiologists.

If one looks at the first column in each group, one can see that the computer sensitivity is the same (68%) for the 9–14.5 mm diameter and 15–19.5 mm diameter cancers, but is less (59%) for the 20–27 mm diameter cancers. Thus, this specific CADe system shows greater sensitivity for the smaller cancers; this is what it was designed to do. The second column in each group represents the radiologists' performance without CADe. One can see that for the 9–14.5 mm diameter cancers, the radiologists did less well (58%) than for the two sets of larger cancers (68% and 78%). If one compares the fourth column to the second column in each group, one can see that the degree of improvement with CADe shows the greatest increase with the smallest nodules, with the degree of benefit from CADe decreasing with the larger nodules. As the radiologists detected more of the cancers without CADe, the CADe provided less benefit to them.

If one compares the third column to the fourth column in each group, one can see that the theoretical best degree of improvement is always greater than the actual improvement. This is discussed in the next section in this chapter.

21.6 COMPARING THE RESULTS: POTENTIAL IMPROVEMENT VS. ACTUAL IMPROVEMENT

One of the interesting perceptual issues in CADe for chest radiographs is that radiologists do not always accept the correct CADe prompts. This is graphically displayed in Figure 21.14. The

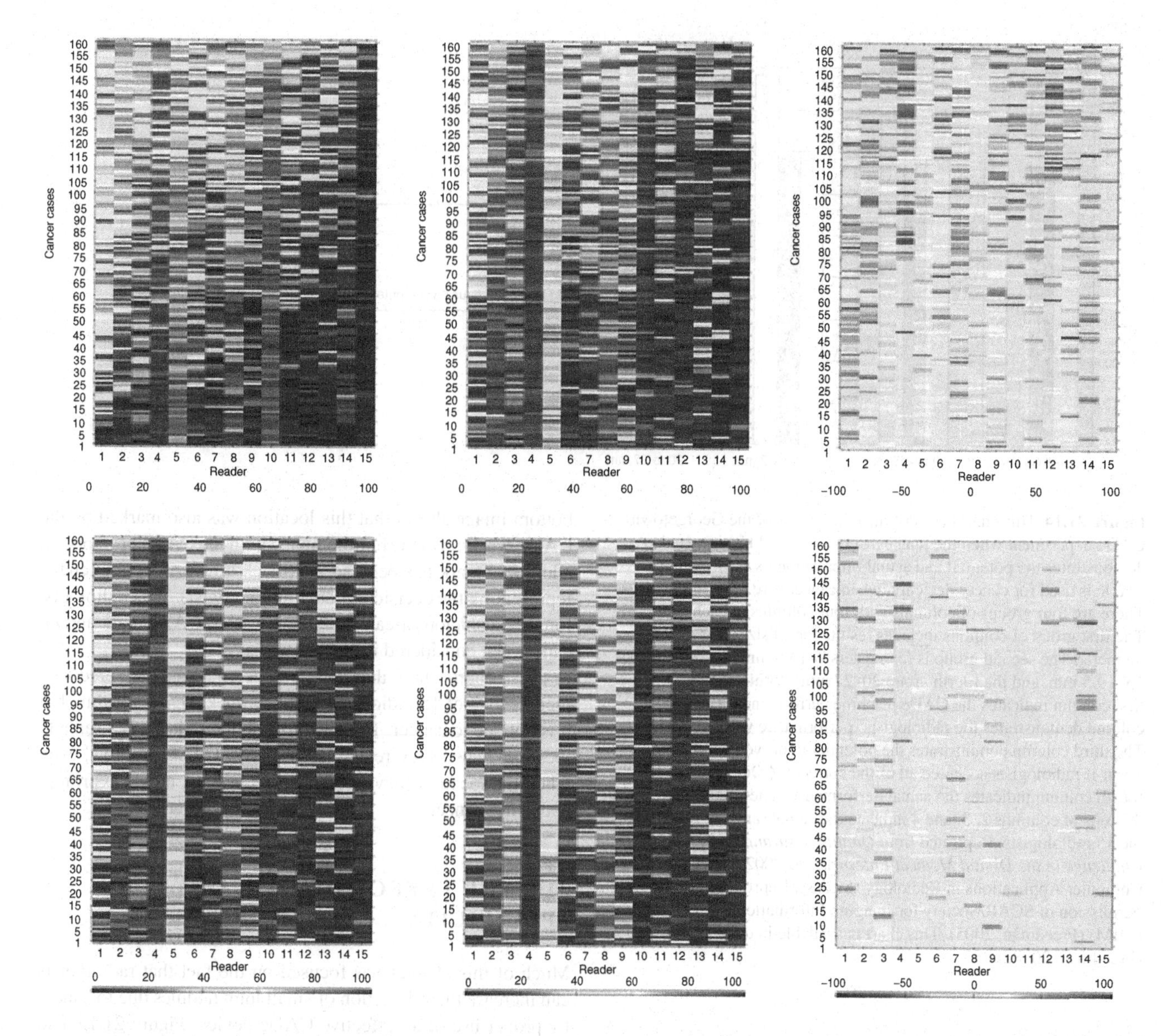

Figure 21.12 (top) and **Figure 21.13** (bottom) These charts contrast the effect of intra-observer variability for 15 radiologists with the effect of CADe for these 15 radiologists. The top charts show the confidence ratings for the two interpretations done without CADe. The y-axis represents the 160 cases without cancer. The x-axis represents the 15 radiologists. The top left chart is the first read without CADe; the top center chart is the second read without CADe. The top right chart is the subtraction showing the changes in confidence ratings between the two readings. The bottom charts demonstrate the effect of CADe. The bottom left chart is of the confidence ratings without CADe. The bottom center chart shows the confidence ratings with CADe. The bottom right chart shows the subtraction to show the change in confidence ratings when the radiologists used CADe (Freedman, 2006). (Reader variability charts are from Freedman, 2008b.) These charts are available in color on the web (Freedman, 2008a).

second column shows the results for the radiologists without CADe. The third column presents the results that could be achieved if the radiologists accepted all of the correct prompts. The fourth column shows what the radiologists actually achieved. When the machine is correct, the radiologists do not always accept this result. The question is "why?"

The answer to this question is uncertain. It involves several factors. For CADe to be effective, the radiologist must (1) see what the CADe is pointing out, (2) consider the finding to be suspicious for cancer, (3) consider this finding to be the most suspicious finding on the chest radiograph, (4) not be misled by the false positives that the CADe produces, and (5) overcome the reluctance that experts have to changing their opinion. This may be why most of the cancers detected by the radiologists when they used CADe were those that they had detected previously (Figure 21.4, read 1), but did not consider suspicious on the second read done without CADe (Figure 21.4, read 2a). Thus, most of the cancers "newly" detected with CADe were those that the radiologist did see before and, therefore, recognized again when they were pointed out by the CADe prompt.

One of the more effective methods of improving the radiologists' results when they read with CADe would be to discover how to persuade the radiologists to accept more of the correct CADe prompts, while rejecting the incorrect ones.

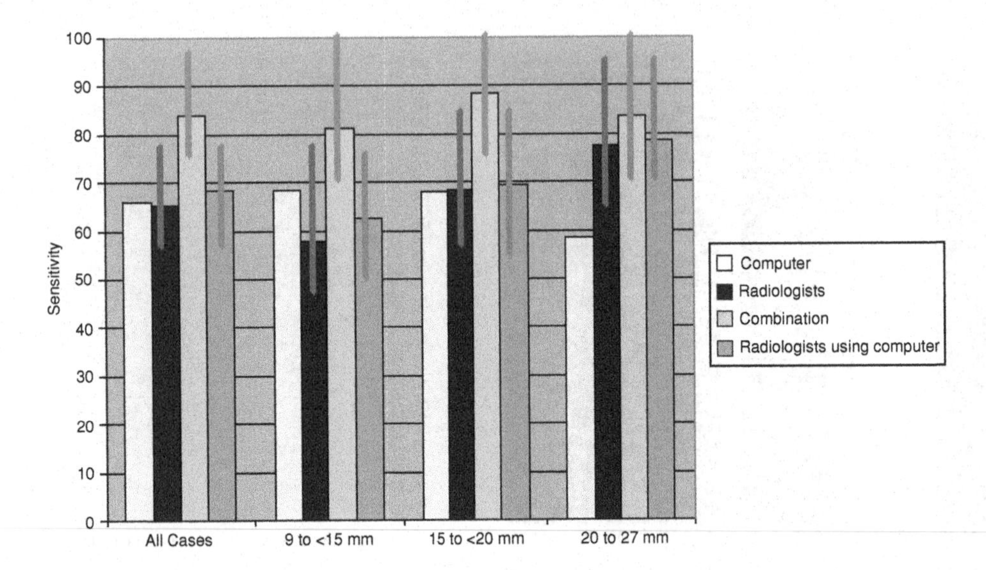

Figure 21.14 This chart demonstrates the results of the Georgetown CADe experiment when the lung cancers are sorted by size. It also demonstrates the potential and actual improvements shown when CADe is used for cancer detection when cancers are of different sizes. There are four groups of columns with four columns in each group. The first group of columns includes results for all sizes 9–27 mm in diameter. The second group is for cancers 9–14.5 mm, the third for 15–19.5 mm, and the fourth group 20–27 mm. Within each group, the first column indicates the CADe machine performance. The second column demonstrates the radiologists' performance without CADe. The third column demonstrates the potential improvement that would occur if radiologists accepted all of the correct CADe prompts. The fourth column indicates the actual performance. The smaller bars on the tops of columns 2, 3, and 4 indicate the actual range of results of the 15 radiologists. Reprinted from *Quality Assurance: Meeting the Challenge in the Digital Medical Enterprise*, © 2002 Society for Computer Applications in Radiology (SCAR). Reproduced with permission of SCAR/Society for Imaging Informatics in Medicine (SIIM) (Freedman, 2005). This chart is available in color on the web (Freedman, 2008a).

21.7 DO RADIOLOGISTS SEE THINGS THAT AREN'T THERE?

Radiologists looking at chest radiographs for evidence of cancer can be confronted with findings that are indeterminate. In those who smoke or have smoked, there are often scars in the lung that could be, by their appearance, small cancers. It should not be surprising that radiologists may incorrectly consider a small scar as a cancer. If many radiologists are looking at the same radiographs, more than one may identify the same scar as suspicious. This finding is displayed in Figure 21.15. In this chart, we record the locations where radiologists have marked something as being suspicious for cancer. There were 15 radiologists in the experiment. Each of these charts contains marks indicating the locations that up to 15 radiologists marked. In Figure 21.15, we have mapped the locations marked by the radiologists on a case that was cancer-free. At the top, the responses without CADe are shown. At the bottom, the responses with CADe are shown. The CADe marks are shown as black hexagons. On the lower left portion of the top chart, we can see that 7 of the 15 radiologists marked the same location as suspicious (left black arrow); the bottom image shows that this location was also marked by the CADe system. On the bottom image with the CADe prompts, an additional mark has been added (black arrow on bottom chart) representing the decision of an 8th radiologist). This radiologist found a suspicious area where there was no CADe mark and no cancer, but considered it suspicious.

These charts show that there can be an area considered suspicious by 7 of 15 radiologists and by the CADe algorithm, but it is not due to cancer. This also shows that radiologists can add locations when they review a case with CADe prompts, even when the location they choose does not have a CADe prompt and is not cancer.

21.8 THE EFFECT OF CADe ON SPECIFICITY

Much of this chapter has focused on the fact that radiologists can increase their detection of small lung nodules due to cancer by proper use of an effective CADe device. Figure 21.15 also has shown that radiologists will identify locations on cancer-free chest radiographs that they consider to be suspicious for cancer and that, with CADe, radiologists will identify new locations on cancer-free cases. When disease is considered to be present on a radiograph without that disease, the result is a decrease in specificity. One of the unwanted detrimental features of a CADe program is that it will result in the identification of locations of presumed disease where disease is not present. On chest radiographs, where scars are more common than cancers, and where even radiologists cannot tell scars from areas suspicious for cancer, it is expected that CADe will result in the identification of small benign lesions that radiologists will consider to be cancers and that a decrease in specificity will result.

In mammography CADe, statistics are often based on whether or not a lesion identified on mammography is considered actionable. In CADe for chest radiographs, the statistics have been based on whether or not a lesion is cancer. This results in two effects: (1) CADe for chest radiographs will have a more detrimental effect by decreasing specificity because small scars that radiologists consider to be actionable will be recommended for workup and will be considered false positives. In studies

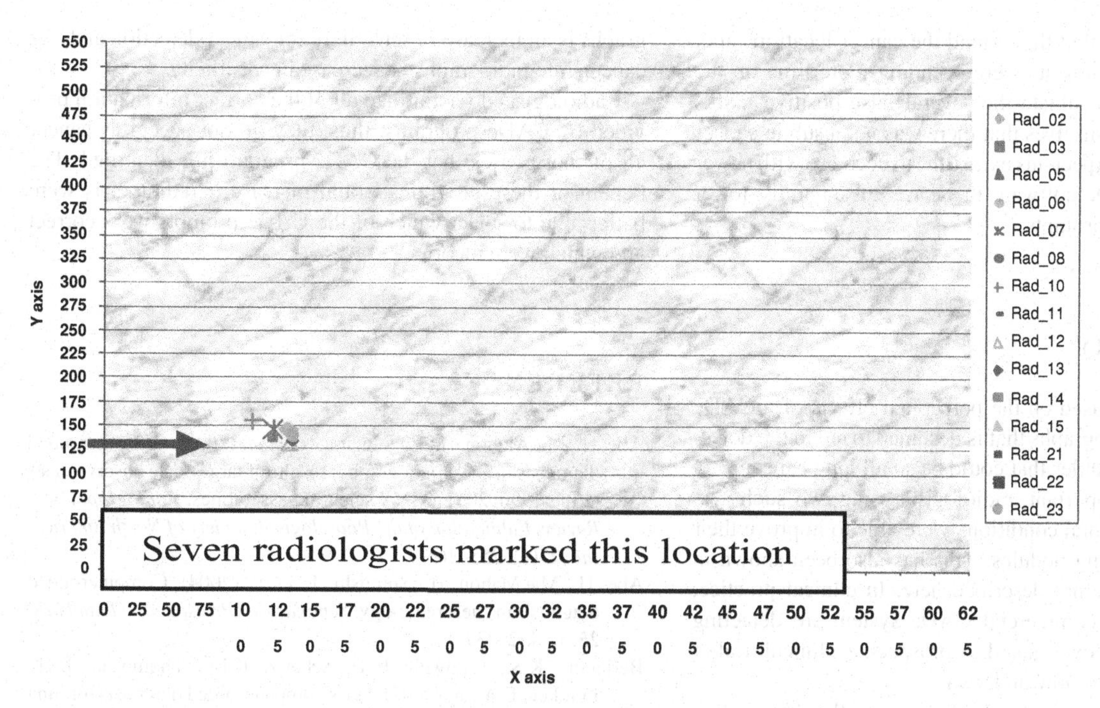

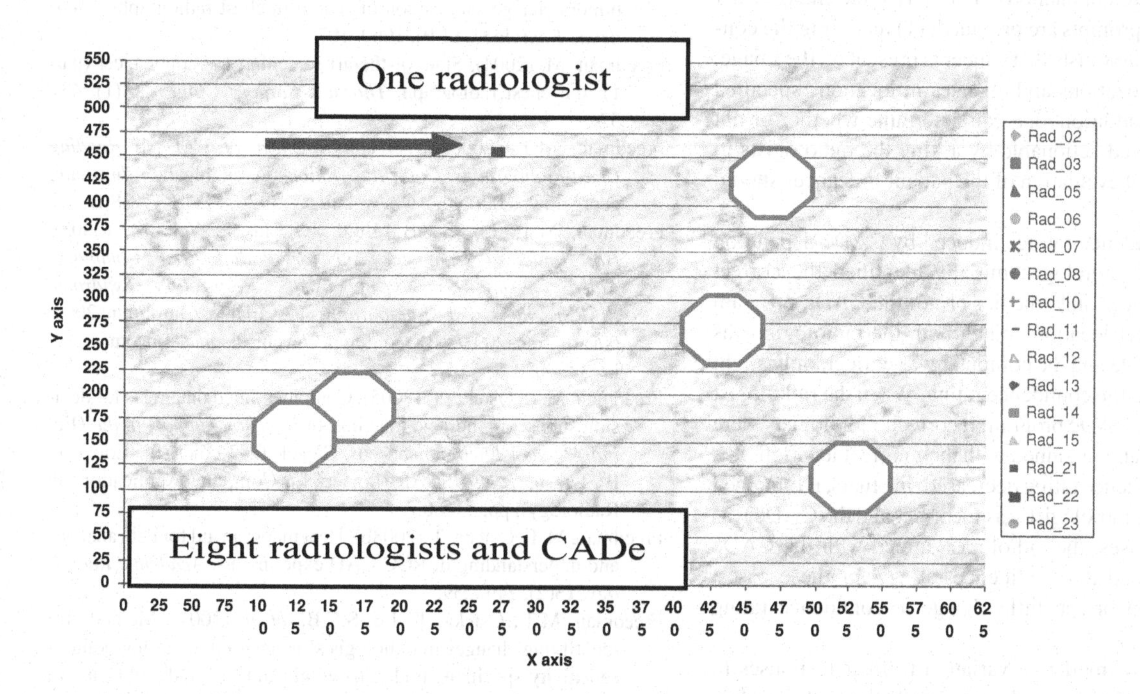

Figure 21.15 This chart demonstrates the locations marked by radiologists in a case without cancer. The top chart shows that 7 of 15 radiologists identified the same location (top chart black arrow). The bottom chart shows the results with CADe. The CADe algorithm marked the same location selected by the 7 radiologists. In addition, a new location was marked (bottom chart black arrow) by an additional radiologist. No CADe prompt is at this location. These charts are available in color on the web (Freedman, 2008a).

of mammography CADe, such results, if the expert panel has accepted them, will be considered true positive findings. CADe for chest radiographs is using a different standard of ground truth.

21.9 THE INCREASE IN TIME REQUIRED BY RADIOLOGISTS WHEN INTERPRETING CHEST RADIOGRAPHS WITH CADe

The radiologists in the first study (Kobayashi, 1996) did not show an increase in the amount of time required to read a case when they used CADe. In the second study, a small increase (an average of 8 seconds) was seen with CADe (Osicka, 2003). The measurements showed that the radiologists, when they knew the CADe results would be presented, used less time, on average, prior to seeing the CADe prompts. Reading without CADe, in the independent test, the radiologists averaged 31 seconds per case; reading in the sequential mode, first without and then with CADe, they averaged 39 seconds per case. The time used for reading with CADe differed for those cases with cancer and those cases without cancer. For those cases with cancer, reading time was, on average, 23.2 seconds without CADe and 22.7 seconds with CADe, not significantly different. For the cancer-free cases, the reading time was 31.3 seconds without CADe and 53.8 seconds with CADe, substantially higher. This suggests, first, that the radiologists spent less time in total when, on cancer cases, they were satisfied they found the cancer locations and, second, that the radiologists were spending more time on the cancer-free cases to evaluate the several false positive marks in order to assure themselves that there was not a subtle cancer present. One would expect this extra time to decrease with newer versions of the CADe software that have substantially fewer false positive CADe prompts.

21.10 SUMMARY

This chapter has focused on the perceptual effects of a CADe system for chest radiographs that is designed to aid in the detection of small lung nodules that could be small lung cancers.

First, and most important, radiologists using two such systems under experimental conditions were able to improve their detection of small lung nodules. This has also been shown in other studies that are not described here. In clinical practice, radiologists using a commercial CADe system are detecting cancers that they had overlooked. A prospective clinical trial of the degree of benefit is now underway.

There are two different methods of observing the changes that occur when CADe prompts are provided: (1) recording the confidence levels of radiologists that cancer is present on the image, and (2) recording locations and changes in locations specified by radiologists. In addition, one can determine whether or not a case was considered actionable by asking the radiologists to indicate if additional evaluation of the patient for cancer should be recommended.

Changes in confidence ratings induced by CADe depend on whether or not the radiologist initially identified the correct location for the cancer and whether or not the CADe program identified the correct location. When both the radiologist was correct without CADe and the computer was correct, only slight changes occurred in the confidence rating. When the radiologist was correct and the CADe program incorrect, radiologists were resistant to change and remained with their correct location. In a few cases, the confidence rating decreased; in others, it remained the same. When the radiologist was incorrect and the CADe was correct, in a few cases, the radiologist agreed with the CADe and greatly increased the confidence ratings for these cases. When both were incorrect, little change in confidence rating occurred.

Radiologists show moderate variation in their responses to subtle radiographs. They can disagree with themselves when they re-read radiographs, a process called intra-observer variability; they can disagree with their colleagues when they have also read the same radiographs, a process called inter-observer variability. The effect of intra- and inter-observer variability is bidirectional, meaning that some benign cases are switched to the cancer category and some cancer cases are switched to benign. In the Georgetown study, this occurred 9–10% of the time.

The effect of CADe, when compared to intra-observer variability, is mainly unidirectional towards greater levels of suspicion (higher confidence of the presence of cancer). This occurs with both cancer-containing radiographs and those without cancer. The effect demonstrated on the cancer-containing radiographs is an increase in sensitivity for cancer detection and, for cancer-free radiographs, a decrease in specificity.

Radiologists do not utilize all of the correct information provided by CADe programs; thus, they do not accept as cancer all of those cases that the CADe program has identified. The benefit of these programs would increase if radiologists were better able to select which of the CADe prompts were correct and which were incorrect.

REFERENCES

Abe, H., MacMahon, H., Engelmann, R., *et al.* (2003). Computer-aided diagnosis in chest radiography: results of large-scale observer tests at the 1996–2001 RSNA scientific assemblies. *Radiographics: a Review Publication of the Radiological Society of North America, Inc*, **23**(1), 255–265.

Abe, H., MacMahon, H., Shiraishi, J., *et al.* (2004). Computer-aided diagnosis in chest radiology. *Seminars in Ultrasound, CT, and MR*, **25**(5), 432–437.

Berbaum, K.S., Caldwell, R.T., Schartz, K.M., Thompson, B.H., Franken, E.A., Jr. (2007). Does computer-aided diagnosis for lung tumors change satisfaction of search in chest radiography? *Academic Radiology*, **14**(9), 1069–1076.

Freedman, M. (2004). State-of-the-art screening for lung cancer (part 1): the chest radiograph. *Thoracic Surgery Clinics*, **14**(1), 43–52.

Freedman, M.T. (2008a). *Perceptual Effect of CAD in Reading Chest Radiographs: Color Illustrations*. Available: http://gushare.georgetown.edu/xythoswfs/webui/_xy-5976759_1-t_y78sSiB9

Freedman, M.T., Lo, S.C.B., Lure, F., Lin, J., Yeh, M. (2003). *Hot Topic: a Computer Aid for Radiologists: Computer-Aided Detection of Severe Acute Respiratory Syndrome (SARS) on Chest Radiography*. Available: http://rsna2003.rsna.org/rsna2003/VBK/conference/event_display.cfm?em_id=3800022

Freedman, M.T., Osicka, T. (2005). Computer aided diagnosis for decision support in thoracic imaging. In *Decision Support in the Digital Medical Environment*, eds. Siegel, E., Reiner, B., Erickson, B., Leesburg, VA: SCAR (Society for Computer Applications in Radiology), pp. 53–68.

Freedman, M.T., Osicka, T. (2008b). Heat maps: an aid for data analysis and understanding of ROC CAD experiments. *Academic Radiology*, **15**(2), 249–259.

Freedman, M.T., Osicka, T., Lo, S.C.B., *et al.* (2001). Methods for identifying changes in radiologists' behavioral operating point of sensitivity-specificity trade-offs within an ROC study of the use of computer aided detection of lung cancer. *SPIE Medical Imaging: Image Processing and Performance*, **4324**, 184–194.

Freedman, M.T., Osicka, T., Lo, S.C.B., *et al.* (2002). Computer aided detection of lung cancer on chest radiographs: effect of machine CAD false positive locations on radiologists' behavior. *Proceedings of SPIE: Image Processing*, **4684**, 1311–1319.

Freedman, M.T., Osicka, T., Zhu, Y., *et al.* (2006). Heat maps as a data visualization and exploration method for multireader-multicase ROC computer-aided detection experiments. *International Journal of Computer Assisted Radiology and Surgery*, **1**(Supp. 1), 348–350.

Giger, M.L., Ahn, N., Doi, K., MacMahon, H., Metz, C.E. (1990). Computerized detection of pulmonary nodules in digital chest images: use of morphological filters in reducing false-positive detections. *Medical Physics*, **17**, 861–865.

Giger, M.L., Doi, K., MacMahon, H. (1988). Image feature analysis and computer-aided diagnosis in digital radiography. 3. Automated detection of nodules in peripheral lung field. *Medical Physics*, **15**, 158–166.

Heelan, R.T., Flehinger, B.J., Melamed, M.R., *et al.* (1984). Non-small-cell lung cancer: results of the New York screening program. *Radiology*, **151**(2), 289–293.

Ishida, T., Katsuragawa, S., Kobayashi, T., MacMahon, H., Doi, K. (1997). Computerized analysis of interstitial disease in chest radiographs: improvement of geometric-pattern feature analysis. *Medical Physics*, **24**(6), 915–924.

Katsuragawa, S., Doi, K., MacMahon, H., *et al.* (1997). Classification of normal and abnormal lungs with interstitial diseases by rule-based method and artificial neural networks. *Journal of Digital Imaging: the Official Journal of the Society for Computer Applications in Radiology*, **10**(3), 108–114.

Kobayashi, T., Xu, X.W., MacMahon, H., Metz, C.E., Doi, K. (1996). Effect of a computer-aided diagnosis scheme on radiologists' performance in detection of lung nodules on radiographs. *Radiology*, **199**(3), 843–848.

Li, F., Engelmann, R., Metz, C.E., Doi, K., MacMahon, H. (2008). Lung cancers missed on chest radiographs: results obtained with a commercial computer-aided detection program. *Radiology*, **246**(1), 273–280.

Lin, J.S., Lo, S.B., Hasegawa, A., Freedman, M.T., Mun, S.K. (1996). Reduction of false positives in lung nodule detection using a two-level neural classification. *IEEE Transactions on Medical Imaging*, **15**(2), 206–217.

Lo, S.B., Lou, S.A., Lin, J.S., *et al.* (1995). Artificial convolution neural network techniques and applications for lung nodule detection. *IEEE Transactions on Medical Imaging*, **14**(4), 711–718.

Muhm, J.R., Miller, W.E., Fontana, R.S., Sanderson, D.R., Uhlenhopp, M.A. (1983). Lung cancer detected during a screening program using four-month chest radiographs. *Radiology*, **148**(3), 609–615.

Oken, M.M., Marcus, P.M., Hu, P., *et al.* (2005). Baseline chest radiograph for lung cancer detection in the randomized prostate, lung, colorectal and ovarian cancer screening trial. *Journal of the National Cancer Institute*, **97**(24), 1832–1839.

Osicka, T., Freedman, M.T., Lo, S.C.B., *et al.* (2003). Computer aided detection of lung cancer on chest radiographs: differences in the interpretation time of radiologists showing vs. not showing improvement with CAD. *Proceedings of SPIE: Image Perception, Observer Performance, and Technology Assessment*, **5034**, 483–494.

Shiraishi, J., Abe, H., Li, F., *et al.* (2006). Computer-aided diagnosis for the detection and classification of lung cancers on chest radiographs: ROC analysis of radiologists' performance. *Academic Radiology*, **13**(8), 995–1003.

Shiraishi, J., Katsuragawa, S., Ikezoe, J., *et al.* (2000). Development of a digital image database for chest radiographs with and without a lung nodule: receiver operating characteristic analysis of radiologists' detection of pulmonary nodules. *AJR American Journal of Roentgenology*, **174**(1), 71–74.

Siegle, R.L., Baram, E.M., Reuter, S.R., *et al.* (1998). Rates of disagreement in imaging interpretation in a group of community hospitals. *Academic Radiology*, **5**(3), 148–154.

Stitik, F.P., Tockman, M.S. (1978). Radiographic screening in the early detection of lung cancer. *Radiologic Clinics of North America*, **16**(3), 347–366.

Stitik, F., Tockman, M., Khouri, N. (1985). Chest radiology. In *Screening for Cancer*, ed. Miller, A.B., New York, NY: Academic Press, pp. 163–191.

Xu, X.W., Doi, K., Kobayashi, T., MacMahon, H., Giger, M.L. (1997). Development of an improved CAD scheme for automated detection of lung nodules in digital chest images. *Medical Physics*, **24**(9), 1395–1403.

Yerushalmy, J. (1969). The statistical assessment of the variability in observer perception and description of roentgenographic pulmonary shadows. *Radiologic Clinics of North America*, **7**(3), 381–392.

Perceptual issues in mammography and CAD

MICHAEL J. ULISSEY

Mammography, as with many other fields in radiology, is imprecise. Some days we notice something on a mammogram and on another day we may fail to notice the same exact finding. If this finding is a cancer, and we failed to initially notice it, we never know why, because too much time has elapsed before the missed cancer is brought to our attention. We cannot go back six months, a year, or two years and remember what we were doing or what we were thinking when we read the original mammogram and failed to notice the footprint of breast cancer. We can suppose, we can theorize, we can try to rationalize why we missed something now deemed important, but we cannot put ourselves back in the same boat to know why we really missed a breast cancer that may seem obvious in retrospect. In 1998 computer aided detection (CAD) was advanced to try to help solve this problem, and over the years some studies have shown it to be beneficial and some have had quite the contrary view. So is CAD an eagle or an albatross?

This chapter will discuss perceptual issues in mammography and review the role CAD can play in an attempt to answer that very question. Please keep in mind that much of the information presented here is the opinion of the author based on his research, experience, and training over the years. This includes research in CAD from both prospective and retrospective standpoints, experience in reading high-volume mammography for ten years, and the most unfortunate but greatest training of all – seeing most of his own misses. In addition, this chapter is written in a more relaxed fashion – intentionally a deviation from normal scholarly formats. I do this so that I can better explain, in clear and easily understandable terms, the facts and figures that have led to my opinions. I will, however, cite scholarly contributions from published literature to tie it all together.

For many years I was in the private practice of breast imaging with my former partner. We had what were essentially solo breast imaging practices in the Dallas Ft. Worth area, where we were reading upward of 20,000 mammograms a year each. One year I read over 24,000 mammograms. As my former partner once said to me, the brutal thing about this type of practice is not the volume of work; it's that you see all your own misses. This woman has been coming to you for years. When you see cancer on a mammogram and you look at the prior study, you don't need to pull the chart to see if your name was on last year's report. It was. To add pressure to the situation, with all our diagnostic examinations we talk to our patients and review their studies with them. To pose a rhetorical question: how would you talk to a patient and how would you dictate a report if you saw a missed cancer on a mammogram and that miss was yours, versus how would you talk to a patient and explain her prior studies if you saw a missed cancer on a mammogram and the miss was your partner's, versus how would you talk to a patient and dictate a report if you saw a missed cancer on a mammogram and the miss was from your competitor's shop down the street – and you don't like him much. It is sometimes difficult for us to face the reality that perceptual issues in mammography affect us on a routine basis.

Let us look at several reasons why this might be so. First, mammograms are black and white images that we often read in batch mode in a screening setting. How often do we stare at black and white images and after a while our eyes begin to see spots? (See Figure 22.1.) Mammograms are black and white images with swirling patterns. How often do we stare at a series of black and white images with swirling patterns and after a while our eyes go bonkers? (See Figure 22.2.) So one issue with respect to perception in mammography is eye fatigue with repetitive similar tasks over time.

Another issue is what I call the vagaries of Monday versus Friday. Look at Figure 22.3 and decide if the first thing you saw was a cocktail glass or two face profiles looking at each other. If it was Monday you might first notice a cocktail glass (white center), but if it were Friday you might first notice two faces looking at each other (black background). It is similar with mammography. On one day a particular finding catches your eye and you take a closer look at it, but on another that same finding, for whatever reason, simply does not catch your eye and you fail to perceive it. We never know why or why not. It is simply one of the vagaries of mammography.

Another issue with respect to perception on a mammogram is boredom with repetitive tasks. We often read batches of 50, 80, 100, or more mammograms at a time, and after a while we lose focus, we lose concentration, we begin to phase out or think of other things. We begin to think about the oil that needs to be changed in our car, or we wonder if we left the garage door open. We think we are actively reading mammograms but we are passively reading. We can't even remember the density of the breast on the last mammogram we read. Sometimes, for whatever unknown reason, we phase out on a series of a few mammograms and fail to notice something important – and we don't find out about this for six months, a year, or longer.

A fourth issue regarding perception is tunnel vision. Some researchers have called it satisfaction of search (Majid, 2003). Look at Figure 22.4. It is easy to notice the large cancer in the right breast and want to jump on it with both feet. It was, after all, a palpable lump and you were seeing the patient for a diagnostic evaluation. You evaluate the patient and eventually feel proud of yourself for getting the workup done quickly and

The Handbook of Medical Image Perception and Techniques, ed. Ehsan Samei and Elizabeth Krupinski. Published by Cambridge University Press.

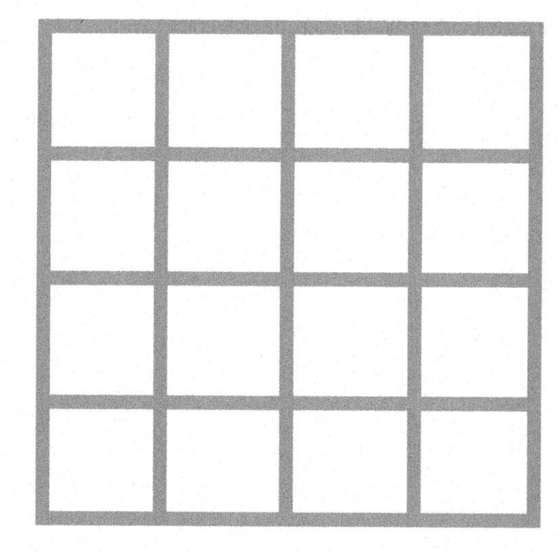

Figures 22.1 and **22.2** Both figures illustrate eye fatigue while reading panels of black and white mammograms.

Figure 22.3 On some days you might notice a cocktail glass in the white center and on other days, two faces looking at each other in the black background.

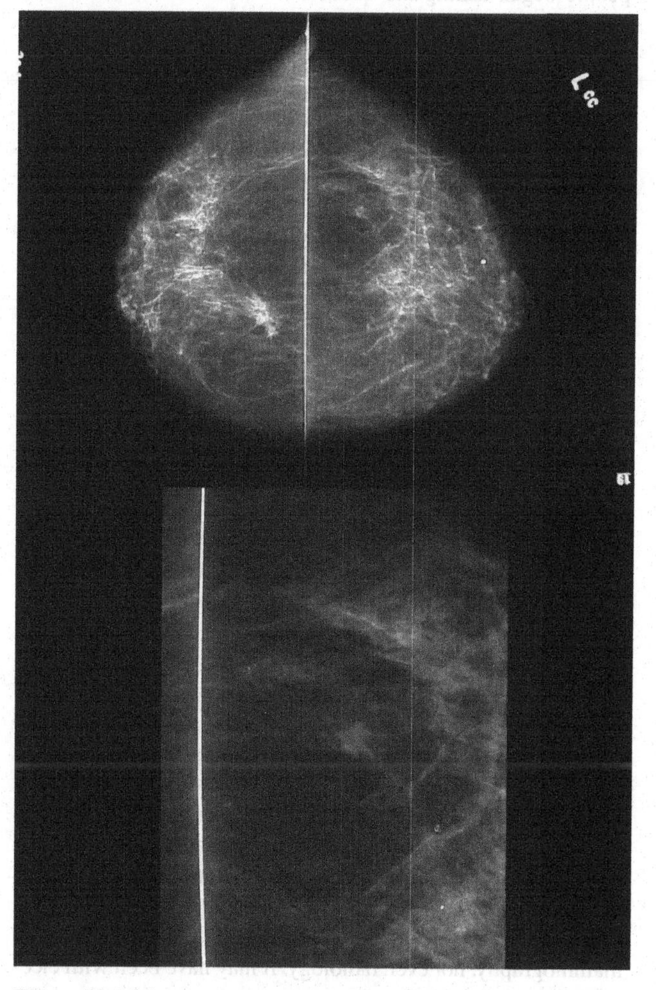

Figure 22.4 It is easy to get tunnel vision on the large right cancer and miss the small left one.

efficiently. She received an ultrasound-guided needle biopsy before the afternoon was over. But you got tunnel vision on the large cancer in the right breast and forgot that the rest of the diagnostic mammogram was also a screening mammogram. You failed to notice the small, maybe 5 mm cancer in the left breast. You got tunnel vision.

Some people have equated the hunt for breast cancer on a mammogram with the notion of *Where's Waldo* (also known as *Where's Wally*). That might be a reasonable analogy, but I think I can make it better. First, is Waldo easier to see because the pictures are in color? Would Waldo be more difficult to see if the photo were in black and white? (See Figure 22.5.) Also, I am going to give you 1000 *Where's Waldo* pictures, but Waldo won't be in every photo. He will only be in 3, 4, or 5 of the 1000. You don't know exactly how many, and you don't know if he will be in more of the early photos, middle photos, later photos, or scattered evenly among them. You have to find every instance of Waldo, and if you miss even one instance you could lose your car, your boat, your house, or your kids' college education. Now let's play *Where's Waldo*!

Figure 22.5 Finding Waldo on a mammogram.

Perhaps one can begin to understand why over a decade ago people began asking the question of whether we could train computers to help us recognize Waldo on a mammogram. Before we address this issue, however, let us first take a look at how computers have been used as detection devices in other areas of our lives that are not related to medicine or mammography.

In their attempt to reduce their costs of exploration by drilling fewer dry holes, the oil industry has used computers to analyze seismic data and guide geologists, engineers, and wildcatters in their search for black gold. Oil explorers used to look at seismic maps and try to interpret where to best drill a hole to explore for oil. They drilled a lot of dry holes, but over time found that if they fed these seismic data into computers, computers could be trained to analyze the data and point out potentially more lucrative areas to drill for oil. (See Figures 22.6–22.8.) Over time, oil companies found these computers to be useful and now they are commonly used in seismic analysis and oil exploration (and oil industry stock seems to have done quite well over time).

In their attempt to better find the missiles of October, intelligence, space, and defense agencies like the CIA, DOD, and NSA have trained computers to analyze satellite photo imagery, filter out the background noise, and home in on important aspects of intelligence information they want to analyze. (See Figures 22.9 and 22.10.) Imagine having to wade through thousands of spy satellite photos and look for the missiles of October. But if computers could be trained to do this, then maybe they could point out important areas for intelligence analysts to focus on, thus making a more efficient and maybe more accurate use of their time. These organizations have created just such computer systems and this is where the technology for CAD in mammography had its origins (Ulissey, 2001).

When CAD first came into the medical field it was not with mammography, nor even radiology. It may have been with electrocardiography (EKGs). For years people were seen in emergency departments with chest pain and the initial EKGs were

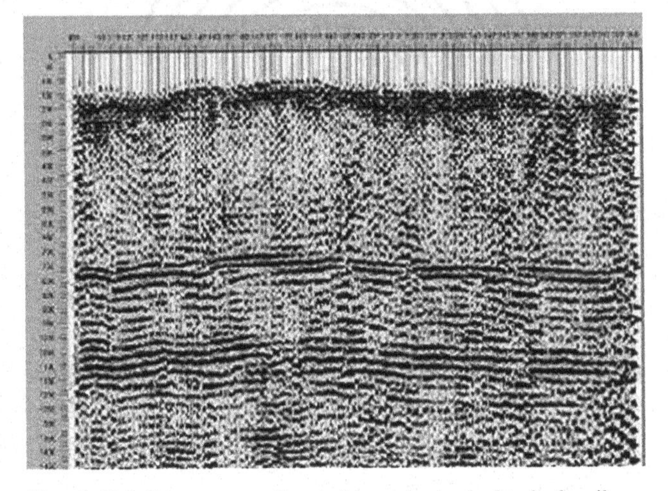

Figure 22.6 Computers can be used to read seismic data in the oil industry.

interpreted as normal. Retrospective review indicated that on some EKGs, the footprint of a heart attack was missed by the ER doc, and this led to some unfortunate patient outcomes (and lawsuits). Over time, EKG machines were outfitted with a computer pre-read to help analyze the patterns of the EKG leads. While the initial programming might have been more crude and less accurate, over time the technology improved and now it is probably difficult if not impossible to buy an EKG machine that does not give a pre-read. (See Figure 22.11.)

In the 1990s, in cooperation with Health and Human Services, the Missiles to Mammograms project was undertaken to come up with a way to use these types of technologies to benefit the breast imaging community and, ultimately, the women we serve (United States Public Health Office, 1996). In 1993 a commercial venture was undertaken to develop the first CAD device for mammography (R2 Technologies, Los Altos, CA – now Hologic), and in June of 1998 that device was

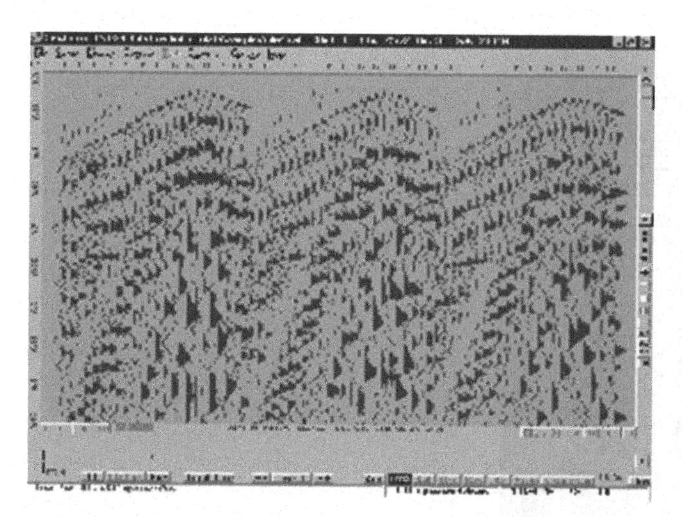

Figure 22.7 The data are fed into computer aided detection algorithms.

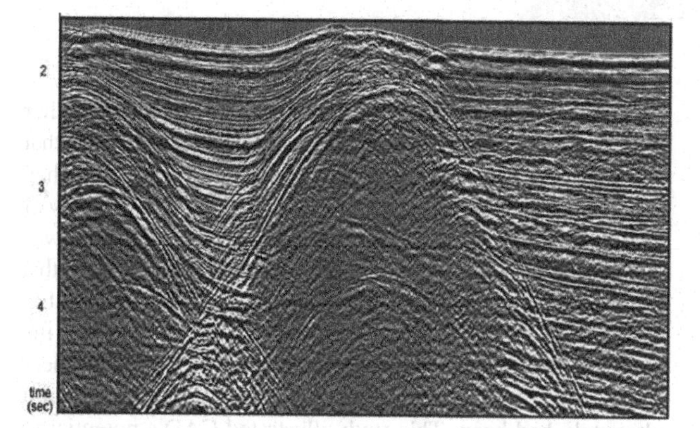

Figure 22.8 Computer aided detection in the oil industry has proven to be beneficial.

Figure 22.9 Computers can analyze satellite imagery and filter out the background noise. Declassified public domain US Government photograph. http://goes.gsfc.nasa.gov/

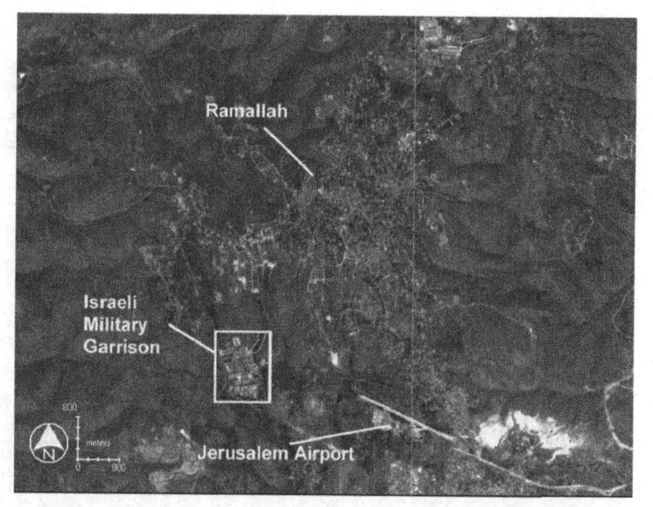

Figure 22.10 Computer aided detection for DOD and other intelligence agencies has proven beneficial. Declassified public domain US Government photograph. http://goes.gsfc.nasa.gov/

approved by the FDA. Since then, multiple similar devices have been approved and CAD for mammography is taking on an increasing role in breast centers worldwide. Over time, some controversies have developed. Some research has indicated that CAD is a useful tool in the hunt for breast cancer (Freer, 2001) and other published papers have indicated the opposite, that it may even be harmful (Fenton, 2007). So is CAD an eagle, or albatross? There has been much published discourse defending and criticizing both sides, and I do not wish to re-hash those controversies. I will submit that CAD is useful to some people and not as useful to others. I cannot believe that it is harmful if used by an honest and trained radiologist.

In order to explain my position I will begin by analyzing two CAD studies that set the foundation for the technology. The first is a retrospective study published by Linda Warren Burhenne (Warren Burhenne, 2000). Although I think this was an excellent study to demonstrate the potential benefits of CAD in mammography, I think the best part of the study is that it may be one of the most comprehensive analyses of our (the radiology community in general) false negative rate in mammography. It waved in our faces the misses we make in mammography, and proposed a way to reduce those unfortunate occurrences by using CAD.

For the study, she and her team of researchers went out and got the mammograms of over 1000 patients from a spectrum of breast centers around the USA. Dr. Warren Burhenne and her team felt that these centers represented a spectrum of the breast imaging that was being done in the USA at the time (personal communication from the lead author). Some were specialty centers, some were not. Some were academic centers, some were not. But all mammograms were read by board-certified MQSA qualified radiologists. Each mammogram was the initial study that indicated the footprint of breast cancer for that specific woman, but it was not that woman's baseline mammogram. The patient had had a prior study maybe six months or a year earlier. The researchers took those 1000-plus mammograms and said "OK, forget about them." They said, "Let's go out and get the most recent prior mammogram that this woman had – the

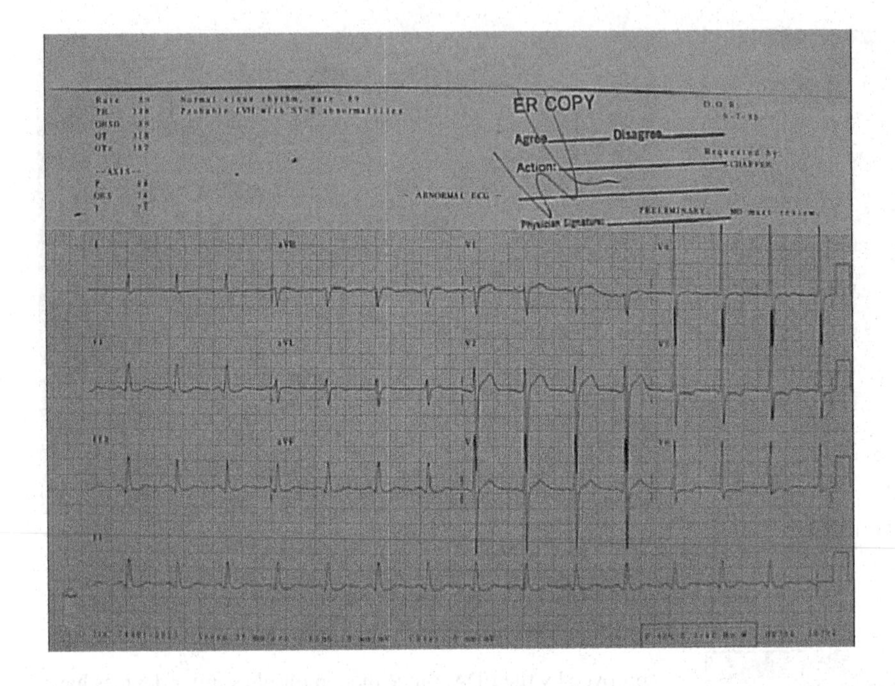

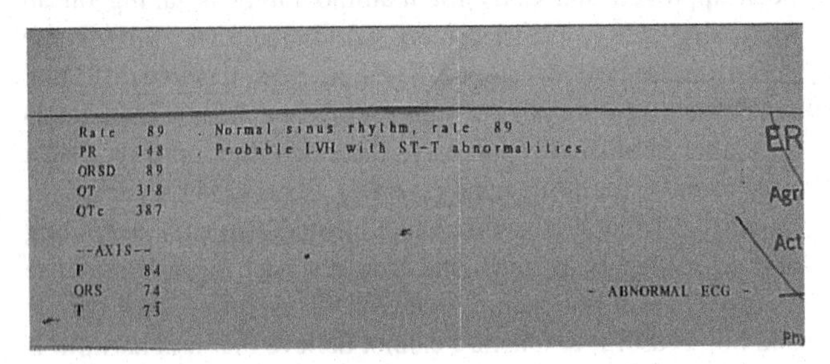

Figure 22.11 Computers analyze EKG information in an effort to assist interpretation.

one that was called *normal* just before this current mammogram showed breast cancer" (poetic license).

The researchers rounded up as many of the most recent prior studies as they could, and focused on those prior mammograms – the so-called *normal* ones. They then had those mammograms analyzed by a team of five independent radiologists to determine if the panel of five could agree as to whether it was OK on the prior film to call it normal, or whether the team would agree that the so-called normal mammogram should have been recalled for further workup.

The results of the study astonished me. Sixty-seven percent of the time (67%), the cancer was on the old film. Albeit in retrospect, it was still there (there is always something on the old film). This is why I now tell residents who train with me that the most important thing to do when you see cancer on a mammogram is not that you should look for the second cancer. No, the most important thing is to pull last year's report and see if your name was on the bottom of it.

The team of reviewers did cut us a break, though. They agreed that only 27% of the time was it a "true miss." In other words, 27% of the time (not 67%), the prior study should not have been called normal. It should have been recalled for further evaluation. This led Dr. Warren Burhenne's team to focus on that group of mammograms – the group in the 27% category. They ran those mammograms through a CAD unit to see how many of the missed cancers would be marked by CAD. The upshot was that had a diligent radiologist read the prior mammogram with a CAD machine, 21% of the time that radiologist would probably have then noticed the cancer and would not have called the mammogram normal. Twenty-one percent of the time, the cancer would probably have been picked up a year or so earlier than it ultimately had been. This study illustrated CAD's potential to help radiologists in their quest for recognizing the footprint of malignancy on a mammogram.

The second study I would like to discuss is a prospective one published by Freer and Ulissey (Freer, 2001). This study was done in a private practice breast center near Dallas (Women's Diagnostic of Texas (WDT)) and began shortly after the first commercial CAD unit for mammography became available. Our centers were essentially solo shops, so double reading was not an option. Therefore, in November of 1998, WDT bought a CAD system in hopes that it would help catch a few extra breast cancers. We then wanted to know how CAD affected us in our daily practice of breast imaging in our community center. It is important to note that we did not set out to publish a research study – rather, we wanted to see if we had made a wise investment. We had paid upwards of $200,000 for the unit and we had ongoing issues of maintenance contracts and upgrades. Our centers were growing so we were also looking at additional equipment purchases. Would it be worth such additional expenditures? Would CAD help us identify additional cancers? If it did help us, what would be the stage of any additional malignancies detected – Stage IV, so no real help for the woman, or Stage I or 0, essentially a speed bump in the woman's life instead of a major disaster. We also wanted to know how CAD would impact our positive predictive rate (PPV) for biopsy

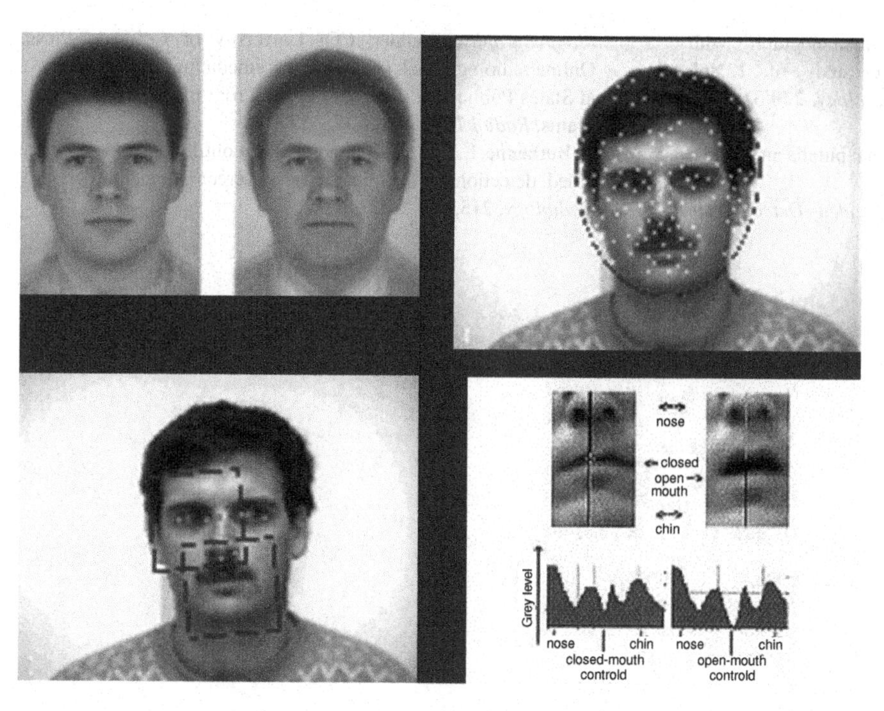

Figure 22.12 Computers can analyze facial features and match elements to databases in the hunt for all sorts of people – good and bad.

and our recall rate. Would CAD prompt us to perform many of the so-called "unnecessary biopsies"? Would CAD double our recall rate? These were questions we needed to answer to allow us to take better care of our patients – and we wanted to analyze this from a prospective standpoint, not a retrospective one.

The study lasted one year and, again, the results of it astonished me. CAD helped us pick up over 19% additional cancers and our recall rate went from only 6.5% to 7.7% – both acceptable numbers in our opinion. Our PPV for biopsy remained the same regardless of whether the cancer was noticed by us or by a CAD prompt, and the best part of the study, to me, was that all additional cancers detected were Stage 0 or Stage I. Good news for the woman.

In the years since publishing that study, I have thought often about what CAD may be doing to me, personally, with respect to perception in mammography. Over the years I have continued to keep and track data. CAD helped me more the first year and less in subsequent years, even though I read more mammograms with it. Some years it helps me only once or twice. A few years it did not help me at all. Did CAD make me a better reader? Maybe. I have always thought of myself as a diligent breast imager – I try the best I can every day with every patient, but I will take all the help I can get. Was it that over time I did not want the computer to "beat me," so I purposely tried to play it at its own game? Maybe. Was there a "first round effect" with CAD – an incidence round versus a prevalence round? Did CAD in the first year of use filter out a set of cancers and the natural effect in subsequent years was for it to "catch fewer ones"? Maybe. That's what I like to believe, rather than my first hypothesis. Maybe CAD has a combined effect – it helps for many different reasons. That is probably the most likely scenario. In any case, I always feel more comfortable reading with CAD rather than without – and I have heard that statement from many of my radiology and breast imaging colleagues.

We have seen, at the beginning of this chapter, how perceptual issues impact breast imaging in the everyday reading of mammograms, and I have tried to proffer several reasons why we as radiologists might miss cancers on a mammogram that in retrospect seem obvious. I have shown you other areas in life where computers have been useful as detection devices, and I have also shown two fundamental published studies that demonstrate both retrospectively and prospectively CAD's ability to assist us in catching some cancers a year or two earlier than they might ultimately be detected on a mammogram. With only a few exceptions, most subsequent CAD studies have also shown it can be helpful with perceptual issues in mammography. But just like CAD did not begin with mammography, it will not end with mammography. Perceptual issues are not limited to radiology or mammography – they face us in everyday life. (See Figure 22.12.) You will be walking through some airport one day, some Olympic village, some shopping center, or some world trade center, and cameras will be scanning your face. They will be extracting facial features and analyzing them. They will be comparing these features to databases of known criminals, missing persons, or other persons of interest. You see, it doesn't matter if we are looking for malignancies on a mammogram, the missiles of October, a missing person, or a bad guy. We are playing the game of perception, the game of hide and seek; and computers, so far, seem to help us win.

REFERENCES

Fenton, J.J., *et al.* (2007). Influence of computer-aided detection on performance of screening mammography. *N Engl J Med*, **356**, 1399–1409.

Freer, T.W., Ulissey, M.J. (2001). Screening mammography with computer-aided detection: a prospective study of 12,860 patients in a community breast center. *Radiology*, **220**(3), 781–786.

Majid, A.S., *et al.* (2003). Missed breast carcinoma: pitfalls and pearls. *Radiographics*, **4**, 881–95.

Ulissey, M.J., Roehrig, J. (2001). *Computer-aided Detection in Mammography*. Boulder, CO: University of Colorado Press. Online radiology book chapter: www.emedicine.com

United States Public Health Office (1996). From missiles to mammograms. *Radiol Technol*, **68**, 175–177.

Warren Burhenne, L.J., *et al.* (2000). Potential contribution of computer-aided detection to the sensitivity of screening mammography. *Radiology*, **215**, 554–562.

How perceptual factors affect the use and accuracy of CAD for interpretation of CT images

RONALD SUMMERS

Radiologic imaging techniques are continually advancing. While not long ago most radiology images were projection images (e.g. chest radiographs), more and more medical images are now tomographic, producing a series of image slices through the patient. Attesting to their value, the growth in tomographic studies such as CT, PET, and MRI far outpaces that of projection images (Maitino, 2003). Techniques that to date have not been tomographic, such as mammography, through recent advances now have tomographic capabilities (Nelson, 1998; Park, 2007b; Rafferty, 2007).

With the advancement of tomography in radiology, physicians have more accurate and precise spatial information about disease processes. However, this comes at the cost of an increase in the number of images. As a consequence, the growth in the number of images that need to be interpreted is outpacing the number of personnel trained to interpret the images. In addition, economic pressures are leading to decreases in reimbursement for tomographic imaging studies, leading to a decrease in the time allotted to interpret the images (Cerdena, 2007). The combination of more images and less time spent interpreting each image has the potential to increase medical error (Robinson, 1997).

Computer-aided detection (CAD) is one potential means of reducing error in diagnostic radiology. However, perceptual factors influence this theoretical benefit, often in unexpected ways that limit CAD's potential. In this review, we will summarize the recent literature on how perceptual factors affect, sometimes adversely, the promise of CAD to improve the accuracy of interpretation of radiology tomographic images. We will focus on the two most common types of CAD for tomographic images: lung nodule and colonic polyp detection on CT images.

23.1 PERCEPTION AND RADIOLOGIST INTERPRETATIVE ERROR

Perception can be defined as the act of observing or achieving understanding. In the context of diagnostic radiology, the understanding takes the form of making a correct diagnosis. Because the perceiving is done by humans and depends on how the diagnostic images are obtained, there is a possibility of making an incorrect diagnosis. Perceptual factors play an important role at several stages of the interpretive process (Figure 23.1).

There have been a number of studies of radiologist error (Bechtold, 1997; Gollub, 1999; Loughrey, 1999; FitzGerald, 2005; Berlin, 2007; Siewert, 2008). In one study, radiologic errors were categorized as technical, active, or a combination of the two (Siewert, 2008). Technical errors include those that relate to image creation, such as scanning parameters and contrast administration. Active errors include those that occur during image interpretation. Interpretative errors may be further classified as those arising from differences between observers (interobserver variability) and differences in interpretation by the same observer interpreting the images at different times (intraobserver variability). These types of variability have been shown to occur across modalities (conventional radiography, mammography, CT, etc.) (Brady, 1994; Bechtold, 1997; Miglioretti, 2007).

23.2 COMPUTER-AIDED DETECTION

CAD may improve the accuracy of radiology interpretation in several ways. For example, by finding abnormalities initially missed by radiologists, CAD may reduce perceptual errors and improve reproducibility. The reduction in missed lesions may particularly help less-trained radiologists achieve accuracies closer to those of expert radiologists, thereby reducing interobserver variability. As we will see, however, these improvements are not consistently achieved in clinical use of CAD.

There are two types of CAD, computer-aided detection and computer-aided diagnosis. In computer-aided detection, the computer determines whether the image is normal or abnormal and identifies possible disease sites. In computer-aided diagnosis, the computer characterizes the nature of the abnormality on an image (e.g. benign or malignant) and outputs the disease type. Most commercial CAD systems for radiology are of the detection rather than diagnosis type. Therefore, in this chapter, we will refer to CAD as computer-aided detection.

23.3 PRINCIPLES OF CAD FOR TOMOGRAPHIC IMAGES

Typical CAD systems for tomographic images, such as those from CT and MRI, identify image features that characterize pathology. Such features include shape, thickness, intensity, edges, and relationships to normal structures. These features, either in isolation or in combination, can help distinguish true from false positives.

A common component of a CAD system is a classifier, typically a mathematical construct that applies weights to the

The Handbook of Medical Image Perception and Techniques, ed. Ehsan Samei and Elizabeth Krupinski. Published by Cambridge University Press.

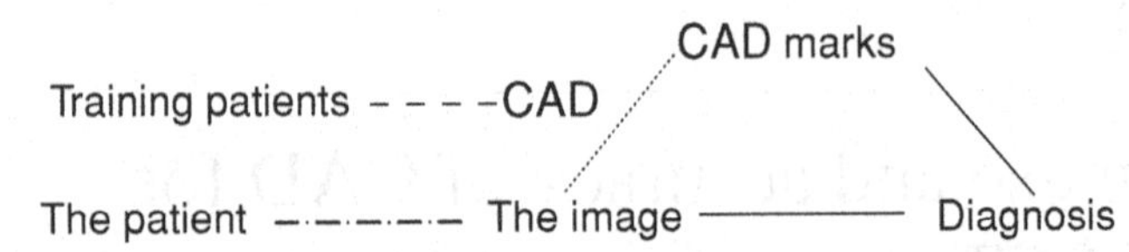

Figure 23.1 The diagnostic pipeline for tomographic images. The lines represent transformations (dashed dotted line = scanning, dotted line = CAD, and solid lines = interpretation). The quality of each stage affects the final outcome because the difference between normal and abnormal is altered (usually degraded) at each stage. Perceptual factors play important roles at all stages after the image is formed. While it may be obvious that human perception is important at the two interpretation stages (solid lines), perception is also a factor during the training stage of the CAD system (dashed line). It is during training that the CAD software is "taught" to distinguish normal from abnormal. In this case, the perceptions of the teacher (the CAD developer) are incorporated into the software.

different features. The classifier combines the features, sometimes in a non-intuitive way, to assign probabilities of disease at specific locations in the image. During development of the CAD system, the classifier is trained on cases with known disease to determine the weights. CAD systems apply false positive reduction, either implicitly through training of the classifier or explicitly by identifying common false positives. For example, the ileocecal valve is a normal structure that is also a common false positive CAD finding on CT colonography (CTC). CAD systems can be trained to identify the ileocecal valve (Summers, 2004; O'Connor, 2006).

Perceptual factors are important during the development of the CAD system. For example, the quality and completeness of the CAD training data set and the measured features, the quality of training of the classifier, and the selection of the operating point (trade-off between sensitivity and false positive rate) are all factors determined in part by human decision-making and therefore affect the performance of the CAD software.

23.4 EXISTING CAD SYSTEMS FOR TOMOGRAPHIC RADIOLOGY IMAGES

The most thoroughly researched areas of CAD for tomographic radiologic imaging are detection of lung nodules on chest CT and colonic polyps on CTC (Summers, 2002a, 2003; Perumpillichira, 2005; Li, 2007; Saba, 2007; Yoshida, 2007). Many other CAD systems for tomographic radiologic imaging have been presented, including automated detection of pulmonary emboli on CT pulmonary angiograms, subcutaneous melanoma metastases on abdominal pelvic CT, and lytic spinal metastases on body CT, but these CAD systems are at an earlier stage of development (Masutani, 2002; Solomon, 2004; O'Connor, 2007).

On MRI, CAD has been used to locate cancer of the breast and prostate (Giger, 2004; Demartini, 2005; Madabhushi, 2005). Cancer detection CAD for MRI analyzes lesion morphology and texture, and kinetic parameters of contrast uptake and washout. Other applications of MRI CAD are to locate cerebral aneurysms on MR angiography and lacunar infarcts on brain MRI (Arimura, 2006; Uchiyama, 2007). Because CAD for MRI is at an earlier stage of investigation compared to CT, we will not discuss it further in this chapter. Nevertheless, many of the perceptual issues for CT CAD are likely to be relevant for MRI CAD.

23.5 LUNG NODULE CAD

Lung cancer is the leading cause of cancer death in the USA. The number of lung cancer cases exceeds that of the next most frequent cancers combined (colon, breast, and prostate cancers). While controversial, early detection of lung cancer by CT screening has been proposed because of the potential benefit of detecting lung cancer at an early stage. Several large clinical studies using helical CT for lung cancer screening including the Early Lung Cancer Action Project (ELCAP), I-ELCAP, and the National Lung Screening Trial (NLST) are either completed or underway (Henschke, 1999, 2006; Clark, 2007; New York Early Lung Cancer Action Project Investigators, 2007).

A number of CAD systems have been developed for chest CT. Such systems incorporate algorithms for distinguishing lung nodules from adjacent blood vessels and detecting lung nodules that lie along the chest wall or pleural surfaces or adjacent to mediastinal structures. As with many CAD systems, lung nodule CAD systems tend to have higher sensitivity and lower false positive rates for detecting larger nodules.

Perceptual factors play an important role in the development and use of lung nodule CAD. During CAD development, perceptual factors affect establishment of the reference standard of truth, as it is sometimes difficult to distinguish pathology from normal anatomic variation. The identification of ground glass opacities is one example where it is sometimes difficult to establish truth. During CAD use, perceptual factors affect accurate measurement of lung nodule size. Perceptual factors also affect the interaction of the radiologist with CAD. Examples include how CAD marks are displayed to radiologists and the differential benefit of CAD according to radiologist skill level.

During the development of the CAD system, it is important to have an accurate training data set in order to properly teach the CAD how to distinguish true from false nodules. Unfortunately, this is easier said than done. For example, consensus panels of multiple radiologists often disagree about whether a finding on chest CT truly represents a nodule (McNitt-Gray, 2007; Ross, 2007). The approach some research groups have taken is to assign probability estimates to nodules and to train CAD systems to detect nodules assigned higher probabilities. The definition of a nodule is somewhat ambiguous due to overlap between potentially cancerous lesions and very common nodular and linear pulmonary scars. The perceptual issues involved in determining the reference standard during CAD development are just beginning to be understood.

An important lung nodule of interest is the ground glass opacity. Challenging to perceive, both by the trained radiologist and the computer, ground glass opacities are relatively low attenuation nodules that can be precursors of cancerous lesions. Lung nodule CAD systems tend to have lower performance for ground glass opacities although the gap in performance may be narrowing (Li, 2008).

Detection of a lung nodule is not the only important task. The size of the nodule is also important as it determines the clinical management of the patient. For example, small nodules may be followed but larger nodules may require immediate biopsy. Unfortunately, there is considerable interobserver and intraobserver variability when measuring nodules either manually or by using computer assistance. It is sometimes difficult to know where to place the electronic calipers when making a measurement. Irregularly shaped nodules pose a particular problem since they are not well described by a single linear measurement. The authors of one study stated that CAD nodule measurement tools are urgently needed (Bogot, 2005).

The clinical utility of CAD is affected not just by whether it detects nodules but also by how it shows the nodules to the radiologist. For example, it has been shown that the method of marking a lung CT CAD finding on the image affects radiologist performance. In one study, 9–17% of CAD true positives were converted to false negatives because the incorrect image was marked and displayed by the CAD system (Armato, 2003).

The utility of lung nodule CAD depends to some extent on the skill level of the user. One study showed that both board-certified radiologists and radiology residents benefited from CAD (Awai, 2004). Another study of 202 observers conducted at the Radiologic Society of North America meeting found that the use of CAD equalized sensitivities between observers with different levels of expertise (Brown, 2005).

23.6 CT COLONOGRAPHY

Colorectal cancer is the second leading cause of cancer mortality in Americans. Most colorectal cancer arises from pre-existing polyps that arise from the wall of the colon. It is known that colorectal cancer screening reduces the incidence of colorectal cancer but insufficient numbers of patients get screened. CTC, also known as virtual colonoscopy, is an emerging technology for colorectal cancer screening and has high sensitivity for polyp detection in some studies (Pickhardt, 2003; Mulhall, 2005; Frentz, 2006).

CTC is typically performed following bowel cleansing and diet restriction to remove fecal residues that could obscure polyps. A typical CTC study consists of 800–1000 images per patient.

23.7 PERCEPTUAL FACTORS FOR CT COLONOGRAPHY

Just as for diagnosis of lung nodules, there are many perceptual issues for finding polyps in the colon, summarized in Figure 23.2. In one study, perceptual errors were a cause of nearly one-half of false negative findings, leading to a potential 20% loss in sensitivity for adenomas and cancers 10 mm or larger (Doshi, 2007). Therefore, one must understand the role of perceptual factors to avoid interpretation errors on CTC. A number of reviews are available to help radiologists learn how to avoid some of the many potential pitfalls (Mang, 2007a).

The perceptual factors may be subdivided into those intrinsic and extrinsic to the colon, although there is some overlap between these two categories. We must first understand these factors before we can understand how CAD can help reduce their adverse consequences. Perceptual factors intrinsic to the colon include those factors relating to the colon itself or its contents. Such factors relate to the colonic and polyp anatomy and the quality of the bowel preparation (Figures 23.3 and 23.4).

The shape of the colon is tortuous and its wall contains over 100 folds that can hide or mimic polyps. For example, optical colonoscopy false negatives revealed by CTC were most commonly on the backside of a fold or near the anus (Pickhardt, 2004). This finding has implications for CTC also. Radiologists must ensure adequate colonic distention and carefully review both sides of haustral folds to avoid missing polyps.

Residual fecal matter within the colonic lumen can mimic or obscure polyps. Poor colonic distention or untagged residual colonic fluid can render parts of the colon uninterpretable, leading to false negative diagnoses (Arnesen, 2005). The bowel preparation must be of high quality and include sufficient cathartic to remove fecal residues. Polyps may be difficult to see if they are relatively small or flat. Flatness is a common characteristic of false negative polyps (MacCarty, 2006). Recent evidence suggests that the visual conspicuity of polyps is determined to a great extent by whether or not the polyp is flat (Summers, 2009).

Just as for lung nodules on chest CT, accurate polyp size measurement is critical, because size is the stratifying variable that determines whether polyps should be reported and immediately resected (Zalis, 2005). Polyp shapes can be complex and it can be hard to determine a "best" size. Variability in polyp size measurement can lead to errors in patient management (Yeshwant, 2006; Park, 2007a; Jeong, 2008).

Perceptual factors extrinsic to the colon include those factors relating to the observation process and the capabilities of interpretation tools. Examples relating to the observation process include the choice and usage of special image displays (MPR, 2D and 3D, window and level settings, virtual dissection, etc.), satisfaction of search, the choice of operating point (level of aggressiveness), and reader skill level and training. Examples relating to tools include the quality and ergonomics of interpretation software and the accuracy of polyp measurement software.

When interpreting CTC images, it is important to use the correct window and level settings. Failure to do so can lead to false negative diagnoses of polyps and cancers that the radiologist does not perceive (Fletcher, 1999). The choice of window and level settings can also affect measurement of polyp size (Burling, 2006). Polyp conspicuity can be affected by the density of contrast-enhanced fluid when fluid and fecal tagging is administered (Slater, 2006). Consequently, in the setting of fluid and fecal tagging, the window and level settings may need to be adjusted.

In satisfaction of search, the observer terminates the image search prematurely after observing only a subset of the abnormalities in the image, leading to false negative diagnoses (Berbaum, 1990). Satisfaction of search errors can occur without or with the use of CAD. For example, by using CAD as a time-saver, an observer could reduce the time spent independently assessing the images, potentially increasing the probability of satisfaction of search errors.

The choice of operating point refers to the trade-off between sensitivity and specificity. The radiologist may adjust their interpretation strategy based on their inherent level of aggressiveness or confidence at interpreting image findings. For example, a

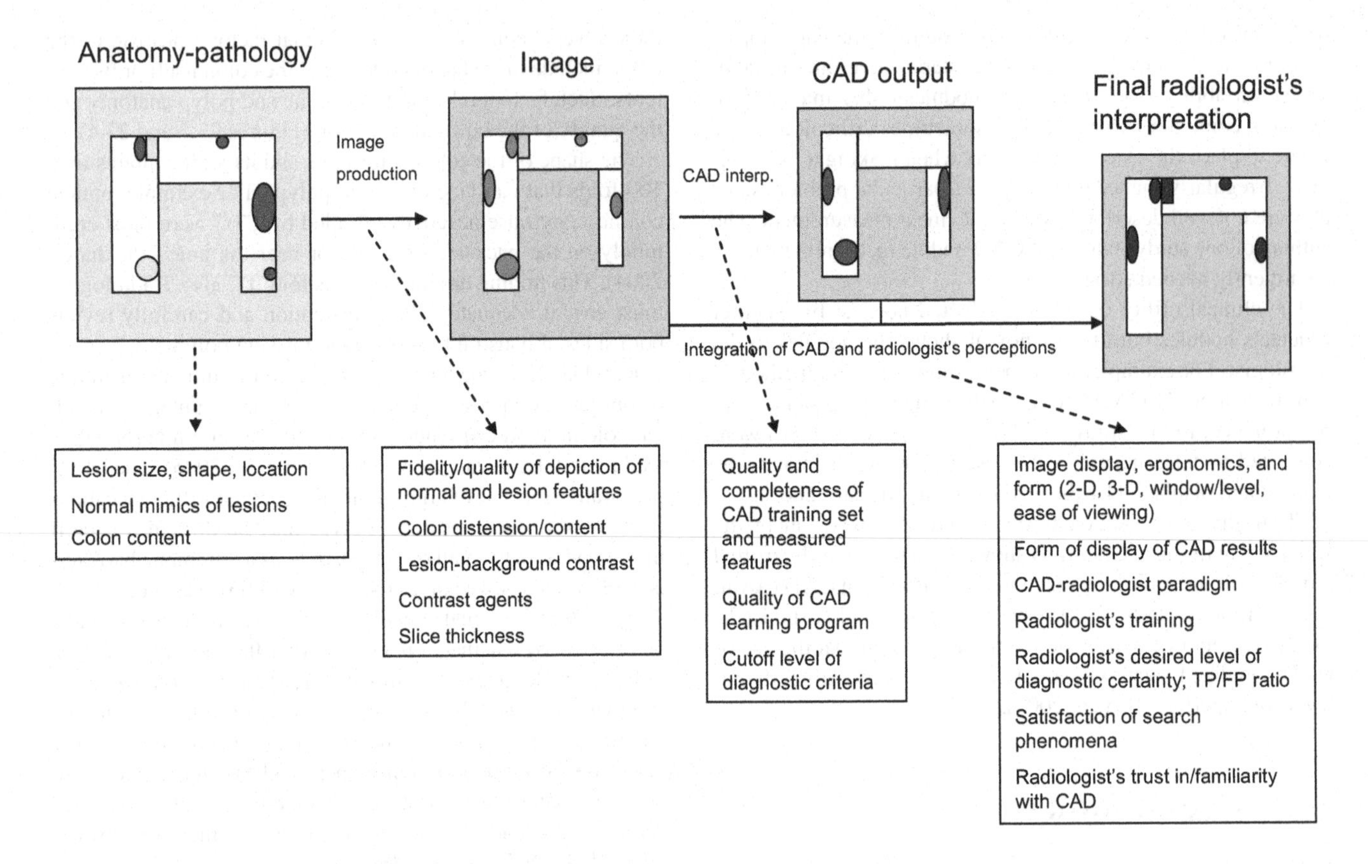

Figure 23.2 Factors of perception of colonic polyps on CT colonography. The final perception depends on factors of component steps. At each step, information may be lost or distorted. The rectangles represent a simplified schematic of colonic segments within a patient or on a CT colonography examination. The circles and ovals within the colon representation indicate polyps of different sizes and locations. For example, a polyp submerged in opacified fluid (lower right of colon in anatomy-pathology box) may be invisible on the image; the size of another polyp may be distorted. The CAD system (third box) missed three polyps (false negatives) and further distorted the size of one polyp. On the final radiologist interpretation (far right box), the radiologist improperly located one polyp (rectangle) in the wrong colonic segment, made a false positive diagnosis, and underestimated the size of one polyp. The radiologist correctly identified the ileocecal valve (lower left of colon in anatomy-pathology box), a normal structure, and did not mention it at the final diagnosis. Figure adapted from one supplied by Andrew Dwyer, MD, NIH.

more aggressive radiologist may seek high sensitivity at the expense of less specificity so as not to miss any cases with disease. Another radiologist may only diagnose an abnormality if they are extremely confident of the diagnosis, raising specificity and potentially lowering sensitivity. These clinical decisions are based in part on the subjective perception of the risk that an abnormality is present. It is not always clear how best to make the trade-off between sensitivity and specificity because low specificity can lead to undesirable interventions in patients subsequently found to be normal.

Reader training and experience have been shown to affect diagnostic performance at interpreting CTC. A multi-observer trial showed large performance gaps between the most and least experienced readers at interpreting CTC (Johnson, 2003). The ESGAR (European Society of Gastrointestinal and Abdominal

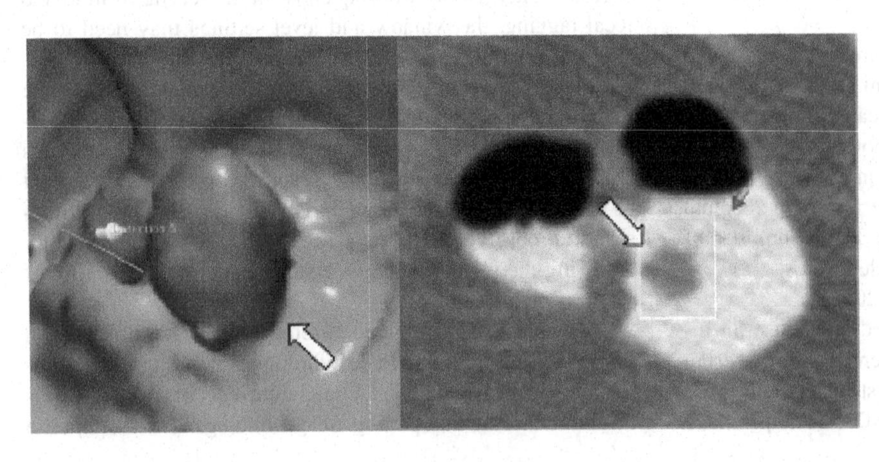

Figure 23.3 CT colonographic images of a 10 mm polyp missed at initial radiologist interpretation but subsequently found with the use of CAD in the second reader mode. The polyp was submerged in opacified colonic fluid, possibly explaining why it was missed. Reprinted from Petrick *et al.* (Petrick, 2008).

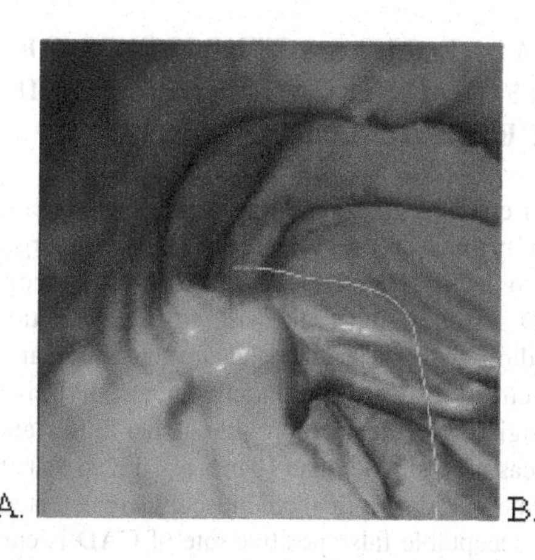

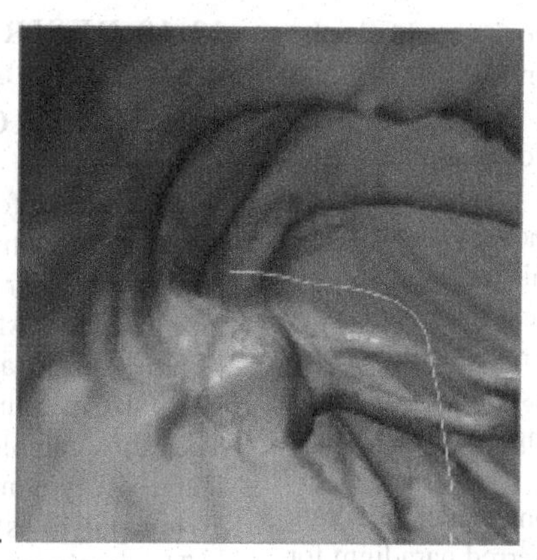

Figure 23.4 The ileocecal valve, a common CT colonography CAD false positive finding. The ileocecal valve can simulate a polyp but is usually easy to recognize. Radiologists need to be able to quickly rule out such false positives. (A) Without and (B) with CAD mark. The line represents the colon centerline.

Radiology) CTC study group showed experienced radiologists performed best, inexperienced radiologists next best, and technologists somewhat lower in performance (ESGAR, 2007). The appropriate amount and type of training are still being determined (Burling, 2007; Rockey, 2007).

23.8 CAD FOR CT COLONOGRAPHY

CAD systems for CTC typically identify polyps by analyzing the shape of the colonic surface. A common method of shape analysis is to compute curvatures of the colonic surface. The curvatures are used to classify local areas of the surface into one of several elemental shapes useful for polyp detection (Summers, 2000, 2003). CAD systems typically further refine initial curvature-based detections to reduce the number of false positives. Such refinements typically take into account such features as the CT attenuation, wall thickness, aspect ratio, and volume of the polyp candidate (Summers, 2000; Yoshida, 2001). More recently reported CAD systems have sensitivities in the range of 80–100% with false positive rates under 10 false positives per patient for detecting 10 mm or larger polyps (Bogoni, 2005; Summers, 2005, 2008a; Fletcher, 2007). For 6–9 mm polyps, sensitivity is typically lower, ranging from 50–90% for false positive rates less than 10 false positives per patient (Summers, 2005, 2008a; Fletcher, 2007; Graser, 2007). Polyps smaller than 6 mm are not typically a target of CAD systems because the sensitivity is generally poor. Typical CAD false positives include ileocecal valve, haustral folds, the rectal tube, and residual fecal matter. Studies have shown that CAD systems can identify polyps missed by radiologists, potentially leading to a complementary or synergistic effect (Summers, 2002b; Taylor, 2006a).

23.9 PERCEPTUAL FACTORS FOR COLONOGRAPHY CAD – READING PARADIGM AND OBSERVER PERFORMANCE STUDIES

The true utility of colonography CAD systems, and the subtle perceptual factors that affect the theoretical benefit of such systems, become apparent in studies that investigate the interaction of the radiologist and the CAD software. Such studies, called observer performance studies, are typically designed to determine whether CAD helps radiologists interpret images better, how radiologists should use CAD, the potential interpretation pitfalls, and whether CAD promotes time efficiency or increases throughput. A number of such studies have shown that CAD improves the performance of radiologists at the task of locating polyps at CTC.

Observer studies can shed light on such perceptual factors as how radiologists alter their diagnostic performance based on their experience with the CAD system or on what they are told about the performance of CTC and CAD. For example, the radiologist may have lower confidence in a CAD system that has either too few or too many false positives and may shift their operating point to compensate. As a consequence, CAD true positives may be ignored or CAD false positives may be misclassified as polyps.

To understand the observer performance studies, it is first necessary to discuss the various reading paradigms for CTC CAD. Reading paradigms describe the order in which the radiologist views the CTC images and the CAD findings. The three reading paradigms proposed for CTC CAD are first reader, concurrent reader, and second reader. There are few studies comparing the various reading paradigms so some of the advantages and disadvantages described in the paragraphs that follow must be viewed as preliminary.

In the "first reader" paradigm, the radiologist only reviews the sites identified on the CTC as potentially abnormal by CAD and does not review the entire colon. This method has the potential advantage of fast interpretation time and high specificity (since the choice of false positives is limited to the CAD findings) but the potential disadvantage of lower sensitivity relative to the concurrent and second reader paradigms. However, radiologists are naturally reluctant to use the first reader paradigm because only the computer reviews the entire image. Despite its limitations, CAD utilized as a first reader may help less experienced readers raise their sensitivity to approach that of more experienced readers (Mani, 2004).

In the "concurrent reader" paradigm, the CAD marks are visible during the radiologist's primary interpretation of the images. The radiologist evaluates the CAD marks as they appear in the image. The potential advantage of this method is reduced interpretation time (Halligan, 2006; Taylor, 2006b, 2008b) but the potential disadvantage is that the CAD marks could distract the radiologist from other findings in the vicinity of a mark, leading to "satisfaction of search" errors. The radiologist could also be lulled into a relaxed search strategy that tends towards a first reader interpretative mode.

In the "second reader" paradigm, the radiologist reviews the images, arrives at a preliminary diagnosis, reviews the CAD findings, and revises the preliminary diagnosis to arrive at a final diagnosis. The potential advantage of this technique is sensitivity higher than either the first or concurrent readers (Taylor, 2008b). The proven disadvantages are longer reading times and a decrease in specificity induced by radiologist misclassification of CAD false positives (Baker, 2007; Mang, 2007b; Petrick, 2008).

In the opinion of the author, since the second reader paradigm leads to the highest sensitivity, it is the preferred paradigm for CAD use by most radiologists. When using CAD in second reader mode, the author recommends that if CAD does not call an abnormality found at the radiologist's preliminary diagnosis, the radiologist should not change the preliminary diagnosis since CAD can make false negative diagnoses. This recommendation potentially prevents the false negative perceptual error arising from a loss in the radiologist's confidence in their initial unaided diagnosis when the CAD system does not confirm the initial diagnosis. The radiologist would inspect each CAD detection and ignore common false positives such as ileocecal valves and most thick folds. An observer study found that the use of 3-D viewing slightly increased reader accuracy in classifying CAD polyp candidates (Shi, 2006). A number of factors were significantly associated with reader accuracy, including polyp size and quality of the examination.

The performance improvements with the use of CAD are usually more pronounced when CAD is used by non-expert or untrained readers than when it is used by expert readers (Baker, 2007; Mang, 2007b). Focused training may improve the interpretation of CTC in combination with CAD (Taylor, 2008a). The authors of this study remarked that in their experience, many novice interpreters of CTC interpret the scans too rapidly, leaving them at risk of making perceptual errors.

A recent observer study showed that polyp conspicuity is determined to a great extent by the height (elevation off the surface) of the polyp (Summers, 2009). Polyp width played a negligible role in determining conspicuity for polyps in the 6–9 mm size category. The CAD system under investigation was less sensitive for polyps that were rated to be less conspicuous. The authors concluded that CAD developers may need to focus on improving the detectability of flatter polyps.

A preliminary report suggests that a CAD system that detects the polyp on both the supine and prone CTC scans may lead to improved polyp detection (Summers, 2008b). Such a CAD system may also lead to improved radiologist confidence of interpretation. When radiologists are unable to locate the potential polyp on both scans, the radiologist may misinterpret the polyp as an artifact or mobile stool. Identification of the polyp candidate on both scans allows the radiologist to confirm that the finding is fixed to the colonic wall and greatly reduces doubt as to the importance of the finding.

In summary, CAD can help overcome some perceptual problems in search and improve sensitivity, particularly for less experienced radiologists, although it may sometimes also lower the specificity.

23.10 DESIRABLE CHARACTERISTICS OF A HIGH-PERFORMING CAD SYSTEM AND THEIR PERCEPTUAL IMPLICATIONS

A CAD system for colonography will typically have a laddered increase in sensitivity that increases for larger polyps since larger polyps are easier to detect. Therefore, radiologists must recognize that the CAD system's sensitivity will vary as a function of polyp size. Radiologists must also be aware that automated polyp size measurement software may underestimate polyp size. Therefore, radiologists may need to provide a size buffer and identify polyps measuring smaller than 10 mm at CTC so that all 10 mm polyps as measured at optical colonoscopy are detected.

The maximum acceptable false positive rate of CAD is currently unknown. However, it is likely that too many false positives will lead to decreased throughput and decreased radiologist confidence in CAD. In the author's opinion, the median false positive rate is desirably less than 10 per patient (total for supine and prone examinations). Nevertheless, a recent study suggests that increasing numbers of false positives do not adversely affect specificity, but do prolong reading times (Taylor, 2008c).

Clinically useful CAD systems will likely have other capabilities that reduce the burden on the radiologist or reduce perceptual difficulties. For example, a CAD system that properly deals with oral contrast for fecal and fluid tagging may enable more confident detection of polyps submerged under fluid or stool or at the air-fluid boundary, a particularly challenging detection task. A CAD system that identifies common false positives such as the ileocecal valve and rectal tube and that links the supine and prone images may lead to improved radiologist performance and time efficiency, but these benefits remain to be proven.

The radiologist should be aware of blind spots in current CAD systems. These include polyps under fluid or at the air-fluid boundary, small sessile or flat polyps, polyps touching or on haustral folds, polyps on the ileocecal valve, and polyps touching the rectal tube.

23.11 OVERCOMING BARRIERS TO ACHIEVE CAD'S FULL POTENTIAL FOR REDUCING PERCEPTUAL ERRORS

The performance of the CAD system is only one factor, albeit an important one, in determining the overall performance of a radiologic test. Other factors such as the quality of integration of CAD into the reading environment (ergonomics), the training and experience of the radiologists, and the application of appropriate quality assurance techniques will be important for helping radiologists achieve CAD's full potential (Figure 23.2). Economic factors may also play a role, for example if the reduced specificity of second-read CAD leads to an unacceptable increase in interventions.

More comprehensive and versatile CAD systems are envisioned for the future. It is likely that the more CAD can do and the more easily it can be used by radiologists, the more useful it will be. CAD plug-ins for PACS may lead to more widespread utilization of CAD.

In summary, while CAD can achieve high performance in the laboratory, perceptual factors strongly influence and may reduce the potential benefit of CAD in the clinic. Better understanding of these factors may lead to improved CAD systems that realize their full benefit. Such understanding can only come if radiologists actively embrace CAD and work with CAD developers to improve the CAD software.

ACKNOWLEDGMENTS

We thank Andrew Dwyer, MD, for critical review of the manuscript and for supplying an early version of Figure 23.2. We thank Perry J. Pickhardt, MD, for supplying CTC data. Viatronix supplied free research software used to produce the endoluminal images. This work was supported by the Intramural Research Program of the National Institutes of Health Clinical Center.

REFERENCES

Arimura, H., Li, Q., Korogi, Y., *et al.* (2006). Computerized detection of intracranial aneurysms for three-dimensional MR angiography: feature extraction of small protrusions based on a shape-based difference image technique. *Med Phys*, **33**, 394–401.

Armato, S.G. (2003). Image annotation for conveying automated lung nodule detection results to radiologists. *Acad Radiol*, **10**, 1000–1007.

Arnesen, R.B., Adamsen, S., Svendsen, L.B., *et al.* (2005). Missed lesions and false-positive findings on computed-tomographic colonography: a controlled prospective analysis. *Endoscopy*, **37**, 937–944.

Awai, K., Murao, K., Ozawa, A., *et al.* (2004). Pulmonary nodules at chest CT: effect of computer-aided diagnosis on radiologists' detection performance. *Radiology*, **230**, 347–352.

Baker, M.E., Bogoni, L., Obuchowski, N.A., *et al.* (2007). Computer-aided detection of colorectal polyps: can it improve sensitivity of less-experienced readers? Preliminary findings. *Radiology*, **245**, 140–149.

Bechtold, R.E., Chen, M.Y., Ott, D.J., *et al.* (1997). Interpretation of abdominal CT: analysis of errors and their causes. *J Comput Assist Tomogr*, **21**, 681–685.

Berbaum, K.S., Franken, E.A., Dorfman, D.D., *et al.* (1990). Satisfaction of search in diagnostic radiology. *Invest Radiol*, **25**, 133–140.

Berlin, L. (2007). Accuracy of diagnostic procedures: has it improved over the past five decades? *Am J Roentgenol*, **188**, 1173–1178.

Bogoni, L., Cathier, P., Dundar, M., *et al.* (2005). Computer-aided detection (CAD) for CT colonography: a tool to address a growing need. *Br J Radiol*, **78**, S57–S62.

Bogot, N.R., Kazerooni, E.A., Kelly, A.M., *et al.* (2005). Interobserver and intraobserver variability in the assessment of pulmonary nodule size on CT using film and computer display methods. *Acad Radiol*, **12**, 948–956.

Brady, A.P., Stevenson, G.W., Stevenson, I. (1994). Colorectal cancer overlooked at barium enema examination and colonoscopy: a continuing perceptual problem. *Radiology*, **192**, 373–378.

Brown, M.S., Goldin, J.G., Rogers, S., *et al.* (2005). Computer-aided lung nodule detection in CT: results of large-scale observer test. *Acad Radiol*, **12**, 681–686.

Burling, D., Halligan, S., Taylor, S., *et al.* (2006). Polyp measurement using CT colonography: agreement with colonoscopy and effect of viewing conditions on interobserver and intraobserver agreement. *Am J Roentgenol*, **186**, 1597–1604.

Burling, D., Moore, A., Taylor, S., La Porte, S., Marshall, M. (2007). Virtual colonoscopy training and accreditation: a national survey of radiologist experience and attitudes in the UK. *Clin Radiol*, **62**, 651–659.

Cerdena, E.A., Corigliano, B.A. (2007). Effective and basic business strategic tools to overcome the DRA impact in outpatient imaging centers. *Radiol Manage*, **29**, 46–53.

Clark, K.W., Gierada, D.S., Moore, S.M., *et al.* (2007). Creation of a CT image library for the lung screening study of the national lung screening trial. *J Digit Imaging*, **20**, 23–31.

Demartini, W.B., Lehman, C.D., Peacock, S., Russell, M.T. (2005). Computer-aided detection applied to breast MRI: assessment of CAD-generated enhancement and tumor sizes in breast cancers before and after neoadjuvant chemotherapy. *Acad Radiol*, **12**, 806–814.

Doshi, T., Rusinak, D., Halvorsen, R.A., *et al.* (2007). CT colonography: false-negative interpretations. *Radiol*, **244**, 165–173.

European Society of Gastrointestinal and Abdominal Radiology (ESGAR) CT Colonography Group Investigators. (2007). Effect of directed training on reader performance for CT colonography: multicenter study. *Radiol*, **242**, 152–161.

FitzGerald, R. (2005). Radiological error: analysis, standard setting, targeted instruction and teamworking. *Eur Radiol*, **15**, 1760–1767.

Fletcher, J.G., Booya, F., Summers, R.M., *et al.* (2007). Comparative performance of two polyp detection systems on CT colonography. *Am J Roentgenol*, **189**, 277–282.

Fletcher, J.G., Johnson, C.D., MacCarty, R.L., *et al.* (1999). CT colonography: potential pitfalls and problem-solving techniques. *Am J Roentgenol*, **172**, 1271–1278.

Frentz, S.M., Summers, R.M. (2006). Current status of CT colonography. *Acad Radiol*, **13**, 1517–1531.

Giger, M.L. (2004). Computerized analysis of images in the detection and diagnosis of breast cancer. *Semin Ultrasound CT MR*, **25**, 411–418.

Gollub, M.J., Panicek, D.M., Bach, A.M., Penalver, A., Castellino, R.A. (1999). Clinical importance of reinterpretation of body CT scans obtained elsewhere in patients referred for care at a tertiary cancer center. *Radiol*, **210**, 109–112.

Graser, A., Kolligs, F.T., Mang, T., *et al.* (2007). Computer-aided detection in CT colonography: initial clinical experience using a prototype system. *Eur Radiol*, **17**, 2608–2615.

Halligan, S., Altman, D.G., Mallett, S., *et al.* (2006). Computed tomographic colonography: assessment of radiologist performance with and without computer-aided detection. *Gastroenterology*, **131**, 1690–1699.

Henschke, C.I., McCauley, D.I., Yankelevitz, D.F., *et al.* (1999). Early Lung Cancer Action Project: overall design and findings from baseline screening. *Lancet*, **354**, 99–105.

Henschke, C.I., Yankelevitz, D.F., Libby, D.M., *et al.* (2006). Survival of patients with stage I lung cancer detected on CT screening. *N Engl J Med*, **355**, 1763–1771.

Jeong, J.Y., Kim, M.J., Kim, S.S. (2008). Manual and automated polyp measurement comparison of CT colonography with optical colonoscopy. *Acad Radiol*, **15**, 231–239.

Johnson, C.D., Toledano, A.Y., Herman, B.A., *et al.* (2003). Computerized tomographic colonography: performance evaluation in a retrospective multicenter setting. *Gastroenterology*, **125**, 688–695.

Li, Q. (2007). Recent progress in computer-aided diagnosis of lung nodules on thin-section CT. *Comput Med Imaging Graph*, **31**, 248–257.

Li, Q., Li, F., Doi, K. (2008). Computerized detection of lung nodules in thin-section CT images by use of selective enhancement filters and an automated rule-based classifier. *Acad Radiol*, **15**, 165–175.

Loughrey, G.J., Carrington, B.M., Anderson, H., Dobson, M.J., Ping, F.L.Y. (1999). The value of specialist oncological radiology review of cross-sectional imaging. *Clin Radiol*, **54**, 149–154.

MacCarty, R.L., Johnson, C.D., Fletcher, J.G., Wilson, L.A. (2006). Occult colorectal polyps on CT colonography: implications for surveillance. *Am J Roentgenol*, **186**, 1380–1383.

Madabhushi, A., Feldman, M.D., Metaxas, D.N., Tomaszeweski, J., Chute, D. (2005). Automated detection of prostatic adenocarcinoma from high-resolution ex vivo MRI. *IEEE Trans Med Imaging*, **24**, 1611–1625.

Maitino, A.J., Levin, D.C., Parker, L., Rao, V.M., Sunshine, J.H. (2003). Nationwide trends in rates of utilization of noninvasive diagnostic imaging among the Medicare population between 1993 and 1999. *Radiol*, **227**, 113–117.

Mang, T., Maier, A., Plank, C., *et al.* (2007a). Pitfalls in multi-detector row CT colonography: a systematic approach. *Radiographics*, **27**, 431–454.

Mang, T., Peloschek, P., Plank, C., *et al.* (2007b). Effect of computer-aided detection as a second reader in multidetector-row CT colonography. *Eur Radiol*, **17**, 2598–2607.

Mani, A., Napel, S., Paik, D.S., *et al.* (2004). Computed tomography colonography – feasibility of computer-aided polyp detection in a "first reader" paradigm. *J Computer Assisted Tomography*, **28**, 318–326.

Masutani, Y., MacMahon, H., Doi, K. (2002). Computerized detection of pulmonary embolism in spiral CT angiography based on volumetric image analysis. *IEEE Trans Med Imaging*, **21**, 1517–1523.

McNitt-Gray, M.F., Armato, S.G., Meyer, C.R., *et al.* (2007). The Lung Image Database Consortium (LIDC) data collection process for nodule detection and annotation. *Acad Radiol*, **14**, 1464–1474.

Miglioretti, D.L., Smith-Bindman, R., Abraham, L., *et al.* (2007). Radiologist characteristics associated with interpretive performance of diagnostic mammography. *Natl Cancer Inst*, **99**, 1854–1863.

Mulhall, B.P., Veerappan, G.R., Jackson, J.L. (2005). Meta-analysis: computed tomographic colonography. *Ann Intern Med*, **142**, 635–650.

Nelson, T.R., Pretorius, D.H. (1998). Three-dimensional ultrasound imaging. *Ultrasound Med Biol*, **24**, 1243–1270.

New York Early Lung Cancer Action Project (ELCAP) Investigators. (2007). CT screening for lung cancer: diagnoses resulting from the New York Early Lung Cancer Action Project. *Radiol*, **243**, 239–249.

O'Connor, S.D., Summers, R.M., Yao, J., Pickhardt, P.J., Choi, J.R. (2006). CT colonography with computer-aided polyp detection: volume and attenuation thresholds to reduce false-positive findings owing to the ileocecal valve. *Radiol*, **241**, 426–432.

O'Connor, S.D., Yao, J., Summers, R.M. (2007). Lytic metastases in thoracolumbar spine: computer-aided detection at CT – preliminary study. *Radiol*, **242**, 811–816.

Park, S.H., Choi, E.K., Lee, S.S., *et al.* (2007a). Polyp measurement reliability, accuracy, and discrepancy: optical colonoscopy versus CT colonography with pig colonic specimens. *Radiol*, **244**, 157–164.

Park, J.M., Franken, E.A., Garg, M., Fajardo, L.L., Niklason, L.T. (2007b). Breast tomosynthesis: present considerations and future applications. *Radiographics*, **27** Suppl 1, S231–240.

Perumpillichira, J.J., Yoshida, H., Sahani, D.V. (2005). Computer-aided detection for virtual colonoscopy. *Cancer Imaging*, **5**, 11–16.

Petrick, N., Haider, M., Summers, R.M., *et al.* (2008). CT colonography and computer-aided detection as a second reader: observer performance study. *Radiol*, **246**, 148–156.

Pickhardt, P.J., Choi, J.R., Hwang, I., *et al.* (2003). Computed tomographic virtual colonoscopy to screen for colorectal neoplasia in asymptomatic adults. *N Engl J Med*, **349**, 2191–2200.

Pickhardt, P.J., Nugent, P.A., Mysliwiec, P.A., Choi, J.R., Schindler, W.R. (2004). Location of adenomas missed by optical colonoscopy. *Ann Intern Med*, **141**, 352–359.

Rafferty, E.A. (2007). Digital mammography: novel applications. *Radiol Clin North Am*, **45**, 831–843.

Robinson, P.J. (1997). Radiology's Achilles' heel: error and variation in the interpretation of the Rontgen image. *Br J Radiol*, **70**, 1085–1098.

Rockey, D.C., Barish, M., Brill, J.V., *et al.* (2007). Standards for gastroenterologists for performing and interpreting diagnostic computed tomographic colonography. *Gastroenterology*, **133**, 1005–1024.

Ross, J.C., Miller, J.V., Turner, W.D., Kelliher, T.P. (2007). An analysis of early studies released by the Lung Imaging Database Consortium (LIDC). *Acad Radiol*, **14**, 1382–1388.

Saba, L., Caddeo, G., Mallarini, G. (2007). Computer-aided detection of pulmonary nodules in computed tomography: analysis and review of the literature. *J Comput Assist Tomogr*, **31**, 611–619.

Shi, R., Schraedley-Desmond, P., Napel, S., *et al.* (2006). CT colonography: influence of 3D viewing and polyp candidate features on interpretation with computer-aided detection. *Radiol*, **239**, 768–776.

Siewert, B., Sosna, J., McNamara, A., Raptopoulos, V., Kruskal, J.B. (2008). Missed lesions at abdominal oncologic CT: lessons learned from quality assurance. *Radiographics*, **28**, 623–638.

Slater, A., Taylor, S.A., Burling, D., *et al.* (2006). Colonic polyps: effect of attenuation of tagged fluid and viewing window on conspicuity and measurement – in vitro experiment with porcine colonic specimen. *Radiol*, **240**, 101–109.

Solomon, J., Mavinkurve, S., Cox, D., Summers, R.M. (2004). Computer-assisted detection of subcutaneous melanomas: feasibility assessment. *Acad Radiol*, **11**, 678–685.

Summers, R.M. (2002a). Challenges for computer-aided diagnosis for CT colonography. *Abdom Imaging*, **27**, 268–274.

Summers, R.M., Beaulieu, C.F., Pusanik, L.M., *et al.* (2000). Automated polyp detector for CT colonography: feasibility study. *Radiol*, **216**, 284–290.

Summers, R.M., Frentz, S., Liu, J., *et al.* (2009). Conspicuity of colorectal polyps at CT colonography: visual assessment, CAD performance, and the important role of polyp height. *Acad Radiol*, **16**, 4–14.

Summers, R.M., Handwerker, L.R., Pickhardt, P.J., *et al.* (2008a). Performance of a previously validated CT colonography computer-aided detection system in a new patient population. *Am J Roentgenol*, **191**, 168–174.

Summers, R.M., Jerebko, A.K., Franaszek, M., Malley, J.D., Johnson, C.D. (2002b). Colonic polyps: complementary role of computer-aided detection in CT colonography. *Radiol*, **225**, 391–399.

Summers, R.M., Liu, J., Rehani, B., *et al.* (2008b). CT colonography computer-aided polyp detection: an observer study showing the benefit of a CAD system that detects the polyp on both the supine and prone scan. *Int J CARS*, **3** (Suppl 1), S189–S190.

Summers, R.M., Yao, J., Johnson, C.D. (2004). CT colonography with computer-aided detection: automated recognition of ileocecal valve to reduce number of false-positive detections. *Radiol*, **233**, 266–272.

Summers, R.M., Yao, J., Pickhardt, P.J., *et al.* (2005). Computed tomographic virtual colonoscopy computer-aided polyp detection in a screening population. *Gastroenterology*, **129**, 1832–1844.

Summers, R.M., Yoshida, H. (2003). Future directions: computer-aided diagnosis. In *Atlas of Virtual Colonoscopy*, ed. Dachman, A.H., New York, NY: Springer, pp. 55–62.

Taylor, S.A., Burling, D., Roddie, M., *et al.* (2008a). Computer-aided detection for CT colonography: incremental benefit of observer training. *Br J Radiol*, **81**, 180–186.

Taylor, S.A., Charman, S.C., Lefere, P., *et al.* (2008b). CT colonography (CTC): investigation of the optimum reader paradigm using computer aided detection software. *Radiol*, **246**, 463–471.

Taylor, S.A., Greenhalgh, R., Ilangovan, R., *et al.* (2008c). CT colonography and computer-aided detection: effect of false-positive results on reader specificity and reading efficiency in a low-prevalence screening population. *Radiol*, **247**, 133–140.

Taylor, S.A., Halligan, S., Burling, D., *et al.* (2006a). Computer-assisted reader software versus expert reviewers for polyp detection on CT colonography. *Am J Roentgenol*, **186**, 696–702.

Taylor, S.A., Halligan, S., Slater, A., *et al.* (2006b). Polyp detection with CT colonography: primary 3D endoluminal analysis versus primary 2D transverse analysis with computer-assisted reader software. *Radiol*, **239**, 759–767.

Uchiyama, Y., Yokoyama, R., Ando, H., *et al.* (2007). Computer-aided diagnosis scheme for detection of lacunar infarcts on MR images. *Acad Radiol*, **14**, 1554–1561.

Yeshwant, S.C., Summers, R.M., Yao, J.H., *et al.* (2006). Polyps: linear and volumetric measurement at CT colonography. *Radiol*, **241**, 802–811.

Yoshida, H., Nappi, J. (2001). Three-dimensional computer-aided diagnosis scheme for detection of colonic polyps. *IEEE Trans Med Imaging*, **20**, 1261–1274.

Yoshida, H., Nappi, J. (2007). CAD in CT colonography without and with oral contrast agents: progress and challenges. *Comput Med Imaging Graph*, **31**, 267–284.

Zalis, M.E., Barish, M.A., Choi, J.R., *et al.* (2005). CT colonography reporting and data system: a consensus proposal. *Radiol*, **236**, 3–9.

24

CAD: risks and benefits for radiologists' decisions

EUGENIO ALBERDI, ANDREY POVYAKALO, LORENZO STRIGINI, AND PETER AYTON

24.1 INTRODUCTION

Computer aids for reading medical images bring obvious potential benefits, as discussed in previous chapters in this handbook. But incorporating them into medical practice also presents risks, which may be less obvious. This chapter discusses some of the things that can go wrong with the use of computer aids, as well as ways of assessing their benefits. Designing computer tools and planning their use needs to take into account how best to balance the potential for gain for certain patients and damage for others. This trade-off also adds complexity to the task of assessing such tools and their impact in medical practice.

Some chapters in this handbook have been written from the point of view of radiologists, or other medical practitioners. To complement this perspective, in this chapter we present the outcomes of interdisciplinary research work, which combined insights from a variety of disciplines: reliability engineering, computing, psychology, human factors, and sociology (Alberdi, 2005).

We document and discuss these issues with reference to the use of computer support for the early detection of cancer in breast screening. In screening for breast cancer, expert clinicians (whom we will call "readers") examine mammograms (sets of X-ray images of a woman's breasts), and decide whether the patient should be recalled for further tests because they suspect cancer.

For over ten years now, since 1998, computer aided detection (CAD)[1] tools have been available to assist the interpretation of mammograms. CAD is designed to alert a reader (usually a radiologist) to areas of a mammogram. Typically the CAD tool processes a digitized version of a mammogram and marks it with "prompts" to highlight mammographic features that the reader should examine. The design goal for CAD is to aid the readers to notice features in a mammogram that might indicate cancer that they may otherwise miss. CAD is not meant to be a diagnostic tool, in the sense that it only marks areas of interest (possibly with additional indications about suspected type or degree of suspicion), which should be subsequently classified by the reader to reach a "recall/no recall" decision. The goal of using CAD is to increase readers' *sensitivity* (the proportion of cancers recalled out of all cancers) without adversely affecting their *specificity* (the proportion of normal cases not recalled, out of all normal cases).

In a typical procedure for using CAD, the reader looks at a mammogram (on film or on a video screen) and interprets it as usual, then activates the tool and looks at a digitized image of the mammogram with the tool's prompts for regions of interest, checks whether he/she has overlooked any features with diagnostic value and then revises his/her original assessment, if appropriate. Figure 24.1 schematically shows the "system" formed by the CAD tool and its human user.

Early evaluations of CAD, based on work by Warren Burhenne and co-workers (Warren Burhenne, 2000), showed the *potential* of this technology to help mammogram readers, and were used to support the approval of CAD by the US Food and Drug Administration (FDA) (FDA, 1998). Subsequent studies have shown not only CAD's potential to help but also factual evidence of significant increases in cancer detection by radiologists when using the tool (Freer, 2001; Cupples, 2005; Gromet, 2008). However, other studies have shown negligible effects of CAD on reader performance (Gur, 2004; Taylor, 2004b), even inferior to the benefits of double reading (Khoo, 2005); that is, separate reading of a patient's mammograms by two human readers. Furthermore, there is evidence of harmful effects of CAD use in certain situations, which manifest themselves either as a decrease in the specificity of CAD-supported readers (Fenton, 2007) or as readers' reduced ability to detect cancers if they are not prompted by CAD (Alberdi, 2004a; Zheng, 2004; Taplin, 2006).

This research naturally calls for a reassessment of the circumstances under which CAD prompts will actually improve readers' decisions. The first requirement for the effectiveness of CAD is of course accuracy of the prompts. It seems clear that, to be of any use, the prompts must be correct "often enough," although no CAD tool can identify all cancers and avoid all false prompts. However, we can expect any approved CAD system to be quite accurate. Ease of use, readability, etc. are also obvious requirements. But research suggests that less obvious aspects of CAD performance are also important. First, there is the issue of whether the CAD tools' help is focused on the real needs of the readers: we will refer to this as the issue of *diversity* between the CAD tools and the readers, because a CAD tool that only offered correct advice in cases that readers would process correctly anyway would not actually reduce reader errors. Second, there are changes in the way readers process information when they are assisted by CAD prompts. The simplest change to envisage is that any user of a computer support system may over time become too reliant on the computer, blindly following its

[1] As is common in the radiological literature, we will also use the abbreviation "CAD" to mean the computer tool, whenever the context does not create ambiguities with its literal meaning "detection activity aided by a computer."

The Handbook of Medical Image Perception and Techniques, ed. Ehsan Samei and Elizabeth Krupinski. Published by Cambridge University Press.

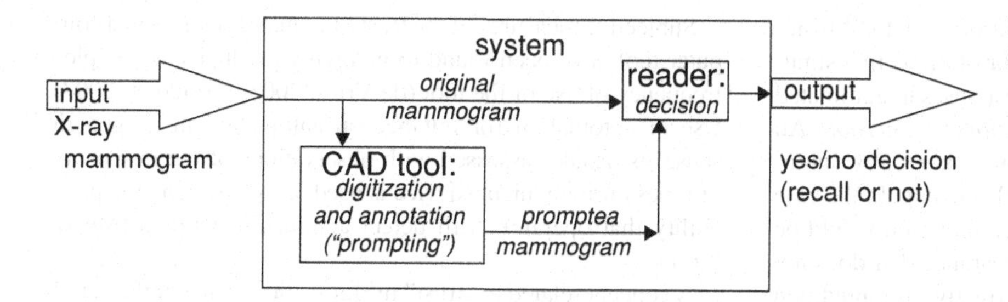

Figure 24.1 The decision system formed by a human "reader" with the CAD tool.

advice, even when it is wrong. One may be tempted to dismiss these problems as only requiring better discipline by the users. However, in reality, there is a family of possible related changes of reader behavior, and some may well be inevitable; rather than trying to avoid them by better discipline, it may be easier or even necessary to tweak the design of CAD to achieve the best overall results given these inevitable changes. We will refer to this set of problems as "automation bias."

This chapter describes studies and analyses that could help us understand some of the disparate results obtained by studies of CAD effectiveness, with pointers to relevant work by the computer engineering and human factors communities. In the last few decades, a growing body of research has shown potential pitfalls of using automated tools to support human decisions. We review, in Section 24.2, some relevant literature on the decision biases associated with using automated decision support tools ("automation bias"). We will also provide evidence (from our previous work and from work by other authors) of automation bias in the use of CAD for mammography (Section 24.3). This will be followed by explanations (conjectured by us and other authors) to account for these "automation bias" phenomena (Section 24.4). Section 24.5 discusses some implications of the work described in this chapter for the evaluation and design of CAD. Section 24.6 contains concluding remarks.

24.2 A BRIEF OVERVIEW OF THE LITERATURE ON AUTOMATION BIAS

Literature on human factors in engineering and computing has long recognized that even well-designed automation may fail to deliver the improvements it promised, due to initially unexpected effects on people. For instance, Bainbridge (Bainbridge, 1983) summarized known "ironies of automation" in industrial control: automated systems that were meant to substitute human operators and deliver more accurate, error-free performance actually relegated operators to narrower but more critical tasks (monitoring of the automated process and emergency intervention), while depriving them of the hands-on practice that these more critical tasks required. So, the task originally performed by a person alone was now performed by a "human-machine system" in which, however, the person's performance was made worse by how automation reshaped human tasks.

A partial solution to this has been seen in automation that assisted human experts rather than supplanting them, and this category of computer systems is actually widespread in "decision support" applications. However, here too there are concerns. We will refer to this set of related concerns by the label "automation bias," used by some researchers during the last decade.

As early as 1985, Sorkin and Woods (Sorkin, 1985), using signal detection theory, concluded that in a monitoring task performed by a human-machine system (a person assisted by a computer), optimizing the computer's performance as though it had to perform the task by itself would not always optimize the performance of the human-computer system. A human-machine system can only be optimized or improved as a whole: improving the automation alone or human training alone may be ineffective or wasteful.

"Automation bias" refers to those situations where a human operator makes more errors when being assisted by a computerized device than when performing the same task without computer assistance (Mosier, 1998; Skitka, 1999). Similar or related concepts are automation-induced "complacency" (Wiener, 1981; Singh, 1993; Azar, 1998) and "over-reliance" on automation (Parasuraman, 1997).

Studies of automation bias have been predominantly conducted in the domain of aviation (and only rarely in medical domains) and typically investigate the behavior of pilots or air traffic controllers assisted by computerized decision aids. Although, at first, the activities conducted by aviation professionals may appear very different from the activities involved in breast screening, they arguably have much in common. They are both "signal detection" tasks (McNicol, 1972), in which the human user is faced with the task of detecting some crucial perceptual pattern (a target) in a complex display. Both missing some targets (false negative error; FN) and identifying a target where none is present (false positive error; FP) carry costs. For example, one aviation task is to identify, in a complex radar image, any airplanes that present dangers of collision. In the case of breast screening, a reader must interpret the appearances in an X-ray to determine the presence of cancer. Automation, in the form of computerized advisory tools, is introduced in both domains to prompt the human operators to help them detect their targets.

In an influential paper, Parasuraman and Riley (Parasuraman, 1997) discuss ways in which this kind of human-computer interaction can go wrong, citing both anecdotal evidence and results from various empirical studies. They talk about three aspects of ineffective human use of automation: *disuse*, i.e. underutilization of automation, where humans ignore automated warning signals; *misuse*, i.e. over-reliance on automation, where humans are more likely to rely on computer advice (even if wrong) than on their own judgment; and *abuse*, when functions are automated "without due regard to the consequences for human... performance and the operator's authority" (Parasuraman, 1997).

The phrase "automation bias" was introduced by Mosier (Mosier, 1998) when studying the behavior of pilots in a simulated flight. They divided inappropriate responses to automated prompts into *errors of commission* and *errors of omission*. An *error of commission* occurs when a decision-maker follows automated advice even when the automated aid is wrong. In the case of CAD for mammography, an error of commission would be to recall a patient for a mammogram appearance that does not indicate cancer but which CAD has incorrectly prompted. On the other hand, an *error of omission* occurs when a decision-maker, given evidence of a problem, fails to react appropriately because the automated tool did not highlight the problem. In the case of mammography, a reader's error of omission would be failing to recall a patient because CAD failed to prompt the area of the mammogram where a cancer is located.

The finding that automation may bias pilot decisions was subsequently replicated in studies using non-pilot samples, namely, student participants in laboratory settings simulating aviation monitoring tasks (Skitka, 1999). Essentially these studies showed that, when the automated tool behaved reliably, participants using it made more correct responses than those without the tool's aid. However, if the automation was imperfect (i.e. unreliable for some of the tasks), participants using it were more likely to make errors than those who performed the same tasks without automated advice. For the monitoring tasks that Skitka and colleagues used in their studies, the participants had access to other (non-automated) sources of information. In the automated condition they were informed that the automated tool was not completely reliable but all other instruments were 100% reliable. Still, many chose to follow the advice of the automated tool even when it was wrong and was contradicted by the other sources of information. The authors concluded that these participants had been biased by automation and interpreted their errors (especially their errors of omission) as a result of complacency or reduction in vigilance.

Individual differences seem to play an important role in human reactions to automation (Skitka, 1999, 2000; Dzindolet, 2003). One would expect that more experienced people would tend to be less susceptible to bias from automation, when performing tasks in their area of expertise. However, as highlighted above, automation bias effects have been reported for both laymen (e.g. student participants) and experts (e.g. air pilots and clinicians). In fact, Galletta and colleagues (Galletta, 2005) have shown (in a verbal skills domain: the use of spelling and grammar checking software) that more skilled and experienced individuals were more likely to be damaged by certain types of computer errors (i.e. FNs) than were the less skilled individuals.

Some researchers have discussed human reliance on automation in terms of "trust" (Bisantz, 2001; Dassonville, 1996; Dzindolet, 2003; Lee, 1994, 2003; Muir, 1987, 1994, 1996; Singh, 1993; Tan, 1996). The idea is that the more human operators trust an automated aid the more likely they are to rely on the aid. So, it is important that this trust be well placed. However, if a human trusts (and thus relies on) an unreliable tool, then automation bias may occur (or misuse of automation as defined above). Similarly if a person does not trust a highly reliable tool, the person may end up "disusing" (as defined above) or under-using the tool, hence not obtaining the full potential benefits of automation.

Subjective measures of the trust of human operators in a computer tool have been found to be highly predictive of people's frequency of use of the tool (de Vries, 2003; Dzindolet, 2003). Use of automation (or reliance on automation in its generic sense) is usually assessed with observations of the proportion of times during which a device is used or by assessing the probability that operators will detect automation failures (Meyer, 2001).

A concept related to "trust" in automation is that of the "credibility" or "believability" of automation (Tseng, 1999). There are indications that people tend to perceive computers as infallible, and may put excessive trust in them (Martin, 1993; Fogg, 1999). Indeed, Dzindolet *et al.* (Dzindolet, 2003) have shown empirically that people have an inclination to trust and rely on an automated aid regardless of its reliability. However, there is also evidence of a pattern of behavior in which, as soon as humans become aware of the errors made by a computer tool, their trust in the tool, and their subsequent reliance on it, decrease sharply (de Vries, 2003; Dzindolet, 2002, 2003). This reduction in trust, in turn, can be attenuated by increasing people's understanding of the computer errors. Dzindolet *et al.* (Dzindolet, 2003) found that people who were told the reasons for the aid's errors were more likely to trust it and follow its advice than those who were not aware of these reasons.

24.3 AUTOMATION BIAS IN CAD FOR MAMMOGRAPHY

The effectiveness of CAD in assisting mammogram readers in breast screening has been the subject of many studies, with variable results between different settings, and some controversy (see reviews by Astley and Gilbert (Astley, 2004) and Bazzochi *et al.* (Bazzochi, 2007)). Here we present evidence of automation bias in the use of CAD, first reporting on a case study of the effects of CAD on the decisions of mammogram readers.

The case study described here was a follow-up to a previous retrospective controlled study of CAD conducted by other researchers for the UK Health Technology Assessment (HTA) program, on a specific CAD tool (Taylor, 2004b), hereon "the HTA study." This study is described in some detail in Section 24.3.1. Our team subjected the raw data from the HTA study to various exploratory statistical analyses, and focused on the interactions between correctness of computer output, difficulty of individual decision problems, and reader skills (Section 24.3.2). We conducted some subsequent experiments to study the effects of incorrect computer output on decisions (Section 24.3.3). We also report the results of more fine-grained analyses, which included previously unexamined details about reader decisions and CAD effects at the level of individual features in mammograms (Section 24.3.4). Finally, we briefly summarize findings from other researchers that report similar automation bias in the use of CAD for mammography (Section 24.3.5).

24.3.1 HTA study

The goal of the HTA study was to assess the impact of a commercial CAD tool, the then market leader in mammography

(R2 ImageChecker M1000) (FDA, 1998). This CAD tool had been shown to have a high sensitivity, that is, to prompt a high proportion of breast cancer appearances. An overall sensitivity of up to 90% had been reported (Castellino, 2000). However, high sensitivity comes at the expense of low specificity, that is, this tool generated many false prompts, with an average of 2.06 prompts per case in one study.

The HTA study was run with 50 readers, experienced in breast screening, who examined 180 cases (a mixture of 60 cancers and 120 normal cases) distributed in 3 sets of 60. All participating readers saw all the cases in two different experimental conditions: (1) with CAD; and (2) without CAD. The order of conditions was randomized across the participants. In both conditions, the participants saw two versions of each case: (1) the mammograms positioned on a standard viewing roller; and (2) a digitized version of the mammograms printed on paper. In the "with CAD" condition, the printouts contained the prompts generated by CAD. Participants were asked to make their decisions as to whether a case should be recalled for further tests as though they were viewing the mammograms as single readers in the UK breast screening program. More details of the procedures can be found in Taylor *et al.* (Taylor, 2004b).

Analysis of the results showed no statistically significant impact of CAD (no improvement and no reduction) on the readers' sensitivity and specificity (Taylor, 2004b).

24.3.2 Interactions between correctness of computer output, difficulty of cases, and human skills

The authors of the HTA study kindly granted us permission to analyze their data in our investigation of the fault-tolerant characteristics of human-machine systems (i.e. the ability of computers and their users to compensate, to some extent, for each other's errors), performed under the Interdisciplinary Research Collaboration on Dependability (DIRC), funded by the UK Engineering and Physical Sciences Research Council. Although conventional statistical analyses (e.g. ANOVA) of the data from the HTA study showed no significant impact of CAD on average, our subsequent exploratory analyses indicated that CAD was actually affecting readers' decision making in systematic ways (Alberdi, 2005).

These supplementary exploratory analyses were informed by probabilistic modeling of the system formed by the CAD machine and the reader (Strigini, 2003), and motivated by previous work on computer systems with diverse redundancy (Littlewood, 2002): systems in which the risk of erroneous results is reduced by having two or more different programs to calculate these results, and comparing their outputs for consistency.

Probabilistic modeling highlighted how variation and covariation in the "difficulty" of input cases (patient mammograms) for CAD and for the reader substantially affect the accuracy of the overall system (i.e. of the final decision about a patient). For example, it showed that it may be misleading to focus on the average probabilities of the failures of either the reader or CAD: reducing either or both of these may not substantially improve the statistics of the accuracy of decisions made. What matters more is how often the CAD tool fails on the same case on which the unaided reader would also fail, so that CAD brings no improvement; and whether, and how often, erroneous behavior by the CAD tool may negatively affect the reader's own decision.

Our analyses focused on the different effects of correct vs. incorrect CAD output, evaluating their interaction with *case difficulty*. The difficulty of a case was defined as the fraction of readers in the study who reached a wrong decision about that case. The correctness of CAD prompting and of human decisions was defined as follows. CAD output is considered to be "incorrect":

- for a cancer, if CAD "misses" the cancer (i.e. it gives an FN), which it can do in two ways: (a) by failing to place any prompt on the mammogram, or (b) by placing prompts in areas other than the actual location of the cancer (all prompts are "false"). CAD is said to provide "correct output" for a cancer (i.e. it processes it correctly; true positive) if it prompts the area on the mammogram where the cancer is located, even if it also prompts other areas of the mammogram.
- for a normal case, if CAD places any prompt on the mammogram ("incorrectly marked" normal case; FP). CAD is said to provide "correct output" for a normal case if it places no prompts ("unmarked" normal; a "true negative").

When referring to readers' decisions, "error by a reader" (or "incorrect decision") means that he/she recalls a normal case (reader's FP) or does not recall a cancer (reader's FN); "correct decision" means that he/she does not recall a normal case or recalls a cancer.

In this chapter we focus on evidence concerning CAD effects on the frequency of FN failures by readers. This will be sufficient to illustrate the possibility of "automation bias" phenomena. In breast screening, an FN is a more severe failure than an FP, although false recalls and the unnecessary investigations (radiology, biopsy) associated with them can contribute to serious psychological and physical damage to healthy patients as well as impose high costs on a medical organization.

The supplementary analyses of the HTA results (Alberdi, 2005) indicated that:

- the correctness of the computer output affected the decisions of the readers. In particular, correct computer prompting was likely to help readers in reaching a correct decision while incorrect prompting hindered their decision making.
- reader and computer errors were strongly correlated; CAD tended to (a) prompt correctly those cancers which readers were also likely to recall without CAD; and (b) process incorrectly most of the cancer cases that readers would also miss unaided; consequently, for those cases that were more difficult for humans to interpret, the computer was less likely to give useful support.

Additionally, we used logistic non-linear regression (a form of interpolation on the raw data) to look for possible general patterns in the effect of CAD. This showed that CAD tended to make cancers that were relatively "easy" (low "difficulty") "easier," and cases which were relatively difficult even more difficult (Povyakalo, 2004).

An abstract representation of these effects is shown in Figure 24.2. The horizontal axis represents case difficulty without computer aid; the vertical axis shows the impact of CAD, as the difference between difficulty of a case when seen with CAD and

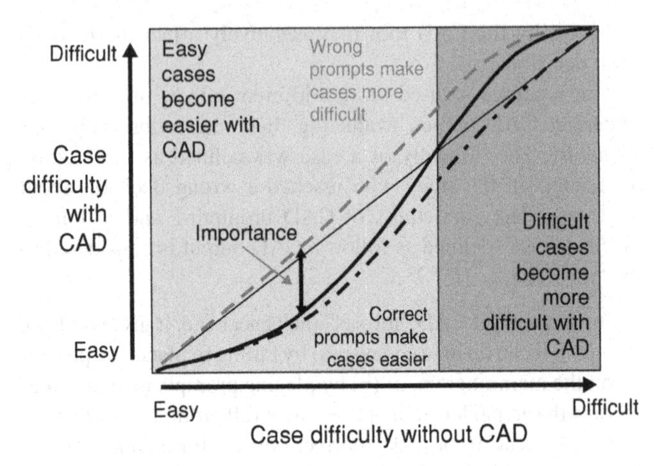

Figure 24.2 Impact of CAD on the HTA results, as a function of "case difficulty."

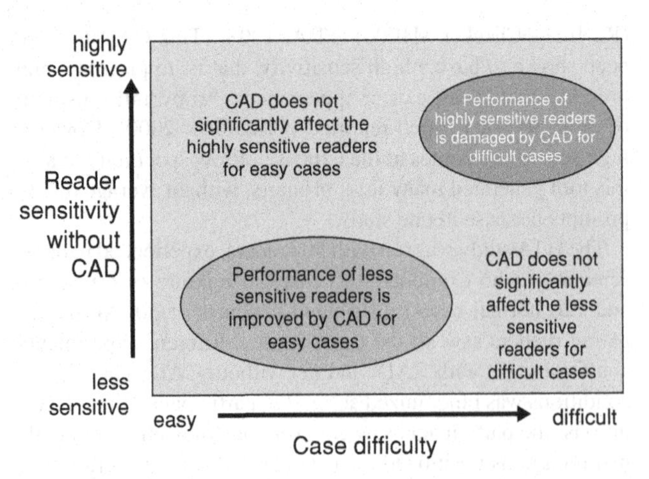

Figure 24.3 The pattern of effects of CAD suggested by the exploratory statistical analyses of the data from the HTA study (schematic representation). Whether the average effect of CAD is positive or negative would depend on the frequencies of the different groups of cases and of readers.

its difficulty without CAD. So, a point below 0 on the y-axis indicates a cancer case for which CAD appears to reduce the rate of FN errors. In other words, readers detect these cancers more often when using CAD than when not using CAD. The curves show the regression estimates for the mean value of the impact of CAD as a function of the difficulty of cases. The dashed curve is obtained from the data for the incorrectly prompted cancers, the dotted-dashed curve from the correctly prompted cancers, and the solid curve from all cancers together.

Figure 24.2 shows that, for "difficult" cases, incorrect decisions were more common with computer aid than without. Since less sensitive readers were very unlikely to recognize "difficult" cancers anyway, this increase in FN errors for "difficult" cases must be due to an adverse effect of the computer aid on the decisions of the more sensitive readers, plausibly caused by incorrect computer prompts (see Figure 24.2, again). Similarly, for the "easy" cases, without computer aid the less sensitive readers made a larger number of incorrect decisions: they were thus more likely to benefit from the correct computer prompts on "easy" cases than on "difficult" ones, and on easy cases they were more likely to benefit than the more sensitive readers.

Thus, the use of computer support is likely to be more beneficial for less sensitive readers and more questionable for the more sensitive ones, especially when these are dealing with the more "difficult" cases. This conjecture is supported by additional (logistic regression) analyses of these data (Povyakalo, 2006). Figure 24.3 shows an abstract representation of the effects of the interaction between case difficulty and unaided readers' sensitivity as estimated by those analyses.

24.3.3 Empirical study of the effects of false negative prompting

We summarize here the results of two follow-up studies, investigating in more detail the effects of FNs from CAD on human decision making (Alberdi, 2004a).

Study 1 used a set of mammograms containing a large proportion of cancers that CAD had missed. All other characteristics of the test set were kept as similar as possible to the sets used in the HTA study, to make sure readers perceived this study as a natural extension of the HTA study (all 20 readers in Study 1 had also participated in the HTA study) and to behave in a comparable way. Study 1 used essentially the same procedures used in the HTA study, except that the readers in Study 1 saw all the cases only once, always with the benefit of CAD. Human and CAD errors were defined as in the previous section.

Our original goal for Study 1 was to collect empirical data to estimate the probability of reader error given an FN error by the CAD tool, rather than to compare the performance of readers with and without CAD. However, the results were unexpected and intriguing: the proportion of correct decisions was very low, and particularly so for cancer cases not prompted by the tool. This suggested that CAD errors may have had a significant negative impact on readers' decisions. To explore this issue further, we designed Study 2, in which a new but equivalent set of readers read the same mammograms without computer aid. Study 2 used the same procedure as in Study 1, except that readers did not see the computer prompts.

Reader sensitivity in Study 2 (without CAD) was significantly higher (73%) than in Study 1 (with CAD: 61%). The measured difference between the with-CAD and without-CAD conditions was more acute for cancer cases that CAD had processed incorrectly and, crucially, even more so for cancer cases where the tool had placed no prompts at all (54% sensitivity without CAD vs. 33% with CAD). See Alberdi *et al.* (Alberdi, 2004a) for details.

These findings strongly suggest that, at least for some categories of cases, FN errors by the CAD tool (especially absence of prompts on cancer cases) had a significant detrimental effect on human decisions. Similar damaging effects of CAD have been reported by Zheng and Taplin (Zheng, 2004; Taplin, 2006) (see Section 24.3.5).

24.3.4 Analysis at the level of individual mammographic features

CAD tools are designed to help in the identification of individual features in a mammogram. However, medical studies most frequently measure CAD's effectiveness in terms of readers' recall of a case seen as a whole. This makes sense for the purpose of determining whether a particular tool helps or not, but not to understand whether its effect on the reader's work is as intended by the designers. Consistently with most studies, and partly for practical reasons, our analyses of the HTA study (and the follow-up experiments, see Section 3.3) also used data about reader decisions collected at the level of *cases* (recall or no-recall decisions), rather than at the level of mammographic *features* (which features the reader judged to be suspicious enough to warrant a recall). Reader decisions were classified simply as recall/no recall decisions, without further details. This omitted much information from the readers' reports that may shed light on how exactly the CAD prompts affected readers' decisions.

For example, as defined earlier, a reader's decision was classified as correct if it entailed recalling a patient who had cancer. However, this classification did not take into account whether the reader had actually detected the area in the mammogram where the cancer was located. To illustrate how an apparently correct recall may be due to the wrong process, we can imagine a mammogram where the cancer features were difficult to identify, and on which CAD would only add "false" prompts, i.e. on non-cancer areas. It is conceivable that a reader who did not recall the patient when not using CAD might then recall them when using CAD, after judging as suspicious some non-cancer area. Although this was seemingly a correct decision, it was, in fact, a recall prompted by the "wrong" mammographic appearance. In other words, the decision appears to be correct at the *case level* because a patient with cancer has been recalled. However, it is a clearly incorrect decision at the *feature level*: the specific area that has triggered the recall contains no cancer.

We outline in this section the results of a study (Alberdi, 2008) funded by Cancer Research UK, whose goal was precisely to conduct detailed feature-level analyses of previously unprocessed data from the HTA study.

Information about the specific features used by readers to make their decisions was available in the mammographic reports that readers filled in during the HTA study. However, it had not been incorporated into the electronic databases used in the earlier analyses; these only contained information about readers' recall decisions, malignancy of the case, and correctness of CAD prompting. In this new study (Alberdi, 2008), the previously unexamined, detailed information from several thousand mammography reports from the HTA study (5,839 mammography reports produced by 50 readers for 59 cancer cases in the previous study) was entered into a new electronic database, including the specific mammographic features that had been identified by the readers. This information was used to elucidate how judgments about features were affected by CAD.

The new electronic database contained information about all the different mammographic features (areas of interest) which had one or more of these characteristics: (a) identified by an expert radiologist as cancer regions, *a posteriori*, based on biopsy reports; (b) marked by CAD ("prompted") with a prompt; and (c) marked by a reader on the printout of the mammogram in the mammography report during the HTA study. Henceforth we will refer to these areas (features) as Marked Areas Of Interest (MAOIs).

Additionally, the new database included other information that readers entered in each mammography report, namely, the degree of suspicion and type of abnormality (mass, calcification, asymmetry) the reader assigned to each MAOI he/she marked, plus other possible notes, and a recall/no recall decision about the whole case.

Using this newly extracted information, the recalls *of cancer cases* were carefully classified into "true-target," if the reader identified at least one cancer feature as suspicious (or at least as the most suspicious in the mammogram), and "false-target" otherwise. About 13.5% of recall decisions were found to be "false-target recalls," that is, they were due to non-cancerous features. This pattern was independent of whether the readers were using CAD or not.

In other analyses, we investigated how readers reacted to features prompted by CAD and how the tool affected readers' identification of cancer features. These analyses showed that the lack of significant effect of CAD use on readers' sensitivity, observed at the level of whole mammograms (recall decisions) in the HTA study (Taylor, 2004b), is also present, *on average*, at the level of individual cancer features. This is the result of significant effects (in opposite directions) on the reader by individual prompts. More specifically, in comparing the decisions made with CAD against those made without, we found that:

- for those lesions that CAD prompted, readers were more likely to examine them when using CAD (i.e. seeing the prompts) than when not using CAD (a desirable effect of CAD).
- for those lesions that CAD prompted, readers were more likely to consider them malignant when using CAD than when not using CAD; in other words, many lesions that readers *did* consider in their decisions even when looking at a case without CAD were treated as *more suspicious* when the case was seen with CAD. This suggests that CAD prompts affect readers' *interpretation* of features and were not used merely as attention cues but also as *diagnostic* cues, going beyond the scope of computer aided *detection*.
- for those cancer features that CAD did not prompt, readers were less likely to detect them, or to consider them malignant, when using CAD than when not using CAD (an undesirable effect of CAD).
- false prompts made non-cancer features more likely to be classified as cancer by readers when using CAD than when not using CAD; this effect outweighed that of true prompts increasing the probability of readers correctly classifying cancer features.

Additionally, similarly to what we described in Section 3.2, exploratory (logistic non-linear) regression analyses were conducted using as factors:

- readers' "*effectiveness*," that is, readers' ability to discriminate between cancers and normal cases, calculated as the difference between a reader's measured sensitivity and 1 minus the reader's measured specificity.
- the "*difficulty*" of the features, that is, the fraction of all readers that failed to detect the feature.

Figure 24.4 shows the results of the regression analysis. The areas marked in the figure represent different degrees of estimated CAD impact. The dashed lines bound areas where the effect is statistically significant. For example, for readers of "effectiveness" 0.50, the estimated CAD impact is between 0.1

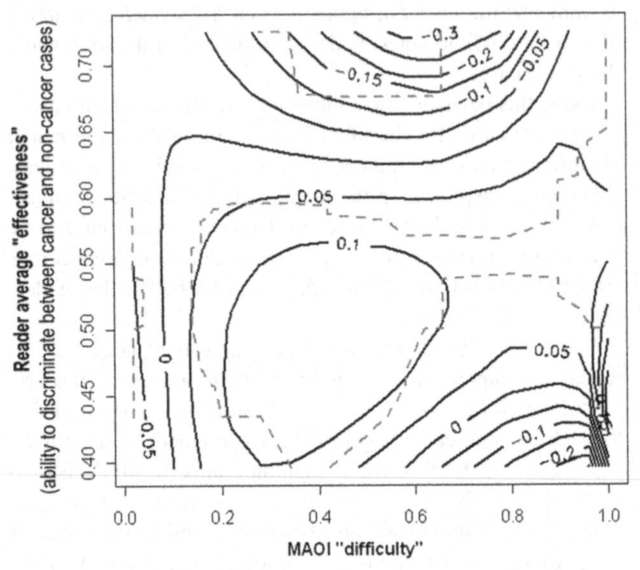

Figure 24.4 "Feature-level" study of the HTA results: exploratory regression analysis of readers' reactions to prompts.

and 0.2 when judging correctly prompted cancer MAOIs of 40% difficulty, that is, they fall within the big closed area in the center. Since this closed area is mostly within a dashed outline, this estimated impact is statistically significant. The estimates suggest that a CAD prompt significantly reduced (by up to 10%), compared to what it was without CAD, the probability of FN error for the less effective readers (those whose discrimination ability is less than 60%; more than two-thirds of our sample), and mostly on MAOIs whose difficulty (measured across all readers) is moderate.

Another statistically significant pattern appears at the top of the figure. For the 10–20% most effective readers there is a significant *negative* estimated effect, i.e. an estimated *increase* in the probability of FN errors about features of moderate difficulty. Again, this supports, at the level of features, the earlier finding that prompting may damage the decisions of more skilled readers for some categories of patients. See (Alberdi, 2008) for more details.

24.3.5 Further evidence of automation bias in the use of CAD

In this section we present evidence collected by other researchers that supports some of the findings reported so far.

Zheng and colleagues (Zheng, 2001, 2004) conducted a retrospective laboratory study of seven radiologists interpreting digitized mammograms in five different conditions. In one condition, the radiologists saw no prompts, whereas the other four conditions were created by manipulating the images, by adding CAD prompts to produce different FN and FP rates of CAD prompting. These authors found that a reduction of prompting sensitivity and specificity significantly increased the radiologists' FN rates in unprompted areas. Their conclusion was that CAD tools have the potential to significantly improve diagnostic performance in mammography, but poor combinations of CAD's FN and FP rates could adversely affect radiologists' performance, for both prompted and unprompted areas.

Similar results were found by Taplin *et al.* (Taplin, 2006) in a retrospective study comparing the sensitivity of 19 radiologists who read a set of 341 mammograms twice, with and without CAD. They found that visible mass cancer lesions that had not been prompted by CAD were less likely to be considered as cancer by radiologists than they were when CAD was not used. Use of CAD was associated with: (a) improved radiologist sensitivity for visible cancers prompted by CAD; and (b) decreased reader sensitivity for unprompted visible cancers.

A different type of bias has been reported by Fenton *et al.* (Fenton, 2007) in one of the largest studies to date, which reviewed decisions, with and without CAD, by a large population of community radiologists (in 43 screening centers) for over 200,000 patients in the course of 4 years. The authors report negligible improvement in cancer detection for readers using CAD and significantly reduced overall accuracy, assessed by use of receiver operating characteristic (ROC) curves. CAD-supported readers exhibited reduced specificity; recall and biopsy rates were significantly increased without an associated rise in sensitivity.

24.4 POSSIBLE EXPLANATIONS FOR AUTOMATION BIAS IN CAD USE

Here we look at cognitive processes or strategies that may be underlying the examples of apparent automation bias described so far. We present these mechanisms in the context of CAD for mammography, but we believe they are relevant for other computer assisted decision environments.

24.4.1 Effects of CAD's false negatives

The results of our analyses and empirical studies summarized earlier in this chapter suggest that incorrect outputs of the CAD tools may have biased the decision making of at least some of the readers for some categories of cases. In particular, the results indicate that, for difficult cancers, FN errors in prompting often led to incorrect reader decisions.

Using Skitka *et al.*'s (Skitka, 1999, 2000) terminology introduced earlier, these incorrect decisions could be classified as "errors of omission": readers interpreted absence of prompts on (an area of) a mammogram as an indication of "no cancer" and therefore failed to take appropriate action (i.e. they failed to recall the patient). There are, however, important differences between Skitka *et al.*'s tasks and mammogram reading as investigated in the HTA study and our follow-up studies. In the former, the participants seemed to use the computer output to replace calculations they could perform otherwise by using alternative, highly reliable, non-computer-mediated indicators. In contrast, uncertainty in mammogram reading is greater. The readers did not have access to any sources of information other than the X-ray films (mammograms) and the output of the CAD tool on digitized versions of the films. It is important to remember that CAD prompts are designed merely as attention cues – and not to replace other sources of information.

With CAD, over-reliance on automation could manifest itself in readers using prompts instead of thoroughly examining the mammograms, that is, readers becoming complacent, or less vigilant, when using automation – as hypothesized by Skitka's team and others to explain errors of omission (Meyer, 2001, 2002, 2003, 2004; Parasuraman, 1997, 2004; Skitka, 1999, 2000).

One could argue that, based on their past experience with the tool, readers tended to assume that the absence of prompts was a strong indication that a case was normal. CAD generates many FPs, which readers claim to find distracting. As a result, the absence of prompts in a mammogram may be seen as more informative than their presence, so readers may have become complacent, paying less attention than necessary to those mammograms that had no computer prompts.

A problem with the term "complacency" (and similar accounts of automation bias) is that it implies value judgments on the human experts. As Moray (Moray, 2003) has recently pointed out, the claim that automation fosters complacency implies that operators are at fault, when the problem often lies in the characteristics of the automated tools.

An alternative (perhaps complementary) way of accounting for the association between incorrect prompting and reader errors contemplates how readers deal with "indeterminate" cases – that is, cases where they have detected anomalies with unclear diagnosis, so that they are uncertain as to whether the cases should be recalled. For example, a reader perhaps noticed an ambiguous abnormality without using CAD; then, when looking at the CAD output, the reader revised his/her decisions for those features that he/she had already detected. In other words, they may have used the absence of prompts as reassurance for a "no recall" decision when dealing with features they found difficult to interpret. Conceivably readers were using any available evidence to resolve uncertainty. The implication is that the CAD tool was being used not only as a detection aid but also as a classification or diagnostic aid – specifically *not* what the tool was designed for. Readers' use of CAD in such a way has been reported before for the CAD tool investigated here (Hartswood, 2003) and for other similar CAD tools (Hartswood, 2000). The following transcript (Hartswood, 2003) is an example of a reader's use of prompts to inform his/her decisions: "This is a case where without the prompt I'd probably let it go ... but seeing the prompt I'll probably recall ... it doesn't look like a mass but she's got quite difficult dense breasts ... I'd probably recall." In other instances, readers were observed using the absence of a prompt as evidence for "no recall" (Hartswood, 2003).

Arguably, this use of CAD violates an explicit warning in the device labeling (1998): "A user should not be dissuaded from working up a finding if the device fails to mark that site." However, such warnings could only be useful if readers were aware they exploited prompts in this fashion.

An alternative conjecture about "indeterminate cases" posits that CAD may alter readers' decision thresholds. The conjecture is as follows: when seeing certain "indeterminate cases" without computer support, readers are likely to recall them for further investigation. However, with computer support, they know that supplementary information will be available in the form of CAD prompts; therefore, their preliminary decisions (before checking the prompts) are more likely to be "no recall" than they would be without CAD. A similar point was previously raised about other studies (Astley, 2004; Taylor, 2004a). In practice, this makes the presence of computer outputs a determinant for the "recall/no recall" decision, despite readers *never* changing a preliminary decision from "recall" to "no recall" based on absence of prompts, or being in violation of the prescribed procedures. For those indeterminate cases that are "difficult" cancers, CAD is likely to produce incorrect output, substantially reducing the chances of the case being recalled. Readers may not change any individual decisions because of absence of prompts, but they may change their decision thresholds for some cases due to the very presence of computer support. It may be very difficult (or even impossible) to prevent this by simple prescription.

24.4.2 Effects of false prompts

In this section we focus on observed and hypothesized effects of false prompts on reader decisions. As noted earlier, CAD tools show high sensitivity at the expense of specificity. A large proportion of the prompts generated by CAD are false. An expected outcome of the low specificity of CAD is that it will affect the specificity of its readers, predictably making readers less specific, as reported by Fenton *et al.* and Astley (Fenton, 2007; Astley, 2005). However, the opposite effects have been reported as well: some studies show that readers are more specific when using CAD than when not using it (Alberdi, 2004a; Taplin, 2006).

Less predictable, perhaps, is the finding that a high FP rate from CAD also reduces readers' sensitivity. We have already noted readers' tendency to over-rely on the absence of prompts as an indication of the absence of cancer. This phenomenon could be partly explained by an excess of false prompts. When a large proportion of the prompts are false, the absence of prompts becomes more informative than their presence.

Readers also report frequently that they find many of the CAD prompts distracting and confusing, possibly because the false prompts cause attentional overload. This distracting effect can manifest itself in different ways. For example, false prompts may distract readers' attention to unimportant areas of the mammogram, making them spend more time than due on normal appearances and leaving them with less time to exhaust their search for true cancer areas, especially if these have not been prompted by CAD.

Another related but different effect is that the presence of "obvious" false prompts may affect how readers interpret true prompts (Astley, 2005). Let us imagine that a reader is assessing a cancer area which is difficult to interpret but which has been correctly prompted by CAD. If the area and the true prompt are surrounded by many false prompts that are easily recognizable as spurious, the value of the true prompt may get diminished in the eyes of the reader. As a result, an area that a reader may have considered as cancer without CAD may be dismissed by the same reader when seeing it with CAD, even if the tool has prompted the area correctly. Analyses of readers' decisions at the level of features have provided evidence of many instances in which a reader performs worse with CAD than without for cancer appearances that CAD has prompted correctly. Similar effects have also been reported by Cupples *et al.* (Cupples, 2005).

24.5 SOME IMPLICATIONS FOR THE EVALUATION AND DESIGN OF CAD

24.5.1 Methodological implications for assessment of CAD

Some of the findings reported earlier in this chapter have obvious methodological implications for the evaluation and study of CAD use. We focus here on two limitations that arguably affect many evaluation studies.

24.5.1.1 Limitations of studying the impact of CAD at the level of case (vs. feature)

We have noted how, when looking at fine-grained analyses of CAD evaluation data (Alberdi, 2008), "false-target recalls" (i.e. recalls of cancer cases without correct identification of the cancerous features) appeared to be rather frequent (about 13.5%). This result should be treated with caution – our criterion for deciding whether a reader had correctly identified the cancer features is just one of several plausible, different criteria – but the evidence that these situations can occur so frequently has important methodological implications for the evaluation of CAD, for two reasons. Although CAD is designed to help in the identification of individual features, a CAD tool's effectiveness is most frequently measured in terms of readers' recalls of cases. This might seem appropriate for deciding how effective a CAD system will be in clinical practice (one of the important goals of studies about CAD). But if "false-target recalls" are frequent, the "case level" sensitivity measures might not represent the probability of the cancer being detected by biopsy, and this estimation error could differ between the with-CAD and without-CAD conditions, giving misleading estimates of CAD's effects.

Imagine, for example, a patient who has cancer on her *right* breast, and this area goes unnoticed by the reader, who is misled by CAD to interpret an area in the *left* breast as the site of cancer. It is difficult to judge during a trial whether such recall decisions would be a detrimental outcome of screening. In the best case, the patient would be properly diagnosed. In the worst case, the further examinations of the patient may not only fail to detect the true cancer but also result in damage to healthy areas of the patient's body.

The need for examining effects at the feature level is even more apparent with respect to the other purpose of research, i.e. to improve the design and use of CAD. It is important to know *how* a reader with CAD achieves a certain sensitivity and specificity at the case level. If this happens through unintended effects of CAD at the feature level, designers may receive misleading signals. For instance, in the above scenario of false-target recall, CAD might be contributing to a recall of a *patient* with cancer *despite not achieving its primary goal* of helping readers to notice cancerous *features*.

In conclusion, although the ultimate goal of CAD is to improve case recall decisions in breast screening, we argue that a fine-grained, feature-level analysis of data is necessary to understand how the tool actually affects the readers' decisions. This is necessary, for example, in order to discriminate among the different conjectures we have outlined in the previous section to explain the detrimental effects observed in specific categories of cases, and allow improvements in the design and use of CAD.

24.5.1.2 Limitations of studying the average effects of CAD on the reader population

Evaluation studies often focus on the effects of automation on readers' *average* performance. But averages may hide substantial variations between sub-populations. Our statistical analyses have proven very useful here. These analyses were motivated by the "diversity modeling" approach (Littlewood, 2002; Strigini, 2003), which focuses on how performance varies across classes of cases.

Whether analyzing data at the level of cases (considering readers' recall decisions) or at the level of features (considering readers' evaluation of specific mammographic areas), we found that CAD had both beneficial and detrimental *systematic* effects, depending on the reader–case (or reader–feature) pairs. In the HTA study, these effects just happened to cancel each other out, due to the composition of the sample of cases used, leading to the original study's result of no significant *average* impact.

If the systematic effects that we detected also occurred in practical clinical use, the net overall effect of CAD on the number of FN errors might still be close to zero, if the sample used in the study were representative of the populations of actual patients and readers. If not, the net overall effect of CAD could be positive, negative, or null, in addition to some possible transfer of risk between categories of patients. Samples used in studies are not necessarily representative of the population for every clinical use, both because these populations may differ, and because studies may use intentionally skewed samples for practical reasons. Then, the observed average effect from the use of CAD in a study is *no guidance* for its effect in clinical practice, nor does it provide precise enough feedback for improving CAD. We therefore argue that the effects of CAD should be analyzed by stratifying the sample of decisions appropriately (as proposed, for example, in Alberdi *et al.* (Alberdi, 2005)), attempting to identify "strata" of readers and cases (or features) within which CAD effects vary less dramatically. Our classifications of cases or features by "difficulty" and of readers by sensitivity or "effectiveness" are just a first step. Ideally, one would classify them by variables that can be estimated both in studies and in clinical use before introducing CAD and that are sufficiently predictive of the effect of CAD. Further studies are necessary to find any such variables. This would give specific feedback to designers about strengths and weaknesses of tools; and stratified measurements would support preliminary estimation of the net effect of a CAD system in any future clinical use from the results in an evaluation study ("preliminary" as there will be other differences between the study and clinical environment; importantly, the magnitude of "automation bias" effects is probably affected by the prevalence of targets, and many study samples have artificially high cancer prevalence). The overall effect on the whole population of patients and readers could be estimated as a weighted sum of the different effects expected for each class of case–reader pairs (or feature–reader pairs), weighted with the relative frequencies of these classes (Strigini, 2003).

24.5.2 Implications for design of the human-CAD system (tool, procedures, training)

As noted earlier, CAD errors in the HTA study were heavily correlated with the errors of readers without CAD (Alberdi, 2005). In other words, there was limited *diversity* between readers and the CAD tool. This may be due to two factors: the same mammograms that are hard to interpret correctly by readers are also hard for the tool to prompt correctly; or, incorrect prompting of a case by the CAD tool makes readers more likely to err in their turn. The data from the HTA study suggest that both factors were present.

Multiple approaches are open to those seeking to improve the effectiveness of CAD. We have pointed out that for some of the "automation bias" effects observed, there are alternative explanations, and research so far has not determined which ones are the real or predominant causes. More refined results would help to finesse approaches to improvement. Nonetheless, we can discuss here some approaches that are feasible, some of which are adopted in practice.

24.5.2.1 Diversity between errors by unaided readers and by CAD tools

An obvious path towards improvement is to improve either the CAD tool or its user, the reader. The tool may be improved by improvements in algorithms. The quality of the reader behavior may be improved by training. However, such improvements (besides being possibly expensive, and subject to a law of diminishing returns) may be ineffective because of lack of "diversity" between the error patterns of the CAD tool and the readers (see the mathematical models in Littlewood *et al.* and Strigini *et al.* (Littlewood, 2002; Strigini, 2003)). For example, if we improve a CAD tool's ability to prompt types of cancers that readers very seldom miss anyway, the final FN rate may not improve.

It may be better to target the interaction and correlation between machine and reader error. Informally, the goal would be to focus the design of CAD tools on correctly prompting those cases where unaided readers would tend to fail. So, improvements in CAD effectiveness could be sought by increasing this diversity, even without improving the average sensitivity or specificity of a CAD tool. The tool could be tuned to be more sensitive for classes of cases on which readers tend to be less effective; these cases are natural candidates for CAD to make a difference. This strategy is feasible, since the designers of an alerting tool have a degree of freedom in choosing the trade-off between its FP and FN rates. Especially if the tool uses a combination of algorithms for identifying situations that require an alert, tuning these multiple algorithms may allow some "targeting" of the peaks and troughs in FN and FP rates to different classes of cases. The overall improvement is determined by the changes achieved for each class, weighted with their relative frequencies (Strigini, 2003). Thus tuning a CAD tool for "diversity" from its readers may improve the latter's performance more than tuning it just to be as good as possible (in terms of sensitivity, specificity, or any weighted combination of the two). An additional desirable effect of this strategy may be the scope for designers to reduce the overall FP rate of the tool (which tends to be high if the main design goal is to ensure a low FN rate for all cases); as discussed earlier, readers find FP prompts annoying, and a high FP rate is plausibly one of the factors contributing to the observed "automation bias" effects. In view of how effects vary among readers, it may even be desirable to tune CAD differently for each individual reader, automatically or manually.

The importance of "diversity" between reader and CAD may have more general implications for designers of these tools (and alerting tools in general). A design philosophy aiming at reproducing in the CAD tool the outward behavior of human experts may limit the effectiveness of these tools: they may tend to help a reader most reliably on those cases where the reader needs less help. They will still be immune to fatigue and random lapses of attention, and thus serve the goal of making reader performance more uniform, if these were (as they are thought to be) important causes of human failure. But, even from this viewpoint, a tool design focused on helping readers with cases where they are least effective could still be the more effective solution (Strigini, 2003).

24.5.2.2 The impact of absence of prompts on reader error

The evidence presented for possible error-inducing effects of CAD points to apparent "automation bias," with readers behaving in "inappropriate" ways, e.g. tending to miss, when using CAD, cancers that they would notice when not using CAD. Defenses against such "inappropriate" behaviors may be devised with procedures and training as well as with changes to CAD tools.

Instructions for CAD use may say that readers must not allow machine prompts (or lack thereof) to change their opinion about a case from "recall" to "do not recall." If readers comply, then a FN error from a CAD tool would not lead to a reader's FN, except on cases on which the reader failed to notice the symptoms to start with. Computer support "could do no harm." But can readers actually comply with such guidelines? Much of an expert's skill is formed by non-explicit pattern recognition rules and heuristics. If experts slowly adapt to relying on the tool's prompts for advice, at least for some types of cases, they may not realize that they are doing so. At least two forms of protection can be pursued.

One protection would be to make it easier for the reader to obey the prescribed procedure, or more difficult to violate it. For instance, the user interface can request readers to mark all features they are going to consider in their decision, before showing the tool's prompts. However, such measures may be cumbersome to enforce, especially with high load on the readers, causing them to take advantage of all available help.

Another possible protection is to make the reader less prone to being influenced in the wrong way by the tool, for instance by practices that remind readers of the possibility of erroneous prompting. Which forms of reminders would be appropriate depends on the specific working practices in a given environment. Examples of erroneous prompting could be given during training, just as, for example, pilot simulator training includes an unrealistically high rate of mechanical failure. A probably more effective, and more controversial, process would be to plant fictitious cases with incorrect CAD output in readers' normal workload, rarely yet frequently enough to refresh their memory of types of possible prompting errors.

One should notice that simply asking readers not to allow the presence or absence of a CAD prompt to affect their judgment about a feature could be "normatively incorrect" – that is, a prescription for non-optimal decision making. Prompts have some informative value: with high-quality tools, absence of a prompt may be indeed a good indication of absence of cancer. It may be desirable to specify a heuristic procedure that readers could be trained to follow, that would give absence of prompts approximately the right weight in decisions. Simply training readers to ignore absence of prompts altogether is a simple, although suboptimal, such heuristic. The question is whether readers could ever succeed in applying it, since their reactions to absence of prompts may be involuntary.

A factor that will certainly affect readers' performance is their level of experience with CAD. Several authors assume, both for cancer screening CAD and for advisory systems in general, that the phenomenon of incorrect prompting causing reader error is limited to users who lack experience with the prompting tool, and disappears with use, as users learn the "strengths and weaknesses" of their tools. This seems a one-sided, over-optimistic statement, since it is not yet supported by decisive empirical evidence:

- experience will help readers to recognize that the tool is fallible, but it will also teach them that in many situations it is normally reliable. CAD tools are tuned to have low FN rate at the cost of a high FP rate, so they are indeed quite reliable for the kinds of cancer signs they target. A natural learned behavior may be to rely on absence of prompts to exclude cancer, even when cancer is present and could have been detected by the unaided reader.
- experience will teach readers to recognize some situations in which the tool tends to fail, but this will be more likely the more the situation is one in which the reader is reliable and so needs the tool's support less.

24.5.2.3 Cognitive overload

A plausible cause for many errors with CAD is simple overload. Readers with CAD have additional work to do compared to readers without CAD: examining prompts, which may be many if the CAD tool has a high FP rate. Reducing these FP rates may thus improve the sensitivity of readers with CAD (in addition to the probable effect of improving their specificity). As we suggested earlier, tuning CAD for lower overall sensitivity but more diversity (from readers) may allow designers to further reduce its FP rates. Other improvements can be sought by changing the information the CAD tool gives readers, for example, by not repeating prompts on features that readers have already noticed and marked.

A likely factor in the readers' cognitive load is their concern to explain the presence or absence of prompts in terms of regular patterns of CAD behavior (Hartswood, 2000; 2003). To reduce this load, the CAD tool could be made to explain its behavior when asked, for example incorporating decision support in the form of knowledge-based diagnostic advice (Alberdi, 2004b). The challenge is to reliably produce the sorts of accounts that would be useful to a reader, which calls for an understanding of the sorts of explanations where confusion may arise (Hartswood, 2003).

24.6 FINAL REMARKS

The main findings we have shown in this chapter are indications that:

- CAD prompts do help readers detect some cancers, which is the main purpose of CAD use.
- the presence of a prompt on a cancer feature makes readers not only more likely to detect the feature but also more likely to *interpret* the feature as cancerous, which goes beyond the scope of computer aided *detection*.
- the absence of prompts on a cancer feature can reduce a reader's ability to detect the abnormality.
- an excessive number of false prompts might damage not only reader specificity but sensitivity as well; it is one of the possible reasons for readers being less likely to notice true cancer features if these have not been prompted by CAD.
- CAD's effects on reader decisions vary in complex ways, being beneficial for some groups of readers and cancer appearances and detrimental for others.
- this combination of systematic positive and negative effects of CAD on reader decisions is likely to explain the lack of average effects of computer prompts on reader decisions reported in some evaluations of CAD.
- any detrimental effects from CAD are not limited to the less accurate readers but, in fact, can be more acute for very skilled readers, at least for some categories of cancer features.

The general picture is thus mixed. In an ideal scenario, CAD would simply help readers to notice cancers that they would have otherwise missed, and thus improve their sensitivity, without excessive reductions to their specificity. However, this benefit is limited by various somewhat subtle factors including: (1) possible "lack of diversity" between CAD tools and readers, meaning that the rare errors by CAD occur in precisely those cases in which a reader would need correct prompting; (2) "automation bias," meaning that sometimes CAD prompting (if incorrect, and sometimes even if correct) causes mistakes that a reader would not have made without CAD, because CAD use changes the readers' mental processes. To design CAD tools and activities, or to assess the benefit to be expected from CAD, requires consideration of all these factors.

We have also suggested some methodological recommendations:

- to assess the benefits and risks of using a CAD tool, statistical analyses need to go beyond average sensitivity and specificity, but take into account the role of readers' skills, the difficulty of features and cases, and the effect of correct or wrong computer prompting on reader decisions.
- to understand better how CAD affects reader performance, it is important to study the decisions of readers at the level of features (vs. case level).

CAD tools are continually evolving and improving. It is tempting to assess a tool by its average FP and FN rates. But we have shown how effectiveness for improving readers' decisions depends on more complex statistical characteristics of a tool's behavior, and on subtle effects on readers' work. Improving a tool's sensitivity and specificity may not be the only (nor, in fact, the best) way for making it more effective. Striving to improve CAD tools' discrimination by improving it for "easy" cases might actually *amplify* risks for difficult cases (when read by more effective readers) without improving performance of the more effective readers. Dedicated studies would be necessary to refute this hypothesis.

More research is certainly desirable towards more detailed understanding of how the technology affects human decisions. Its results would help to resolve the controversies about the effectiveness of CAD tools as well as to improve their design and assessment.

ACKNOWLEDGMENTS

The work described in this paper was supported in part by: Cancer Research UK (Population & Behavioural Science Committee) via grant C22515/A7339; the UK Engineering and Physical Sciences Research Council via the Interdisciplinary Collaboration on the Dependability of Computer Based Systems (DIRC), and via project INDEED, "Interdisciplinary Design and Evaluation of Dependability" (EP/E000517/1); and the European Union Framework Programme 6 via the ReSIST (Resilience for the Information Society) Network of Excellence, contract IST-4–026764-NOE.

We would like to thank: Rosalind Given-Wilson for her collaborations and expert radiological advice; Paul Taylor for his help in making available the HTA data and for several discussions of this set of studies; Carla Nunes, who contributed to designing the data entry procedures and performed the data entry in the Cancer Research UK project; and our colleagues in the DIRC project for early discussions of these topics.

REFERENCES

Alberdi, E., Povyakalo, A.A., Strigini, L., Ayton, P. (2004a) Effects of incorrect CAD output on human decision making in mammography. *Acad Radiol*, **11**, 909–918.

Alberdi, E., Taylor, P., Lee, R. (2004b) Elicitation and representation of expert knowledge for computer aided diagnosis in mammography. *Methods Inf Med*, **43**, 239–246.

Alberdi, E., Povyakalo, A.A., Strigini, L., *et al.* (2005) Use of computer-aided detection (CAD) tools in screening mammography: a multidisciplinary investigation. *Br J Radiol*, **78**, S31–S40.

Alberdi, E., Povyakalo, A.A., Strigini, L., Ayton, P., Given-Wilson, R. (2008) CAD in mammography: lesion-level versus case-level analysis of the effects of prompts on human decisions. *Int J CARS*, **3**, 115–122.

Astley, S.M., Gilbert, F.J. (2004) Computer-aided detection in mammography. *Clin Radiol*, **59**, 390–399.

Astley, S.M. (2005) Evaluation of computer-aided detection (CAD) prompting techniques for mammography. *Br J Radiol*, **78**, S20–S25.

Azar, B. (1998) Danger of automation: it makes us complacent. *APA Monitor*, **29**, 3.

Bainbridge, L. (1983) Ironies of automation. *Automatica*, **19**, 775–779.

Bazzocchi, M., Mazzarella, F., Del Frate, C., Girometti, R., Zuiani, C. (2007). CAD systems for mammography: a real opportunity? A review of the literature. *Radiologia Medica*, **112**, 329–353.

Bisantz, A.M., Seong, Y. (2001) Assessment of operator trust in and utilization of automated decision-aids under different framing conditions. *International Journal of Industrial Ergonomics*, **28**, 85–97.

Castellino, R., Roehrig, J., Zhang, W. (2000) Improved computer aided detection (CAD) algorithms for screening mammography. *Radiology*, **217**, 400.

Cupples, T.E., Cunningham, J.E., Reynolds, J.C. (2005) Impact of computer-aided detection in a regional screening mammography program. *Am J Roentgenol*, **185**, 944–950.

Dassonville, I., Jolly, D., Desodt, A.M. (1996) Trust between man and machine in a teleoperation system. *Reliability Engineering & System Safety (Safety of Robotic Systems)*, **53**, 319–325.

de Vries, P., Midden, C., Bouwhuis, D. (2003) The effects of errors on system trust, self-confidence, and the allocation of control in route planning. *International Journal of Human-Computer Studies*, **58**, 719–735.

Dzindolet, M.T., Pierce, L.G., Beck, H.P., Dawe, L.A. (2002) The perceived utility of human and automated aids in a visual detection task. *Human Factors*, **44**, 79–94.

Dzindolet, M.T., Peterson, S.A., Pomranky, R.A., Pierce, L.G., Beck, H.P. (2003) The role of trust in automation reliance. *International Journal of Human-Computer Studies*, **58**, 697–718.

FDA (1998) Pre-market approval decision. Application P970058 – http://www.fda.gov/cdrh/pdf/p970058.pdf. US Food and Drug Administration.

Fenton, J.J., Taplin, S.H., Carney, P.A., *et al.* (2007) Influence of computer-aided detection on performance of screening mammography. *N Engl J Med*, **356**, 1399–1409.

Fogg, B.J., Hsiang, T. (1999) The elements of computer credibility. In Altom, M.W., Williams, M.G. (eds.) *Proceedings of the ACM CHI99 Human Factors in Computing Systems Conference*. Pittsburgh, PA: ACM Press.

Freer, T.W., Ulissey, M.J. (2001) Screening mammography with computer-aided detection: prospective study of 12,860 patients in a community breast center. *Radiology*, **220**, 781–786.

Galletta, D.F., Durcikova, A., Everard, A., Jones, B.M. (2005) Does spell-checking software need a warning label? *Commun ACM*, **48**, 82–86.

Gromet, M. (2008) Comparison of computer-aided detection to double reading of screening mammograms: review of 231,221 mammograms. *Am J Roentgenol*, **190**, 854–859.

Gur, D., Sumkin, J.H., Rockette, H.E., *et al.* (2004) Changes in breast cancer detection and mammography recall rates after the introduction of a computer-aided detection system. *J Natl Cancer Inst*, **96**, 185–190.

Hartswood, M., Procter, R. (2000) Computer-aided mammography: a case study of error management in a skilled decision-making task. *Journal of Topics in Health Information Management*, **20**, 38–54.

Hartswood, M., Procter, R., Rouncefield, M., *et al.* (2003) "Repairing" the machine: a case study of the evaluation of computer-aided detection tools in breast screening. In Kuutti, K., Karsten, E.H., Fitzpatrick, G., Dourish, P., Schmidt, K. (eds.) *Eighth European Conference on Computer Supported Cooperative Work (ECSCW 2003)*. Helsinki, Finland: Springer.

Khoo, L.A.L., Taylor, P., Given-Wilson, R.M. (2005) Computer-aided detection in the United Kingdom National Breast Screening Programme: prospective study. *Radiology*, **237**, 444–449.

Lee, J.D., Moray, N. (1994) Trust, self-confidence, and operators' adaptation to automation. *International Journal of Human-Computer Studies*, **40**, 153–184.

Lee, J.D., See, K.A. (2003) Trust in computer technology. Designing for appropriate reliance. *Human Factors*, **46**, 50–80.

Littlewood, B., Popov, P., Strigini, L. (2002) Modelling software design diversity – a review. *ACM Computing Surveys*, **33**, 177–208.

Martin, C.D. (1993) The myth of the awesome thinking machine. *Commun ACM*, **36**, 39–44.

McNicol, D. (1972) *A Primer of Signal Detection Theory*. London: George Allen & Unwin.

Meyer, J. (2001) Effects of warning validity and proximity on responses to warnings. *Human Factors*, **43**, 563–572.

Meyer, J., Bitan, Y. (2002) Why better operators receive worse warnings. *Human Factors*, **44**, 343–353.

Meyer, J., Feinshreiber, L., Parmet, Y. (2003) Levels of automation in a simulated failure detection task. *IEEE International Conference on Systems, Man and Cybernetics, 2003*. Washington, DC.

Meyer, J. (2004) Conceptual issues in the study of dynamic hazard warnings. *Human Factors*, **46**, 196–204.

Moray, N. (2003) Monitoring, complacency, scepticism and eutactic behaviour. *International Journal of Industrial Ergonomics*, **31**, 175–178.

Mosier, K.L., Skitka, L.J., Heers, S., Burdick, M. (1998) Automation bias: Decision making and performance in high-tech cockpits. *International Journal of Aviation Psychology*, **8**, 47–63.

Muir, B.M. (1987) Trust between humans and machines, and the design of decision aids. *International Journal of Man-Machine Studies*, **27**, 527–539.

Muir, B.M. (1994) Trust in automation: part I. Theoretical issues in the study of trust and human intervention in automated systems. *Ergonomics*, **37**, 1905–1922.

Muir, B.M., Moray, N. (1996) Trust in automation: part II. Experimental studies of trust and human intervention in a process control simulation. *Ergonomics*, **39**, 429–460.

Parasuraman, R., Riley, V. (1997) Humans and automation: use, misuse, disuse, abuse. *Human Factors*, **39**, 230–253.

Parasuraman, R., Miller, C.A. (2004) Trust and etiquette in high-criticality automated systems. *Communications of the ACM*, **47**, 51–55.

Povyakalo, A.A., Alberdi, E., Strigini, L., Ayton, P. (2004) Evaluating "human + advisory computer" systems: a case study. In Watts, A.D. (ed.) *HCI2004, 18th British HCI Group Annual Conference*. Leeds, UK: British HCI Group.

Povyakalo, A.A., Alberdi, E., Strigini, L., Ayton, P. (2006) *Divergent effects of computer prompting on the sensitivity of mammogram readers*. London, UK: Technical Report, Centre for Software Reliability, City University.

Singh, I.L., Molloy, R., Parasuraman, R. (1993) Automation-induced "complacency": development of the complacency-potential rating scale. *International Journal of Aviation Psychology*, **3**, 111–122.

Skitka, L.J., Mosier, K., Burdick, M.D. (1999) Does automation bias decision making? *International Journal of Human-Computer Studies*, **51**, 991–1006.

Skitka, L.J., Mosier, K., Burdick, M.D. (2000) Accountability and automation bias. *International Journal of Human-Computer Studies*, **52**, 701–717.

Sorkin, R.D., Woods, D.D. (1985) Systems with human monitors: A signal detection analysis. *Human-Computer Interaction*, **1**, 49–75.

Strigini, L., Povyakalo, A.A., Alberdi, E. (2003) Human-machine diversity in the use of computerised advisory systems: a case study. *2003 International Conference on Dependable Systems and Networks (DSN03)*. San Francisco, CA: IEEE.

Tan, G., Lewandowsky, S. (1996) A comparison of operator trust in humans versus machines. *First International Cyberspace Conference on Ergonomics*.

Taplin, S.H., Rutter, C.M., Lehman, C.D. (2006) Testing the effect of computer-assisted detection on interpretive performance in screening mammography. *Am J Roentgenol*, **187**, 1475–1482.

Taylor, P., Given-Wilson, R., Champness, J., Potts, H.W., Johnston, K. (2004a) Assessing the impact of CAD on the sensitivity and specificity of film readers. *Clin Radiol*, **59**, 1099–1105.

Taylor, P.M., Champness, J., Given-Wilson, R.M., Potts, H.W.E., Johnston, K. (2004b) An evaluation of the impact of computer-based prompts on screen readers' interpretation of mammograms. *Brit J Radiol*, **77**, 21–27.

Tseng, S., Fogg, B.J. (1999) Credibility and computing technology. *Communications of the ACM*, **42**, 39–44.

Warren Burhenne, L.J., Wood, S.A., D'Orsi, C.J., *et al.* (2000) Potential contribution of computer-aided detection to the sensitivity of screening mammography. *Radiology*, **215**, 554–562.

Wiener, E.L. (1981) Complacency: is the term useful for air safety? *26th Corporate Aviation Safety Seminar*. Denver, CO: Flight Safety Foundation, Inc.

Zheng, B., Ganott, M.A., Britton, C.A., *et al.* (2001) Soft-copy mammographic readings with different computer-assisted detection cuing environments: preliminary findings. *Radiology*, **221**, 633–640.

Zheng, B., Richard, G.S., Sara, G., *et al.* (2004) Detection and classification performance levels of mammographic masses under different computer-aided detection cueing environments. *Acad Radiol*, **11**, 398–406.

PART V

OPTIMIZATION AND PRACTICAL ISSUES

25

Optimization of 2D and 3D radiographic imaging systems

JEFFREY H. SIEWERDSEN

This chapter provides an overview of technologies, applications, and methods of imaging performance optimization for two-dimensional (2D) and three-dimensional (3D) radiographic imaging modalities. The 2D radiographic modalities include various large-area X-ray projection imaging modalities, including radiography, mammography, fluoroscopy (arguably a 3D spatio-temporal imaging modality), and dual-energy imaging. Emphasis therein is upon digital X-ray imaging technologies, most notably active matrix flat-panel detectors (FPDs). The 3D radiographic modalities include volumetric imaging systems that form 3D tomographic reconstructions from large-area 2D projections, with emphasis upon tomosynthesis and cone-beam CT (CBCT) using FPDs.

Section 25.1 provides an overview of 2D and 3D radiographic imaging modalities. Section 25.2 briefly summarizes metrics of imaging performance, with emphasis on spatial-frequency-dependent, Fourier-based metrics of spatial resolution and noise. Section 25.3 presents the detectability index as an objective function for system optimization that combines such Fourier metrics with a spatial-frequency description of the imaging task. Finally, Section 25.4 presents a variety of examples of performance evaluation and optimization in 2D and 3D radiographic imaging.

25.1 2D AND 3D RADIOGRAPHIC IMAGING MODALITIES

25.1.1 Detectors for 2D and 3D radiographic imaging

A brief description is provided below of the various technologies for 2D and 3D radiographic imaging. Details of each abound in textbooks and the scientific literature (Bushberg, 2002; Hendee, 2002). Below, the basic principles, strengths, and limitations of these technologies are very briefly summarized.

25.1.1.1 Screen-film

The most prevalent conventional means of 2D X-ray radiographic image capture is a screen-film cassette – typically a silver halide film on a flexible plastic support in combination with an intensifying phosphor screen (or pair of screens – e.g. Gd_2O_2S:Tb). Such systems are relatively inexpensive, offer sub-millimeter spatial resolution sufficient for a broad range of examinations, and have represented the mainstay of radiographic examination through most of the previous century. While the technology is extremely mature and has benefited from decades of research, development, and optimization, it faces a variety of fundamental limitations. Images are not recorded in real time, and compromises must be made to simultaneously optimize image capture (i.e. the screen-film combination) and display (typically a light box). Screen-film systems often exhibit a fairly narrow latitude (exposure range over which the detector is sensitive). Such limitations have motivated the development of digital X-ray detectors.

25.1.1.2 Computed radiography (CR)

The development of CR systems over the last 20–30 years represents a significant advance. Such systems consist of a photostimulable storage phosphor (e.g. BaFBr, etc.) that records a "latent image" upon exposure, which is subsequently read out by laser scanning. For example, a red laser is scanned across the phosphor, thereby de-exciting metastable states in the phosphor to stimulate emission of (green-blue) light that is collected by a photomultiplier. Such systems provide digital image readout that separates image capture from display and, therefore, permits a variety of sophisticated image processing techniques. CR is prevalent across the spectrum of radiographic imaging applications, including chest, breast, and extremities. The limitation of non-real-time readout remains.

25.1.1.3 X-ray image intensifiers (XRIIs)

XRIIs typically consist of a large-area scintillator (e.g. CsI:Na) coupled to a photocathode at the input to a vacuum tube. Electrons emitted from the photocathode are accelerated across the vacuum tube and focused onto a smaller output phosphor, which in turn is coupled to a video camera. Such systems offer high gain (i.e. can operate at very low exposure rate) and real-time readout and have been the primary means of fluoroscopic imaging over the last ~50 years. Limitations include size (bulky vacuum tube), veiling glare, distortion, and susceptibility to magnetic fields.

25.1.1.4 Scanning slot detectors

Charge-coupled devices (CCDs) coupled to efficient scintillators present a variety of detector configurations that offer to overcome many of the limitations mentioned above. Large-area 2D images may be acquired by configuring a row of CCD detectors as a slit (or many rows of CCD detectors as a slot) and scanning the assembly across the field of view in coordination with the X-ray collimators. A major advantage of such scanning systems

The Handbook of Medical Image Perception and Techniques, ed. Ehsan Samei and Elizabeth Krupinski. Published by Cambridge University Press.

Table 25.1 *A few commercially available FPD systems. The table shows a partial listing of detectors developed by various manufacturers for a broad range of applications. Note: product names and specifications are subject to change.*

Example application	Pixel format	Area (cm^2)	Pixel pitch (μm)	Fill factor	Converter material	Model	Manufacturer
Radiography	2022×2022	41×41	200	~0.8	CsI:Tl	Definium 5000/8000	General Electric
	3001×3001	43×43	143	~0.7	CsI:Tl	Pixium 4600	Trixell
	2688×2688	43×43	160	NA	Gd_2O_2S:Tb or CsI:Tl	CXDI-40G/C	Canon
Fluoroscopy	2048×1536	40×30	194	~0.7	CsI:Tl	PaxScan 4030A	Varian
	2022×2022	41×41	200	~0.8	CsI:Tl	Innova 4100	General Electric
	2880×2880	43×43	150	>0.9 effective	a-Se	Safire	Shimadzu
	2881×2881	43×43	148	NA	CsI:Tl	Pixium 4343RF	Trixell
Mammography	3584×4096	25×29	70	>0.9 effective	a-Se	Selenia	Hologic
	2048×2816	18×24	85	>0.9 effective	a-Se	LMAM	Anrad
	2400×3100	24×31	100	NA	CsI:Tl	Senographe Essential	General Electric
MV Portal Imaging	1024×1024	41×41	400	~0.8	Gd_2O_2S:Tb or CsI:Tl	RID 1640-A	PerkinElmer

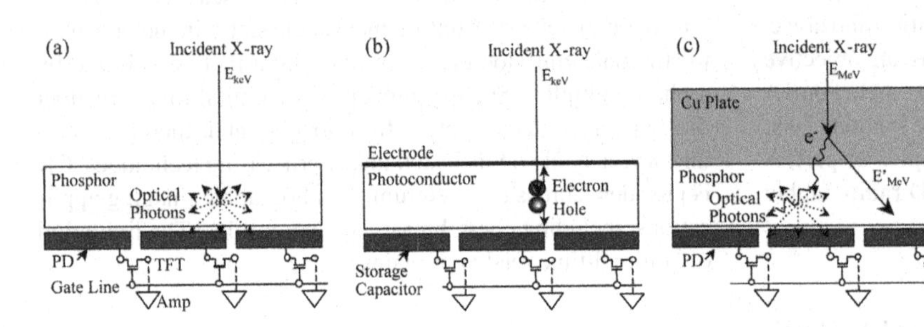

Figure 25.1 Cross-sectional illustration of (a) indirect-detection FPD, (b) direct-detection FPD, and (c) megavoltage (indirect-detection) FPD.

is the reduction in X-ray scatter inherent to the narrower-beam (slot) geometry, motivating research and development for chest and breast imaging. Limitations of such systems are the acquisition time, high heat loading of the X-ray tube, and susceptibility to patient motion.

25.1.1.5 Large-area CCD-based detectors

Producing large-area CCD-based detectors appropriate to human anatomy requires either tiling of CCDs or coupling of a large-area scintillator to the CCD via minifying optics. Such systems have been produced for applications ranging from chest imaging to mammography (Roehrig, 1994). These systems face limitations of cost, readout speed, and efficiency (loss of optical photons in the coupling optics).

25.1.1.6 Flat-panel detectors (FPDs)

The development of large-area X-ray detectors that are efficient, distortionless, real-time, and cost-effective posed a considerable materials science challenge over the course of the 1980s and 1990s. Hydrogenated amorphous silicon (a-Si:H) offered a semiconductor that could be fabricated in large areas (e.g. solar panels), is resistant to radiation damage (by virtue of its amorphous, non-crystalline nature), and could be used to form thin-film transistors (TFTs) in active matrix arrays (e.g. in liquid crystal displays) as well as photodiodes. Such forms the base technology for active matrix readout of both "indirect-detection" FPDs (for which the sensitive pixel elements typically consist of a scintillator – e.g. CsI:Tl – coupled to an a-Si:H photodiode) (Antonuk, 1997) and "direct-detection" FPDs (for which X-rays are converted to charge directly in a photoconductor, such as a-Se or PbI_2) (Rowlands, 1996). The process of converting X-rays to electronic charge is illustrated in Figure 25.1. FPDs have become a prominent, commercially available technology for numerous applications, several addressed in the following section. Systems have been developed with area up to 43 × 43 cm^2 and pixel pitch typically ~100–400 μm. A summary of example systems and specifications is shown in Table 25.1.

25.1.1.7 Photon counting detectors

In contrast to energy-integrating detectors – including both direct- and indirect-detection FPDs, for which signal intensity is proportional to the total energy deposited – photon counting detectors count the number of X-ray interactions per pixel during a given time interval. Photon counting can be achieved via several methods. For example, gas-based photon counters incorporate a gas layer between electrodes, which undergoes ionization created by the incident X-ray. The resulting electrons further ionize the gas in avalanche multiplication. Each X-ray interaction is therefore amplified many-fold, and a threshold can be set to discriminate X-ray interactions according to their energy (Maidment, 2006). Another photon counting technology typically employs a specialized pixel readout chip (del Risco Norrlid, 2005) that increments the photon count for each X-ray interaction above a specified energy threshold. Advantages of photon counting detectors include the near absence of electronic readout noise and high detective quantum efficiency (Cahn, 1999). Limitations of detector size and paralyzation at high exposure rates are actively investigated in ongoing research.

25.1.1.8 Future 2D and 3D detector technologies

The development of novel high-performance detector technologies presents a vibrant area of medical imaging research that promises to advance detector capabilities and open new areas of application. While FPDs present an important base technology for large-area X-ray imaging, current research offers to improve upon the efficiency, gain, noise, lag, and physical implementation of such technology. New pixel architectures incorporating multiple diode or TFT components as on-pixel amplifiers offer the potential for improved imaging performance at extremely low radiation dose by virtue of increased gain (Sawant, 2006). Similarly, high-efficiency, high-gain materials for direct-detection FPDs, including PbI_2, CdZnTe, and others offer potentially dramatic improvement in low-dose imaging performance (Antonuk, 2004), as do systems offering avalanche gain (Hunt, 2002). New readout architectures offer higher frame rates and reduced image lag (Boyce, 2005). Advances in photon counting detectors, briefly mentioned above, point to exciting new capabilities altogether based on the ability to discriminate X-ray photon energies, including digital dual-energy decomposition and scatter rejection. Finally, flexible detector substrates and the potential for "ink jet" style deposition of pixel components point to exciting new physical embodiments and reduction in cost.

Furthermore, it is worth noting a convergence of conventionally "radiographic" area detectors and conventionally "tomographic" detectors. Over the last two decades, X-ray projection imaging has advanced from screen-film to large-area active matrix FPDs that define the state of modern digital radiography. In the same time frame, CT detectors have advanced from single-row gas detectors to multi-row solid-state detectors that define the state of modern CT. The distinction between detectors for radiography and tomography has begun to fade and can be expected to diminish further in the coming decades.

25.1.2 Applications of 2D and 3D radiographic imaging

Below, a brief description of the various applications of radiographic detectors for 2D and 3D imaging is offered, summarizing basic applications, breakthroughs, and challenges.

25.1.2.1 2D radiography

The term "radiography" refers to a broad scope of applications ranging from general radiography, chest imaging, extremities imaging, etc., to more specialized exams such as mammography (discussed below). Herein, we use the term to refer to X-ray imaging that produces a static 2D projection.

Projection images are obtained from the differential attenuation of X-rays through the object being imaged as described by the Beer-Lambert Law:

$$N = N_0 \exp(-\mu t) \tag{25.1a}$$

where N denotes the transmitted number of X-rays, N_0 denotes the number of X-rays incident on the object, μ denotes the linear attenuation coefficient of the material, and t denotes the distance traveled through the material. Allowing for an energy-dependent spectrum of incident X-rays:

$$N(E) = N_0(E) \exp(-\mu(E)t) \tag{25.1b}$$

Further allowing for a heterogeneous object in which μ varies in depth (z):

$$N(E) = N_0(E) \exp\left(-\int_0^t \mu(z;E)\,dz\right) \tag{25.1c}$$

Considering an energy-integrating 2D detector, therefore, the radiographic image is

$$I(x,y) = \int_0^{E_{max}} N_0(x,y;E) \exp\left(-\int_0^t \mu(z;E)\,dz\right) dE \tag{25.1d}$$

Two limitations of 2D radiography are apparent. Consider a structure of interest (e.g. a soft tissue tumor) embedded within a surrounding (uniform or heterogeneous) medium. First, the contrast (i.e. signal difference) between the structure of interest and the background in a radiograph is much less than the intrinsic difference between their attenuation coefficients. That is, the difference between line integrals over $\mu(z)$ is much less than the difference in μ. Second, the integral over depth means that overlying anatomical structures are superimposed in the 2D radiograph, resulting in an anatomical background clutter that may decrease the conspicuity of subtle underlying features. The lack of spatial and material discrimination is an important motivation for the development of 3D radiographic imaging systems as described below.

Radiographic imaging systems vary widely in form and components, depending on the application; however, most radiographic systems employ a tungsten anode X-ray tube operating in the range 50–150 kVp. Imaging technique considerations include selection of kVp, added filtration, tube current, exposure time, geometry of the X-ray tube and detector (e.g. air gap and magnification), and thickness and position of the patient (e.g. posterior-anterior versus lateral). A typical technique for a chest exam employs a 120 kVp X-ray beam with short X-ray exposure (~30 ms) sufficient to freeze anatomical motion. An example chest radiograph acquired using an FPD is shown in Figure 25.2(a). Imaging tasks vary widely from application to application. In chest radiography, for example, tasks include detection

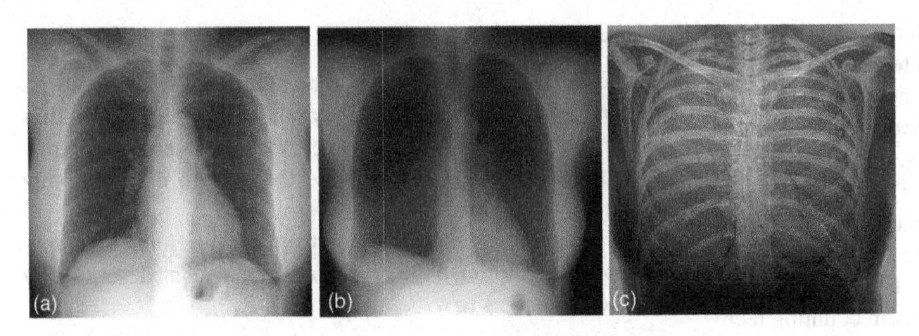

Figure 25.2 Example 2D radiographic images. (a) DR chest radiograph. (b) DE soft-tissue image. (c) DE bone image.

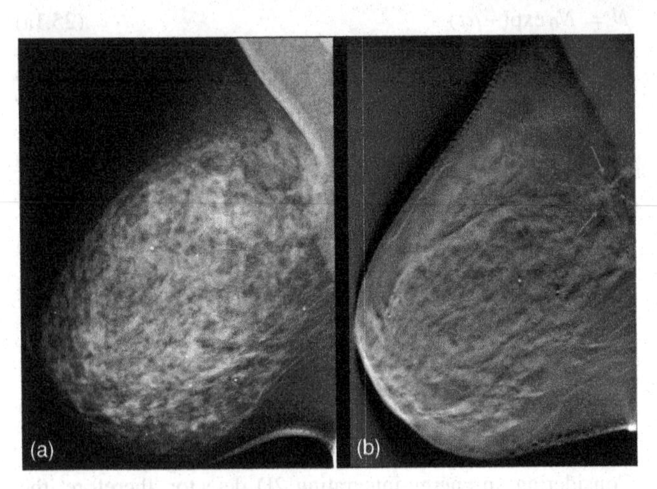

Figure 25.3 Example of 2D and 3D mammographic images. (a) DR mammogram. (b) Breast tomosynthesis image. (Images courtesy of Duke University.)

and diagnosis of lung nodules (early stage lung cancer), pneumonia, fracture, pulmonary embolism, and pneumothorax.

25.1.2.2 2D breast imaging (mammography)

Mammography refers to 2D static imaging of the breast – similar to radiography, except the X-ray source, X-ray energy, system geometry, antiscatter grid, detector, image processing, etc. are all tuned (optimized) to the specific task of detecting and/or diagnosing breast disease. The X-ray source typically incorporates Mo, Rh, and/or W targets, with X-ray energies in the range 20–40 kVp. System geometry typically incorporates breast compression paddles, near-contact geometry, and antiscatter grids. Modern digital mammography systems incorporate 2D scanning slot, CR, or FPD-based detection systems. An example mammogram is shown in Figure 25.3(a). In breast cancer screening, the imaging tasks include detection of abnormal lesions, discrimination of masses from benign cysts, and detection of microcalcifications that are often associated with malignant disease.

25.1.2.3 2D megavoltage imaging (portal imaging)

Radiation therapy relies on the precise delivery of tumoricidal doses of radiation to a prescribed target volume. To ensure localization of the target with respect to the treatment beam, 2D images can be acquired using the megavoltage beam itself. Such images are conventionally referred to as "portal" images in that they provide a beam's eye view through the "portal" of a given treatment field as shaped by cerrobend moulded blocks or multi-leaf collimators. The energy of the treatment beam is typically 10–100 times greater than diagnostic-quality X-rays – e.g. as low as ~1 MeV for treatment with Co^{60} and 6–18 MeV typical in modern linear accelerators. This dramatically affects the detector design considerations, image quality characteristics, and radiation dose compared to diagnostic imaging applications discussed above. As illustrated in Figure 25.1(c), imaging detectors often incorporate a metal sheet in combination with a scintillator to boost quantum detection efficiency, with high-energy X-rays interacting in the metal sheet to produce scattered electrons which, in turn, interact in the scintillator to produce optical photons. Imaging tasks vary considerably among treatment sites – e.g. delineation of bony ridges in the pelvis – with the common goal of localizing the target within the treatment portal.

25.1.2.4 Dual-energy (DE) radiography

Discrimination of material content within a 2D radiograph can be achieved by decomposing images obtained at separate energies. In DE imaging, low- and high-energy images are combined to cancel a given material – e.g. the extinction of bone yielding a "soft-tissue" image and the extinction of soft-tissue yielding a "bone" image. The approach exploits the linear independence of photoelectric and Compton cross-sections in the attenuation coefficient, and several algorithms for image decomposition have been described (Warp, 2003). The simplest is weighted log subtraction, which yields soft-tissue and bone images through the selection of a tissue cancellation factor:

$$\ln\left(I_{soft}^{DE}\right) = \ln\left(I^H\right) - w_s \ln\left(I^L\right) \tag{25.2a}$$

$$\ln\left(I_{bone}^{DE}\right) = -\ln\left(I^H\right) + w_b \ln\left(I^L\right) \tag{25.2b}$$

The factor w_s is selected to cancel bone and is given by the ratio of bone attenuation coefficients at high and low energy. Similarly, w_b cancels soft-tissue and is given by the ratio of soft-tissue coefficients at high and low energy. DE imaging has found utility in a broad range of clinical applications, including chest, breast, and cardiac imaging. In each case, the ability to cancel specific materials reduces the presence of overlying anatomical structures in the resulting 2D image and thereby improves the conspicuity of an underlying structure of interest. Imaging tasks are correspondingly broad, including: lung nodule detection, discrimination of rib fracture from bony metastasis, visualization of iodine contrast enhancement, and detection of calcified plaque. Example DE images of the chest are shown in Figure 25.2(b–c).

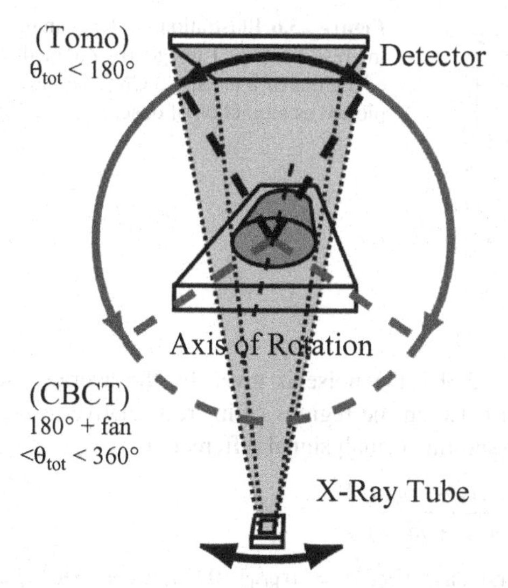

Figure 25.4 Illustration of tomosynthesis and CBCT (circular) source-detector orbit.

25.1.2.5 Real-time fluoroscopy

Fluoroscopy provides dynamic 2D images that can be displayed and recorded as they change in real time. Such technology may involve an XRII with a CCD camera or an FPD read and displayed at a high frame rate (e.g. ~30 frames per second (fps)). The applications of real-time fluoroscopy are numerous in image-guided interventions, including the placement of catheters, stents, and other interventional equipment in real time.

25.1.2.6 3D tomosynthesis

Both tomosynthesis and CBCT (below) reconstruct 3D images of the patient through the acquisition of multiple 2D radiographs at various angles about the patient. While CT and CBCT are typically characterized by hundreds or thousands of projections acquired in circular orbits of the source and detector through at least 180° about the patient, tomosynthesis is characterized by a smaller plurality of images (e.g. 10–50) acquired across a limited arc (as in Figure 25.4), circle, or linear source-detector trajectory. Systems for chest tomosynthesis are becoming commercially available, acquiring approximately 20–60 images as the source and detector translate linearly across ~16° in the superior-inferior direction (Dobbins, 2003). Conversely, breast tomosynthesis acquires ~11 images as the X-ray tube traverses a ~50° arc (with the detector stationary) (Rafferty, 2007). Tomosynthesis on a C-arm gantry (Tutar, 2003; Bachar, 2007), on the other hand, utilizes ~45 images acquired in a ~45° source-detector circular arc. Tomosynthesis images are characterized by high in-plane spatial resolution (determined essentially by the detector system) and depth resolution on the order of ~10 mm, depending on the system geometry, acquisition technique (number of views and orbital extent), and reconstruction algorithm. Filtered backprojection and iterative algebraic reconstruction techniques are the most prevalent, the former offering speed and ease of implementation and the latter offering potentially improved noise performance. The level of depth discrimination achieved in tomosynthesis significantly reduces the influence of overlying anatomical clutter within a slice and thereby improves the conspicuity of subtle underlying lesions. Example images are shown in Figure 25.3(b).

25.1.2.7 3D cone-beam CT

Some principles of CT, including helical CT, multi-detector CT, etc., are addressed in Chapter 26. CBCT operates upon similar principles to CT but is characterized by a large-area detector (e.g. a flat-panel detector), source-detector trajectories that are not necessarily bound to a circular orbit (Figure 25.4), and account of the divergent X-ray beam in 3D reconstruction. Methods of CBCT reconstruction have been an active area of research since the original algorithm by Feldkamp *et al.* (Feldkamp, 1984) and include a variety of filtered-backprojection, iterative, and more sophisticated techniques. Use of 2D radiographic imaging systems for 3D CBCT became a prevalent area of research and clinical application with the broad availability of high-performance FPDs. Applications include image-guided interventions (e.g. image-guided radiation therapy (Jaffray, 2002), surgery (Siewerdsen, 2005; Rafferty, 2006), and vascular interventions (Fahrig, 2000)) and diagnostic imaging (e.g. 3D breast (Boone, 2006) and maxillofacial (Bartling, 2007) imaging). CBCT image quality is characterized by soft-tissue detectability in combination with nearly isotropic, sub-millimeter spatial resolution. While image quality in FPD-based CBCT is typically less than in diagnostic CT (due, for example, to reduced detector efficiency, increased electronic noise, and high levels of X-ray scatter), the technology is well suited to a broad host of applications requiring a simpler mechanical platform (as opposed to a slip ring gantry), open geometry (e.g. a C-arm), and/or combined CBCT + fluoroscopy capability. Example images are shown in Figure 25.5.

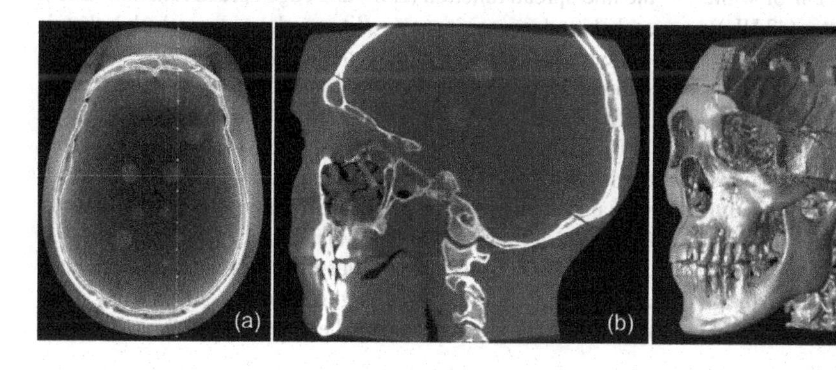

Figure 25.5 Example 3D CBCT images. (a) Axial, (b) sagittal, and (c) surface rendering of an anthropomorphic skull phantom.

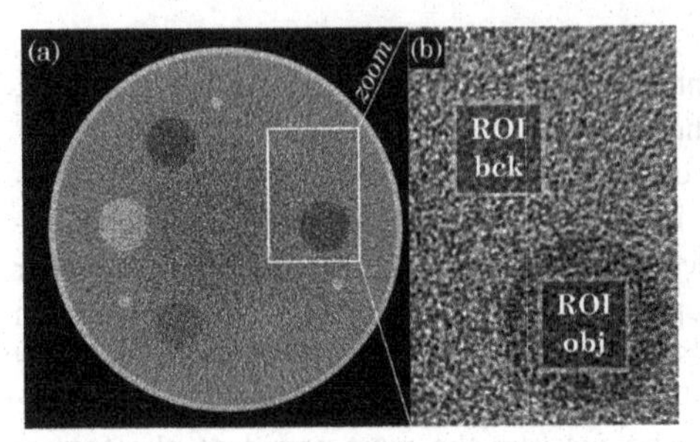

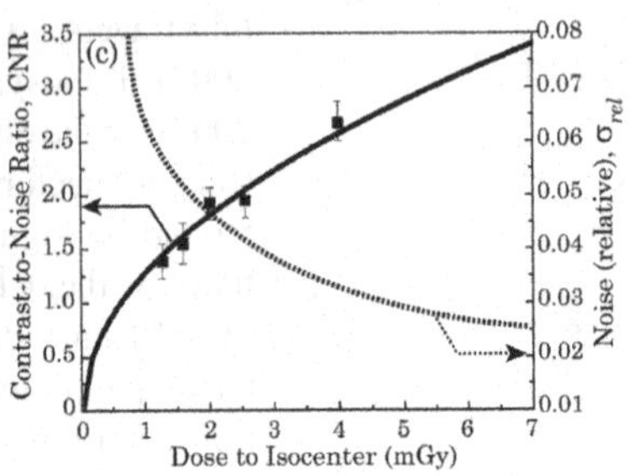

Figure 25.6 Illustration. (a) CNR in regions of interest in a CBCT image. (b) Zoomed-in view of regions of interest. (c) CNR and noise plotted as a function of dose.

25.2 IMAGING PERFORMANCE

25.2.1 Metrics of imaging performance

Metrics of imaging performance are described in Chapter 11. A brief description of imaging performance metrics in the context of 2D and 3D imaging optimization is provided below. The term "imaging performance" herein refers to quantitative characteristics of the imaging system aside from the observer. The term therefore describes characteristics of the detector (e.g. spatial resolution and noise) and – to an extent – the object (e.g. "air" or "water" for analysis based on uniform input or "soft-tissue" for analysis of contrast and contrast-to-noise ratio). The term "image quality," on the other hand, is reserved for descriptors that include assessment by a human observer (e.g. receiver operating characteristic (ROC) analysis).

25.2.1.1 Contrast, noise, and CNR

Contrast and noise refer to large-area transfer characteristics that may be evaluated in the spatial domain of the image data. The term "contrast" refers to the difference in image signal between regions corresponding to an object of interest and the background:

$$C_{rel} = \frac{|\overline{m_{obj}} - \overline{m_{bck}}|}{(\overline{m_{obj}} + \overline{m_{bck}})/2} \quad (25.3a)$$

Example regions of "object" and background are illustrated in the CBCT image of Figure 25.6(a–b). The contrast defined in Eq. (25.3a) describes the signal difference relative to the mean signal. For example, in chest radiography, one might describe the contrast of a lung nodule as 1%. Alternatively, the absolute contrast (*C*) may be defined in terms of the numerator alone (carrying the same units as the signal). For example, in CT, one might describe the contrast of muscle as 65 HU (equal to the difference between the CT # of muscle relative to water) or the contrast of white matter as 3 HU (equal to the CT # of white matter (46 HU) minus that of neighboring gray matter (43 HU)).

Noise refers to the standard deviation in pixel values over a given region of interest (ROI). This reasonably characterizes signal variability, provided that the ROI is sufficiently large to capture the signal variations of interest (i.e. the ROI size is much greater than the correlation length of signal fluctuations) and that the variations are stochastic in nature (e.g. related to X-ray quantum noise rather than image artifacts, anatomical clutter, etc.). Within a given ROI, the noise is given simply by the standard deviation within the ROI. With respect to the two ROIs from which contrast is evaluated as in Figure 25.6(b) and Eq. (25.3b), the noise is given by the average standard deviation between the regions of interest relative to the mean signal (assuming a small signal difference):

$$\sigma_{rel} = \frac{(\sigma_{obj} + \sigma_{bck})/2}{(\overline{m_{obj}} + \overline{m_{bck}})/2} \quad (25.3b)$$

In the first case (i.e. $\sigma = \sigma_{ROI}$), the noise carries the same units as the signal. For example, in CT the noise as measured in a uniform water cylinder might be described as 5 HU. The relative noise, on the other hand, describes the noise in relation to the mean signal about two regions.

The contrast-to-noise ratio, CNR, combines the large-area characteristics of C and σ (or C_{rel} and σ_{rel}) as:

$$CNR = \frac{C}{\sigma} = \frac{C_{rel}}{\sigma_{rel}} = \frac{|\overline{m_{obj}} - \overline{m_{bck}}|}{(\sigma_{obj} + \sigma_{bck})/2} \quad (25.3c)$$

CNR provides a coarse figure of merit for detectability, with classical perception models indicating a minimum CNR of ~2–5 for confident detection under certain assumptions of signal size and background (Rose, 1948). As shown in Figure 25.6(c) for CBCT, noise reduces as the inverse square root of dose, and CNR improves as the square root of dose. Note that "contrast" (i.e. mean signal difference) is independent of dose.

25.2.1.2 Modulation transfer function (MTF)

The spatial frequency-dependent signal transfer characteristics of the imaging system are described by the modulation transfer function (MTF), analogous to the 1D transfer function ubiquitous to temporal frequency analysis in digital signal processing and taken as a measure of spatial resolution. The MTF is given by the Fourier transform of the impulse response function, as estimated from the point-spread function (PSF), line-spread function (LSF), or edge-spread function (ESF). As described in Section 25.2.2, below, common experimental methods of MTF estimation in digital imaging systems include measurement of the line-spread function (LSF) and edge-spread function (ESF), each based on an input (a slit or edge, respectively) that is angled with respect to the sampling matrix to give an oversampled response function as described by Fujita *et al.* (Fujita, 1992).

The magnitude of the MTF describes the extent to which the imaging system modulates (attenuates) a signal of a given spatial frequency in transferring information from the input to the output. Typical MTFs for (#1) nominal, (#2) blurry, and (#3) edge-enhancing systems are illustrated in Figure 25.7. In each case, MTF(0) is unity, implying that the system does not affect the mean (DC) signal value. The "ideal" MTF is unity at all

Table 25.2 *Example analytical forms of the MTF for three special cases of impulse response*

Description	Response function	MTF
Rectangular aperture	$psf(x, y) = \frac{1}{A_x A_y}\Pi(x/A_x)\Pi(y/A_y)$	$MTF(f_x, f_y) = \text{sinc}(\pi A_x f_x)\,\text{sinc}(\pi A_y f_y)$
Exponential spread	$lsf(x) = \frac{1}{2k}\exp(-\lvert x\rvert/k)$	$MTF(f_x, f_y) = \frac{1}{1 + 4\pi^2 k^2 (f_x^2 + f_y^2)}$
Gaussian spread	$psf(x, y) = \frac{1}{2\pi k^2}\exp[-(x^2 + y^2)/2k^2]$	$MTF(f_x, f_y) = \exp(-\pi^2 k^2 (f_x^2 + f_y^2))$

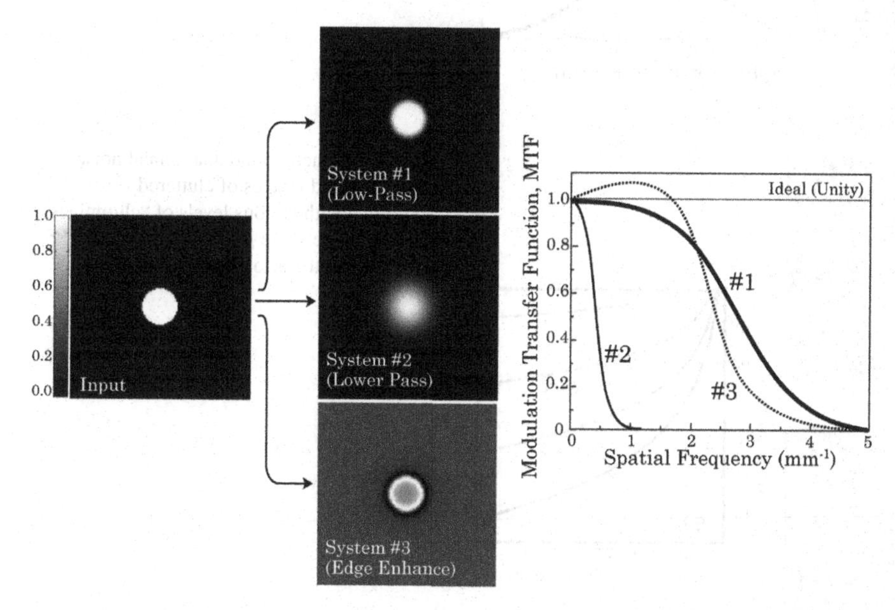

Figure 25.7 Illustration of input, output, and MTF characteristics.

frequencies, meaning the system transfers signals at all spatial frequencies without attenuation of the signal magnitude. A few analytical examples are listed in Table 25.2.

25.2.1.3 Noise-power spectrum (NPS)

The magnitude and spatial frequency dependence of image fluctuations are characterized by the noise-power spectrum (NPS), which can be understood as the spectral density of image fluctuations, or as the variance of the Fourier coefficients of image fluctuations. The NPS is experimentally determined from the Fourier transform of "noise-only" images as follows (for a 2D image):

$$NPS(f_x, f_y) = \frac{A_x A_y}{N_x N_y}\left\langle \left| FFT\left[I(x, y) - \overline{I}\right]\right|^2 \right\rangle \tag{25.4}$$

where $I(x, y)$ denotes a uniform ("noise-only") image with mean signal $\overline{I}$. FFT denotes the 2D Fourier transform, | | the modulus, and < > the ensemble average. The terms A_i and N_i denote the pixel size and number of pixels along direction i of each realization in the ensemble. Note that the NPS carries units of (signal)^2 times dimensions of image domains – e.g. (signal^2)(mm^2) for a 2D image and (signal^2)(mm^3) for a 3D image. The normalized NPS, denoted NNPS, is given by the NPS divided by the mean signal squared.

Requirements of NPS analysis include stationarity (i.e. spatial invariance of the signal mean and noise). Note that whereas the MTF is a characteristic of the system (e.g. "The system has an MTF of . . . ") the NPS is a characteristic of the image (e.g. "The image has an NPS of . . . "). Figure 25.8 illustrates various noise realizations and the corresponding NPS, including uncorrelated ("white") noise as is characteristic of perfectly pixelated detectors, low-pass noise as is characteristic of a blurry detector, mid-pass noise as is characteristic of CT, and high-frequency noise as might be associated with a very sharp filter.

In some circumstances, the anatomical background may be described in terms of NPS characteristics, termed background anatomical noise (NPS_B). Such a description is subject to the same requirements as the NPS described above for stochastic quantum noise fluctuations (namely, stationarity in the mean and variance) and has proven useful in describing the performance in, for example, 2D breast imaging (Burgess, 2001). Such noise characteristics are typically described in empirical terms as power-law noise according to a $1/f$ characteristic, as is common in a fascinatingly broad variety of statistical phenomena (e.g. variations in the periodicity of sunspot activity and precession of the Earth's axis). The background NPS is typically characterized by empirical fits of the form:

$$NPS_B(f) = \frac{K}{f^\beta} \tag{25.5}$$

where f is radial frequency, K characterizes the magnitude of variations, and β describes the steepness of falloff in the frequency domain – e.g. $\beta = 0$ for uncorrelated noise, and β increases for greater degrees of "clumpiness" as illustrated in

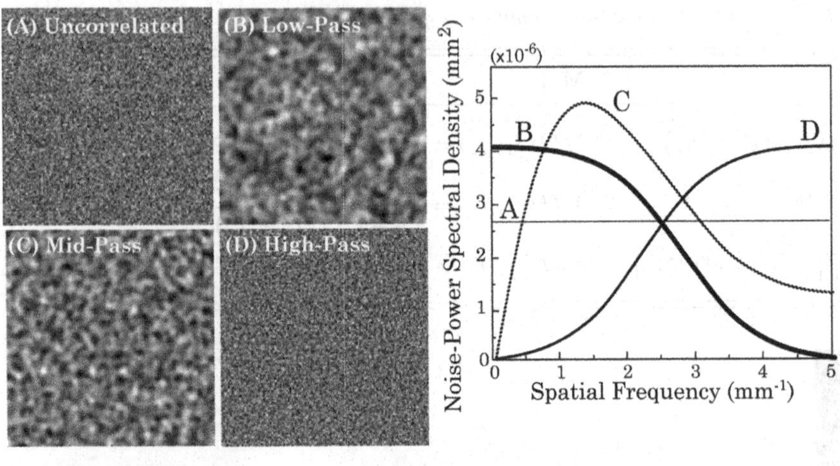

Figure 25.8 Illustration of noise realizations and NPS.

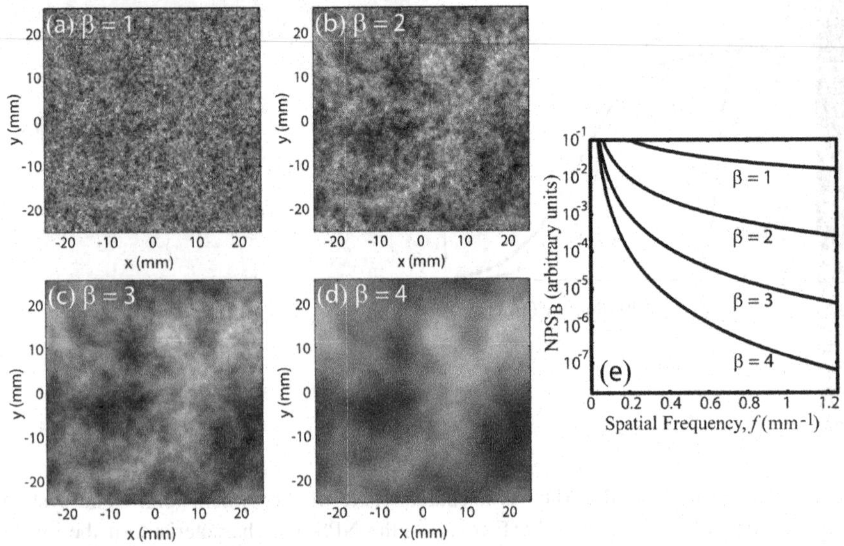

Figure 25.9 Background anatomical noise. (a–d) Simulated images of cluttered background with various levels of "clumpiness" attributed to the value β in Eq. (25.5). (e) NPS corresponding to various levels of β.

Figure 25.9. For 2D mammography, values of β in the range 2.6–3.1 have been reported (Burgess, 2001). For 2D chest imaging, clutter associated with ribs, bronchial structures, and vessels correspond to a value of $\beta \sim 3.75$ (Richard, 2005b). Evaluation of anatomical clutter in this manner is underway in 3D imaging as well (Metheany, 2007; Nishikawa, 2007).

25.2.1.4 Noise-equivalent quanta (NEQ) and detective quantum efficiency (DQE)

As described in Chapter 11, the noise-equivalent quanta (NEQ) describes the spatial frequency-dependent signal-to-noise ratio as follows:

$$NEQ(f_x, f_y) = \frac{MTF^2(f_x, f_y)}{NNPS(f_x, f_y)} \tag{25.6a}$$

giving the effective number (fluence) of quanta used by the system in forming an image. The effective number of quanta is necessarily less than the actual number of quanta incident on the detector, $\bar{q}_0$, with the efficiency of utilization given by the detective quantum efficiency:

$$DQE(f_x, f_y) = \frac{MTF^2(f_x, f_y)}{\bar{q}_0 NNPS(f_x, f_y)} \tag{25.6b}$$

NEQ and DQE have become prevalent metrics of imaging performance in the development of digital X-ray detectors. Experimental methods for their determination are becoming increasingly standardized within the medical imaging physics community, and theoretical methods for their estimation continue to represent an active area of research. Example measurement and prediction (cascaded systems analysis (see below)) of the NEQ and DQE for a 2D radiographic imaging system are illustrated in Figure 25.10. Note that the NEQ scales in proportion to the incident exposure, whereas the DQE is nearly independent of incident exposure (governed in this respect by the signal in proportion to the electronics noise). The low-frequency DQE is given by the quantum detection efficiency of the system times the Swank factor, a term related to variations in the conversion of X-rays to secondary (optical or electronic) quanta (Swank, 1973). The frequency dependence of the NEQ and DQE are governed by the system MTF, electronics noise, and sampling (aliasing).

Note that the denominator in the NEQ and DQE is the normalized NPS (i.e. the absolute NPS divided by the mean signal squared). The "generalized" NEQ and DQE, denoted GNEQ and GDQE, respectively, can be alternatively defined to also include the anatomical background NPS_B in the denominator (ICRU, 1996; Siewerdsen, 2004b; Richard, 2005a).

Table 25.3 *Example RQA beams for quality assurance tests and measurement of imaging performance*

Beam quality	kVp (approx)	Added filtration (mm Al)	mR/mAs (at 100 cm)	Mean energy (keV)	HVL (mm Al)	TVL (mm Al)	Fluence/exposure (X-rays/mR/mm²) (approx)
RQA 3	50	10	0.13	39.5	3.9	13.7	192,840
RQA 5	70	21	0.17	52.6	6.8	24.4	265,390
RQA 7	90	30	0.26	63.3	9.3	32.2	284,140
RQA 9	120	40	0.43	76.4	11.6	40.0	273,150
RQA 10	150	45	0.68	85.4	12.8	44.4	258,020

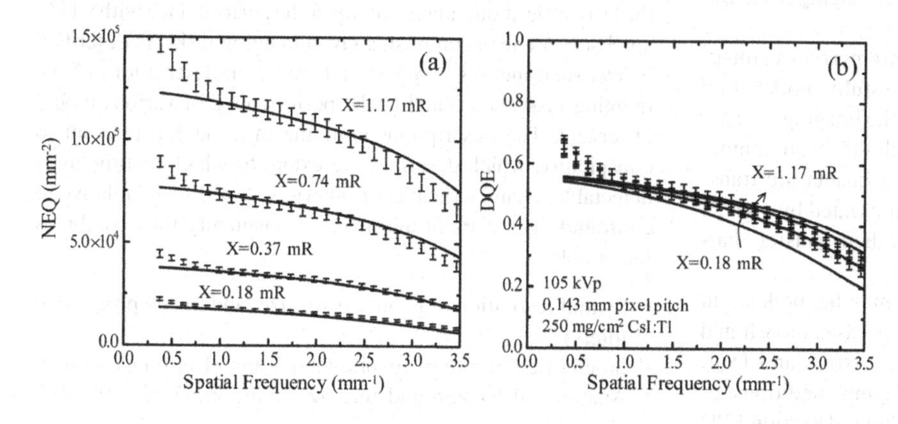

Figure 25.10 Example (a) NEQ and (b) DQE measured (symbols) and predicted (curves) at various exposure levels. The system considered is the Pixium 4600 of Table 25.1.

25.2.2 Experimental and theoretical methods of imaging performance characterization

25.2.2.1 Measurement of MTF, NPS, DQE, and NEQ

Experimental methods for measurement of MTF, NPS, DQE, and NEQ have become increasingly standardized over the last decade. An essential first consideration in such measurements is the quality of the X-ray beam, which should be selected with kVp and added filtration appropriate to a given application. For example, Table 25.3 lists a variety of standard beam qualities (IEC Committee 62B, 2003).

The MTF is measured in terms of the Fourier transform of the impulse response function, where the impulse may be a point, line, or edge input. In 2D radiography, measurement of the PSF is uncommon, since the fluence (and therefore the image signal) through a narrow pinhole is so low that measurements are dominated by noise. More common are measurements of the LSF or ESF to obtain a 1D "slice" of the 2D MTF along a given direction. In each case, the line or edge is angled with respect to the detector matrix as described by Fujita *et al.* (Fujita, 1992) to effect oversampling of the response function. The LSF (or ESF) is typically measured as the image of a narrow slit (or a strongly attenuating edge) placed in contact with the detector. The derivative of the (oversampled) ESF yields the LSF. The 1D Fourier transform of the (area-normalized) LSF gives a slice of the MTF in the direction perpendicular to the line. For 3D imaging (e.g. CBCT) the MTF has been measured in a manner drawn from axial CT, in which a wire oriented longitudinally in the image presents a PSF in axial planes, the Radon transform of which yields the LSF, and the 1D FFT of which yields a slice of the MTF. Positioning the wire at a slight angle with respect to the longitudinal axis allows measurement of an oversampled LSF analogous to the Fujita angled slit (Silverman, 2009). More recently, a 3D ESF technique has been applied to CBCT in which the image of a large (several cm), high-contrast sphere provides the edge response function in any direction.

The NPS may be measured in terms of the Fourier transform of "noise-only" images as discussed above in Eq. (25.4). Important considerations in such measurement include: detrending of the otherwise uniform image data; the number of realizations forming the ensemble; and the size of each realization. Standardization of these and other factors is underway. More recently, the 3D NPS of tomosynthesis and CBCT have been described (Siewerdsen, 2000a), determined in a manner analogous to Eq. (25.4) and demonstrating a number of non-trivial noise characteristics that are distinct from conventional CT (Riederer, 1978; Hanson, 1979).

A knowledge of the MTF and NPS allows determination of the NEQ (Eq. (25.6a)) and DQE (Eq. (25.6b)), provided we have an estimate of the X-ray fluence ($\bar{q}_0$) incident on the X-ray detector. The fluence per unit exposure ($\bar{q}_0/X$) is estimated from the area-normalized incident X-ray spectrum ($q_{rel}(E)$) as

$$\frac{\overline{q_0}}{X} = \int_0^{E_{max}} k \frac{q_{rel}(E)}{E\left[\mu_{ab}(E)/\rho\right]_{air}} dE \tag{25.7}$$

where k is a constant (5.45×10^8 eV/g/mR) determined by the definition of exposure in air, E has units of eV, and μ_{ab}/ρ is the energy absorption coefficient for air. The spectrum $q(E)$ may be obtained from measurements (Fewell, 1977) or, more commonly, from a variety of X-ray spectrum models, such as those of Tucker-Barnes (Tucker, 1991), the TASMIP algorithm (Boone, 1997), or the SPEKTR toolkit (Siewerdsen, 2004c).

25.2.2.2 Theoretical analysis of MTF, NPS, DQE, and NEQ

The ability to describe and predict the image signal and noise characteristics either analytically (e.g. cascaded systems analysis) or numerically (e.g. Monte Carlo simulation) provides an important understanding of the fundamental factors governing imaging performance and a guide to system optimization. Such approaches have been useful in the design of 2D and 3D radiographic systems – for example: understanding the tradeoffs in MTF and NPS in direct and indirect detection FPDs (Siewerdsen, 1997; Zhao, 1997); examining the low-dose fluoroscopic performance of FPDs and XRIIs (Cunningham, 1994; Antonuk, 2000); understanding system design in mammography (Vedantham, 2004) and angiography (Ganguly, 2003); and selecting the low and high kVp in dual-energy imaging (Richard, 2007).

Under the assumptions of linearity and shift invariance, imaging systems such as FPDs have been successfully modeled by cascaded systems analysis (CSA). In CSA, the imaging chain is conceptualized as a series of stages in which the mean number and spatial frequency distribution of image quanta are transferred according to the physical process represented by a given stage. The basic types of stages along with their transfer characteristics are summarized in Table 25.4.

For example, an indirect-detection FPD may be modeled in simplest terms as in Table 25.5. Such analysis has demonstrated good agreement with measurement of NPS, NEQ, and DQE for a broad variety of FPD designs and imaging conditions, as illustrated in Figure 25.10. Modeling of a direct-detection FPD is similar, understanding that the secondary quanta generated at Stage 2 are electron-hole pairs (not optical photons), and the spread of such quanta at Stage 3 is small (i.e. T_3 is close to 1) compared to diffuse scatter of optical photons in a scintillator. An important modification to the serial cascade involves a parallel cascade of stages associated with K-fluorescence at Stage 2, as described by Yao and Cunningham (Yao, 2001). Such analysis has been applied in 2D radiography (Richard, 2005b), fluoroscopy (Vedantham, 2004), mammography, and portal imaging (Bissonnette, 1997), and has been extended more recently to describe the signal and noise characteristics of 2D DE imaging (Richard, 2005a) as well as 3D tomosynthesis and CBCT (Siewerdsen, 2003).

Extension of such descriptions to include additional factors related to the "system" includes incorporation of focal spot blur and X-ray scatter in measurement and/or theoretical analysis of the DQE. For example, X-ray scatter may be incorporated in the DQE as a source of "additive quantum noise" (Siewerdsen, 2000b). More complicated factors may be addressed in numerical simulation – e.g. Monte Carlo studies that examine factors that do not lend well to a linear analytical approach. For example, X-ray scatter is recognized as a significant factor in image quality in both 2D and 3D radiographic imaging. Analysis of the X-ray scatter sensitivity of various detectors (Spies, 2001) has proven useful in examining the intrinsic scatter response characteristics of phosphors, scintillators, and photoconductors. For 3D imaging, the X-ray scatter magnitude and distribution associated with various system geometries have also been examined using Monte Carlo simulation, providing a possible means of scatter artifact correction (Jarry, 2006). The effect of oblique X-ray incidence on the detector spatial resolution has been similarly examined, suggesting nonstationarity in the MTF (i.e. degradation in MTF with more oblique incidence) (Badano, 2007). Further, Monte Carlo simulations of the entire imaging chain have been developed as useful means of system optimization (Gallas, 2004).

25.3 IMAGING TASK AND DETECTABILITY

25.3.1 Conspicuity

The various signal and noise characteristics detailed above describe the fidelity with which an imaging system transfers the signal and noise from its input to its output. By themselves, they say little about image quality or the performance with which an observer can accomplish a given imaging task. As described below, such metrics may be combined with descriptions of the imaging task and related to the performance of various model observers. The descriptions combine in a manner relevant to conspicuity, which describes the extent to which a stimulus is detectable, characterizable, or otherwise "well seen" relative to confounding factors in the image. Conspicuity thus combines factors of:

- Spatial resolution (e.g. blur, pixel size, and image processing filters)
- Stochastic noise (e.g. quantum noise and electronics noise)
- Anatomical background (e.g. overlying bone or soft-tissue clutter)
- Artifacts (e.g. shading or streaks)
- Stimulus magnitude (contrast, or signal power)
- Stimulus extent (size or other spatial characteristics)

The first two factors are described well by the Fourier performance metrics (MTF and NPS). The second two relate to more deterministic effects that may or may not be amenable to a simple Fourier-based statistical approach (e.g. in the anatomical NPS_B). The last two factors relate to the imaging task, described in greater detail below.

Among the various 2D and 3D modalities summarized in Section 25.1, it is interesting to note the characteristics of the "advanced" modalities (i.e. DE imaging, tomosynthesis, and CBCT) in relation to conspicuity. Each represents a means of enhancing conspicuity over that presented in conventional 2D radiographs. In each case, there is a tradeoff between reduced anatomical background and increased quantum noise. For example, DE imaging increases the quantum noise through a combination of (statistically independent) low- and high-energy images. Since anatomical background is typically the dominant factor in conspicuity (by an order of magnitude or more), the tradeoff is usually worthwhile in terms of overall conspicuity. Thus, DE imaging can be viewed as improving conspicuity by means of material discrimination in canceling overlying anatomical structure from the 2D image. Similarly, tomosynthesis and CBCT improve conspicuity by eliminating the superposition of overlying anatomical structure within any particular "slice" of a 3D volume image.

25.3.2 Imaging task

The performance of an imaging system should ultimately be evaluated with respect to a given imaging task (or combination of tasks). While the term "task" invites qualitative interpretations

Table 25.4 *Summary of types of stages and the associated signal and noise transfer relations in cascaded systems analysis*

Stage/physical process	Description	Parameter	Signal and noise transfer relations
Gain	Change in the mean number of image quanta:	$\bar{g}_i$: mean gain $\bar{g}_i > 1$ (amplification) $\bar{g}_i < 1$ (attenuation)	$\bar{q}_i = \bar{g}_i \bar{q}_{i-1}$ $S_i(f) = \bar{g}_i^2 S_{i-1}(f) + \sigma_{g_i}^2 + S_{add_i}(f)$
Stochastic spreading	Random scattering of image quanta (blur)	$T_i(f)$: MTF of the scattering process	$\bar{q}_i = \bar{q}_{i-1}$ $S_i(f) = [S_{i-1}(f) - \bar{q}_{i-1}] T_i^2(f) + \bar{q}_{i-1}$
Deterministic spreading	Integration of image quanta over an aperture	$T_i(f)$: MTF of the integrating aperture a: width of integrating aperture	$\bar{q}_i = a^2 \bar{q}_{i-1}$ $S_i(f) = a^4 S_{i-1}(f_v) T_i^2(f)$
Sampling	Discrete sampling of continuous image quanta at specific locations	$III_i(f)$: Comb function	$\bar{q}_i = \bar{q}_{i-1}$ $S_i(f) = S_{i-1}(f) ^{**} III_i(f)$

Table 25.5 *Example serial cascade for CSA modeling of signal and noise for an indirect-detection FPD*

Stage	Physical process	Parameter	Mean signal and NPS
0	Fluence incident on detector	$\bar{q}_0$	$\bar{q}_0$ $S_0(f) = \bar{q}_0$
1	Interaction of incident X-rays in scintillator	$\bar{g}_1$	$\bar{q}_1 = \bar{q}_0 \bar{g}_1$ $S_1(f) = \bar{q}_0 \bar{g}_1$
2	Conversion of incident X-rays to secondary quanta (optical photons)	$\bar{g}_2 \varepsilon_{g_2}$	$\bar{q}_2 = \bar{q}_0 \bar{g}_1 \bar{g}_2$ $S_2(f) = \bar{q}_0 \bar{g}_1 \bar{g}_2 (\bar{g}_2 + 1 + \varepsilon_{g_2})$
3	Stochastic spread of secondary quanta	$T_3(f)$	$\bar{q}_3 = \bar{q}_0 \bar{g}_1 \bar{g}_2$ $S_3(f) = \bar{q}_0 \bar{g}_1 \bar{g}_2 [1 + (\bar{g}_2 + \varepsilon_{g_2}) T_3^2]$
4	Coupling of secondary quanta to photodiode	$\bar{g}_4$	$\bar{q}_4 = \bar{q}_0 \bar{g}_1 \bar{g}_2 \bar{g}_4$ $S_4(f) = \bar{q}_0 \bar{g}_1 \bar{g}_2 \bar{g}_4 [1 + \bar{g}_4 (\bar{g}_2 + \varepsilon_{g_2}) T_3^2]$
5	Integration of secondary quanta over (photodiode) aperture	$T_5(f)$	$\bar{q}_5 = a^4 \bar{q}_0 \bar{g}_1 \bar{g}_2 \bar{g}_4$ $S_5(f) = a^4 \bar{q}_0 \bar{g}_1 \bar{g}_2 \bar{g}_4 [1 + \bar{g}_4 (\bar{g}_2 + \varepsilon_{g_2}) T_3^2] T_5^2$
6	Sampling of 2D pixel matrix	$III_6(f)$	$\bar{q}_6 = a^4 \bar{q}_0 \bar{g}_1 \bar{g}_2 \bar{g}_4$ $S_6(f) = a^4 \bar{q}_0 \bar{g}_1 \bar{g}_2 \bar{g}_4 [1 + \bar{g}_4 (\bar{g}_2 + \varepsilon_{g_2}) T_3^2] T_5^{2**} III_6$
7	Readout with electronics noise	$S_{add7}(f)$	$\bar{q}_7 = a^4 \bar{q}_0 \bar{g}_1 \bar{g}_2 \bar{g}_4$ $S_7(f) = a^4 \bar{q}_0 \bar{g}_1 \bar{g}_2 \bar{g}_4 [1 + \bar{g}_4 (\bar{g}_2 + \varepsilon_{g_2}) T_3^2] T_5^{2**} III_6 + S_{add7}$

Table 25.6 *Terms and symbols related to definition of spatial frequency-dependent task functions*

	Spatial domain	Fourier domain
Hypothesis 1	$h_1(x)$	$H_1(f)$
Hypothesis 2	$h_2(x)$	$H_2(f)$
Task function	$W_{task}(x)$ $= h_1(x) - h_2(x)$	$W_{task}(f)$ $= \lvert H_2(f) - H_1(f) \rvert$

ranging from the overall clinical objective (e.g. achieve cure) to more specific diagnostic objectives (e.g. determine the nature of an ailment), herein the term is used in a manner that quantifies characteristics associated with the object being imaged and the interpretation of it by a model observer with respect to various hypotheses. Task functions (templates) are described below in terms of the spatial frequencies of interest for a given imaging task and may be used to relate spatial-frequency-dependent physical performance metrics (e.g. NEQ) to the model observer SNR, which in turn may be cast in terms of the area under an ROC curve or the proportion correct in an alternative forced-choice (AFC) test. The magnitude of a task function relates to the contrast (or more specifically, the signal power) associated with the task, and the spatial frequency dependence depends on the frequency content of the structure(s) of interest.

Herein, we consider binary classification tasks, where the observer decides which of two specific hypotheses is more likely represented by the image. As illustrated in various examples below, the task function is described by the Fourier transform of the object functions corresponding to the two hypotheses. Terms and symbols related to the hypotheses and task function are in Table 25.6.

Binary classification is obviously an idealized approach and is fairly limited in terms of the scope of realistic tasks that can be described. Representing higher-order classification tasks (e.g. deciding among three or more hypotheses) is the subject of ongoing research. Similarly, such an approach does not apply directly to a continuously variable estimation task (e.g. size estimation or

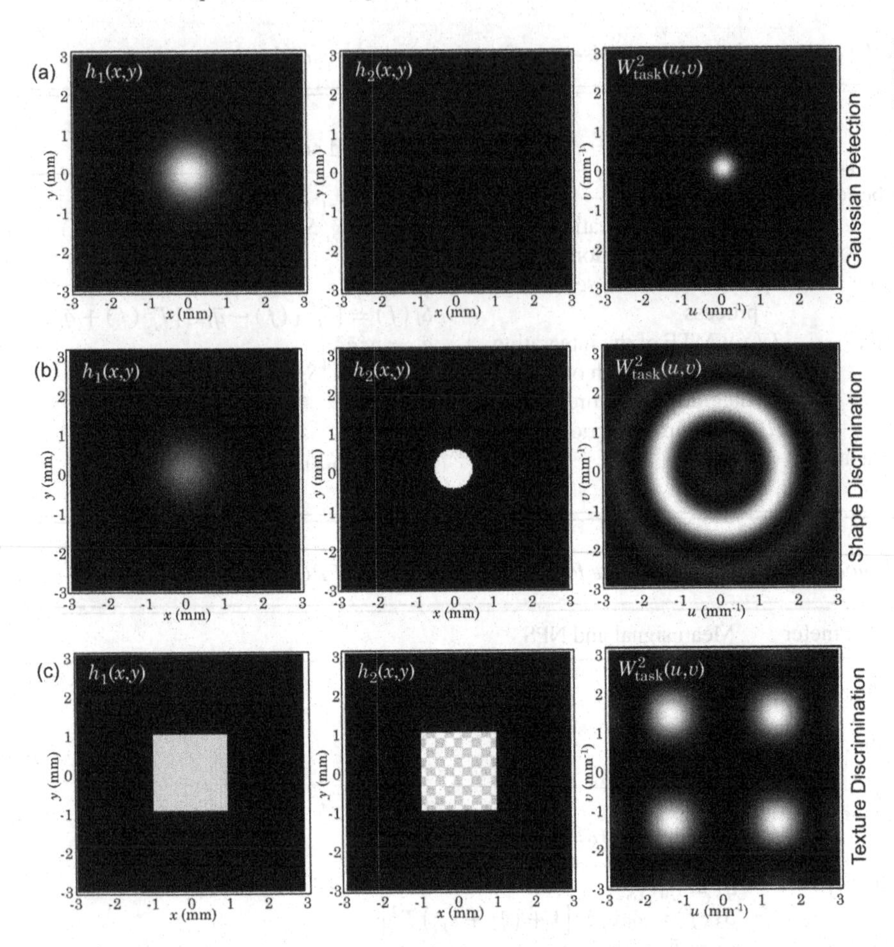

Figure 25.11 Summary of three example task functions. (a) Gaussian detection. (b) Shape discrimination. (c) Texture discrimination.

stimulus localization), except that it could be idealized in a simple extreme as a decision between two hypotheses – e.g. size estimation of "A" or "B," or location estimation as "here" or "there." Further, the task functions below do not explicitly account for possible variability in one or both hypotheses, and both hypotheses need to be defined. For example, "A or Not A" (e.g. a 3 mm lesion or some other lesion) cannot be directly cast in terms of a simple Fourier difference of two hypotheses. Still, quantifying the spatial frequency dependence of idealized imaging tasks – even for simple, abstract tasks – provides a starting point for bridging the gap between detector performance (MTF, NPS, and NEQ) and observer performance. As shown in later sections, such simple models can offer predictive power of detectability in real human observers.

In the sections below, three idealized imaging tasks are presented analytically as examples of binary hypothesis-testing classification tasks. Each case involves two hypothesis object functions (h_1 and h_2), the Fourier transforms of which determine a task function corresponding to the spatial frequencies of interest in accomplishing the task. Example idealized tasks are illustrated in Figure 25.11, including: (a) detection on a uniform background; (b) shape discrimination; and (c) texture discrimination.

25.3.2.1 Detection on a uniform background

The simplest idealized task is detection of an object against a uniform background. For example, the object to be detected could be an extremely fine detail (e.g. a point). In this case, h_1 is modeled as a delta function, with Fourier transform equal to a constant. The alternative hypothesis, h_2, and its Fourier transform corresponding to a uniform (zero) background, are zero, such that the difference $(H_1 - H_2)$ and therefore the task function, W_{task}, are simply a constant. Thus, detection of a point against a uniform background corresponds to all spatial frequencies equally weighted in the task. As shown below, the detectability index for this task is given by the integral over the NEQ.

Figure 25.11(a) illustrates the example in which the object to be detected is a Gaussian cloud, for which h_1 is therefore also a Gaussian. We write:

$$h_1(x, y) = A_1 \exp\left(\frac{-2\left(x^2 + y^2\right)}{w_1^2}\right) \tag{25.8a}$$

$$H_1(u, v) = \frac{\pi}{2} A_1 w_1^2 \exp\left(\frac{-\pi^2 w_1^2 \left(u^2 + v^2\right)}{2}\right) \tag{25.8b}$$

where A_1 designates the peak signal amplitude of the stimulus, and w_1 is its Gaussian "width" (equal to two standard deviations). Taking h_2 again as a uniform (zero) background, the task function is therefore also a Gaussian:

$$W^2_{task}(u, v) = \frac{\pi^2}{4} A_1^2 w_1^4 \exp\left(-\pi^2 w_1^2 \left(u^2 + v^2\right)\right) \tag{25.8c}$$

Hence, detection of a Gaussian object corresponds to a preferential weighting of low spatial frequencies, with the magnitude of the task function related to the size and contrast of the object, and breadth of the task function inversely related to the size of the object.

25.3.2.2 Shape discrimination

Figure 25.11(b) illustrates the somewhat more interesting case in which the task is to classify the shape of the object. This can be modeled as discrimination between two possible hypotheses – e.g. an object with a diffuse boundary (a Gaussian) versus an object with a sharp boundary (a disk). Taking these as h_1 and h_2, respectively, we have

$$h_1(x, y) = A_1 \exp\left(\frac{-2\left(x^2+y^2\right)}{a_1^2}\right) \tag{25.9a}$$

$$H_1(u, v) = \frac{\pi}{2} A_1 w_1^2 \exp\left(\frac{-\pi^2 w_1^2\left(u^2+v^2\right)}{2}\right) \tag{25.9b}$$

and

$$h_2(x, y) = A_2, \text{if} \sqrt{x^2+y^2} \leq w_2/2 \\ = 0, \text{otherwise} \tag{25.9c}$$

$$H_2(u, v) = \frac{\pi}{2} A_2 w_2^2 \frac{J_1\left(\pi w_2 \sqrt{u^2+v^2}\right)}{\pi a_2 \sqrt{u^2+v^2}} \tag{25.9d}$$

where A_2 is the amplitude of the disk signal, w_2 is the diameter of the disk, and J_1 is a first-order Bessel function. The task function for shape discrimination is therefore

$$W_{task}^2(u, v) = \frac{\pi^2}{4}\left[A_1 w_1^2 \exp\left(\frac{-\pi^2 w_1^2\left(u^2+v^2\right)}{2}\right) - A_2 w_2^2 \frac{J_1\left(\pi w_2 \sqrt{u^2+v^2}\right)}{\pi w_2 \sqrt{u^2+v^2}}\right]^2 \tag{25.9e}$$

Note that the shape discrimination task varies as the square of the amplitude and fourth power of the object width. It becomes evident that the shape of the task function is intimately related to the size and contrast of the two hypothesized object functions; in fact, the task function can assume any variety of shapes depending on A_1, w_1, A_2, and w_2. For example, taking $A_1 w_1^2 >> A_2 w_2^2$ (such that the Gaussian dominates over the disk), the task function is essentially a Gaussian – i.e. the discrimination task function has the same form as a detection task function. On the other hand, for the case

$$A_1 w_1^2 = \frac{A_2 w_2^2}{2} \tag{25.9f}$$

then $W_{task}(0, 0) = 0$ (noting that $\lim_{x\to 0} \frac{J_1(x)}{x} = \frac{1}{2}$), and the task function is quite different from a simple Gaussian. Middle and high frequencies are most heavily weighted, quite different from those associated with Gaussian detection. From a perception standpoint, use of these frequencies corresponds to the observer's ability to discriminate the sharpness of the edge.

25.3.2.3 Texture discrimination

Figure 25.11(c) illustrates the case in which the task is to classify the texture of the object as either "smooth" or "textured." This can be modeled as discrimination between two possible hypotheses – e.g. a smooth rect function versus a textured checkerboard. Taking these as h_1 and h_2, respectively, we have

$$h_1(x, y) = A_1 \Pi(\frac{x}{w_1})\Pi(\frac{y}{w_1}) \tag{25.10a}$$

$$H_1(x, y) = A_1 w_1^2 \sin c\,(\pi w_1 u) \sin c\,(\pi w_1 v) \tag{25.10b}$$

and

$$h_2(x, y) = h_1(x, y) + \Pi(\frac{x}{w_1})\Pi(\frac{y}{w_1}) A_2 S(2\pi x/w_2) S(2\pi y/w) \tag{25.10c}$$

$$H_2(u, v) = A_1 w_1^2 \sin c\,(\pi w_1 u) \sin c\,(\pi w_1 v) + \ldots \\ A_2 w_1^2 \left\{\sum_{n=-\infty}^{\infty} \frac{2}{\pi} \frac{(-1)^n}{2n+1} \sin c\left[\pi w_1 \left(u - \frac{2n+1}{w_2}\right)\right]\right\} \\ \times \left\{\sum_{m=-\infty}^{\infty} \frac{2}{\pi} \frac{(-1)^m}{2m+1} \sin c\left[\pi w_1 \left(v - \frac{2m+1}{w_2}\right)\right]\right\} \tag{25.10d}$$

where A_1 and w_1 are the amplitude and width of the rect, respectively, A_2 is the amplitude of the individual crenellae (checkerboards), and w_2 is the width of each crenella. The notation S denotes a square wave, and the coefficients in the summation are simply those for the Fourier series of such a wave. The task function is therefore:

$$W_{task}^2(u, v) \\ = A_2^2 w_1^4 \left\{\sum_{n=-\infty}^{\infty} \frac{2}{\pi} \frac{(-1)^n}{2n+1} \sin c\left[\pi w_1 \left(u - \frac{2n+1}{w_2}\right)\right]\right\}^2 \\ \times \left\{\sum_{m=-\infty}^{\infty} \frac{2}{\pi} \frac{(-1)^m}{2m+1} \sin c\left[\pi w_1 \left(v - \frac{2m+1}{w_2}\right)\right]\right\}^2 \tag{25.10e}$$

The example in Figure 25.11(c) illustrates a richness of non-zero, off-axis spatial frequencies associated with this task. Such task functions are most easily computed numerically, rather than in analytical forms as above.

25.3.3 Detectability index

As described above, the detectability combines the spatial-frequency-dependent detector performance (NEQ) with the spatial frequencies of interest (W_{task}). In its most generic form, the detectability index is given by the integral over the NEQ and Fourier transform of the image:

$$d'^2 = \int NEQ(f)\, |\Delta H(f)|^2\, df \tag{25.11a}$$

A number of models may be derived from this basic form that treat the imaging task and model observer somewhat differently. See Chapter 17 for a more complete discussion of observer models beyond the Fourier task-based forms considered here.

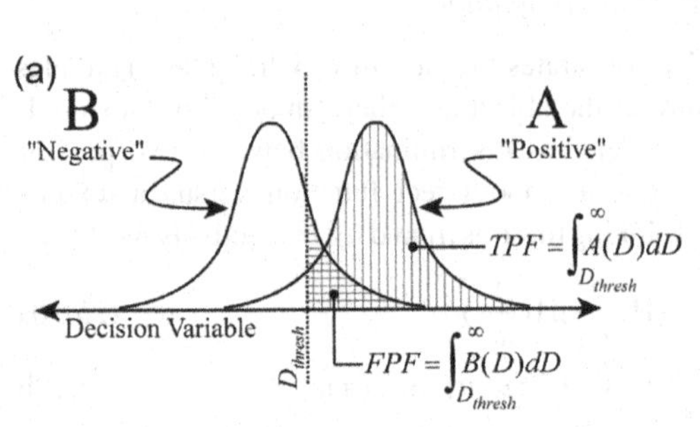

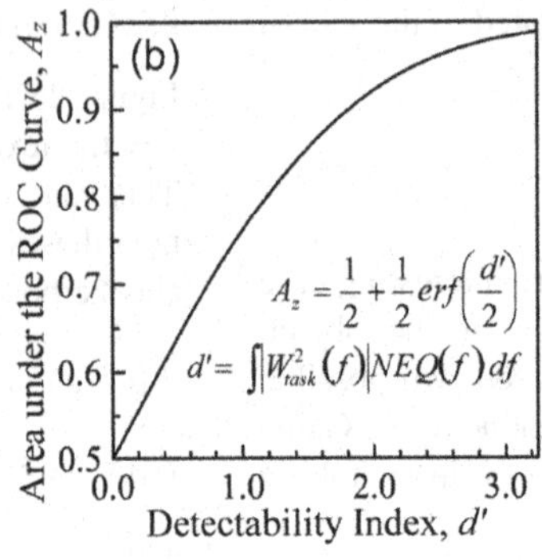

Figure 25.12 (a) Decision functions for cases A and B, with corresponding definitions of TPF and FPF. (b) Relationship between the detectability index and the area under the ROC curve.

Perhaps the simplest model weighs the NEQ by the imaging task as

$$d'^2 = \int \frac{MTF^2(f)W^2_{task}(f)}{NNPS(f)} df \tag{25.11b}$$

corresponding to the prewhitening (or Fisher-Hotelling) model observer. This can be modified in a straightforward manner to include the transfer characteristic of the eye (Barten, 1999) along with stochastic variations associated with the visual system (Burgess, 1994):

$$d'^2 = \int \frac{MTF^2(f)W^2_{task}(f)E^2(f)}{NNPS(f)E^2(u,v) + N_{int}} df \tag{25.11c}$$

where $E(f)$ is the eye transfer function, and N_{int} is the internal noise of the observer. Note that the eye function applies separately to the signal and the noise but does not cancel out of the detectability index due to the internal noise term.

An alternative form corresponds to a non-prewhitening model observer in which the task function applies separately to the signal and noise transfer characteristics of the detector, giving:

$$d'^2 = \frac{\left[\int MTF^2(f)W^2_{task}(f)df\right]^2}{\int MTF^2(f)W^2_{task}(f)NNPS(f)df} \tag{25.11d}$$

This corresponds to an observer which is unable to account for correlations in the noise. Similarly, this model may be modified to include the eye filter and internal noise as:

$$d'^2 = \frac{\left[\int MTF^2(f)W^2_{task}(f)E^2(f)df\right]^2}{\int \left[MTF^2(f)W^2_{task}(f)NNPS(f)E^4(f) + MTF^2(f)W^2_{task}(f)N_{int}\right]df} \tag{25.11e}$$

The extent to which such models agree with the performance of human observers has been a subject of considerable investigation in perception science over the last several decades. In the context of 2D and 3D radiographic imaging with FPDs, as considered in this chapter, there is growing evidence for such models providing a reasonable description of real observer performance to an extent that detectability index may be taken as a valid, meaningful objective function in the design and optimization of such imaging systems.

25.3.4 Detectability index in relation to human observer performance

With knowledge of system characteristics (MTF and NPS) and the imaging task (denoted W^2_{task}), the detectability index can be calculated for any number of idealized observers and corresponds to an information-theoretic description of image quality. This framework is useful conceptually, as it provides a means of decoupling the description of the detector, imaging task, and observer. However, assumptions associated with this framework (e.g. stationarity and the simplicity of idealized tasks) represent challenges in relating d' to human observer performance.

As detailed in other chapters, a common technique for characterization of observer response involves determination of the ROC (Metz, 1978) which typically measures an observer's performance in discriminating between two cases, A and B (e.g. positive/negative, signal present/absent, abnormal/normal, etc.). As illustrated in Figure 25.12, the two cases are assumed to occupy distributions as a function of a decision variable used by the observer in forming a hypothesis, with the width of each distribution related to image noise and object variability, and the separation between the distributions relative to their widths related to the observer SNR. Given an arbitrary decision threshold, observer performance is evaluated by the fraction of actually-positive cases correctly identified (the true-positive fraction (TPF)) versus the fraction of actually-negative cases incorrectly identified (the false-positive fraction (FPF)). The ROC curve is formed by varying the decision threshold, with the area under the curve, A_z, providing a summary measure of performance – e.g. $A_z = 0.5$ or 1.0 in the worst (random response) or best (perfect response) cases, respectively.

The relationship between detectability index and area under the ROC curve is straightforward under the assumption that the distributions are normal and symmetric (ICRU Report 54, 1996):

$$A_z = \frac{1}{2} + \frac{1}{2}erf\left(\frac{d'}{2}\right) = \frac{1}{2} + \frac{1}{\sqrt{\pi}}\int_0^{d'/2} e^{-x^2}dx \tag{25.12a}$$

$$d'^2 = 4\, inverf^2\left[2\left(A_z - {}^1\!/_2\right)\right] \tag{25.12b}$$

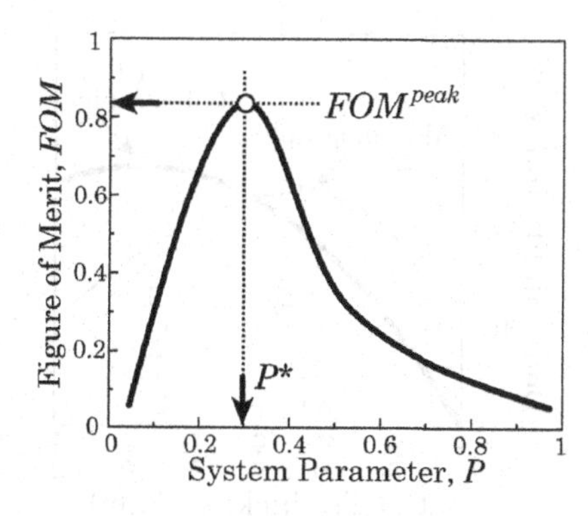

Figure 25.13 Optimization. Generic performance metric (FOM) versus generic system parameter (P).

Further, the relation to the proportion correct, P_C, in an alternative forced-choice (AFC) test involving M alternatives is given by

$$P_C(d', M) = \frac{1}{\sqrt{2\pi}} \int_{-\infty}^{\infty} e^{-\left(\frac{(x-d')^2}{2}\right)} [\Phi(x)]^{M-1} dx \qquad (25.12c)$$

where $\Phi(x)$ is a cumulative normal distribution function. These relations provide the common ground between the Fourier-based approach (in which NEQ is evaluated experimentally or theoretically) and the observer-based approach (in which A_z is evaluated by ROC or AFC), with detectability index, d', providing the bridge.

25.4 OPTIMIZATION

25.4.1 Maximizing performance with respect to the imaging task

The term "optimization" refers to the maximization of a given imaging performance or image quality metric with respect to one or more system parameters. For example, a given figure of merit (*FOM*) for imaging performance measured or calculated as a function of a given system parameter (*P*) is illustrated in Figure 25.13. The maximum value of the *FOM* is denoted FOM^{peak}, and the corresponding value of the parameter that optimizes performance is denoted P^*. As discussed above, the FOM is evaluated with respect to a given imaging task – e.g. as in the detectability index evaluated from the spatial frequencies of interest in testing the two hypotheses.

Often, an imaging system is used to accomplish more than one imaging task – e.g. nodule detection *and* shape discrimination. If different optimal parameters are identified for each imaging task, a dilemma may be presented in identifying a single optimal value, particularly if the optimum involves a system design parameter (e.g. scintillator thickness) or technique factor (e.g. kVp) that cannot be varied after image acquisition. Multi-function optimization helps to address this problem by minimizing the total reduction in the FOM for all tasks considered. As a simple example, let $FOM_i(P)$ be the FOM for the ith task, where P denotes a given system parameter and P_i^* denotes the value which maximizes $FOM_i(P)$. Further, let a_i denote an optional weight proportional to the relative "importance" of the ith task with respect to all other tasks. Such weighting may be subjectively determined, quantitatively evaluated by independent cost-benefit analysis, selected according to the frequency with which a task may occur, or simply taken equal for all tasks. The effective optimum, denoted P^*_{eff}, may be evaluated by numerous methods – e.g. in a form that minimizes the total loss in FOM:

$$\sum_i a_i \left[FOM_i(P_i^*) - FOM_i(P_{eff}^*)\right]$$

$$= \min_p \left(\sum_i a_i \left[FOM_i(P_i^*) - FOM_i(P)\right] \right) \qquad (25.13)$$

This equation balances tradeoffs among all tasks, where for the ith task, P^*_{eff} gives performance that is maintained in proportion to a_i. If the a_i terms are defined as relative frequencies of occurrence, this definition is equivalent to maximizing the expectation value (i.e. averaged across many cases) of the FOM. For all tasks weighted equally ($a_i = 1 \; \forall \; i$), P^*_{eff} gives performance that minimizes the total loss in FOM across each task. Conversely, P^*_{eff} can be considered to maximize the total FOM across each task.

For example, Figure 25.14(a) shows a hypothetical system for which the peak FOM with respect to Task A is achieved at P^*_A, whereas the peak for Task B is achieved at P^*_B. The effective optimum lies somewhere between P^*_A and P^*_B such that the total loss in FOM is minimized. Note that depending on the shape and magnitude of FOM(P) for each task, the multi-task optimum value P^*_{eff} may significantly reduce the ability to perform any particular task; however, it is the best choice in light of all tasks considered (even if no task is performed particularly well!). For example, Figure 25.14(b) illustrates a case of widely disparate task-specific optima and steeply varying

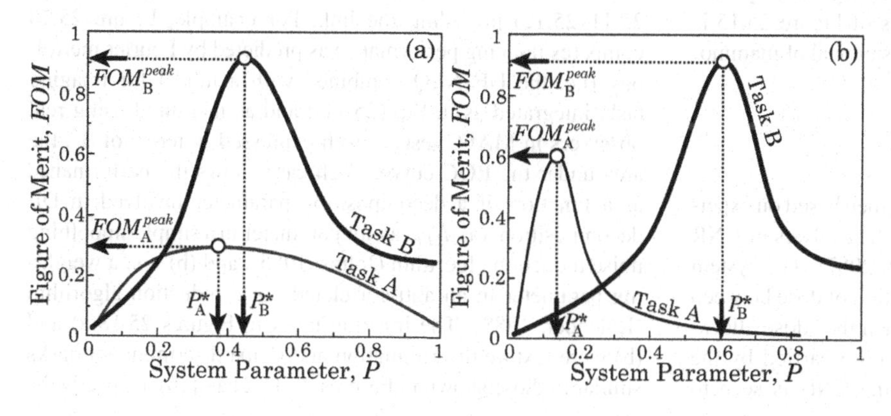

Figure 25.14 Optimization relative to multiple tasks.

FOMs, yielding an effective optimum that may equate to a fairly low FOM. Thus, the term "optimal" does not necessarily equate to "good" or "ideal" for each imaging task; rather, it indicates the best the system can perform given competing objectives and represents the best compromise. Therefore some consideration must clearly be applied in cases where competing tasks suggest vastly different individual optima.

Several examples in which the various imaging performance metrics described above are applied to the evaluation and optimization of 2D and 3D radiographic/tomographic imaging systems are described below. Examples were chosen to demonstrate the breadth of imaging technologies that have benefited from such analysis (from 2D radiography to 3D CBCT) as well as the utilization of various performance metrics (from CNR to detectability index).

25.4.2 Example: 2D projection imaging (radiography and mammography)

There has been tremendous work in the evaluation and optimization of 2D radiography and mammography systems based on FPDs since the inception of the technology in the 1990s and 2000s. An early example of system optimization in terms of detectability index is in the analysis of performance as a function of scintillator thickness. This was (and continues to be) an important system design parameter, since the converter thickness has a strong effect on the efficiency and spatial resolution characteristics of the detector. The desire for high quantum detection efficiency (achieved by using a thicker converter) must be balanced against increased image blur, and the tradeoffs should be considered quantitatively in relation to the imaging task.

For example, univariate optimization of CsI:Tl thickness is illustrated for the case of chest radiography in Figure 25.15 (Siewerdsen, 1998). In this case, the imaging task was a simple 1 mm Gaussian detection task, and all other system parameters were fixed at values typical of clinical chest imaging systems. Detectability is seen to improve with increased screen thickness up to ~475 μm by virtue of improved quantum detection efficiency. Beyond the optimal thickness, detectability degrades due to system blur. The optimum implied by such early analysis is consistent with the ~500 μm CsI:Tl thickness typical of FPDs developed more recently for chest imaging. Similarly in mammography, univariate analysis of scintillator thickness suggests an optimum below which task performance is degraded by noise, and above which task performance is degraded by blur. The optimum implied in the early analysis of Figure 25.15 is roughly consistent with the CsI:Tl thickness typical of mammographic FPD systems.

25.4.3 Example: dual-energy imaging

System optimization can be similarly performed based on experimentally measured figures of merit, as in the analysis of CNR for DE imaging in Figure 25.16 (Shkumat, 2007). The system parameter under consideration is the allocation of dose between the low- and high-energy projections, termed the "dose allocation" and defined as the fraction of total dose carried by the low-energy image. For fixed total dose, the CNR is seen to improve with allocation up to A ~0.30, beyond which CNR gradually declines. The results pertain to the visualization of the simulated soft-tissue nodule illustrated in Figure 25.16 and demonstrate optimal performance for conditions in which ~30% of the dose is imparted by the low-energy image, and ~70% of the dose is imparted by the high-energy image. The latter requires a higher proportion of dose to compensate for the lower DQE for the high-energy image (reduced quantum detection efficiency).

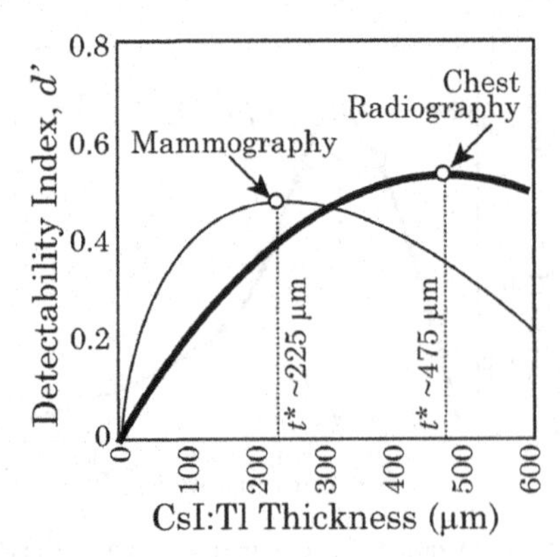

Figure 25.15 Univariate optimization of scintillator thickness in FPDs for 2D projection imaging.

DE imaging can be similarly evaluated in terms of theoretical analysis of detectability index (Richard, 2007). As shown in Figure 25.17 (for a 1 mm Gaussian detection task), d' may be determined in multi-variate analysis of imaging performance as a function of tissue cancellation factor (w_s, providing optimal bone cancellation), allocation (A, as in the previous example), and selection of kVp pair. Figure 25.17(a) shows that for a fixed dose and kVp pair, detectability index is optimized at w_s ~0.4 and A ~0.3 (consistent with Figure 25.16). Varying the kVp pair across the range 60–150 kVp (but keeping the total dose fixed) and re-evaluating the optimal w_s and A* at each point, Figure 25.17(b) suggests maximum performance at a kVp pair of 60 kVp (low) and 150 kVp (high) – i.e. at maximum deliverable energy separation. At other dose levels and for other imaging tasks, the optimal kVp was shown to vary, in each case demonstrating a strong dependence on the low-kVp and a weak dependence on the high-kVp, as in Figure 25.17(b).

Recent work has helped to bridge the relation between Fourier metrics of imaging performance in DE imaging (e.g. the DE NEQ) and the performance of real human observers, with models of imaging task (Section 25.3.4) and detectability index (Eqs. 25.11–25.12) providing the link. For example, Figure 25.18 compares imaging performance as predicted by Fourier metrology (i.e. the DE NEQ combined with a model for imaging task, integrated as in Eq. (25.11)) and as measured using real observers in MAFC tests – each expressed in terms of A_z, the area under the ROC curve. Each case shows the performance as a function of a decomposition parameter involved in DE decomposition: (a) d_{LPF}, a blur parameter in a simple smoothing noise reduction algorithm (Johns, 1985); and (b) w_N, a weighting parameter in an anti-correlated noise reduction algorithm (Kalender, 1988). The imaging tasks in Figures 25.18(a) and (b) were texture discrimination and shape discrimination tasks similar to those shown in Figures 25.11(c) and (b), respectively.

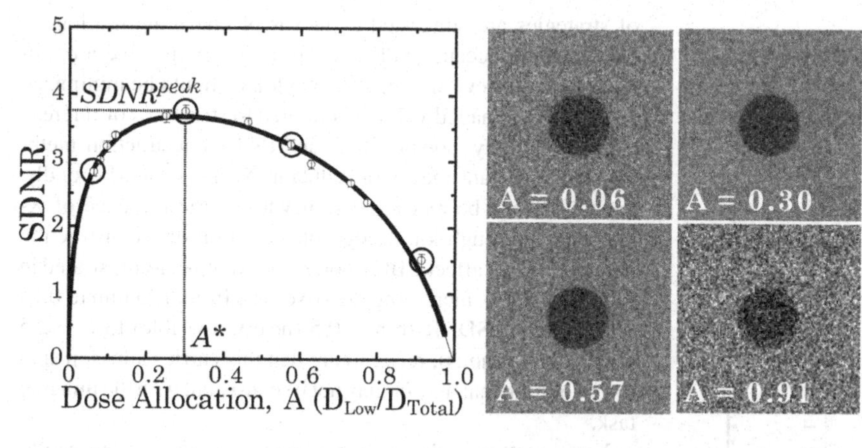

Figure 25.16 DE CNR (or SDNR) optimization (kVp pair).

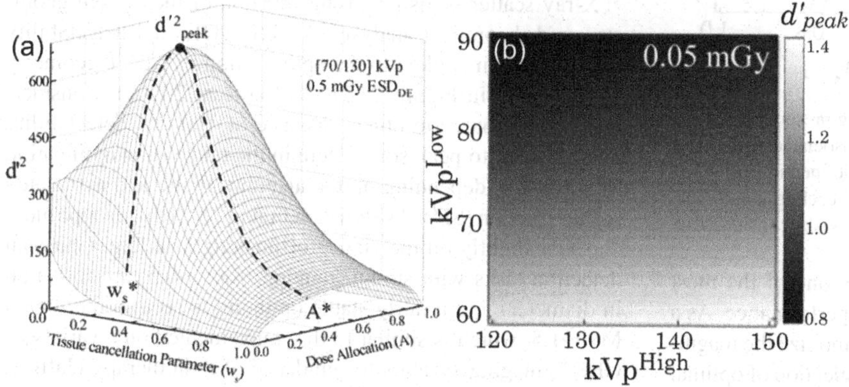

Figure 25.17 Theoretical optima of DE parameters (kVp pair, dose allocation, and noise reduction parameters). (a) Detectability computed as a function of tissue cancellation parameter and dose allocation. (b) Peak detectability (from (a)) computed as a function of low- and high-kVp. (Figure adapted from Richard *et al.* (Richard, 2007) with permission from the publisher.)

In each case, reasonable correspondence is observed between purely theoretical and measured performance, suggesting that Fourier-based metrics of imaging performance can provide realistic descriptions of human observer performance for simple tasks can provide and, therefore, a valid basis for system optimization.

DE imaging also presents an example in which multiple tasks can present competing optima. For example, considering lung nodule detection in DE images of the chest, Richard *et al.* (Richard, 2005a) showed the optimal dose allocation (i.e. fraction of total dose imparted by the low-energy image) to be A* ~0.60 for the DE soft-tissue image. As illustrated in Figure 25.19, however, the dose allocation that maximized the detectability index in the DE bone image was A* ~0.92. Taking Eq. (25.13) and considering each task as equally important (i.e., $a_1 = a_2$), the resulting effective optimum is A* ~0.79, which imparts a small tradeoff in detectability for each task. Thus, the effective optimum results in performance that is maximized for neither the soft-tissue nor bone image, reflecting the notion discussed above that optimization relates not simply to performance maximization, but performance maximization with respect to the given task (or tasks).

25.4.4 Example: 3D imaging (tomosynthesis and cone-beam CT)

Theoretical and experimental performance evaluation has proven similarly important in the development of 3D radiographic imaging systems such as tomosynthesis and CBCT. In the development of such systems using FPDs, it was recognized

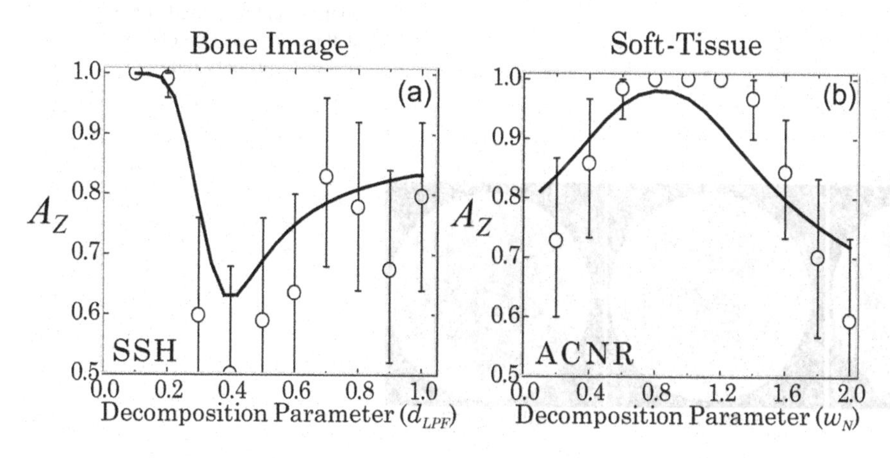

Figure 25.18 Measured and theoretical performance in DE imaging. (a) Texture discrimination task in the DE soft-tissue image. (b) Shape discrimination in the DE soft-tissue image. The correspondence of theory and measurement demonstrate that Fourier metrics can be related to real observer performance and provide a meaningful basis for system optimization.

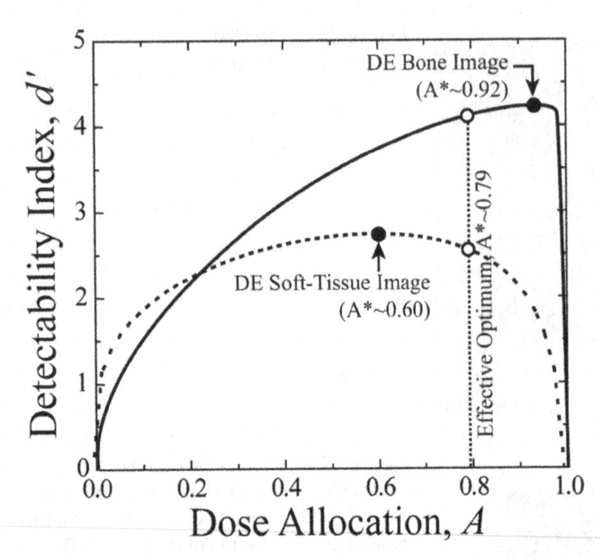

Figure 25.19 Multi-task optimization. DE imaging presents an example in which the competing optima may be associated with different imaging tasks. The effective optimum – defined in Eq. (25.13) – minimizes the loss in detectability across each task and thereby maximizes performance with respect to all tasks considered.

from an early stage that X-ray scatter poses one of the most significant physical factors limiting imaging performance. As a result, there has been considerable effort to minimize the magnitude of X-ray scatter in CBCT (e.g. through selection of optimal geometry and antiscatter grids) and to reduce scatter artifacts (e.g. through estimation of scatter fluence) (Siewerdsen, 2001, 2004a; Jarry, 2006).

The imaging performance metrics described in Section 25.2.2 have provided an important basis for CBCT performance evaluation and optimization with respect to X-ray scatter. For example, measurements of soft-tissue CNR (alternatively termed SDNR, the signal-difference-to-noise ratio) as a function of scatter-to-primary ratio (SPR) are shown in Figure 25.20. Such work shows the intrinsic reduction in soft-tissue contrast imparted by X-ray scatter (Johns, 1983; Siewerdsen, 2001) as well as the magnitude of scatter artifacts (e.g. strong cupping) associated with the typically high SPR (e.g. SPR ~100%) in CBCT. A variety of strategies are suggested as means of managing the loss in SDNR. First, because SPR is a strong function of the longitudinal field of view (i.e. the cone angle), it should be minimized such that the imaged volume is limited to structures of interest. Correspondingly, bowtie filters and ROI reconstruction methods are important means of reducing X-ray scatter. Secondly, the SDNR can be recovered somewhat in the presence of X-ray scatter through knowledgeable selection of reconstruction parameters that affect CBCT noise; for example, as illustrated in Figure 25.20(a), increasing the voxel size from 0.25 mm to 0.75 mm improves SDNR from ~0.5 (nearly invisible) to $>$ ~2.5 (fairly conspicuous), recognizing that this carries a loss in spatial resolution that may or may not be consistent with the imaging task.

X-ray scatter is also a strong function of the system geometry, and theoretical analysis of NPS, DQE, and detectability index has been applied to the optimization of CBCT geometry as illustrated in Figure 25.21 (Siewerdsen, 2000b). Considering a fixed source-to-object distance of 100 cm, detectability index is seen to peak somewhere in the range of magnification M ~1.3–2.0, depending on the anatomical site and associated scatter fraction (SF′). As shown in Figure 25.21(b), the optimum depends slightly on the choice of imaging task (e.g. Gaussian detection tasks with stimuli ranging from ~0.5 mm to 4 mm in diameter), but a fairly stable optimum is seen in the region M* ~1.5, which is similar to that implemented on a variety of CBCT imaging systems for guidance of radiotherapy (Jaffray, 2002).

Finally, the optimal filters associated with 3D cone-beam CT reconstruction may be evaluated in terms of detectability index with respect to a given imaging task, as illustrated in Figure 25.22. The parameter of interest in this case is the width of the (cosine) apodization filter, denoted h_{win}, which varies from h_{win} = 0.5 (equivalent to a smooth Hann filter) to 1.0 (equivalent to a sharp Ram-Lak filter). The results are fairly intuitive: a low-frequency detection task is optimized with a smoother apodization filter, while a higher-frequency task (delta-function detection) requires a sharper filter. The approach allows optimization therein with respect to various imaging conditions (e.g. radiation dose) and multiple (possibly competing) imaging tasks.

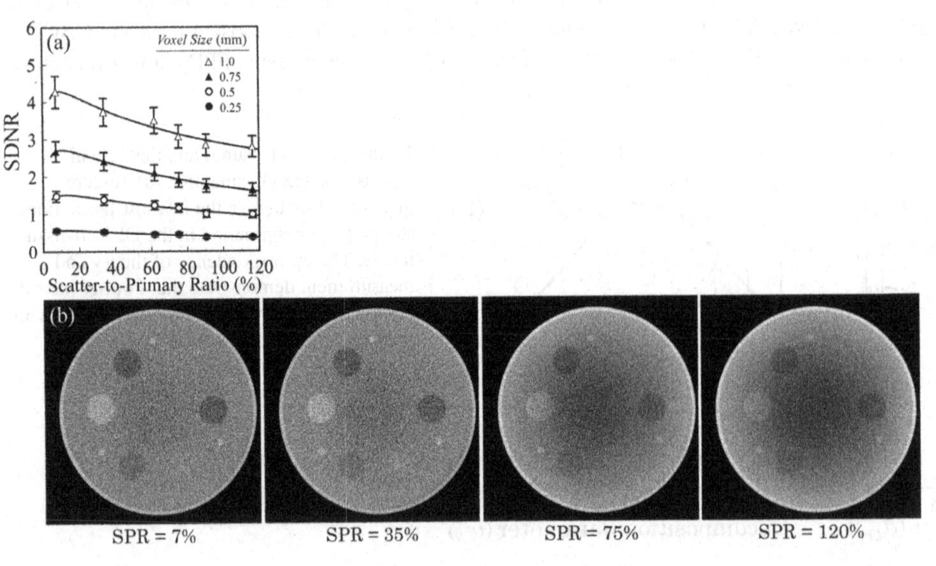

Figure 25.20 X-ray scatter in CBCT. (a) Degradation of SDNR due to X-ray scatter. (b) Illustration of reduced soft-tissue contrast and increased cupping artifact as a function of scatter-to-primary ratio.

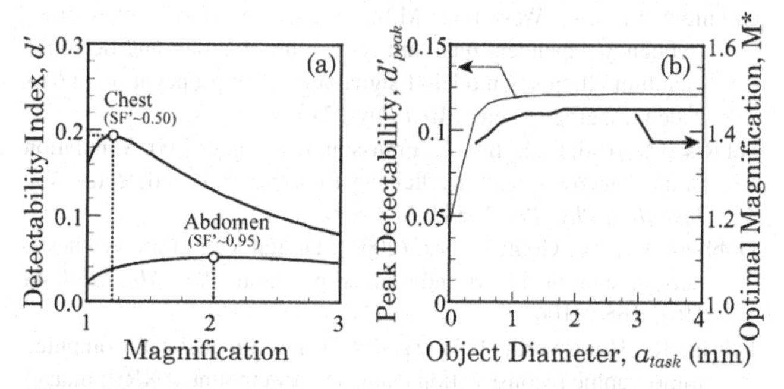

Figure 25.21 Optimization of system geometry in CBCT. (a) Detectability index computed as a function of magnification. (b) Peak detectability and optimal magnification (from (a)) computed as a function of object diameter in a Gaussian detection task.

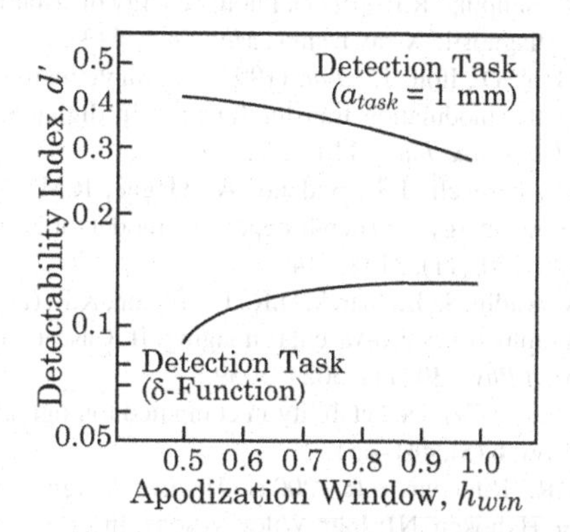

Figure 25.22 Optimization of CBCT reconstruction filter. (Figure adapted from Siewerdsen *et al.* (Siewerdsen, 2004b) with permission from the publisher.)

25.5 SUMMARY AND DISCUSSION

This chapter has attempted to provide a basic overview of both the technologies associated with 2D and 3D radiography (i.e. projection radiography, mammography, fluoroscopy, DE imaging, tomosynthesis, and cone-beam CT) as well as the principles and methods of imaging performance evaluation and optimization. Each of these presents a moving target. First, imaging technologies are undergoing continuous evolution and improvement, and the emphasis above on FPDs is consistent with their role as the most prevalent base technology for 2D and 3D radiography at present. However, there are numerous emerging technologies undergoing rapid development for future applications – e.g. complementary metal-oxide semiconductor (CMOS) detectors, various systems exhibiting avalanche gain, and photon counting detectors. Second, the methods of imaging performance characterization are still subjects of research and development. Standardization of measurement methods, refinements to cascaded systems analysis, quantitation of imaging task, and connection between model and human observers are all areas of activity throughout the medical imaging community.

As discussed in Section 25.1, the development of detector technologies presents an active area of ongoing research, and the distinction between 2D and 3D radiographic technologies has begun to fade. Correspondingly, one may expect the capabilities of imaging systems to broaden – for example, as evident in modern C-arms (once serving as dedicated fluoroscopes) now capable of fluoroscopy, tomosynthesis, and CBCT. One may expect similar expansion of functionality in radiographic systems – e.g. chest imaging systems capable of radiography, motion cine, dual-energy imaging, and tomosynthesis – as well as CT scanners – e.g. interventional CT scanners providing fluoroscopy and volumetric imaging. As detector technologies advance and mature, one may expect a rapidly expanding spectrum of 2D and 3D X-ray imaging capabilities. While the first century of the X-ray's application in medicine was marked by the projection "shadowgram," the next will be marked by capabilities for temporal, material, and spatial discrimination that are adaptable to a broad spectrum of imaging tasks, making the "new X-ray" an indispensible component of diagnostic and interventional medicine.

The basic approach to performance optimization outlined above rests on Fourier-based metrics of imaging performance (MTF, NPS, DQE, and NEQ) combined with a Fourier description of the imaging task (e.g. detection and discrimination tasks). These quantities integrated according to a given observer model (e.g. Fisher-Hotelling or non-prewhitening observers) yield a detectability index that may be evaluated as a figure of merit (i.e. an objective function) in optimizing performance with respect to one or more image acquisition or reconstruction parameters. Optimization as considered in this chapter regards maximization of imaging performance, and it should certainly be recognized that there are numerous other aspects of system optimization that are not addressed here, including logistics, workflow, and usability. These belong more to the disciplines of systems engineering and human factors analysis and are certainly essential to the successful implementation of medical imaging systems.

Consideration of the imaging task is central to meaningful optimization of imaging performance. After all, an imaging system is always designed for a given task – or more commonly, a spectrum of tasks. Description of simple detection and discrimination tasks as spatial-frequency templates, as described above, provides an important starting point for the quantitative incorporation of tasks in system optimization. Of course, this is an abstract, idealized approach and cannot pretend to describe the complexity of factors affecting a diagnostic decision. As suggested by Figure 25.18, however, the approach can demonstrate reasonable correspondence between the detectability index as derived from first principles of image science and the performance of human observers. More sophisticated imaging task models and performance descriptors represent an important area of future research. Such an understanding is key to evolution from the traditional diagnostic imaging paradigm (in which the highest quality image is acquired everywhere all the time) to a truly task-based paradigm in which an image is acquired for the purpose of a specific task, and the acquisition, reconstruction, and processing techniques are knowledgeably tuned for that purpose.

ACKNOWLEDGMENTS

Acknowledgment and gratitude are owed to the colleagues who made this chapter possible. Samuel Richard (University of Toronto) and Daniel J. Tward (Ontario Cancer Institute) assisted with the figures, references, and editing throughout.

Dr. John Yorkston (Carestream Health Inc.) provided valuable conversations regarding digital X-ray imaging technologies. Dr. Ian A. Cunningham (University of Western Ontario) provided essential guidance regarding cascaded systems analysis. Much of the work was performed by members of the Image-Guided Therapy (IGTx) Lab at the Ontario Cancer Institute in years of fruitful collaboration between the author and Dr. David Jaffray (Princess Margaret Hospital).

REFERENCES

Antonuk, L.E., El-Mohri, Y., Siewerdsen, J.H., *et al.* (1997). Empirical investigation of the signal performance of a high-resolution, indirect detection, active matrix flat-panel imager (AMFPI) for fluoroscopic and radiographic operation, *Med Phys*, **24** (1), 51–70.

Antonuk, L.E., Jee, K.W., El-Mohri, Y., *et al.* (2000). Strategies to improve the signal and noise performance of active matrix, flat-panel imagers for diagnostic x-ray applications, *Med Phys*, **27** (2), 289–306.

Antonuk, L.E., Zhao, Q., Su, Z., *et al.* (2004). Systematic development of input-quantum-limited fluoroscopic imagers based on active-matrix flat-panel technology, *Proc SPIE Med Imag*, **5368**, 127–138.

Bachar, G., Siewerdsen, J.H., Daly, M.J., Jaffray, D.A., Irish, J.C. (2007). Image quality and localization accuracy in C-arm tomosynthesis-guided head and neck surgery, *Med Phys*, **34** (12), 4664–4677.

Badano, A., Kyprianou, I.S., Jennings, R.J., Sempau, J. (2007). Anisotropic imaging performance in breast tomosynthesis, *Med Phys*, **34** (11), 4076–4091.

Barten, P.G.J. (1999). *Contrast Sensitivity of the Human Eye and its Effect on Image Quality*, Bellingham, WA: SPIE Publications.

Bartling, S.H., Majdani, O., Gupta, R., *et al.* (2007). Large scan field, high spatial resolution flat-panel detector based volumetric CT of the whole human skull base and for maxillofacial imaging, *Dentomaxillofac Radiol*, **36** (6), 317–327.

Bissonnette, J.P., Cunningham, I.A., Jaffray, D.A., Fenster, A., Munro, P. (1997). A quantum accounting and detective quantum efficiency analysis for video-based portal imaging, *Med Phys*, **24** (6), 815–826.

Boone, J.M., Lindfors, K.K. (2006). Breast CT: potential for breast cancer screening and diagnosis, *Future Oncol*, **2** (3), 351–356.

Boone, J.M., Seibert, J.A. (1997). An accurate method for computer-generating tungsten anode x-ray spectra from 30 to 140 kV, *Med Phys*, **24** (11), 1661–1670.

Boyce, S.J., Chawla, A., Samei, E. (2005). Physical evaluation of a high frame rate, extended dynamic range flat panel detector for real-time cone beam computed tomography applications, *Proc SPIE Med Imag*, **5745**, 591–598.

Burgess, A.E. (1994). Statistically defined backgrounds: performance of a modified nonprewhitening observer model, *J Opt Soc Am A*, **11** (4), 1237–1242.

Burgess, A.E., Jacobson, F.L., Judy, P.F. (2001). Human observer detection experiments with mammograms and power-law noise, *Med Phys*, **28** (4), 419–437.

Bushberg, J.T., Seibert, J.A., Leidholdt, E.M., Boone, J.M. (2002). *The Essential Physics of Medical Imaging*, Hagerstown, MD: Lippincott Williams & Wilkins.

Cahn, R.N., Cederstrom, B., Danielsson, M., *et al.* (1999). Detective quantum efficiency dependence on x-ray energy weighting in mammography, *Med Phys*, **26** (12), 2680–2683.

Cunningham, I.A., Westmore, M.S., Fenster, A. (1994). A spatial-frequency dependent quantum accounting diagram and detective quantum efficiency model of signal and noise propagation in cascaded imaging systems, *Med Phys*, **21** (3), 417–427.

del Risco Norrlid, L., Edling, F., Fransson, K., *et al.* (2005). Simulation of the detective quantum efficiency for a hybrid pixel detector, *Nuc Inst Meth Phys Res A*, **543**, 528–536.

Dobbins, J.T., III, Godfrey, D.J. (2003). Digital x-ray tomosynthesis: current state of the art and clinical potential, *Phys Med Biol*, **48** (19), R65–R106.

Fahrig, R., Holdsworth, D.W. (2000). Three-dimensional computed tomographic reconstruction using a C-arm mounted XRII: image-based correction of gantry motion nonidealities, *Med Phys*, **27** (1), 30–38.

Feldkamp, L.A., Davis, L.C., Kress, J.W. (1984). Practical cone-beam algorithm, *J Opt Soc Am A*, **1**, 612–619.

Fewell, T.R., Shuping, R.E. (1977). Photon energy distribution of some typical diagnostic X-ray beams, *Med Phys*, **4** (3), 187–197.

Fujita, H., Tsai, D., Itoh, T., *et al.* (1992). A simple method for determining the modulation transfer function in digital radiography, *IEEE Trans Med Imag*, **11** (1), 34–39.

Gallas, B.D., Boswell, J.S., Badano, A., Gagne, R.M., Myers, K.J. (2004). An energy- and depth-dependent model for X-ray imaging, *Med Phys*, **31** (11), 3132–3149.

Ganguly, A., Rudin, S., Bednarek, D.R., Hoffmann, K.R. (2003). Micro-angiography for neuro-vascular imaging. II. Cascade model analysis, *Med Phys*, **30** (11), 3029–3039.

Hanson, K.M. (1979). Detectability in computed tomographic images, *Med Phys*, **6** (5), 441–451.

Hendee, W.R., Ritenour, E.R. (2002). *Medical Imaging Physics, 4th edition*, Hoboken, NJ: John Wiley & Sons, Inc.

Hunt, D.C., Kirby, S.S., Rowlands, J.A. (2002). X-ray imaging with amorphous selenium: X-ray to charge conversion gain and avalanche multiplication gain, *Med Phys*, **29** (11), 2464–2471.

ICRU (1996). ICRU Report 54. *Medical Imaging – the Assessment of Image Quality*. Bethesda, MD: International Commission on Radiation Units and Measurements.

IEC Committee 62B (2003). *IEC 62220–1 Medical Electrical Equipment Characteristics of Digital X-Ray Imaging Devices: Part 1. Determination of the Detective Quantum Efficiency*. Geneva, Switzerland: IEC.

Jaffray, D.A., Siewerdsen, J.H., Wong, J.W., Martinez, A.A. (2002). Flat-panel cone-beam computed tomography for image-guided radiation therapy, *Int J Radiat Oncol Biol Phys*, **53** (5), 1337–1349.

Jarry, G., Graham, S.A., Moseley, D.J., *et al.* (2006). Characterization of scattered radiation in kV CBCT images using Monte Carlo simulations, *Med Phys*, **33** (11), 4320–4329.

Johns, H.E., Cunningham, J.R. (1983). *The Physics of Radiology, 4th edition*, Springfield, IL: Charles C. Thomas.

Johns, P.C., Yaffe, M.J. (1985). Theoretical optimization of dual-energy x-ray imaging with application to mammography, *Med Phys*, **12** (3), 289–296.

Kalender, W., Klotz, E., Kostaridou, L. (1988). An algorithm for noise suppression in dual-energy CT material density images, *IEEE Trans Med Imag*, **7**, 218–224.

Maidment, A.D.A., Ullberg, C., Lindman, K., *et al.* (2006). Evaluation of a photon-counting breast tomosynthesis imaging system, *Proc SPIE Phys Med Imag*, **6142**, 61420B-1–61420B-11.

Metheany, K., Boone, J.M., Abbey, C.K., Packard, N. (2007). A comparison of anatomical noise properties between breast CT and projection breast imaging, *Med Phys*, **34** (6), 2563.

Metz, C.E. (1978). Basic principles of ROC analysis, *Semin Nucl Med*, **8** (4), 283–298.

Nishikawa, R.M., Engstrom, E., Reiser, I. (2007). Comparison of the breast tissue power spectrum for mammograms, tomosynthesis projection images, and tomosynthesis reconstruction images. Annual meeting of the Radiological Society of North America (RSNA).

Rafferty, E.A. (2007). Digital mammography: novel applications, *Radiol Clin N Am*, **45** (5), 831–843, vii.

Rafferty, M.A., Siewerdsen, J.H., Chan, Y., *et al.* (2006). Intraoperative cone-beam CT for guidance of temporal bone surgery, *Otolaryngol Head Neck Surg*, **134** (5), 801–808.

Richard, S., Siewerdsen, J.H. (2007). Optimization of dual-energy imaging systems using generalized NEQ and imaging task, *Med Phys*, **34** (1), 127–139.

Richard, S., Siewerdsen, J.H., Jaffray, D., Moseley, D.J., Bakhtiar, B. (2005a). Generalized DQE analysis of radiographic and dual-energy imaging using flat-panel detectors, *Med Phys*, **32**, 1397–1413.

Richard, S., Siewerdsen, J.H., Jaffray, D.A., Moseley, D.J., Bakhtiar, B. (2005b). Generalized DQE analysis of dual-energy imaging using flat-panel detectors, *Proc SPIE Phys Med Imag*, **5745**, 519–528.

Riederer, S.J., Pelc, N.J., Chesler, D.A. (1978). The noise power spectrum in computed X-ray tomography, *Phys Med Biol*, **23** (3), 446–454.

Roehrig, H., Fajardo, L.L., Yu, T., Schempp, W.S. (1994). Signal, noise and detective quantum efficiency in CCD based X-ray imaging systems for use in mammography, *Proc SPIE Phys Med Imag*, **2163**, 320–332.

Rose, A. (1948). The sensitivity performance of the human eye on an absolute scale, *J Opt Soc Am*, **38**, 196–208.

Rowlands, J.A. (1996). Digital X-ray systems based on amorphous selenium, *AJR Am J Roentgenol*, **167** (2), 409–411.

Sawant, A., Antonuk, L.E., El-Mohri, Y., *et al.* (2006). Segmented crystalline scintillators: empirical and theoretical investigation of a high quantum efficiency EPID based on an initial engineering prototype CsI(Tl) detector, *Med Phys*, **33**, 1053–1066.

Shkumat, N.A., Siewerdsen, J.H., Dhanantwari, A.C., *et al.* (2007). Optimization of image acquisition techniques for dual-energy imaging of the chest, *Med Phys*, **34** (10), 586–601.

Siewerdsen, J.H., Antonuk, L.E. (1998). DQE and system optimization for indirect-detection flat-panel imagers in diagnostic radiology, *Proc SPIE Phys Med Imag*, **3336**, 546–555.

Siewerdsen, J.H., Antonuk, L.E., El-Mohri, Y., *et al.* (1997). Empirical and theoretical investigation of the noise performance of indirect detection, active matrix flat-panel imagers (AMFPIs) for diagnostic radiology, *Med Phys*, **24** (1), 71–89.

Siewerdsen, J.H., Jaffray, D.A. (2000a). Cone-beam CT with a flat-panel imager: noise considerations for fully 3D computed tomography, *Proc SPIE Phys Med Imag*, **3977**, 408–416.

Siewerdsen, J.H., Jaffray, D.A. (2000b). Optimization of x-ray imaging geometry (with specific application to flat-panel cone-beam computed tomography), *Med Phys*, **27** (8), 1903–1914.

Siewerdsen, J.H., Jaffray, D.A. (2001). Cone-beam computed tomography with a flat-panel imager: magnitude and effects of X-ray scatter, *Med Phys*, **28** (2), 220–231.

Siewerdsen, J.H., Jaffray, D.A. (2003). Three-dimensional NEQ transfer characteristics of volume CT using direct and indirect-detection flat-panel imagers, *Proc SPIE Phys Med Imag*, **29** (11), 2655–2671.

Siewerdsen, J.H., Moseley, D.J., Bakhtiar, B., Richard, S., Jaffray, D.A. (2004a). The influence of antiscatter grids on soft-tissue detectability in cone-beam computed tomography with flat-panel detectors, *Med Phys*, **31** (12), 3506–3520.

Siewerdsen, J.H., Moseley, D.J., Burch, S., *et al.* (2005). Volume CT with a flat-panel detector on a mobile, isocentric C-arm: pre-clinical investigation in guidance of minimally invasive surgery, *Med Phys*, **32** (1), 241–254.

Siewerdsen, J.H., Moseley, D.J., Jaffray, D.A. (2004b). Incorporation of task in 3D imaging performance evaluation: the impact of asymmetric NPS on detectability, *Proc SPIE Phys Med Imag*, **5368**, 89–97.

Siewerdsen, J.H., Waese, A.M., Moseley, D.J., Richard, S., Jaffray, D.A. (2004c). Spektr: a computational tool for x-ray spectral analysis and imaging system optimization, *Med Phys*, **31** (11), 3057–3067.

Silverman, J.D., Paul, N.S., Siewerdsen, J.H. (2009). Investigation of lung nodule detectability in low-dose 320-slice computed tomography. *Med Phys*, **26**, 1700–1710.

Spies, L., Ebert, M., Groh, B.A., Hesse, B.M., Bortfeld, T. (2001). Correction of scatter in megavoltage cone-beam CT, *Phys Med Biol*, **46** (3), 821–833.

Swank, R.K. (1973). Absorption and noise in x-ray phosphors, *J Appl Phys*, **44** (9), 4199–4203.

Tucker, D.M., Barnes, G.T., Chakraborty, D.P. (1991). Semiempirical model for generating tungsten target x-ray spectra, *Med Phys*, **18** (2), 211–218.

Tutar, I.B., Managuli, R., Shamdasani, V., *et al.* (2003). Tomosynthesis-based localization of radioactive seeds in prostate brachytherapy, *Med Phys*, **30** (12), 3135–3142.

Vedantham, S., Karellas, A., Suryanarayanan, S. (2004). Solid-state fluoroscopic imager for high-resolution angiography: parallel-cascaded linear systems analysis, *Med Phys*, **31** (5), 1258–1268.

Warp, R.J., Dobbins, J.T., III (2003). Quantitative evaluation of noise reduction strategies in dual-energy imaging, *Med Phys*, **30** (2), 190–198.

Yao, J., Cunningham, I.A. (2001). Parallel cascades: new ways to describe noise transfer in medical imaging systems, *Med Phys*, **28** (10), 2020–2038.

Zhao, W., Rowlands, J.A. (1997). Digital radiology using active matrix readout of amorphous selenium: theoretical analysis of detective quantum efficiency, *Med Phys*, **24** (12), 1819–1833.

26

Applications of AFC methodology in optimization of CT imaging systems

KENT OGDEN AND WALTER HUDA

26.1 INTRODUCTION

Maximizing the quality of medical images is a primary goal in radiological imaging. For digital modalities, such as computed tomography (CT), this task is complex because of the large number of independent factors that contribute to the final reconstructed images. The task of quantifying image quality in itself has always been difficult, and a variety of metrics have been employed (Tapiovaara, 2005). However, simple image quality measures such as pixel standard deviation (to quantify noise) or modulation transfer function (to quantify resolution) are not able to accurately predict the visibility or conspicuity of lesions. For example, it has been shown that the texture in the CT image plays a role in lesion signal to noise ratio (SNR) for ideal observers (Boedeker, 2007) and for CAD systems (Ochs, 2006), and this texture is dependent on the selected reconstruction filter. It is because of such difficulties that observer studies are performed, as they give direct information on observer performance as technical parameters are changed.

In CT imaging, the operator must select values for multiple technical factors when acquiring a patient scan. There are many more parameters available for digital modalities than there were for film-based imaging, and the choices the operator makes are not governed by the need to reach a target image density as they were for film. The user therefore has increased freedom in selecting from the range of parameters available. There are constraints on the available choices, in that the techniques must be achievable by the CT system (e.g. tube heating and generator power limitations), as well as the need to keep the exposure time to a minimum value to minimize motion artifacts. Nonetheless, a range of values for many of the available parameters will yield images of diagnostic quality.

The choice of techniques in CT will not only have an impact on the image quality, but will also affect the dose delivered to the patient. For example, increasing the quantity of radiation used (e.g. increased tube current) and decreasing the beam energy (tube accelerating potential, kV) will generally improve the image quality and observer performance (Huda, 2007). However, the resultant dose to the patient may increase by a factor that does not justify the increase in image quality. The task of optimizing CT, then, is one of choosing an appropriate set of technical parameters such that the image quality is in some way optimal for the diagnostic task, while keeping the patient dose as low as possible.

26.1.1 CT parameters

CT image acquisition presents the user with many parameters that could be subjected to an optimization process. We group this (non-exhaustive) list of parameters into three categories and give an overview of each category.

26.1.1.1 Beam characteristics

Physical beam characteristics would include beam *quantity* as well as beam *quality*. Beam quantity will be primarily determined by tube current (mA) and tube rotation time (s), and is denoted as the current-exposure time product (mAs). Increasing the mAs value will increase the patient dose proportionally, and will decrease the image noise. If the system is quantum noise limited (as CT generally is in normal operation), the noise will decrease such that the contrast to noise ratio (CNR) will increase proportional to $\text{mAs}^{1/2}$. The beam quantity is also related to the beam shaping filter used (i.e. head or body filter) and the selected pitch, if the scan is acquired in helical scan mode. The total amount of radiation used to acquire a scan is inversely proportional to selected pitch (all else being constant), though the effect on image noise as pitch changes is dependent on the scanner design and the reconstruction algorithm used (Primak, 2006).

Beam quality is primarily determined by the tube accelerating potential (kV), as well as the beam shaping filter. Modern scanners generally have a limited number of tube voltage settings (80, 100, 120, and 140 kV are typical). However, this limited number of settings represents a wide range of beam qualities. Increasing the kV will increase the fraction of the beam that penetrates the patient and therefore the amount of detected radiation, which will reduce noise. Higher beam energies, however, will decrease the contrast of all tissues, and may result in low-contrast lesions being harder to detect if image noise is not decreased to compensate. Also, the tube output will increase (radiation intensity per unit mAs) as kV increases, and this needs to be considered when selecting kV.

26.1.1.2 Scan geometry

Scan parameters related to the acquisition geometry include image thickness, field of view (which affects pixel size and therefore resolution), focal spot size, and scan pitch, if acquired in helical mode.

The Handbook of Medical Image Perception and Techniques, ed. Ehsan Samei and Elizabeth Krupinski. Published by Cambridge University Press.

Reducing the reconstructed slice thickness with all other parameters fixed will result in increased image noise, as fewer detected photons will be contributing to each image. Countering this increase in image noise, however, is the reduction in partial volume artifact, in which objects thinner than the image slice thickness are "blended" with adjacent tissues and can lose contrast or become undetectable.

Scan pitch, as mentioned above, can affect the total amount of radiation used, and may affect the image noise. Helical scanning may also introduce specific artifacts, since there may be incomplete data available to reconstruct an image at any given slice location. Data from adjacent projections may be interpolated to create a complete data set to reconstruct a given slice location. This results in reduced longitudinal resolution, as measured by the slice sensitivity profile, as well as other helical artifacts (Hu, 1996; Hsieh, 1997, 2003).

Changing the focal spot size can affect system resolution. However, the focal spot size may not be selectable for a given scan, as the tube current may require the use of a large spot size to prevent damage to the tube anode. Small focal spot sizes may be available only for specific scans.

26.1.1.3 Reconstruction algorithm and display parameters

All modern CT scanners offer a choice of reconstruction algorithms tailored to specific diagnostic purposes. It is normal in many cases to reconstruct multiple image sets using different algorithms for a single scan acquisition. In general, these algorithms affect the spatial frequency content of the resulting images, which will affect both image noise and spatial resolution. These algorithms are not standardized and different manufacturers use various naming conventions to denote the algorithms available on their scanners.

Display of CT images represents the endpoint of the imaging chain, and the final displayed image quality can be influenced by a large number of parameters. These could include printer, film, viewbox characteristics, ambient conditions for hardcopy display, monitor calibration characteristics, and the associated display parameters for softcopy display. Window/level settings are adjusted by the radiologist in virtually all CT scans as they are being interpreted in softcopy mode, and will have a dramatic impact on the ability to visualize pathology. In this chapter, we will not explore the variables associated with image display, but we mention them here for completeness.

There are other parameters involved in CT scanning that are not addressed here. For example, in dynamic studies such as CT angiography, timing is very important to ensure that the appropriate anatomy is scanned during the maximum opacification of the vessels of interest. The delay is typically adjusted for each patient, using, for example, a "timing bolus" to determine the precise delay before scan initiation for that patient. Another example is the total beam width (collimation) used during a scan. In general, the maximum collimation available will be used for most scans. However, certain scanners may require the use of smaller beam widths to achieve their maximum z-axis resolution (i.e. minimum slice thickness). We generally assume in this chapter that the maximum beam width is used.

As mentioned earlier, the goal in optimizing CT is to determine the techniques or processing that provide the appropriate diagnostic information, while delivering the least possible dose to the patient. This implies an ability to obtain some metric of the information content of an image (i.e. the "image quality") as well as an ability to evaluate an appropriate dose metric, such as effective dose, to quantify the risk. We will not discuss dose issues in this chapter, instead focusing on determining some measure of image quality.

It is important to recognize that "image quality" is dependent on the diagnostic task at hand. Therefore, any "optimal" set of scan parameters is likely only appropriate for a specific diagnostic task or range of tasks. In the case where the diagnostic task is not specifically known in advance (i.e. a "general" diagnostic scan), a compromise must be reached that will provide good overall image quality.

26.2 OPTIMIZING CT SCAN PARAMETERS

There are multiple approaches to the problem of optimizing medical image quality. Perhaps the simplest is to use traditional analytical measures such as image noise and contrast, which may be used to calculate a CNR for a given set of scan parameters. There are difficulties that arise immediately from this approach, however. Contrast itself is a straightforward concept, and relatively easy to measure. "Noise" is more complicated, but is usually taken to mean quantum, electronic, or other sources of random fluctuations in image pixel values that impede the ability to detect subtle features.

Simplistic measures of noise include measuring the pixel standard deviation in an otherwise uniform region of interest. It has long been recognized that this simple noise metric does not completely determine the likelihood of detecting subtle pathology. This is because the "texture" of the noise (its visual structure, determined by its frequency content) can play an important role in the perception and detection of pathology in an image. In CT, noise texture is strongly affected by the reconstruction algorithm and may change radically when different algorithms are used on the same scan data. It is possible to have images with equal pixel standard deviations that have very different visual appearance due to the differences in the texture of the noise.

Another difficulty when using random noise measurements to predict diagnostic usefulness of an image is that the image signal variations due to patient anatomy, sometimes termed "structure" or "anatomic" noise, can mask pathology more effectively than the random noise. This is because the variations in intensity due to anatomic differences are usually many times larger than the random fluctuations, as well as the fact that the typical sizes of anatomic structures may be similar to those of the pathology of interest. In the presence of significant anatomical structure and for typical clinical techniques, random (quantum) noise may have little effect on the ability to detect lesions, and any increases in technique to reduce random noise will achieve little or no benefit, while increasing patient dose.

In the work described in this chapter, we do not try to determine an image quality metric such as CNR directly. Instead, we look at the effects changing acquisition parameters have on an observer's performance for a specific detection task. By doing this, there is no need to make assumptions about the effect changing an imaging parameter will have on lesion detection, as it is measured directly. This is an advantage over metrics such as CNR, which may not directly predict observer performance in detecting lesions.

Measuring observer performance may be accomplished in multiple ways as described in other chapters of this book. We have chosen to use the alternative forced-choice (AFC) methodology, and our typical diagnostic task is to locate nodular lesions in various regions of the anatomy. Although this is a limited task, it is appropriate for CT imaging as nodular lesions of various pathologic origins may be found throughout the body.

In the next section, we provide a detailed description of the AFC software we have developed. This description should allow the interested reader with appropriate programming skills to recreate this methodology.

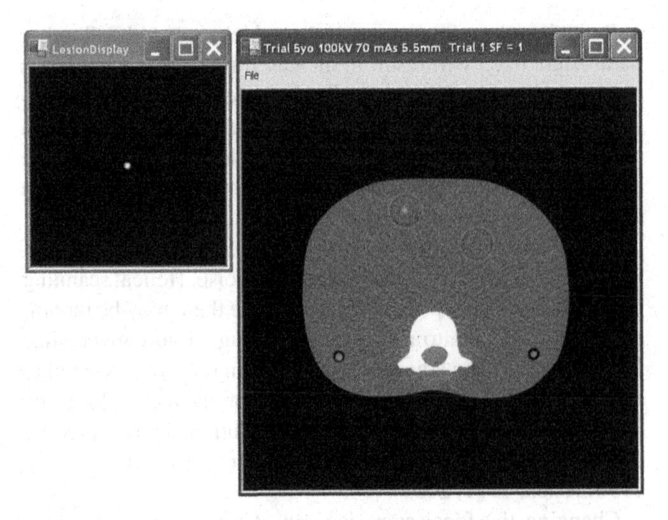

Figure 26.1 Graphical user interface of the AFC software. The smaller window on the left shows the observer the current lesion. The contrast of the lesion in this display will track changes in the contrast of the embedded lesion, but is at a higher level for easy visualization. The contrast of the cue circles shown here is higher than in an actual experiment for printing considerations.

26.3 DESCRIPTION OF AFC METHODOLOGY

The AFC approach to measuring the effect of changing CT acquisition parameters on observer performance is straightforward in principle. Given a parameter to be tested (e.g. the mAs used to acquire a scan), multiple image sets are collected for an appropriate rang of values of that parameter. Each set of images will then be used in AFC experiments to measure the observers' performance at that particular parameter setting. The results of all the experiments may then be compared to determine the effect on observer performance as changes are made to that parameter.

AFC studies are either performed using lesions that are "real" (taken from patient exams), or simulated mathematically. An important aspect of these studies is that the observers' performance will change as the lesion size is changed. To systematically investigate this effect, it is necessary to have a "library" of lesions available in different sizes, which may present a problem when using patient scan-derived lesions. Mathematically generated lesions permit any lesion size to be generated, and this is the approach taken in the examples presented in this chapter.

In an observing session (denoted an "experiment") the observer is presented with multiple choices for the possible location of a lesion embedded in a CT image. The observer must select (forced-choice) which of the locations (alternatives) actually contains the lesion. There is no subjectivity involved in determining the observer's accuracy. The observer is either right or wrong, and does not have to provide a "confidence" level that the lesion is actually present, as is commonly done with ROC-based experimental methods.

26.3.1 AFC implementation

The interface of the software developed by the authors is shown in Figure 26.1. A detailed description of the software design goals and logical flow follows.

Historically, AFC experiments have been conducted using multiple images for each presentation to the observer. The images used were typically regions extracted from radiographic images, not complete images. This was done out of practical considerations, since entire digitized images require more storage and time to load into memory, and full images may be larger than the available display resolution allows without minification. We have chosen a different approach, in which a single, complete image is used and which contains multiple possible (non-overlapping) locations for the embedded lesion. The locations are indicated with low-contrast cue circles as in Figure 26.1. The user must decide which of the locations contains the lesion, and click on it using the computer mouse. This is a more natural presentation of the images than using multiple (possibly sub-region) images, although not as realistic as a methodology requiring the user to search for possible lesion locations.

26.3.2 Experiment logical flow

Initially, the lesion is embedded in the background image using the full lesion intensity. In other words, the pixel values stored in the lesion image files are used "as is." The lesion pixel values are chosen (since the lesion is generated mathematically) to result in a high contrast when inserted at the original intensity, such as in Figure 26.1. After clicking on the (correct) cue circle, the software will display the next image with its possible lesion locations, but the lesion intensity will be decreased by multiplying all lesion pixel values by a factor less than 1.0 (referred to here as the "scale factor"). This results in a decreased contrast in the display of the embedded lesion. Scale factors are chosen in fixed steps such that every eight steps equal a factor of two. That is, there are eight steps between 1.0 and 0.5, eight steps between 0.5 and 0.25, etc. This geometric progression gives finer control of lesion intensity at low scale factors without requiring a large number of steps at higher scale factors.

At the beginning of the experiment, the software logic reduces the lesion intensity rapidly, using three steps per presentation while the observer continues to make correct selections.

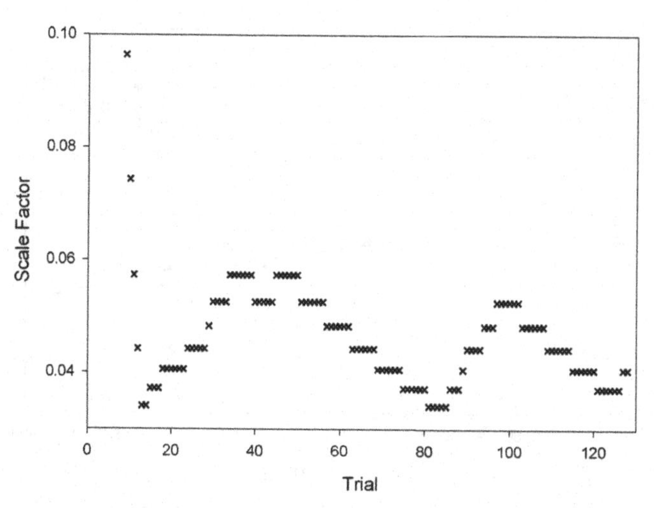

Figure 26.2 Graph showing lesion contrast values as experiment progresses.

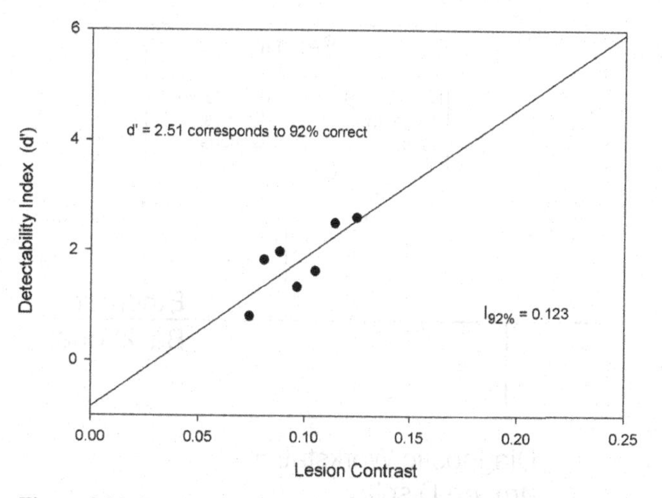

Figure 26.3 Graph showing typical experimental d′ versus contrast results.

Eventually, the observer will be unable to determine the location of the lesion and will be forced to guess. When a mistake is made, the software moves into a second mode of operation. After the mistake, the lesion intensity is increased by one step. If the observer makes correct selections at this scale factor for six consecutive presentations, the lesion intensity will be dropped by one step. If a mistake is made, the intensity increases again by one step and counting restarts. The experiment continues in this fashion until the total number of presentations determined in the experiment parameters is reached. At this point the experiment is complete and the software analyzes the results. A typical experiment contrast progression is shown in Figure 26.2.

An important feature of the software is the use of an intensity "band" that limits the maximum and minimum scale factors used. The maximum limit is ignored at the beginning of the experiment, but controls the maximum scale factor used after the observer makes their initial mistake. This prevents the contrast from rising above a set limit in the event that several mistakes are made consecutively. If this happened, it could take many trials to reduce the contrast to a reasonable range.

The minimum lesion scale factor is always enforced during the experiment. This is important, as it is conceivable that the observer will guess correctly for several successive presentations even when the contrast is lower than the detection threshold. At the beginning of the experiment, when larger steps in intensity are being taken, this could result in reaching a very low lesion contrast, and many mistakes would be required to increase contrast to a perceptible level. During that period of time there would also likely be some correct guesses, which would further delay the contrast returning to a perceptible level.

26.3.3 Analysis of experimental results

During the experiment, the software is recording the observers' performance at each contrast (scale factor) level. The total number of trials at each of the discrete levels is counted, as is the number of correct selections at those levels. These data are used in the final analysis of performance. The first step of this analysis is to calculate the percentage correct at each scale factor. Some of the scale factors will be unused, and these will be ignored. Also, the observer will score 100% correct at some of the levels, and these also will be ignored in the analysis.

The percentage correct values are then converted into d′ (equivalent to SNR at that contrast level) values using a lookup table. The purpose of this step is to linearize the data, such that simple linear regression methods may be used to fit the resulting values when plotted versus the lesion contrast (scale factor). A typical set of experimental results is shown in Figure 26.3.

After conversion to d′ values (see Green and Swets for a discussion of d′ values in AFC experiments) (Green, 1988), the results are fit by a simple weighted linear regression. The weighting at each contrast level is simply the total number of presentations at that level. This provides more robust results and limits the effect from contrast levels where there were few (as little as two is possible) presentations. The resulting curve fit is used to estimate the lesion contrast required for the observer to achieve 92% correct ($I_{92\%}$). This number encapsulates the observers' performance for that particular experiment.

The AFC software has a client-server architecture as shown in Figure 26.4. This architecture allows a distributed network of computers to be used to perform AFC experiments, with the data needed to complete the experiment (images, lesions, and experiment parameters) as well as the experimental results being stored on a central server.

26.4 OBSERVER STUDIES IN CT

In the following sections we will present some experimental results for studies that investigate three important parameters associated with CT imaging: (1) radiation quantity (i.e. mAs) used to acquire a scan, (2) radiation quality (i.e. kVp) selected for a scan, and (3) choice of reconstruction algorithm.

26.4.1 Data acquisition for observer studies

For these studies, a four-slice LightSpeed CT scanner (GE Medical Systems, Milwaukee, WI) was used to acquire axial images of the abdomen of an adult male Rando phantom (The Phantom Laboratory, Salem NY). All images were reconstructed

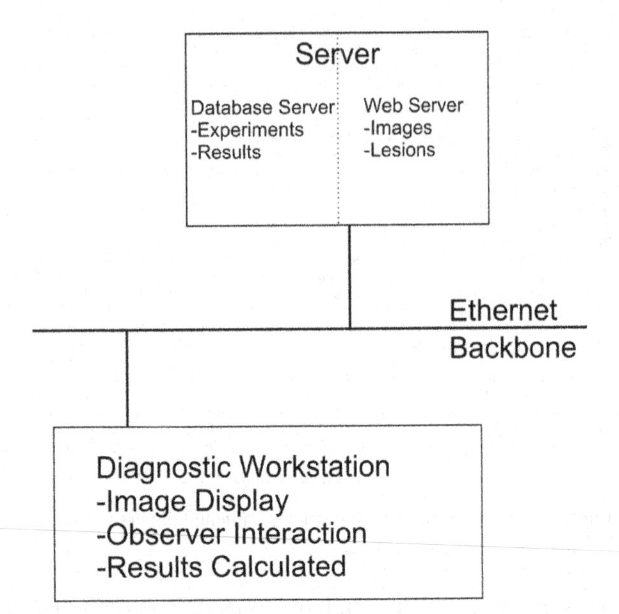

Figure 26.4 Block diagram of client-server architecture for AFC software.

with a slice thickness of 1.25 mm and a 30 cm display field of view. Observer studies were performed using a diagnostic quality monitor that had been calibrated in accordance with the DICOM Grayscale Standard Display Function.

The Rando phantom is composed of a skeleton and plastic tissue equivalent materials, and is widely used for measuring patient doses in diagnostic radiology and radiation oncology. The phantom is taken to correspond to a 1.75 m tall adult male who weighs 73.5 kg. There are 35 sections in the phantom that are each 2.5 cm thick. Images of the Rando phantom have greater anatomical realism than, for example, cylindrical phantoms of acrylic, but do not contain all of the anatomical details that are normally found in clinical head and body images.

As mentioned earlier, synthetic lesions were obtained mathematically by projecting a sphere with a diameter of 100 pixels of arbitrary size, which were blurred by a Gaussian with a standard deviation of 1 pixel. The lesion was scaled to the appropriate diameter, with adjustments made to relate lesion diameter in pixels to the corresponding image pixel size. Lesion sizes are reported in mm, and lesion intensity values were scaled so that the maximum pixel value was 100. This is a convenient value that will have a high contrast when inserted at full intensity into a CT image.

26.4.2 Investigating the effect of changing X-ray tube output (mAs)

The CT output is defined as the product of the X-ray tube current, and the X-ray tube rotation time (360°). In these experiments, all images were obtained in axial mode and processed using the Standard reconstruction filter and a 2 second scan time. The display level was 40 and the display window width was 450, which are typical for abdominal CT imaging. Experiments were performed at a constant X-ray tube voltage of 120 kV, and X-ray tube outputs ranged from 90 mAs to 360 mAs.

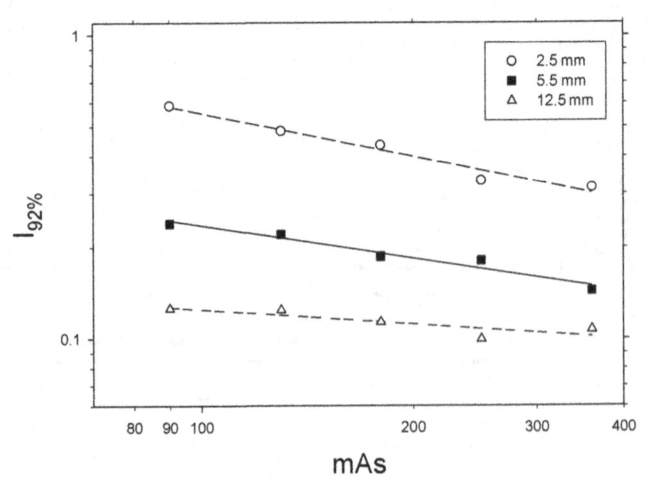

Figure 26.5 Experimental results of $I_{92\%}$ versus mAs curves for three lesion sizes.

Table 26.1 *Effect of lesion size and mAs value on lesion detection performance*

Lesion size (mm)	Slope of the $I_{92\%}$ vs mAs curve (Figure 26.5)	($I_{92\%}$ at 90 mAs)/($I_{92\%}$ at 360 mAs)
2.5	−0.47	1.87
5.5	−0.35	1.66
12.5	−0.16	1.19

Figure 26.5 shows how changing the X-ray technique (mAs) affects lesion detection for three lesion sizes (i.e., 2.5 mm; 5.5 mm; and 12.5 mm). Table 26.1 shows how changing the mAs affected the slope of the curve for each of these lesions, as well as the change in $I_{92\%}$ when the mAs increased by a factor of four (i.e., 90 mAs to 360 mAs). The Rose model of lesion detection, where random (quantum) noise is the only factor that influences lesion detection, predicts a slope of −0.5 when $\log(I_{92\%})$ is plotted as a function of log(mAs).

The data in Table 26.1 show that the Rose model predicted observer performance reasonably well for the 2.5 mm lesion, where the measured slope of −0.47 is in good agreement with the value of −0.5 predicted by the Rose model. However, the data in Table 26.1 also show that the slopes get progressively shallower as lesion size increases, which suggests that factors other than quantum mottle play a role in determining observer performance. Another way of illustrating the same point is to note that quadrupling the mAs would double detection performance for the Rose model, but the data in the last column of Table 26.1 show that it is only the 2.5 mm lesion where this is found to be the case.

26.4.3 Effect of changing X-ray tube voltage on observer performance

We investigated the effects of changes in the X-ray tube voltage (80 to 140 kV) by generating a set of Rando images at a constant X-ray tube output (180 mAs). Since we add lesions *after* image acquisition, our experiments will only quantify the effects of

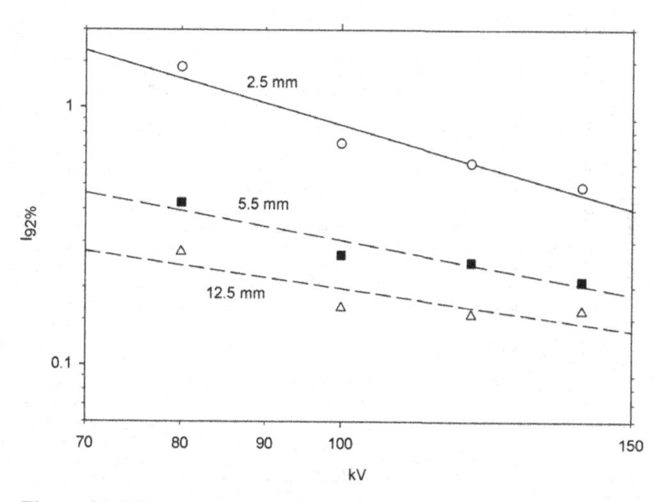

Figure 26.6 Experimental results of $I_{92\%}$ versus kV curves for three lesion sizes.

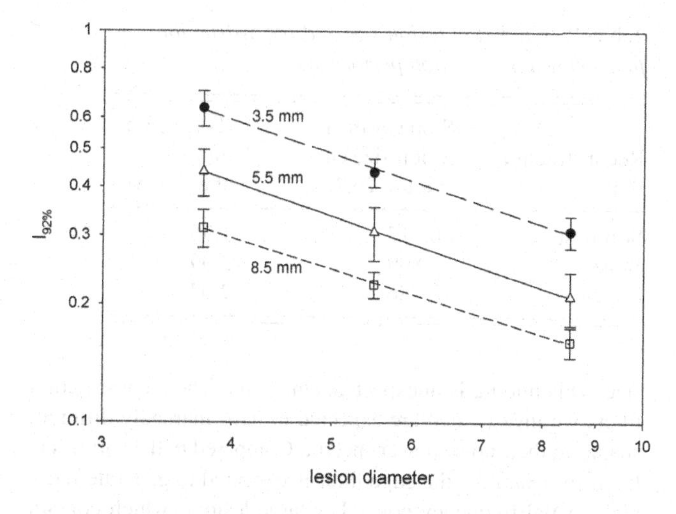

Figure 26.7 Experimental contrast-detail curves obtained for three CT reconstruction filters.

Table 26.2 *Effect of lesion size and kV on lesion detection performance*

Lesion size (mm)	Slope of the $I_{92\%}$ vs kV curve (Figure 26.6)	($I_{92\%}$ at 80 kV)/($I_{92\%}$ at 140 kV)
2.5	−1.85	2.92
5.5	−1.18	2.01
12.5	−0.93	1.70

changes in X-ray tube output and the corresponding reductions in image noise. It is important to note that our experimental paradigm has no effect on lesion contrast *per se*; the experimental results are thus deemed to be relevant for soft tissue lesions, where contrast differences over the clinical range of X-ray tube voltages will be very small. For high atomic number lesions (e.g. iodinated contrast), changes in X-ray tube voltage would have a major impact on lesion contrast, and our experiments are clearly not directly relevant for such diagnostic tasks.

Figure 26.6 shows how changing the X-ray tube voltage (kV) affects detection for three lesion sizes (i.e. 2.5 mm, 5.5 mm, and 12.5 mm). Table 26.2 shows how changing the kV affected the slope of the curve for each of these lesions, as well as the changes in $I_{92\%}$ when the X-ray tube voltage increased from 80 to 140 kV. Increasing the X-ray tube voltage from 80 to 140 kV improves lesion detection performance substantially; for 2.5 mm lesions, the improvement is close to a factor of three, and for 12.5 mm lesions, the improvement is a factor of 1.7. Qualitatively, the trends depicted in Figure 26.6 are similar to those observed in Figure 26.5.

Results for varying mAs are shown in Figure 26.5/Table 26.1 and can be compared to the results obtained with varying kV as shown in Figure 26.6/Table 26.2. Doubling the mAs for detection of 2.5 mm lesions would improve detection performance by a factor of 1.4, whereas the corresponding improvement in lesion detection from a doubling of X-ray tube voltage would be a factor of 3.6. For 12.5 mm lesions, doubling the mAs would improve detectability by a factor of 1.12, whereas a doubling of the X-ray tube voltage for these larger lesions would be expected to improve lesion detection by a factor of 1.9. It is therefore evident that lesion detection is much more sensitive to X-ray tube voltage (at fixed mAs) than to X-ray tube output at all lesion sizes investigated. This finding is not unexpected, given that the X-ray tube output increases linearly with mAs, but supra-linearly with kV. It is of interest, however, to note that these differences are much more important for smaller lesions.

26.4.4 Effect of reconstruction filter on observer performance

The same projection data set can be reconstructed using different filters; filters that have good spatial resolution characteristics will have high levels of image noise and vice versa. We investigated the effects of reconstruction filter on lesion detection by performing a single acquisition (120 kV; 180 mA; tube rotation time 1 second), and reconstructing images in the same manner with a slice thickness of 1.25 mm, a fixed 30 cm display field of view, and viewed using a display level of 40 and a display window width of 440. We studied images reconstructed using three filters: (1) Soft Tissue with the lowest level of noise, with a normalized noise value of 1; (2) Bone with a relative noise of 4.5; and (3) Edge with a relative noise of 7.7.

Figure 26.7 shows the results obtained for $I_{92\%}$ versus lesion size (3.5 mm, 5.5 mm, and 8.5 mm), and where the vertical uncertainties are the standard error obtained by averaging results from four observers. Table 26.3 shows how changing the lesion size affected the slope of the curve in Figure 26.7, as well as the changes in $I_{92\%}$ when the lesion size increased from 3.5 mm to 8.5 mm (i.e. by a factor of 2.3). The computed slopes of straight line fit (i.e. $\log(I_{92\%})$ versus log(diameter)) are all in reasonable agreement with the value of −1.0 that is predicted by the Rose model. The Rose model is a slightly better predictor for the filters that result in more noise; the Bone and Edge filters have an averaged slope of −0.90, whereas the Soft Tissue and detail filters have an average slope of −0.81.

The results presented in Figure 26.7 all show similar trends in the change of detection performance as a function of lesion

Table 26.3 *Effect of lesion size and reconstruction filter on lesion detection performance*

Reconstruction filter	Slope of the $I_{92\%}$ vs lesion size curve (Figure 26.7)	($I_{92\%}$ at 3.5 mm)/($I_{92\%}$ at 8.5 mm)
Soft tissue	–0.817	1.95
Bone	–0.899	2.09
Edge	–0.886	2.07

size. This finding is unexpected, since all of the reconstruction filters we investigated are expected to have markedly different shapes in their noise power spectra. Compared to the soft tissue filter, for example, the edge filter is expected to generate more high spatial frequency noise. For large lesions, which contain signals at lower spatial frequencies, the relative $I_{92\%}$ values for the edge and soft filters should therefore differ from the relative values for the smaller lesions whose signal will be located at higher spatial frequencies. The plots in Figure 26.7 indicate that the relative differences are approximately equal across all lesion sizes.

It is also of interest to investigate the quantitative differences in lesion detectability with the measured differences in image noise, where the latter are quantified by the standard deviation of pixel values in a region of interest in images of a uniform (water) phantom. Figure 26.8 shows the measured relationship between changes in detection performance ($I_{92\%}$) and the corresponding changes in image noise. All data in Figure 26.8 were normalized to unity for the lowest noise reconstruction filter (soft tissue), the increases in $I_{92\%}$ for the bone/edge filters were then determined for each of three lesion sizes, and the data shown in Figure 26.8 are the resultant averages (± standard deviation). These data show that increases in noise by a factor of ~8 reduced lesion detection performance by only a factor of ~2. Although increases in image noise qualitatively predict changes in low-contrast lesion detectability, this parameter is clearly a poor quantitative predictor of detection performance, as mentioned earlier. This is presumably due to the differences in the texture (power spectrum) of the noise produced by the three reconstruction filters.

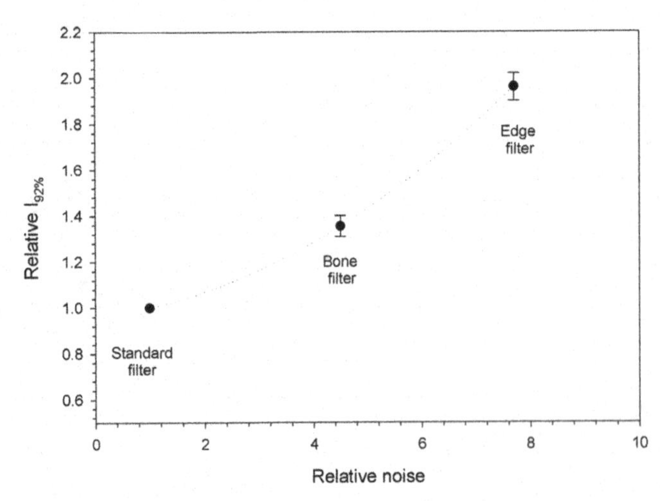

Figure 26.8 Changes in $I_{92\%}$ (i.e. detection performance) versus the corresponding changes in image noise (i.e., standard deviation of pixel value in a uniform phantom).

26.5 CONCLUSION

Since the "optimal" image quality depends on the specific diagnostic task, it is not possible for equipment manufacturers to provide equipment that is "pre-tuned" to produce the best image quality for all imaging applications. Instead, equipment is designed to provide flexibility to the user so that imaging study parameters may be adjusted to meet a variety of diagnostic needs. The operators of the equipment must decide how best to select image acquisition parameters. The selection of acquisition parameters will have a direct impact on the perception of different pathologies in CT images, and therefore on the efficiency with which they are detected. To address these perceptual issues, information regarding observer performance is helpful as it allows changes in detection performance to be quantified.

AFC techniques can provide useful quantitative data that allow informed selection of techniques when acquiring CT scans. There are several limitations of the described method that should be noted, however.

- This methodology does not simulate the physics of acquisition, and so the effect on lesion contrast as kV is changed is not being modeled. This is expected to have a small effect on the detection of soft tissue lesions in non-contrast enhanced studies, however.
- The image reconstruction algorithm selected would have an effect on lesion characteristics as well as background characteristics, and this effect is also not modeled by the AFC software.
- Diagnostic image quality is task-dependent. The methodology described here presents the observer with a simple detection task that may not indicate how observer performance will change for other tasks.

In this chapter, we have outlined a useful AFC methodology and some of the results we have obtained. These results show that changes in reconstruction parameters, such as the filtered back-projection filter used, can have a drastic impact on the detection of nodular lesions in CT images. This emphasizes the need for this type of information, and the importance of having an understanding of the effects that scan acquisition parameters can have on detection performance.

REFERENCES

Boedeker, K.L., McNitt-Gray, M.F. (2007). Application of the noise power spectrum in modern diagnostic MDCT: part II. Noise power spectra and signal to noise. *Phys Med Biol* **52** (14), 4047–4061.

Green, D.M., Swets, J.A. (1988). *Signal Detection Theory and Psychophysics*. Los Altos Hills, CA: Peninsula Publishing, pp. 505.

Hsieh, J. (1997). Nonstationary noise characteristics of the helical scan and its impact on image quality and artifacts. *Med Phys* **24** (9), 1375–1384.

Hsieh, J. (2003). *Computed Tomography: Principles, Design, Artifacts, and Recent Advances*. Bellingham, WA: SPIE, p. 230.

Hu, H., Fox, S.H. (1996). The effect of helical pitch and beam collimation on the lesion contrast and slice profile in helical CT imaging. *Med Phys* **23** (12), 1943–1954.

Huda, W., *et al.* (2007). In Hsieh, J., Flynn, M.J., eds., *Medical Imaging 2007: Physics of Medical Imaging*. Bellingham, WA: SPIE, pp. 65104P:1–12.

Ochs, R., *et al.* (2006). In Reinhardt, J.M., Pluim, J.P.W., eds., *Medical Imaging 2006: Image Processing*. Bellingham, WA: SPIE.

Primak, A.N., *et al.* (2006). Relationship between noise, dose, and pitch in cardiac multi-detector row CT. *Radiographics* **26** (6), 1785–1794.

Tapiovaara, M. (2005). Image quality measurements in radiology. *Radiat Prot Dosimetry* **117** (1–3), 116–119.

27

Perceptual issues in reading mammograms

MARGARITA ZULEY

27.1 INTRODUCTION

Considering perception in mammography is a multifaceted discussion that encompasses all of the variables in image production and interpretation. Even though there is a widely documented mortality reduction of 25–30% in populations that have undergone yearly interval screening mammography, still 20–40% of breast cancers are missed when they are initially present on screening images. False negative readings may occur because the cancer is truly occult on the study, it is obscured by overlapping dense breast tissue, the technical factors of acquisition are poor, display conditions may not be optimized, or because the radiologist might not correctly detect or classify the findings. All of these possibilities together account for the miss rate of the test. Not only are some cancers missed on mammography, but also there exists a significant false positive rate. On average, in the USA, 11% of all screened patients are recalled for additional workup. The vast majority of these recalled patients do not have breast cancer. All cases recalled that do not lead to a diagnosis of breast cancer are termed false positive studies. The intention of this chapter is to discuss, in detail, all of the technical and human factors that lead to these mentioned statistics, with an eye toward the principles of perception that are involved.

27.1.1 Anatomy, physiology, and pathology

An understanding of the anatomy and physiology of the body is critical to the radiologist, for one must understand what normal is and why it is normal before abnormalities can be correctly identified and diagnosed. Further, an understanding of how and why abnormalities grow and how they affect the organ aids in interpretation. The breast is a modified sweat gland that develops from a breast bud. Growth starts as young girls progress through puberty. Estrogen and progesterone are two hormones produced by the ovaries that are the main stimulants to the breast. The cells of the breast tissue have receptors that bind these hormones as they circulate through the blood stream. When the hormones are bound, they cause the cells to function. Therefore, the gland's function varies with a woman's hormonal status. The breast is composed of an intricate network of ducts that start at the nipple and extend into the tissue, branching and eventually ending in what is called the terminal duct lobular unit. Eight to ten of these ductal segments are present in the breast. Surrounding the ducts is the parenchymal tissue of the breast, also called fibroglandular tissue. This is admixed with some amount of adipose tissue. Running through this stroma are lymphatic channels, arteries, and veins. Finally, there is a fine network of suspensory ligaments that support the organ, called Cooper's ligaments. The breast tissue extends into the axilla, termed the tail of Spence, and lies immediately on top of the deep pectoral fascia of the chest wall. Typical lymphatic drainage of the breast is to the ipsilateral axillary lymph nodes.

One of the greatest difficulties in interpreting mammography is that every person's mammogram is unique. This is unlike looking at the brain, liver, uterus or almost any other organ. Figure 27.1 shows the variability in the normal breast between patients. In addition, the normal appearance of the organ changes as the hormonal conditions of the patient change. In order to aid in the interpretation of mammography, almost every facility captures information from the patient on each visit to facilitate understanding of the conditions in which the breast is imaged each time. For example, we ask when the last menstrual cycle began and ask about exogenous hormone use, menopausal status, and so forth.

Because the breast is a hormonally active gland, it is very frequently hyperactive. There is a common benign condition affecting the breast called fibrocystic breast disease, which is, essentially, an overreaction to hormonal stimuli. In this condition, patients present with symptoms like breast pain and lumps. In turn, radiologists see solid masses, collections of ductal fluid, and sometimes calcifications on the mammograms and ultrasound. This is one of the more common reasons for false positive findings. The radiologist must understand this entity, its typical age of presentation, and its appearance to avoid confusing it with breast cancer. Not only does this condition lead to many recalls from the screening population, to false positive biopsies and even to surgery, but it also can obscure malignant findings or distract the reader, causing them to overlook a malignancy. Figure 27.2 shows fibrocystic changes.

Breast cancer is the most commonly diagnosed cancer in women, with approximately 184,000 new cancers found annually in the USA alone. Over a woman's lifetime, the chances of developing a breast cancer are one in seven to eight, with the incidence of breast cancer increasing as women age. Women who are at higher risk than the general population, due to a genetic mutation, family history, or other reasons, have even worse statistics. Ninety percent of breast cancer is called ductal carcinoma. It starts, typically, in the terminal duct lobular unit of the breast. While *in situ*, it may calcify and therefore small or new clusters of calcification are an important finding on a mammogram. Once the cancer grows, it will either create a mass and/or distort the local architecture of the adjacent normal tissue. At this point, the cancer has typically grown through the

The Handbook of Medical Image Perception and Techniques, ed. Ehsan Samei and Elizabeth Krupinski. Published by Cambridge University Press.

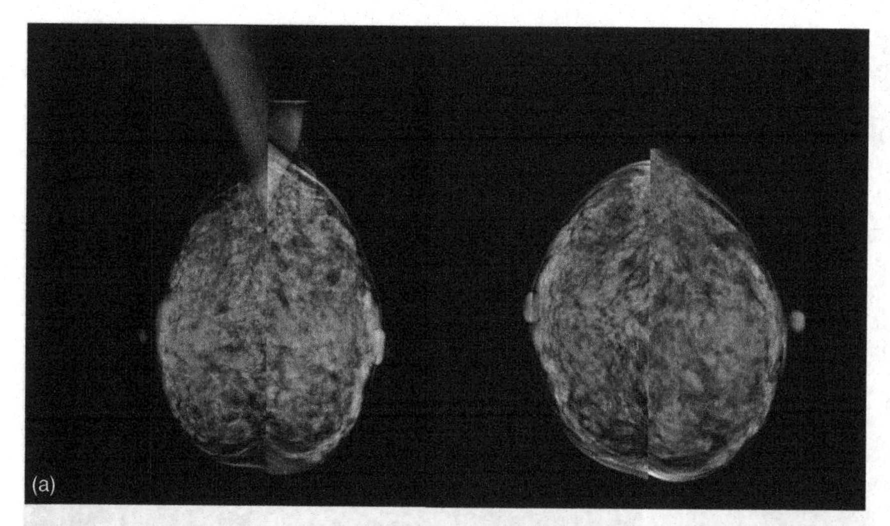

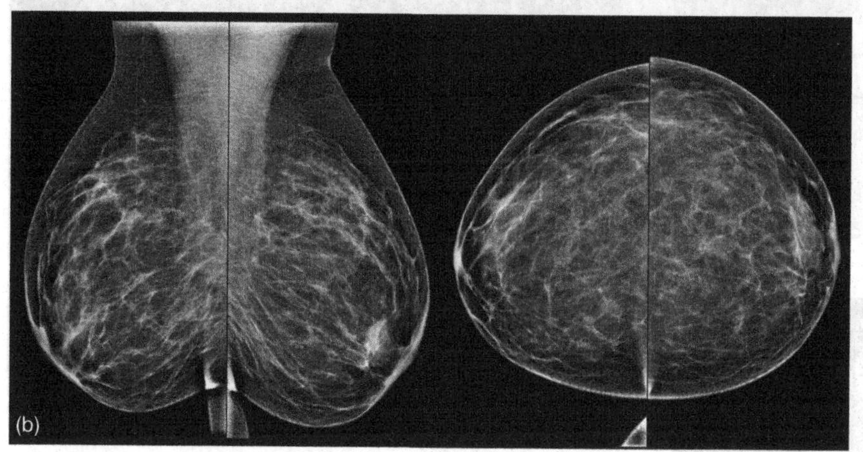

Figure 27.1 The variability of what constitutes normal breast tissue is exemplified in these two mammograms. The first mammogram (a) contains much more dense breast tissue and the second (b) contains more fatty tissue.

duct wall and invaded the adjacent fibroglandular and adipose tissue. Most commonly, the mass has spiculated margins, as it may be schirrous or cause a reaction in the adjacent normal tissues. However, if the tumor is fast growing, it may expand so rapidly that the margins are more rounded than irregular. This makes it look more benign. Even harder to detect, some cancers will appear as only a new area of what looks like breast tissue on the mammogram. If left unchecked, the disease will invade the lymphatic channels in the breast and be carried to the axillary lymph nodes and then out of the breast to other organs such as the brain, liver, and bone. Figure 27.3 shows some typical appearances of breast cancer. It is this understanding of the normal anatomy and physiology, and of both benign and malignant processes, that is the foundation of perception for the radiologist.

27.1.2 Evolution of mammography

Next, let us consider the history of mammography, because it is unique in medicine and its history drives the current state of the art. Mammography has been studied for decades with large randomized controlled trials in the USA, Europe, and Canada. The first such trial in the USA, called the Health Insurance Program of New York Project (HIP), occurred in the 1960s. In this trial, more than 60,000 women were enrolled. Patients either had four consecutive years of screening mammography or had the standard care of the time. The patients who underwent yearly screening mammography had a 29% mortality reduction at 9 years that persisted as a 23% mortality reduction at 18 years. Several European trials looking at results of national screening

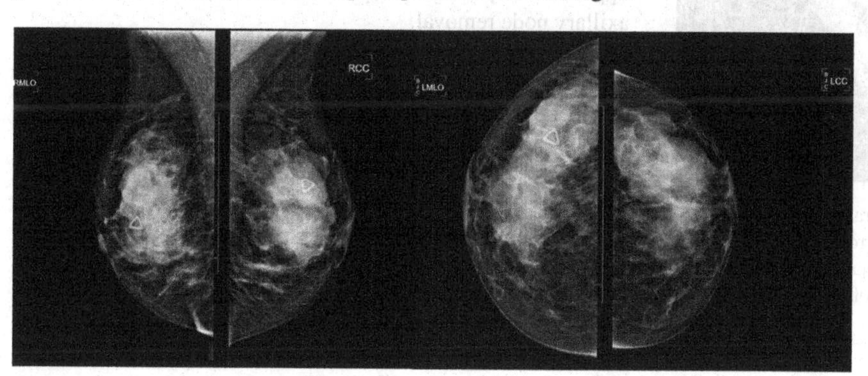

Figure 27.2 This patient has fibrocystic changes. There are rounded masses in both breasts that proved to be harmless pockets of fluid, called cysts. The triangular markers on the images denote palpable masses. All of these changes can distract the radiologist from identifying a possible cancer or may obscure a cancer.

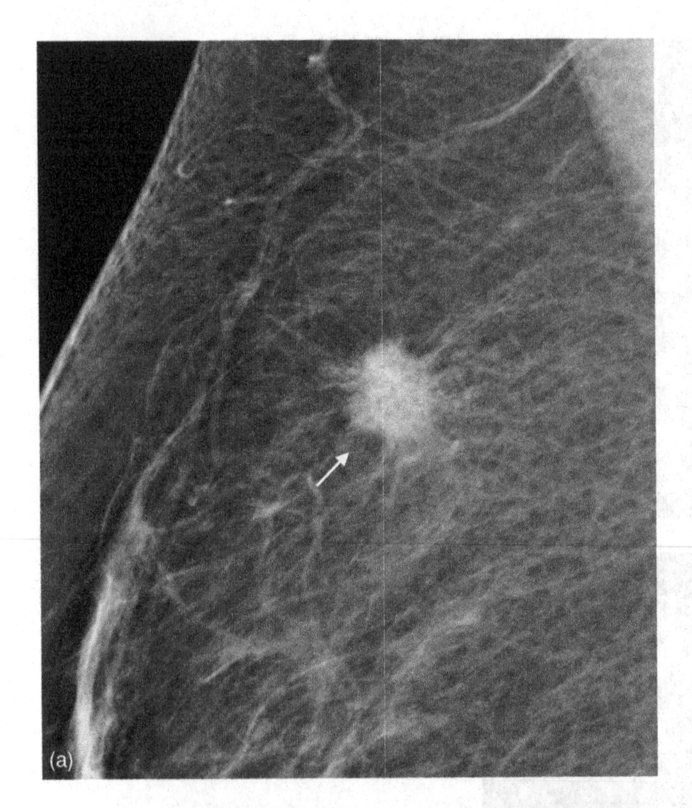

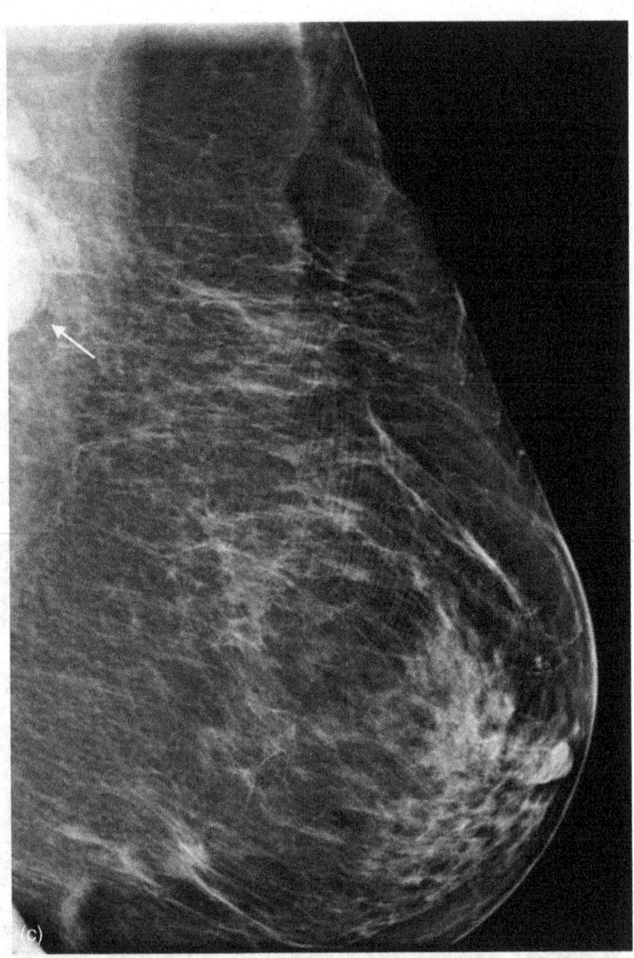

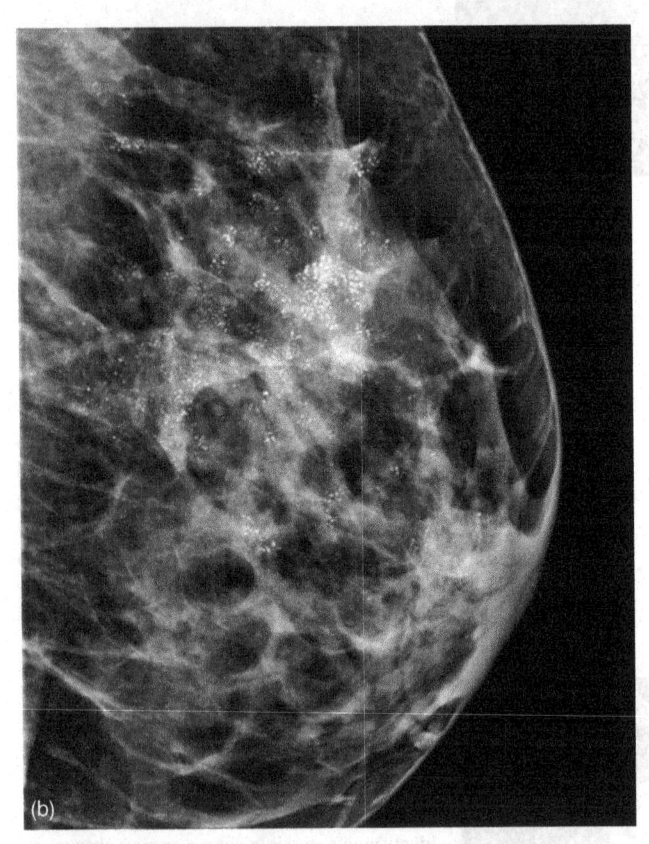

Figure 27.3 Three different and common appearances of breast cancer: in (a) there is a spiculated mass (white arrow), in (b) there are malignant calcifications, and in (c) there is an enlarged, cancerous lymph node partially included on the back edge of the image (white arrow).

programs yielded similar results. In addition to decreasing mortality, the morbidity associated with the diagnosis of breast cancer is reduced in the screened populations. Cancers detected with mammography are typically smaller than those detected by clinical exam and more frequently do not involve the axillary lymph nodes. This earlier detection translates into less aggressive therapy and, hence, reduced morbidity. One such trial showed that 50% of screen detected cancers were stage 0 or 1 as opposed to only 19% in the clinically detected group (Tabar, 2000). The earlier stage of disease at diagnosis leads to a reduced need for adjuvant chemotherapy, and allows more women to undergo breast-conserving surgery and limited axillary node sampling as opposed to requiring a complete mastectomy and more extensive axillary node removal.

Early systems had higher patient doses than are present today and, very early on, the technology was similar to what was used in general radiography. In fact, the first mammograms were performed on general purpose X-ray units that were also used for chest X-rays, bone films, abdominal films, etc. Quickly, it was discovered that mammography was a special case that required dedicated equipment in order to be able to perceive abnormalities in the breast. At this early point, mammography separated from general radiology and has remained relatively separated ever since. Early work in product development showed that the X-ray source and receptor required specialized equipment because of the relatively little differences in X-ray beam attenuation between the soft tissue structures in the breast, as well as the small size of breast lesions. This is in contrast to the remainder of the body, where differences between structures as varied as bone, soft tissues in organs, and air in the lungs is necessary. As a result, molybdenum and rhodium targets and filters were developed for mammography, replacing the tungsten/aluminum combinations in general radiography equipment. The characteristics of these softer beams allowed for better tissue delineation in the breast. In addition, specialized screen film combinations and extended processing were developed in order to maximize the fine detail of the organ.

27.1.3 MQSA

Mammography is the only imaging study regulated by an act of congress. This law, the Mammography Quality Standards Act (MQSA), was passed in 1992 to ensure that patients receive at least a minimum quality of imaging and interpretation. The Food and Drug Administration has been charged with enforcing this law and, in turn, the American College of Radiology (along with a few states) has been the primary organization that has ensured that facilities performing mammography and radiologists interpreting it in the United States adhere to the requirements of MQSA. As a result, every detail of image production and interpretation has been scrutinized and studied in order to clearly define boundaries of acceptability. The MQSA regulations are intended to ensure that the quality of the images, the conditions in which the images are interpreted, and the knowledge of the reader are all adequate to render an accurate diagnosis.

When MQSA guidelines were published, phantoms that evaluate a system's ability to see fine specks, small masses, and linear structures further helped to improve image quality. Radiologists, medical physicists, and technologists quickly became intimately involved in assessing the quality of the images produced from systems, and learned to recognize when any part of the system fell out of an acceptable range. There are still daily, weekly, monthly, and yearly quality checks that go on in the mammography department as a result of this history.

Every aspect of image acquisition, film screen processing, and reading conditions is specified in the MQSA guidelines. These are longstanding and well mapped for screen film mammography and under active development for digital mammography. Facilities must undergo acceptance testing yearly by an accredited medical physicist as well as maintain daily quality assurance logs for every aspect of the imaging chain. Radiologists are required to interpret a minimum number of studies and maintain a certain level of continuing medical education specific to mammography to qualify as a reader. Every single acquisition unit, technologist, and radiologist must meet the requirements before a facility is certified for mammography. If the facility does not meet the regulations, then the facility is closed.

In screen film, because the radiologist is so accustomed to the quality control issues of the imaging chain, the effects that aberrations in any component have on the final image are well understood. For example, one can identify artifacts that result from problems with the temperature of the developer chemicals versus defects in the screen in which the films are placed. Problems with the exposure and filtering system of the mammogram machine are also readily detectable. This intimate knowledge allows the radiologist to be able to understand the impact of the analog imaging chain on the final image quality, thus improving the quality of interpretation.

27.1.4 Viewing conditions

Not only did the acquisition units evolve as specialized systems from the rest of radiology, but also the conditions in which the images are read diverged. The images are read on light boxes, or alternators that have a very high luminance, on the order of 3000 cd/m^2. This is a much higher luminance than light boxes used in general radiology. Just as the X-ray beam, filter, exposure, and screen film combination were designed specifically to allow perception of small structures in areas of relatively low contrast, the increased luminance also serves the same purpose. In addition, limitation of extraneous light from any source is mandated, including light passing through the film around the breast, light leaking from the view box around the films, reflected light onto the images, and ambient lighting in the room. Therefore, special reading rooms have been designed in many departments expressly for mammography interpretation. This attention to detail in the reading conditions allows the radiologist to have the ability to perceive the small lesions and subtle findings typically seen on mammograms. Over the last several decades, all of the independent variables have been studied to ascertain their impact on the final image perception and, consequently, the accuracy of interpretation. It has been shown time and time again that improved perception of lesions is possible when all of the conditions of the reading environment are optimized.

After all of that regulation and tight control, one would think that perception of real findings in mammography is as close to a sure thing as is possible in radiology. This is far from the truth. Though screening mammography has stood the test of time and is the only radiology study that is still widely used as a screening tool for the general population of women over the age of 40, it is not a perfect test. Many variables are still present so that the sensitivity of the test varies from as high as 80–90% to as low as 60% for different patient subgroups. Patients who have more dense breast tissue and less adipose tissue in their breasts have harder mammograms to interpret and thus have an overall lower sensitivity of detection.

The popular media has publicized the success of screening mammography, extolling the real mortality and morbidity reductions that this test has achieved in population based studies. The media attention to some degree has created the misconception among the lay public that a mammogram reported to be normal is tantamount to not having breast cancer. The benefits of mammography are real, but in fact mammography does not find all breast cancers. Part of the limitation is that a mammogram is a two-dimensional, flat display of a conical, three-dimensional organ. Typically two views are obtained of each breast: the cranial-caudal view and the medio-lateral oblique view. These views are named for the direction in which the X-ray beam leaves the tube, moves through the tissue, and eventually is incident on the detector or screen film surface. The reader looks at these two views and, in their mind, triangulates the images into a three-dimensional structure. There are, therefore, four mammographic views in the standard mammogram. Of course, many exceptions exist for this, including extra views for patients with implants, extra views for patients presenting with a diagnostic problem and so forth.

27.1.5 The radiologist

Radiologist themselves are another part of the puzzle of perception. In order to interpret mammograms in the USA, the radiologist must be accredited under MQSA. The qualifications for this include at least three months of training in mammography during radiology residency, at least board eligibility for diagnostic radiology, interpretation of at least 240 mammograms per year, and ongoing continuing medical education credits specific to mammography.

In the process of viewing these images, the reader evaluates the tissue itself for abnormalities. As was discussed earlier, the breast is a modified sweat gland that is composed of a mixture of breast tissue and adipose tissue. Each person's mammogram has a characteristic appearance, which changes over time as the organ is exposed to different hormonal states. The possible findings that one might see in a breast include masses, areas of architectural distortion, calcifications, areas of asymmetry, and developing densities (the last being a possible mass visible on only one of the two standard views).

With this background in mind, considering how perception plays into the interpretation of the mammogram is a relatively logical series of considerations. By starting at the beginning of the imaging chain and working through to the final interpretation, one can consider the perceptual factors along the way.

27.2 SCREEN FILM MAMMOGRAPHY

27.2.1 Acquisition parameters

In screen film mammography, a film that is coated on one side with a fine powder of silver bromide particles is placed in a cassette that contains a screen. When the powder on the film is exposed to the X-rays, it changes from a grayish-white opaque surface to black. The more radiation that reaches the film, the darker the film becomes. So, areas that contain more dense breast tissue will be lighter, or more white, than areas that contain fat or air. It is in these shades of gray that all abnormalities are perceived.

During acquisition, several key factors are involved in producing an image that maximizes the information and minimizes the noise. The first of these factors is adequate physical compression of the breast tissue. For each image obtained, the gland is compressed between a clear plastic paddle and the detector so that the tissues are separated as much as possible and the X-ray beam passes through as little tissue as possible. This configuration not only helps to identify abnormalities and spread out structures, but also helps to limit the radiation dose to the skin and breast tissue. If a patient has limited tolerance for compression or has extremely dense breast tissue, the time of exposure can exceed the amount of time that the patient can remain motionless. Motion on the image creates blurring of structures and loss of detail, thus limiting the ability to perceive subtle findings such as the spicules of a cancer or small clusters of calcium.

The second technical factor in producing a quality image is limiting the scatter of the radiation beam. The X-ray beam comes from a point source and as it travels through space, it diverges. The parts of the beam that spread and hit the tissue at an angle, or bounce off at an angle, are called scattered radiation. Scattered radiation creates noise in the image, therefore reducing the quality of the image. Most systems have two mechanisms to reduce scatter: a built-in grid system that blocks the scattered radiation, and the ability to cone the field of view.

The third technical factor important in image quality is the exposure parameters themselves. All systems have an automatic exposure control (AEC) that, in most instances, will acquire an image with the appropriate parameters so as to produce the best possible image within the acceptable radiation exposure limits. These parameters determine both the energy of the beam and the amount of photons that are emitted for any given exposure. The technologist is able to override the AEC when appropriate in order to manually obtain an image.

Above all, the most important facet of the acquisition is the technologist's skill. A good technologist makes the patient feel comfortable so that she is relaxed. This allows the technologist to pull as much tissue into the field of view as possible and also allows the technologist to compress the tissue adequately in order to improve spatial resolution and reduce motion. After each image is acquired, the technologist inspects the image for quality prior to sending it to the radiologist to read.

In analog imaging, the parts of the film that have no attenuation of the beam by tissue are maximally exposed to the X-ray beam and are black. This black background around the breast tissue is ideal because light cannot pass through this part of the film during reading. One downside of screen film imaging is that the tissues that contain mostly adipose tissue, like the subcutaneous fat at the edges of the breast, absorb very little of the X-ray beam and so can be very dark. In many instances, the radiologist has to use a high-luminance light, called a hot light or bright light, to see these parts of the image. This is most common when the breast is thick and contains a majority of dense breast tissue, because the AEC gives more doses to penetrate the tissue.

After the images are acquired, the technologist, in the analog environment, takes them to a darkroom and processes the images in a specially designed processor that runs the films through wet chemicals and then dries them. He or she then makes sure that the images are satisfactory for interpretation before sending them to the radiologist.

27.2.2 Interpretation

The next phase of the process is the interpretation of the images. The job of the radiologist is to assess the image quality and to render an opinion. While viewing the images, the reader searches for abnormalities and determines the significance of any findings or changes. From a perception standpoint, the radiologist has two tasks: to detect possible abnormalities and to classify the findings. Based on these perceptions, the radiologist will provide an assessment and a recommendation. The process of perceiving these abnormalities evolves as the reader accumulates experience.

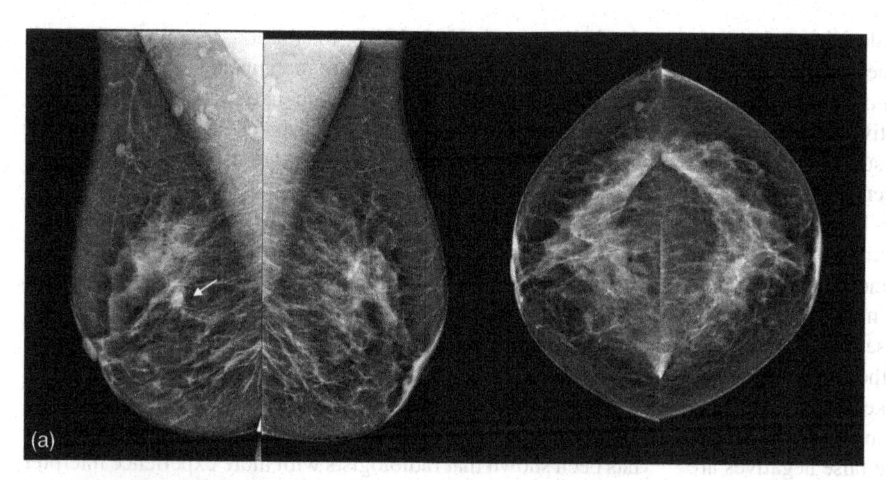

Figure 27.4 This patient was recalled from her screening visit for a possible mass on the right (white arrow in Figure (a)). The additional view (b) taken on her return shows that the area was normal and due to overlapped tissue. This is an example of a false positive screening interpretation.

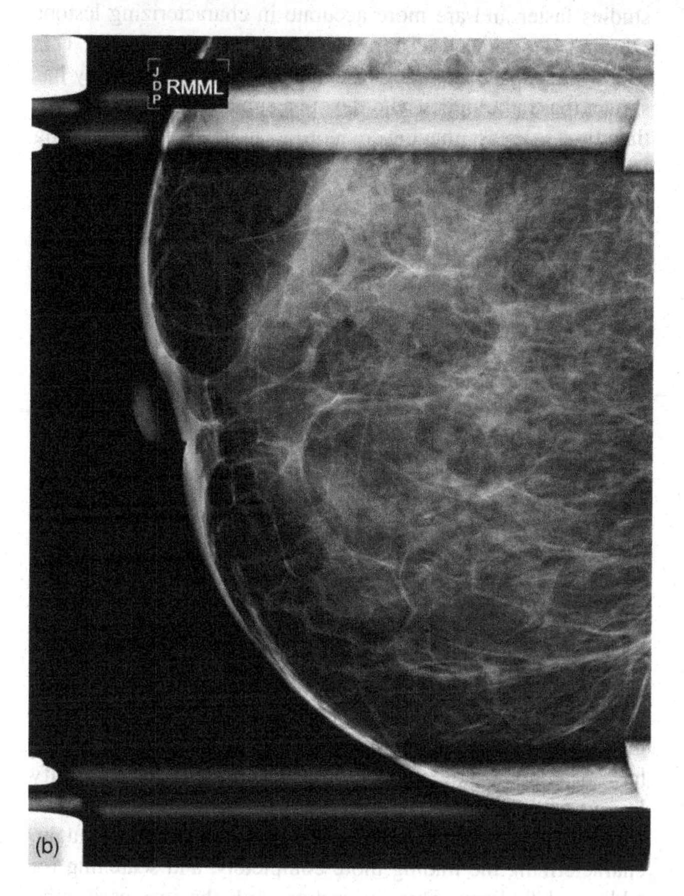

27.2.2.1 Eye tracking research

Many studies have been done tracking eye movements of radiologists and time to interpretation, working toward understanding exactly how we perceive findings (Nodine, 2001, 2002; Krupinski, 2005a; Mello-Thoms, 2005). The basis to eye tracking is that when the fovea fixes on a point, some cognitive process is occurring. Therefore, when a radiologist's eye dwells on an area, the perception research suggests that the radiologist has detected a possible finding and is deciding if it is a normal perturbation of breast tissue, a benign finding, or a possible malignancy. This research has shown that there are several points in reading where perception mistakes occur (Saunders, 2006). The first type of error that occurs in reading mammograms is that the area is never seen (as evidenced by the fovea not fixing on a finding); in other words, there is a search error. The second type of error is termed a recognition error. These occur when a lesion is fixed upon but then quickly dismissed. In this type of perception error, the eye does not fix again on that spot. These first two types of misperceptions produce false negative decisions. The third type of error is a decision making error. In this error, an area is identified as potentially abnormal and the eye fixes on it. Instead of dismissing it quickly, however, the eye dwells on it for a period of time, or will repeatedly return to it before a decision is made. This last type of misperception will cause either false positive or false negative decisions. For false negative decisions generated from either a recognition or decision making error, the radiologist identifies the finding but decides that the area is attributable to the normal variability of the breast tissue or to a benign process like fibrocystic change. For false positive results that occur from a decision error, an area is reported as potentially abnormal but in fact is either benign or normal. Many recalled patients from screening mammography are misperceptions of decision making that result in a false positive finding. Figure 27.4 shows a false positive finding from a screening mammogram. As one may imagine, the more benign lesions in the breast and the more dense breast tissue that there is, the harder the perception tasks become. With experience, it has been shown that the false positive rates of radiologists decline. The thought is that with experience, benign lesions and normal tissue perturbations are more accurately classified because the radiologist has seen them enough times before that they learn to dismiss them.

27.2.2.2 Speed/accuracy research

One of the ways to define expertise in radiology is to evaluate the accuracy and speed at which an abnormality is identified on an image. The second type of perception research looks at decision making time. It has been shown that decision speed and accuracy of decisions both improve with experience. The concept is that mammographers over time accumulate a knowledge bank based on years of interpretations. Reading thousands of studies trains them on what are normal variations in breast tissue, benign findings, and cancers; this fund of knowledge translates into rapid and accurate perceptions. The research in this area is correlative to the eye tracking research.

Several studies have shown that incorrect decisions take more time than correct decisions. This holds true for both positive and negative errors. In other words, true positives are identified faster than false positives and true negatives are determined faster than false negatives. This research suggests that experienced radiologists correctly identify cancers even within two seconds of looking at a mammogram (Mello-Thoms, 2005). The false positives and false negatives then occur as the radiologist incorrectly dismisses subtle real findings and incorrectly classifies benign or normal areas as potentially malignant (Nodine, 1999, 2002; Mello-Thoms, 2005). The false positives that are identified are typically findings that cause the eye to dwell and cause hesitation in decision making, unlike the true positive findings, which are typically seen quickly and acted upon with decisiveness by the experienced reader. The false negatives are either search errors or findings that are identified, but dismissed (a misclassification).

27.2.2.3 Search patterns

Over time, each radiologist develops a pattern of reading. With this search pattern the reader looks at every case the same way, especially in the screening environment. Some first compare current and prior images to look for changes over time. Others look for symmetry of the left and right breasts. Most then go on to look for subtle findings such as microcalcifications, masses that are partially obscured by overlying dense breast tissue, and lesions that are at the edges of the film. Many readers with significant experience do a global initial analysis, not focusing on any one point of the image. After the initial evaluation, the reader then spends the rest of the viewing time confirming their original perception and making sure that nothing was overlooked. Conversely, radiologists with much less experience have a search pattern that uses a point-to-point evaluation of the images.

27.2.2.4 Inter observer variability

Many studies have suggested that there is a significant amount of variability between radiologists in the interpretation of mammography. This is termed inter observer variability. The majority of the studies looking at this topic have been reader studies that contained an enriched population of cancer cases and were conducted using just a two-view mammogram without clinical history, additional views, or older images for comparison (Ciccone, 1992; Elmore, 1994; Kerlikowske, 1998; Sirovich, 1999; Berg, 2000). The radiologist had to provide an assessment and a recommendation based on these two views alone. This is an artificial situation and not the way that decisions are made in the clinical setting. In the normal course of patient care, pertinent medical history is on hand and prior images, when available, are compared with the current study at the time of interpretation. If an abnormality is detected, then additional, more detailed images are obtained, and then possibly ancillary tests such as ultrasound are performed to render a final assessment and recommendation.

Several authors have shown that having prior images, having pertinent clinical history, and having the information of the additional diagnostic workup improve accuracy (Burnside, 2002; Houssami, 2003, 2004). In addition, in some of the studies, assessments of "suggestive of malignancy" and "highly suggestive of malignancy" (two common assessment categories in mammography) were considered disparate, despite that fact that both of these assessments would lead to the same outcome of a biopsy and correct diagnosis in the clinical setting. One study by Berg *et al.* (Berg, 2000) did provide this additional information. Even though there was variability in the descriptors used for lesions, for the five readers in that study, nearly all of cancers in the cohort were correctly identified by all of the readers. This suggests that there is more consistency between radiologists' perception than is reported in the literature.

Some published studies have looked at radiologist characteristics in an attempt to understand inter observer variability. It has been shown that radiologists with more experience interpret studies faster and are more accurate in characterizing lesions (Nodine, 1999; Mello-Thoms, 2005; Miglioretti, 2007). This follows the speed accuracy research. More than one study has shown that radiologists who devote a substantial proportion of their time to breast imaging are more accurate than those reading fewer mammograms, both in the screening and diagnostic environments (Sickles, 2002; Leung, 2007; Miglioretti, 2007). This follows the concept that accurate perception on mammography is in part an acquired skill based on years of interpretations.

Alternatively, the more subtle the findings, the greater the inter observer variability and the higher the false negative rate (Kerlikowske, 1998; Krupinski, 2005a; Mello-Thoms, 2005). This is a logical statement, since one would expect that as the perception task becomes harder, the chances of missing the finding increase. There is significant overlap between inter observer variability generated by the subtle nature of the lesion, or at least its conspicuity at display, and the radiologist's experience and skill.

27.2.3 Screening

In the screen film environment, typically the radiologist spends less than a minute and at most two minutes per case. So, the viewing time is relatively short. As was discussed previously, studies have been done showing that experienced radiologists identify lesions within seconds of seeing a mammogram, and then spend the rest of the viewing time confirming their initial thought, characterizing the finding more completely, and searching for additional findings. This interaction with the images is relatively quick when reading screen film mammography because the screen film image obtained is the full resolution of the acquisition. The films are hung on specially designed light boxes that cut out extraneous light around the films and the four typical views of each breast can be hung immediately adjacent to the prior images so that the tasks of comparison and lesion identification can be done quickly, and essentially simultaneously. The light box luminance has been optimized for viewing of mammography as well. If more light control is needed, or the physician wants to magnify the image, they simply pick up the magnifier or light coner that sits at the viewing station. In this environment, the radiologist's eyes never leave the mammogram when reading, so full attention is given to the work of perception and the viewing conditions have been optimized for the job.

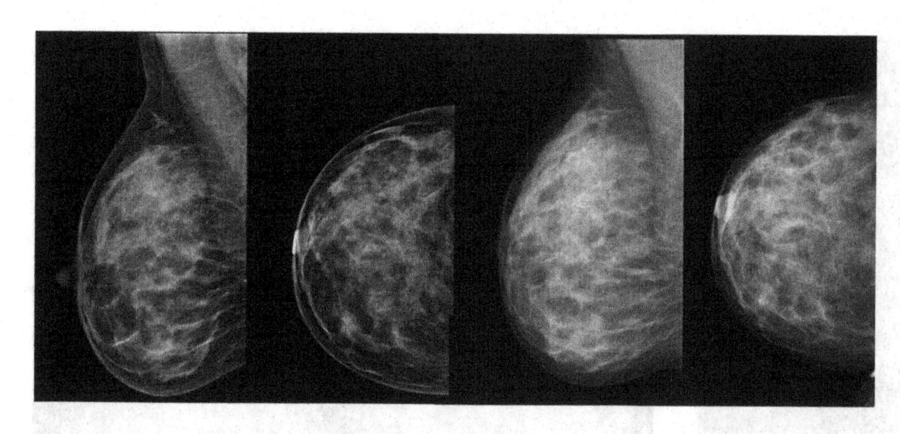

Figure 27.5 This patient was imaged in consecutive years on two different systems. The images that appear more black and white were acquired on a system that uses a molybdenum/molybdenum combination and the images that appear more gray were made using a tungsten/aluminum combination.

27.3 DIGITAL MAMMOGRAPHY

27.3.1 Image acquisition

With the advent of digital mammography systems, all of the knowledge gained from the analog years was transferred to digital. Several of the systems still use the specialized molybdenum and rhodium combinations, but interestingly, because of the ability of digital systems to apply processing and because of detector characteristics, many manufacturers also have tungsten X-ray tubes and silver, aluminum, or even copper filters.

Several vendors have FDA-approved systems for digital mammography. Each vendor has utilized different technology for the detector, with some using amorphous selenium or silicon, some using silicon crystals, and some using phosphorescent plates in a CR format. These different technologies, as one would expect, produce different signal to noise ratios and have different detective quantum efficiencies (DQEs). The DQE of a system describes, in essence, how efficient the system is at using the X-ray beam. The higher the DQE, the more efficient the system, and the less exposure needed to obtain quality images. With the advent of these efficient digital units, for the first time in years, vendors are able to start to consider how to lower radiation exposure to patients while, at the same time, not impacting the ability to perceive abnormalities on the images. Add to that the fact that some systems operate with a tungsten tube and an aluminum filter, and others use a molybdenum or rhodium tube with either a molybdenum or rhodium filter, and quickly one realizes that even though all of the units meet the FDA specifications, the images produced from the units will be different.

For instance, an image produced with a tungsten tube and an aluminum filter will be, typically, softer, or more gray than the same patient's image acquired with a molybdenum/molybdenum combination. This is simply the case because tungsten is a harder beam and so penetrates the tissues more than molybdenum. These technical differences create variation in the images, which the radiologist has to take into consideration when interpreting. From a perceptual standpoint, the variability in characteristics of the image due to acquisition parameters is noise that the radiologist must learn to mentally subtract in order to perceive any changes in the image that are actually caused by a lesion. Specifically, consider the situation where a tungsten/aluminum combination is used to produce a mammogram one year and the next year the same patient has her mammogram on another unit that uses a molybdenum/molybdenum combination. The mammogram done with the molybdenum/molybdenum combination will have a more contrasting look. Figure 27.5 shows such variability. This could be perceived as the patient having more breast tissue than the year before. Since one of the cornerstones of reading mammography is assessing change over time, this sort of acquisition variability can cause a misperception.

27.3.2 Acquisition matrix

Not only do the detector and anode/cathode technologies vary, but the pixel size (pitch) and matrix size of the detectors are different between vendors. As pixel size decreases, both spatial resolution and noise increase. Perceptually, then, lesion margins and subtle calcifications potentially will be more clearly defined with smaller pixels, if the noise of the image is offset by proper dose and processing adjustments. If, however, the added noise of the smaller pixel is overriding, then the image will actually be degraded. So the pixel pitch of the system can help or hinder lesion conspicuity and perception. Figure 27.6 shows an example.

27.3.2.1 Fit to viewport display

Most facilities are now reading digital mammography on soft copy review workstations. One of the main ways that the images are hung on the workstation is in what is termed "fit to viewport" mode. In this mode, multiple images can be displayed at once, for direct comparison. Fit to viewport means that all of the acquired pixels from the detector (or at least all of the pixels that are inside the skin line) are assigned to a portion of a monitor. So the larger the matrix of acquisition, the smaller the image will appear at display, and the more down sampling or interpolation of acquired pixels will occur to fit the image into that viewport. As has already been mentioned, a mainstay of reading mammograms is comparing current and prior studies for temporal change. If a patient has mammograms performed on units of different matrix size, then, in the fit to viewport mode, the breast will display as a different size. Consider the case where a patient has two sequential mammograms done, with the original study done on a smaller matrix detector. If that patient has a mass in the breast, the mass will appear larger on the current study simply because of the smaller matrix of the current study (Figure 27.7). The radiologist learns to take into account this variability when assessing size changes. In order to facilitate

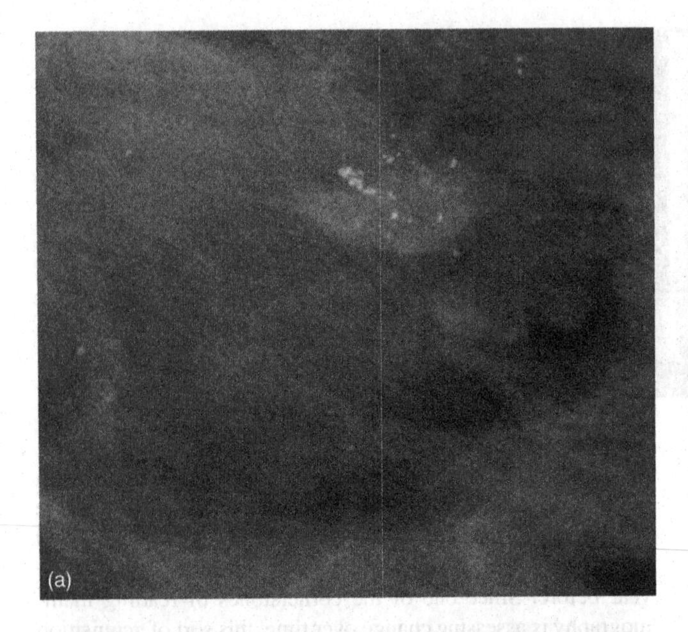

Figure 27.6 In Figure (a) the image is acquired with a smaller pixel size than in Figure (b), but there is more noise in the first image so overall image (b) is superior to (a) in its ability to see fine calcifications.

accurate perception and diagnosis, advanced workstations now accommodate for these variable pixel pitches and matrices by scaling images appropriately in the fit to viewport mode so that all of the images appear to be the same size.

27.3.3 Processing algorithms

Even given the significant differences in matrix size as well as detector and anode/cathode components, the single largest variable in digital mammographic image appearance is a result of image processing. Two sets of images are produced from any digital mammography unit: "for processing" images and "for presentation" images. Understanding how the images are generated leads to an understanding of how perception in interpretation is affected by these technical details. Digital images

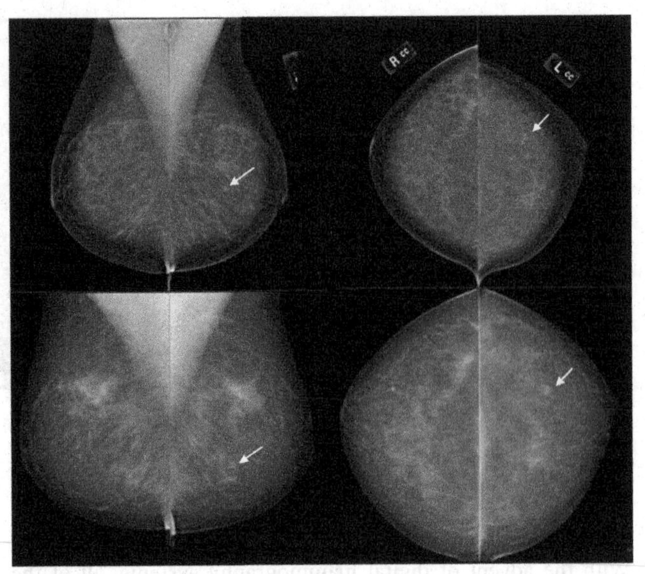

Figure 27.7 The same patient imaged on two systems with different sized detectors, and displayed in the "fit to viewport" mode on the workstation. The radiologist must decide if the nodule in the left breast (white arrow) has grown or just appears larger because of the scaling problem of the display.

are created by the anode exposing the detector to a dose of radiation. The resultant amount of energy that is absorbed by the detector in any one pixel is called an analog to digital unit (ADU). Hence, after an exposure, an ADU map of the matrix is available. The ADU count will be the highest in pixels where there was no attenuation of the X-ray, namely where there is only air. The more dense the tissue that the X-ray beam passes through, the more attenuated the beam, and thus the lower the ADU count in that pixel. So, the ADU map is a reflection of the density of the breast tissue at any pixel location. Those ADUs are then converted into a gray scale image and then processing algorithms are applied to the data to correct for detector inhomogeneities, to accommodate for thickness differences at the edges of the breast, as well as to enhance the differences between fatty tissue and breast parenchyma. "For processing" images are partially processed images that have had all detector corrections made. These are used by computer aided diagnostic (CAD) algorithms but not interpreted. The "for presentation" images are further processed images that are meant for diagnostic interpretation by the radiologist. This set of images then has window width and window level look-up tables applied to them for proper display.

Every mammography vendor has created one or more looks for their images by using image processing. Typically the user has little or no knowledge as to what type of processing is applied to the data, and has little or no ability to manipulate the data, save for changing preset window width and window level settings on some images.

Since some vendors process their data to correct for low modulation transfer functions and some process to enhance edges, for example, the resultant images that emerge from different units are drastically different; this difference is far greater than the differences between screen film systems. This has created a significant perception problem for the radiologist, because

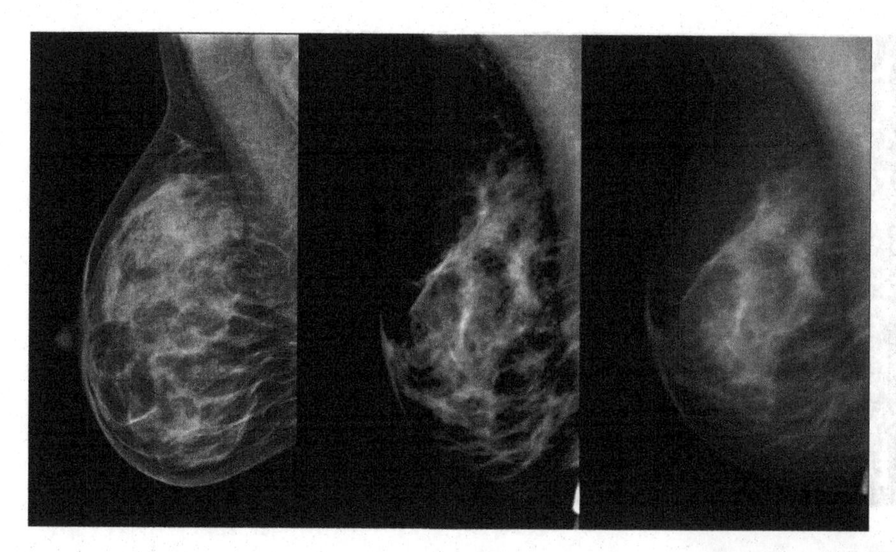

Figure 27.8 Example images from different digital mammography systems show the variability in the look of the mammogram across equipment. The radiologist must adapt to this additional variability during interpretation.

changes in the appearance of the tissue are possibly due to processing differences and not a real change in the patient's tissue. Figure 27.8 shows how different digital mammograms may look. The knowledge fund that the radiologist accumulated in screen film reading is only partially transferable to digital mammography, in part because of the new normal look of images created with these algorithms.

27.3.4 Interpretation in the digital environment

27.3.4.1 Change over time

If we consider how a radiologist interprets a mammogram, and how abnormalities are perceived, then the potential impact of the technical variations on interpretation will be more evident. Because the breast is a functional, hormonally stimulated organ, the normal anatomy of the breast not only varies from one patient to another but also varies in any given patient over time. For instance, the general distribution of breast tissue on a person's mammogram is somewhat analogous to their fingerprint in that it is uniquely theirs. We rely heavily on this attribute to look for a developing cancer because there may be a focal change in the appearance of the breast tissue when old images are compared with current ones. For example, there may be a mass or an area of asymmetry that is new. Figure 27.9 shows an example. Once the radiologist perceives the change, then the area is more closely analyzed to characterize it. At this point, the radiologist is trying to determine if the perceived change is due to positioning of the patient, a technical difference in the images, a focal abnormality, or part of a more global change in the mammogram due to a hormonal shift in the patient. So how the change is perceived by the radiologist drives the diagnosis. For masses, margin assessment is important because benign and malignant masses often look different. For calcifications, distribution, symmetry, and morphology are all evaluated.

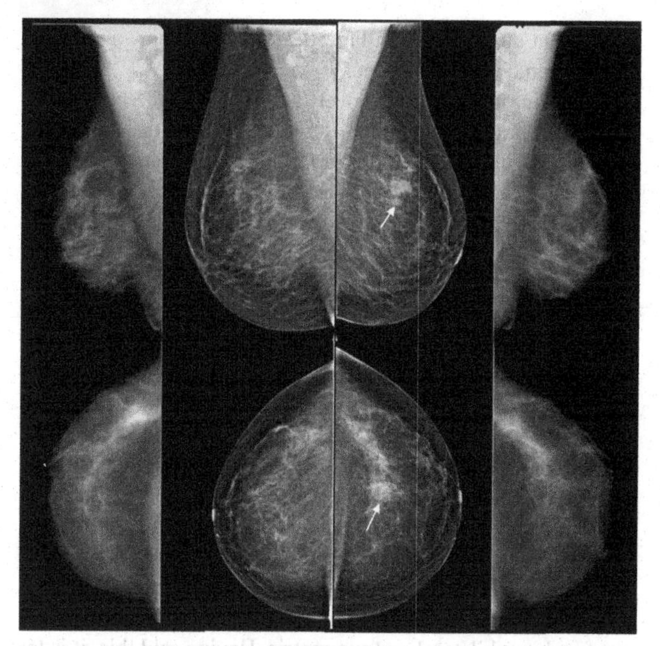

Figure 27.9 Example of a screen detected mass in the left breast. The central four views are the current study and the peripheral four views are the prior. The mass represents a change over time, and is thus perceived by the radiologist so that the patient is appropriately recalled for additional evaluation.

27.3.4.2 Physiologic changes

Mammography is a two-dimensional image of a three-dimensional, functional, hormonally influenced gland. As such, the normal variation of the breast tissue over time in any given patient causes the mammogram to change. In younger, premenopausal patients, the gland is active and varies with the monthly cycle. With pregnancy and lactation, the glandular component hypertrophies and the breast takes on a characteristic appearance. Then, with the cessation of lactation, the breast returns to its pre-pregnancy state. Benign findings, like cysts, which are collections of normal ductal fluid, are common, especially in patients in their forties. After menopause, the amount of estrogen in the body is minimal and so the breast tissue involutes and the mammogram becomes more fatty replaced. Also, as the tissue ages, calcifications develop in many women. These typical physiologic changes happen in the majority of women and are predictable changes that the radiologist learns to identify when reading mammograms. Still, these changes add variability that must be factored into the perception of the status of

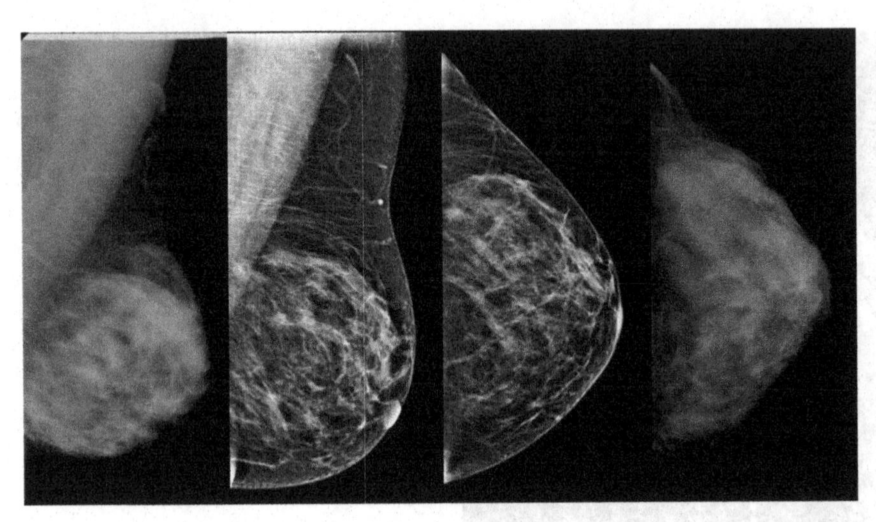

Figure 27.10 These images are all from the same patient prior to (outside images) and following (central images) menopause. Notice how the amount of white, glandular tissue has diminished with the hormonal change of menopause. Also, over this period of time the mammogram has changed from screen film to digital technology, adding even more variability into the images.

the patient during reading. Figure 27.10 shows how a patient's images can change over time.

Observing these changes over time is one of the basic tasks of interpreting mammography. It is the job of the mammographer to decipher what is normal, expected change and then to perceive what change is caused by an abnormality. As was discussed, in digital mammography, processing algorithms are applied to the data to create the final images for interpretation. These algorithms vary not only between vendors, but between different versions of the same vendors' equipment. Some vendors even offer more than one algorithm for interpretation of any individual study. So in the digital environment, the radiologist has a more challenging job of determining if a perceived change is due to a normal hormonal variation in the patient, due to processing of the image, or due to a lesion that requires action. This additional variable makes perception harder. One of the primary ways that the reader deals with this is to look at the images globally to assess if the difference affects all of the breast tissue equally and symmetrically or if the change is a more focal process. Most typically, findings that require action are focal, not bilateral and symmetric. Having said this, it is far easier to overlook a focal abnormality when the background is altered, because in the mind of the reader, the noise of the new background pattern, whether from a physiologic perturbation or a technical shift in processing of the image, is more difficult to read through. Even with this additional variable, radiologists have been shown to be highly adaptable and are able to accommodate for this additional variability.

In a large prospective multi-institutional study done by the American College of Radiology Investigational Network (ACRIN) called the Digital Mammography Imaging Screening Trial (DMIST), readers were found to be more sensitive for the detection of cancer in certain patient subsets using digital mammography than screen film, even though this study was performed while readers had little accumulated experience with the digital technique (Pisano, 2005). In this same study, the overall accuracy of digital mammography and screen film mammography was equivalent. Several other studies, performed both in the USA and abroad, have shown similar accuracy for screen film and digital mammography (Lewin, 2001, 2002; Skanne, 2003, 2004; Cole, 2004; Del Turco, 2007). What is interesting about all of these studies, from a perception standpoint, is that radiologists were able to accommodate to the new look of the digital mammograms and still were able to be at least as accurate in diagnosing breast cancer as they were with the screen film images – images that they had viewed for years and had a good deal of comfort interpreting.

27.3.5 Soft copy display

27.3.5.1 Screening

Perception in mammography is also influenced by the conditions in which the images are viewed. First, let us consider the screening environment. Even though breast cancer is the most commonly diagnosed cancer in women, the incidence of a cancer in a screening population is approximately 3 in 1000 cases read or less. Therefore, the radiologist views many normal cases for each cancer seen. In addition, often, screening mammograms are read in batches, where in a busy facility several hundred screening mammograms are performed and interpreted daily. In such an environment, it is important that the reader stays alert for the relatively rare positive case. In the same reading session, on average, the radiologist will recall 11% of cases for additional evaluation because of a possible lesion. Most of these recalled cases turn out to be perceptual misclassifications due to normal overlap of tissue or benign fibrocystic disease. Figure 27.11 shows a patient recalled from a screen mammogram for fibrocystic changes. Given the large work load, short viewing time, and rare actionable findings, the screening reading environment must be optimized in order to minimize physical and mental fatigue, as well as to reduce misperceptions.

27.3.5.2 Hanging protocols

Earlier, search patterns were discussed. These develop for each person over time and improve interpretation speed and accuracy. In screen film interpretation, most radiologists developed search patterns that combined the tasks of comparison and individual image interrogation, because all of the information could be presented simultaneously. In fact experienced radiologists see the important findings within seconds of looking at the images. In the digital mammography department, viewing of the images

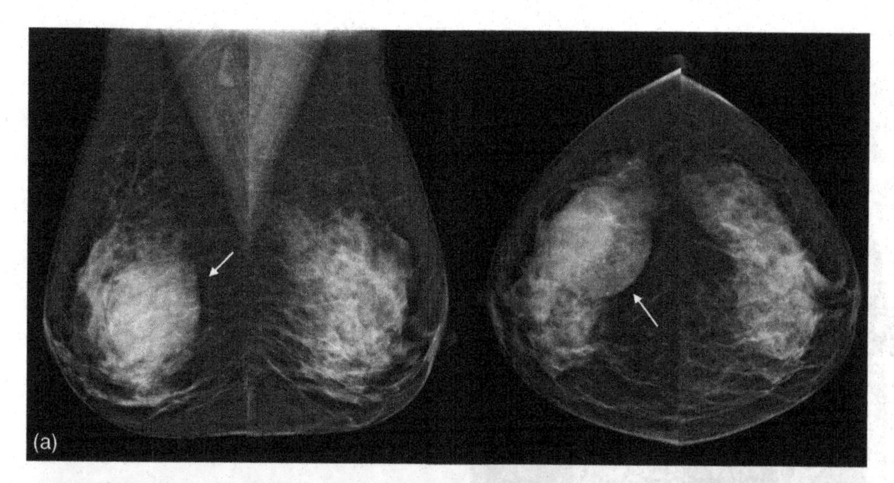

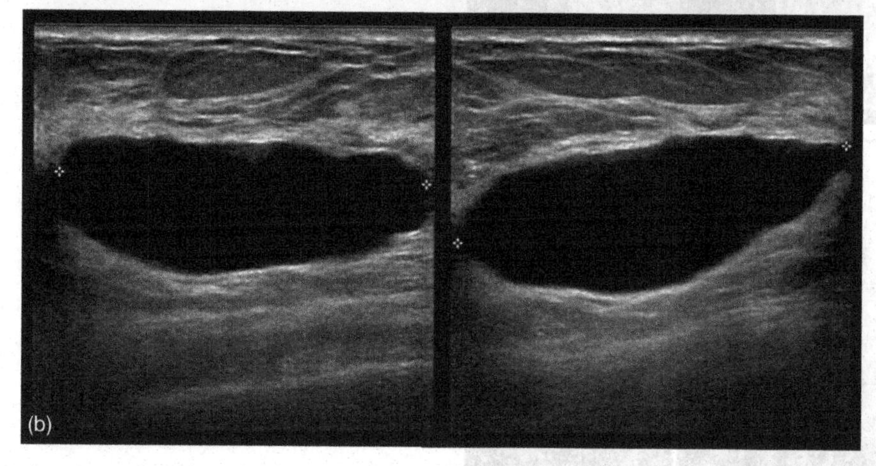

Figure 27.11 This patient was recalled from a screening mammogram (a) for a mass in the right breast (white arrows). On return, ultrasound (b) showed that this mass was a cyst, which is part of the spectrum of benign fibrocystic disease.

is more difficult because all of the image data is not typically displayed simultaneously. Because some microcalcifications in the breast are on the order of 150 microns in diameter, digital mammography detectors have been designed with very small pixel pitches. Current detectors on the market vary in pixel pitch from 50 microns to 100 microns. Depending on the size of the detector, that translates to an acquisition matrix of approximately 8 million to up to 30 million pixels for each view. The monitors that are currently used at diagnostic workstations have 5 million pixels. In order to see a mammographic image at full resolution, the breast is often larger than the monitor and so the radiologist must navigate through the image either by using the pan function or by using a vendor-supplied sequential hanging of the image at full resolution. As mentioned earlier, there are multiple reading tasks in image interpretation: comparison of current and prior images for change over time, comparison of right and left images for symmetry, and evaluation of any image for a subtle lesion or characterization of a finding detected from the comparison tasks. In digital mammography, those tasks have been separated, to a large degree, because of the large matrix of acquisition and the relatively small matrix of display. Not only does this cause a shift in reading patterns, but it also means that the reader is interacting with a computer interface in order to change the display as they read. Therefore, part of their focus is on the interface, not the mammogram. As a result, there is a potential for perception errors. Consider also the fact that, at full resolution, less than one-fifth of the entire image may be available to view and so loss of orientation is common when reading digital studies. When this happens the reader has to break from the viewing and change the display of the image to regain orientation. That is a new problem created with digital mammography. All of these minor changes add up to a very different reading experience and necessitates that the reader develops a new search pattern. Figure 27.12 shows an example of a screening hanging protocol.

27.3.5.3 Display modes

One more interesting twist on perception is directly tied to the matrix size of acquisition and display. In soft copy display, there are three primary ways that images are viewed: fit to viewport, true size, and full resolution. Fit to viewport means that all of the pixels acquired at the detector are displayed by a portion of the monitor that is prescribed by the hanging protocol. For the comparison task this typically means that, if prior soft copy images are available, eight images are displayed on the monitor pair at one time; this is illustrated in Figure 27.12. In order to accomplish this, the workstation does some form of interpolation of the data such that an average ADU is displayed for a group of pixels. Therefore, all of the acquired information is not actually displayed to the radiologist in this mode. If the prior and current studies happen to be from different systems, or the prior is a digitized analog study and the current is a digital mammogram, then very possibly the pixel pitch will be different between the two studies. So, in the fit to viewport mode, the breast might appear to be different in size, if the workstation cannot scale

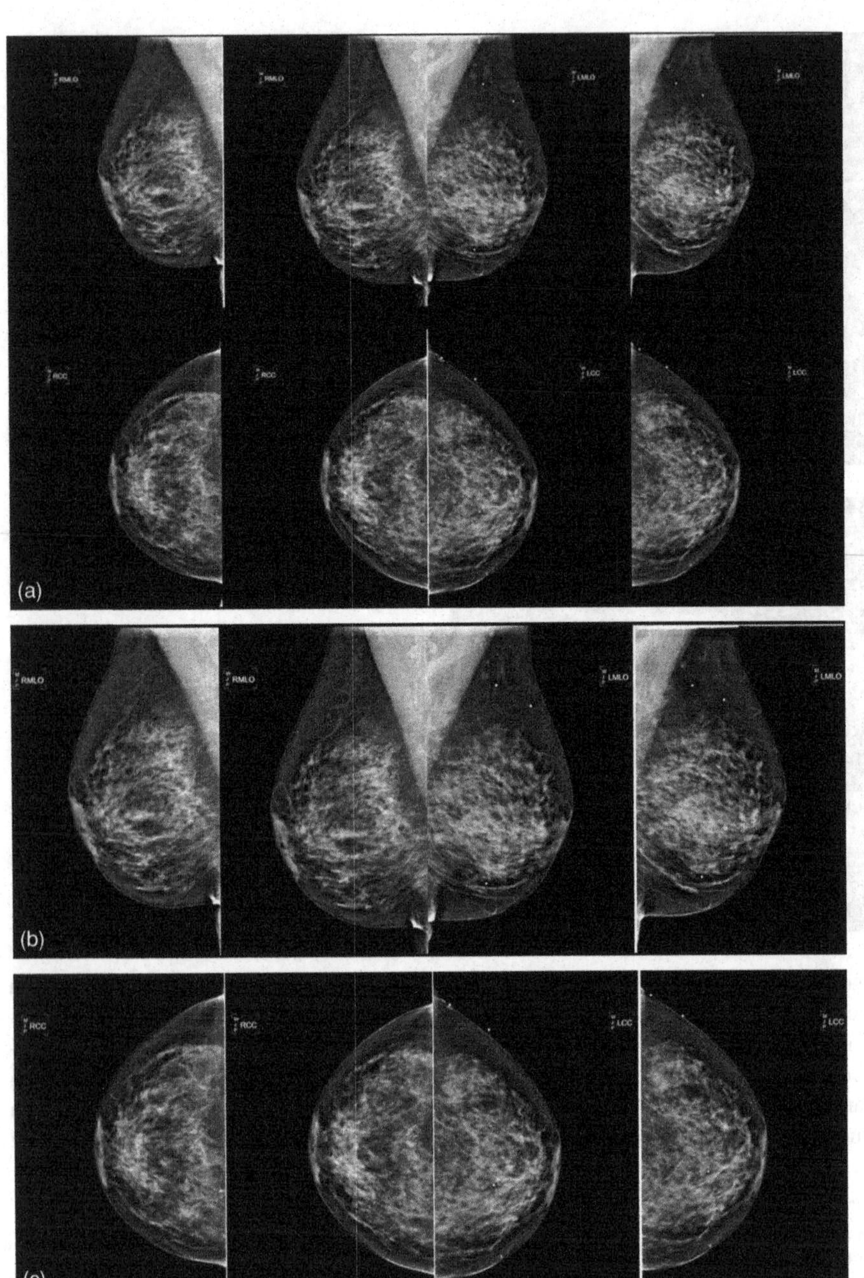

Figure 27.12 Images (a–e) each show the display of one step in a screening hanging protocol. The radiologist would step through each of these displays to read this patient's mammogram. Notice that the steps that allow comparison of current and prior images are separate from the steps that allow inspection of fine detail, and so these perception tasks are forcibly separated as well.

the display to the different pixel pitches. Analyzing if an area of asymmetry is larger or a mass has grown requires that the radiologist take into account this variable and either learn how to negate that effect or use a measuring tool to check sizes. As soon as this situation arises, the radiologist is distracted from the task at hand (namely comparison of the breast tissue) and is now focused on interpreting the "noise" of the variable displayed breast sizes. More sophisticated workstations are just beginning to appear on the market that account for these variable sized images and scale the display of different matrices so that the breast looks the same size in the fit to viewport mode.

27.3.5.4 Monitor technology

Monitor technology, calibration, positioning, and ambient lighting also have an important role in perception in digital mammography. First, let us consider the technology options of monitors. The FDA has required that all digital mammography be read on at least 5 MP monitors that can display at least 8 bits of information and are specifically approved for mammography use by the FDA. On the market currently are CRT and LCD monitors with variable bit depth capability. CRT monitors are older technology, but preferred by some users. The light source on these monitors is transmitted through glass in the system and the light beam diverges as it travels through space. Because of this, the resultant image is slightly smoother than the same image displayed on an LCD with the same bit depth. LCD monitors, conversely, are an array of adjacent thin film transistors. These transistors are either on or off, depending on what is prescribed by the video card that drives the monitors. There is virtually no spreading of the light source and so the image is crisper. Structured noise is introduced into the image by the

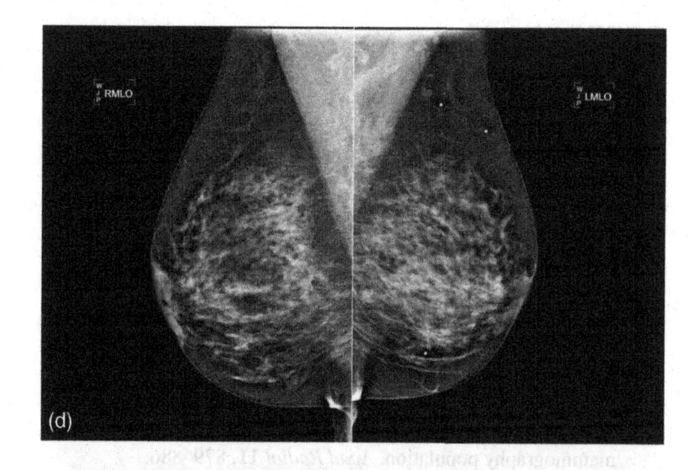

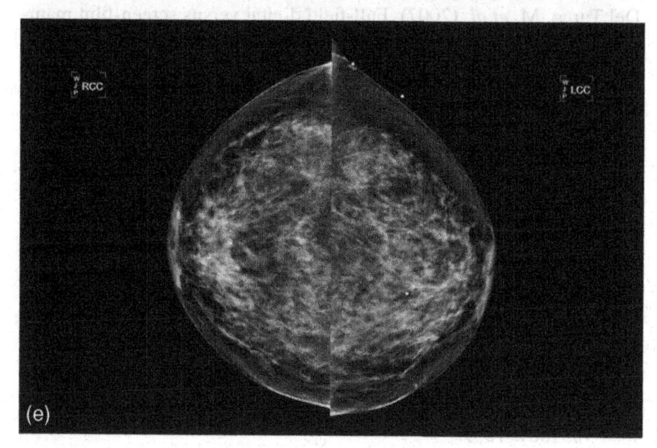

Figure 27.12 (*cont.*)

nature of this design. Even given these technical differences, it has been shown that radiologists' performance is equivalent with the two monitor technologies (Zuley, 2006). The CRTs also refresh approximately 30,000 times each minute, which has been shown to cause eye fatigue. LCDs do not refresh so eye strain is less.

27.3.5.5 Monitor calibration

For years, it was thought that the human eye could not perceive more than 256 shades of gray (or 8 bits of data). Very likely this conclusion was drawn because technology at the time that research was done was not what we have available today. With the higher luminance of the LCDs, this has been shown to not be true. In fact, radiologists can perceive more than 1000 shades of gray in the right environment. The perception of shades of gray is in part driven by the luminance of the monitors. On average, the maximum luminance of a CRT is 250 cd/m^2 and the maximum luminance of an LCD is 600 cd/m^2. All medical monitors must adhere to certain specifications which state that there is a fixed relationship of the maximum and minimum luminance output of any monitor pair. The higher the maximum luminance, conceivably, the better the ability of the reader to detect more shades of gray, but also then the minimum luminance has to be set higher to follow these required specifications. The minimum luminance controls how black the dark areas are and the maximum luminance controls how white the bright areas appear. So, if the maximum luminance is set very high, the whites can be so bright as to almost blind the radiologist to the adjacent structures and the background becomes more gray than black. If the maximum luminance is set too low, then the system performance and the ability of the reader to perceive the displayed anatomy are compromised, because it is harder for the human eye to see variability in the very dark areas. From a perception standpoint, one can imagine how important monitor calibration is in order to ensure that the radiologist is able to visualize different shades of gray in the whole image from the brightest to the darkest areas. It is recommended that facilities calibrate all monitor pairs in the facility to the same maximum and minimum luminance so that the display of the images on all workstations in any facility is constant.

27.3.5.6 Ambient lighting

Even when the monitors are calibrated properly, the conditions in which the interpretation occurs influence perception. Optimal ambient lighting is different for CRTs and LCDs. For CRTs, because the maximum luminance is limited, there should be virtually no ambient light in the room at the time of reading. If one considers that in the reading room, there is often an alternator or light box for viewing of prior analog films, and workstations on which tasks like keyboarding are done, then the ambient light issue is problematic. For instance, for every mammogram that has an analog prior, if the original images are viewed, the radiologist selects the current digital case from the worklist on the workstation, then has to move the alternator to the board where the prior images are hanging and turn on the light boxes. Then, the light boxes should be turned off and the current soft copy digital images viewed. Not only does the radiologist have to remember what the prior images looked like after the lights are off so that he or she can make a comparison with them, but his or her eyes have to adjust to the different lighting conditions before reading the current soft copy study. Imagine reading like this in a batch reading mode where potentially hundreds of cases are read.

For reading on LCD monitors, it is recommended that there is a low level of background ambient lighting, a different requirement than the conditions for CRT interpretation. Of course, no matter what the monitor technology, reflected light on the monitor screen will decrease the quality of display. LCDs also suffer from loss of luminance with increasing viewing angles. Typically manufacturers recommend that the monitors be turned toward each other a bit so that the viewing angle is reduced. As was discussed earlier, the lower the available luminance, the fewer shades of gray the human eye can see. Thus, the harder it is to perceive subtle changes in these areas. Couple this with the fact that mammograms are typically displayed with the chest wall sides of the mammograms abutting each other and the skin sides closer to the lateral sides of the monitor pair. The area on the mammogram that has the darkest display and thus the lowest contrast is in the subcutaneous tissues under the skin. Therefore, this is the area that is inherently one of the harder to perceive changes, even with optimized viewing. Adding image degradation from loss of luminance due to off-axis viewing makes perception of changes here harder.

27.3.5.7 Other hardware

The diagnostic workstation is not just monitors; there is also typically a keyboard, a hard drive, a mouse, and/or a special keypad that has been designed for reading mammography. In addition, because digital mammography is still relatively new, and because all aspects are tightly regulated by the FDA, many mammography workstations are not able to be integrated into the system-wide PACS. Therefore, radiologists are very frequently reading mammograms on stand-alone workstations. While these workstations are typically good at displaying at least screening mammography, they usually have limited workflow functionality. Every time the reader has to interact with all of these components, the reading times increase and perception may be hampered, because they cause the reader to stop and start their perceptual tasks. The longer reading times increase fatigue for the reader, especially in the batch screening setting.

27.3.6 Ergonomics

Proper ergonomics of the reading area decreases both mental and physical fatigue. Many of the factors that affect proper display of the images also have ergonomic implications. How the monitors are positioned not only impacts perceived luminance and thus ability to detect subtle differences in the tissues, but also impacts on the fatigue of the radiologist and the time it takes to interpret studies. When the face of each monitor in the pair is placed at 180 degrees to the other, the radiologist either has to lean his/her body or move his/her head to see all of the image well. In essence, the reader is using their body to compensate for poor positioning of the monitors. This causes strain on the neck and eyes and increases fatigue. One study has shown that as luminance of the display drops, reading times increase and performance decreases (Krupinski, 1999, 2005b). So, all of these factors are intertwined in the perception of the study. Similarly, if the radiologist is sitting improperly, without back and arm support, muscle fatigue becomes a problem, especially when reading large groups of studies, such as batch reading of screening mammograms. Not only does the ambient light and reflected light affect perception of findings, but again, wrong lighting causes the radiologist to try to compensate with their eyes and body, which in turn increases strain and fatigue.

27.4 CONCLUSION

In conclusion, perception in mammography is influenced by a complex interaction of technical and human factors. Essentially, all of the technical components of acquisition, processing, and display of the images impact on the ability of the radiologist to interpret the images correctly. The human tasks of perception, including experience, ergonomics, and fatigue, also play an important role. Many of the skills and knowledge gained in screen film mammography have been transferred to the digital environment, but the tasks of perception are even more challenging in this environment due to new, additional variables. Understanding some of the variables involved in mammography perception helps in product development, reading room design and staffing, and education.

REFERENCES

Berg, W., Campassi, C., Langenberg, P., Sexton, M. (2000). Breast imaging reporting and data system: inter- and intraobserver variability in feature analysis and final assessment, *AJR Am J Roentgenol* **174**, 1769–1777.

Burnside, E., Sickles, E., Sohlich, R., Dee, K. (2002). Differential value of comparison with previous examinations in diagnostic versus screening mammography, *AJR Am J Roentgenol* **179**, 1173–1177.

Ciccone, G., Vineis, P., Frigerio, A., Segnan, N. (1992). Inter-observer and intra-observer variability of mammogram interpretation: a field study, *Eur J Cancer* **28A**, 1054–1058.

Cole, E. *et al.* (2004). Diagnostic accuracy of Fischer Senoscan Digital Mammography versus screen-film mammography in a diagnostic mammography population, *Acad Radiol* **11**, 879–886.

Del Turco, M. *et al.* (2007). Full-field digital versus screen-film mammography: comparative accuracy in concurrent screening cohorts, *AJR Am J Roentgenol* **189**, 860–866.

Elmore, J., Wells, C., Lee, C., Howard, D., Feinstein, A. (1994). Variability in radiologists' interpretations of mammograms, *N Engl J Med* **331**, 1493–1499.

Houssami, N. *et al.* (2003). The contribution of work-up or additional views to the accuracy of diagnostic mammography. *The Breast* **12**, 270–275.

Houssami, N. *et al.* (2004). The influence of clinical information on the accuracy of diagnostic mammography, *Breast Cancer Res Treat* **85**, 223–228.

Kerlikowske, K. *et al.* (1998). Variability and accuracy in mammographic interpretation using the American College of Radiology Breast Imaging Reporting and Data System, *J Natl Cancer Inst* **90**, 1801–1809.

Krupinski, E., Roehrig, H., Furukawa, T. (1999). Influence of film and monitor display luminance on observer performance and visual search, *Acad Radiol* **6**, 411–418.

Krupinski, E. (2005a). Visual search of mammographic images: influence of lesion subtlety, *Acad Radiol* **12 (8)**, 965–969.

Krupinski, E., Johnson, J., Roehrig, H., Nafziger, J., Lubin, J. (2005b). On-axis and off-axis viewing of images on CRT displays and LCDs: observer performance and vision model predictions, *Acad Radiol* **12 (8)**, 957–964.

Leung, J. *et al.* (2007). Performance parameters for screening and diagnostic mammography in a community practice: are there differences between specialists and general radiologists? *AJR Am J Roentgenol* **188**, 236–241.

Lewin, J. *et al.* (2001). Comparison of full-field digital mammography with screen-film mammography for cancer detection: results of 4,945 paired examinations, *Radiology* **218 (3)**, 873–880.

Lewin, J. *et al.* (2002). Comparison of full-field digital mammography and screen-film mammography for detection of breast cancer, *AJR Am J Roentgenol* **179 (3)**, 671–677.

Mello-Thoms, C. *et al.* (2005). Effects of lesion conspicuity on visual search in mammogram reading, *Acad Radiol* **12 (7)**, 830–840.

Miglioretti, D. *et al.* (2007). Radiologist characteristics associated with interpretive performance of diagnostic mammography, *J Natl Cancer Inst* **99**, 1854–1863.

Nodine, C. *et al.* (1999). How experience and training influence mammography expertise, *Acad Radiol* **6**, 575–585.

Nodine, C. *et al.* (2001). Blinded review of retrospectively visible unreported breast cancers: an eye-position analysis, *Radiology* **221 (1)**, 122–129.

Nodine, C., Mello-Thoms, C., Kundel, H., Weinstein, S. (2002). Time course of perception and decision making during mammographic interpretation, *AJR Am J Roentgenol* **179 (4)**, 917–923.

Pisano, E. *et al.* (2005). Diagnostic performance of digital versus film mammography for breast-cancer screening, *N Engl J Med* **353**, 1773–1783.

Saunders, R., Samei, E. (2006). Improving mammographic decision accuracy by incorporating observer ratings with interpretation time, *Br J Radiol* **79**, 117–122.

Sickles, E., Wolverton, D., Dee, K. (2002). Performance parameters for screening and diagnostic mammography: specialist and general radiologists, *Radiology* **224**, 861–869.

Sirovich, B., Sox, H. (1999). Breast cancer screening, *Surg Clin North Am* **79**, 961–991.

Skanne, P., Young, K., Skjennald, A. (2003). Population-based mammography screening: comparison of screen-film and full-field digital mammography with soft-copy reading – Oslo I study, *Radiology* **229**, 877–884.

Skanne, P., Skjennald, A. (2004). Screen-film mammography versus full-field digital mammography with soft-copy reading: randomized trial in a population-based screening program – the Oslo II study. *Radiology* **232**, 197–204.

Tabar, L. *et al.* (2000). The Swedish two-county trial twenty years later. Updated mortality results and new insights from long-term follow-up, *Radiol Clin North Am* **38** (**4**), 625–652.

Zuley, M. *et al.* (2006). Full-field digital mammography on LCD versus CRT monitors, *AJR Am J Roentgenol* **187** (**6**), 1492–1498.

28

Perceptual optimization of display processing techniques

RICHARD VANMETTER

28.1 INTRODUCTION

Medical image processing encompasses a wide range of digital processing techniques that are applied to medical images for many diverse purposes. The goal, however, is always to maximize the useful diagnostic information conveyed to highly-trained human observers. For some modalities, CT and MRI for example, image reconstruction processing is necessary to produce images that can be correlated with patient anatomy. Optimization of these techniques is often complicated by trade-offs among image resolution, noise, and susceptibility to artifacts. In these and other three-dimensional modalities, further image processing is then applied to the reconstructed image data. For example, image segmentation and surface rendering can be used to enhance the visualization of specific structures. Two-dimensional, projection imaging modalities also utilize a wide variety of image processing techniques, commonly including segmentation, a wide range of automated detection and measurement techniques, texture recognition as well as display enhancement techniques informed by the needs of the human visual system. These examples give some idea of the scope encompassed by the large and rapidly expanding field of medical image processing. It can be better appreciated by perusing the available literature as exemplified by edited volumes (Sonka, 2000) and ongoing conference series (Pluim, 2007).

This chapter will focus on display processing of projection radiographic images. This application draws upon a small subset of the vast collection of image processing techniques. The goal is to optimize the relationship between an acquired set of digital image data and the image that is displayed for clinical interpretation. The choice of techniques is based on an understanding of the diagnostic task, the anticipated clinically relevant features, and the characteristics of the human visual system related to visual perception. To appreciate the diagnostic task and the perceptual characteristics of the human visual system, the reader is referred to other chapters of this volume and references therein.

Focusing on display processing of projection radiographic imaging is not as limiting as might first appear. Many of the display processing techniques that will be discussed in this chapter are applicable to other sources of digital image data. Therefore, the choice of projection radiography is intended to be illustrative. This is particularly true of the role of perceptual requirements for optimal image display, which although task-dependent are largely independent of imaging modality.

28.2 CHARACTERISTICS OF DIGITAL IMAGE DATA

Image processing techniques are used to transform the pixel values of a RAW IMAGE that is acquired by a digital projection imaging device to the pixel values of an ORIGINAL IMAGE that is suitable for display processing and finally to the pixel values of a DISPLAY-READY IMAGE. Each of these is a digital image represented as a two-dimensional array of unitless pixel values. Because digital images have no units or other physical attributes, it is important to distinguish data files that represent RAW IMAGES, ORIGINAL IMAGES, and DISPLAY-READY IMAGES. First some definitions:

RAW IMAGE – is the uncorrected output of a digital imaging device. The RAW IMAGE is not calibrated and may contain artifacts associated with the acquisition device. As such it is not directly suitable for display processing.

ORIGINAL IMAGE – is created from the RAW IMAGE by calibrating and correcting artifacts caused by the acquisition device. An ORIGINAL IMAGE is suitable for display processing.

DISPLAY-READY IMAGE – is created from the ORIGINAL IMAGE by the display processing. Pixel values in the DISPLAY-READY IMAGE are suitable for interpretation with a standard display.

Display processing depends critically on a correct understanding of the physical correlates of the pixel values in the ORIGINAL IMAGE and the DISPLAY-READY IMAGE. The pixel values of the ORIGINAL IMAGE must be related in some way to the X-ray fluence reaching the image detector. Likewise, the pixel values of the DISPLAY-READY IMAGE must be created with a full understanding of the optical image that will be produced by the display device. Figure 28.1 shows an overview of the imaging chain from acquisition to display.

The first step is digital acquisition in which the incident X-ray fluence, $E_A(x,y)$, interacts with the digital detector which produces a RAW IMAGE. The acquisition system then does the necessary calibration and artifact correction to convert the RAW IMAGE to a calibrated, artifact-free ORIGINAL IMAGE represented as an array of digital pixel values $\{x_A(i,j)\}$. In this process the incident X-ray fluence is encoded according to the function f(E). Subsequent image processing must understand this encoding in order to correctly interpret the meaning of the input pixel values. This is made explicit in the source image decoding step in which the inverse of the encoding function, f^{-1}, is applied. At this point, the display processing that is the central subject of this chapter is applied to create an array of

The Handbook of Medical Image Perception and Techniques, ed. Ehsan Samei and Elizabeth Krupinski. Published by Cambridge University Press.

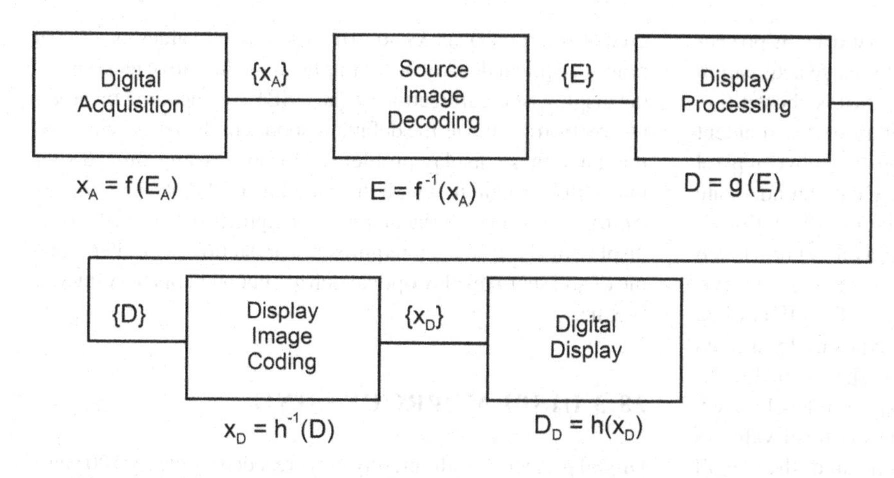

Figure 28.1 Digital projection radiography imaging chain.

display values, D(i,j). These values express the physical attributes of the displayed pixel, such as luminance, log-luminance, and optical density. They must then be translated to pixel-values of a DISPLAY-READY IMAGE, $x_D(i,j)$, appropriate for driving a display device to the desired luminance levels for viewing the image. This is done by applying $h^{-1}(D)$, the inverse of the display function that characterizes the display device. Finally, the display device displays the pixel values of the image in terms of its designed response characteristics, h(x). Before discussing the display processing in detail it is appropriate to briefly review the characteristics of the acquisition and display devices.

28.2.1 Image acquisition

There are several technologies for acquiring digital projection radiographic image data. These include photostimulable-phosphor-based computed radiography (Rowlands, 2002), flat-panel digital radiography (Rowlands, 2000), and charge-coupled-device-based digital radiography (Hejazi, 1997). Each of these technologies creates a 'RAW IMAGE' or uncorrected digital representation of the incident X-rays.

28.2.1.1 Computed radiography (CR)

Computed radiography uses a planar detector containing a storage phosphor that stores part of the incident X-ray energy. This image-wise energy distribution can be liberated as emitted light by exciting the phosphor with "actinic" light of a different wavelength. The resulting RAW IMAGE is acquired either by scanning the storage phosphor with a laser focused to a small spot while collecting and detecting the emitted light; or by scanning the storage phosphor with a line of light-emitting diodes and collecting the emitted light with a corresponding line of photo-diodes. In either case, the signal is digitized to form a representation of the spatial distribution of the emitted light, a RAW IMAGE.

28.2.1.2 Flat-panel digital radiography (FP-DR)

Flat-panel digital radiography is based on full-sized arrays of thin-film transistors that are similar to those used for image displays. The signal at each pixel is produced either by interaction of the incident X-rays with an X-ray photoconductor or with an X-ray phosphor optically coupled to a photoconductor sensitive to the light emitted by the phosphor. An electrical charge is collected at each pixel that is proportional to the X-ray energy deposited. Switching circuitry allows electrical charge from each pixel to be digitized to form a RAW IMAGE.

28.2.1.3 CCD-based digital radiography (CCD-DR)

Charge coupled device (CCD)-based digital radiography uses a full-size X-ray phosphor together with minifying optics and a charge coupled device. The charge coupled device has a sensitive area that is much smaller than the X-ray image area, typically 5 cm or smaller in linear dimension. The minifying optics reduce the size of the optical image produced by the phosphor to that of the CCD. The CCD provides a method of digitizing the incident optical image to create the RAW IMAGE.

28.2.1.4 Image representation

All of the image acquisition technologies are fundamentally linear in that the signal produced by the detector is linearly related to the incident X-ray exposure. However, it is important to recognize that the signal produced by a detector is affected by the energy spectrum and the angular distribution of the incident X-rays. Verifying linearity therefore assumes carefully controlled experimental conditions that are generally not realized in clinical practice.

The RAW IMAGE from each of the image acquisition technologies is affected by imaging artifacts that can be mitigated with appropriate image pre-processing. This is often the first digital image processing that is done and is commonly hidden from the user. Examples of the image artifacts in the RAW IMAGE that can be corrected by image processing include variability in gain and offset among the pixels, dead pixels and lines, as well as distortions caused by scanning or minification optics.

Pre-processing of the RAW IMAGE produces an ORIGINAL IMAGE in which artifacts have been mitigated and the digital value at each pixel is mapped in a well-defined relationship to the incident X-ray exposure under controlled conditions. The choice of this relationship may differ among manufacturers. There is no current standard for encoding ORIGINAL IMAGES in projection radiography. In many cases, although the image acquisition is intrinsically linear, a LOGARITHMIC ENCODING is chosen (i.e. pixel values are proportional to the logarithm of the incident X-ray exposure). In others, the linear relationship between pixel values and incident exposure is preserved. Given an adequate number of digitization levels (bit depth), the representation of the ORIGINAL IMAGE is unimportant as long as it is known and utilized by the subsequent processing. LOGARITHMIC ENCODING has some conceptual and practical advantages. The effect of an object in the X-ray beam is to reduce the exposure by a fixed percentage. This corresponds to a constant change in log X-ray air kerma, independent of the X-ray exposure level. When LOGARITHMIC ENCODING is used, the change in pixel value in the ORIGINAL IMAGE will be constant independent of the overall X-ray exposure level. This means that equal changes in pixel value anywhere in an image correspond to occlusion of the X-ray beam by an equivalent object. A direct consequence is that a change in overall exposure level will only result in a shift in the pixel value histogram in the ORIGINAL IMAGE, while the shape of the histogram will be unchanged. This property facilitates analysis of the ORIGINAL IMAGE and application of display processing. The LOGARITHMIC ENCODING representation for the ORIGINAL IMAGE will be used for the illustrations in this chapter.

28.2.2 Image display

Modern image displays interpret pixel values to produce display luminances that follow a linear range of the Digital Imaging and Communications in Medicine (DICOM) Grayscale Standard Display Function (GSDF) values. The GSDF values are described in detail elsewhere in this volume as well as in part 14 of the DICOM standard. The purpose of the GSDF values is to provide a set of luminance levels that represent equal changes in perceived brightness under specified viewing conditions. This perceptual linearity of the GSDF values is intended to accommodate displays with different luminance ranges while retaining the relative perceptual differences among pixels in the image. Because use of the GSDF is nearly universal, the pixel values of DISPLAY-READY IMAGES output by display processing are most often intended to represent GSDF values over a specified luminance range. However, it must be recognized that GSDF is not normally the native response of display devices. Therefore, the pixel values of the DISPLAY-READY IMAGE must be recast by the diagnostic workstation to accommodate the response function of the associated display device. Modern medical imaging workstation software can be assumed to provide this functionality.

Prior to the adoption of part 14 of the DICOM standard, a variety of display pixel interpretations were encoded and a variety of display responses realized by different vendors. This led to the very real possibility of poor displayed image quality when the encoding of the DISPLAY-READY IMAGE and the response of the display were poorly matched. Some examples of now obsolete DISPLAY-READY IMAGE encoding included a linear range of optical densities and matching the luminance characteristics of "typical" CRT displays used in medical imaging. Likewise, workstations and film printers might be configured to interpret DISPLAY-READY IMAGE pixel values as a linear range of optical density or to simply allow the native response of the display. Standardization on the GSDF for encoding DISPLAY-READY IMAGES and as the default response of display devices has mitigated these display problems. However, some clinical sites may still be found in which the encoding of the DISPLAY-READY IMAGES represents a linear range of optical densities. Correct display of these images requires a workstation or printer capable of and set to display optical density encoded DISPLAY-READY IMAGES.

28.3 DISPLAY PROCESSING

Digital projection radiography was preceded by nearly 100 years of SCREEN-FILM-based projection radiography. Therefore, digital radiography represents a replacement technology for a wide range of clinical examinations. It is expected to replicate the diagnostic capabilities of the predicate technology and to provide opportunities to improve patient care. Chapters in this volume have described how radiologists employ pattern recognition skills established during intensive training and honed by years of practical experience. Therefore, an important initial goal for display processing of digitally acquired images is to replicate the overall image appearance of SCREEN-FILM images. This maximizes the value of the radiologists' training and experience when transitioning to digital radiography.

Once digital radiography is established in a medical practice, the flexibility of display processing and display can be used to enhance the image in valuable ways. This is particularly true with the widespread availability of high-quality displays, which have replaced film as the display media of choice for digital radiography. Techniques that go beyond the replication of SCREEN-FILM imaging encompass the broad range of image processing and display techniques that are now widely accepted in clinical practice. Such enhanced display processing techniques continue to develop, therefore the specific techniques cited in this chapter are offered as illustrative examples.

28.3.1 Gray scale rendering

The most important characteristic of SCREEN-FILM technology is the relationship between X-ray exposure and the resulting darkening of the film. This is commonly described by plotting the optical density of the film as a function of the logarithm of the incident air kerma. OPTICAL DENSITY, the logarithm of the reciprocal of the optical transmittance, was adopted by early workers because they found it to be more closely associated with the perceived darkness of the film than the optical transmittance itself. The luminance image that the viewer sees results from the back lighting of the radiograph with a source of uniform luminance. The log of X-ray incident air kerma is a convenient choice because the effect of an object in the X-ray beam is to reduce the exposure to the SCREEN-FILM by a fixed percentage. This corresponds to a constant change in log X-ray air kerma independent of the X-ray exposure level.

Figure 28.2 shows the response function for several SCREEN-FILM systems. OPTICAL DENSITY is plotted as a function of log incident air kerma. SCREEN-FILM images are negatives in

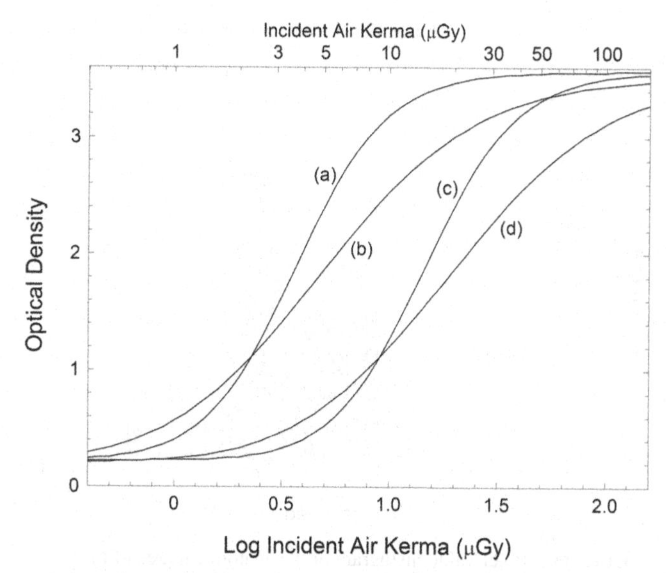

Figure 28.2 Screen-film characteristic curves showing optical density of the developed film as a function of the air kerma incident upon the screen-film system; (a) high-speed screen with high-contrast film, (b) high-speed screen with low-contrast film, (c) low-speed screen with high-contrast film, (d) low-speed screen with low-contrast film.

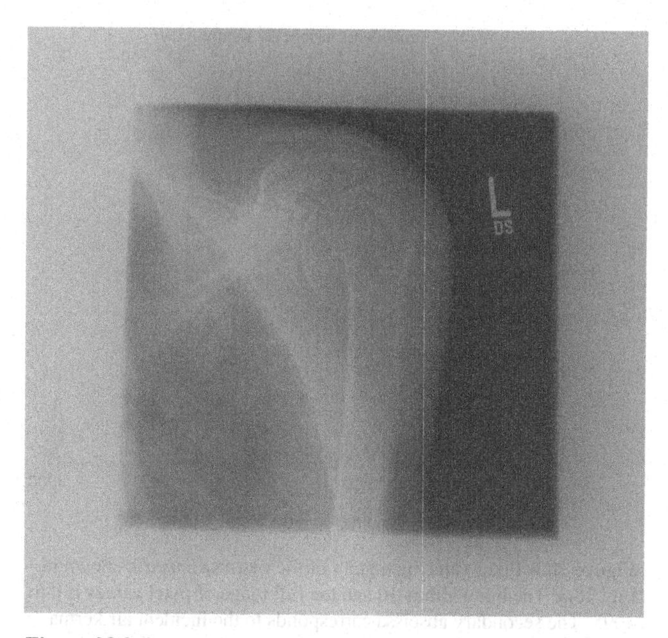

Figure 28.3 ORIGINAL IMAGE of the anterio-posterior projection of a shoulder.

that the images are darker in areas of higher X-ray exposure. The speed of a SCREEN-FILM system is characterized by the incident air kerma needed to produce a fixed response (OPTICAL DENSITY). The slope of the curves indicates the "CONTRAST" (i.e. the change in OPTICAL DENSITY that will be caused by a fixed change in log exposure). CONTRAST affects the ability to visualize subtle features. Higher contrast is desirable up to the point that noise becomes visible in the image. The range of log air kerma over which suitable CONTRAST and OPTICAL DENSITY are produced is described as the LATITUDE. A wide LATITUDE is desirable to optimally visualize the full range of X-ray transmittances in a clinical image. It should be clear that increased LATITUDE comes at the expense of reduced CONTRAST, and vice versa.

Each SCREEN-FILM combination is characterized by a fixed curve shape and a fixed speed. The curve shape, and hence the CONTRAST, is determined by the film. In Figure 28.2 two film types are illustrated. These differ in CONTRAST and therefore in LATITUDE, the range of log-exposure over which useful images can be obtained. The speed is determined by both the screen and the film. In this example, two screens of different speeds are illustrated. Curves (a) and (c) result from a high-contrast film used with a fast and a slow screen, respectively. Likewise, curves (b) and (d) show the response of a low-contrast film with the same two screens. An important limitation of SCREEN-FILM is that the curve shape and speed cannot be adjusted. Therefore, if the range of exposures for an image does not fall within the range for which the SCREEN-FILM provides adequate CONTRAST and OPTICAL DENSITIES, the image often needs to be re-acquired. Such mis-exposures waste time and materials and have the adverse effect of increasing patient radiation dose.

To simply replicate the gray scale display characteristics of SCREEN-FILM, the pixel values of a digital ORIGINAL IMAGE could be remapped to DISPLAY-READY IMAGE pixel-values that would replicate one of the fixed curves shown in Figure 28.2. However, this would unnecessarily re-introduce the fixed-speed and fixed-contrast limitations of SCREEN-FILM technology. Instead, most manufacturers of digital projection radiography systems provide for the adjustment of the display processing after the image data are acquired to compensate for a range of potential mis-exposures and for variations in subject contrast. This is a substantial advantage.

Since manual adjustment of the display-processing parameters for each image would be unacceptably time-consuming, various methods of automated display processing are provided by most manufacturers. This requires analysis of the ORIGINAL IMAGE to determine the appropriate adjustment of display processing for each acquired image. A variety of methods are used for this analysis. The most common methods are based on analysis of the pixel-value histogram for all or part of the ORIGINAL IMAGE. More sophisticated and robust methods depend on a more detailed analysis of the image content.

A more complete understanding of the role of display processing can be obtained by considering a concrete example. An AP shoulder examination has been selected as an example that is intended to illustrate some of the challenges and opportunities for display processing. In particular, the ORIGINAL IMAGE illustrated in Figure 28.3 was chosen to include two noteworthy features relevant to robust display processing. First, there is a significant collimated area in the image, typical of many widely used radiographic examinations. We denote this as FOREGROUND. Second, there is a smaller but still significant area of direct X-ray exposure, also typical of many radiographic examinations. We denote this as BACKGROUND. These areas have a profound influence on the histogram of the ORIGINAL IMAGE, shown in Figure 28.4. Here the number of pixels having a particular range of values is plotted as a function of pixel value. In this example, the pixel values are proportional to the logarithm of the incident air kerma. This acquisition system is capable of responding to incident air kerma from about 0.1 μGy

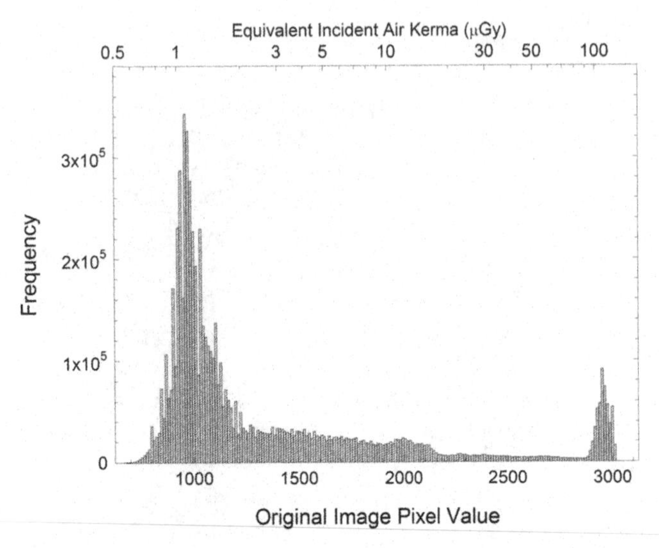

Figure 28.4 Pixel-value histogram of the ORIGINAL IMAGE shown in Fig. 28.3. The bin width is 16 and the full range of pixel values is 0 to 4095. The secondary abscissa corresponds to the incident air kerma required to produce the pixel values shown on the primary abscissa under prescribed X-ray spectrum (calibration) conditions.

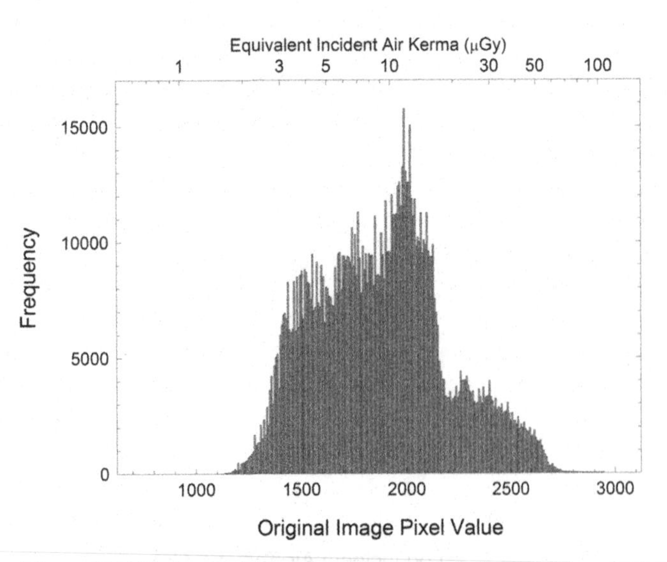

Figure 28.5 Pixel-value histogram of the anatomical area of the ORIGINAL IMAGE shown in Fig 28.3. The bin width is 16 and the full range of pixel values is 0 to 4095. The secondary abscissa corresponds to the incident air kerma required to produce the pixel values shown on the primary abscissa under prescribed X-ray spectrum (calibration) conditions.

to 1000 μGy, producing pixel values from 0 to 4095. The histogram bins in Figure 28.4 are each 16 pixel-values wide. Thus the number of pixels with values from 0 to 15 are counted and plotted as a bar located at pixel-value 8. This process is continued for each subsequent range of 16 pixel values. The histogram in Figure 28.4 is calculated from the entire ORIGINAL IMAGE, including the collimated (FOREGROUND) and direct-exposure (BACKGROUND) areas. The histogram is dominated by two peaks at opposite ends corresponding to the FOREGROUND and BACKGROUND areas. It is instructive to compare this histogram with that shown in Figure 28.5, which is calculated from only the anatomical areas of the ORIGINAL IMAGE. First, note that the ordinate scale has changed reflecting the smaller number of pixels in the anatomical area compared to the entire image. Also note that the prominent peaks near pixel values 1000 and 3000 have completely disappeared. A prominent broad peak now appears near pixel value 2000. This histogram more closely conveys the range of pixel values that need to be optimally displayed to produce a useful image of the clinically important areas. The difference between the histograms in Figures 28.4 and 28.5 illustrates the importance of image segmentation prior to histogram analysis for optimal display processing.

The histograms in Figures 28.4 and 28.5 have been plotted over the same range of air kerma as the SCREEN-FILM response plot in Figure 28.2 to facilitate comparison. Even when non-anatomical areas are eliminated (Figure 28.5), the histogram covers a substantial range of exposures; in this case, pixel values from ~1200 to ~2800 correspond to equivalent incident air kerma of ~1.4 to ~55 μGy. While this range is well encompassed by a low-contrast film, it would be challenging to optimally image on a high-contrast film. It has inadequate LATITUDE. The trade-off between CONTRAST and LATITUDE will be addressed below as one of the limitations of SCREEN-FILM radiography that can be mitigated by digital acquisition and optimal display processing.

28.3.2 Image analysis

The goal of image analysis is to provide the necessary information to automatically place a gray scale rendering curve relative to the pixel values of each ORIGINAL IMAGE. The first step is commonly a segmentation of the clinically significant areas of the ORIGINAL IMAGE. The simplest approach would be to use the full image area. However, as shown above, the resulting histogram would be difficult to interpret because of the variability in the extent and pixel values of collimated and direct exposure areas. A more sophisticated approach might be based on the expected location of the anatomical area of the image. For example, a central area of the image might be used, which for this image would work reasonably well. However, this approach is not robust to changes in positioning or to variations in anatomy. Clinical images are not always centered on the acquisition device and even when centered, the central part of the image may not be representative of the full anatomical area. For example, consider the case of two views imaged on a single CR imaging plate such as an AP and lateral hand. Likewise, consider the central portion of a PA chest image which would consist primarily of the low-penetration mediastinum. This would poorly represent the pixel values in the more important lung areas. Better approaches segment the anatomical area based on image-wise characteristics of the ORIGINAL IMAGE.

Background detection. Identifying BACKGROUND is important because of the wide pixel-value range of the ORIGINAL IMAGE (Figure 28.4) relative to the more limited pixel-value range of the anatomical area (Figure 28.5). One does not want to waste luminance range in the DISPLAY-READY IMAGE attempting to optimally render BACKGROUND areas. It is desirable for the BACKGROUND areas of the image to be rendered at the lowest luminance levels.

BACKGROUND typically corresponds to the highest pixel values in the ORIGINAL IMAGE where the unobstructed radiation hits the imaging plate. Several factors make the detection of these regions a challenge. First, the radiation field across the image may be non-uniform due to the orientation of the X-ray source relative to the imaging device. The effect of scatter from thicker anatomical regions compounds this problem. Also, for some exams, multiple exposures may be done on a single receptor, resulting in multiple BACKGROUND signal levels. Finally, an ORIGINAL IMAGE may or may not contain an area of direct radiation exposure. The method of BACKGROUND detection must be able to robustly detect the areas of direct radiation as well as the absence of direct radiation.

One successful method of detecting background begins by characterizing direct exposure regions based upon the characteristic transition from anatomy to BACKGROUND and from FOREGROUND to BACKGROUND. In this method, each line of the image is analyzed to detect such transitions, which are characterized as continuously increasing or decreasing line segments. Segments of each line that are BACKGROUND candidates are then evaluated to determine whether the variation and texture are typical of direct exposure areas. Candidate BACKGROUND pixels for each line are thereby accumulated. The histogram of background pixel values invariably contains a well-defined peak (Barski, 1997). However, it is important to allow for multiple exposure fields that may result in multiple background peaks. From this histogram, a threshold for background pixels is established and the contiguous areas of BACKGROUND delineated. This approach has been shown to successfully deal with the problems of non-uniform background and multiple exposures.

Foreground detection. Like BACKGROUND, identifying FOREGROUND is essential for several reasons. As with BACKGROUND areas, one does not want to waste luminance range in the DISPLAY-READY IMAGE attempting to optimally render non-clinical areas behind the collimation blades. More importantly, FOREGROUND areas having low exposure are normally displayed near-white. When viewing the FOREGROUND regions in an image, the brightness of these areas can cause viewing flare, which reduces the contrast sensitivity of the eye to darker (anatomical) areas of the image, causes eye strain, and hinders optimal clinical interpretation.

Several approaches to FOREGROUND detection have been discussed. The simplest, based on histogram analysis, which do not depend on explicit detection of the collimation boundaries, do not generally perform as well as more sophisticated algorithms that identify FOREGROUND using *a priori* knowledge of the collimation process. We know a good deal about collimation. For example, we normally expect the collimator blades to be either orthogonal or parallel to one another. We also expect certain characteristics of the edge transition between FOREGROUND and other regions. The first step to detecting collimation blades is therefore to identify pixels in the image coinciding with edge transitions. There are many methods to detect edges in digital images. The detailed methods can be found in general image processing texts. The output of edge detection algorithms is an image in which only edge transitions have high pixel values. Such an edge detection image will find the collimation blades along with many other edges in the image unrelated to the collimation blades. For example, bones generally have clearly delineated edges. HOUGH TRANSFORMS have the useful property of mapping straight lines into a single point in the transform space. Since most collimators have straight edges, this can be a powerful tool to eliminate from further consideration the vast majority of non-collimator features detected in the edge detection image. This greatly simplifies and speeds up the search for collimator blades. It also has the advantage over "edge-walking" techniques of being much less sensitive to gaps or irregularities in the edge profile of the collimation boundary. However, one disadvantage of the HOUGH TRANSFORM approach is that it precludes a search for collimators with non-straight edges. For example, round collimators have sometimes been used for a small number of examinations. A simple solution is for radiographers to use only linear collimation.

The general result of the above analysis is to identify a superset of candidate collimation boundaries that includes the actual collimation blades as well as some number of "false positives." Having a set of candidate collimation boundaries, there are three approaches commonly used to select the best estimate of the actual collimator blade configuration: (1) rule-based, (2) statistics-based, and (3) figure-of-merit-based. The rule-based approach employs a set of fixed rules based on an *a priori* knowledge of the possible configurations of actual collimator blades. These can include the number, location, and relative orientation of the collimator blades. However, a strictly rule-based approach can lack robustness because of the wide variability in clinical radiographs and the complexity of coordinating a large set of rules. In statistics-based approaches, the distribution of a likelihood function needs to be estimated from a set of training data. Success depends critically on the training set. Gathering a representative training set and training the algorithm can be a time-consuming process. The robustness of the result will depend heavily on the correlation between the training set and the set of images encountered in practice. The more robust figure-of-merit-based approach shares the same principles as the statistics-based approach in that a likelihood function is utilized to yield estimates while incorporating the *a priori* knowledge of the rule-based approach. In particular, this hybrid process uses rule-based decision and fuzzy-logic score-based evidence accumulation, which offers a more robust reasoning process than either alone. The characteristics of collimation that can be utilized in this approach include geometric properties such as orthogonality, parallelism, convexity, and aspect ratio, as well as spatial properties such as centrality, occupancy, perimeter, and boundary contrast. The evidence accumulation process lends itself well to the computation of a continuously valued figure-of-merit associated with a particular combination of blades, minimizing the number of binary thresholds and heuristics that must be determined during creation, tuning, and updating of the algorithm. Like other statistics-based approaches, the figure-of-merit must be refined by validation against image datasets.

In practice, figure-of-merit-based collimation detection has been demonstrated to be accurate in 97% of a set of images studied. This result was based on a study of 7700 actual clinical images not used for algorithm development or training (Luo, 1997). Two kinds of errors occur: false negatives and false positives. False negative errors occur when collimation blades in the image are undetected, resulting in collimated areas of the image that may be interpreted as part of the clinically important part of the image and which may not be masked in the DISPLAY-READY IMAGE. False negatives were found to occur in 2.5% of the images studied. False positives correspond to the false detection of features in the image which are not collimation blades. This can have the serious consequence of occluding diagnostically important areas of the image when masking is applied to the DISPLAY-READY IMAGE. This was found to occur in 0.5% of the 7700 images studied. Improvements in these techniques continue to be made. It is likely that the performance of current algorithms exceeds that reported in this somewhat dated study.

Since there will always be the possibility of FOREGROUND detection errors, solutions must be provided to deal with collimation detection errors as part of the quality assurance by the radiographic technologist. These should include both the ability to remove collimation masking from the DISPLAY-READY IMAGE as well as allowing collimation masking to be added as needed to occlude bright collimated areas in the DISPLAY-READY IMAGE. The implementation of these solutions must be intuitive and easy since they are expected to be used infrequently.

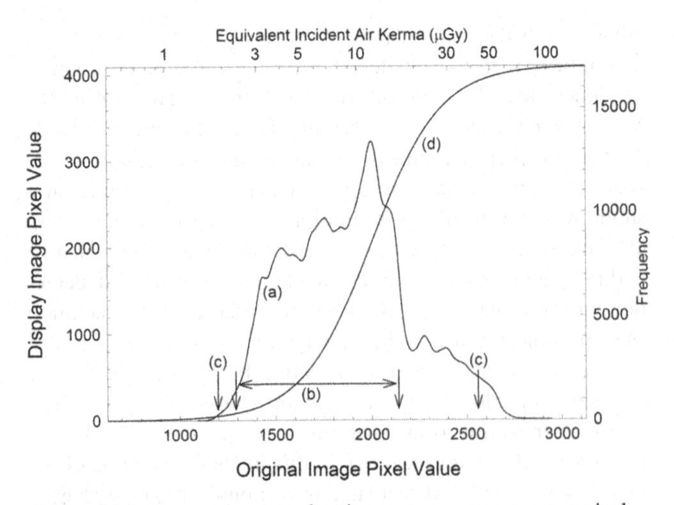

Figure 28.6 Tone-scale curve showing DISPLAY-READY IMAGE pixel value as a function of ORIGINAL IMAGE pixel value.
(a) smoothed histogram (second ordinate) of the ORIGINAL IMAGE pixel values used for image analysis, (b) primary range of pixel values within which maximum contrast is desired, (c) extensions of the pixel range for which reduced contrast is acceptable, (d) tone-scale curve based on the four reference pixel values shown.

28.3.2.1 Histogram analysis

By accurately detecting BACKGROUND and FOREGROUND areas of the ORIGINAL IMAGE, the remaining areas can be considered as clinically relevant for histogram analysis. Depending on the acquisition technology, there may be one or more such areas. Multiple exposures are sometimes done on CR imaging plates. For example, AP, lateral, and oblique projections of a finger, toe, wrist, or ankle may be exposed on a single CR imaging plate with each exposure having its own collimation borders which may or may not overlap. Multiple exposures are generally not possible in FP-DR or CCD-DR because the detector is read out and reset after each exposure. While special software has been described that detects background, foreground, and the clinical areas of multiple exposures, I will confine the following discussion of histogram analysis to the case of single-exposure images.

The goal of histogram analysis is to identify the range of ORIGINAL-IMAGE pixel values that should be used to produce the best possible rendering for each radiographic image. This is most simply described in terms of fiducial points describing the upper and lower limits of that range (i.e. two pixel-values in the ORIGINAL IMAGE). Some algorithms produce four fiducial points corresponding to a range in which maximum display contrast is to be applied, as well as an extended range over which reduced contrast is acceptable. These fiducial points must be determined for each image. The analysis to do this is often one of the most proprietary aspects of display processing. Methods are commonly kept as trade secrets. Therefore, little has been written or patented about the detailed methods. Nonetheless, general methods have been described.

Methods for histogram analysis are largely heuristic, being selected for accuracy and robustness over an available set of images. Histograms computed directly from the pixel values are often jagged in appearance with many minor peaks and troughs. It is often useful to smooth histograms before further analysis. This helps to identify and interpret histogram peaks more robustly. A smoothed histogram for the non-BACKGROUND and non-FOREGROUND areas of the ORIGINAL IMAGE shown in Figure 28.3 is shown in Figure 28.6. Comparison of Figures 28.5 and 28.6 shows the effects of histogram smoothing. Comparison to Figure 28.4 suggests how difficult analysis of the histogram of the entire image would be. In addition to pixel-value histograms, more sophisticated histograms are sometimes used. For example, "level-crossing histograms" are used, which are sensitive to structure in the image. Level-crossing histograms evaluate images one line at a time, counting the number of times the pixel values of a line cross each pixel value. To count as a level-crossing, minimal deviations below or above the value must have occurred. Level-crossing histograms discriminate against large uniform areas in an image, such as prostheses or foreground and background areas that were not already detected and eliminated from consideration (Lee, 1997a). Redundancies in removing non-clinical areas from consideration when determining the range of pixel values used for optimal rendering may seem unnecessary; however, they can significantly increase the robustness of the overall algorithms.

The shapes of ORIGINAL-IMAGE histograms vary considerably, particularly among different body parts and projections. Therefore, body part and projection dependent heuristic rules are often used to compute the fiducial points. The histogram characteristics commonly used may include, for example, percentages of the cumulative histogram, location relative to histogram peaks, histogram values relative to the largest peak, as well as points of inflection or maximum slope. Four fiducial points are shown in Figure 28.6, corresponding to the range (b) in which maximum display contrast is to be applied, as well as an extended range, points (c), over which reduced contrast is acceptable. These points are descriptive of this particular image and are passed to subsequent algorithms for determining the tone-scale curve, edge-enhancement, and equalization.

28.3.2.2 Tone-scale development and placement

It has been found that the best perceptual quality is obtained by providing a variety of shapes for the tone-scale that maps ORIGINAL IMAGE pixel values to DISPLAY-READY pixel values. This is a substantial advantage of digital acquisition over the fixed curve shapes and speed of SCREEN-FILM radiography. Tone-scale rendering may be based on pre-defined curve shapes, on curves with shaping parameters, or on a combination of the two. The final selection of the optimal curve shape that will be used for each body part and projection is often done with the assistance of the customer.

The choice of curve shape determines the way in which the available perceptual contrast will be allocated over the range of pixel values in the ORIGINAL IMAGE that has been identified by histogram analysis as the most important for each image. A simple example is a curve that provides equal perceptual contrast over a specified range with reduced perceptual contrast beyond that range. This approach has been successfully applied in commercially available CR and FP-DR systems.

Optimal rendering of digital images requires a tone-scale that properly incorporates the characteristics of the human visual system while identifying the most important pixel-value range of the ORIGINAL IMAGE. A very useful way of quantifying a range of luminance values in a displayed image is in terms of a scale of luminance values that correspond to equal changes in perceived brightness. Such scales have been developed, but are somewhat dependent on the details of the experimental technique. As mentioned above, a particular such scale has been applied to the standardization of medical imaging displays by DICOM and has been reviewed (Hemminger, 1995; Blume, 1996). Medical imaging display devices are now standardized (DICOM part 14) to produce a range of luminance levels that are spaced so that they produce equal perceptual brightness intervals (Pizer, 1980; Sezan, 1987). If the pixel values of the DISPLAY-READY IMAGE are linearly interpolated into the available range of pixel values for such displays, a perceptually linear response is obtained. The interpolation is necessary because the number of pixel values used to encode the DISPLAY-READY IMAGE may be different from the number of display levels available on the display. For example, DISPLAY-READY IMAGES are often represented as 12 bits (0–4095) while displays are often limited to 8 or 10 bits (0–255 or 0–1023).

If the DISPLAY-READY IMAGE is to be displayed by a DICOM part 14 compliant display, a simple linear relationship between the ORIGINAL IMAGE and DISPLAY-READY IMAGE pixel values will provide a perceptually optimal tone-scale rendering of pixel value differences in the selected range of the ORIGINAL IMAGE (Lee, 1997b). For this to be the optimal tonal rendering, two things are necessary. First, it is necessary that equal changes in pixel value of the ORIGINAL IMAGE are of equal importance. Recall that encoding of the ORIGINAL IMAGE is not yet standardized. However, we have argued that the same feature will correspond to equal changes in pixel value when LOGARITHMIC ENCODING of the ORIGINAL IMAGE is chosen. Second, it is necessary that the correct range of pixel values in the ORIGINAL IMAGE has been selected. However, a strictly linear relationship between the ORIGINAL IMAGE and the DISPLAY-READY IMAGE would result in the abrupt truncation of the linear mapping at the upper and lower pixel value limits in the DISPLAY-READY IMAGE. Pixel values in the ORIGINAL IMAGE that are above the upper limit would be rendered at the darkest pixel value of the DISPLAY-READY IMAGE and those below the lower limit would be rendered at the brightest pixel value of the DISPLAY-READY IMAGE. This produces an unnecessary image artifact known as clipping, particularly if the pixel-value range is too small. A much more pleasing and useful image is produced by gradually reducing the perceptual contrast outside the selected range of the ORIGINAL IMAGE. While this has the disadvantage of reducing contrast within the selected range, it has the advantage of preserving the visibility of pixel values beyond this range, albeit at a still lower contrast. This choice produces a more robust tone-scale choice that is tolerant of small errors in the selection of the optimal display range in the ORIGINAL IMAGE.

28.3.3 Edge restoration and enhancement

There are several processes in the acquisition of projection radiographs that lead to a loss of sharpness or edge detail in the resulting images (Sprawls, 1995). The first is the blurring associated with the finite size of the focal spot of the X-ray tube and the distance between the object being radiographed and the imaging detector. A second loss of sharpness is the blurring associated with many of the most common imaging detectors. This loss depends significantly on the detector technology. All phosphor-based technologies introduce blurring of the image due to lateral light spread in the phosphor. Some phosphors, particularly those having a needle-like structure (evaporated layers of CsI and CsBr), exhibit less lateral light spread than coated powder phosphors consisting of powdered phosphor in a transparent binder, such as BaFBrI commonly used for CR. Therefore, depending on the acquisition technology, the need for edge restoration will vary. It must be remembered that in SCREEN-FILM imaging, which uses a coated powder phosphor as the primary detector, no edge-restoration was possible. Therefore, radiologists trained on SCREEN-FILM may prefer less edge restoration than those trained on digital technologies. Thus, optimal edge restoration and enhancement is likely to be a matter of individual preference that can be influenced by training. It is important to recognize that great care must be exercised in selecting edge restoration and enhancement as they can influence diagnostic interpretation.

28.3.3.1 Unsharp masking

Edge enhancement can be achieved in several ways. Perhaps the simplest conceptually is UNSHARP MASKING. Before digital imaging, UNSHARP MASKING was done photographically by creating an unsharp (intentionally blurred) negative (inverted gray scale) copy of an ORIGINAL IMAGE. A print was then made by superimposing the ORIGINAL IMAGE and the mask (the unsharp negative copy) onto a high-contrast positive (copy) film.

In digital imaging, the unsharp copy of the ORIGINAL IMAGE can be created by averaging the value of a range of pixels surrounding each pixel in the image. More sophisticated methods weight the pixel values as they are averaged depending on their distance from the original pixel. In any case, the resulting image is an unsharp version of the original. Edge enhancement

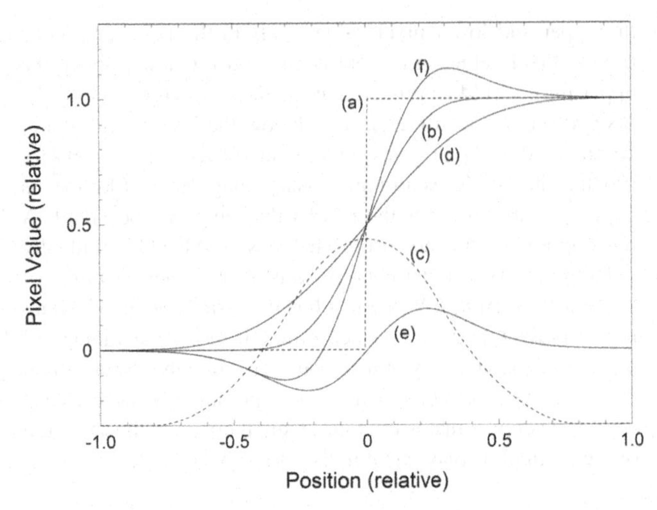

Figure 28.7 One-dimensional illustration of unsharp masking. Relative pixel value as a function of position. (a) ideal edge image, (b) ORIGINAL IMAGE, (c) BLURRING KERNEL used to create the UNSHARP MASK IMAGE, (d) UNSHARP MASK IMAGE, (e) ORIGINAL IMAGE minus the UNSHARP MASK IMAGE, (f) EDGE-ENHANCED IMAGE.

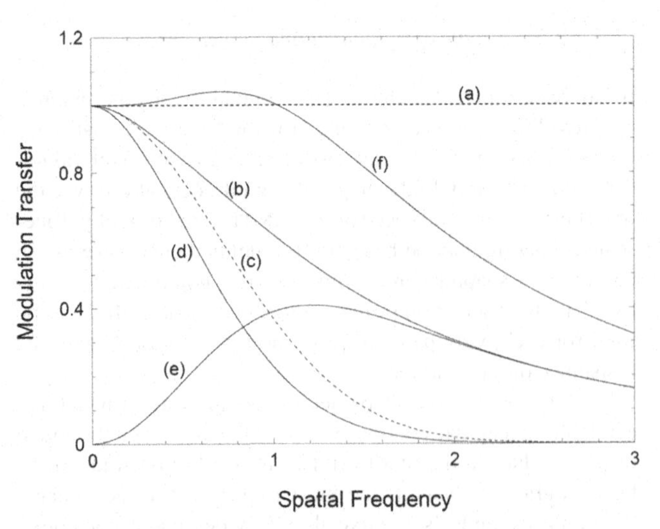

Figure 28.8 One-dimensional illustration of unsharp masking. Relative modulation or MTF as a function of spatial frequency. (a) relative modulation of an ideal edge image, (b) MTF for the ORIGINAL IMAGE, (c) BLURRING KERNEL used to create the UNSHARP MASK IMAGE, (d) MTF for the UNSHARP MASK IMAGE, (e) MTF for the ORIGINAL IMAGE minus the UNSHARP MASK IMAGE, (f) MTF for the EDGE-ENHANCED IMAGE.

is achieved by first subtracting the pixel values of the UNSHARP MASK IMAGE from the ORIGINAL IMAGE to produce an EDGE-ONLY IMAGE. A multiple of the EDGE-ONLY IMAGE is then added to the ORIGINAL IMAGE to enhance edges. This can best be illustrated with a one-dimensional example. For the purposes of this illustration, pixel values will be replaced by continuous curves which are more easily interpreted.

Consider the ideal image of a one-dimensional object that contains an abrupt edge as represented by curve (a) in Figure 28.7. Because the image acquisition process is generally characterized by several sources of image blur, the ORIGINAL IMAGE might appear as illustrated by curve (b). The purpose of edge restoration and enhancement is to create an image that is closer to the ideal image. The first step in digital UNSHARP MASKING is to define a BLURRING KERNEL. An example of a BLURRING KERNEL is shown as curve (c) in Figure 28.7. The BLURRING KERNEL (c) is convolved with the ORIGINAL IMAGE (b) to create a further blurred version of the ORIGINAL IMAGE, the UNSHARP MASK IMAGE, shown as curve (d). If this UNSHARP MASK IMAGE (d) is subtracted from the ORIGINAL IMAGE (b), the resulting EDGE-ONLY IMAGE (e) contains only the edges of the ORIGINAL IMAGE (b). This zero-mean image can be weighted (multiplied by a constant, in this case 1.0) and added to the ORIGINAL IMAGE (b) to produce an EDGE-ENHANCED IMAGE illustrated as curve (f). The EDGE-ENHANCED IMAGE (f) is characterized by an edge gradient that is higher than that of the ORIGINAL IMAGE (b) and closer to that of the ideal image (a). It is also characterized by undershoot and overshoot artifacts. These are the places where curve (f) lies below curve (a) to the left of the edge and where curve (f) lies above curve (a) to the right of the edge. These artifacts are characteristic of UNSHARP MASK edge enhancement. The useful amount of edge enhancement is limited by this artifact.

It is useful to also consider this example in terms of the spatial frequency characteristics of each of the images, shown in Figure 28.8. The ideal image would have required an image acquisition system with ideal frequency response that would be characterized by a modulation transfer function (MTF) of one at all spatial frequencies, curve (a). The sources of image blur in the actual acquisition system result in an ORIGINAL IMAGE characterized by a lower MTF, particularly at the higher frequencies, curve (b). The BLURRING KERNEL has a spatial-frequency response shown as curve (c). Convolution of the BLURRING KERNEL with the ORIGINAL IMAGE corresponds to the multiplication of the corresponding MTF curves in the spatial-frequency domain. The product of curves (b) and (c) is the MTF of the UNSHARP MASK IMAGE and is shown as curve (d). As expected the UNSHARP MASK IMAGE is characterized by a lower MTF than that of the ORIGINAL IMAGE. The MTF of the EDGE-ONLY IMAGE is obtained by subtracting that of the UNSHARP MASK IMAGE (d) from that of the ORIGINAL IMAGE (b). This is shown as curve (e). The EDGE-ONLY IMAGE is characterized by having suppressed low frequencies, approaching zero at zero frequency. Finally, when the EDGE-ONLY IMAGE is added to the ORIGINAL IMAGE, the MTF values (e) and (b) add to produce the MTF of the EDGE-ENHANCED IMAGE (f). Comparing the MTF of the EDGE-ENHANCED IMAGE (f) to that of the ORIGINAL IMAGE (b) shows that the response at higher frequencies is increased. It also shows that the MTF first increases with increasing frequency leading to a peak in the MTF curve at a non-zero frequency. This peak in the MTF curve is directly related to the undershoot and overshoot artifacts in the spatial domain shown in curve (f) of Figure 28.7.

A radiographic imaging example of UNSHARP MASKING is shown in Figure 28.9. This is the central part of an AP projection of a knee. In practice, the ideal image is unavailable and the blur associated with image acquisition is not completely characterized. Therefore, we begin with the ORIGINAL IMAGE, shown in Figure 28.9(a). Window and level have been adjusted to give a pleasing presentation of the image data. The UNSHARP MASK image, Figure 28.9(b), is produced by blurring this image with a square kernel. The size of the kernel is indicated by the

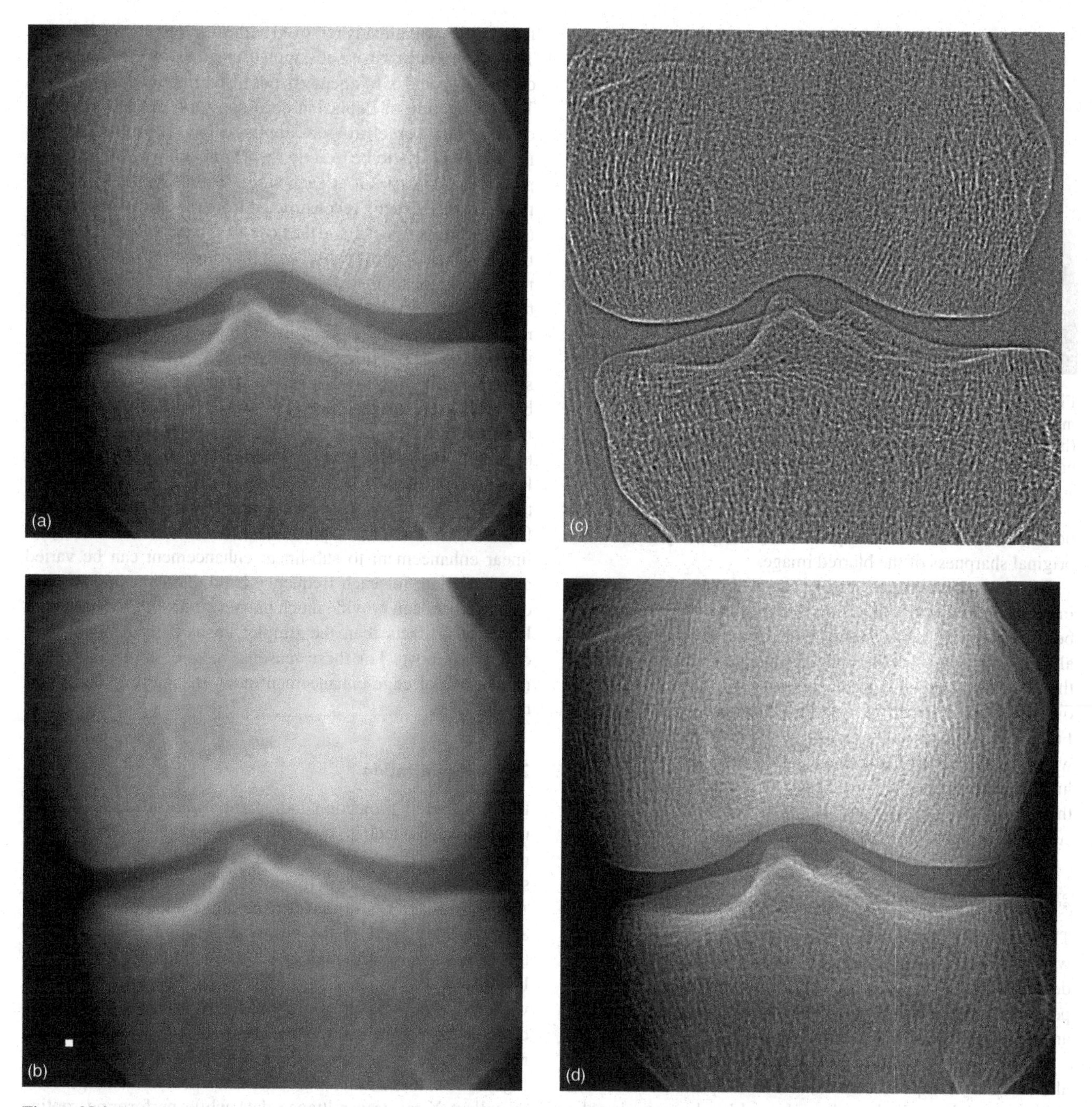

Figure 28.9 Radiographic image illustration of edge enhancement. (a) ORIGINAL IMAGE, (b) UNSHARP MASK IMAGE with the size of the masking kernel illustrated by the white square, (c) EDGE-ONLY IMAGE obtained by subtracting UNSHARP MASK IMAGE from ORIGINAL IMAGE, (d) EDGE-ENHANCED IMAGE.

white square in the corner of Figure 28.9(b). The window and level settings are the same as for Figure 28.9(a). Subtracting the UNSHARP MASK IMAGE from the ORIGINAL IMAGE produces an EDGE-ONLY IMAGE, Figure 28.9(c). This zero-mean image has been offset and enhanced in contrast for display. Finally, pixel values in the zero-mean EDGE-ONLY IMAGE are doubled and added to the ORIGINAL IMAGE to produce the EDGE-ENHANCED IMAGE shown as Figure 28.9(d). The window and level settings for Figure 28.9(d) are the same as for the ORIGINAL IMAGE and the UNSHARP MASK IMAGE. Edge enhancement has increased the edge contrast and hence the apparent sharpness of edges in Figure 28.9(d). Keeping the window and level settings the same for the images in Figures 28.9(a), (b), and (d) is important to illustrate the influence of edge contrast on the perception of overall contrast in these images. Edge enhancement has increased the visibility of the bone trabecula and subtle structures while leaving the large-area (macro) contrast of the image unchanged. This kind of edge enhancement is useful when the visibility of subtle details is limited by the contrast sensitivity of the viewer rather than by image noise. The ability to restore edges in blurred images is often limited by the loss of signal-to-noise ratio (SNR) intrinsic to many blurring processes. If there were

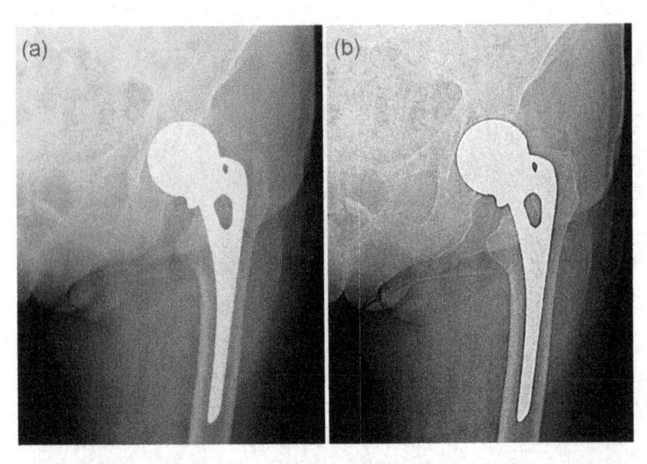

Figure 28.10 Overshoot/undershoot artifact associated with unsharp masking. (a) ORIGINAL IMAGE free of artifact, (b) EDGE-ENHANCED IMAGE showing dark-band artifact bordering hip prosthesis resulting from excessive edge enhancement with UNSHARP MASKING.

no loss of SNR, it would be possible to completely restore the original sharpness of the blurred image.

From a perceptual perspective, the visibility of some noise in an image is a good indication that further enhancement will be of diminishing value. The undershoot and overshoot artifacts already mentioned can also limit the amount of unsharp masking that should be applied to medical images. Visible undershoot or overshoot artifacts can have serious diagnostic consequences. Figure 28.10 shows an example of a hip joint prosthesis in which an undershoot artifact has caused a dark (apparently radiolucent) area around the prosthesis. This can be mistaken for the clinical finding of a gap between the prosthesis and the surrounding bone, a serious medical condition.

28.3.3.2 Advanced frequency enhancement techniques

There are several ways in which edge enhancement achieved with UNSHARP MASKING can be improved. For example, the degree of edge enhancement can be pixel value dependent. This permits reduced edge enhancement where it is desirable, such as areas of higher noise or where edge enhancement is less needed.

A more sophisticated method for edge enhancement which allows more complete control of the frequencies at which enhancement is applied can be achieved by decomposing the image into a set of images each containing information from a discrete band of frequencies (Couwenhoven, 2004). One well-known method for doing this, Laplacian decomposition, employs a weighted 5×5 blurring kernel that is applied to the ORIGINAL IMAGE to produce a LOW-PASS IMAGE. This image is subtracted from the ORIGINAL IMAGE to create a HIGH-PASS IMAGE. At this point, the ORIGINAL IMAGE can be exactly reconstructed by simply adding the LOW-PASS IMAGE and HIGH-PASS IMAGE. To continue the frequency decomposition, the LOW-PASS IMAGE is then subsampled to reduce the number of pixels in each dimension by a factor of two. The same procedure that was applied to the ORIGINAL IMAGE is then repeated on the subsampled image. This produces a new pair of LOW-PASS and HIGH-PASS IMAGES. The resulting HIGH-PASS IMAGE is a BAND-PASS IMAGE of the ORIGINAL IMAGE because the highest band of frequencies was already removed in the first decomposition step. The LOW-PASS IMAGE is subsampled, and the process repeated to create a pyramid of frequency bands that differ by a factor of two. The key feature of Laplacian decomposition is that the choice of smoothing kernel and subsampling allows the full-resolution LOW-PASS IMAGE to be exactly reconstructed from the subsampled LOW-PASS IMAGE at each stage. Therefore, the ORIGINAL IMAGE can be exactly reconstructed from the HIGH-PASS IMAGE, the BAND-PASS IMAGES, and the LOW-PASS IMAGE. The HIGH-PASS IMAGE and each of the subsequent BAND-PASS IMAGES have zero mean. Only the final LOW-PASS IMAGE retains information about the mean pixel values of the ORIGINAL IMAGE. Edge enhancement is achieved by selectively enhancing the HIGH-PASS and BAND-PASS IMAGES prior to reconstruction. This can be done by simply multiplying pixel values in selected BAND-PASS IMAGES by a factor greater than one (Vuylsteke, 1994). Edge enhancement can be further refined by applying non-linear enhancement to pixel values of the HIGH-PASS and BAND-PASS IMAGES. Non-linear enhancement allows low-contrast (difficult to see) features to be selectively enhanced, while high-contrast (easily visible) features are not. The amplitude for the transition from super-linear enhancement to sub-linear enhancement can be varied as appropriate for each frequency band. This method of edge enhancement can provide much greater control as well as lower levels of artifacts than the simpler unsharp masking method described above. For these reasons, most commercial implementations of edge enhancement now use multiple frequency bands.

28.3.4 Equalization

Prior to digital projection radiography, the difficult compromise between adequate CONTRAST and adequate LATITUDE for many examinations was accepted as an intrinsic limitation of SCREEN-FILM technology. Images that provide adequate CONTRAST for diagnosis in mid-density areas often fail to provide adequate CONTRAST in the under- and over-exposed areas of the same image. Although practitioners appreciated that the limited LATITUDE of X-ray films could not encompass the wide range of exposures in many X-ray examinations, it was difficult to acquire quantitative data on the exposure latitude requirements for a representative population of patients. As digital detectors came into clinical use it became possible to collect X-ray transmittance data while performing routine examinations. For example, the required LATITUDE for chest radiography was systematically studied (Van Metter, 1992). This study quantified the range of X-ray transmittances showing that for many patients diagnostically important areas of the image exceeded the LATITUDE of commonly used SCREEN-FILM systems, confirming that not all areas of the chest could be optimally displayed in a single image. The clinical implications of this result were reported by Slone *et al.* (Slone, 1996), who studied the effects of varying the placement of the tone-scale curves on the clinical utility of chest radiographs. This research showed that the tone-scale curves of a commonly used SCREEN-FILM combination were unable to provide optimal display of all areas of interest in chest radiographs simultaneously. More specifically, they showed that the position of the tone-scale curve needed to be varied to yield optimal imaging of different clinical areas (unobstructed lungs, retrocardiac, mediastinum, and subdiaphragm).

The need for some form of exposure equalization was therefore widely recognized. A variety of mechanical approaches were applied with limited success and practicality. X-ray compensation filters were devised that reduced the range of X-ray transmittance of the body parts imaged in common radiographic examinations. Filters placed near the X-ray source were demonstrated (Edholm, 1971) and were commercially available for a range of radiographic examinations including chest, scoliosis, extremities, lumbar spine, and cervical spine. However, the disadvantage of fixed mechanical filters was that they were made to compensate a single well-defined anatomical shape and were sub-optimal when the anatomy under examination deviated from that shape. A further disadvantage was that they could be inadvertently left in the X-ray beam when body parts for which they were not intended were imaged.

Some of the limitations of fixed compensation filters can be mitigated by the use of filters that are tailored to the anatomical features of each examination. To this end, electro-mechanical devices to address the CONTRAST/LATITUDE trade-off, by modulating large-area exposure as a function of X-ray attenuation, were successfully introduced to practice by several groups in the early 1980s. These approaches can be usefully classified according to whether the exposing beam was the full area of the image (Hasegawa, 1984), a fan beam (Vlasbloem, 1988), or a scanning spot (Plewes, 1982). While many laboratory prototypes were investigated, the only successful commercial product was the Advanced Multiple-Beam Equalization Radiography System (AMBER, Oldelft, The Netherlands). It used a fan beam within which 22 individual attenuators achieved regional control of the scanning spot. This approach had the image quality advantages of raster modulation with the tube loading advantages of a fan beam. The image quality advantages of AMBER chest radiographs were demonstrated for simulated interstitial disease (Schultz Kool, 1994) and lung nodules (Schultz Kool, 1988). The success of pre-exposure X-ray equalization therefore naturally pointed toward the opportunity for display equalization in digital projection radiography, where the wide latitude of many digital detectors could be exploited.

Without equalization, a tone-scale mapping, $\rho(x)$, is applied to an ORIGINAL IMAGE, or an edge-enhanced ORIGINAL IMAGE. The output image can then be expressed as

$$x_D(i,j) = \rho(x_A(i,j)) \tag{28.1}$$

where $x_A(i,j)$ is the pixel value of the ORIGINAL IMAGE or the edge-enhanced ORIGINAL IMAGE at location i,j and $x_D(i,j)$ is the DISPLAY-READY IMAGE pixel value at the same pixel.

Equalization image processing is accomplished by dividing the image into very-low-frequency and residual (high-frequency) component images. The contrast of the very-low-frequency image is reduced, thereby decreasing the latitude (range of exposures) of the image. The contrast of the high-frequency image is maintained or slightly enhanced to preserve the visual appearance of image detail. Finally, the very-low- and high-frequency images are recombined and the tone-scale mapping applied. The pixel values $x_A(i,j)$ are replaced by the EQUALIZED-IMAGE pixel values

$$x_A'(i,j) = a \bullet (x_A(i,j) \otimes K) + (1-\alpha) \bullet x_{mid} + \beta \bullet \{x_A(i,j) - (x_A(i,j) \otimes K)\} \tag{28.2}$$

where K is a smoothing kernel, $\otimes$ is the convolution operator, α is the very-low-frequency gain, β is the high-frequency gain, and x_{mid} is the pixel value in the ORIGINAL IMAGE or edge-enhanced ORIGINAL IMAGE that will remain invariant under the transformation. Typical values for α and β are 0.5 and 1.1, respectively. The DISPLAY-READY IMAGE pixel values are then obtained as

$$x_D'(i,j) = \rho(x_A'(i,j)) \tag{28.3}$$

where $\rho(x)$ is the same tone-scale defined above (Van Metter, 1999).

Figure 28.11 illustrates the effect of equalization on the lateral projection image of the cervical spine. Figure 28.11(a) shows the DISPLAY-READY IMAGE without equalization. There are seven cervical vertebrae (C1–C7) which need to be seen in this examination. It is also important to see the first thoracic vertebra (T1). In this image only the first five cervical vertebrae are well visualized. The lower cervical vertebrae and the first thoracic vertebra are obscured by the very low X-ray transmittance of the shoulder and thorax. However, because of the intrinsically wide latitude of the digital capture technology, the required information is present in the ORIGINAL IMAGE but poorly displayed. Shifting the tone-scale mapping, $\rho(x)$, would reveal significant detail in this area, but render the rest of the image too dark. Figure 28.11(b) shows the result of equalization. The contrast of details in the already well visualized areas of the image is preserved; however, the areas of the image which previously appeared underexposed (too light) are now well visualized. In particular, the details of the lower cervical vertebrae and the first thoracic vertebra are now evident. It is important to emphasize that these details were present in the ORIGINAL IMAGE but that equalization has enabled them to be appreciated in the DISPLAY-READY IMAGE. This is particularly important since only the DISPLAY-READY IMAGE is retained for long-term storage in most picture archiving and communication systems (PACS). Therefore, once the ORIGINAL IMAGE is purged from the acquisition device, only the DISPLAY-READY IMAGE is accessible. With inadequate display processing, information inherent in the ORIGINAL IMAGE can be permanently lost. Finally, equalized images have been shown to reduce the time needed by radiologists for diagnostic interpretation of clinical chest images (Krupinski, 2001).

28.3.5 Surround masking

Having detected the collimation configuration for each image, it is useful to apply masking to the collimated areas of the image. The purpose of masking is to darken the non-clinical areas of the image so the amount of unnecessary light reaching the observer's eye is reduced. This has the advantage of reducing flare in the viewer's optical system and of allowing the adaptation level to be more appropriate to the areas of clinical interest in the image. Masking of the surround has been done with both replacement with black (minimal luminance) display values and with gray display values intended to approximate the

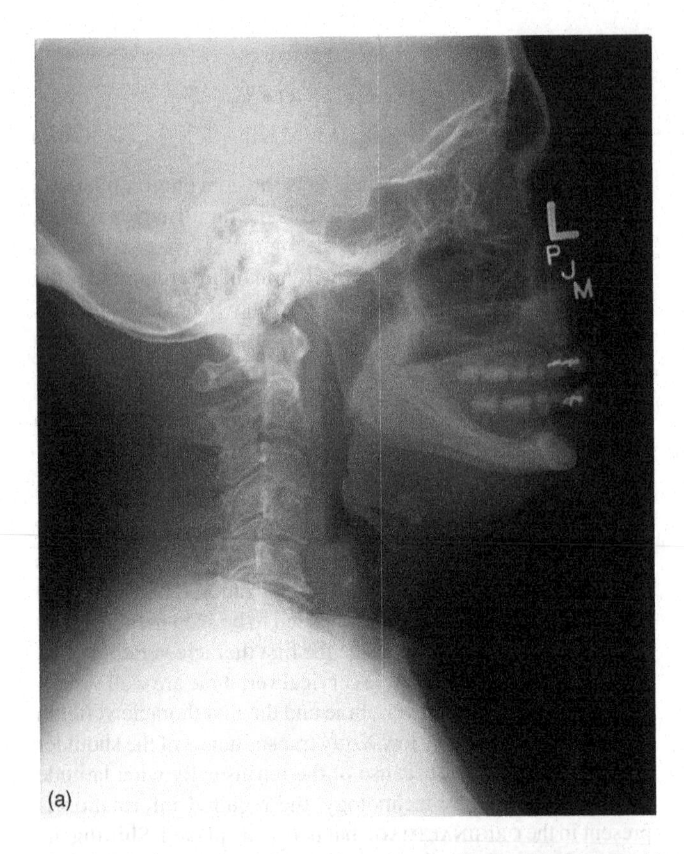

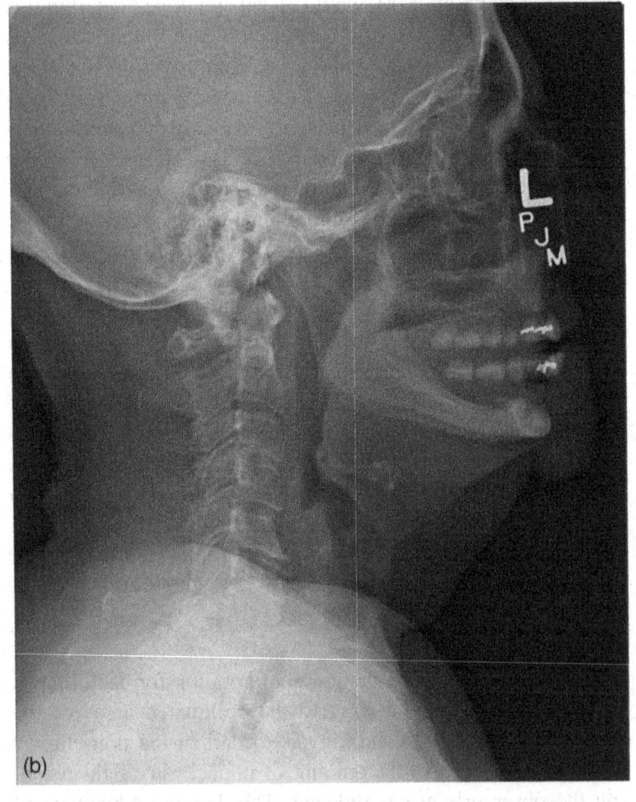

Figure 28.11 Equalization processing. (a) DISPLAY-READY IMAGE without equalization processing, (b) DISPLAY-READY IMAGE with equalization processing shows increased visibility of vertebrae and skin line without loss of image detail.

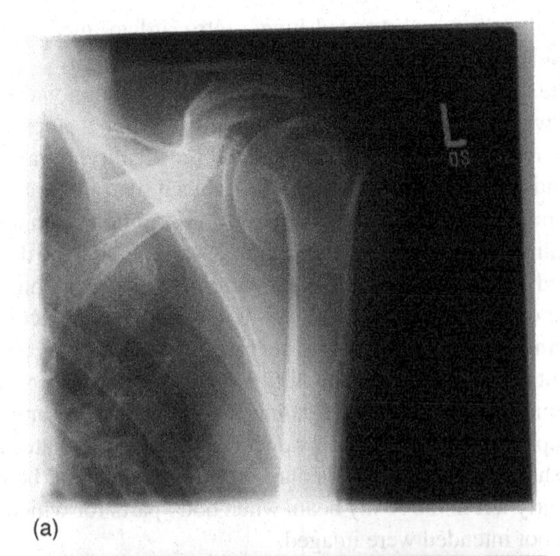

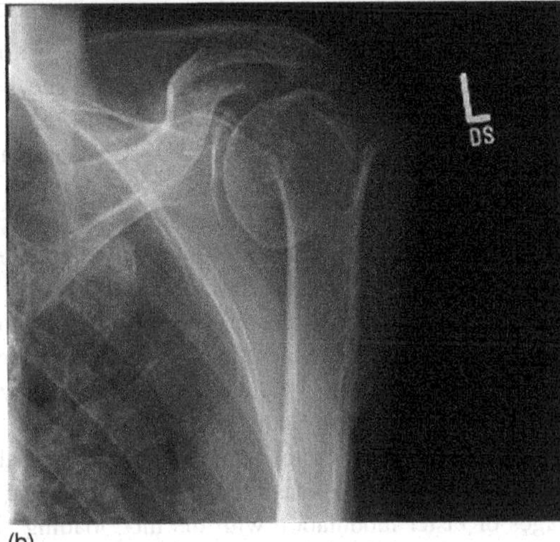

Figure 28.12 Effect of surround masking on image quality. (a) DISPLAY-READY IMAGE without surround masking, (b) DISPLAY-READY IMAGE with black surround mask.

average luminance of the clinical areas of the displayed image. Both approaches dramatically improve the perceptual quality of the display for many images. The effect is so intuitively obvious that no definitive experiments have been done to compare them.

Figure 28.12 compares a masked and unmasked display of the same image. The apparent contrast of details in the darker parts of the image is reduced in the unmasked image because of viewing flare associated with the bright surround. The effect of masking is less dramatic on the printed page than for images displayed for diagnostic viewing, where the maximum luminance of the display device is higher and the ambient illuminance is lower than a normal reading environment. Failure to control flare light caused by non-diagnostic areas of the image or from other sources of unnecessary light in the viewing environment can have an overwhelming impact on the quality of the displayed image.

28.3.6 Grid line removal

When stationary grids are used to reduce X-ray scatter, the grid is placed between the patient and the image acquisition device. As a result, the individual septa of the grid are sometimes imaged. Grid lines caused by stationary grids can often be easily seen in SCREEN-FILM images. They do not normally create problems for diagnostic interpretation because they appear as a high-frequency, regular pattern unlike anatomical or pathological details of interest to radiologists. The situation can be different for digital projection radiography because the regular pattern of grid lines can interact with the regularly spaced pixels to produce quasi-periodic artifacts at much lower spatial frequencies. This effect is known as a Moire pattern (Belykh, 2001).

To avoid Moire artifacts, it is necessary to carefully choose the frequency of stationary grids so they are either well below the limiting "NYQUIST" FREQUENCY (½ the sampling frequency) of the acquisition device or well above the spatial frequencies to which the device has appreciable response. In the former case, the grid frequency is below the NYQUIST FREQUENCY of the acquisition device and is correctly imaged. In the latter case, the frequency of the grid is high enough that the image of the grid is adequately blurred so that the residual Moire artifact is below the visual contrast threshold.

However, even when the grid is at a low frequency that is correctly imaged by the acquisition device, a secondary problem can occur when the image is displayed at reduced pixel resolution. Many display devices subsample digital images for lower-resolution display without filtering the frequency content of the images. When this is done, a grid which was below the NYQUIST FREQUENCY of the acquisition device can be above the NYQUIST FREQUENCY of the displayed image. Moire artifacts can occur, unless the image is filtered to eliminate frequency components above the NYQUIST FREQUENCY of the display. The purpose of grid line removal is to remove the grid lines from an image so that regardless of the display resolution used, no artifact will appear.

Grid line removal is effected by first detecting the presence of grid lines in the image data and then applying a notch filter to remove those spatial frequencies (Belykh, 2001). While some anatomical information is removed from the image, notch filters are not observed to cause objectionable image artifacts.

28.4 FUTURE OPPORTUNITIES AND DIRECTIONS

Image processing is a critical component of all digital projection image acquisition devices. High-quality image processing must automatically provide DISPLAY-READY IMAGES that are optimized for diagnostic interpretation and do so reliably and robustly. In addition to the ability to produce a high-quality diagnostic image, the most important measure of image processing is the fraction of images that require manual adjustment prior to acceptance for diagnostic viewing. Minimizing the fraction of images requiring manual adjustment and facilitating that adjustment when needed are two important areas of continuing research.

The most common current practice is to complete image processing within the acquisition system and to transmit a DISPLAY-READY IMAGE in the form of a DICOM "for display" image for diagnosis and archiving. In this process, the ORIGINAL IMAGE is discarded, which precludes reprocessing and quantitative use of the ORIGINAL IMAGE data. As increased computational power becomes ubiquitous, it is likely that image processing for display may move to the diagnostic workstation. This has several advantages:

- The ORIGINAL IMAGE would be preserved, providing the ability for alternate processing and for subsequent quantitative use of the ORIGINAL IMAGE.
- The image processing preferences for each viewer could be applied at the time images are displayed.
- When comparisons of images acquired at different times are needed, image processing can be applied in a manner that facilitates intercomparison. This may be as simple as ensuring that the images have similar contrast and brightness, but also could include difference-image and image overlay techniques.

Such a change would also force a re-examination of the attributes of the DICOM "for processing" image. When each acquisition system is processing the ORIGINAL IMAGE, the interpretation of the pixel values in that image need only be known internally. Wider utilization of the ORIGINAL IMAGE will require some method of pixel interpretation standardization. This may best be done by extension of the DICOM standard, perhaps in the same way that DICOM part 14 provided a standard pixel interpretation for display devices. As was the case for the part 14 standard, the particular standard selected will be less important than having a commonly agreed upon standard.

Another area of opportunity is the image processing and display of multi-component image sets. A simple example is dual-energy image in which two images are acquired with X-ray beams having different average X-ray energies. From such a pair of images, image processing can produce images in which bone or soft tissue is eliminated. Optimal image processing will require that the gray scale relationships between the resulting images be maintained. Utilization of these images will be facilitated by optimal display configurations to preserve the anatomical context between the component images. The promise of multi-component imaging techniques will be facilitated by the development of distributed image processing on standardized image sets.

In summary, the image processing of projection images is still in its infancy. The goal of restoring the image characteristics of SCREEN-FILM technology has largely been achieved. These characteristics include providing the expected tonal rendering, restoring the expected sharpness, compensating for the detector characteristics, and accommodating the image display characteristics. These are crucial to preserve the years of training already invested by radiologists that are needed for diagnostic interpretation. The first steps to go beyond replicating the familiar have also been taken. Among these, perhaps the most important include automatically compensating for exposure variability and changes in subject contrast, enhancing images specifically for the human visual system, and increasing the range of exposures that can be visualized in a single image without loss of detail contrast. Clearly, other challenges and opportunities lie ahead for the full potential of digital projection imaging to be realized.

REFERENCES

Barski, L., Senn, R. (1997). Determination of direct x-ray exposure regions in digital medical imaging, *US Patent* 5,606,587.

Belykh, I.N., Cornelius, C.W. (2001). Antiscatter stationary grid artifacts automated detection and removal in projection radiography images, *Proc SPIE*, **4322**, 1162–1166.

Blume, H. (1996). The ACR/NEMA proposal for a grey-scale display function standard, *Proc SPIE*, **2707**, 344–360.

Couwenhoven, M., Senn, R., Foos, D. (2004). Enhancement method that provides direct and independent control of fundamental attributes of image quality for radiographic imagery, *Proc SPIE*, **5367**, 474–481.

Edholm, P.R., Jacobson, B. (1971). Primary x-ray dodging, *Radiol*, **99**, 694–696.

Hasegawa, B.H. *et al.* (1984). Application of a digital beam attenuator to chest radiography, *Proc SPIE*, **486**, 2–7.

Hejazi, S., Trauernicht, D.P. (1997). System considerations in CCD-based x-ray imaging for digital chest radiography and digital mammography, *Med Phys*, **24**, 287–297.

Hemminger, B.M., Johnston, R.E., Rolland, J.P., Muller, K.E. (1995). Introduction to the perceptual linearization of video display systems for medical image presentation, *J Digital Imaging*, **8**, 21–34.

Krupinski, E.A. *et al.* (2001). Enhanced visualization processing: effect on workflow, *Acad Radiol*, **8**, 1127–1133.

Lee, H., Barski, L., Senn, R. (1997a). Automatic tone scale adjustment using image activity measures, *US Patent* 5,633,511.

Lee, H.-C., Daly, S., Van Metter, R. (1997b). Visual optimization of radiographic tone scale, *Proc SPIE*, **3036**, 118–129.

Luo, J., Senn, R. (1997). Collimation detection for digital radiography, *Proc SPIE*, **3034**, 74–85.

Pizer, S.M., Chan, F.H. (1980). Evaluation of the number of discernible levels produced by a display. In *Information Processing in Medical Imaging*, DiPaola, R., Kahn, E., eds., Paris: Editions INSERM, pp. 561–580.

Plewes, D.B., Wandtke, J.C. (1982). A scanning equalization system for improved chest radiography, *Radiol*, **142**, 765–768.

Pluim, J.P.W., Reinhardt, J.M. (2007). *Medical Imaging 2007: Image Processing*. Bellingham, WA: SPIE Press.

Rowlands, J.A., Yorkston, J. (2000). Flat panel detectors for digital radiography. In *Handbook of Medical Imaging: Volume 1. Physics and Psychophysics*, Beutel, J., Kundel, H.L., Van Metter, R.L., eds., Bellingham, WA: SPIE Press, pp. 225–328.

Rowlands, J.A. (2002). The physics of computed radiography, *Phys Med Biol*, **47**, R123–R166.

Schultz Kool, L.J. *et al.* (1988). Advanced multiple-beam equalization radiography in chest radiology: a simulated nodule detection study, *Radiol*, **169**, 35–39.

Schultz Kool, L.J. *et al.* (1994). Optimization of chest films of equalization radiography (advanced multiple beam equalization radiography). Comparison by means of a receiver operating characteristic study of simulated nodular interstitial disease, *Invest Radiol*, **29**, 1020–1025.

Sezan, M.I., Yip, K.-L., Daly, S.J. (1987). Uniform perceptual quantization: applications to digital radiography, *IEEE Trans Syst Man Cybern*, SMC-**17**, 622–634.

Slone, R.M., Van Metter, R., Senol, E., Muka, E., Pilgrim, T.K. (1996). Effect of exposure variation on the clinical utility of chest radiographs, *Radiol*, **199**, 497–504.

Sonka, M., Fitzpatrick, J.M. (2000). *Handbook of Medical Imaging: Volume 2. Medical Image Processing and Analysis*. Bellingham, WA: SPIE Press.

Sprawls, P. (1995). *Physical Principles of Medical Imaging*. Madison, WI: Medical Physics Publishing, pp. 267–281.

Van Metter, R., Lemmers, H., Schultz Kool, L. (1992). Exposure latitude for thoracic radiography, *Proc SPIE*, **1651**, 52–61.

Van Metter, R., Foos, D. (1999). Enhanced latitude for digital projection radiography, *Proc SPIE*, **3658**, 468–483.

Vlasbloem, H., Schultz Kool, L.J. (1988). AMBER: a scanning multiple-beam equalization system for chest radiography, *Radiol*, **169**, 29–34.

Vuylsteke, P., Schoeters, E. (1994). Multi-scale image contrast amplification (MUSICA), *Proc SPIE*, **2167**, 551–560.

29

Optimization of display systems

ELIZABETH KRUPINSKI AND HANS ROEHRIG

29.1 INTRODUCTION

The radiology reading room has changed quite dramatically since the introduction of digital images and Picture Archiving and Communications Systems (PACS), and these changes have the potential to impact significantly the perception of radiographic images and thus the diagnostic interpretation process. Similar changes have occurred in other specialties as well. Pathology has always been dependent on the traditional light microscope to view glass slides but with the advent of digital scanners to create virtual slides (Weinstein, 2004), display technology has become a very hot topic. Telemedicine has also created the need for increased attention to display technologies, as many specialties such as dermatology (Krupinski, 2008) and ophthalmology (Taylor, 2007) use digital photographs to acquire and display diagnostic images.

Even though many aspects of the digital imaging chain have become simpler, there have also been complexities introduced. In radiology in particular, PACS has improved workflow in many ways as images are now easier to store and transmit within and between hospitals as compared to film. One consequence appears to be that radiologist productivity has improved – by more than 50% in some studies (Siegel, 2002). Reiner *et al.* found that reading CT exams from softcopy (a PACS workstation) resulted in a 16.2% reduction in overall interpretation time compared to hardcopy film reading (Reiner, 2001). The time savings differed for each radiologist, but they all experienced a general decrease in reading time with softcopy display. This may in part be related to increased experience with digital displays over the years. For example, in 1994 Krupinski *et al.* measured reading times of radiologists as they searched chest images for various lesions on hardcopy versus softcopy versions (Krupinski, 1994), and found that reading time with the softcopy display was about one minute longer per image than with film images.

Improved user interfaces and increased experience with softcopy displays over the years have probably contributed as well to decreased softcopy viewing times. It is not clear, however, that all softcopy reading times have decreased since technology may have outpaced the reader to some extent. Helical CT produces thousands of thin slice images that the radiologist must scroll through and MRI likewise produces hundreds more images than it did with film. It simply takes more time to read all these images with a potentially negative impact on workflow. Computer-aided detection and/or discrimination (CAD) may help in reducing reading times, but recent studies have revealed mixed findings, with both increases and decreases in reading times being reported (Beyer, 2007; Kakeda, 2008; Mang, 2007).

There is also some evidence that PACS itself increases the number of images per study in some cases, although it is difficult to isolate the exact influence PACS has on volume to the exclusion of all other factors (Horii, 2002). Teleradiology has also increased significantly in recent years, adding a significant number of studies, often without a commensurate increase in staff (Krupinski, 2003). At the University of Arizona we have provided teleradiology services since 1996 and by the end of 2007 had interpreted over 650,000 cases in addition to the in-house cases (see Figure 29.1). Although there are benefits associated with increased case volume, such as additional income and variety in the types of cases received to be interpreted (also adding to the quality of the educational experience of residents), there are also increased burdens being placed on the radiologists' time as well as their perceptual and cognitive resources.

Due to the increase in number and type of images being viewed by clinicians today, there are a number of important issues that need to be addressed. One of the main issues is how to optimize the display of images, especially as the images continue to change as technology evolves. Monochrome displays are currently the most common type of display found in radiology reading rooms. However, the number of color displays used to present both grayscale and color clinical images is increasing rapidly and display manufacturers are now manufacturing medical-grade color displays for image presentation in radiology and other clinical areas as well. True-color images are not common in radiology, but pseudo-color images are becoming much more prevalent. Multi-modality imaging techniques such as PET-CT use colors to fuse images acquired from different acquisition modalities. 3-D surface and volume rendering techniques use colors to facilitate the perception of depth and distinguish overlaying structures better. The use of pseudo-colors may increase the conspicuity of features in the images, and in turn improve diagnostic performance.

Other medical specialties that utilize color images in the diagnostic process are starting to purchase medical-grade color displays. In pathology and telemedicine (e.g. dermatology), color often serves as the basis for the diagnosis. How color images are reproduced on color displays, the accuracy of color reproduction by the displays, and the consistency of the color reproduction among color displays can all affect clinical practice. It is necessary to ensure that displays are set up and calibrated equally to prevent luminance and chrominance differences between displays from affecting the diagnosis.

The Handbook of Medical Image Perception and Techniques, ed. Ehsan Samei and Elizabeth Krupinski. Published by Cambridge University Press.

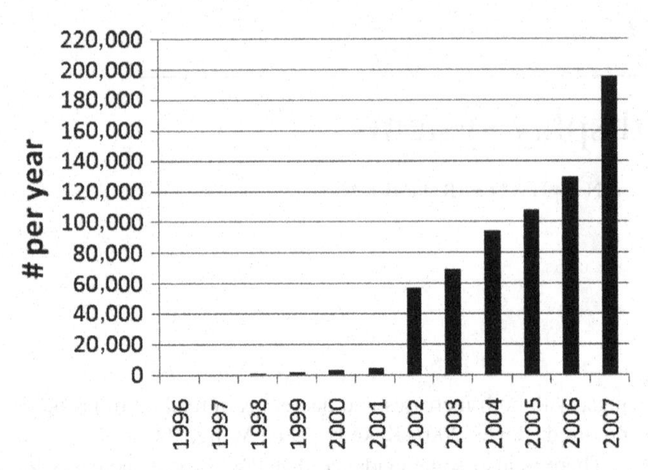

Figure 29.1 Number of teleradiology cases interpreted at the University of Arizona since the inception of the teleradiology program.

29.2 PHYSICAL CHARACTERIZATION FOR OPTIMIZATION OF DISPLAYS

Before discussing the relationship between medical image perception and displays, it is worth spending a little time discussing the most common types of displays being used to present images to the clinician.

29.2.1 The cathode ray tube (CRT) display

Digital imaging sensors and softcopy displays have replaced the film-screen sensor and the film light-box display in radiology. As a result the detector and the display are independent. They can be optimized separately and the images can be presented to the human observer at optimum information transfer. An early candidate for softcopy displays was the CRT. More recently, LCDs have become widespread as the primary diagnostic display.

Figure 29.2 is a schematic of a typical CRT. Details of the construction are found in the literature (Roehrig, 2000, 2002a, 2003). Since Ferdinand Braun invented the CRT in 1896, many types have been developed, but, remarkably, today's CRTs are very similar in design to Braun's invention.

The CRT consists of an evacuated glass tube, flaring from a narrow neck to a practically rectangular glass faceplate of the size of the expected image. The faceplate is covered on the vacuum side with a phosphor screen. This tube contains, at the narrow tube neck, an electron gun and beam-forming electrodes. The resulting electron beam writes the image on the phosphor screen making use of cathodo-luminescence. There are two types of phosphor screens: P-45 and P-104. They differ in the color as well as in the granular appearance. The P-45 has a slight blue tone and exhibits almost an order of magnitude less spatial noise than the P-104, while the P-104 has an almost white tone. A particularly welcome feature is the fact that CRTs are Lambertian radiators.

Despite all the digital controls of modern CRTs, the CRT is still an analog device needing analog video signals as the input. The video signals are generated by the display controller, the CRT's interface to the computer. The conversion from digital signals to analog signals is done with the aid of digital-to-analog converters (DACs). The CRT is no longer the most mature electronic display available and it is no longer the primary candidate to display digital radiographs. In fact its performance falls short in many areas. While most high-performance CRTs can display images with matrix sizes of 2048 × 2560, the modulation transfer function (MTF) and veiling glare are severely limited. They cannot compare to the almost perfect MTFs and the hardly noticeable veiling glare of active matrix LCDs (AM-LCDs).

29.2.2 The active matrix LCD

LCDs are based on two properties of liquid crystals: (1) they have a specific plain of polarization, and (2) they align themselves with an applied electrical field (Roehrig, 2002b, 2003). Consequently liquid crystals permit the plane of polarization to be rotated with the aid of an electrical field. Figure 29.3 (left) illustrates how the transmission of light through a cell filled with liquid crystal can be controlled by an electrical field when the cell with the liquid crystal and the electric field is

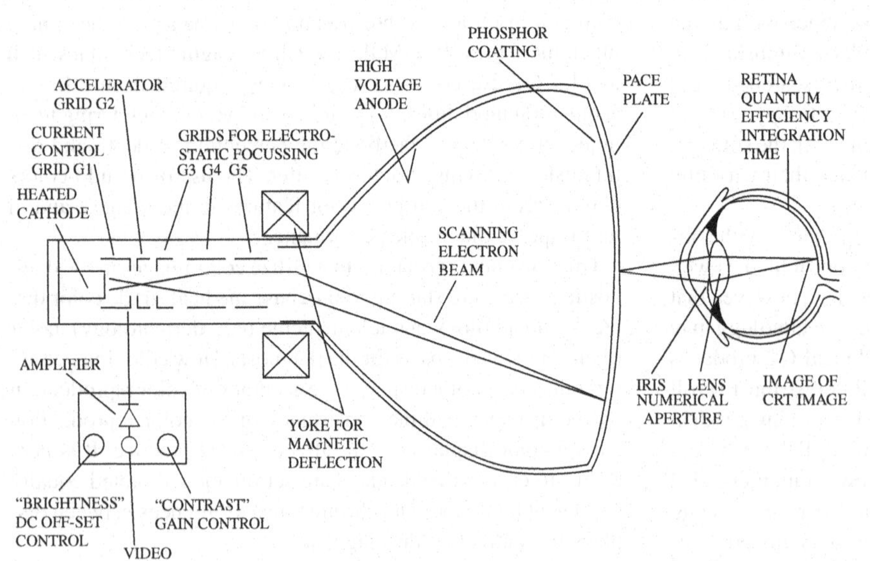

Figure 29.2 Schematic of a typical CRT display with electrostatic focus and magnetic deflection. The essential imaging components of the eye of the human observer (the lens with its numerical aperture and the retina with its quantum efficiency) are included to remind the reader that the human observer is an integral part of the display.

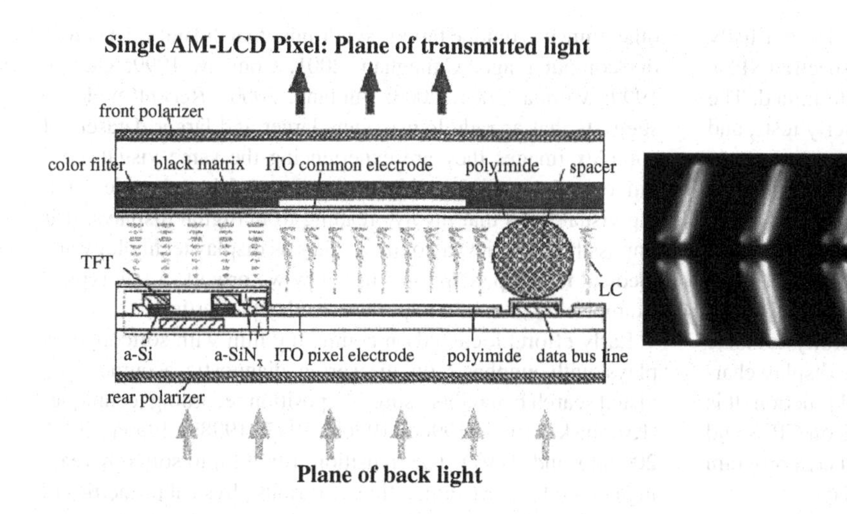

Figure 29.3 Left: Single AM-LCD pixel as it would be in a complete liquid crystal display. **Right:** Microscopic view of a single LCD pixel with three sub-pixels (dual-domain in-plane switching).

placed in between two polarizers, the polarization planes of which are oriented at 90 degrees with respect to each other. Twisted nematic (TN) and super-twisted nematic (STN) are frequently used types of liquid crystal. Figure 29.3 also identifies the thin-film-transistor (TFT) switch, which is why this type of LCD is called an AM-LCD. We mention in passing that most of the LCDs use the STN type of liquid crystal (Roehrig, 2002b). Figure 29.3 (right) is a microscopic view of a single LCD with three sub-pixels (dual-domain in-plane switching).

Much information on the properties of AM-LCDs has been presented recently (Blume, 2002, 2003; Roehrig, 2002a, 2002b, 2003). In general it appears that in many respects AM-LCDs have image quality superior to that of CRTs. LCDs can be controlled digitally or by analog controls, even though ultimately pixel addressing is done digitally. A major problem of LCDs is the dependence of both contrast and luminance for viewing off-axis (Blume, 2002, 2003; Martin, 2001).

Since the operation of the LCD is based on the transmission of the light from the back light, maximum and minimum luminance frequently are not controlled separately. Consequently there is usually only one control. The minimum luminance or black level of the LCD is then defined by the maximum optical density of the crossed polarizers and the perfection of the alignment of the liquid crystals in the black state. The inherent display function is related to the sensitivity of the rotation of the plane of polarization with the applied electrical field. Unfortunately this curve is very steep and cannot be used for display of a grayscale. Reference voltages are applied to reduce the inherent steepness, and generate a display function with a gamma of about 2.2 but at the cost of producing edges and kinks (Blume, 2001; Roehrig, 2002b).

29.2.3 Characterizing and optimizing displays

Quality control (QC) is playing an increasingly important role in medical imaging, since perception and diagnostic accuracy can be affected significantly by the quality of the image. For years, the American College of Radiology has offered guidelines on image quality (ACR, 2008), and other image-based specialties have begun to do so as well (ATA, 2004; Krupinski, 2008). Additionally a number of organizations have developed and evaluated guidelines for display performance and image quality testing (Deutsches Institut fuer Normung, 2001; IEC, 2008; Japan Industries Association of Radiological Systems, 2005; VESA, 2007). Most notable are those formulated by SMPTE (SMPTE, 1991) and the Digital Imaging and Communications in Medicine (DICOM, 2000) committee. The SMPTE pattern was for many years the dominant test pattern but it has been replaced by one of the TG18 patterns. The most visible part of DICOM is the DICOM 14 Grayscale Standard Display Function (GSDF) which determines the display function. The specific display function selected by DICOM is based on the Barten model and offers the additional advantage of perceptual linearization (Blume, 1996). Now it is inconceivable that electronic displays are used for diagnostic applications without proper DICOM calibration. It also is important to note that the display function generated with the German DIN system is different from the DICOM 14 GSDF at low luminance values.

The American Association of Physicists in Medicine (AAPM) through its Task Group 18 (TG18) has generated a thorough QC program titled "Assessment of Display Performance for Medical Imaging Systems" (Samei, 2005). The TG18 recommendations cover two types of tests: (1) visual tests, often called qualitative or basic tests, where a human observer views the display and makes a decision on the presence or absence of a test object; (2) quantitative or advanced tests, where an instrument like a photometer or a CCD camera is used to make a measurement and provide quantitative data.

The TG18 tests include methods for assessing the luminance and contrast performance of medical displays, taking into consideration the psychophysical factors discussed earlier. The guidelines also suggest minimum expected performance values. Similarly, further tests and criteria include display resolution and noise as well as other display characteristics such as angular response, reflection, glare, distortion, color tint, and artifacts. The guidelines also recommend a strategy to ascertain the maximum allowable illumination in the reading room by using the reflection and luminance characteristics of the display. Most of the recommended tests use newly developed test patterns.

The visual tests are usually based on some sort of a contrast pattern where the threshold contrast varies as the display image quality changes. These visual tests are performed as part of the image quality control in the reading room. The quantitative or advanced tests are done with one or several fairly high-performance CCD cameras in a laboratory. Here usually performance parameters like MTF, noise power spectra (NPS), and signal-to-noise ratio per display pixel are determined. The TG18 program recommends monthly tests, quarterly tests, and annual tests.

An important part of the TG18 recommendations are the display specifications. Display specifications are critical to the ultimate quality of the images displayed by the device. Some of the important engineering specifications of display devices are described in Table 1 of Samei *et al.* (Samei, 2005). When acquiring a display system, the user should carefully evaluate the specifications of the device to ensure that the display characteristics meet or exceed the needs of the desired function. It is important to note that the emphasis on displays is on CRTs and not on LCDs, as at the time of preparation of the TG18 program CRTs were the most widely used electronic displays.

29.2.4 Color displays

With respect to color displays, the tri-chromatic theory holds that the retina of the eye consists of a mosaic of three different receptors. Each responds to specific wavelengths corresponding to blue, green, and red. These three elements, which overlap considerably in response, are separately connected through nerves to the brain where the sensation of color is derived by the brain's analysis of the relative stimuli (Bass, 1995; Malacara, 2002). The operation of color displays is based on the fact that the mixture of three primary colors in suitable quantities can produce many color sensations. The three primaries may be, but are not necessarily, monochromatic. According to the CIE, colors may be specified by points on a 2-D chromaticity diagram (CIE, 1991).

The CIE 1976 u',v' chromaticity diagram provides a perceptually uniform color spacing for colors at approximately the same luminance. Colors can also be specified in terms of color temperature. This statement holds for those colors emitted by sources following Planck's radiation law. Accordingly, color fidelity is achieved if the color coordinates of the object to be imaged and the color coordinates of the images on the color display are identical. Color coordinates and temperatures can be measured with colorimeters or spectroradiometers. Colorimeters are photometers with three or more detectors, calibrated to provide chromaticity and luminance information within the field of view of these detectors. There is increasing use of color images and color displays in many clinical specialties. However, unlike the initial digital foray into electronic displays for radiology, there is no clear set of tools and guidelines to calibrate medical-grade color displays.

29.3 PERCEPTUAL PERFORMANCE IN THE OPTIMIZATION OF DISPLAYS

Once a display has been physically characterized and optimized, the question is whether it matters. Is diagnostic accuracy or reader efficiency improved? We include reading efficiency in addition to accuracy because human factors research has shown that in virtually all viewing environments and tasks, visual displays tax users' vision, cognition, and posture, leading among other things to unique fatigue syndromes that did not exist before the computer age (Callaghan, 2001; Conway, 1999; Gomzi, 1999; Murata, 2001, 2003; Yunfang, 2000). Recent evidence suggests that as radiologists view larger and larger datasets of complex images they are experiencing the same visual, cognitive, and postural problems encountered in other demanding visual tasks that involve reading from digital displays. It is impossible to review all of the studies on display technology and medical image perception, but below we provide a few typical examples on the more important display properties.

Early efforts focussed on comparing film with softcopy displays with emphasis on measuring diagnostic accuracy and visual search behaviors using eye-position recording techniques (Krupinski, 1994, 1996a, 1996b, 1997, 1998a, 1999a, 2000, 2002a; Lund, 1997). The transition from film to softcopy reading resulted in the examination of various physical properties of monitors to determine whether they influence diagnostic accuracy and visual search efficiency. It was found for example that increased display luminance (Krupinski, 1999a) and a perceptually linearized (i.e. with the DICOM standard versus a non-perceptually linearized) display (Krupinski, 2000) did indeed lead to better diagnostic accuracy and more efficient visual search. The effectiveness and use of image processing (Krupinski, 1998a) and reader experience have also been found to play important roles in the interpretation of medical images (Krupinski, 1996a, 1996b, 2005, 2006).

Even recent studies continue to assess image quality of CRT and LCD technologies. Saunders compared five commercial displays and confirmed that LCDs offer higher MTFs than CRT displays (Saunders, 2006a). Yet, the resolution advantages of LCDs must be considered in light of their noise properties. Physically, the CRT displays showed a lower MTF, but also demonstrated lower noise. Saunders also examined radiologists' performance with the displays in a mammography task. Radiologists had similar overall classification accuracy (LCD: 0.83 ± 0.01, CRT: 0.82 ± 0.01) and lesion detection accuracy (LCD: 0.87 ± 0.01, CRT: 0.85 ± 0.01) on both displays. Overall, the two displays did not exhibit any statistically significant difference (Saunders, 2006b).

Samei also compared a CRT and LCD for detection of masses and microcalcifications in mammograms (Samei, 2007). The detection rate in terms of the area under the ROC curve (Az) showed a 2% increase and a 4% decrease from CRT to LCD respectively, but the differences were not statistically significant. They also looked at viewing angle and the data showed better microcalcification detection but lower mass detection at 30° viewing orientation. The overall results varied notably from observer to observer yielding no statistically discernible trends across all observers, suggesting that within the 0–50° viewing angle range and in a controlled observer experiment, the variation in the contrast response of the LCD has little or no impact on the detection of mammographic lesions.

Saunders examined the effects of different resolution and noise levels on task performance in digital mammography (Saunders, 2007). Human observer results showed decreasing display resolution had little effect on overall classification accuracy and individual diagnostic task performance, but increasing noise caused overall classification accuracy to decrease by a statistically significant 21% as the breast dose went to one quarter of its normal clinical value. The noise effects were most prominent for the tasks of microcalcification detection and mass discrimination. The primary conclusion is that quantum noise appears to be the dominant image quality factor in mammography.

Samei also looked at display matrix sizes (3–9 Mpixel) with CRTs and LCDs and effects on performance. While these devices exhibit variations in their physical characteristics, they do not appear to translate into notable variations in the inherent contrast detectability performance of the display devices (Samei, 2008). This study, based on a contrast-detail paradigm, found that 5 and 3 Mpixel monochrome LCD displays perform slightly better than 5 Mpixel CRTs and a 9 Mpixel color LCD. The 5 Mpixel LCD performed slightly better than the 3 Mpixel LCD as well. The relative superiority of different displays was subtle and may not necessarily translate into discernible performance differences in clinical use and thus needs to be placed in perspective with other implementation considerations such as clinical application and cost.

Another important question that radiologists are currently asking is what bit depth is required in a display. Most commercial and medical-grade monitors manufactured today display only 8 bits (256 gray levels) of data. This is sufficient for some medical image interpretation tasks in which the acquired data are 8 bits or less, but many medical images are acquired at higher bit depths (e.g. 12–16 bits or 4096–6553 gray levels) (Chunn, 2000). Because of this disparity, all acquired gray levels cannot be displayed at once even with mammography displays with 1024 levels of gray. The result is a potentially significant loss of information during the diagnostic interpretation process when window/level is not utilized and the potential for artifacts to be introduced due to down-sampling images to 8 bits. Additionally, the uses of window and level to manipulate the displayed gray levels can slow down the interpretation process, adversely affecting workflow.

Perceptually speaking, higher bit-depth displays may or may not improve performance. Clinicians take in a huge amount of information during the initial view of an image (i.e. the Gestalt or global percept). The more information or gray levels available in the initial view, the more efficient and informative the initial impression is going to be (Nodine, 1987a, 1987b). More information may reduce the need for excessive windowing and leveling, reducing the time needed by the clinician to render a diagnosis. Evidence, however, indicates that the human visual system can only detect about 1000 gray levels (far below 12–16 bits or 4096–6553 gray levels) at luminance levels currently used in medical grade monitors, and consequently displaying more gray levels may not be useful (Barten, 1992, 1999).

One study that examined bit depth used three sets of 8-bit and 11-bit, 3 Mpixel, monochrome, portrait-mode, medical-grade LCD monitors (Krupinski, 2007a). All were calibrated to the DICOM GSDF. One hundred direct digital radiography (DR) chest images (General Electric Revolution XQ/I System) were used; 50 nodule-free cases and 50 cases with subtle solitary pulmonary nodules (verified by CT). Three sites participated, each with six radiologist observers. Each radiologist viewed all 100 images twice – once on the 8-bit and once on the 11-bit monitor. They viewed each image and decided whether a nodule was present or absent, then provided their confidence in that decision. The confidence data were analyzed using the multi-reader multi-case (MRMC) ROC technique (Dorfman, 1992). Use of window/level processing function during interpretation was recorded, as was total viewing time. Visual search efficiency was also measured on a sub-set of images at one site using the 4000SU Eye-Tracker. The eye-position data characterized time to first fixate a lesion, total search time, and dwells associated with each decision type (true and false, positive and negative) (Hu, 1994; Krupinski, 1996a, 1996b, 1997, 1999a, 2000, 2002a, 2004; Nodine, 1992).

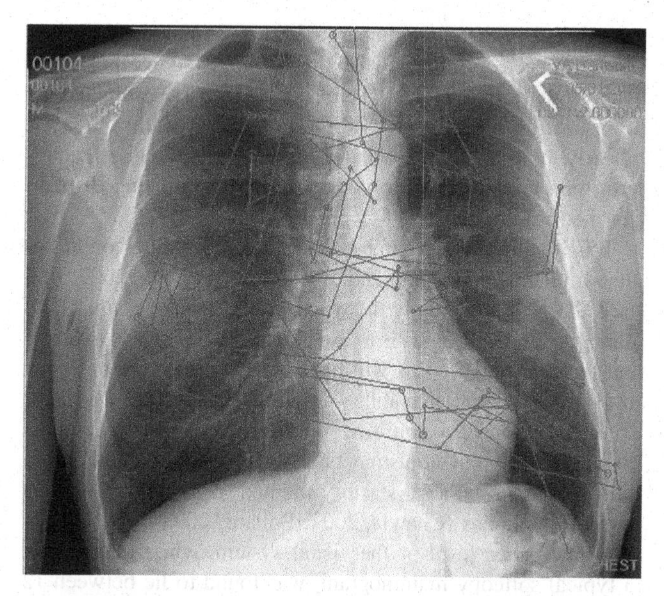

Figure 29.4 Typical example of an eye-tracking pattern of a radiologist searching for nodules in a chest image. The small circles represent locations where the eye lands with high-resolution foveal gaze and the lines represent the order in which they were generated.

The results revealed that there was no statistically significant difference in area under the curve (Az) performance as a function of 8 vs 11 bit-depth. Average Az for the 8-bit display was 0.8284 and average performance for the 11-bit display was 0.8253. There were no statistically significant differences between the 8-bit and 11-bit displays for any of the three systems. There were no differences in the percentage of cases on which window/level was used. Preference for window/level seemed to be an individual trait – some readers used it and some did not, but an individual reader used it about the same with both displays.

Figure 29.4 shows a typical scan pattern from the eye-position part of the study. Total viewing times were significantly shorter for the 11-bit than the 8-bit displays. Time to first hit the nodules during search revealed that for true positive and false negative decisions the time to first hit was shorter with the 11-bit than the 8-bit display although the differences did not reach statistical significance.

Also examined were cumulative dwell times for true positive, false negative, false positive, and true negative decisions. For the 11-bit display, cumulative dwell times for each decision category were lower than for the 8-bit display, the differences for true negative decisions reaching significance. Although search efficiencies may not seem important when one or two images are concerned, inefficiency adds up over an entire day's worth of reading numerous images and the end result is fewer images being read in the same amount of time compared to displays that have been optimized to the user's visual system capabilities. This study represents just one of many that have evaluated the optimization of displays for the interpretation of medical images. Clearly, both diagnostic accuracy and interpretation efficiency are important variables to consider when optimizing displays for medical imaging.

It is important to remember that ambient lighting conditions also influence perception and display use (Chawla, 2007; Pollard, 2008). Perceptually the way to avoid negative impacts of changing ambient lights is to use a controlled increase of ambient lighting because this will likely not degrade and may even improve an observer's ability to detect low-contrast objects. The utilization of a luminance-induced pupil response model to minimize pupil adjustments during radiograph interpretation appears to be a satisfactory method of improving reading room ergonomics (Chawla, 2007; Pollard, 2008). The adaptation luminance level of the visual system, while interpreting a typical softcopy mammogram, was found to lie between 12 and 20 cd/m^2. Increasing room illuminance to approximately 50–80 lux may maximize clinician comfort without sacrificing diagnostic performance. However, this increase should be accompanied by the proper calibration of the display, taking into consideration the exact level of ambient lighting and the reflective properties of both the display and materials within the reading room.

29.4 COLOR DISPLAYS

As already noted, color displays are clearly becoming an important modality both in radiology and many other clinical specialties. In a recently completed radiology study we compared diagnostic accuracy for detecting nodules in DR chest images on a medical-grade color display with a commercial off-the-shelf (COTS) display (Krupinski, 2009). Both were calibrated using the DICOM GSDF, but there was no special color calibration used. Overall there was a significant difference between the medical-grade color display (mean ROC Az = 0.9101) and the COTS color display (mean ROC Az = 0.8424). This difference could occur for a number of reasons, but recall that this was exactly the reason why the DICOM calibration was initially developed for monochrome displays – the same images on two different monitors that were presumably comparable in most other respects did not look the same to viewers and often yielded different diagnostic performance. That is precisely what we observed here with two different color displays – the same images displayed on different color monitors did not look the same and yielded different diagnostic accuracy levels. A method to calibrate color displays in much the same way that DICOM calibrates monochrome could potentially reduce these differences.

Color displays are crucially important from a perceptual perspective in clinical specialties other than radiology. For example, we conducted a teledermatology study using 308 cases, each diagnosed in person and then via digitally acquired images displayed on a color monitor (Krupinski, 1999b). We found that there was overall 85% concordance between in-person and digital interpretation. The dermatologists also rated image color and generally rated it as being excellent or good. There was also a clear relationship between rated image color quality and performance. For those cases rated as having only fair or poor color quality on the display monitor, there were significantly more differences in diagnostic accuracy for the digital images compared to the gold standard in-person diagnoses.

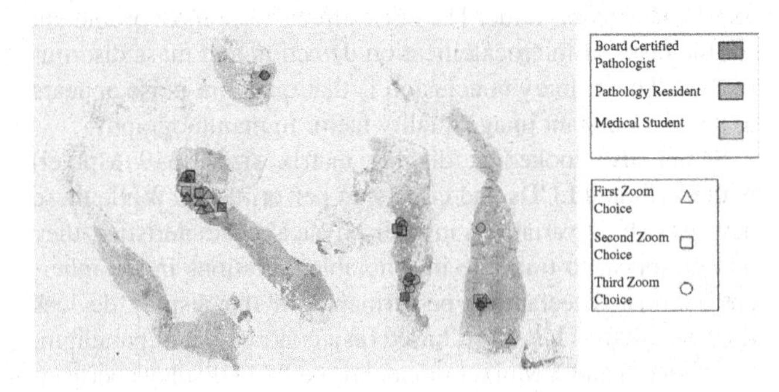

Figure 29.5 Locations selected for zooming by experienced pathologists, pathology residents, and medical students. Sporadic locations are those selected by only one reader and common ones are those selected by more than one.

In pathology, display considerations involve another unique aspect of medical image perception. Digitized pathology slides, or virtual slides, are very large. A single image can require as much as 1 GB of storage space depending on the size of the scanned area and the challenge is to display all of this information to the pathologist in an efficient manner. In a recent perception study (Krupinski, 2006), we used eye-tracking to characterize where pathologists initially look at a virtual slide – is search random or are the eyes attracted initially to regions of diagnostic interest? The goal was to determine if there is a way to pre-select diagnostically relevant regions of interest for initial display, leaving the rest of the image off the display unless actively accessed.

A set of 20 breast core biopsy surgical pathology cases was digitized using the Dmetrix DX-40 virtual slide processor (Weinstein, 2004). Low-magnification images (i.e. the full slide not zoomed to any particular region, average 39.55 × 23.4 cm) were displayed on an IBM T221 9 Mpixel color LCD. Three pathologists, three pathology residents, and three medical students served as observers, viewing the slides while their eye-position was recorded. The readers had to select the top three locations that they would zoom onto if they were going to view the image in greater detail to render a diagnostic decision.

The locations marked by only one person were considered "sporadic" and those by more than one "common" (see Figure 29.5). There were significantly more common locations marked per image than sporadic. On average there were 4.40 common locations and 1.45 sporadic locations per virtual slide. The pathologists, residents, and medical students selected 20%, 43%, and 37% of the sporadic locations, respectively. To determine if the preferred zoom locations were clinically meaningful, a senior pathologist reviewed each location to determine whether it contained information that could contribute to a diagnostic decision. Of all locations, 92% contained diagnostically relevant information (85% of the malignant and 95% of the benign lesions). For areas that did not contain relevant information,

55% were selected by medical students, 36% by residents, and 10% by pathologists.

The eye-position data revealed that the pathologists (mean = 4.471 sec) viewed each virtual slide image for significantly less total time than the residents (mean = 7.148 sec) or the medical students (mean = 11.861 sec). The mean total viewing time per virtual slide for each of the readers differed for each experience level group as well. It is clear from the eye-position data that the zoom locations are identified very quickly. On average, the experienced pathologists selected all three locations within five seconds – enough information was extracted in the initial global impression and through peripheral vision to reduce significantly the need for examining all tissue in foveal vision.

The pathologists generated fewer eye fixations and made longer saccades between fixations compared to the residents and medical students. However, given the high number of common locations, the inexperienced observers made comparable decisions to the more experienced pathologists, although their scanning strategies were less efficient.

29.5 DISPLAY INTERFACE

Image quality is not the only factor that impacts the perceptual and cognitive systems of clinicians. The user interface is the backbone of the workstation and studies have shown that it should be fast, intuitive, and user friendly (Krupinski, 2007b). It has already been noted that from a perceptual point of view, the quality of the default image presentation is extremely important. Many acquisition devices actually preprocess images to improve appearance, providing the potential to reduce viewing times and the number of image manipulations (e.g. window/level or zooming) needed (Krupinski, 2001; Thrall, 2005). The hanging protocol is also a critical perceptual element and default settings are widely used to save precious interpretation time as a preset hanging protocol brings up the most important images in a preferred format.

Other factors that often influence cognitive and perceptual performance with digital displays include: short and long-term memory capacity, decision-making strategies, attention span or scope, search and scanning strategies, time perception, arousal and vigilance levels, fatigue, perceptual load, monotony and boredom, aging, and even circadian rhythms (Shneiderman, 1998). The perception of time is something that people rarely consider. Computers are getting faster and image retrieval times have decreased over the years, but it still seems to take forever for the desired image to appear. Clinicians are generally not aware of fatigue, boredom, and perceptual load as factors that affect their performance, but they do and they can even vary as a function of time of day. Gale examined circadian rhythms in radiologists and found that the detection of pulmonary nodules decreased significantly after lunch compared to reading in the morning or early evening (Gale, 1984). Factors such as personality, gender, and cultural differences can affect the ways that people interact with digital displays and the information presented on them (Shneiderman, 1998).

Key perceptual principles for organizing information on digital displays include: consistency in data display, efficient information assimilation by the user, minimal memory load on the user, and flexibility for user control of data display (Shneiderman, 1998). Providing an interface that balances automation and human control so that the user can devote as much perceptual and cognitive resource to the primary task of rendering a diagnostic decision, rather than figuring out how to navigate through and visualize the data needed to complete the task, is very important. Moise demonstrated this when he examined the way that readers interacted with two image-hanging formats (Moise, 2005). The readers were first presented with a pair of chest images (a PA and lateral) side-by-side representing the current case to be diagnosed. They had to determine if there was a change in status from the prior interpretation. The interface tools to access and display the prior images were of two types and in both cases the prior images were shown as thumbnails above the current full-resolution images. In scenario one, the readers had to select each image individually and remove the one in the current display that they wanted to replace the thumbnail with. To achieve this, multiple clicks of the mouse were needed. In scenario two, the thumbnail images were paired, so a single click of the mouse brought up the current and prior AP images side-by-side automatically. Response time, accuracy (in selecting the desired images), and number of interaction steps were measured, as was eye-position. The "one click does it all" scenario yielded significantly shorter response times, higher accuracy (i.e. fewer incorrect selections), and a 9% reduction in the time spent looking at the control areas compared to scenario one. The design of the interface obviously affected both perceptual and cognitive performance.

Krupinski and Lund also showed the need for simplicity of interface design (Krupinski, 1997). Softcopy and hardcopy bone images with fractures were compared while recording eye-position. Twenty percent of the fixation clusters on softcopy display were on the menu and other toolbar functions rather than on the diagnostic parts of the image. Most menu/toolbar scanning (87%) took place during the first ten seconds, but the rest occurred sporadically during search of the anatomic regions. This took visual attention away from the diagnostic task. Voice-activated commands or other alternative means of interacting with the computer may be useful as displays become more complex and more information needs to be displayed on more monitors. Visual attention should not be drawn away from the diagnostic image by non-diagnostic information.

29.6 DISPLAY CONFIGURATIONS AND PERCEPTION

One solution to viewing more images is to use more displays, and there is some evidence that this influences interpretation performance. In radiology when film images were shown on alternators it was easy for the radiologist to perceive a number of them as a continuous display. Individual films could be displayed side-by-side without any physical gaps. With softcopy each display is independent, even when running them as a single desktop. There are significant physical gaps (the monitor casing or frame) between the displays that the user's visual and attentional systems must deal with. No studies have been done to date to determine whether these "gaps" between images influence search or detection behaviors, but display manufacturers are starting to make displays with much larger sizes so that more image real estate is available on a single display.

It is possible, however, that there is an optimal number of displays per workstation. Siegel and Reiner examined a variety of configurations (one, two, or four monitors) and how they influenced workflow and performance (Siegel, 2002). Differences were found between one- and two-monitor workstations but not between two and four. Reading speed was 25% faster with the four-monitor workstation than with one, and radiologists were less fatigued. Users rated 57% of studies as low fatigue and 1% as high with the four-monitor workstation, and rated 39% with the one-monitor workstation as low fatigue and 24% as high. In a related study, four monitors were set up in a 2 × 2 arrangement, with two on the bottom and two above them instead of side-by-side (Krupinski, 1998b). The radiologists found this to be quite inconvenient since the images on the top monitors were hard to see without moving their chair back, causing significant neck and shoulder strain.

A potentially significant problem with multiple displays is navigation. Users seem to "get lost" perceptually in large or multiple display configurations (Andre, 1992; Murray, 1996; Obradovich, 1996; Watts-Perotti, 1999). The problem is magnified when the displays contain lots of disparate pieces of information that need to be attended to when making a decision. Studies have looked at aircraft cockpit display designs for example, to determine the best way to provide pilots with the varied sources of information they need to fly a plane safely (Andre, 1992; Hofer, 1997). Grouping compatible or related information sources onto individual displays within the larger layout tends to improve search and performance. Perceptual grouping may improve the signal-to-noise ratio of the relevant information and create a flow of information that draws the viewer around the display rather than forcing them to actively sort through display items to determine how they are related (Andre, 1992; Ellis, 1968; Hofer, 1997; Murray, 1996; Obradovich, 1996; Pomplun, 2003; Watts-Perotti, 1999).

Navigation strategies for radiology workstations have also been studied. One study compared a viewing strategy that grouped images according to body area and examination view (associated with a sketch of the human body in what they termed an anthropomorphic display) versus a simple linear image organization (Roger, 1989). The study did not examine diagnostic performance, but subjectively the radiologists preferred the anthropomorphic navigation strategy to having to scroll through a list of ungrouped images.

Another perceptual issue is whether multiple images should be shown simultaneously or sequentially. Gur compared simultaneous versus sequential display of chest images (Gur, 1992) and found that diagnostic performance and perceived image quality were equivalent. Two readers had short viewing times with sequential viewing and two had short times with simultaneous, and there was no preference or comfort level for one display versus the other. CT and MRI exams, where sequential versus simultaneous display of images is also a choice, have been studied as well. For both, stack mode or cine viewing is generally more efficient perceptually because it eliminates the need for the radiologist to scan through the images and skip from image to image while trying to keep the individual images in sequence. Organizing different sets of images (e.g. pre- and post-contrast) into consecutive displays may help improve performance even more (Tazawa, 2001).

Perceptually, one of the main goals for display is to organize and simplify the information as much as possible to create a balance between automation and human control so that perceptual resources can be devoted to the interpretation task rather than the navigation task. Another way to accomplish this is with image-processing techniques. Reiner *et al.* showed a series of CR musculoskeletal images to radiologists with four processing algorithms (Reiner, 2003). It was found that edge enhancement and multi-frequency processing algorithms improved sensitivity and specificity for non-displaced fractures. In a study of various processing algorithms for CR chest images, however, Krupinski *et al.* found that image processing was just as likely to change decisions from false negative to true positive as from true positive to false negative and changes in confidence were just as likely to increase as decrease (Krupinski, 1998a).

Another useful strategy for displaying medical image data is the use of 3-D and stereoscopic displays. All forms of imaging today are 2-D representations of 3-D anatomic structures (even in pathology virtual slides since no scanner as of yet scans more than a single plane). Clinicians have to visualize in their minds the 3-D relationships between the various 2-D representations rendered on images. Scrolling through a series of stack-mode image slices as is done in radiology facilitates the perception of 3-D, but it is not truly a 3-D display. Volume rendering takes it a step further and with the proper software clinicians can virtually manipulate volumetric images to visualize structures in 3-D (Rosenbaum, 2000). One technique tried using rotating video displays of CT, MR, and digital subtraction angiography (DSA) datasets and reconstructed them for display with one- or two-degree angular views. Radiologists viewed them with polarized crystal display glasses for the stereoscopic effect. Image findings changed in 61% of the cases after viewing in 3-D, and decision confidence increased in 73% of the cases.

Getty has evaluated the use of stereoscopic viewing for mammographic interpretation (Getty, 2007). In a series of studies, readers have read mammography cases twice – once on film or with traditional viewing methods and once stereoscopically on a stereo display workstation. ROC analysis has repeatedly shown that performance increases significantly with stereoscopic viewing. Readers also appear to be able to detect a number of likely lesions with stereoscopy compared to traditional viewing methods. Clearly, the addition of 3-D stereoscopic viewing has immense potential. The only possible drawback is that the best stereoscopic viewing is obtained with special glasses and displays, but if their utility can be demonstrated clearly stereoscopic viewing of large datasets may become commonplace in the future.

29.7 CONCLUSIONS

The future of displays in medical imaging is difficult to predict. New types of information such as that from molecular imaging are continually being added to the clinician's arsenal in the detection and diagnosis of disease processes (Bremer, 2003; Hillman, 2002; Rollo, 2003). This continually necessitates the re-evaluation of how medical images should be displayed for optimal perception and interpretation. This chapter has hopefully illustrated the need to keep two important aspects of the interpretation process in mind. The first is that at its most basic level, medical image interpretation is a perceptual process. The display environment needs to be optimized to account for the strengths and weaknesses of the human visual system. Pertinent information needs to be readily acquired visually from images and the easier it is for clinicians to do this, the easier it will be to render an accurate diagnostic decision. Secondly, the display optimization process must account for the degree to which the clinician interacts with the display to obtain and visually process diagnostic information. Clinicians need to devote as much of their perceptual resources to the diagnostic task as possible rather than navigating through the display figuring out how to find and process relevant diagnostic information.

REFERENCES

American College of Radiology. (2008). Guidelines and standards. http://www.acr.org/SecondaryMainMenuCategories/quality_safety/guidelines.aspx Last accessed May 12, 2009.

American Telemedicine Association. (2004). Telehealth practice recommendations for diabetic retinopathy. *Telemed J e-Health*, **10**, 469–482.

Andre, A.D., Wickens, C.D. (1992). Compatibility and consistency in display-control systems: implications for aircraft decision aid design. *Hum Fact*, **34**, 639–653.

Barten, O.G.J. (1992). Physical model for contrast sensitivity of the human eye. *Proc SPIE Med Imag*, **1666**, 57–72.

Barten, P.G.J. (1999). *Contrast Sensitivity of the Human Eye and its Effects on Image Quality*. Bellingham, WA: SPIE Press.

Bass, M. (1995). *Handbook of Optics*, Volume I. New York, NY: McGraw-Hill.

Beyer, F., Zierott, L., Fallenberg, E.M., *et al.* (2007). Comparison of sensitivity and reading time for the use of computer-aided detection (CAD) of pulmonary nodules at MDCT as concurrent or second reader. *Eur Radiol*, **17**, 2941–2947.

Blume, H. (1996). The ACR/NEMA proposal for a gray-scale display function standard. *Proc SPIE Med Imag*, **2707**, 344–360.

Blume, H., Ho, A.M.K., Stevens, F., Steven, P.M. (2001). Practical aspects of grayscale calibration of display systems. *Proc SPIE Med Imag*, **4323**, 28–41.

Blume, H., Steven, P., Cobb, M., *et al.* (2002). Characterization of high-resolution liquid crystal displays for medical images, Part I. *Proc SPIE Med Imag*, **4681**, 271–292.

Blume, H., Steven, P., Ho, A., *et al.* (2003). Characterization of liquid-crystal displays for medical images, Part 2. *Proc SPIE Med Imag*, **5029**, 449–473.

Bremer, C., Ntziachristos, V., Weissleder, R. (2003). Optical-based molecular imaging: contrast agents and potential medical applications. *Eur Radiol*, **13**, 231–243.

Callaghan, J.P., McGill, S.M. (2001). Low back joint loading and kinematics during standing and unsupported sitting. *Ergonomics*, **44**, 280–294.

Chawla, A., Samei, E. (2007). Ambient illumination revisited: a new adaptation-based approach for optimizing medical imaging reading environments. *Med Phys*, **34**, 81–90.

Chunn, T., Honeyman, J. (2000). Storage and database. In *Handbook of Medical Imaging Volume 3: Display and PACS*, ed. Kim, Y., Horii, S.C. Bellingham, WA: SPIE Press, pp. 365–401.

Commission Internationale de l'Eclairage. (1991). CIE standard illuminants for colorimetry. http://www.cie.co.at/div2/TCs/d2publ.html Last accessed May 12, 2009.

Conway, F.T. (1999). Psychological mood state, psychosocial aspects of work, and musculoskeletal discomfort in intensive video display terminal (VDT) work. *Intl J Human-Computer Interaction*, **11**, 95–107.

Digital Imaging and Communications in Medicine (DICOM) Part 14: Grayscale Standard Display Function. (2000). NEMA PS 3.14. Rosslyn, VA: National Electrical Manufacturers Association. http://medical.nema.org/ Last accessed May 12, 2009.

DIN-6868–57. (2001). Image quality assurance in x-ray diagnostics, acceptance testing for image display devices. Berlin, Germany: Deutsches Institut fuer Normung.

Dorfman, D.D., Berbaum, K.S., Metz, C.E. (1992). Receiver operating characteristic rating analysis: generalization to the population of readers and patients with the jackknife method. *Invest Radiol*, **27**, 723–731.

Ellis, K. (1968). Methods of scanning a large visual display. *Occupational Psych*, **42**, 181–188.

Gale, A.G., Murray, D., Millar, K., Worthington, B.S. (1984). Circadian variation in radiology. In *Theoretical and Applied Aspects of Eye-Movement Research*, ed. Gale, A.G., Johnson, F. London, England: Elsevier Science Publishers, pp. 313–321.

Getty, D.J. (2007). Improved accuracy of lesion detection in breast cancer screening with stereoscopic digital mammography. Paper presented at the 93rd Annual Meeting of the Radiological Society of North America, Chicago, IL.

Gomzi, M., Bobic, J., Ugrenovic, Z., Goldoni, J. (1999). Personality characteristics of VDT operators and computer-related work conditions. *Studia Psychologica*, **41**, 15–21.

Gur, D., Good, W.F., King, J.L. (1992). Simultaneous and sequential display of ICU AP-chest images. *Proc SPIE Med Imag*, **1653**, 159–163.

Hillman, B.J., Neiman, H.L. (2002). Translating molecular imaging research into radiologic practice: summary of the proceedings of the American College of Radiology Colloquium, April 22–24, 2001. *Radiol*, **222**, 19–24.

Hofer, E.F. (1997). Bridging the gap between theory and application: an evaluation of the display issues of map clutter, rotation and integration. *Dissertation Abstracts International: Section B: The Sciences & Engineering*, pp. 57–7259.

Horii, S., Nisenbaum, H., Farn, J. (2002). Does use of a PACS increase the number of images per study? A case study in ultrasound. *J Dig Imag*, **15**, 27–33.

Hu, C.H., Kundel, H.L., Nodine, C.F., Krupinski, E.A., Toto, L.C. (1994). Searching for bone fractures: a comparison with pulmonary nodule search. *Acad Radiol*, **1**, 25–32.

International Electrotechnical Commission. (2008). http://www.iec.org Last accessed May 12, 2009.

Japan Industries Association of Radiological Systems. (2005). JESRAX-0093–2005. http://www.jira-net.or.jp/e/index.htm Last accessed May 12, 2009.

Kakeda, S., Korogi, Y., Arimura, H., *et al.* (2008). Diagnostic accuracy and reading time to detect intracranial aneurysms on MR angiography using a computer-aided diagnosis system. *Am J Roentgenol*, **190**, 459–465.

Krupinski, E.A. (1996a). Visual scanning patterns of radiologists searching mammograms. *Acad Radiol*, **3**, 137–144.

Krupinski, E.A. (2009). Medical grade vs off-the-shelf color displays: influence on observer performance and visual search. *J Dig Imag*. In press.

Krupinski, E.A., Berger, W., Dallas, W., Roehrig, H. (2004). Pulmonary nodule detection: what features attract attention? *Proc SPIE Med Imag*, **5372**, 122–127.

Krupinski, E.A., Burdick, A., Pak, H., *et al.* (2008). American Telemedicine Association's practice guidelines for teledermatology. *Telemed J e-Health*, **14**, 289–301.

Krupinski, E.A., Evanoff, M., Ovitt, T., *et al.* (1998a). The influence of image processing on chest radiograph interpretation and decision changes. *Acad Radiol*, **5**, 79–85.

Krupinski, E.A., Kallergi, M. (2007b). Choosing a radiology workstation: technical and clinical considerations. *Radiol*, **242**, 671–682.

Krupinski, E.A., LeSueur, B., Ellsworth, L., *et al.* (1999b). Diagnostic accuracy and image quality using a digital camera for teledermatology. *Telemed J*, **5**, 257–263.

Krupinski, E.A., Lund, P.J. (1997). Differences in time to interpretation for evaluation of bone radiographs with monitor and film viewing. *Acad Radiol*, **4**, 177–182.

Krupinski, E.A., Maloney, K., Bessen, S.C., *et al.* (1994). Receiver operating characteristic evaluation of computer display of adult portable chest radiographs. *Invest Radiol*, **29**, 141–146.

Krupinski, E.A., McNeill, K.M. (1998b). *Aurora Technologies Sunrise diagnostic radiology workstation evaluation*. Technical Report RRL-TR1998.8.17.

Krupinski, E.A., McNeill, K.M., Haber, K., Ovitt, T.W. (2003). High-volume teleradiology service: focus on radiologist satisfaction. *J Dig Imag*, **16**, 34–36.

Krupinski, E.A., Radvany, M., Levy, A., *et al.* (2001). Enhanced visualization processing: effect on workflow. *Acad Radiol*, **8**, 1127–1133.

Krupinski, E.A., Roehrig, H. (2000). The influence of a perceptually linearized display on observer performance and visual search. *Acad Radiol*, **7**, 8–13.

Krupinski, E.A., Roehrig, H. (2002a). Pulmonary nodule detection and visual search: P45 and P104 monochrome monitors versus color monitor displays. *Acad Radiol*, **9**, 638–645.

Krupinski, E.A., Roehrig, H., Dallas, W., Fan, J. (2005). Differential use of image enhancement techniques by experienced and inexperienced observers. *J Dig Imag*, **18**, 311–315.

Krupinski, E.A., Roehrig, H., Furukawa, T. (1999a). Influence of film and monitor display luminance on observer performance and visual search. *Acad Radiol*, **6**, 411–418.

Krupinski, E.A., Siddiqui, K., Siegel, E., *et al.* (2007a). Influence of 8-bit vs. 11-bit digital displays on observer performance and visual search: a multi-center evaluation. *J Soc Inf Disp*, **15**, 385–390.

Krupinski, E.A., Tillack, A.A., Richter, L., *et al.* (2006). Eye-movement study and human performance using telepathology virtual slides. Implications for medical education and differences with experience. *Hum Path*, **37**, 1543–1556.

Krupinski, E.A., Weinstein, R.S., Rozek, L.S. (1996b). Experience-related differences in diagnosis from medical images displayed on monitors. *Telemed J*, **2**, 101–108.

Lund, P.J., Krupinski, E.A., Pereles, S., Mockbee, B. (1997). Comparison of conventional and computed radiography: assessment of image quality and reader performance in skeletal extremity trauma. *Acad Radiol*, **4**, 570–576.

Malacara, D. (2002). *Color Vision and Colorimetry: Theory and Applications*. Bellingham, WA: SPIE Press.

Mang, T., Peloschek, P., Plank, C., *et al.* (2007). Effect of computer-aided detection as a second reader in multidetector-row CT colonography. *Eur Radiol*, **17**, 2598–2607.

Martin, S., Badano, A., Kanicki, J. (2001). Characterization of a high quality monochrome AM-LCD monitor for digital radiology. *Proc SPIE Med Imag*, **4681**, 293–304.

Moise, A., Atkins, S. (2005). Designing better radiology workstations: impact of two user interfaces on interpretation errors and user satisfaction. *J Dig Imag*, **18**, 109–115.

Murata, A., Uetake, A., Matsumoto, S., Takasawa, Y. (2003). Evaluation of shoulder muscular fatigue induced during VDT tasks. *Intl J Human-Computer Interaction*, **15**, 407–417.

Murata, A., Uetake, A., Otsuka, M., Takasawa, Y. (2001). Proposal of an index to evaluate visual fatigue induced during visual display terminal tasks. *Intl J Human-Computer Interaction*, **13**, 305–321.

Murray, S.A., Caldwell, B.S. (1996). Human performance and control of multiple systems. *Hum Factors*, **38**, 323–329.

Nodine, C.F., Kundel, H.L. (1987a). The cognitive side of visual search. In *Eye Movements: From Physiology to Cognition*, ed. O'Regan, J.K., Levy-Schoen, A. Amsterdam, Netherlands: Elsevier Publishers, pp. 573–582.

Nodine, C.F., Kundel, H.L., Polikoff, J.B., Toto, L.C. (1987b). Using eye movements to study decision making of radiologists. In *Eye Movement Research: Physiological and Psychological Aspects*, ed. Luer, G., Lass, U., Shallo-Hoffman, J. Lewiston, NY: CJ Hogrefe Publishers, pp. 349–363.

Nodine, C.F., Kundel, H.K., Toto, L.C., Krupinski, E.A. (1992). Recording and analyzing eye-position data using a microcomputer workstation. *Behav Res Meth Instrum Comput*, **24**, 475–485.

Obradovich, J.H., Woods, D.D. (1996). Users as designers: how people cope with poor HCI design in computer-based medical devices. *Hum Factors*, **38**, 574–592.

Pollard, B.J., Chawla, A.S., Delong, D.M., Hashimoto, N., Samei, E. (2008). Object detectability at increased ambient lighting conditions. *Med Phys*, **35**, 2204–2213.

Pomplun, M., Reingold, E.M., Shen, J. (2003). Area activation: a computational model of saccadic selectivity in visual search. *Cog Sci*, **27**, 299–312.

Reiner, B.I., Siegel, E.L., Hooper, F.J., *et al.* (2001). Radiologists' productivity in the interpretation of CT scans: a comparison with conventional film. *Am J Roentgenol* **176**, 861–864.

Reiner, B., Siegel, E., Moffitt, R., Brower, S. (2003). Utility of advanced computed radiography image-processing algorithms in the soft-copy interpretation of musculoskeletal trauma. *J Dig Imag*, **16**, 1–2.

Roehrig, H. (2000). The monochrome cathode ray tube display and its performance. In *Handbook of Medical Imaging Volume 3: Display and PACS*, ed. Kim, Y., Horii, S.C. Bellingham, WA: SPIE Press, pp. 155–220.

Roehrig, H., Chawla, A., Krupinski, E.A., Fan, J., Gandhi, K. (2003). Why should you calibrate your display? *Proc SPIE Med Imag*, **5199**, 181–192.

Roehrig, H., Fan, J., Chawla, A., Gandhi, K. (2002a). The liquid crystal display (LCD) for medical imaging in comparison with the cathode ray tube display (CRT). *Proc SPIE Pen Rad Sys Apps IV*, **4786**, 114–131.

Roehrig, H., Fan J., Furukawa, T., *et al.* (2002b). Performance evaluation of LCD displays. In *Proceedings CARS-2002*, ed. Lemke, H.U., Vannier, M.W., Innamura, K. Berlin, Germany: Springer Publishers, pp. 461–466.

Roger, E., Goldberg, M. (1989). Image organization and navigation strategies for a radiological workstation. *Proc SPIE Med Imag*, **1091**, 315–324.

Rollo, F.D. (2003). Molecular imaging: an overview and clinical applications. *Radiol Mngt*, **25**, 28–32.

Rosenbaum, A.E., Huda, W., Lieberman, K.A., Caruso, R.D. (2000). Binocular three-dimensional perception through stereoscopic generation from rotating images. *Acad Radiol*, **7**, 21–26.

Samei, E., Badano, A., Chakraborty, D., *et al.* (2005). Assessment of display performance for medical imaging systems. Report of the American Association of Physicists in Medicine (AAPM) Task Group 18. Madison, WI: Medical Physics Publishing. AAPM on-line Report No. 03, http://deckard.duhs.duke.edu/~samei/tg18.htm Last accessed May 12, 2009.

Samei, E., Poolla, A., Ulissey, M.J., Lewin, J. (2007). Digital mammography: comparative performance of LCD and CRT displays. *Acad Radiol*, **14**, 539–546.

Samei, E., Ranger, N.T., Delong, D.M. (2008). A comparative contrast-detail study of five medical displays. *Med Phys*, **35**, 1358–1364.

Saunders, R.S., Baker, J.A., Delong, D.M., Johnson, J.P., Samei, E. (2007). Does image quality matter? Impact of resolution and noise on mammographic task performance. *Med Phys*, **34**, 3971–3981.

Saunders, R.S., Samei, E. (2006a). Resolution and noise measurements of selected commercial medical displays. *Med Phys*, **33**, 308–319.

Saunders, R.S., Samei, E., Baker, J.A., *et al.* (2006b). Comparison of LCD and CRT displays based on utility for mammographic tasks. *Acad Radiol*, **13**, 1317–1326.

Shneiderman, B. (1998). *Designing the User Interface*. Reading, MA: Addison-Wesley, pp. 20–21.

Siegel, E.L., Reiner, B. (2002). Image workflow. In *PACS: a Guide to the Digital Revolution*, ed. Dreyer, K.J., Mehta, A., Thrall, J.H. New York, NY: Springer-Verlag, pp. 161–190.

SMPTE (1991). Specifications for medical diagnostic imaging test pattern for television monitors and hard-copy recording cameras. SMPTE RP 133. White Plains, NY: Society of Motion Picture and Television Engineers.

Taylor, C.R., Merin, L.M., Salunga, A.M., *et al.* (2007). Improving diabetic retinopathy screening ratios using telemedicine-based digital retinal imaging technology: the Vine Hill study. *Diabetes Care*, **30**, 574–578.

Tazawa, S., Gotoh, Y., Takahashi, S., *et al.* (2001). Cine viewing of abdominal CT. *Comp Meth Prog Biomed*, **66**, 105–110.

Thrall, J.H. (2005). Reinventing radiology in the digital age, Part II. New directions and new stakeholder value. *Radiol*, **237**, 15–18.

Video Electronics Standards Association. (2008). http://www.VESA.org Last accessed May 12, 2009.

Watts-Perotti, J., Woods, D.D. (1999). How experienced users avoid getting lost in large display networks. *Intl J Human-Computer Interaction*, **11**, 269–299.

Weinstein, R.S., Descour, M.R., Liang, C., *et al.* (2004). An array microscope for ultrarapid virtual slide processing and telepathology. Design, fabrication, and validation study. *Hum Path*, **35**, 1303–1314.

Yunfang, L., Wenjing, W., Bingshuang, H., Changji, L., Chenglie, Z. (2000). Visual strain and working capacity in computer operators. *Homeostasis Health Dis*, **40**, 27–29.

30

Ergonomic radiologist workspaces in the PACS environment

CARL ZYLAK

Given the continuing increase in the complexity of imaging studies, particularly cross-sectional ones, radiologists must meet the difficult challenge of image management and interpretation. Currently, interpretation of images is largely performed in the soft copy environment. The multiplicity of images per study precludes using hard copy.

Radiologists have traditionally found themselves interpreting images in an environment that is often less than ideal. Most workers in any other industry would not tolerate the often abysmal environment in which we find radiologist workplaces. Without meaningful radiologist consultation, including expert interpretations and recommendations for further workup, radiology's economic enterprise would fail. Without us, the enterprise cannot move forward. We must create an environment in which attention is paid to ergonomic issues that will result in maximizing efficiency, optimizing productivity, and reducing fatigue and discomfort of the radiology community.

Ergonomics has been defined by the International Ergonomics Association as the application of scientific information concerning objects, systems, and environment for human use. The term ergonomics, derived from the Greek words *ergon* (work) and *nomos* (natural laws) first entered the modern lexicon when Wojciech Jastrzebowski used the word in his 1857 article, "The outline of ergonomics, i.e. science of work, based on the truths taken from the natural science" (http://en.wikipedia.org/wiki/Ergonomics). Ergonomics is about the fit between people, their work activities, equipment, work systems, and environment to ensure that workplaces are safe, comfortable, and efficient, and that productivity is not compromised (www.powerhomebiz.com/Glossary).

Developing an ergonomic design for the workplace requires development of a plan, which will encompass several diverse important considerations. The pathway to radiologists' workplace design can take various routes (Krupinski, 2007; Nagy, 2003; Prabhu, 2005; Ratib, 2000). One approach is an iterative one whereby the current environment is assessed and opinions are sought from users with continuous reassessment and revision.

At the outset, one must identify issues related to the current environment. These issues can be assessed by observation and extended surveys, of not only the radiologists but of all users, including clinicians.

Issues can relate to existing poor workspace design (Figure 30.1). Placement of computers and monitors on existing desks and tables can result in the monitor being at a fixed height, which is too high for some and too low for others. Ghost images representing reflected light on the monitors from adjacent light sources such as adjacent monitors, view boxes, and other light sources that are positioned directly opposite interfere with the presentation of the images.

Ambient lighting and noise must be assessed. Inadequate or poorly directed lighting can result in eyestrain, burning or itchy eyes, headaches, and blurred or double vision. Excessive noise from adjacent alternator motors and lack of workspace partitioning can result in loss of concentration and fatigue. Workspaces may be crowded with poor visibility for teaching and consultation. Air circulation and adequate temperature controls are often overlooked.

Chairs, at a minimum, need to provide horizontal support for the arm, to reduce stress on the shoulder. Extended periods with an outstretched arm can result in elbow and shoulder pain. Neck strain can result from static loading of the shoulder and neck muscles during periods of extended flexion or extension.

Repetitive stress injuries (www.cdc.gov/niosh/docs/2002–122) occur with poorly designed workspaces. Repetitive stress injury results from overuse or misuse of muscles, tendons, and nerves, associated with performance of repetitive tasks. This can lead to cumulative muscle loading, particularly if working in an awkward or fixed posture with insufficient rest time. Cumulative trauma disorders (www.multimediagroup.com) may develop slowly over time.

Repetitive stress injury can affect fingers, hands, wrists, elbows, arms, shoulders, back, and neck. Symptoms such as aching, tenderness, swelling, pain, numbness, and loss of joint movement can be the result. Static muscle loading tenses the muscle, which squeezes the blood vessels, decreasing the flow of blood through the muscles. Muscle switches to an insufficient form of energy supply. Lactic acid builds up, which results in pain and fatigue.

Common injuries relate to inflammation of tendons when muscles and tendons are repeatedly tense. The tendon may fray or thicken, resulting in decreased mobility, which may be permanent. Tenosynovitis represents inflammation of the synovial sheath caused by repetitive motion. This restricts tendon movement. The sheath may scar, resulting in limited mobility. Synovia in the hands, wrists, elbows, and shoulders are most often affected.

In order to resolve the issues, various approaches can be taken. One approach is to form a users' group, consisting of radiologists, residents, physicists, and administrators, to address the issues. It is useful to have someone champion the effort, particularly if that person has influence in the budgetary decision making process. Often in picture archiving and communication system development, reading room dollars get shortchanged.

The Handbook of Medical Image Perception and Techniques, ed. Ehsan Samei and Elizabeth Krupinski. Published by Cambridge University Press.

Figure 30.1 Workspace before redesign. Note unsupported arm extension, and necessity to hold hard copy images on lap due to absence of a work bench. The radiologist in the immediate background has learned to avoid stress on the shoulder and neck by placing the mouse on a book supported by his thigh. Adjacent to him, a multiviewer adds unwanted illumination.

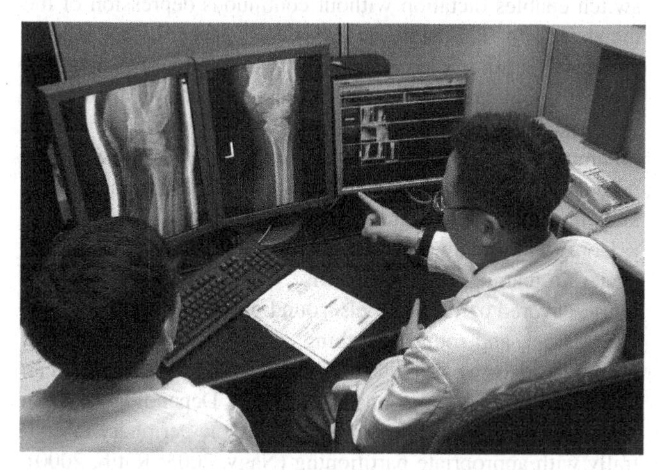

Figure 30.2 Workplace after redesign. This is an example of a two-person workspace for teaching and consultations. A full-width keyboard bench (Anthro) provides a working surface for both persons. A side table for the telephone, support for indirect lighting, hard copy images, etc., is conveniently located next to the work bench. A two-box film viewer is attached to the 4 × 6 foot sound-absorbing partition panels.

Solutions pursued include important consideration of the design of the workstation platform (desk) (Harisinghani, 2004). Various manufacturers take different approaches: www.radricktechnologies.com, www.anthro.com, and www.afcindustries are but a few examples (Figure 30.2).

Important considerations include the ability of independent height adjustment of the table and the viewing monitors. This can take the form of motorized split benches with independent height controls. Other approaches include motorized height control of a single platform used in conjunction with monitors with independent arm adjustment. An important often-overlooked issue relates to poor management of the multiple wires required to run a PACS workstation. These should be bundled and dropped from the ceiling within a single protective conduit. A separate

Figure 30.3 Example of a motorized split bench (AFC Industries).

portal should be considered for personal computers. Insertion of flash drives should be simple!

The ability for independent adjustment of the table and monitors offers the potential for avoiding cumulative static muscle loading of the shoulder. Shoulder impingement injuries relate to irritation or damage to the rotator cuff tendons. This is particularly associated with repetitive upward and outward shoulder motion where the elbow is unsupported. This can result in inflammation of the bursa with partial tears of the tendons, particularly the supraspinatus. Independent adjustment of the worktable and monitors facilitates working in either the sitting or standing positions. When choosing to work in the standing position, low back strain is reduced by the use of a footstool. Recall the brass foot rail in the saloons shown in the old Western movies! Elevation is also useful for teaching and consultation with small groups (Figure 30.3).

Keyboards should be placed in a flat or slightly negative tilt position (Hedge, 1995). Retracting the feet of the standard keyboard can facilitate this. Smaller, flatter, and thinner keyboards, particularly those with a negative slope, are preferred. If anything, the keyboard and hands should be slightly below the supported upper arm. Wrist posture should be neutral with the hands, wrists, and the anterior aspect of the forearm in a relatively straight line avoiding ulnar or radial deviation, excessive extension, and flexion. The elbow should be held close to the side of the body with no twisting or distortion. If wrist rests are to be used, it is important to avoid having the rests immediately under the carpal tunnel as this will add to the pressure and contribute to, rather than alleviate, the problem.

Mouse designs have improved (Aaras, 2002; Gustafsson, 2003; Jensen, 2002). The use of a wheel as opposed to a tapping motion reduces stress, which can result in trigger finger. In trigger finger, pain is present when bending the finger, particularly the index finger. Swelling and thickening of the flexor tendons of the digit are caused by repetitive flexion of the digit. A small nodule can develop on the synovial sheath that results in the index finger becoming "stuck" in the flexed position. To overcome the resistance produced by the nodule, manual assistance is required to extend the digit.

Other problems associated with mouse usage relate to lateral epicondylitis. This results from overuse of extensor muscles, extensor carpi radialis longus and brevis, which attach at the lateral epicondyle of the humerus. Healing causes tendonosis or thickening of the tendon. Bending the wrist back with static loading with the arm extended and tapping of the index finger all contribute to lateral epicondylitis.

The carpal tunnel syndrome (Ruess, 2003) is associated with pain and numbness in the hand and fingers along the course of the median nerve. The median nerve gives sensation to the thumb, index finger, long finger, and half of the ring figer. Symptoms include gradual tingling and numbness in the area supplied by the nerve. This is followed by dull, vague pain in the same distribution. This can result from any repetitive hand motion or vibration. The flexor tendons enclosed in the space bounded anteriorly by the flexor retinaculum and posteriorly by the carpal bones become inflamed, which increases pressure in the carpal tunnel, squeezing the median nerve. The use of padded wrist rests often adds to the problem by increasing the pressure on the median nerve within the carpal tunnel.

Guyon's canal syndrome consists of pain in the little finger and half of the ring finger. This is due to overuse of the wrist, particularly bending the wrist down and out with constant pressure on the palm. The ulnar nerve is compressed, resulting in discomfort.

Considering seating, one size certainly does not fit all. Chairs must have easy adjustments to accommodate differing body habitus. Adjustment of chair height, arm height and width, back tension, and stop position with lumbar support and a pad tilt are minimal requirements. To ensure stability, five supporting legs are required. The chair should be able to swivel to allow turning while in the seated position. One must be careful to avoid pressure on the ulnar nerve as it crosses the elbow when resting on the armrest. This can result in numbness along the ring and little finger with hand pain and muscle weakness. Prolonged direct pressure on the olecranon can lead to bursitis.

Regarding monitors, LCD monitors as illustrated in Figures 30.2 and 30.3 seem to be the choice in terms of brightness and ease of usage. They provide high brightness and resolution with a thin light cabinet. Ambient light reflection is significantly reduced.

The top of the monitor screen should be at eye level (Sommerich, 2001). The viewing area of the monitor should be 15 to 20 degrees below eye level with the screen tilted away from the observer. The eye-to-screen monitor distance should be about 66 centimeters or 26 inches, which is about an arm's length. The use of dedicated reading glasses has been helpful in reducing neck strain which is commonly associated with tilting of the head when reading with bifocal, etc., glasses.

Tungsten canister lights should replace overhead fluorescent lights. If a side table is required, a hooded fluorescent table light provides adequate illumination for reading documents while minimizing monitor light reflection. An LED placed on the desk is a good alternative. As a general rule, the level of ambient room light should approximate the background brightness of the monitor. Wall color is also to be considered, with a bluish-gray-purple combination favored. During the transition phase, a view box will be required for viewing hard copy images. In the transition, multiviewers should be removed in their entirety if possible. However, often a few may be required during the transition phase. A two-bank film viewer is generally sufficient.

Voice recognition systems are now commonplace. These can be awkward to use while rendering an interpretation and maintaining visual focus on the displayed images. Excessive thumb loading with poor arm and elbow position is often a consequence. It is important to recognize that most of these have options for left hand or right hand control. A toggle on-off switch enables dictation without continuous depression of the dictate button. This prevents excessive pressure by the thumb, which can result in discomfort in the MP joint during dictation. This gripping pressure can also produce pain on the side of the wrist and forearm, above the thumb. The pain results from inflammation of the tenosynovium and tendons of the abductor pollicis brevis and extensor pollicis longus. This syndrome is known as DeQuervain's tenosynovitis.

Considerable attention needs to be paid to acoustics. A variety of design features can reduce noise. Common among these are fabric-covered partition panels, sound-absorbing wall covering, floor carpet, and sound-absorbing ceiling tiles. Acoustic foam tiles are said to be particularly effective. White noise devices can be helpful in reducing sound/noise levels. Depending on room size, workstations can be arranged along the periphery or centrally with appropriate partitioning (Nagy, 2003; Ratib, 2000). Conversion of a traditional ballroom space to accommodate 20 soft copy workstations is shown in Figure 30.4. An example of a recent design for 8 workplaces is shown in Figures 30.5 and 30.6.

The benefits of an ergonomic workplace can be negated by poor posture, poor arm position, and insufficient work breaks. Users need to be taught and reminded how to use the components of the workspace. Posture instructions include forming a 90-degree angle at the elbow with the arms hanging downward at the sides. The shoulder should remain relaxed and in a lower position during keyboard use. Seat height should be adjusted so that the femur is horizontal with the feet flat and knees at 90-plus degrees. If necessary, a footrest can be used if the chair does not allow the feet to be flat on the floor. The chair back support should be adjusted to enable the individual to sit comfortably with appropriate lumbar support. Various studies have demonstrated that pressure in the lower lumbar disc changes with varying seated and standing positions (Andersson, 1974). Contrary to what we learned from our mothers, it is best not to sit erect. Lumbar disc pressure is lowest for a supported reclined posture of anywhere from 105 to 130 degrees (Bashir, 2006).

Figure 30.4 Initial design of conversion of a ballroom from hard copy to soft copy reporting. The multiviewers have since been removed. Inferior quality sound-absorbing materials in the ceiling and the lack of baffles hung from the ceiling allow voices to reflect from the workstations, resulting in significant distraction and irritation.

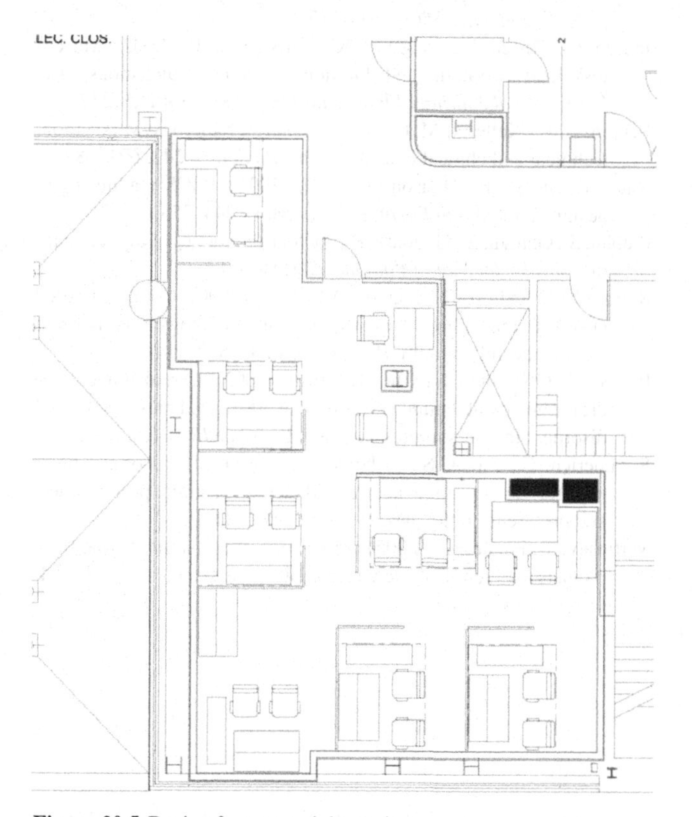

Figure 30.5 Design for a new eight-station reporting room. The arrangement prevents glare from adjacent workplaces.

It is important to take frequent work breaks. Repetitive stress injuries occur during stationary exertion or continual loading of muscles for long periods. Changing the sitting position every 10 minutes together with an active break every 30 minutes is very helpful, particularly when at a workstation for 2 to 3 hours.

Eye fatigue and blurriness result from long periods of image interpretation. Use the 20–20–20 rule, meaning: look away from the monitor at least every 20 minutes and focus on something over 20 feet away for 20 seconds. A recent study of prevalence of eyestrain among radiologists (Vertinsky, 2005) reported

Figure 30.6 Foam-absorbing 2 × 2 foot, 2-inch thick Sonex harmoni ceiling tiles and carpeting on the floor and walls has significantly reduced the reflected noise levels. The noise reduction coefficient of the tiles is 0.9, particularly effective in the normal low voice range. Lighting is done with wall units that illuminate the ceiling in an indirect manner. Canister lights illuminate walkways without casting light on the monitors.

that there was a prevalence of eyestrain in 36% of the radiologists surveyed. Eyestrain results from fatigue of the ciliary and extraocular muscles due to prolonged accommodation inversions required for near vision work. Looking straight ahead increases the exposed corneal surface, and the blink rate is decreased, because of mental concentration. Both factors contribute to dryness of the eyes. This is why it is important to tilt the monitor screen away from the observer. This decreases the amount of cornea exposed to air, which will decrease the dryness factor. In this study, increased eyestrain was associated with a variety of factors, including a longer workday that is more than six hours, associated with fewer work breaks. Intense viewing work associated with CT screening was also a factor.

An in-house survey (Hedge, A., *et al.*, 2007 (unpublished)) of radiologists at one institution, the Henry Ford Health System, Detroit, MI, found concerns regarding discomfort of the neck and lower back, which the radiologists thought related to monitor position, difficulties with the hand employed to use the mouse, and neck difficulties associated with workstation chair design. The suggestion coming from this informal survey was a need for continuing education of the radiologists regarding proper adjustments of the various components and emphasizing the necessity of frequency of rest breaks.

PACS requires new methods for case conference presentations. Large format computer displays offer opportunities to improve the teaching/learning environment (Figure 30.7).

In summary, ergonomic success relates to designing a workplace that maximizes workflow and efficiency and minimizes physical and mental strain. All of the issues discussed are important. A major area with potential for considerable negative impact on radiologist productivity relates to noise control. Lastly, ergonomic success in part relates to having an ergonomic advocate who can facilitate administrative and budgetary support. Ongoing financial commitment must be available for assessment and application of evolving

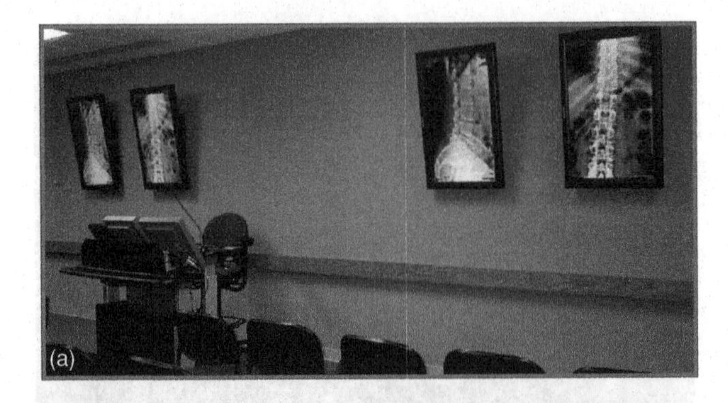

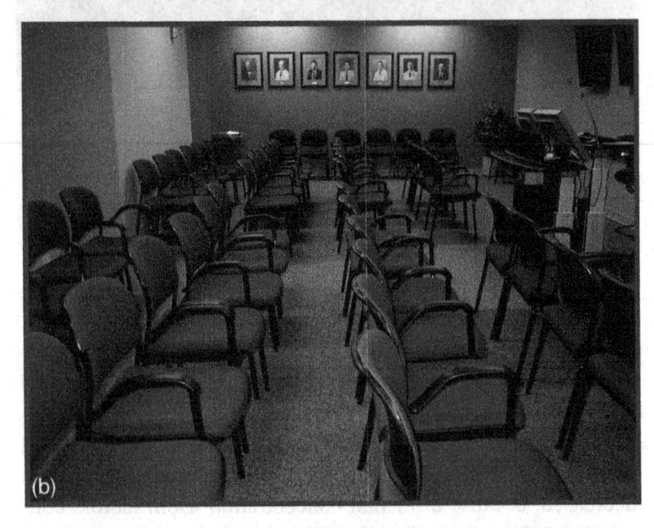

Figure 30.7 (a). Conference room utilizing dual portrait 40″ LCD monitors. Cases are retrieved online via the PACS workstation. For PowerPoint presentations (PACS system off), screens drop from the ceiling with the images projected from ceiling-mounted dual LCD projectors. (b). Dual monitors and screens enable orientation of seating providing all participants with ideal viewing opportunity.

technological and related equipment innovations. Continuing vigilance and education regarding the radiologists' workplace will result in fewer workplace-related injuries, greater job satisfaction, and an increase in productivity.

REFERENCES

Aaras, A., Dainoff, M., Ro, O., Thoresen, M. (2002). Can a more neutral position of the forearm when operating a computer mouse reduce the pain level for VDU operators? *International Journal of Industrial Ergonomics* **30**: 307–324.

Andersson, B., Örtengren, R. (1974). Lumbar disc pressure and myoelectric back muscle activity during sitting. *Scand J Rehab Med* **6**: 115–121.

Bashir, W., Torio, T., Smith, F., Takahashi, K., Pope, M. (2006). Alteration of water content in lumbar intervertebral discs related to variable sitting postures using whole-body positional MR imaging. Abstract, RSNA.

Gustafsson, E., Hagberg, M. (2003). Computer mouse use in two different hand positions: exposure, comfort, exertion and productivity. *Applied Ergonomics* **34**: 107–113.

Harisinghani, M., Blake, M., Saksena, M., *et al.* (2004). Importance and effects of altered workplace ergonomics in modern radiology suites. *Radiographics* **24**: 615–627.

Hedge, A., Powers, J. (1995). Wrist postures while keyboarding: effects of a negative slope keyboard system and full motion forearm supports. *Ergonomics* **38**(3): 508–517.

Jensen, C., Finsen, L., Søggaard, K., Christensen, H. (2002). Musculoskeletal symptoms and duration of computer and mouse use. *International Journal of Industrial Ergonomics* **30**: 265–275.

Krupinski, E., Kallergi, M. (2007). Choosing a radiology workstation: technical and clinical considerations. *Radiology* **242**: 671–682.

Nagy, P., Siegel, E., Hanson, T., *et al.* (2003). PACS reading room design. *Seminars in Roentgenology* **38**(3), 244–255.

Prabhu, S., Gandhi, S., Goddard, P. (2005). Ergonomics of digital imaging. *British Journal of Radiology* **78**, 582–586.

Ratib, O., Valentino, D., McCoy, M., *et al.* (2000). Computer-aided design and modeling of workstations and radiology reading room for the new millennium. *Radiographs* **20**: 1807–1816.

Ruess, L., O'Connor, S., Cho, K., *et al.* (2003). Carpal tunnel syndrome and cubital tunnel syndrome: work-related musculoskeletal disorders in four symptomatic radiologists. *AJR* **181**: 37–42.

Sommerich, C., Joines, S., Psihogios, J. (2001). Effects of computer monitor viewing angle and related factors on strain, performance, and preference outcomes. *Human Factors* **43**(1): 39–55.

Vertinsky, T., Forster, B. (2005). Prevalence of eye strain among radiologists: influence of viewing variables on symptoms. *AJR* **184**: 681–686.

PART VI

EPILOGUE

31

Future of medical image perception

ELIZABETH KRUPINSKI AND EHSAN SAMEI

At the 2007 Radiologic Society of North America Meeting, Elias Zerhouni, MD (then Director of the National Institutes of Health) gave the Eugene P. Pendergrass New Horizons Lecture on "Major Trends in the Imaging Sciences" (Zerhouni, 2008). His vision for the future of medical imaging was inspiring, noting that medicine in general is going through a transition from what he called the "hardware" phase of discoveries to the "software" phase. In the "hardware" phase the life sciences were geared towards understanding specific elements, while in the "software" phase we are now trying to understand how elements interact and work together in both normal and abnormal processes. The role of imaging in both phases is clear, as imaging has the potential to reveal structure, function, and metabolism in vivo. Because of this, medicine is entering a new era in which medicine will be predictive, personalized, pre-emptive, and participatory.

No matter where the future takes medicine and the imaging sciences, at the core of all image-based clinical specialties, especially radiology, is the interaction of the clinician's eye-brain system with visually presented medical image data. Any changes in the ways that images are acquired and displayed, and the information that they contain, will undoubtedly impact the clinical decision. As this book has hopefully demonstrated, our continued investigation into the mechanisms underlying this complex interpretation process will help ensure that the information presented to the clinician is done in an optimized and ergonomic fashion so that the detection and correct classification of abnormalities and diseases continue to improve.

As Dr. Zerhouni noted, the clinician practicing medicine ten years from now will, in all likelihood, be working in a very different environment than today. In radiology, analog films are already becoming a rare commodity, as are light microscopes in pathology. Clinicians in other specialties such as dermatology and ophthalmology are increasingly using images to render diagnoses and this is likely to grow even more rapidly as telemedicine reimbursement increases. There will be new types of digital images, acquired using different technologies than exist today, and processed in ways that have yet to be conceived. Molecular imaging will surely have a huge impact on the types of studies being done and the types of images being interpreted. The exchange of images and data from one specialty to another to enhance the decision-making process will also increase as hospital information systems become more sophisticated and integrated. As we move into the future and look further into the human body with advanced imaging techniques, we need to include the human interpreter of the data in our investigations and technology developments.

These goals are not new. In November of 1994 the National Institutes of Health (NIH), the National Cancer Institute (NCI), and the Conjoint Committee on Diagnostic Radiology sponsored a conference on Long-Term Plans for Imaging Research (Kundel, 1995). Five priorities for medical image perception research were identified: (a) develop psychophysical models for the detection of abnormalities in medical images; (b) improve understanding of the mechanisms of perception as they apply to medical images; (c) develop aids for enhancing perception that provide interactions between vision and displays; (d) study perceptually acceptable alternatives to sequential sections for viewing cross-sectional data; and (e) perform methodological research aimed at improving the evaluation of medical imaging systems.

After a series of "Far West Image Perception Conferences" that began in 1985, the Medical Image Perception Society (MIPS) was organized in 1997 to further promote these research goals and education in medical image perception, and to provide a forum for the discussion of perceptual, psychophysical, and cognitive issues. At that time, two articles were written by MIPS to update the goals of medical image perception research (Krupinski, 1998a, 1998b). Those goals broadly included (a) mathematical modeling of the detection task; (b) understanding visual search; (c) understanding the nature of expertise; (d) developing perceptually based standards for image quality; (e) developing computer-based aids to image perception; and (f) developing quantitative methods for describing natural images and for measuring human detection and recognition performance.

In 2001 the Far West Image Perception Conference was renamed the Medical Image Perception Society Conference. The most recent meeting was in 2007 at the University of Iowa. During this meeting, five panels were held discussing the current state of medical image perception research and its future. Summarized here are the three major points from each of the panels, and it is clear that many of these questions and issues have been raised and considered in this text. It is also clear that these same themes in medical image perception research have been expressed previously in one form or another. Many of the chapters in the current text incorporated and expanded upon these themes, clearly demonstrating the importance of medical image perception in today's practice of medicine as well as the future's.

The Handbook of Medical Image Perception and Techniques, ed. Ehsan Samei and Elizabeth Krupinski. Published by Cambridge University Press.

31.1 BASIC MEDICAL IMAGE PERCEPTION SCIENCE

Even though computer-based image processing and analysis tools will become increasingly prevalent in the future as described in some of the chapters in this text, clinicians will continue to interpret medical images for years to come and thus they will remain a central consideration in performance of diagnostic systems. We need to understand the cognitive and perceptual mechanisms underlying reports of abnormalities and use this information to help guide the development of better acquisition and display devices, improving patient care and treatment. By studying how the observer samples parts of an imaging study and allocates attention and perceptual resources, we may inform the training of new readers and the development of effective machine readers. Some topics for future research in this area include the following.

(1) Clinicians rarely get feedback about misses in clinical practice. Will this change with the Integrated Health Enterprise (IHE) and will it actually impact how clinicians perceive/interpret images?
(2) Imaging clinicians must read increasing numbers of cases and images resulting in perceptual and cognitive overload. Computer-aided detection (CAD) schemes and other image processing and analysis tools can help, but they do not seem to help everyone and not in all interpretation tasks. Perception science can help better understand clinicians' limitations as well as strengths to better guide the development of perceptual and cognitive interpretation aids.
(3) Molecular imaging is on the rise and will present a whole new set of challenges to clinicians and so those studying medical image perception need to start looking at this area of imaging.
(4) Increasingly, clinical decisions are made based on multiple types of images. Medical image perception science can address how, from a cognitive and psychophysical standpoint, the information can be integrated towards best outcome.

31.2 DISPLAYS AND TOOLS

Ergonomic and human factors issues are becoming more important as new technologies emerge and the clinical work environment changes to a more computer-based one. In today's clinical environment the issues of the display and the display interface are quite important. We need to explore the interaction between ergonomics and cognition, perception, and performance in the ways noted below.

(1) Does the wide variability of vendor products used to present images impact the perception of images? This seems likely and it has implications not only for clinical practice but research as well. Perceptionists can help understand these differences and better optimize the imaging display, navigation tools, and reading environment.
(2) Can standardized lexicons help reduce reader variability? If we give the same tools to everyone it could be quite useful but would it simply take too long (if ever) to get everyone to the same level?
(3) Is it worth developing training sets that can help with reducing disparities? This could be very difficult and time-consuming and one worries about time delays and relevance.

31.3 ROC METHODOLOGY

Assessment is crucial to any research venue and as technologies become more complicated, so must the assessment methods themselves change and advance as well. We have seen a number of improvements and variations made to the standard receiver operating characteristic (ROC) methods over the years, and in the future we will see even more improvements. Some questions to guide that research include the following.

(1) Do we need more complicated ROC designs – split-plot, unbalanced designs, and those that deal with missing data? Most people still seem to be using basic ROC techniques developed years ago even though much more advanced techniques are out there with validated software. Is it a publicity issue or lack of understanding?
(2) Detection of single abnormalities is still the core application for ROC, but there is the multiple lesion issue, and the problem with truth or the gold standard. More concentrated efforts to define better ways of ascertaining truth are warranted.
(3) Ideally, the way we measure diagnostic accuracy should emulate the interpretation tasks performed by the clinicians. Future research should focus on methods by which we can assess diagnostic accuracy that are most relevant in that way.

31.4 VISUAL SEARCH

Experienced imaging-based clinicians (particularly radiologists) process a huge amount of diagnostic information in a very short amount of time (the initial global or gist view), but even the best clinician needs to scan the eyes over the images to thoroughly evaluate them. For years, many image perceptionists have used eye-tracking technology to understand the visual search process. If we can understand the nature of visual search of medical images, we may be better able to understand why errors of diagnosis occur and what we can do to improve performance. We also may be better able to understand how expertise develops and how we can better and more efficiently train residents to become expert radiologists. As we move to the future, the following questions should be addressed.

(1) Is the eye-tracking technology good enough yet for routine use in the clinic – could we use it as a feedback tool?
(2) Can we move visual search to more complex data – scrolling through CT slices and even looking at 3D images?
(3) Can CAD be improved with eye-position recording? CAD misses some things and so does the human – if we

combine them with eye-tracking then we may do even better at catching things.

31.5 OBSERVER MODELS

Observer models can serve as the basis for evaluating performance of image acquisition assuming an "ideal observer," i.e. one who could theoretically take advantage of all of the information contained in an image. Greater generalizability of these techniques to clinical radiology has been a recent goal of many researchers. Observer modeling can be used for a variety of tasks, including assessment of image quality throughout the imaging chain, and prediction of observer performance as a function of modality or some other imaging parameter. Some user avenues for future research include the following.

(1) Are observer models useful for clinicians or are they just theoretical tools for researchers?
(2) Can we use the models for training? If we show what the ideal is and what is possibly preventing the observer from achieving the ideal, maybe perceptionists can help bridge the gap.
(3) Most models are still very basic – simulated signals in simulated/real backgrounds without much visual search incorporated. Medical images are dynamic and much more complicated – we need to improve the complexity of the models being developed and used.
(4) Observer models should be extended to enable processing image data from a variety of sources (e.g. multiple modalities, or temporal images), similar to the way clinical practice is increasingly performed.

31.6 IMAGE QUALITY

Image quality is obviously an important aspect of medical images. Image quality has many contributing components and many metrics by which it can be defined. The imaging chain is a complex sequence of events during which the image is susceptible to a wide array of potentially degrading factors that imaging scientists devote their careers to eliminating. The chain actually begins with the object being imaged (the patient) and proceeds through acquisition, transmission, display, and finally processing by the human eye-brain system. Since the images may need to be reviewed in the future, storage and retrieval also become important steps in the chain.

When dealing with image quality, it is tempting to think that the only thing of importance is the final "beautiful" image that the radiologist, pathologist, or other clinician views. The ultimate "quality," however, should be judged not by our ability to provide the clinicians with the most "beautiful" image, but with the image that helps them render an accurate interpretation in the most efficient and timely manner. The real question in medical image perception is not "is the image pleasing to look at?" but rather "can the information required to render an accurate and confident diagnostic interpretation be visualized by the human observer?"

One challenge in medical imaging, and medical image perception in particular, is that the definition and hence the quantitative and objective measurement of image quality differs depending on how the question is posed and to whom. If you take two images that you want to compare in terms of image quality and ask engineers "which one is better?" they are likely to suggest that you look at the raw image data and calculate some figure of merit such as mean square error or signal-to-noise ratio. If you ask the same question of medical physicists they will suggest measuring something like noise equivalent quanta or detective quantum efficiency. Image analysis scientists might argue that the "better" image is the one that a computer has analyzed or manipulated to provide CAD and diagnosis information to the human observer. Ask vision scientists and they tell you to measure perceived differences using figures of merit such as just-noticeable differences or to measure observer performance using ROC techniques. The clinician will say that the only valid metric is diagnostic accuracy and the ability to incorporate this into a treatment recommendation. No single approach is the ultimate answer as the chapters in this book have shown in the wide variety of approaches described to understand the medical image perception process. Medical image perception research needs to take into account all of these perspectives in order to better understand the interaction between the image and the eye-brain system of the person interpreting the image.

Future research on image quality may focus on the following areas.

(1) It is important to understand and correlate the various measures of image quality so that the results of different studies would not lead to conflicting conclusions.
(2) We need to have better quantitative metrics of image quality that are more closely and directly related to diagnostic accuracy.
(3) It is important to substantiate the impact of image quality on aspects of medical interpretations other than diagnostic accuracy. Those include timing efficiency, fatigue, and ease of interpretation.
(4) For many imaging modalities, image quality (determined by the combination of image acquisition and processing protocols) are set based on aesthetic reasons. Those, for example, include CT and digital radiography. Future research should focus on optimizing image quality based on quantitative metrics that are correlated with diagnostic accuracy.
(5) Given the multi-component nature of medical imaging technology, it is important to characterize and optimize image quality in an integrated fashion, such that the relative contribution of each component is substantiated and the "bottleneck" components are identified for closer examination and adjustment.

It should be clear that there are many things to consider when trying to optimize observer performance in the medical imaging environment. One must take the display, the room, the images, and especially the clinician into account to get a complete picture of the image interpretation process. Understanding even something as simple as how often a clinician should take a break from examining images to avoid fatigue and the probability of making errors is important. Even more important is to realize that our understanding of these factors and how the clinician performs is not a static question. Technology has changed and will continue to change and these changes will present new challenges to the clinician's perceptual and cognitive systems. Our continued exploration into medical image perception and the way the clinician interacts with medical images will always be important.

REFERENCES

Krupinski, E.A., Kundel, H.L. (1998a). Update on long-term goals for medical image perception research. *Acad Radiol*, **5**, 629–633.

Krupinski, E.A., Kundel, H.L., Judy, P.F., Nodine, C.F. (1998b). The Medical Image Perception Society: key issues for image perception research. *Radiol*, **209**, 611–612.

Kundel, H.L. (1995). Medical image perception. *Acad Radiol*, **2**, S108–S110.

Zerhouni, E.A. (2008). Major trends in imaging sciences: 2007 Eugene P. Pendergrass New Horizons Lecture. *Radiol*, **249**, 403–409.

INDEX

Printed in the United States
By Bookmasters